Surveys in Differential Geometry

Vol. 1: Lectures given in 1990
edited by S.-T. Yau and H. Blaine Lawson

Vol. 2: Lectures given in 1993
edited by C.C. Hsiung and S.-T. Yau

Vol. 3: Lectures given in 1996
edited by C.C. Hsiung and S.-T. Yau

Vol. 4: Integrable systems
edited by Chuu Lian Terng and Karen Uhlenbeck

Vol. 5: Differential geometry inspired by string theory
edited by S.-T. Yau

Vol. 6: Essay on Einstein manifolds
edited by Claude LeBrun and McKenzie Wang

Vol. 7: Papers dedicated to Atiyah, Bott, Hirzebruch, and Singer
edited by S.-T. Yau

Vol. 8: Papers in honor of Calabi, Lawson, Siu, and Uhlenbeck
edited by S.-T. Yau

Vol. 9: Eigenvalues of Laplacians and other geometric operators
edited by A. Grigor'yan and S-T. Yau

Vol. 10: Essays in geometry in memory of S.-S. Chern
edited by S.-T. Yau

Vol. 11: Metric and comparison geometry
edited by Jeffrey Cheeger and Karsten Grove

Vol. 12: Geometric flows
edited by Huai-Dong Cao and S.-T. Yau

Vol. 13: Geometry, analysis, and algebraic geometry
edited by Huai-Dong Cao and S.-T.Yau

Vol. 14: Geometry of Riemann surfaces and their moduli spaces
edited by Lizhen Ji, Scott A. Wolpert, and S.-T. Yau

Volume IX

Surveys in Differential Geometry

Eigenvalues of Laplacians and other geometric operators

edited by

Alexander Grigor'yan and Shing-Tung Yau

International Press
www.intlpress.com

Surveys in Differential Geometry, Vol. 9

Editors:
Alexander Grigor'yan, Bielefeld University, Germany
Shing-Tung Yau, Harvard University

ISBN 978-1-57146-180-3

Paperback reissue 2010. Previously published in 2004 under ISBN 1-57146-115-9 (clothbound).

Typeset using the LaTeX system.

Preface

This year, we have decided to concentrate on an important topic in Geometry: the eigenvalues of the Laplacian and other geometric operators. This has been one of the most fundamental subjects in the area. The works on Hodge theory and the heat equation proof of the Hodge theorem certainly show its importance. One of the most dramatic and influential papers was the one by Mark Kac on "How to hear the shape of the drum", which intimately connected the geometry of a manifold with the spectrum of the Laplacian on it.

A strong understanding of spectral properties of differential operators has deep consequences in Geometry, Physics, Number Theory, Probability Theory, etc. Analytic methods have penetrated also into Discrete Mathematics, where the study of the spectral properties of difference operators leads to deeper understandings of the combinatorial questions on graphs.

This volume is an attempt to survey the diversity of directions unified under the above headings. Although the subject is too big to be covered by a single book, we still believe that the present collection does introduce a substantial part of the subject, and we hope it will be beneficial to geometers and experts from the related areas.

Alexander Grigor'yan and Shing-Tung Yau
April 2004

Contents

Surveys in Differential Geometry
IX, International Press

Anomalous diffusion and stability of Harnack inequalities

Martin T. Barlow

CONTENTS

1. Introduction.

Let (M, g) be a complete Riemannian manifold, with metric $d(x, y)$, and let $\Delta = \text{div}\nabla$ be the Laplace-Beltrami operator on M. Let $V(x, r)$ denote the volume of the ball $B(x, r)$ with centre x and radius r. In this paper we survey the stability of elliptic and parabolic Harnack inequalities on M.

A function $u = u(x)$ is *harmonic* in a domain $D \subset M$ if it is a solution of the Laplace equation:

$$\Delta u(x) = 0, \qquad x \in D.$$

M satisfies the *elliptic Harnack inequality* (EHI) if there exists a constant C_E such that, for any ball $B(x, R)$, whenever u is a non-negative harmonic function on $B(x, R)$ then

$$\sup_{B(x,R/2)} u \leq C_E \inf_{B(x,R/2)} u.$$

The parabolic Harnack inequality (PHI) is a little more complicated to state. Let $x \in M$, $R > 0$, $T = R^2$, $\widetilde{D}(x_0, R) = (0, 4T) \times B(x_0, 2R)$, and

$$Q_- = (T, 2T) \times B(x_0, R), \qquad Q_+ = (3T, 4T) \times B(x_0, R).$$

M satisfies the PHI if there exists C_P such that, for any x_0 and R, if u is a non-negative solution of

$$\frac{\partial}{\partial t} u(x, t) = \Delta u(x, t), \qquad (x, t) \in \widetilde{D}(x_0, R) \tag{1.1}$$

2000 *Mathematics Subject Classification.* Primary 60J27; Secondary 60J35.
Research partially supported by a NSERC (Canada) grant.

then

$$\sup_{Q_-} u \le C_P \inf_{Q_+} u.$$

Since any harmonic function solves (1.1), the PHI immediately implies the EHI.

DEFINITION 1.1. (a) M satisfies volume doubling (VD) if there exists a constant C_1 such that

$$V(x,2R) \le C_1 V(x,R) \quad \text{for all } x \in M,\ R \ge 0. \tag{1.2}$$

(b) M satisfies a Poincaré inequality, denoted PI(2), if there exists a constant C_2 such that, for any $B(x,R)$ and $f \in C^\infty(B(x,R))$,

$$\int_{B(x,R)} |f - \overline{f}_B|^2 d\mu \le C_2 R^2 \int_{B(x,R)} |\nabla f|^2 d\mu. \tag{1.3}$$

(c) M satisfies HK(2) if the heat kernel $p_t(x,y)$ on M satisfies the two-sided Gaussian bound

$$\frac{c_1}{V(x,t^{1/2})} e^{-c_2 d(x,y)^2/t} \le p_t(x,y) \le \frac{c_3}{V(x,t^{1/2})} e^{-c_4 d(x,y)^2/t}. \tag{1.4}$$

In [**47**] Li-Yau proved that if M has non-negative Ricci curvature then M satisfies HK(2). This result was refined by subsequent work of Grigoryan [**22**] and Saloff-Coste [**54**], who proved that PHI is equivalent to two conditions on M: volume doubling, and a family of Poincaré inequalities. The relation with HK(2) comes from [**44**].

THEOREM 1.2. *(*[**22**],[**54**],[**20**]*). Let M be a complete Riemannian manifold. The following conditions are equivalent:*
(a) The heat kernel $p_t(x,y)$ on M satisfies the two-sided Gaussian bound (1.4).
(b) M satisfies the parabolic Harnack inequality.
(c) M satisfies VD and PI(2).

The arguments used to prove Theorem 1.2 are quite general, and can be translated into other contexts: see [**18**] for graphs, and [**57**] for general metric spaces. Since the condition (c) is stable under rough isometries (see Section 5), it follows that HK(2) and the PHI are also stable under rough isometries (with suitable 'side conditions'). Theorem 1.2 gave a fairly complete characterization of the PHI, but left open two significant questions concerning the EHI:

1. Is the EHI equivalent to the PHI?
2. Is the EHI stable under rough isometries?

Of these, (2) is still open, while the (negative) answer to (1) had been for some time implicit in work on diffusions on fractals, before being made explicit in [**6**], following a conversation between the author and A. Grigoryan at MSRI in 1997.

In the early 1980s mathematical physicists became interested in the properties of random structures, such as percolation clusters, at criticality. Let us recall the definition of percolation on the Euclidean lattice $\mathbb{Z}^d$: one regards the edges $\{x,y\}$ (with $|x-y|=1$) as *bonds*, and each bond is *open* with probability $p \in [0,1]$, independently of all other bonds – see [**28**]. The *open clusters* are the collections of points connected by paths consisting of open bonds. For small p all the open clusters are finite, while for p close to 1 there is a single giant open cluster with small holes. At a critical value $p_c = p_c(d) \in (0,1)$ there is a *phase transition*; all clusters are finite if $p < p_c$, while if $p > p_c$ there is a unique infinite cluster. (All statements of this kind are *almost sure*, that is, disregarding a set of probability zero.) At the critical value $p = p_c$ it is believed that all clusters are finite (this is known if $d = 2$ or $d \geq 19$), but nevertheless there are many large clusters – in fact (see [**14**]) a cube of side n will contain, with high probability, clusters of diameter n. In some cases (see [**41**], [**34**], [**35**]) it is possible to define an *incipient infinite cluster* $\widetilde{\mathcal{C}}$ – essentially the cluster containing 0 conditioned to be infinite. The local structure of $\widetilde{\mathcal{C}}$ should then be similar to that of large finite critical clusters.

Figure 1.1. Percolation at criticality; the points in the largest cluster are marked with bullets.

The physicist's conjecture (confirmed in some cases by theorems in these papers) is that $\widetilde{\mathcal{C}}$ has a fractal structure. Physicists are interested in what are called *transport properties* of percolation clusters – that is, to mathematicians, the behaviour of solutions to the Laplace, heat and wave equations. In view of the connections between the heat equation and Markov processes, this motivated the study of random walks on percolation clusters: the 'ant in the labyrinth' of De Gennes [**17**]. (The wave equation is much harder – one reason being the difficulty in making a useful probabilistic connection.)

Since it is hard make exact calculations on sets such as $\widetilde{\mathcal{C}}$, physicists therefore proposed (see [**1**], [**53**]) that one should look random walks on regular, deterministic fractals such as the *Sierpinski gasket*. This idea has proved fruitful – not only is the study of random walks and diffusions on such sets interesting in its own right, but recent work indicates that random walks on fractals and critical percolation clusters have similar behaviour.

Figure 1.2. The first and fourth stages in the construction of the Sierpinski gasket.

For a brief survey of heat kernels and fractal sets see [**4**], and for more details on the various families of regular fractals which have been considered see the book [**42**], and lectures [**2**]. See also the survey [**56**], and for a review of the physics literature [**31**].

The limit of the construction illustrated in Figure 1.2 is a compact subset $F_{SG} \subset \mathbb{R}^2$. (We take the origin as the lower left hand corner of F_{SG}.) One can define an unbounded set $\widetilde{F}_{SG}$ with a local structure similar to F_{SG} by setting

$$\widetilde{F}_{SG} = \bigcup_{n=0}^{\infty} 2^n F_{SG}.$$

On certain sets $\widetilde{F}$ of this type, including the Sierpinski gasket, (see the references above for other examples) one can define a Laplacian operator $\mathcal{L}$, and the associated heat kernel $p_t(x, y)$ satisfies

$$\begin{aligned} c_1 t^{-\alpha/\beta} \exp(-c_2 (\frac{d(x,y)^\beta}{t})^{1/(\beta-1)}) &\le p_t(x,y) \\ &\le c_3 t^{-\alpha/\beta} \exp(-c_4 (\frac{d(x,y)^\beta}{t})^{1/(\beta-1)}), \qquad x, y \in \widetilde{F},\ t > 0. \end{aligned} \tag{1.5}$$

Here α and β depend on the fractal F; α is the Hausdorff dimension of F while $\beta \ge 2$ is a number (called the *walk dimension* of F), which gives the space-time scaling of the heat equation on F. One finds that $2 \le \beta \le 1 + \alpha$, and that these are the only constraints on α and β – see [**3**], [**33**], [**23**]. While fractal sets with $\beta = 2$ are known (see [**15**], [**46**]), the main families of regular fractals have $\beta > 2$: for example the Sierpinski gasket has $\alpha_{SG} = \log 3/\log 2$ and $\beta_{SG} = \log 5/\log 2$. Let X be the diffusion process associated with $p_t(x, y)$; then (1.5) leads easily to the bound

$$\mathbb{E}^x d(x, X_t)^2 \asymp t^{2/\beta}, \qquad t > 0. \tag{1.6}$$

The case $\beta \ne 2$ is called by physicists *anomalous diffusion*. The intuitive explanation for (1.6) for the Sierpinski gasket is that the motion of X is impeded by a sequence of successively larger obstacles.

It might be thought that this unusual scaling is due to the local fractal structure of $\widetilde{F}_{SG}$. However, $\widetilde{F}_{SG}$ is self-similar, and the large time behaviour of $p_t(x, y)$ is governed by the large scale structure of $\widetilde{F}_{SG}$. In fact the bound (1.5) (for $t \geq 1$ and $d(x, y) \leq t$) holds for various 'classical' sets with a large scale fractal structure. Thus Jones [**38**] proved that the random walk on the graphical Sierpinski gasket G_{SG} (see Figure 1.3) satisfies this bound, while Bass and I in [**7**] obtained an analogous result for graphical Sierpinski carpets. (These graphs are sometimes called 'pre-fractals'.)

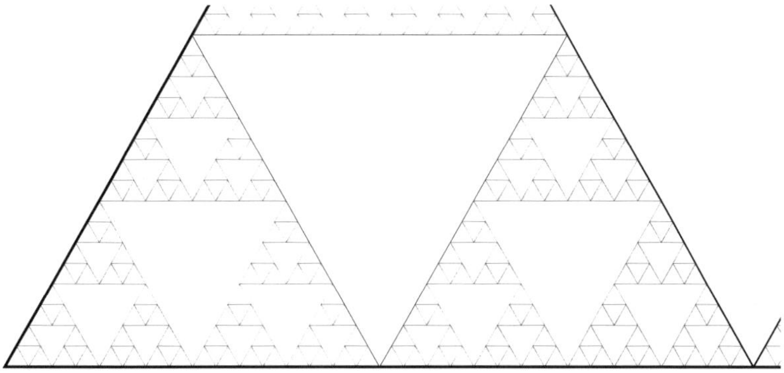

Figure 1.3. The graphical Sierpinski gasket G_{SG}: the small triangles have side 1.

One can also look at 'pre-fractal' domains in $\mathbb{R}^d$. These are open domains $D \subset \mathbb{R}^d$ with a large scale structure similar to some unbounded regular fractal – see [**6**], [**8**] for the pre-Sierpinski carpet. (Pre-fractal manifolds have also been considered in [**45**], [**11**].)

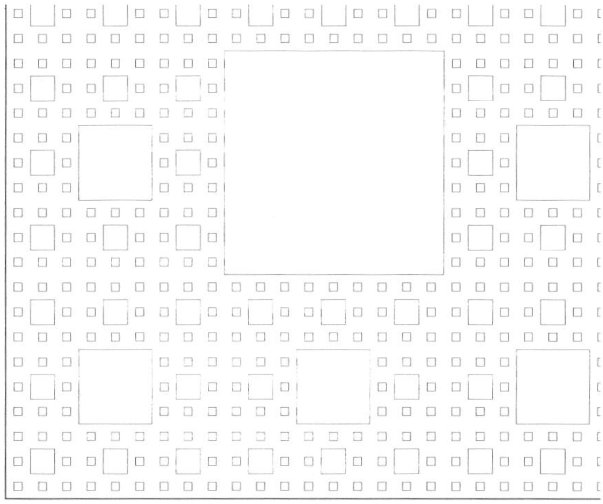

Figure 1.4. The pre-Sierpinski carpet: the small squares have side 1.

These sets satisfy the elliptic Harnack inequality. For the graphical Sierpinski gasket there is a very easy direct proof – see Theorem 2.6 of [**4**]. The proof for pre-Sierpinski carpets in dimensions $d \geq 3$ is quite long, and uses probabilistic coupling – see [**6**]. If $\beta > 2$ it is easy to deduce from (1.5) that PHI fails for these sets. The underlying reason is that in (1.5) the space-time scaling is given by the 'anomalous diffusion' coefficient β, rather than 2; the EHI contains no information about this scaling while the PHI does.

Given this, it is clear how the PHI should be generalized. Given $\beta \geq 2$ one replaces $T = R^2$ in the definition above by $T = R^\beta$; denote this condition PHI(β). It is immediate that PHI(β) implies EHI.

The argument in [**20**] used to prove the equivalence of PHI and (1.4) extends to the case $\beta > 2$: see Section 5 of [**32**]. Thus it follows that PHI(β) (with the appropriate β) holds for the Sierpinski gasket, carpet, and other families of regular fractals. One can also prove quite easily from (1.5) that these spaces satisfy a Poincaré inequality with anomalous scaling, denoted PI(β), and obtained by replacing R^2 by R^β in (1.3).

The proofs of (1.5) in the literature all use very strongly various symmetry properties of the spaces. In transferring results from one kind of pre-fractal object to another (for example, from the graphical Sierpinski gasket to a manifold made from it in the same way the 'jungle gym' is from the Euclidean lattice – see [**52**], [**40**]) one would wish for the same kind of stability in the case $\beta > 2$ as is given by Theorem 1.2 if $\beta = 2$.

An initial guess that Theorem 1.2 holds if one just replaces 2 by β is easily seen to be false – for example the product graph $G = \mathbb{Z} \times G_{SG}$ satisfies VD and PI(β_{SG}), but fails EHI – see the proof of Theorem 6 of [**3**]. (The reason is that the different space-time scaling in the two directions means that a random walk will with high probability leave a ball in the $\mathbb{Z}$ direction before it has moved very far in the G_{SG} direction.)

In [**9**] Bass and I, in the graph case, gave an additional condition, denoted CS(β), which, with VD and PI(β) is equivalent to PHI(β). This condition, which is unfortunately quite complicated, is described in Section 2 below. Given open sets $U_1 \subset \overline{U}_1 \subset U_2$, we will say a function f is a *cutoff function for* $U_1 \subset U_2$ if $f \geq 1$ on U_1 and $f = 0$ on U_2^c.

In Euclidean space (or a manifold satisfying VD) if we look for the lowest energy cutoff function f for $B(x,R) \subset B(x,2R)$ then this has energy $\mathcal{E}(f,f) = \int |\nabla f|^2 \asymp R^{-2}V(x,r)$. On a pre-fractal domain D, such as the pre-Sierpinski carpet (Figure 1.4), one can do better, and for the optimal f one obtains $\mathcal{E}(f,f) \asymp R^{-\beta}V(x,r)$. (One takes $|\nabla f|$ higher on shells where the set D is relatively thin.) The condition CS(β) is that there exists a large family of well-behaved cutoff functions for balls $B(x,R) \subset B(x,2R)$, with energy of order $R^{-\beta}V(x,R)$. Balls with radius less than 1 in a graph are trivial, but for manifolds we need to be able to treat balls of any size. Since a manifold is locally Euclidean, one expects the usual R^2 behaviour for small R. One therefore needs to introduce the function

$$\Psi(r) = \begin{cases} r^2 & \text{if} \quad r < 1, \\ r^\beta & \text{if} \quad r \geq 1. \end{cases}$$

and discuss PI(Ψ), PHI(Ψ) and CS(Ψ).

We have the following:

THEOREM 1.3. *Let M be a complete smooth Riemannian manifold. The following are equivalent:*
(a) M satisfies PHI(Ψ).

(b) M satisfies VD, PI(Ψ), and CS(Ψ).
(c) The heat kernel $p_t(x,y)$ on M satisfies a two-sided bound denoted $HK(\Psi)$.

A similar theorem for graphs was the main result of [**9**], and was based on methods used in [**8**] to study divergence from operators on the pre-Sierpinski carpet. The proof of Theorem 1.3 is very similar, and we will only sketch the main arguments. Full details will appear in [**10**], which will treat the general case of a measure metric space with a Dirichlet form: this covers pre-fractal domains and manifolds, as well as true fractals.

The conditions VD, PI(Ψ), and CS(Ψ) are all stable under rough isometries, given suitable 'side conditions' on the families of sets considered. (Since rough isometries only preserve global properties, it is clear that some conditions of this kind are necessary.)

While CS(Ψ) (with VD, PI(Ψ)) is therefore a necessary and sufficient for PHI(Ψ)), of course it may not be the simplest such condition. In fact, in the 'strongly recurrent' case ($\alpha < \beta$) a much simpler characterization of PHI(Ψ) is possible, just using VD and estimates on the electrical resistance: this will appear in [**12**]. (See also [**43**].)

One might hope that these stability results for PHI(Ψ) would lead to the stability of EHI. However, this still seems a hard problem. Delmotte [**19**] has shown, by joining two different graphs by one edge, that EHI can hold even if VD fails. Let G_i be pre-fractal graphs satisfying EHI with indices α_i and β_i, such that $\beta_2 - \alpha_2 = \beta_1 - \alpha_1 = \zeta > 0$. Then (see [**3**]) joining G_1 and G_2 by one edge gives a graph G satisfying EHI but with different space-time scaling in different regions. Thus any heat equation approach to the stability of the EHI would appear to need to deal with spaces with rather irregular properties. However, in this example the electrical resistance between points x and y decays as $d(x,y)^{-\zeta}$, so that, in terms of electrical resistance, the graph G is quite regular. This suggests that it may be possible to characterize EHI in terms of electrical resistance – see the open problem mentioned at the end of [**5**].

The structure of the remainder of this survey is as follows. In Section 2 we introduce precisely the main concepts and give our main results. In Section 3 we sketch the proof that PHI(Ψ) implies CS(Ψ); the argument uses Green's functions to build a suitable family of cutoff functions. Section 4 deals with the implication (b) $\Rightarrow$ (a). We begin by showing how Moser's iteration argument breaks down in the case $\beta > 2$ if we try to use standard cutoff functions. We then sketch how the difficulty can be overcome using a weighted Sobolev inequality derived from CS(Ψ). This proves that VD, PI(Ψ) and CS(Ψ) imply EHI; PHI(Ψ) then follows easily using the scaling information in PI(Ψ). The stability of CS(Ψ) under rough isometries is proved in Section 5.

We write c_i to denote positive constants which are constant within each argument, and $f \asymp g$ to mean there exist positive constants c_i such that $c_1 f \le g \le c_2 f$.

2. Stability theorems

Let M be a complete smooth non-compact Riemannian manifold, $\Delta = \mathrm{div}\nabla$ be the Laplace-Beltrami operator on M, d be the Riemannian metric and μ be the volume. We write $B(x,r)$ for open balls in M, and set $V(x,r) = \mu(B(x,r))$. We define the Dirichlet form $\mathcal{E}(f,f)$ on $L^2(M)$ with core $C_0^\infty(M)$ by taking

$$\mathcal{E}(f,f) = \int_M |\nabla f|^2 d\mu, \qquad f \in C_0^\infty(M).$$

(See [**21**] for more details.) Note (see [**54**]) that we have

$$d(x,y) = \sup\{|f(x) - f(y)| : f \in C_0^\infty(M), \, |\nabla f| \le 1\}.$$

We write $W = (W_t, t \ge 0, \mathbb{P}^x, x \in M)$ for the Brownian motion on M.

We now give a number of conditions M may or may not satisfy. Let $\beta \ge 2$ and define Ψ as in (1.7). Recall from Section 1 the definitions of volume doubling VD and EHI. It is easy to deduce from VD that there exists $\alpha < \infty$ such that if $x, y \in M$ and $0 < r < R$ then

$$\frac{V(x,R)}{V(y,r)} \le c_1 \Big(\frac{d(x,y) + R}{r} \Big)^\alpha. \tag{2.1}$$

Definition 2.1. M satisfies the *Poincaré inequality* PI(Ψ), if there exists a constant c_1 such that for any ball $B = B(x,R) \subset M$ and $f \in C^\infty(B)$

$$\int_B (f(x) - \overline{f}_B)^2 d\mu(x) \le c_1 \Psi(R) \int_B |\nabla f|^2 d\mu. \tag{2.2}$$

Here $\overline{f}_B = \mu(B)^{-1} \int_B f(x) d\mu(x)$.

Definition 2.2. M satisfies the *parabolic Harnack inequality* (PHI(Ψ)), if there exists a constant c_P such that the following holds. Let $x_0 \in M$, $R > 0$, $T = \Psi(R)$ and $u = u(t,x)$ be a non-negative solution of the heat equation $\partial_t u = \Delta u$ in $(0,4T) \times B(x_0, 2R)$. Let $Q_- = (T,2T) \times B(x_0,R)$ and $Q_+ = (3T,4T) \times B(x_0,r)$; then

$$\sup_{Q_-} u \le c_P \inf_{Q_+} u. \tag{2.3}$$

Definition 2.3. Let A, B be disjoint subsets of M. We define the effective resistance $R_{\text{eff}}(A,B)$ by

$$R_{\text{eff}}(A,B)^{-1} = \inf\Big\{ \int_M |\nabla f|^2 : f = 0 \text{ on } A \text{ and } f = 1 \text{ on B}, \, f \in C^\infty(M) \Big\}.$$

M satisfies the condition RES(Ψ) if for any $x \in M$, $R \ge 0$,

$$R_{\text{eff}}(B(x,R), B(x,2R)^c) \asymp \frac{\Psi(R)}{V(x,R)}. \tag{2.4}$$

We have by the same arguments as in Lemma 5.1 of [**9**]

Lemma 2.4. *Let M satisfy VD, EHI, and PI(Ψ). Then M satisfies RES(Ψ).*

Let $p_t(x,y)$ be the fundamental solution of the heat equation on M. Recall that we have a 'crossover' from the classical $t = r^2$ scaling when $r < 1$ to the anomalous scaling $t = r^\beta$ for $r \ge 1$. For $(t,r) \in (0,\infty) \times [0,\infty)$ we define the regions:

$$\Lambda_1 = \{(t,r) : t \le 1 \vee r\}, \qquad \Lambda_2 = \{(t,r) : t \ge 1 \vee r\}. \tag{2.5}$$

Let $\theta \ge 2$ and

$$h_\theta(r,t) = \exp(-(r^\theta/t)^{1/(\theta-1)}).$$

DEFINITION 2.5. M satisfies the heat kernel bounds HK(Ψ), if, writing $r = d(x, y)$,

$$c_1 V(x, t^{1/2})^{-1} h_2(c_2 r, t) \le p_t(x, y) \le c_3 V(x, t^{1/2})^{-1} h_2(c_4 r, t) \tag{2.6}$$

for $x, y \in M$ and $t \in (0, \infty)$ with $(t, r) \in \Lambda_1$, and

$$c_1 V(x, t^{1/\beta})^{-1} h_\beta(c_2 r, t) \le p_t(x, y) \le c_3 V(x, t^{1/\beta})^{-1} h_\beta(c_4 r, t) \tag{2.7}$$

for $x, y \in M$ and $t \in (0, \infty)$ with $(t, r) \in \Lambda_2$.

REMARKS 2.6.
1. In [**6**] it was proved that the pre-Sierpinski carpet satisfies HK(Ψ). We will see below that this also holds on sufficiently regular pre-fractal manifolds.
2. If $r > t$ then $h_\beta(r, t) \ge h_2(r, t)$.
3. To understand why the crossover in HK(Ψ) takes the form it does, it is useful to consider the contribution to $p_t(x, y)$ from various types of path in M. First, if $0 < t \le 1$ and $d(x, y) < 1$ then the behaviour is essentially local, and the locally Euclidean structure of M gives the usual type of bound (2.6).

If $r > t$ then we are in the 'large deviations' regime: the the main contribution to $p_t(x, y)$ is from those paths of the Brownian motion W which are within a distance $O(t/r)$ of a geodesic from x to y. So, once the length of the geodesic is given, only the local structure of M plays a role. Note that in this case the term in the exponential is smaller than e^{-ct}, so that the volume term $V(x, t^{1/2})^{-1}$ could be absorbed into the exponential with a suitable modification of the constants c_2 and c_4.

Finally, if $t > 1$ and $r < t$, then the paths which contribute to $p_t(x, y)$ fill out a much larger part of M: those which lie in $B(x, t^{1/\beta})$ if $r < t^{1/\beta}$, and those which are within a distance $O(t/r^{\beta-1})$ of a geodesic from x to y in the case when $t^{1/\beta} \le r \le t$.

We will also want to discuss local versions of these conditions. We say M satisfies $\mathrm{VD}_{\mathrm{loc}}$ if (1.2) holds for $x \in M$, $0 < R \le 1$. Similarly we define $\mathrm{PI}(\Psi)_{\mathrm{loc}}$, $\mathrm{EHI}_{\mathrm{loc}}$ and $\mathrm{PHI}(\Psi)_{\mathrm{loc}}$ by requiring the conditions only for $0 < R \le 1$. For $\mathrm{HK}_{\mathrm{loc}}$ we require the bounds only for $t \in (0, 1)$ – so only (2.6) is involved. Note that these local conditions are all independent of the the parameter β. The value 1 here is just for simplicity: each of the local conditions implies an analogous local condition for $0 < R \le R_0$ for any (fixed) $R_0 > 1$ – see Section 2 of [**32**].

The following result of Hebisch and Saloff-Coste localizes Theorem 1.2.

THEOREM 2.7. *([**32**], Theorem 2.7). The following are equivalent:*
(a) M satisfies VD_{loc} and $PI(\Psi)_{loc}$.
(b) M satisfies $PHI(\Psi)_{loc}$.
(c) M satisfies $HK(\Psi)_{loc}$.
If any of (a)–(c) hold then M satisfies EHI_{loc}, and $p_t(x, y)$ is continuous on $(0, \infty) \times M \times M$.

In addition we will need the following:

THEOREM 2.8. *(See [**32**], Theorem 5.3, [**26**]). The following are equivalent:*
(a) M satisfies PHI(Ψ).
(b) M satisfies HK(Ψ).
(c) M satisfies VD, EHI and RES(Ψ).

PROOF. The equivalence of (a) and (b) is given in [**32**]; and that these are equivalent to (c) is proved in [**26**]. (See [**25**] for the graph case.) □

We now introduce the condition CS(Ψ).

DEFINITION 2.9. *M* satisfies CS(Ψ) for $\beta \geq 2$ if there exists $\theta \in (0,1]$ and constants c_1, c_2 such that the following holds. For every $x_0 \in M$, $R > 0$ there exists a cutoff function $\varphi (= \varphi_{x_0,R})$ with the following properties:
(a) $\varphi(x) \geq 1$ for $x \in B(x_0, R/2)$.
(b) $\varphi(x) = 0$ for $x \in B(x_0, R)^c$.
(c) $|\varphi(x) - \varphi(y)| \leq c_1 (d(x,y)/R)^\theta$ for all x,y.
(d) For any ball $B(x,s)$ with $0 < s \leq R$ and $f : B(x,2s) \to \mathbb{R}$

$$\int_{B(x,s)} f^2 |\nabla \varphi|^2 d\mu \leq c_2 (s/R)^{2\theta} \Big(\int_{B(x,2s)} |\nabla f|^2 d\mu + \Psi(s)^{-1} \int_{B(x,2s)} f^2 d\mu \Big). \tag{2.8}$$

REMARKS 2.10.
1. We call (2.8) a weighted Sobolev inequality. It is clear that to prove (2.8) it is enough to consider nonnegative f.
2. Suppose CS(Ψ) holds for M, but with (a) above replaced by

$$\varphi(x) \geq 1 \text{ for } x \in B(x_0, \delta R), \tag{2.9}$$

for some $\delta < \frac{1}{2}$. Then an easy covering argument (using VD) gives CS(Ψ) with $\delta = \frac{1}{2}$.
3. Let $\lambda > 1$. Suppose that CS(Ψ) holds, except that instead of (2.8) we have

$$\int_{B(x,s)} f^2 |\nabla \varphi|^2 d\mu \leq c_2 (s/R)^{2\theta} \Big(\int_{B(x,\lambda s)} |\nabla f|^2 d\mu + \Psi(s)^{-1} \int_{B(x,\lambda s)} f^2 d\mu \Big). \tag{2.10}$$

Then once again it is easy to obtain CS(Ψ) with $\lambda = 2$ by a covering argument.
4. Any operation on φ which reduces $|\nabla \varphi|$ while keeping properties (a), (b) and (c) of Definition 2.9 will generate a new cutoff function which still satisfies (2.8). We can therefore assume that any cutoff function φ satisfies the following:
(a) $0 \leq \varphi \leq 1$.
(b) For each $t \in (0,1)$ the set $\{x : \varphi(x) > t\}$ is connected and contains $B(x_0, R/2)$.
(c) Each connected component A of $\{x : \varphi(x) < t\}$ intersects $B(x_0, R)^c$.
5. CS(2) always holds since one can take $\varphi(x) = (2/R) d(x, B(x_0, R)^c)$. Then $|\nabla \varphi| \leq 2/R$ and (2.8) (with $\theta = 1$) follows easily.
6. Let $2 \leq \beta \leq \beta'$, and write Ψ, Ψ' for the functions given by (1.7). Then PI(Ψ) implies PI(Ψ'), while CS(Ψ') implies CS(Ψ). Further (see Lemma 2.14 below) if PI(Ψ) holds, then CS(Ψ') cannot hold for any $\beta' > \beta$.
7. If M_i, $i = 1, 2$, are manifolds satisfying PHI(Ψ_i), respectively, with $\beta_1 < \beta_2$, then the product $M = M_1 \times M_2$ satisfies PI(Ψ_2). However, since M does not satisfy PHI(Ψ_2) it cannot satisfy CS(Ψ_2). Thus the conditions PI(Ψ) and CS(Ψ) are independent.

The following theorem gives a characterization of PHI(Ψ) in terms of conditions which have good stability properties.

THEOREM 2.11. *The following are equivalent:*
(a) M satisfies VD, PI(Ψ) and CS(Ψ).
(b) M satisfies PHI(Ψ).

This was proved in the graph case in [**9**]. The extension to manifolds is sketched below in Sections 3 and 4. Full details, and the extension to a general class of measure metric spaces will be given in [**10**].

We will see in Section 5 that the conditions VD, PI(Ψ) and CS(Ψ) are stable under rough isometries between manifolds and graphs, provided these have the appropriate local regularity.

THEOREM 2.12. *Let G be a graph with bounded vertex degree. Let M be a manifold with Ricci curvature bounded below, and positive injectivity radius, which is roughly isometric to G. The following are equivalent:*
(a) G satisfies PHI(Ψ).
(b) M satisfies PHI(Ψ).

See [**40**] for the definition of injectivity radius.

EXAMPLES 2.13. Given a graph G with bounded vertex degree one can create a manifold M satisfying the conditions of Theorem 2.12 by replacing the edges of G by tubes of length 1, (and diameter say 1/10) and gluing these together smoothly at the vertices.

Using Theorem 2.12 with Theorems 2.8 and 2.11 we see that the pre-Sierpinksi gasket manifolds defined in [**45**] satisfy HK(Ψ) with $\beta = \log 5/\log 2$. We also deduce that the manifolds based on the family of Vicsek fractals studied in [**11**] satisfy HK(Ψ), where $\Psi(r) = r^2 \vee r^\beta$, and β is the 'walk dimension' of the associated graph.

We conclude this section by discussing the relation of PI(Ψ) and CS(Ψ) with the spectral gap of balls. Let $B = B(x,R)$ be an open ball in M. Let $\mathcal{M}(B) = \{f \in C^\infty(B) : \int_B f = 0, ||f1_B||_2 \neq 0\}$; then the smallest non-zero Neumann eigenvalue of $-\Delta$ on B is given by

$$\lambda_1(B) = \inf_{f \in \mathcal{M}(B)} \frac{\int_B |\nabla f|^2}{\int_B |f|^2}. \tag{2.11}$$

LEMMA 2.14.
(a) If M satisfies PI(Ψ) then

$$\lambda_1(B(x,R)) \geq c\Psi(R)^{-1}, \qquad x \in M,\ R > 0.$$

(a) If M satisfies VD and CS(Ψ) then

$$\lambda_1(B(x,R)) \leq c\Psi(R)^{-1}, \qquad x \in M,\ R > 0.$$

PROOF. (a) is immediate from the definition of PI(Ψ) and (2.11).
(b) Let γ be a geodesic from x_0 to $y \in \partial B$ and $x_1 \in \gamma$ with $d(x_0, x_1) = 2R/3$. For $i = 0, 1$ let $B_i = B(x_i, R/8)$ and $B_i^* = B(x_i, R/4)$. Using VD we have

$$c_1 \mu(B_1^*) \leq \mu(B_2) \leq \mu(B_2^*) \leq c_2 \mu(B_1).$$

By CS(Ψ) there exist cutoff functions φ_i for $B_i \subset B_i^*$. Let $g = \varphi_0 - \varphi_1 - c_3$, where c_3 is chosen so that $\int_B g = 0$. It is easy to check that

$$\int_B g^2 d\mu \geq c\mu(B).$$

By (2.8) applied to the constant function 1

$$\int_{B_i^*} |\nabla \varphi_i|^2 d\mu \le c_2 \Psi(R/4)^{-1} \int_{B(x_i,R/2)} d\mu.$$

Hence

$$\int_B |\nabla g|^2 \le c\Psi(R)^{-1}\mu(B) \le c'\Psi(R)^{-1} \int_B g^2,$$

so that $\lambda_1(B) \le c'\Psi(R)^{-1}$. □

3. Construction of cutoff functions

Throughout this section we assume that M satisfies HK(Ψ) (and hence VD and PHI(Ψ)). We will sketch the proof that M satisfies CS(Ψ); the argument, which runs along the same lines as that in [**9**], uses the Greens functions $g_\lambda(x,y)$, given by

$$g_\lambda(x,y) = \int_0^\infty e^{-\lambda t} p_t(x,y)dt,$$

to build a cutoff function φ. The main difference here is that as [**9**] used $g(x,y)$ rather than $g_\lambda(x,y)$, a strong transience condition (called (FVG)) was needed in the initial arguments. (This was then removed using a standard trick with product spaces.)

LEMMA 3.1. *Let $x_0 \in M$, $R \ge 1$. Then there exists $\delta > 0$ such that if $\lambda = cR^{-\beta}$*

$$g_\lambda(x_0,y) \le c_2 \frac{R^\beta}{V(x_0,R)}, \qquad y \in B(x_0,R)^c, \tag{3.1}$$

$$g_\lambda(x_0,y) \ge 2c_2 \frac{R^\beta}{V(x_0,R)}, \qquad y \in B(x_0,2\delta R). \tag{3.2}$$

PROOF. This follows easily from HK(Ψ) by integration. □

LEMMA 3.2. *Let $x_0 \notin B(x_1,r)$. Then there exists $\theta > 0$ such that*

$$|g_\lambda(x_0,x) - g_\lambda(x_0,y)| \le (\frac{d(x,y)}{r})^\theta \sup_{B(x_1,r)} g_\lambda(x_0,.). \tag{3.3}$$

PROOF. The Hölder continuity of p_t is given by PHI(Ψ). Integrating we obtain (3.3). □

We begin the proof of CS(Ψ) with the special case when we only require the weighted Sobolev inequality for $I = B$.

PROPOSITION 3.3. *Let $x_1 \in M$, $r > 0$. There exists $\delta > 0$ and a cutoff function φ for $B(x_1,\delta r) \subset B(x_1,r)$ such that, writing $B' = B(x_1,\delta r)$, $B = B(x_1,r)$,*

$$\int_{B'} f^2 |\nabla \varphi|^2 d\mu \le c \int_B |\nabla f|^2 d\mu + c\Psi(r)^{-1} \int_{B'} f^2. \tag{3.4}$$

PROOF. If $r \le c$ then we can take φ to be a local cutoff function for $B' \subset B$, so suppose $r > c$.

Let $D = B(x_1, r-\varepsilon)$ where $\varepsilon < r/10$, and $\lambda > 0$. Let G_λ^D be the resolvent associated with the process W killed on exiting D, that is,

$$G_\lambda^D f(x) = \mathbb{E}^x \int_0^{\tau_D} e^{-\lambda t} f(W_t)dt,$$

for bounded measurable f, where

$$\tau_D = \inf\{t : W_t \in M - D\}.$$

Let $h = 1_{B'}$, and let $v = G_\lambda^D h$. Using the estimates on the heat kernel of W, as in Lemma 4.7 of [**9**], we can take λ small enough so that

$$v(x) \geq c_1 \Psi(r), \qquad x \in B',$$

and

$$v(x) \leq c_2 \Psi(r), \qquad x \in D.$$

Let $f \in C_0^\infty(M)$. Then

$$\int_B f^2 |\nabla v|^2 d\mu \leq \int_M f^2 |\nabla v|^2 d\mu = \int_M (\nabla(f^2 v).\nabla v) d\mu - \int_M 2fv\nabla f.\nabla v d\mu.$$

By Gauss-Green

$$\int_M (\nabla(f^2 v).\nabla v) d\mu = -\int_M f^2 v \Delta v d\mu \leq c\Psi(r) \int_{B'} f^2 d\mu.$$

Using Cauchy-Schwarz we obtain

$$|\int_M 2fv(\nabla f.\nabla v) d\mu| \leq c\Big(\int_M v^2 |\nabla f|^2 d\mu\Big)^{1/2} \Big(\int_M f^2 |\nabla v|^2 d\mu\Big)^{1/2}$$
$$\leq c\Psi(r) \Big(\int_B |\nabla f|^2 d\mu\Big)^{1/2} \Big(\int_M f^2 |\nabla v|^2 d\mu\Big)^{1/2}.$$

So, writing $I = \int_M f^2 |\nabla v|^2 d\mu$, $J = \int_B |\nabla f|^2 d\mu$, $K = \int_B f^2 d\mu$, we have

$$I \leq c\Psi(r) K + c\Psi(r) J^{1/2} I^{1/2},$$

from which it follows that $I \leq c\Psi(r)K + c\Psi(r)^2 J$. Setting $\varphi = 1 \wedge (c\Psi(r)^{-1} v)$, where c is chosen so that φ is a cutoff function for $B' \subset B$, (3.4) follows. □

Now fix $x_0 \in M$ and $R \geq 1$. Let $\lambda = cR^{-\beta}$, δ, c_2 be as in Lemma 3.1, and

$$h = c_2 R^\beta / V(x_0, R).$$

We now define

$$Q(b) = Q(x_0, b) = \{y : g_\lambda(x_0, y) > b\}.$$

As in [**9**] we can approximate $Q(b)$ by balls. Let ρ be an approximate identity with support $B(x_0, \delta^2 R)$, and

$$\omega_0(x) = G_\lambda \rho(x), \qquad \omega(x) = (2h \wedge \omega(x) - h)^+.$$

Thus $h^{-1}\omega$ is a cutoff function for $B = B(x_0, \delta R) \subset B(x_0, R)$.

PROPOSITION 3.4. *Let $x_0 \in M$, $R > 0$, ω be as above. Let $I = B(x_1, \delta s)$, with $s \leq R$, and $I^* = B(x_1, s)$. Suppose that either*

$$I^* \subset Q(2h) \tag{3.5}$$

or

$$I^* \cap B(x_0, \delta R) = \emptyset. \tag{3.6}$$

Let $f \in C^\infty(M)$. There exists $c_1 < \infty$ such that

$$h^{-2} \int_I f^2 |\nabla \omega|^2 \leq c_1 (s/R)^{2\theta} \Big(\int_{I^*} |\nabla f|^2 d\mu + c\Psi(r)^{-1} \int_I f^2\Big). \tag{3.7}$$

PROOF. The argument follows the lines of [**9**], Proposition 4.10. As ω is constant on $Q(2h)$ it is enough to consider the case when (3.6) holds. Let v be a cutoff function for $I \subset I^*$ given by Proposition 3.3. Let $\omega_1(x) = \omega_0(x) - c_3$, where c_3 is chosen so that $\omega_1 \geq 0$ on I^* and

$$\omega_1(x) \leq ch(s/R)^\theta = L, \qquad x \in I^*;$$

this is possible by Lemma 3.2. Let

$$\begin{aligned} A &= \int_I f^2 |\nabla \omega|^2 d\mu, \\ D &= \int_{I^*} |\nabla f|^2 d\mu + \Psi(r)^{-1} \int_I f^2 d\mu, \\ F &= \int_{I^*} f^2 v^2 |\nabla \omega_1|^2. \end{aligned}$$

Now

$$A \leq F = \int_{I^*} f^2 v^2 \nabla \omega_1 . \nabla \omega_0 = \int_{I^*} \nabla (f^2 v^2 \omega_1) . \nabla \omega_0 - \int_{I^*} \omega_1 \nabla (f^2 v^2) . \nabla \omega_0. \tag{3.8}$$

For the first term in (3.8), by Gauss-Green

$$\begin{aligned} \int_{I^*} \nabla (f^2 v^2 \omega_1) . \nabla \omega_0 &= - \int_M (f^2 v^2 \omega_1) \Delta \omega_0 \\ &= - \int_M (f^2 v^2 \omega_1)(\lambda \omega_0 - \rho) \\ &\leq \int_M (f^2 v^2 \omega_1) \rho = 0. \end{aligned}$$

Here we used the fact that $\omega_1 \geq 0$ on I^*, that v has support I^*, and that v and ρ have disjoint supports. The second term in (3.8) is handled exactly as in [**8**] and [**9**]. That is, using Cauchy-Schwarz,

$$\begin{aligned} &| \int_{I^*} \omega_1 \nabla (f^2 v^2) . \nabla \omega_0 | \\ &\quad \leq c \Big(\big(\int_{I^*} v^2 |\nabla f|^2 d\mu \big)^{1/2} + \big(\int_{I^*} f^2 |\nabla v|^2 d\mu \big)^{1/2} \Big) \Big(\int_{I^*} \omega_1^2 f^2 v^2 |\nabla \omega_0|^2 d\mu \Big)^{1/2} \\ &\quad \leq c L D^{1/2} F^{1/2}, \end{aligned}$$

where we used Proposition 3.3 in the final line. Thus we obtain $F \leq cL^2 D$. □

PROPOSITION 3.5. *Let M satisfy PHI(Ψ). Then M satisfies VD, PI(Ψ) and CS(Ψ).*

PROOF. The arguments that M satisfies VD and PI(Ψ) are standard. CS(Ψ) follows from Proposition 3.4 and an easy covering argument just as in Corollary 4.11 of [**9**]. □

4. Proof of Harnack inequalities

We begin by explaining the necessity of CS(Ψ) in the anomalous diffusion case. Let M be a manifold satisfying VD, PI(Ψ) and having regular volume growth

$$V(x,r) \asymp r^\alpha, \qquad x \in M, \ r \geq 1.$$

Since M need not satisfy CS(Ψ), any argument to prove EHI must fail, but it is still instructive to see where the problem arises. Let us try to follow Moser's proof of the EHI in [**49**]; a similar obstacle would arise if one tried other approaches, such as that in [**20**]. The difficulty is with the first ('easy') part of Moser's argument.

Write

$$\fint_B f = \mu(B)^{-1} \int_B f d\mu.$$

From PI(Ψ) one obtains (see [**54**], [**55**], Section 5.2) the Sobolev inequality

$$\left(\fint_B |f|^{2\kappa}\right)^{1/\kappa} \le cR^\beta \fint_B |\nabla f|^2, \tag{4.1}$$

for $f \in C_0^\infty(B)$, where B has radius $R \ge 1$ and $\kappa > 1$.

Let $u > 0$ be harmonic in $B = B(x,R)$; as M is locally regular we have that u is continuous. Let $\frac{1}{2} < a_2 < a_1 < 1$, $B_i = B(x, a_i R)$, $p > 2$, $\psi \in C_0^\infty(B_1)$ be a cutoff function for $B_2 \subset B_1$ and $v = u^p$. Moser's argument (see [**55**], p. 121) gives

$$\int_{B_1} |\nabla(\psi v)|^2 \le c||\nabla\psi||_\infty^2 \int_{B_1} v^2. \tag{4.2}$$

By (4.1) applied to $f = v\psi = u^p\psi$

$$\left(\fint_{B_2} u^{2\kappa p}\right)^{1/\kappa} \le \left(\fint_{B_1} (v\psi)^{2\kappa}\right)^{1/\kappa} \le cR^\beta \fint_{B_1} |\nabla(v\psi)|^2. \tag{4.3}$$

Since we can find ψ such that $||\nabla\psi||_\infty \le 2R^{-2}(a_1 - a_2)^{-1}$ it follows that

$$\left(\fint_{B_2} u^{2\kappa p}\right)^{1/\kappa} \le cR^{\beta-2}(a_1 - a_2)^{-2} \fint_{B_1} u^{2p}. \tag{4.4}$$

Now let $a_k = \frac{1}{2}(1 + 2^{-k})$, $p_k = p_0\kappa^k$ where $p_0 > 2$, and $B_k = B(x, a_k R)$. Then, if

$$I_k = \left(\fint_{B_k} u^{2p_k}\right)^{1/2p_k},$$

(4.4) implies that

$$I_{k+1} \le (cR^{\beta-2}2^{2k})^{1/2p_k} I_k, \quad k \ge 0. \tag{4.5}$$

If $\beta = 2$ this leads, by iteration as in [**49**], to the bound

$$u(y) \le C \fint_B u^{2p_0}, \qquad y \in B(x, \tfrac{1}{2}R). \tag{4.6}$$

If $\beta > 2$ one still obtains an L^∞ bound on u in $B(x, \frac{1}{2}R)$, but the constant C now depends on R, so that the final constant in the EHI will also depend on R.

Inspecting the argument above, the crucial loss is in using the bound (4.2) to go from (4.3) to (4.4); one needs a cutoff function ψ such that the final term in (4.3) can be controlled by a term of order $R^{-\beta}$.

We shall now see how CS(Ψ) enables one to do this. As the arguments in Section 5 of [**9**] can be repeated in this more general context with minor changes, we only sketch the main ideas. Full details will be given in [**10**].

Fix $x_0 \in M$, let $R \ge 0$, and φ be a cutoff function for $B(x_0, R/2) \subset B(x_0, R)$ given by CS(Ψ). We regularize φ so that it satisfies conditions (a)–(c) of Remark 2.10.4. We define the measure γ by

$$d\gamma = d\mu + R^\beta |\nabla\varphi|^2 d\mu.$$

We do not know if this measure satisfies volume doubling, but using CS(Ψ) we do have $\gamma(B(x_0,R)) \le cV(x_0,R)$.

This first step is to use CS(Ψ) (and PI(Ψ)) to obtain the following weighted Sobolev inequality, which will replace (4.1) in the iteration argument. We write $J^{(s)} = \{x : d(x,J) \le s\}$.

PROPOSITION 4.1. *(See Theorem 5.4 of* [**9**].*) Let $s \le R$ and $J \subset B(x_0,R)$ be a finite union of balls of radius s. There exist $\kappa > 1$ and $c_1 < \infty$ such that*

$$\Big(\mu(J)^{-1}\int_J |f|^{2\kappa} d\gamma\Big)^{1/\kappa} \le c_1\Big(R^\beta \mu(J)^{-1}\int_{J^s} |\nabla f|^2 d\mu + (s/R)^{-2\theta}\mu(J)^{-1}\int_J f^2 d\gamma\Big). \tag{4.7}$$

Now let u be harmonic in a domain $D \subset X$. By Theorem 2.7 u satisfies a local Harnack inequality, so is Hölder continuous. As in [**49**] we have

LEMMA 4.2. *Let D be a domain in M, let u be positive and harmonic in D, $v = u^k$, where $k \in \mathbb{R}$, $k \ne \frac{1}{2}$, and let $\eta \in C_0^\infty(D)$. Then*

$$\int_D \eta^2 |\nabla v|^2 d\mu \le c_1\Big(\frac{2k}{2k-1}\Big)^2 \int_D v^2 |\nabla \eta|^2 d\mu.$$

Now let $u > 0$ be harmonic in $B(x_0,R)$, and γ be as above.

PROPOSITION 4.3. *Let v be either u or u^{-1}. There exists c_1, $\delta > 0$ such that if $0 < q < 2$, then*

$$\sup_{B(x,\delta R)} v^{2q} \le c_1 V(x,R)^{-1}\int_{B(x,R)} (R^\beta |\nabla v^q|^2 + v^{2q}) d\mu. \tag{4.8}$$

PROOF. If $R < 1$ this follows from the local Harnack inequality, so suppose $R \ge c$. Let φ, γ be as above. Let $h_n = 1 - 2^{-n}$, $0 \le n \le \infty$, so that $0 = h_0 < h_\infty = 1$. For $k \ge 0$ set

$$\varphi_k(x) = (\varphi(x) - h_k)^+, \qquad A_k = \{x : \varphi(x) > h_k\},$$

and note that $B(x,R/2) \subset A_n \subset A_0 \subset B(x,R)$ for every $n \ge 0$. We therefore have, writing $V = V(x,R)$,

$$c_2 V \le \mu(A_k) \le V, \quad k \ge 0.$$

The Hölder condition on φ given by CS(Ψ) implies that if $x \in A_{k+1}$ and $y \in A_k^c$, then $d(x,y) \ge c_3 2^{-k/\theta} R$. Set $s_k = \frac{1}{2} c_3 2^{-k/\theta} R$, and note that $\varphi_k > c_4 2^{-k}$ on $A_{k+1}^{(s_k)}$. Let $\{B_i\}$ be a cover of A_{k+1} by balls of radius $s_k/2$, and let $J_{k+1} = \cup B_i$. Write $J'_{k+1} = J_{k+1}^{(s_k/2)}$ $A'_{k+1} = A_{k+1}^{(s_k)}$, and note that $A_{k+1} \subset J_{k+1} \subset J'_{k+1} \subset A'_{k+1} \subset A_k$.

From Proposition 4.1 with $f = v^p$ and s replaced by $s_k/2$,

$$\begin{aligned} \Big(V^{-1}\int_{A_{k+1}} f^{2\kappa} d\gamma\Big)^{1/\kappa} &\le \Big(V^{-1}\int_{J_{k+1}} f^{2\kappa} d\gamma\Big)^{1/\kappa} \\ &\le c_5 V^{-1}\Big[R^\beta \int_{J'_{k+1}} |\nabla f|^2 d\mu + (R/s_k)^{2\theta}\int_{J'_{k+1}} f^2 d\gamma\Big] \\ &\le c_6 V^{-1}\Big[R^\beta \int_{A'_{k+1}} |\nabla f|^2 d\mu + 2^{2k}\int_{A_k} f^2 d\gamma\Big]. \end{aligned} \tag{4.9}$$

The first term in (4.9) is controlled by Lemma 4.2:

$$\begin{aligned} R^\beta \int_{A'_{k+1}} |\nabla f|^2 &\le R^\beta (c_7 2^{-k})^{-2} \int_{A'_{k+1}} \varphi_k^2 |\nabla f|^2 \\ &\le c_8 2^{2k} R^\beta \int_{A_k} \varphi_k^2 |\nabla f|^2 \\ &\le c_9 2^{2k} R^\beta \Big(\frac{2p}{2p-1}\Big)^2 \int_{A_k} f^2 |\nabla \varphi_k|^2 \\ &\le c_{10} 2^{2k} \Big(\frac{2p}{2p-1}\Big)^2 \int_{A_k} f^2 d\gamma. \end{aligned}$$

We therefore deduce that

$$\Big(V^{-1}\int_{A_{k+1}} f^{2\kappa} d\gamma\Big)^{1/\kappa} \le c_{11}\Big(\frac{2p}{2p-1}\Big)^2 2^{2k} V^{-1} \int_{A_k} f^2 d\gamma. \tag{4.10}$$

Given this, the usual iteration argument, with $p_n = q_0 \kappa^n$ for an appropriate q_0, leads to

$$\sup_{B(x,r/2)} v \le c\Big(V^{-1}\int_{B(x,r)} v^{2q_0} d\gamma\Big)^{1/(2q_0)}. \tag{4.11}$$

(4.8) now follows using Hölder's inequality and CS(Ψ) – see [**9**] for details. □

While the right hand side of (4.8) is a little different from that in (4.6), one can still use the ideas of [**13**] and [**51**] to complete the proof of the EHI – see [**9**].

THEOREM 4.4. *Let M satisfy VD, PI(Ψ) and CS(Ψ). Then M satisfies EHI.*

It is possible that the same arguments could also be used to prove PHI(Ψ) directly. But, in view of Theorem 2.8, and Lemma 2.4 the EHI is enough: VD plus EHI plus PI(Ψ) implies RES(Ψ), and hence M satisfies PHI(Ψ).

5. Stability under rough isometries.

We will need to consider two types of space: weighted graphs, and manifolds.

DEFINITION 5.1. Let (G, E) be an infinite locally finite connected graph. Define edge weights (conductances) $\nu_{xy} = \nu_{yx} \ge 0$, $x, y \in G$, and assume that ν is adapted to the graph structure by requiring that $\nu_{xy} > 0$ if and only if $x \sim y$. Let $\nu_x = \sum_y \nu_{xy}$, and extend to a measure ν on G. We call (G, ν) a *weighted graph.*

We write $d(x, y)$ for the graph distance, and define the balls

$$B(x, r) = \{y : d(x, y) < r\}.$$

Given $A \subset G$ write $\partial A = \{y \in A^c : d(x, y) = 1 \text{ for some } x \in A\}$ for the exterior boundary of A, and let $\overline{A} = A \cup \partial A$.

DEFINITION 5.2. A weighted graph (G, ν) has *controlled weights* if there exists $p_0 > 0$ such that for all $x \in G$

$$\frac{\nu_{xy}}{\nu_x} \ge p_0, \qquad x \sim y. \tag{5.1}$$

This was called the p_0-condition in [**25**].

The Laplacian is defined on (G, ν) by

$$\Delta f(x) = \frac{1}{\nu_x} \sum_y \nu_{xy}(f(y) - f(x)).$$

We also define a Dirichlet form $(\mathcal{E}, \mathcal{F})$ by taking $\mathcal{F} = L^2(G, \nu)$, and

$$\mathcal{E}(f,g) = \frac{1}{2} \sum_x \sum_y (f(x) - f(y))(g(x) - g(y))\nu_{xy}, \quad f, g \in \mathcal{F}.$$

If $f \in \mathcal{F}$ we define the measure $\Gamma(f, f)$ on G by setting

$$\Gamma(f,f)(x) = \sum_{y \sim x} (f(x) - f(y))^2 \nu_{xy}. \tag{5.2}$$

The conditions VD, EHI and PHI(Ψ) for graphs are defined in exactly the same way as for manifolds. The definitions of PI(Ψ) and RES(Ψ) are also the same if we replace $|\nabla f|^2 d\mu$ by $d\Gamma(f, f)$ in (2.2) and (2.4). For the bound HK(Ψ) we only require (2.7). The condition CS(Ψ) is also the same; the weighted Sobolev inequality (2.8) takes the form

$$\begin{aligned} &\sum_{x \in B(x_1,s)} f(x)^2 \Gamma(\varphi, \varphi)(x) \\ &\qquad \le c_2 \Big(\frac{s}{R}\Big)^{2\theta} \Big(\sum_{x \in B(x_1,2s)} \Gamma(f,f)(x) + s^{-\beta} \sum_{x \in B(x_1,2s)} \nu_x f(x)^2 \Big). \end{aligned} \tag{5.3}$$

DEFINITION 5.3. Let (X_i, d_i, μ_i), $i = 1, 2$ be complete measure metric spaces; that is each (X_i, d_i) is a complete metric space and μ_i is a measure such that $\mu_i(B) \in (0, \infty)$ for each ball B in X_i. A map $\varphi : X_1 \to X_2$ is a *rough isometry* if there exist constants $C_1 - C_4$ such that

$$C_1^{-1}(d_1(x,y) - C_2) \le d_2(\varphi(x), \varphi(y)) \le C_1(d_1(x,y) + C_2), \tag{5.4}$$

$$\bigcup_{x \in X_1} B_{d_2}(\varphi(x), C_2) = X_2, \tag{5.5}$$

$$C_3^{-1}\mu_1(B_1(x, C_2)) \le \mu_2(B_{d_2}(\varphi(x), C_2)) \le C_3 \mu_1(B_1(x, C_2)). \tag{5.6}$$

If there exists a rough isometry between two spaces they are said to be *roughly isometric*. (One can check this is an equivalence relation.)

This concept was introduced (for manifolds) by Kanai in [**39**]–[**40**], but without the condition (5.6). In those papers both manifolds were assumed to have Ricci curvature bounded below and positive injectivity radius; this leads to volume bounds which imply (5.6) – see p. 394 of [**40**].

A rough isometry between X_1 and X_2 means that the global structure of the two spaces is the same. For example, it is easy to prove that VD is stable under rough isometries. However, to have stability of Harnack inequalities, we also require some control over the local structure. In the case of graphs it is enough to have controlled weights, but for general measure metric spaces more regularity is needed.

Our main stability result is the following.

THEOREM 5.4. *For $i = 1, 2$ let X_i be either a manifold satisfying VD_{loc} and PI_{loc} or a graph with controlled weights. Suppose there exists a rough isometry $\varphi : X_1 \to X_2$.*
(a) If X_1 satisfies VD and PI(Ψ) then X_2 satisfies VD and PI(Ψ).
(b) If X_1 satisfies VD and CS(Ψ) then X_2 satisfies VD and CS(Ψ).

This result is proved in the graph case in [**29**]. Since rough isometry is an equivalence relation, to prove Theorem 5.4 it is enough to prove that if M is a manifold (satisfying $\mathrm{VD}_{\mathrm{loc}}$ and $\mathrm{PI}_{\mathrm{loc}}$) and G is a graph constructed by taking an appropriate net of M, (so that M and G are roughly isometric), then CS(Ψ) (resp. PI(Ψ)) holds for M if and only if it holds for G.

We will only prove (b) here. But note that since balls in the two metrics are deformations of each other, the initial argument for (a) only gives stability of weak Poincaré inequalities. An argument such as that in Jerison [**37**] (see also [**30**] for a more general formulation) is then needed to derive the (strong) Poincaré inequality from VD and the weak PI.

Let M be a manifold satisfying $\mathrm{VD}_{\mathrm{loc}}$. Let $G \subset M$ be a maximal set such that

$$d(x,y) \geq 1 \text{ for } x, y \in G, x \neq y.$$

Thus $B(x, \frac{1}{2})$, $x \in G$ are disjoint, and $\cup_{x\in G} B(x,1) = M$. Give G a graph structure by letting $x \sim y$ if $d(x,y) \leq 3$. Let d_G be the usual graph distance on G, and write $B_G(x,r)$ for balls in G. It is straightforward to check that G is connected, and that

$$\tfrac{1}{3}d(x,y) \leq d_G(x,y) \leq d(x,y) + 1, \quad x, y \in G. \tag{5.7}$$

Since M satisfies $\mathrm{VD}_{\mathrm{loc}}$ we have, as in Lemma 2.3 of [**39**], that the vertex degree in G is uniformly bounded.

For each $x \sim y$ in G let z_{xy} be the midpoint of a geodesic connecting x and y, and $A_{xy} = B(z_{xy}, 5/2)$, so that $B(x,1) \subset A_{xy} \subset B(x,4)$. Let $\nu_{xy} = 0$ if $x \not\sim y$, and if $x \sim y$ let

$$\nu_{xy} = \mu(A_{xy}).$$

As usual we set $\nu_x = \sum_{y\sim x} \nu_{xy}$. Write $A_x = \cup_{y\sim x} A_{xy}$. Since M satisfies $\mathrm{VD}_{\mathrm{loc}}$, we have

$$\mu(B(x,1)) \leq \nu_x \leq c\mu(A_x) \leq c\mu(B(x,4)) \leq c'\mu(B(x,1)), \tag{5.8}$$

and using (2.1) it is easy to verify that (G,ν) has controlled weights.

Define $\imath : G \to M$ by $\imath(x) = x$. We have

PROPOSITION 5.5. *Let M be a manifold satisfying VD_{loc}. Then the weighted graph (G,ν) has controlled weights and $\imath$ is a rough isometry.*

To prove Theorem 5.4 we will need to transfer functions between $C(G, \mathbb{R}_+)$ and $C(M, \mathbb{R}_+)$. Let $f \in C(M, \mathbb{R}_+)$. Define

$$\widehat{f}(x) = \mu(B(x,1))^{-1} \int_{B(x,1)} f d\mu, \quad x \in G. \tag{5.9}$$

The transfer in the other direction requires a bit more care. Using the fact that M satisfies $\mathrm{VD}_{\mathrm{loc}}$ we can find a partition of unity (ψ_x), $x \in G$, with the following properties:

(i) $\psi_x(z) = 1$ for $z \in B(x, \frac{1}{4})$,
(ii) $\psi_x(z) = 0$ for $z \in B(x, \frac{3}{2})^c$,

(iii) $\psi_x \in C^\infty(M)$ and $|\nabla \psi_x| \le C_1$ for each $x \in G$.

Now if $g : G \to \mathbb{R}_+$ define $\widetilde{g} \in C^\infty(M)$ by

$$\widetilde{g}(z) = \sum_{x \in G} g(x)\psi_x(z), \quad z \in M. \tag{5.10}$$

Set also, if $f : G \to \mathbb{R}$, $k \in \mathbb{N}$,

$$V_k f(x) = \sup_{z: d_G(z,x) \le k} |f(x) - f(z)|, \quad x \in G.$$

LEMMA 5.6. *Let M satisfy VD_{loc}. Let $f : G \to \mathbb{R}_+$, and $x \in G$.*
(a) If $\psi_z(w) > 0$ for some $w \in B(x_0, 1)$ then $d(x,z) < 3$ and $x \sim z$.
(b) If $\psi_z(w) > 0$ for some $w \in A_x$ then $d_G(x,z) \le 4$, and

$$|f(x) - \widetilde{f}(w)| \le V_4 f(x), \quad w \in A_x.$$

(c) Let $A \subset G$, and $A' = \{y : d_G(y, A) \le 4\}$. Then

$$\sum_{z \in A} V_4 f(z)^2 \nu_z \le c \sum_{y,z \in A'} (f(y) - f(z))^2 \nu_{yz}.$$

(d)

$$\int_{B(x,r)} \widetilde{f}(w)^2 d\mu(w) \le \sum_{y \in G \cap B(x,r+2)} f(y)^2 \nu_y.$$

(e) On $B(x,1)$, $|\nabla \widetilde{f}|^2 \le c_1 V_1 f(x)^2$.

PROOF. (a) If $w \in B(x,1)$ and $\psi_z(w) > 0$ then $d(x,z) < 1 + \frac{3}{2} < 3$, so $x \sim z$.
(b) Suppose $w \in A_{xy}$ and $\psi_z(w) > 0$. Then $w \in B(x', 1)$ for some $x' \in G$, and $d(x, x') < 5$. Then $x' \sim z$ and it is easy to check that $d_G(x, x') \le 3$. Since

$$\widetilde{f}(w) = f(x) + \sum_z (f(z) - f(x))\psi_z(w),$$

and only those z with $d(z,x) \le 4$ contribute to the sum, the second part is immediate.
(c) For $x \in G$ we can, by (b), find a path $y_i(x)$, $0 \le i \le k(x) \le 4$, such that $x = y_0(x)$, and $V_4 f(x) = |f(x) - f(y_{k(x)})(x)|$. Then

$$\sum_{x \in A} V_4 f(x)^2 \nu_x \le 4 \sum_{x \in A} \sum_{i=1}^{k(x)} (f(y_i(x)) - f(y_{i-1}(x)))^2 \nu_x,$$

and using the fact that (G, ν) has controlled weights, (c) now follows.
(d) Since $\widetilde{f}(w)^2 \le \sum_z f(z)^2 \psi_z(w)$,

$$\int_{B(x_0,r)} \widetilde{f}(w)^2 d\mu \le \sum_z f(z)^2 \int_B \psi_z(w) d\mu$$
$$\le c \sum_{z \in G \cap B(x_0, r+2)} f(z)^2 \nu_z.$$

(e) By (a) we can write, for $w \in B(x,1)$,

$$\widetilde{f}(w) = \widetilde{f}(x) + \sum_{z \sim x} \psi_z(w)(f(z) - f(x)).$$

Hence

$$\begin{aligned}|\nabla \widetilde{f}|^2 &= \sum_{z\sim x}\sum_{z'\sim x}(f(z)-f(x))(f(z')-f(x))(\nabla\psi_z.\nabla\psi_{z'})\\ &\le \sum_{z\sim x}(f(z)-f(x))^2|\nabla\psi_z|^2\\ &\le C_1^2\sum_{z\sim x}V_1f(x)^2 \le c'V_1f(x)^2.\end{aligned}$$

□

LEMMA 5.7. *Let M satisfy VD_{loc} and PI_{loc}. Let $g: X\to\mathbb{R}_+$, $x\in G$, and $y\sim x$. Then*

$$(\widehat{g}(x)-\widehat{g}(y))^2\nu_{xy}\le c\int_{A_{xy}}|\nabla g|^2d\mu.$$

PROOF. Write

$$\overline{g}=\mu(A_{xy})^{-1}\int_{A_{xy}}g(w)d\mu(w).$$

Then we have $(\widehat{g}(x)-\widehat{g}(y))^2\le 2(\widehat{g}(x)-\overline{g})^2+2(\widehat{g}(y)-\overline{g})^2$. It is enough to bound the first term:

$$\begin{aligned}(\widehat{g}(x)-\overline{g})^2\nu_{xy} &= \frac{\mu(A_{xy})}{\mu(B(y,1))}\int_{B(x,1)}(\widehat{g}(x)-\overline{g})^2d\mu\\ &\le c\int_{B(x,1)}(g(w)-\overline{g})^2d\mu(w)\\ &\le c\int_{A_{xy}}(g(w)-\overline{g})^2d\mu(w)\le c\int_{A_{xy}}|\nabla g|^2d\mu,\end{aligned}$$

where we used $\mathrm{PI}_{\mathrm{loc}}$ in the final line. □

In the arguments that follow, we will use the fact, given in Remarks 2.10, that to verify CS(Ψ) it is enough to do so for any $\delta>0$ in (2.9) and $\lambda>0$ in (2.10).

PROPOSITION 5.8. *Let M satisfy VD_{loc} and PI_{loc}. Suppose that M satisfies VD and CS(Ψ). Then (G,ν) satisfies VD and CS(Ψ).*

PROOF. Let $B_G(x_0,R)$ be a ball in G; we need to construct a cutoff function $\widehat{\varphi}$ satisfying (a)–(d) of Definition 2.9. If $R\le c$ then it is easy to check that we can take $\widehat{\varphi}(x)$ be the indicator of $B_G(x_0,R/2)$.

So assume $R>c$. We can find a constant c_1 such that

$$B_G(x_0,c_1R)\subset G\cap B(x_0,R/8-6)\subset G\cap B(x_0,R/4+6)\subset B_G(x_0,R).$$

It is enough to construct a cutoff function $\widehat{\varphi}$ for $B_G(x_0,c_1R)\subset B_G(x_0,R)$. Let φ be a cutoff function for $B(x_0,R/8)\subset B(x_0,R/4)$, and let $\widehat{\varphi}$ be given by (5.9). Properties (a)–(c) of Definition 2.9 are easily checked, and it remains to verify the weighted Sobolev inequality (5.3).

Let $x_1\in G$, $1\le s\le R$, and $A_G=B_G(x_1,s)$. Choose c_2,c_3 so that

$$A_G\subset B(x_1,c_2s-6)\cap G\subset B(x_1,2c_2s)\cap G\subset B_G(x_1,c_3s-6).$$

Write $A'_G=B(x_1,c_3s)$, and let $f: A'_G\to\mathbb{R}_+$. We extend f to G by taking f to be zero outside A'_G, and define $\widetilde{f}$ by (5.10).

Let $x \in G$, and $y \sim x$. Then by Lemma 5.6(b) and Lemma 5.7

$$\begin{aligned} f(x)^2(\widehat{\varphi}(x) - \widehat{\varphi}(y))^2 \nu_{xy} &\le c \int_{A_{xy}} f(x)^2 |\nabla \varphi|^2 d\mu(w) \\ &\le 2c \int_{A_{xy}} \widetilde{f}(w)^2 |\nabla \varphi|^2 d\mu(w) + 2c \int_{A_{xy}} V_4 f(x)^2 |\nabla \varphi|^2 d\mu(w). \end{aligned}$$

Therefore

(5.11)

$$\begin{aligned} &\sum_{x \in A_G} \sum_{y \sim x} f(x)^2 (\widehat{\varphi}(x) - \widehat{\varphi}(y))^2 \nu_{xy} \\ &\quad \le c \sum_{x \in A_G} \sum_{y \sim x} \int_{A_{xy}} \widetilde{f}(w)^2 |\nabla \varphi|^2 d\mu(w) + c \sum_{x \in A_G} \sum_{y \sim x} \int_{A_{xy}} V_4 f(x)^2 |\nabla \varphi|^2 d\mu(w) \\ &\quad \le c \int_{B(x_1, c_2 s)} \widetilde{f}(w)^2 |\nabla \varphi|^2 d\mu(w) + c \sum_{x \in A_G} \sum_{y \sim x} V_4 f(x)^2 \int_{A_{xy}} |\nabla \varphi|^2 d\mu(w). \end{aligned}$$

Applying (2.8) to the ball A_{xy} gives

$$\int_{A_{xy}} |\nabla \varphi|^2 d\mu \le c R^{-2\theta} \mu(B(z_{xy}, 5)) \le c' R^{-2\theta} \nu_{xy}. \tag{5.12}$$

Therefore, using Lemma 5.6(c), the second term in the final line of (5.11) is bounded by

$$c R^{-2\theta} \sum_{x \in A_G} \sum_{y \sim x} V_4 f(x)^2 \nu_{xy} \le c R^{-2\theta} \sum_{x \in A'_G} \Gamma(f,f)(x). \tag{5.13}$$

Using (2.8) again

$$\begin{aligned} &\int_{B(x_1, c_2 s)} \widetilde{f}(w)^2 |\nabla \varphi|^2 d\mu(w) \\ &\qquad \le c (s/R)^{2\theta} \Big(\int_{B(x_1, 2c_2 s)} |\nabla \widetilde{f}|^2 + \Psi(s)^{-1} \int_{B(x_1, 2c_2 s)} \widetilde{f}^2 d\mu \Big). \end{aligned} \tag{5.14}$$

By Lemma 5.6(e)

$$\begin{aligned} \int_{B(x_1, 2c_2 s)} |\nabla \widetilde{f}|^2 &\le \sum_{x \in G \cap B(x_1, 2c_2 s + 1)} V_1 f(x)^2 \mu(B(x,1)) \\ &\le \sum_{x, y \in A'_G} (f(x) - f(y))^2 \nu_{xy}, \end{aligned} \tag{5.15}$$

while by Lemma 5.6(d)

$$\int_{B(x_1, 2c_2 s)} \widetilde{f}(w)^2 d\mu(w) \le \sum_{x \in B_G(x_1, c_3 s)} f(x)^2 \nu_x. \tag{5.16}$$

Combining the estimates (5.11)–(5.16) completes the proof. □

PROPOSITION 5.9. *Let M satisfy VD_{loc} and PI_{loc}. Suppose that (G, ν) satisfies VD and CS(Ψ). Then M satisfies VD and CS(Ψ).*

PROOF. Let $B = B(x_0, R)$ be a ball in M. If $R \le c_1$ then we can use the local regularity to construct a cutoff function φ for B. So assume $R \ge c_1$. We can therefore assume that $x_0 \in G$.

Given $A \subset G$ write $A^{(1)} = \cup_{x\in A} B(x,1)$. We can find c_i such that

$$B(x_0, c_1 R) \subset B_G(x_0, c_2 R - 6)^{(1)} \subset B_G(x_0, 2c_2 R + 6)^{(1)} \subset B(x_0, R).$$

Let φ_G be a cutoff function for $B_G(x_0, c_2 R) \subset B_G(x_0, 2c_2 R)$, and let

$$\varphi(w) = \widetilde{\varphi}_G(w) = \sum_{z\in G} f(x)\psi_z(w).$$

Properties (a)–(c) of φ follow easily from those of φ_G, and it remains to verify (2.8).

Let $B_1 = B(x_1, s)$ with $s \in (0, R)$. If $s \le c_3$ then, as $V_1\varphi(x) \le cR^{-2\theta}$,

$$\int_{B(x_1,s)} g^2 |\nabla\varphi|^2 d\mu \le cR^{-2\theta} \int_{B(x_1,s)} g^2 d\mu.$$

Now suppose $s \ge c_3$. Then we can assume $x_1 \in G$, and there exist c_i so that

$$B(x_1, s) \subset B_G(x_1, c_4 s - 6)^{(1)} \subset B_G(x_1, 2c_4 s + 6)^{(1)} \subset B(x_1, c_5 s - 6).$$

Let $g : B(x_1, c_5 s) \to \mathbb{R}_+$. Define $\widehat{g}$ on $B_G(x_1, 2c_4 s + 6)$ by (5.9). Then

$$\begin{aligned} \int_{B_1} g^2|\nabla\varphi|^2 d\mu &\le \sum_{x\in B_G(x_1,c_4 s)} \int_{B(x,1)} g(w)^2 |\nabla\varphi|^2 d\mu(w) \\ &\le 2 \sum_{x\in B_G(x_1,c_4 s)} \int_{B(x,1)} (g(w) - \widehat{g}(x))^2 |\nabla\varphi|^2 d\mu(w) \qquad (5.17) \\ &\quad + 2 \sum_{x\in B_G(x_1,c_4 s)} \int_{B(x,1)} \widehat{g}(x)^2 |\nabla\varphi|^2 d\mu(w). \end{aligned}$$

By Lemma 5.6(e) the first term above is bounded by

$$cR^{-2\theta} \sum_{x\in B_G(x_1,c_4 s)} \int_{B(x,1)} (g(w) - \widehat{g}(x))^2 d\mu,$$

and using $\mathrm{PI}_{\mathrm{loc}}$ this is bounded by

$$cR^{-2\theta} \sum_{x\in B_G(x_1,c_4 s)} \int_{B(x,1)} |\nabla g|^2 d\mu \le c' R^{-2\theta} \int_{B(x_1,c_5 s)} |\nabla g|^2 d\mu. \qquad (5.18)$$

For the final term in (5.17), by Lemma 5.6(e) and (2.8) for φ_G,

$$\begin{aligned} &\sum_{x\in B_G(x_1,c_4 s)} \widehat{g}(x)^2 \int_{B(x,1)} |\nabla\varphi|^2 d\mu(w) \\ &\quad\le \sum_{x\in B_G(x_1,c_4 s)} \widehat{g}(x)^2 V_1\varphi_G(x)^2 \mu(B(x,1)) \\ &\quad\le c \sum_{x\in B_G(x_1,c_4 s)} \widehat{g}(x)^2 \Gamma(\varphi_G, \varphi_G)(x). \\ &\quad\le c(s/R)^{2\theta} \Big(\sum_{x\in B_G(x_1,2c_4 s)} \Gamma(\widehat{g},\widehat{g})(x) + \Psi(s)^{-1} \sum_{x\in B_G(x_1,2c_4 s)} \widehat{g}(x)^2 \nu_x \Big) \end{aligned}$$

Using Lemma 5.7 for the first term, and an easy bound for the second, (2.8) now follows. □

References

[1] S. Alexander and R. Orbach, Density of states on fractals: "fractons ". *J. Physique (Paris) Lett.* **43** (1982) L625–L631.

[2] M.T. Barlow. Diffusions on fractals. In: *Lectures on Probability Theory and Statistics, Ecole d'Éte de Probabilités de Saint-Flour XXV - 1995*, 1-121. Lect. Notes Math. **1690**, Springer 1998.

[3] M.T. Barlow. Which values of the volume growth and escape time exponent are possible for a graph? To appear *Revista Math. Iberoamericana.*

[4] M.T. Barlow. Heat kernels and sets with fractal structure. To appear *Contemp. Math.*

[5] M.T. Barlow. Some remarks on the elliptic Harnack inequality. Preprint.

[6] M.T. Barlow and R.F. Bass. Brownian motion and harmonic analysis on Sierpinski carpets. *Canad. J. Math.* **51** (1999) 673-744.

[7] M.T. Barlow and R.F. Bass. Random walks on graphical Sierpinski carpets. In: *Random walks and discrete potential theory*, ed. M. Piccardello, W. Woess, *Symposia Mathematica XXXIX* Cambridge Univ. Press, Cambridge, 1999.

[8] M.T. Barlow, R.F. Bass. Divergence form operators on fractal-like domains. *J. Funct. Analysis* **175** (2000), 214-247.

[9] M.T. Barlow, R.F. Bass. Stability of parabolic Harnack inequalities. To appear *Trans. A.M.S.*

[10] M. T. Barlow, R.F. Bass, T. Kumagai. Stability of parabolic Harnack inequalities on measure metric spaces. In preparation.

[11] M. Barlow, T. Coulhon, A. Grigor'yan. Manifolds and graphs with slow heat kernel decay. *Invent. Math.* **144** (2001), 609-649.

[12] M. T. Barlow, T. Coulhon, T. Kumagai. In preparation.

[13] E. Bombieri, E. Giusti. Harnack's inequality for elliptic differential equations on minimal surfaces. *Invent. Math.* **15** (1972), 24–46.

[14] C. Borgs, J.T. Chayes, H. Kesten, J. Spencer. Uniform boundedness of critical crossing probabilities implies hyperscaling. Statistical physics methods in discrete probability, combinatorics, and theoretical computer science (Princeton, NJ, 1997). *Random Structures Algorithms* **15** (1999), no. 3-4, 368–413.

[15] M. Bourdon, H. Pajot. Poincaré inequalities and quasiconformal structure on the boundary of some hyperbolic buildings. *Proc. A.M.S.* **127** (1999), 2315–2324.

[16] E.B. Davies. Large deviations for heat kernels on graphs. *J. London Math. Soc.(2)* **47** (1993), 65–72.

[17] P.G. de Gennes. La percolation: un concept unificateur. *La Recherche* **7** (1976), 919–927.

[18] T. Delmotte. Parabolic Harnack inequality and estimates of Markov chains on graphs. *Rev. Math. Iberoamericana* **15** (1999), 181-232.

[19] T. Delmotte. Graphs between the elliptic and parabolic Harnack inequalities. *Potential Anal.* **16** (2002), 151–168.

[20] E.B. Fabes and D.W. Stroock. A new proof of Moser's parabolic Harnack inequality via the old ideas of Nash. *Arch. Mech. Rat. Anal.* **96** (1986) 327–338.

[21] M. Fukushima, Y. Oshima, and M. Takeda, *Dirichlet Forms and Symmetric Markov Processes.* de Gruyter, Berlin, 1994.

[22] A. Grigor'yan. The heat equation on noncompact Riemannian manifolds. *Math. USSR Sbornik* **72** (1992) 47-77.

[23] A. Grigor'yan. Heat kernels and function theory on metric measure spaces. To appear *Contemp. Math.*

[24] A. Grigor'yan, A. Telcs. Sub-Gaussian estimates of heat kernels on infinite graphs. *Duke Math. J.* **109**, (2001) 452-510.

[25] A. Grigor'yan, A. Telcs. Harnack inequalities and sub-Gaussian estimates for random walks. *Math. Annalen* **324** (2002) no. 3, 521-556.

[26] A. Grigor'yan, A. Telcs. In preparation.

[27] A. Grigor'yan, J. Hu, K-S Lau. Heat kernels on metric-measure spaces and an application to semi-linear elliptic equations. *Trans. A.M.S.* **355** (2003) no.5, 2065-2095.

[28] G.R. Grimmett. *Percolation.* (2nd edition). Springer 1999.

[29] B.M. Hambly, T. Kumagai. Heat kernel estimates for symmetric random walks on a class of fractal graphs and stability under rough isometries. To appear *Proc. of Symposia in Pure Math.*

[30] P. Hajłasz, P. Koskela. Sobolev met Poincaré. *Mem. Amer. Math. Soc.* **145** (2000).

[31] S. Havlin and D. Ben-Avraham: Diffusion in disordered media, *Adv. Phys.* **36**, (1987) 695-798.

[32] W. Hebisch, L. Saloff-Coste. On the relation between elliptic and parabolic Harnack inequalities. *Ann. Inst. Fourier, Grenoble* **51** (2001) 1437-1481.

[33] M. Hino. On short time asymptotic behavior of some symmetric diffusions on general state spaces. *Potential Anal.* **16** (2002), no. 3, 249–264.

[34] R. van der Hofstad, F. den Hollander, G. Slade. Construction of the incipient infinite cluster for spread-out oriented percolation above 4+1 dimensions. *Commum. Math. Phys.* **231** (2002) 435-461.

[35] R. van der Hofstad, A. A. Járai. The incipient infinite cluster for high dimensional unoriented percolation. Preprint 2003.

[36] J.E. Hutchinson. Fractals and self-similarity. *Indiana J. Math.* **30**, 713-747 (1981).

[37] D. Jerison. The weighted Poincaré inequality for vector fields satisfying Hörmander's condition. *Duke Math. J.* **53** (1986), 503–523.

[38] O.D. Jones. Transition probabilities for the simple random walk on the Sierpinski graph. *Stoch. Proc. Appl.* **61** (1996), 45–69.

[39] M. Kanai. Rough isometries and combinatorial approximations of geometries of non-compact reimannian manifolds. *J. Math. Soc. Japan* **37** (1985), 391-413.

[40] M. Kanai. Analytic inequalities, and rough isometries between non-compact reimannian manifolds. *Lect. Notes Math.* **1201**, Springer 1986, 122-137.

[41] H. Kesten. The incipient infinite cluster in two-dimensional percolation. *Probab. Theory Related Fields* **73** (1986), 369–394.

[42] J. Kigami. *Analysis on fractals.* Cambridge Tracts in Mathematics, **143**, Cambridge University Press, Cambridge, 2001.

[43] T. Kumagai. Heat kernel estimates and parabolic Harnack inequalities on resistance forms. To appear *Publ. RIMS Kyoto Univ.*

[44] S. Kusuoka, D.W. Stroock. Applications of the Malliavin calculus. III. *J. Fac. Sci. Univ. Tokyo Sect. IA Math.* **34** (1987), no. 2, 391–442.

[45] S. Kusuoka and X.Y. Zhou. Waves on fractal-like manifolds and effective energy propagation. *Probab. Theory Related Fields* **110** (1998), no. 4, 473–495.

[46] T.J. Laakso. Ahlfors Q-regular spaces with arbitrary $Q > 1$ admitting weak Poincaré inequality. *Geom. Funct. Anal.* **10** (2000), 111–123.

[47] P. Li, S.-T. Yau. On the parabolic kernel of the Schrödinger operator. *Acta Math.* **156** (1986), no. 3-4, 153–201.

[48] P.A.P. Moran. Additive functions of intervals and Hausdorff measure. *Proc. Camb. Phil. Soc.* **42** , 15-23 (1946).

[49] J. Moser. On Harnack's inequality for elliptic differential equations. *Comm. Pure Appl. Math.* **14**, (1961) 577–591.

[50] J. Moser. On Harnack's inequality for parabolic differential equations. *Comm. Pure Appl. Math.* **17**, (1964) 101–134.

[51] J. Moser. On a pointwise estimate for parabolic differential equations. *Comm. Pure Appl. Math.* **24**, (1971) 727–740.

[52] A. Phillips, D. Sullivan. Geometry of leaves. *Topology* **20**, (1981) 209–218.

[53] R. Rammal and G. Toulouse. Random walks on fractal structures and percolation clusters, *J. Physique Lettres* **44**, L13–L22 (1983).

[54] L. Saloff-Coste. A note on Poincaré, Sobolev, and Harnack inequalities. *Inter. Math. Res. Notices* **2** (1992) 27–38.

[55] L. Saloff-Coste. *Aspects of Sobolev-Type inequalities.* Lond. Math. Soc. Lect. Notes **289**, Cambridge Univ. Press 2002.

[56] R.S. Strichartz. Analysis on fractals. *Notices A.M.S.* **46**, (1999) no. 10, 1199-1208.

[57] K.-T. Sturm. Analysis on local Dirichlet spaces III. The parabolic Harnack inequality. *J. Math. Pures. Appl. (9)* **75** (1996), 273-297.

Department of Mathematics, University of British Columbia, Vancouver V6T 1Z2, Canada

E-mail address: barlow@math.ubc.ca

Surveys in Differential Geometry
IX, International Press

From isoperimetric inequalities to heat kernels via symmetrisation

Gérard Besson

This paper is dedicated to M. Berger for his teaching.

ABSTRACT. We briefly describe the series of work that led from the Paul Lévy-Gromov isoperimetric inequality to the estimates on the heat kernel obtained by the technique of symmetrisation. An interpretation in terms of a spectral precompactness theorem is given at the end.

CONTENTS

This text is a brief account of the series of work which led from isoperimetric inequalities to estimates on the heat kernel of a Riemannian manifold. The two basic ingredients are the Paul Lévy-Gromov isoperimetric inequality and the symmetrisation process which is a device allowing to translate geometric inequalities into analytic ones. Beyond estimating the heat kernel we shall describe a spectral version of Gromov's compactness theorem. This survey certainly cannot prevent from reading the deep results on isoperimetric inequalities described in the book [**45**]. The reader should also consult [**15**]. Further general references will be given

2000 *Mathematics Subject Classification.* Primary 54C21, 58J50; Secondary 53C20, 58J35.
Key words and phrases. Differential geometry, algebraic geometry.
The author is partially supported by the European network E.D.G.E. HPRN-CT-2000-00101.

in the sequel. The author apologizes for the restrictive choice made here and for possible omissions.

It is our pleasure to thank V. Bayle, D. Cordero and S. Gallot for many helpful conversations.

1. The classical isoperimetric inequality

The paradigmatic example of an isoperimetric inequality is the so-called classical one. It seems that it always has been known; indeed it is cited in Aristotle and is usually known as the solution to the Dido problem. In modern terms it is stated as follows: let D be a domain in $\mathbb{R}^2$ with smooth boundary and D^* be a ball, say centred at the origin, whose area is equal to the area of D; then,

$$L(\partial D) \geq L(\partial D^*),$$

where L denotes the length of the boundary of the domains D and D^*. Furthermore, the equality in the above inequality implies that D is congruent (isometric) to D^*. Another way of stating it, without refereeing to the ball, is

$$L^2 \geq 4\pi A$$

where L (resp. A) denotes again the length of the boundary of D (resp. the area of D). In the first form, strictly speaking, it should rather be called an isovolumic inequality and its extension to higher dimension is obvious.

Various questions are raised by this inequality. One of them is the possible extension to domains whose boundary is not anymore smooth: what notion of "length of the boundary" can be used? answers can be found in the references [**27**], [**22**] or [**6**] where notions such as the perimeter according to Cacciopoli and De Giorgi or the exterior Minkowski content are discussed. The best way to avoid considering such issues is to use the following form,

THEOREM 1.1 (Isoperimetric inequality in $\mathbb{R}^n$). *Let D be a domain in $\mathbb{R}^n$ and D^* a ball which has the same volume. We denote by D_r (resp.D_r^*) the r-tubular neighbourhood of D (resp. D^*), then*

$$\mathrm{vol}(D_r) \geq \mathrm{vol}(D_r^*)\,.$$

Although the inequality is natural, correct and complete proofs were only produced quite recently. Let us for example mention the proof, for $n = 2$, by A. Hurwitz refined by H. Lebesgue (see [**49**], [**50**] and [**57**]), at the undergraduate level, using a Fourier series expansion of the map from $\mathbf{S}^1$ to $\mathbf{C}$ defining the boundary of D. It does give an easy proof when the domain is simply-connected together with a stability result. We shall here give a proof using the mass transportation.

SKETCH OF PROOF. It relies on the idea of optimal mass transportation that was kindly communicated to us by D. Cordero (see [**35**]). It uses the following result proved by Y. Brenier [**21**] and extended by R. McCann [**61**],

THEOREM 1.2 ([**61**]). *Let μ and ν be two probability measures on $\mathbb{R}^n$ and assume that μ is absolutely continuous with respect to the Lebesgue measure; then there exists a convex function ϕ such that $(\nabla\phi)_*\mu = \nu$. Furthermore, $\nabla\phi$ is unique μ-ae.*

Here $\nabla\phi$ is a vector field, thus a map from $\mathbb{R}^n$ to itself; then $(\nabla\phi)_*\mu$ is the pushed forward measure by this map.

Now let $\mu = \frac{1}{\mathrm{vol}(D)}\mathbf{1}_D$ and $\nu = \frac{1}{\mathrm{vol}(B)}\mathbf{1}_B$, where B is the unit ball. The above theorem asserts that there exists a convex function ϕ such that $(\nabla\phi)_*\mu = \nu$. The map $\nabla\phi$ sends D into B and thus, on D, $|\nabla\phi| \leq 1$. Furthermore, the equality between the two measures yields, by change of variables, a Monge-Ampère equation, *i.e.*,

$$\det(\mathrm{hess}\phi(x)) = \frac{\mathrm{vol}(B)}{\mathrm{vol}(D)}, \quad \forall x \in D$$

Since ϕ is a convex function its hessian is non-negative and thus, for all $x \in D$,

$$(\det(\mathrm{hess}\phi(x)))^{1/n} = \Big(\frac{\mathrm{vol}(B)}{\mathrm{vol}(D)}\Big)^{1/n} \leq \frac{1}{n}(\mathrm{trace}(\mathrm{hess}\phi(x))) = -\frac{1}{n}\Delta\phi(x)$$

where Δ denotes the Laplacian as a positive operator. Then,

$$\Big(\frac{\mathrm{vol}(B)}{\mathrm{vol}(D)}\Big)^{1/n}\mathrm{vol}(D) \leq \frac{1}{n}\int_D(-\Delta\phi(x))dx = \frac{1}{n}\int_{\partial D}\nabla\phi.\vec{n}$$

where $\vec{n}$ is the exterior unit normal to ∂D. We conclude by recalling that $|\nabla\phi(x)| \leq 1$ for $x \in D$. This yields,

$$\frac{1}{n}\mathrm{vol}_{n-1}(\partial D) \geq \mathrm{vol}(B)^{1/n}\,\mathrm{vol}(D)^{1-1/n}$$

□

Remarks 1.3. (1) The article [**21**] contains interesting comments concerning the analogy between the result described in Theorem 1.2 and the polar decomposition of matrices. The analogy with the rearrangement of functions is also described.

(2) It is rather a sketch of the proof since in order to make the change of variables yielding the Monge-Ampère equation we would need to know that the map ϕ is at least C^2. This is obtained in this situation by the regularity results proved by L. Caffarelli (see [**23**],[**24**] and [**25**]).

(3) A similar proof gives the Brunn-Minkowski theorem. It is an easy exercise left to the reader. A very good survey of the inequalities related to the Brunn-Minkowski theory is [**43**] (and the references herein).

(4) The extension to manifolds, even with constant curvature, seems completely open at this stage.

(5) A proof of the isoperimetric inequality by mass transportation is not new. Indeed, M. Gromov used the so-called Knothe transport. The map $\nabla\phi$ is replaced by a vector field whose Jacobian matrix is triangular and whose construction is explicit; the proof also extends to finite dimensional normed spaces. See the appendix I of [**62**].

(6) In the article [**35**], the authors prove sharp Sobolev type inequalities using the mass transportation.

(7) A general reference for the theory of optimal transportation is [**78**].

2. The symmetrisations

The symmetrisations are natural processes used to replace a set by a more symmetric one; it has been invented by Jacob Steiner. When it is applied to the level sets of a function it allows to "rearrange" its values in order to produce a more symmetric function (invariant by some symmetries of the space). This technique is quite powerful and produce nice and sharp inequalities. The basic references are [**71**] and [**6**] in which a whole bunch of analytic inequalities are treated using various symmetrisations. Other good references for rearrangement are [**46**] and [**22**] (where several symmetrisations are discussed from a geometric point of view).

In this section we shall present in a very elementary way two symmetrisations that we shall need in the sequel.

2.1. Symmetrisation. Let D be a domain in $\mathbb{R}^n$ and V a k-dimensional affine subspace (where $k \leq n$), which we shall call a vertical subspace. For any $n-k$-dimensional affine subspace H, orthogonal to V, let us associate to $D \cap H$ a ball in H of the same $n-k$-volume centred at the point $H \cap V$. When H varies the union of these balls designed a domain which we shall also call D^*. Classically, the case $k = n-1$ is called the *Steiner symmetrisation* and the case $k = 1$ is called the *Schwarz symmetrisation* (see [**71**] p.5 and [**22**] p.78). We shall be interested mainly in the two following cases: $k = n-1$ and $k = n$. The main feature of these elementary constructions is the

PROPOSITION 2.1. *In both cases,*

1. $\mathrm{vol}(D) = \mathrm{vol}(D^*)$
2. $\mathrm{vol}_{n-1}(\partial D) \geq \mathrm{vol}_{n-1}(\partial D^*)$

PROOF. For the $k = n$ symmetrisation, the equality (1) is in the definition of D^* and the inequality (2) is the isoperimetric inequality.

For the Steiner and Schwarz symmetrisations, (1) is a corollary of Fubini's theorem. The inequality (2) can be proved by approximation (see [**71**] chapter 7 and [**22**] p. 79 for the spherical Steiner symmetrisation). In dimension 2, for example, if V is a vertical line, one can approximate D by trapezoids whose basis are horizontal. The inequality amounts then to compare the perimeter of such a trapezoid with a symmetric one (with respect to V). □

3. The Faber-Krahn inequality

The Faber-Krahn inequality is the analytical translation of the isoperimetric inequality. It was conjecture by Lord Rayleigh in his book on the theory of sound ([**77**] p. 339) and proved independently by G. Faber ([**38**]) and E. Krahn ([**55**]). Let us consider a measurable function u defined on a domain D of $\mathbb{R}^n$. We assume furthermore that u is positive on D and vanishes on its boundary; it is sufficient to consider such functions for the following theorem. Let D^* be a ball centred at the origin and having the same area as D.

Definition 3.1. If $\mathcal{E}$ denotes the domain in $\mathbb{R}^{n+1}$ below the graph of u and above D, we define $\mathcal{E}^*$ to be the Schwarz-symmetrised domain. Its boundary consists of D^* and the graph of a function from D^* to $\mathbb{R}$ which we call u^*.

One can check that u^* could also be defined as follows: let $D(\mu) = \{x \in D : u(x) \geq \mu\}$, then, for $x \in D^*$, we define $u^*(x) = \sup\{\mu : x \in D(\mu)^*\}$ (see [**6**] chapter II). The two functions u and u^* are thus seen to be equimeasurable, *i.e* the volumes of $D(\mu)$ and $D^*(\mu)$ (with an obvious definition) are equal. An important issue discussed in [**6**] is the regularity of u^*; it is, in particular, proved that if u is a Lipschitz function so is u^*. In the sequel, we shall disregard these questions and the reader is referred to the above mentioned book.

PROPOSITION 3.2. *Let D, D^*, u and u^* be as above, then*

1. *If u is a Lipschitz function so is u^*.*

2. *For all* $p \geq 0$, $\int_D u^p = \int_{D^*} u^{*p}$.
3. $\int_D |du|^2 \geq \int_{D^*} |du^*|^2$.

SKETCH OF PROOF OF (3). We shall assume that all expressions below make sense. The key ingredient is the co-area formula (see [**22**] p. 103 or [**39**] section 3.2): let h be a continuous function, then

$$\int_D h(x)dx = \int_0^{sup(u)} \Big(\int_{\Gamma_t} h|du|^{-1}d\sigma_t\Big)dt$$

where $\Gamma_t = u^{-1}(t)$ and $d\sigma_t$ is the volume element on Γ_t. Let us apply this formula to the set $\Omega_t = \{x \,|\, 0 \leq u(x) < t\} \subset D$ and $h \equiv 1$, this gives

$$\mathrm{vol}(\Omega_t) = \int_0^t \Big(\int_{\Gamma_s} |du|^{-1}d\sigma_s\Big)ds$$

yielding $\frac{d}{dt}(\mathrm{vol}(\Omega_t)) = \int_{\Gamma_t} |du|^{-1}d\sigma_t$. Now

$$\int_D u^p = \int_0^{\sup(u)} t^p \frac{d}{dt}(\mathrm{vol}\,\Omega_t)) = p\int_0^{\sup(u)} t^{p-1}\,\mathrm{vol}(\Omega_t)dt$$

The equimeasurability of u and u^* and the last formula prove (2). We also get from the co-area formula,

$$\int_D |du|^2 = \int_0^{\sup(u)} \Big(\int_{\Gamma_t} |du|d\sigma_t\Big)dt$$

and from the Cauchy-Schwarz inequality,

$$\int_{\Gamma_t} |du|d\sigma_t \geq \Big(\int_{\Gamma_t} d\sigma_t\Big)^2 \Big/ \Big(\int_{\Gamma_t} |du|^{-1}d\sigma_t\Big).$$

If we define $v_{n-1}(t) = \int_{\Gamma_t} d\sigma_t = \mathrm{vol}_{n-1}(\Gamma_t)$ and $v_n(t) = \mathrm{vol}(\Omega_t)$, we then get the inequality,

$$\int_{\Gamma_t} |du|d\sigma_t \geq \frac{(v_{n-1}(t))^2}{v_n'(t)}.$$

For the function u^* we get, with obvious notations,

$$\int_{\Gamma_t} |du^*|d\sigma_t = \frac{(v_{n-1}^*(t))^2}{v_n^{*'}(t)}.$$

But, on the one hand by equimeasurability, we have $v_n(t) = v_n^*(t) \Rightarrow v_n'(t) = v_n^{*'}(t)$ and on the other hand the isoperimetric inequality gives $v_{n-1}(t) \geq v_{n-1}^*(t)$. Finally,

$$\int_{\Gamma_t} |du|d\sigma_t \geq \int_{\Gamma_t^*} |du^*|d\sigma_t\,.$$

The inequality (3) is then obtained by integration. □

Let us give a slightly different point of view, which is more geometric.

SKETCH OF A SECOND PROOF. (see [**6**] chapter II)

Let $\mathcal{G} \subset \mathbb{R}^{n+1}$ (resp. $\mathcal{G}^*$) be the graph of u (resp. u^*), the n-volume of $\mathcal{G}$ is given by,

$$\mathrm{vol}_n(\mathcal{G}) = \int_D (1 + |du|^2)^{1/2}.$$

Now the definition of $\mathcal{G}^*$ and the properties of the Schwarz symmetrisation give,

$$\mathrm{vol}_n(\mathcal{G}) \geq \mathrm{vol}_n(\mathcal{G}^*) .$$

We define $u_\epsilon = \sqrt{\epsilon}u$ for $\epsilon > 0$. Clearly $(\sqrt{\epsilon}u)^* = \sqrt{\epsilon}u^*$, so that

$$\int_D (1 + |du_\epsilon|^2)^{1/2} \geq \int_{D^*} (1 + |du_\epsilon^*|^2)^{1/2} .$$

Now letting ϵ go to 0, expanding in ϵ and using the equality between the volumes of D and D^* give the inequality (3). □

Remark 3.3. Once again let us emphasize that the co-area formula as used in the above "proof" needs some regularity on the function u. If u is smooth, one can choose t to be a regular value of u thanks to Sard's theorem. One other possibility could be to approximate u in a suitable topology by Morse functions with finitely many non degenerate critical points in the interior of D, an argument introduced by Th.Aubin (see [**1**] p. 40 or [**14**] lemma 10bis p. 519 and the appendix of [**9**]). This important issue is also discussed in [**6**].

We now proceed to the Faber-Krahn inequality. Let us denote by $\lambda_1(D)$ the first eigenvalue of the Laplace operator with Dirichlet boundary conditions on D and u a corresponding eigenfunction. We recall that u has constant sign in the interior of D (see [**37**]); we thus can choose it to be positive. The key feature is the variational characterisation of the eigenvalues, also called the min-max principle, through the so-called Rayleigh quotient (see [**6**], [**16**], [**9**] and [**37**]). Precisely for $\lambda_1(D)$ one has,

$$\lambda_1(D) = \inf \left\{ \frac{\int_D |du|^2}{\int_D u^2} \,|\, u \in C^1(D),\ u = 0 \,\mathrm{on}\, \partial D \right\} .$$

THEOREM 3.4 (The Faber-Krahn inequality). *With the above notation,*

$$\lambda_1(D) \geq \lambda_1(D^*),$$

with equality if and only if D is a round ball.

PROOF. The inequality is an immediate consequence of the Proposition 3.2. The equality case is more involved and the reader is referred to [**6**] or [**71**]. □

Remarks 3.5. (1) This inequality is the analytic translation of the isoperimetric inequality.

(2) The same proof also applies to some Sobolev constants as noticed by G. Talenti and Th. Aubin, see [**76**] and [**1**].

(3) In the article [**36**], dated from 1918, R. Courant proved that if $D \subset \mathbb{R}^2$ has smooth boundary and D^{**}, is a ball with the same perimeter, then

$$\lambda_1(D) \geq \lambda_1(D^{**}) .$$

Then, G. Faber (1923) and E. Krahn (1925) published independently the inequality which bears there names. One can easily check that $\lambda_1(D^*) \geq \lambda_1(D^{**})$ (it requires the monotonicity principle on Dirichlet eigenvalues described in [**37**] and [**6**]) which shows that Courant's inequality is weaker than the one by Faber and Krahn. Finally, let us mention that the proof of E. Krahn is simpler than the proof of G. Faber, in the spirit of the above sketch.

It seems that R. Courant suggested the problem to G. Faber giving the impulse in the gestation of this fundamental result. The article [**36**] was shown to us by M. Berger.

4. The Paul Lévy-Gromov isoperimetric inequality

In this section we shall show how to extend the previous estimates on Riemannian manifolds. The core of it which is also the core of this text is the Paul Lévy-Gromov isoperimetric inequality. The basic reference is [**45**]. We shall use here the extension given in [**11**]; the reader can also consult [**41**], [**10**]. For an n-dimensional closed Riemannian manifold (M, g) and $0 \leq \beta \leq 1$, we let $h_{(M,g)}$ be the function defined by

$$h_{(M,g)}(\beta) = \inf \{ \frac{\mathrm{vol}_{n-1}(\partial\Omega)}{\mathrm{vol}(M)} \mid \text{for } \Omega \subset M \text{ with smooth boundary and } \frac{\mathrm{vol}(\Omega)}{\mathrm{vol}(M)} = \beta \} .$$

Definition 4.1. The function $h_{(M,g)}$ is called the *isoperimetric profile* of M.

The isoperimetric profile captures the essence of the isoperimetry on M. Dividing by $\mathrm{vol}(M)$ amounts to normalising the Riemannian measure on M to be a probability measure. This will facilitate comparisons between different Riemannian manifolds.

Example 4.2. On the n-sphere, endowed with its canonical metric, for a given value of β, the infimum in the definition of the isoperimetric profile is achieved for round balls (spherical caps) thanks to the isoperimetric inequality proved by E. Schmidt (see [**74**] and [**22**] p. 83 section 9.7). One can then compute the values of $h_{(S^n, can)}(\beta)$ for each β.

For example, when $n = 2$, we have $h_{(S^2,\mathrm{can})}(\beta) = \sqrt{\beta(1-\beta)}$.

Remarks 4.3. (1) For a given value of β, different from 0 and 1, a minimizing domain for the above infimum is easily seen to have constant mean curvature (where it is defined) and thus the computation of the function h, even on standard manifolds, requires the knowledge of constant mean curvature hypersurfaces. As a consequence there are very few explicit examples of isoperimetric profile. At this stage the isoperimetric profile of the 3-dimensional flat tori are not known.

(2) For the sake of simplicity we shall sometimes omit the reference to the metric g in the isoperimetric profile when there is no ambiguity.

(3) The function h_M satisfies obvious properties, such as the behaviour for β going to 0, the symmetry with respect to 1/2, etc... The reader can consult [**9**] p. 85, [**14**], [**11**] or [**41**].

We now define

$$R_{min}(x) = \inf \{ \mathrm{Ricci}_M(u, u) \mid u \text{ is a unit tangent vector at } x \},$$

where Ricci_M denotes the Ricci curvature of M as a field of bilinear forms and

$$r_{min} = \inf\{R_{min}(x) \mid x \in M\}.$$

We denote by $\mathrm{diam}(M)$ the diameter of M.

THEOREM 4.4 ([**11**]). *Let (M, g) be an n-dimensional closed Riemannian manifold satisfying*

$$r_{min}(M)\,\mathrm{diam}(M)^2 \geq \epsilon(n-1)\alpha^2$$

for $\epsilon \in \{-1, 0, 1\}$ and $\alpha > 0$. Then there exists a positive number $a(n, \epsilon, \alpha)$ such that, for all β in $[0, 1]$,

$$h_{(M,g)}(\beta) \geq h_{(S^n(R),\mathrm{can})}(\beta) ,$$

where $S^n(R)$ is the sphere of radius $R = \mathrm{diam}(M)/a(n,\epsilon,\alpha)$ with the induced (thus canonical) metric from $\mathbb{R}^{n+1}$. The number $a(n,\epsilon,\alpha)$ is explicit (see [**11**] *for its value).*

Remarks 4.5. (1) In [**45**] appendix C, M. Gromov states his result, namely when $r_{min} > 0$. The above theorem is an improvement of the final step in the proof of Gromov's result, the general scheme of the proof being the same. Introducing the diameter allows to get rid of the positive curvature assumption. However we always compare to a round sphere.

(2) The inequality is sharp in the positive curvature case ($\epsilon = 1$).

Let us explain the different steps of the proof on a very simple case: $n = 2$ and M a convex surface in $\mathbb{R}^2$ whose curvature is not smaller than 1. This is the original situation considered by Paul Lévy (see [**58**] Third part, chapter IV). This example contains all the ideas of the proof (see [**42**] Chapter IV section H).

SKETCH OF PROOF. We proceed in three steps.

1st step: Let us fix β different from 0 and 1. The geometric measure theory (see [**39**]) asserts that there exists in M an hypersurface H (a curve in the case under consideration) bounding a domain Ω, of relative volume β, such that $\frac{\mathrm{vol}_{n-1}(H)}{\mathrm{vol}(M)} = h(\beta)$. The existence is indeed easy since it boils down to proving a compactness theorem for certain rectifiable currents. More involved is the regularity theory in this situation. It is known that there may be singularities but they form a set of codimension at least 7. In our (easy) case the curve H is smooth and the mean curvature of H is constant as can be seen by a standard variational technique. For the unit sphere the corresponding hypersurface $\partial\Omega_r^*$ is a circle around the north pole whose radius r is computable in term of β.

2nd step: The main geometric ingredient is the Heintze-Karcher comparison theorem ([**47**]). More precisely H is a two sided hypersurface and we choose a unit normal; furthermore any point in one side (Ω) or the other (Ω^c) is attained by the normal exponential to H. The volumes of Ω and Ω^c can be estimated if one has a control of the Jacobian of the normal exponential map, this is given in [**47**]; precisely, let η be the (constant) mean curvature of H, then

$$\mathrm{vol}(\Omega) \leq \mathrm{vol}_{n-1}(H) \int_0^r (\cos(t) - \eta\sin(t))dt\,,$$

where r is the first point t where $\cos(t) - \eta\sin(t) = 0$, thus $\eta = \cot(r)$. The comparison theorem shows that if starting from H we go in the normal direction in Ω up to distance r we cover all of Ω.

Now we can apply the same argument to Ω^c. The mean curvature, with respect to the same normal is now $-\eta$, thus

$$\mathrm{vol}(\Omega^c) \leq \mathrm{vol}_{n-1}(H) \int_0^{\pi-r} (\cos(t) + \eta\sin(t))dt\,.$$

The furthest point from H in Ω^c is at most at distance $\mathrm{diam}(M) - r$, but by Myers' theorem (see [**42**]) and since the curvature of M has been assumed to be not greater than 1 in this case, one has $\mathrm{diam}(M) \leq \pi$; so if we replace $\mathrm{diam}(M) - r$ by $\pi - r$ we just increase the integral on the right hand side since the integrand is non-negative on the interval $[0, \pi - r]$.

On the model space, the unit sphere, if Ω_r^* denotes the geodesic ball of radius r around the north pole, one has

$$\frac{\mathrm{vol}(\Omega_r^*)}{\mathrm{vol}_{n-1}(\partial\Omega_r^*)} = \int_0^r (\cos(t) - \eta\sin(t))dt = \frac{1}{a(r)}$$

indeed $\eta = \cot(r)$ is the mean curvature of $\partial\Omega_r^*$. In the same way,

$$\frac{\mathrm{vol}(\Omega_r^{*c})}{\mathrm{vol}_{n-1}(\partial\Omega_r^{*c})} = \frac{1}{a(\pi - r)}\,.$$

The conclusion is,

$$\begin{aligned} h_M(\beta) = \frac{\mathrm{vol}_{n-1}(H)}{\mathrm{vol}(M)} &\geq \max\left\{a(r)\,\mathrm{vol}(\Omega), a(\pi - r)\,\mathrm{vol}(\Omega^c)\right\}/\,\mathrm{vol}(M) \\ &\geq \max\left\{\beta a(r), (1-\beta)a(\pi - r)\right\} \\ &\geq \inf_{0\leq t\leq\pi} \max\left\{\beta a(t), (1-\beta)a(\pi - t)\right\}. \end{aligned}$$

3rd step: The function $a(r)$ is explicitly computable (in this case) and is non decreasing. So the inf-max is achieved for t_0 such that

$$\begin{aligned} \beta a(t_0) &= (1-\beta)a(\pi - t_0) \\ \beta\frac{\mathrm{vol}_{n-1}(\partial\Omega_{t_0}^*)}{\mathrm{vol}(\Omega_{t_0}^*)} &= (1-\beta)\frac{\mathrm{vol}_{n-1}(\partial\Omega_{t_0}^*)}{\mathrm{vol}(\Omega_{t_0}^{*c})} \end{aligned}$$

which means that $\Omega_{t_0}^*$ satisfies $\mathrm{vol}(\Omega_{t_0}^*) = \beta\,\mathrm{vol}(S^2)$. This proves the inequality,

$$h_{(M,g)}(\beta) \geq h_{(S^2,\mathrm{can})}(\beta)\,.$$

□

Remark 4.6. In the positive curvature case ($\epsilon > 0$) the sketch of proof given above shows that there is an equality case. Indeed one can show that if $h_{(M,g)}(\beta) = h_{(S^n,\mathrm{can})}(\beta)$ for one value of $\beta \in (0, 1)$ then (M, g) is isometric to (S^n, can). When the curvature is allowed to be non-positive, no such equality can be obtained.

Comments 4.7. One can try to extend the Theorem 4.4 in various directions. One of them is to try to understand more in depth the relationships between the curvature and h. This is the approach developed in the article by F. Morgan and D. Johnson in [**63**] and recently by V. Bayle in [**8**] where it is shown that $h_{(M,g)}$ satisfies a differential inequality in β which yields a different proof of 4.4. More precisely, one can show the following

THEOREM 4.8 ([**8**]). *Let (M, g) be a closed n-dimensional Riemannian manifold satisfying* Ricci$(g) \geq (n-1)\delta g$*, for $\delta \in \mathbb{R}$. Then the function $y = h_{(M,g)}^{n/(n-1)}$ satisfies the differential inequality*

$$y'' \leq -n\delta y^{(2-n)/n}\,,$$

where the second derivative is taken in the sense of distributions.

The basic idea is to study the parallel deformations of the boundary of an optimal domain for a given value of β. The main issue is to deal with the singularities of H. This theorem gives a new proof of Paul Lévy-Gromov's isoperimetric inequality. In this proof the Heintze-Karcher comparison theorem is not used any more. One conclusion of all these works that should be emphasized is that rather than comparing our manifold with a sphere, we are indeed comparing it with an one dimensional model, consisting of an interval (or a union of intervals) with a relevant density; this was developed in details and with a series of examples in [**40**]. This is not that surprising since all the classical geometric comparison theorems (Bishop's, Bishop-Gromov's, etc...) produce comparisons with constant curvature space which are rotationally symmetric, and thus, in spirit, one-dimensional. It is a question whether there exists a (reasonable) comparison geometry! with, sa! y the complex projective or hyperbolic space (for a survey of comparison geometry see [**66**]).

A second point of view is to try to replace L^∞ norms on the Ricci curvature (lower bound of Ricci curvature as above) by L^p norms. In this context this trend was initiated by S. Gallot in the paper [**41**]. It is worth citing the following result.

THEOREM 4.9 ([**41**]). *Let α and D be positive constants and $q \in (n, +\infty)$. Let (M, g) be a closed n-dimensional Riemannian manifold whose diameter is bounded above by D and whose Ricci curvature satisfies*

$$\frac{1}{\mathrm{vol}(M)} \int_M \left(\frac{r_-}{(n-1)\alpha^2} - 1 \right)_+^{q/2} dv_g \leq \frac{1}{2}(e^{B(q)\alpha D} - 1)^{-1},$$

where $r_-(x) = \max(0, -R_{min}(x))$ and $(.)_+$ denotes the positive part. Then, for every domain $\Omega \subset M$, we have

$$\frac{\mathrm{vol}(\partial\Omega)}{\mathrm{vol}(M)} \geq \gamma(\alpha, D) \min \left(\frac{\mathrm{vol}(\Omega)}{\mathrm{vol}(M)}, \frac{\mathrm{vol}(\Omega^c)}{\mathrm{vol}(M)} \right)^{1-\frac{1}{q}}.$$

The functions $B(q)$ and $\gamma(\alpha, D) > 0$ are computed explicitly.

This result is quite optimal in the sense that the hypothesis cannot be relaxed too much as is shown in [**41**]. For example such estimates with $p = n$ are shown to be impossible.

Further interesting developments toward a geometry under integral bounds of curvature have been made, in particular in the articles [**68**] and [**70**] (and the references therein). Concerning the isoperimetric inequalities the most recent results in this direction are given in [**2**] by E. Aubry where the hypothesis on the diameter is dropped. More precisely, one gets a Myers's type theorem as follows.

THEOREM 4.10 ([**2**]). *For any $q > n/2$ there exist two explicitly computable positive constants $C(q,n)$ and $\alpha(q,n)$ such that the following is true:*

1. *If $0 < R \leq 6\pi$, then any n-dimensional complete Riemannian manifold (M, g) such that there exists a point $x_0 \in M$ satisfying*

   $$\|(R_{min} - (n-1))_-\|_{L^q(B(x_0,R))} \leq \epsilon \leq \alpha(q,n),$$

   has its diameter bounded by $\pi(1+C(q,n)\epsilon^{\frac{q}{2q-1}})$. In particular, M is compact.
2. *If $R > 6\pi$, then the same conclusions hold under the more restrictive hypothesis*

   $$R^2\|(R_{min} - (n-1))_-\|_{L^q(B(x_0,R))} \leq \epsilon \leq \alpha(q,n).$$

   Here $B(x_0, R)$ denotes the geodesic ball centred at x_0 and of radius R.

A similar result appeared earlier in [**69**]; in [**2**] p. 134 E. Aubry suggests that the proof given in [**69**] is not complete; this is a point yet to be checked. Now, together with the first part of Theorem 4.9 this yields the isoperimetric inequality that appeared in the second statement in 4.9 with these (weaker) hypothesis. It is to be noticed that under the the above hypotheses the curvature is almost positive (in integral sense); however, the Ricci curvature may take very negative values on a small set. It is not yet a version of Paul Lévy-Gromov isoperimetric inequality under integral bounds on the Ricci curvature, but according to the author ([**2**] p.144) such a statement is attainable and is in progress. For a review of convergence and compactness theorems under integral bounds of the curvature see [**66**].

Finally another important approach is developed in the works by D. Bakry and M. Ledoux ([**3**] and also [**4**]). Here an abstract version of Lévy-Gromov's isoperimetric inequality is proved for an infinite dimensional diffusion generator involving comparison with Gaussian isoperimetric profile, which can be seen as a limit of profiles of suitably renormalised spheres whose dimensions tend to infinity ([**7**]).

5. Estimating eigenvalues

As in section 3 we shall use the symmetrisation procedure to translate into analytic inequalities the isoperimetric inequality 4.4. The symmetrisation consists now to associate to a domain Ω in a closed manifold M a geodesic ball Ω^* on the comparison n-sphere. More precisely, if (M,g) satisfies $r_{min}(M)\,\mathrm{diam}(M) \geq \epsilon(n-1)\alpha^2$ as before, Ω^* is defined by,

$$\frac{\mathrm{vol}(\Omega)}{\mathrm{vol}(M)} = \frac{\mathrm{vol}(\Omega^*)}{\mathrm{vol}(S^n(R))}$$

where R is the explicit radius given in the Theorem 4.4. We shall now consider Riemannian Laplacian on (M,g). The first eigenvalue is always zero and by abuse of notation we shall call $\lambda_1(M)$ the first non-zero eigenvalue. It can be a multiple eigenvalue, and an eigenfunction corresponding to this eigenvalue changes sign; it indeed has exactly two nodal domains (see [**9**]). The same procedure than the one described in section 3 leads to the following theorem:

THEOREM 5.1 ([**11**]). *i) Under the hypotheses of Theorem 4.4, one has*

$$\lambda_1(M) \geq \lambda_1(S^n(R)) = n\left(\frac{a(n,\epsilon,\alpha)}{\mathrm{diam}(M)}\right)^2.$$

ii) Under the same hypotheses, if in addition $r_{min}(M) \geq (n-1)$, then

$$\lambda_1(M) \geq n\left(\frac{\int_0^{\pi/2}(\cos t)^{n-1}dt}{\int_0^{d/2}(\cos t)^{n-1}dt}\right)^{2/n} \geq n. \tag{5.1}$$

Moreover, the equality in the first inequality in (5.1) implies that (M,g) is isometric to $(S^n(R), \mathrm{can})$

The explicit expression that appears in ii) comes from the explicit value for R (see [**11**]). The second part of this theorem is an improvement of the Lichnérowicz-Obata result showing that with the assumption ii) one has $\lambda_1(M) \geq \lambda_1(S^n) = n$ and the equality case ([**60**] and [**64**], see also [**9**]).

COMMENTS ON THE PROOF. The only difference with the proof of the Theorem 3.4 is that a first eigenfunction takes positive and negative values. Let u be such an eigenfunction, we write $u = u^+ - u^-$ where $u^\pm = \sup\{\pm u, 0\}$. We now symmetrise u^+ into a function $(u^+)^*_N$ with respect to the north pole and u^-

into $(u^-)^*_S$ with respect to the south pole and define $u^* = (u^+)^*_N - (u^-)^*_S$. The Proposition 3.2 is still valid together with the fact that,

$$\int_M u = \int_{S^n(R)} u^* = 0\,.$$

The conclusion is obtained by applying the min-max principle for the first non zero eigenvalue on a closed Riemannian manifold, namely

$$\lambda_1(M) = \inf\left\{\frac{\int_M |du|^2 dv_g}{\int_M u^2 dv_g} \mid u \in C^1(M) \text{ and } \int_M u dv_g = 0\right\}.$$

□

Once again this technique also allows to control the Sobolev constants as was first noticed independently by Th. Aubin and G. Talenti (see [**1**], [**76**], [**41**] and [**10**]).

Comments 5.2. The abstract approach developed by D. Bakry and Z. Qian (see [**5**]) extends to a large family of operators previous results obtained by P. Kröger ([**56**]) for the Laplacian on Riemannian manifolds. It yields estimate on the first non zero eigenvalue without going through symmetrisation and also improves the Lichnérowicz-Obata result. Furthermore it gives a sharp bound in the case when $r_{min} \geq 0$ improving a result due to J.Q. Zhong and H.C. Yang ([**79**]) which asserts that if (M, g) is a compact Riemannian manifold with non-negative Ricci curvature, then $\lambda_1(M) \geq \frac{\pi^2}{\mathrm{diam}(M)^2}$. This last statement is optimal as can be seen by considering long thin tori. Theorem 5.1 does not provide a proof of this result. P. Kröger and later D. Bakry and Z. Qian put these two results in the same framework, which consists in comparing the first eigenvalue of the Riemannian manifold to those of some operators on an interval or a! union of intervals. It goes through interesting comparison theorems on the range and on the gradient of eigenfunctions (see also [**59**]). The method of D. Bakry and Z. Quian uses an abstract version of the notion of lower bound for the Ricci curvature and of the Bochner technique (see [**10**]). The fact that the model space is one-dimensional gives more flexibility and this point of view is pushed far away in these works as in [**40**]. At this stage, the approach of [**56**] extended by [**5**] is one of the most powerful for estimating the first eigenvalue; it is however not so easy to draw an explicit lower bound for $\lambda_1(M)$, and it would be interested to see if it can be adapted to integral bounds on the curvature as below. It remains also to interpret V. Bayle's results in the abstract framework of [**5**], if possible. For an overview the reader is refered to the introduction of [**5**].

Concerning this last point, *i.e.* integral bounds on the curvature, one can use the symmetrisation (see [**40**] and [**11**]) to give a lower bound on λ_1. More precisely, combining the Theorem 5.2 in [**69**] and the Theorem 4.10 above we obtain the following result (see [**2**] section 4.3.4.).

THEOREM 5.3 ([**2**]). *Under the hypothesis of Theorem 4.10 above one gets,*

$$\lambda_1(M) \geq n(1 - C'(q, n))$$

where $C'(q, n) \in (0, 1)$ is an explicitly computable constant.

This is an almost sharp and explicit version of Lichnérowicz-Obata's result.

6. Estimating the heat kernel

The heat kernel of a domain or a Riemannian manifold encodes the eigenfunctions and the spectrum of the space under consideration. We shall present here a brief account of estimates obtained in the spirit of the preceding section. Various other points of view concerning the heat equation can be developed, some of which are described in this issue. We shall concentrate here on the symmetrisation technique. General references for this section are [**16**] and [**9**].

6.1. The heat kernel. For the sake of simplicity let us consider the case of a closed Riemannian manifold (M, g). The Laplacian Δ gives rise, by standard functional calculus, to another operator $e^{-t\Delta}$ which is infinitely regularizing. It thus has a kernel which we shall denote by $p_M(t; x, y)$, where $t > 0$ and x and y are points in M. This function is the fundamental solution of the heat equation, *i.e.* it satisfies the parabolic equation,

$$\begin{cases} (\frac{\partial}{\partial t} + \Delta) p_M(t; x, .) & = 0 \\ \lim_{t \to 0^+} p_M(t; x, y) dv_g(y) & = \delta_x \,, \end{cases}$$

where δ_x is the Dirac measure at $x \in M$. The kernel has an (obvious) expression in terms of the eigenfunctions and eigenvalues of the Laplacian. Let $\lambda_i(M)$ be the i-th eigenvalue of the Laplacian of (M, g), counted with multiplicity $(0 \leq i)$; we shall write λ_i when there is no ambiguity. Let $(\phi_i)_{0 \leq i}$ be an orthonormal Hilbert basis of eigenfunctions of $L^2(M, dv_g)$; then

$$p_M(t; x, y) = \sum_{i=0}^{+\infty} e^{-\lambda_i t} \phi_i(x) \phi_i(y) \,.$$

Recall that on any compact manifold the eigenvalues form an increasing sequence of numbers going to $+\infty$ and that their multiplicities are finite. The operator $e^{-t\Delta}$ is of the trace class and its trace is given by the following formulae:

$$Z_M(t) = \text{trace}(e^{-t\Delta}) = \int_M p_M(t; x, x) dv_g(x) = \sum_{i=0}^{+\infty} e^{-t\lambda_i} \,.$$

The semi-group property, *i.e.* $e^{-(t+s)\Delta} = e^{-t\Delta} e^{-s\Delta}$ is translated to the identity

$$p_M(t + s; x, y) = \int_M p_M(t; x, z) p_M(s; z, y) dv_g(z) \,.$$

The important feature, which links the heat kernel to the geometry of M more tightly is the Minakshisundaram-Pleijel asymptotic expansion (see [**16**] p. 204). Namely, there exist C^∞ functions u_i on $M \times M$ such that for any integer q and for all (x, y) in $M \times M$ near the diagonal,

$$p_M(t; x, y) = \frac{1}{(4\pi t)^{n/2}} e^{-r^2(x,y)/4t} (u_0(x, y) + t u_1(x, y) + \cdots + t^q u_q(x, y) + O(t^{q+1})) \,,$$

where $r(x, y)$ is the Riemannian distance between x and y in M, and $t \to 0$. It follows from the proof that this expansion can be differentiated term by term as many times as needed (see [**16**] p.213, [**26**] p.154 and [**44**]). The functions u_i can be expressed as integrals of local quantities computable using the metric and its derivatives.

6.2. Estimating the heat kernel. We shall explain how to symmetrise solutions of parabolic equations or equivalently families of functions; the idea is to compare a symmetrised solution to the solution of a symmetrised problem. The reader can refer to the book [**6**] chapter IV section 3. In the context of Riemannian geometry it first appeared in [**13**], see also [**40**] and [**9**]. Let us quote the comparison theorem proved in [**11**] using the method of [**13**].

THEOREM 6.1 ([**13**] and [**11**]). *Let (M, g) be an n-dimensional closed Riemannian manifold satisfying*

$$r_{min} \operatorname{diam}(M)^2 \geq (n-1)\epsilon\alpha^2$$

for some $\epsilon \in \{-1, 0, 1\}$ and some positive number α. Then, for all $x \in M$ and $t > 0$,

$$\operatorname{vol}(M)p_M(t; x, x) \leq \operatorname{vol}(S^n(R))p_{S^n(R)}(t; p, p) = Z_{S^n(R)}(t) = Z_{S^n(1)}(t/R^2),$$

where p is any point on the sphere $S^n(R)$ of radius $R = R(n, \epsilon, \alpha)$ (here R is the same as in Theorem 4.4).

The last equality comes from the invariance of $p_{S^n(R)}$ on the two-point homogeneous manifold $S^n(R)$. An immediate and trivial corollary is the inequality

$$Z_M(t) \leq Z_{S^n(R)}(t) = Z_{S^n(1)}(t/R^2).$$

SKETCH OF PROOF, SEE [**6**], [**13**], [**9**], [**10**] AND [**40**]. For the sake of simplicity let us put $V = \operatorname{vol}(M)$ and $V^* = \operatorname{vol}(S^n(R))$. For $f \in C^\infty(M)$, a non-negative function on M, we define $D(s) = \{x \in M \mid f(x) > s\}$ and $a(s) = \frac{\operatorname{vol}(D(s))}{V}$ which is a non-increasing function. Let us define,

$$\overline{f}(\beta) = \inf\{s\,,\, a(s) < \beta\},$$

we then have $\overline{f}(a(s)) = s$ for all regular values s. As in the proof of Proposition 3.2, and with the same notations, at a regular value of f, we have

$$a'(s) = -\frac{1}{V}\int_{\partial D(s)} |df|^{-1} d\sigma_s\,.$$

We define now $F(\beta) = \int_{D(\overline{f}(\beta))} f(x)dv_g(x)$ and the co-area formula gives,

$$F(\beta) = \int_{\overline{f}(\beta)}^{\overline{f}(0)} \Big(\int_{\partial D(s)} f|df|^{-1}d\sigma_s\Big)ds = -V\int_{\overline{f}(\beta)}^{\overline{f}(0)} sa'(s)dt = V\int_0^\beta \overline{f}(u)du\,.$$

The last equality is obtained by the change of variable $u = a(s)$. We deduce from Green's formula that

$$\int_{D(\overline{f}(\beta))} \Delta_M f = \int_{\partial D(\overline{f}(\beta))} |df|d\sigma_s \geq \operatorname{vol}_{n-1}(\partial D(\overline{f}(\beta)))^2\Big(\int_{\partial D(\overline{f}(\beta))} |df|^{-1}d\sigma_s\Big)^{-1}.$$

On the other hand,

$$F''(\beta) = V\overline{f}'(\beta) = V(a'(\overline{f}(\beta)))^{-1} \leq 0$$

since a is non-increasing; we thus get,

$$\int_{D(\overline{f}(\beta))} \Delta_M f \geq -\Big(\frac{\operatorname{vol}_{n-1}(\partial D)}{V}\Big)^2 F''(\beta)\,.$$

On $S^n(R)$, if f^* is a radial function, then the level sets $D^*(s)$ are geodesic balls and $|df^*|$ is constant on $\partial D^*(s)$. We then get equality in all the above inequalities, so that, with $F^*(\beta) = \int_{D^*(\overline{f}^*(\beta))} f^*$, we have

$$\int_{D^*(\overline{f}^*(\beta))} \Delta_{S^n(R)} f^* = -\Big(\frac{\mathrm{vol}_{n-1}(\partial D^*)}{V^*}\Big)^2 F^{*''}(\beta)\,.$$

We now apply these computations to $f_t = p_M(t;x,.)$ and $f_t^* = p_{S^n(R)}(t;p,.)$ where p is any chosen point on the sphere. we define, as before $F(t,\beta) = \int_{\{f_t > \overline{f}_t(\beta)\}} f_t$ and, with a similar formula, F^*. With these choices the following lemma is proved in [**6**] (lemma 4.13, p.212),

LEMMA 6.2.

$$\frac{\partial F}{\partial t} = \int_{\{f_t > \overline{f}_t(\beta)\}} \frac{\partial}{\partial t} f_t = -\int_{\{f_t > \overline{f}_t(\beta)\}} \Delta_M f_t$$

and a similar formula is true with F^*.

Combining the above lemma, the definition of the isoperimetric profile and the fact that F'' is non-positive, we get

$$\begin{aligned}\frac{\partial F}{\partial t} &\leq h_M^2(\beta) F''(\beta)\\ \frac{\partial F^*}{\partial t} &= h_{S^n(R)}^2(\beta) F^{*''}(\beta)\end{aligned}$$

Notice that the two functions F and F^* are defined on the same two-dimensional space $(0,+\infty)\times(0,1)$. It is now possible to compare them. We define $G = F - F^*$, the Paul Lévy-Gromov isoperimetric inequality implies,

$$\frac{\partial G}{\partial t} - h_{S^n(R)}^2(\beta)\frac{\partial^2 G}{\partial \beta^2} \leq 0\,.$$

The function G satisfies $G(t,0) = G(t,1) = 0$ for all $t > 0$ and $\lim_{t\to 0^+} G(t,\beta) = 0$ for all $\beta \in [0,1]$, then a parabolic maximum principle (see [**72**]) gives $F \leq F^*$, that is,

$$V\int_0^a \overline{f}_t(\beta)d\beta \leq V^* \int_0^a \overline{f}_t^*(\beta)d\beta \quad \forall a \in [0,1]\,.$$

The following lemma is proved in [**6**] p.174,

LEMMA 6.3.

$$V^2 \int_0^1 \overline{f}_t(\beta)^2 d\beta \leq V^{*2} \int_0^1 \overline{f}_t^*(\beta)^2 d\beta$$

Now the co-area formula yields,

$$V^2 \int_M (p_M(t;x,y))^2 dv_g(y) \leq V^{*2} \int_{S^n(R)} (p_{S^n(R)}(t;p,q))^2 dq$$

and the semi-group property,

$$V p_M(2t;x,x) \leq V^* p_{S^n(R)}(2t;p,p)$$

here p is again any point on the sphere. This is the desired inequality. □

Comments 6.4. As for the Theorem 4.4, the above inequality is sharp for positive curvature ($\epsilon = 1$). If one allows non sharp results the same proof works using the Theorem 4.9 above. Precisely,

THEOREM 6.5 ([**41**], [**40**]). *Under the hypothesis of Theorem 4.9, one has*

$$\mathrm{vol}(M)p_M(t;x,x) \leq p^*(b(q,\alpha,D)t)\,,$$

where $b(q,\alpha,D)$ is a computable function and p^* is the heat kernel of a double cone computed at one vertex of this cone.

The double cone is just a convenient and geometric way of describing the one-dimensional model considered (see [**40**] for the details). The non sharpness of this comparison is seen in the following upper bound,

$$p^*(t) \leq C(q)t^{-q/2}\,.$$

Now, one can also obtain a similar result under the hypothesis of the Theorem 4.10, that is the almost positive curvature condition. We leave to the reader to state such a theorem.

In [**9**] appendix A, an abstract point of view on symmetrisation is presented. More precisely, one considers two Hilbert spaces $\mathcal{H}$ and $\mathcal{K}$ and a cone $\mathcal{K}^+ \subset \mathcal{K}$ satisfying few properties. A symmetrisation S is a map from $\mathcal{H}$ onto $\mathcal{K}^+$ fulfilling two conditions (see [**9**] p.163). Two examples are given:

1. the map $S(f) = |f|$ from the space of L^2 sections of a hermitian vector bundle on a manifold M (for example complex-valued functions on M) to the cone of non-negative functions on M.
2. The second example is the symmetrisation of functions as used above, where the cone is the space of radial functions on the model.

The inequalities on the heat kernel are interpreted as dominations of operators (the two heat operators in the second example). A criterion for domination in term of the quadratic form associated to the operators is given and checked in both examples. In the first one, this is an extension of the so-called Beurling-Deny criterion for positivity, namely the positivity of the heat kernel is equivalent to the following inequality,

$$\int_M |d|u||^2 \leq \int_M |du|^2$$

where u is a C^1-function on M. In other words the Dirichlet integral is reduced when taking the absolute value of the function. For the general case of a section it is related to the so-called Kato inequality as treated in [**48**]. In the second example the criterion is checked in details in [**9**] appendix A.

7. Spectral precompactness theorems

In this section we present an interpretation of the inequalities on the heat kernel in term of a precompactness theorem. The basic example is Gromov's precompactness theorem. Let

$$\mathcal{M}_{n,k,D} = \Big\{(M,g) \mid \dim M = n,\ \mathrm{Ricci}(g) \geq (n-1)kg,\ \mathrm{diam}(M,g) \leq D\Big\}.$$

M. Gromov defined a distance d_{GH} on the family of Riemannian manifold, now called the Gromov-Hausdorff distance, and proved the following (see [**45**]):

THEOREM 7.1 ([**45**]). *The metric space $(\mathcal{M}_{n,k,D}, d_{GH})$ is precompact.*

We are going to define a distance between manifolds using the heat kernel. Before that, we need some consequences of Theorem 6.1.

COROLLARY 7.2 ([**12**]). *There exist explicit functions $A(n,k,D)$, $B(n,k,D)$ and $C(n,k,D)$ such that, for all $(M,g) \in \mathcal{M}_{n,k,D}$,*

1. $\forall j \geq 0, \quad \lambda_j(M) \geq A(n,k,D) j^{2/n}$
2. $N_M(\lambda) = \sharp\{j \,|\, \lambda_j(M) \leq \lambda\} \leq 1 + B(n,k,D)\lambda^{n/2}$
3. $\forall x \in M, \forall \alpha \geq 0,$

$$\sum_{j\geq 1} \lambda_j^\alpha(M) e^{-t\lambda_j(M)} \phi_j^2(x) \leq \frac{C(n,k,D)}{\mathrm{vol}(M)}(\alpha+1)t^{-(n+2\alpha)/2}.$$

Remark 7.3. The first point is easy for a fixed Riemannian manifold thanks to Weyl's asymptotic formula (see [**9**]). What is important is the uniformity on the space $\mathcal{M}_{n,k,D}$. The third point is more interesting since it amounts to an estimate on the time derivative of "order" α of the heat kernel. Let us sketch the (easy) proof of this fact.

SKETCH OF PROOF OF (3). For $x \in M$ let us define the measure μ_x on $\mathbb{R}_+$ by $d\mu_x = \sum_{j\geq 1} \phi_j^2(x)\delta_{\lambda_j}$. Then

$$\sum_{j\geq 1} \lambda_j^\alpha e^{-t\lambda_j} \phi_j^2(x) = \int_{\mathbb{R}} \lambda^\alpha e^{-t\lambda} d\mu_x(\lambda).$$

An integration by parts yields,

$$0 \leq \int_{\mathbb{R}} \lambda^\alpha e^{-t\lambda} d\mu_x(\lambda) \leq \int_{\mathbb{R}} \lambda^{\alpha-1}(\lambda t + \alpha) e^{-t\lambda} \mu_x([0,\lambda]) d\lambda$$

but $\mu_x([0,\lambda]) = \sum_{0<\lambda_j\leq\lambda} \phi_j^2(x)$, thus

$$\sum_{0<\lambda_j\leq\lambda} \phi_j^2(x) \leq e \sum_{0<\lambda_j\leq\lambda} e^{-\lambda_j/\lambda}\phi_j^2(x) \leq e \sup_{x\in M} \left(p_M(\frac{1}{\lambda}; x, x) - \frac{1}{\mathrm{vol}(M)}\right)$$

$$\sum_{0<\lambda_j\leq\lambda} \phi_j^2(x) \leq \frac{e}{\mathrm{vol}(M)}\left(Z_{S^n(1)}(\frac{1}{\lambda R^2}) - 1\right) \leq E(n,k,D)\frac{\lambda^{n/2}}{\mathrm{vol}(M)}$$

and we conclude by a simple integration. □

7.1. Construction of some embeddings. We shall use the Hilbert space of sequences defined by

$$l^2 = \left\{ (a_j)_{j\geq 1}, a_i \in \mathbb{C} \;\middle|\; \sum_{j\geq 1} |a_j|^2 < +\infty \right\}.$$

For a Riemannian manifold (M,g), $\mathcal{A}$ will denote an orthonormal basis of $L^2(M, dv_g)$ consisting of eigenfunctions denoted by $\phi_j^{\mathcal{A}}$ (corresponding to the eigenvalue λ_j). There are plenty of possible choices, even if all eigenvalues are simple, since, in this case, one can choose $\phi_j^{\mathcal{A}}$ or $-\phi_j^{\mathcal{A}}$.

Definition 7.4. Let (M,g) and $\mathcal{A}$ be as above. For any $t > 0$, define the mapping $\Psi_t^{\mathcal{A}}$ by

$$\begin{aligned} \Psi_t^{\mathcal{A}} : M &\longrightarrow l^2 \\ x &\longmapsto \sqrt{2}(4\pi)^{n/4} t^{(n+2)/4} \{e^{-\lambda_j t/2} \phi_j^{\mathcal{A}}(x)\}_{j\geq 1} \end{aligned}$$

The constants that appear are just useful normalisations. In spirit this is just the map $x \longmapsto p_M(t;x,.)$, but we need the target space to be independent of (M,g), that is why the eigenfunctions are used as coordinates in an abstract Hilbert space. We now can prove the following elementary theorem (see [**12**]):

THEOREM 7.5 ([**12**]). *Let (M,g) and $\mathcal{A}$ be fixed.*

1. *For all $t>0$, the mapping $\Psi_t^{\mathcal{A}}$ is an embedding of M into l^2.*
2. *Let* can *be the canonical (Euclidean) metric in l^2 and* $\mathrm{scal}(g)$ *be the scalar curvature of g. Then*

$$(\Psi_t^{\mathcal{A}})^*(\mathrm{can}) = g + t/3(\frac{1}{2}\,\mathrm{scal}(g)g - \mathrm{Ricci}(g)) + O(t^2)\,,$$

when $t \to 0$.

Remarks 7.6. It is clear that $\Psi_t^{\mathcal{A}}$ is equivariant with respect to the action of the isometry group of M (if any) on M and on l^2 by permuting the eigenfunctions in the same eigenspace. Consequently, when (M,g) is homogeneous irreducible, one can check that $(\Psi_t^{\mathcal{A}})^*(\mathrm{can}) = C(t)g$, *i.e.* the embedding is isometric up to a constant. In general it is asymptotically isometric (up to a constant). The tensor that appears in the expansion of the pulled-back metric is geometrically significant however it is still open to draw nice geometric conclusion from this fact.

REMARKS ON THE PROOF. The proof is straightforward. The first part is a consequence of the fact that the eigenfunctions separate the points and of a similar property on their differentials. The second part consists in differentiating the Minakshisundaram-Pleijel asymptotic expansion. In fact one just have to differentiate the distance r and the function u_0. Let us recall that u_0 is related to the volume element at y computed in polar normal coordinates around x, and thus by differentiation the Ricci curvature appears. The metric comes from differentiating the distance and finally $u_1(x,x) = 1/6\,\mathrm{scal}(g)(x)$. The details are in [**12**]. □

Comments 7.7. The embedding is very similar to the case of minimal submanifolds of the sphere S^{n-1} in which the coordinates of $\mathbb{R}^n$ are eigenfunctions when restricted to the submanifold (endowed with the induced metric). Such an embedding is also used in the study of compact harmonic manifolds (see Besse's nice embedding in [**18**]). The idea was indeed to mimic these cases.

7.2. Spectral distances. The idea is now to use the Hausdorff distance in l^2 to define a distance between manifolds. One minor difficulty is to deal with the ambiguity brought by the choices of the basis of $L^2(M,dv_g)$. This ambiguity is given by the product of the orthogonal groups of the eigenspaces; since they are finite dimensional, by compactness of M, the product group is compact. The set of such basis will be called $\mathfrak{B}(M)$. Let us call E_j the eigenspace corresponding to λ_j and μ_j the j-th eigenvalue but counted without multiplicity. For a finite dimensional space E we define the distance between two orthonormal basis to be the Euclidean distance between the identity matrix and the transition matrix between these two basis and we call it d_E. Finally, for a basis $\mathcal{A}$ of $L^2(M,dv_g)$ of eigenfunctions we denote by $\mathcal{A}_{|E_j}$ the basis induced on E_j.

Definition 7.8. 1. For any two orthonormal basis $\mathcal{A}$ and $\mathcal{B}$ of $L^2(M, dv_g)$ of eigenfunctions of the Laplacian, we set

$$d(\mathcal{A},\mathcal{B})^2 = \sum_{j=1}^{+\infty} \mu_j^{-N} d_{E_j}(\mathcal{A}_{|E_j}, \mathcal{B}_{|E_j})^2$$

where $N > n/2$.

2. We define the mapping $I_t^{\mathcal{A}}$ by

$$\begin{array}{rcl} I_t^{\mathcal{A}} : M & \longrightarrow & l^2 \\ x & \longmapsto & \sqrt{\mathrm{vol}(M)}\{e^{-\lambda_j t/2}\phi_j^{\mathcal{A}}(x)\}_{j\geq 1} \end{array}$$

3. Denoting by d_H the Hausdorff distance between compact sets of l^2, we define, for two Riemannian manifolds (M,g) and (M',g'),

$$d_t(M,M') = \max\left\{ \sup_{\mathcal{A}} \inf_{\mathcal{A}'} d_H(I_t^{\mathcal{A}}(M), I_t^{\mathcal{A}'}(M'))\,,\ \sup_{\mathcal{A}'} \inf_{\mathcal{A}} d_H(I_t^{\mathcal{A}}(M), I_t^{\mathcal{A}'}(M')) \right\}$$

where $\mathcal{A} \in \mathfrak{B}(M)$ and $\mathcal{A}' \in \mathfrak{B}(M')$ (in the notation $d_t(M,M')$ we omit the reference to the metrics).

The following proposition is easy to check,

PROPOSITION 7.9. *For any $t > 0$, d_t is a distance on the isometry classes of Riemannian manifolds. In particular $d_t(M,M') = 0 \Rightarrow M$ and M' are isometric.*

7.3. Precompactness. We can then prove

THEOREM 7.10 ([**12**]). *The metric space $(\mathcal{M}_{n,k,D}, d_t)$ is precompact for all $t > 0$.*

This closes up this circle of ideas; indeed, the manifolds in the space $\mathcal{M}_{n,k,D}$, which is a precompact space when endowed with the Gromov-Hausdorff distance, satisfy an isoperimetric inequality, which implies estimates on the heat kernel, which in turn yields a spectral precompactness.

PROOF. Let h^1 be the following Sobolev space of sequences:

$$h^1 = \left\{ (a_j)_{j\geq 1}, a_j \in \mathbb{C} \,\middle|\, \sum_{j\geq 1}(1+j^{2/n})|a_j|^2 < +\infty \right\}.$$

The Rellich theorem (see [**73**]) shows that the obvious injection $h^1 \hookrightarrow l^2$ is compact, that is, the image of a bounded set in h^1 is relatively compact in l^2. Since the family of compact subsets of a compact metric space endowed with the Hausdorff distance is precompact, it suffices to show that the map $I_t^{\mathcal{A}}$ sends the manifolds of $\mathcal{M}_{n,k,D}$ into a bounded subset of h^1. This is given by the following estimates, for $x \in M \subset \mathcal{M}_{n,k,D}$:

$$\begin{aligned} \|I_t^{\mathcal{A}}(x)\|_{h^1}^2 &= \mathrm{vol}(M)\textstyle\sum_{j\geq 1}(1+j^{n/2})e^{-\lambda_j t}\phi_j^2(x) \\ &\leq E(n,k,D)\,\mathrm{vol}(M)\textstyle\sum_{j\geq 1}(1+\lambda_j)e^{-\lambda_j t}\phi_j^2(x) \\ &\leq F(n,k,D)t^{-n/2}(1+t^{-1})\,. \end{aligned}$$

Here we have used the Corollary 7.2 and the fact that $Z_{S^n(1)}(t) - 1 \leq C(n)t^{-n/2}$, for some constant $C(n)$. □

7.4. Some improvements. The first improvement can be made using Theorem 6.5. Corollary 7.2 above can be proved by means of the estimate of Theorem 6.5, and then a precompactness result can be obtained. The only difference is that the heat kernel of the double cone satisfies now $p^*(t) \leq C(q)t^{-q/2}$ for t small, which is irrelevant in the above proof since t is fixed. We thus define

$$\mathcal{M}^*(n,q,\alpha,D) = \Big\{(M,g) \mid \quad \dim M = n, \operatorname{diam}(M) \leq D \text{ and } \frac{1}{\operatorname{vol}(M)} \int_M \left(\frac{r_-}{(n-1)\alpha^2} - 1\right)_+^{q/2} dv_g \leq \tfrac{1}{2}(e^{B(q)\alpha D} - 1)^{-1}\Big\}.$$

THEOREM 7.11. *The metric space* $(\mathcal{M}^*(n,q,\alpha,D), d_t)$ *is precompact for all* $t > 0$.

It is also probably possible to use the result [**2**] to get rid of the restriction on the diameter in some instances. A geometric precompactness theorem with integral norms on the Ricci curvature is proved by P. Petersen and G. Wei in [**70**].

The second improvement of the above technique is due to A. Kasue and H. Kumura ([**53**], [**54**]) and more recently to A. Kasue ([**52**]). The above embeddings are used in a more systematic way, precisely the space $C^0([0,+\infty], l^2)$ is considered together with the map

$$I_{\mathcal{A}}[x](t) = (e^{-(t+1/t)/2} e^{-\lambda_j t/2} \phi_j(x))_{j \geq 0},$$

for $x \in M$. Then the distance in $C^0([0,+\infty], l^2)$ is taken to be

$$\Theta(\gamma,\sigma) = \sup\{\|\gamma(t) - \sigma(t)\|_{l^2} \mid t \in [0,+\infty]\}.$$

An easy computation shows that, for x and x' in M:

$$\Theta(I_{\mathcal{A}}[x], I_{\mathcal{A}}[x'])^2 = \sup_{t>0} e^{(t+1/t)}(p_M(t;x,x) + p_M(t;x',x') - 2p_M(t;x,x')).$$

So, if one defines a new distance on M by,

$$d_M^{\mathrm{spec}}(x,x')^2 = \sup_{t>0} e^{(t+1/t)}(p_M(t;x,x) + p_M(t;x',x') - 2p_M(t;x,x')),$$

then $I_{\mathcal{A}}$ is a distance preserving mapping between (M, d_M^{spec}) and $C^0([0,+\infty], l^2)$.

Then they consider in [**54**] a family $\mathcal{F}$ of compact connected Riemannian manifolds satisfying the following condition: there exist positive constants ν and C_U, such that for any $M \in \mathcal{F}$,

$$p_M(t;x,x) \leq \frac{C_U}{t^{\nu/2}}, \quad \text{for } t \in (0,1],\ x \in M.$$

This condition is implied by our assumptions on the Ricci curvature. The result is as follows.

THEOREM 7.12 ([**54**]). *The families* $\mathcal{F}^{spec} = \{(M, d_M^{spec})\}$ *and* $\mathcal{F}^{geo} = \{(M,g)\}$ *are precompact for, respectively, the spectral distance in* $C^0([0,+\infty], l^2)$ *and the Gromov-Hausdorff distance.*

Possible limits of manifolds in the above metric spaces have nice properties; they are shown to carry a Dirichlet form, a Radon measure, and a heat semi-group; the eigenvalues are shown to converge and the eigenfunctions too (in a certain sense, see [**54**] and the introduction of [**52**]). These are spaces on which one can do analysis; see [**52**] for more details on the limit spaces.

8. Miscellaneous comments and questions

We have left aside other cases of interest, for example, the case of manifolds with boundary and the case of non-compact manifolds. A problem, which also would deserve a survey, is the question of stability of the eigenvalues. Briefly, on the n-sphere the first eigenvalue has multiplicity $n+1$ and one can ask to describe the manifolds with Ricci $\geq (n-1)g$ and such that the $n+1$ (or fewer) first eigenvalues are close to the minimal possible value given by Lichnérowicz-Obata's theorem. This is closely related to the deep works of T. Colding (see [**33**], [**32**] and [**34**]). Recent results have been obtained in [**67**], [**2**] and [**17**].

For the non-compact case we can add the following comments. Let X be a compact manifold which carries an hyperbolic metric and let $\widetilde{X}$ be its universal cover. The heat kernel of the hyperbolic metric on $\widetilde{X}$ goes to zero uniformly on compact sets when t goes to $+\infty$. However the measure $p_{\widetilde{X}}(t;x,.)dv_g$ converges, when t goes to infinity, to the hyperbolic harmonic measure $\overline{p}_{\widetilde{X}}(x,\theta)d\theta$, where $\overline{p}$ denotes here the hyperbolic Poisson kernel, θ is a point on $\partial\widetilde{X} = S^{n-1}$ and $d\theta$ is the canonical measure on $\partial\widetilde{X} = S^{n-1}$ (here one has to choose an origin on $\widetilde{X}$). In [**20**] we use a family of measures on $\partial\widetilde{X}$, indexed by the points in $\widetilde{X}$ and associated to a metric on $\widetilde{X}$ pulled-back from X, to embed $\widetilde{X}$ into an infinite dimensional sphere. We then show that! among all such embeddings (when the metric varies) the embedding using the hyperbolic harmonic measure is absolutely minimizing for the volume of a fundamental domain. This is exactly a non-compact version of Besse's nice embedding (see [**20**] for the details); we do prove that a compact locally harmonic and negatively curved manifold is locally symmetric which is the analogue of the result obtained by Z. I. Szabo ([**75**]) for compact simply-connected harmonic manifolds.

In the compact case no such geometric result has been proved yet. One possibility could be to use a variation of the above theme given in [**12**]. Let us consider the embedding of a Riemannian manifold M into the unit sphere of l^2 defined by:

$$K_t^{\mathcal{A}}(x) = \frac{1}{\left(\sum_{j\geq 1} e^{-\lambda_j t}\phi_j^2(x)\right)^2}\left\{e^{-\lambda_j t/2}\phi_j(x)\right\}_{j\geq 1};$$

then the pulled back metric satisfies

$$(K_t^{\mathcal{A}})^*(\mathrm{can}) = 1/2t[g - \frac{t}{3}\,\mathrm{Ricci}(g) + O(t^2)].$$

QUESTION 1. *Is $K_t^{\mathcal{A}}$ asymptotically minimal?*

Another question is the study of limit spaces for the spectral distance. Although the results of [**52**], [**53**] and [**54**] are very interesting, the link between these limits and those obtained in the works of J. Cheeger and T. Colding (see [**28**], [**29**], [**30**], [**31**]) is not done. Let us call them the geometric limits. One result given by J. Cheeger and T. Colding (see [**29**]) is that when the limit of a sequence of compact Riemannian manifolds M_n with Ricci curvature bounded below (by the same constant) is a compact manifold M_∞, then, for n large, M_n is diffeomorphic to M_∞. It would be interesting to give another proof of this theorem.

QUESTION 2. *Can geometric limits be embedded in l^2, by a reasonable version of the heat kernel, onto the spectral limits?*

The embeddings presented in this text are related to the spectrum of the Laplacian, in particular when two Riemannian manifolds (M, g) and (N, h) are isospectral the embeddings satisfy

$$Z_M(t) = \frac{1}{\mathrm{vol}(M)} \int_M \|I_t^{\mathcal{A}}(x)\|_{l^2}^2 dv_g(x) = \frac{1}{\mathrm{vol}(N)} \int_N \|I_t^{\mathcal{B}}(x)\|_{l^2}^2 dv_h(x) = Z_N(t),$$

for all $t > 0$. Let us recall that the spectrum of the Laplacian determines the volume of the manifold. This means that the image manifolds in l^2 have, for all $t > 0$, the same moment of inertia.

QUESTION 3. *Can this technique say something about the isospectral problem?*

Another important question is to know what can be obtained by using other operators. Possible candidates are the Yamabe operator (see [**19**]) and the Dirac operator (see [**51**]). One can also study perturbed heat kernel, *i.e.* instead of defining the kernel by the parabolic equation considered above one could use the equation associated to an operator of the type:

$$\frac{\partial}{\partial t} + \Delta_{g(t)}$$

the difference being here that the metric varies with t, for example $g(t)$ could be the variation of metrics obtained from the Ricci flow. This kind of situation appears in G. Perel'man's recent works ([**65**]). A wild question can be:

QUESTION 4. *Can one see the Ricci flow as a family of submanifolds of l^2?*

References

[1] Th. AUBIN. *Non linear analysis on Manifolds. Monge-Ampère equations*, volume 252 of *Grundlehren der mathematischen Wissenschaften*. Springer-Verlag, 1982.

[2] E. AUBRY. *Variétés de coubure de Ricci minorée : inégalités géométriques optimales et stabilité des variétés extrémales*. PhD thesis, Université Joseph Fourier (Grenoble I), December 2003.

[3] D. BAKRY and M. LEDOUX. Lévy-Gromov's isoperimetric inequality for an infinite dimensional diffusion generator. *Invent. math.*, 123:259–281, 1996.

[4] D. BAKRY and M. LEDOUX. Sobolev inequalities and myers's diameter theorem for an abstract Markov generator. *Duke Math. Journal*, 85(1):253–270, 1996.

[5] D. BAKRY and Z. QUIAN. Some new results on eigenvectors via dimension, diameter and Ricci curvature. *Advances in Math.*, 155:98–153, 2000.

[6] C. BANDLE. *Isoperimetric Inequality and Applications*, volume 7 of *Monograph and Studies in Math.* Pitman Press, 1980.

[7] F. BARTHE. Log-concave and spherical models in isoperimetry. *Geom. and Funct. Anal.*, 12:32–55, 2002.

[8] V. BAYLE. A differential inequality for the isoperimetric profile. *International Math. Res. Not.*, 7:311–342, 2004.

[9] P. BÉRARD. *Spectral Geometry: Direct and Inverse Problems*, volume 1207 of *Lecture Notes in Mathematics*. Springer-Verlag, 1986.

[10] P. BÉRARD. From vanishing theorems to estimating theorems: the Bochner technique revisited. *Bulletin of the A.M.S.*, 19(2):371–406, 1988.

[11] P. BÉRARD, G. BESSON, and S. GALLOT. Sur une inégalité isopérimétrique qui généralise celle de Paul Lévy-Gromov. *Invent. Math.*, 80:295–308, 1985.

[12] P. BÉRARD, G. BESSON, and S. GALLOT. Embedding Riemannian manifolds by their heat kernel. *G.A.F.A.*, 4(4):373–398, 1994.

[13] P. BÉRARD and S. GALLOT. Inégalités isopérimétriques pour l'équation de la chaleur et applications à l'estimation de quelques invariants. In *Séminaire Goulaouic-Meyer-Schwartz*, Éxposé no. 15. École Polytechnique, 1983-84.

[14] P. Bérard and D. Meyer. Inégalités isopérimétriques et applications. *Ann. Sci. E.N.S. Paris*, 15:513–542, 1982.

[15] M. Berger. *Riemannian geometry during the second half of the century*. Number 17 in University Lecture Series. American Mathematical Society, 2000.

[16] M. Berger, P. Gauduchon, and E. Mazet. *Le spectre d'une variété riemannienne*, volume 194 of *Lecture notes in Math.* Springer-Verlag, 1971.

[17] J. Bertrand. *Pincement spectral en courbure positive*. PhD thesis, Université Paris-Sud, Orsay, Octobre 2003.

[18] A.L. Besse. *Manifolds all of whose geodesics are closed*. Number 93 in Ergebnisse der mathematik und ihrer grenzgebiete. Springer-Verlag, 1978.

[19] A.L. Besse. *Einstein Manifolds*. Number 10 in Ergebnisse der mathematik und ihrer grenzgebiete. 3 folge. Springer-Verlag, 1987.

[20] G. Besson, G. Courtois, and S. Gallot. Volume et entropie minimales des variétés localement symétriques. *G.A.F.A.*, 5(5):731–799, 1995.

[21] Y. Brenier. Polar factorization and monotone rearrangement of vector-valued functions. *Comm. Pure Appl. Math.*, 44:375–417, 1991.

[22] Yu.D. Burago and V.A. Zalgaller. *Geometric Inequalities*, volume 194 of *Grundlehren der Math. Wiss.* Springer-Verlag, 1985.

[23] L. Caffarelli. Boundary regularity of maps with convex potentials. *Comm. Pure Appl. Math.*, 45(9):1141–1151, 1992.

[24] L. Caffarelli. The regularity of mappings with a convex potential. *J. Amer. Math. Soc.*, 5(1):99–104, 1992.

[25] L. Caffarelli. Boundary regularity of maps with convex potentials ii. *Ann of Math.(2)*, 144(3):453–496, 1996.

[26] I. Chavel. *Eigenvalues in Riemannian geometry. With a chapter by B. Randol. With an appendix by J. Dodziuk*. Number 115 in Pure and Applied Mathematics. Inc, 1984.

[27] I. Chavel. *Isoperimetric inequalities*. Number 145. Cambridge University Press, 2001.

[28] J. Cheeger and T. Colding. Lower bounds on Ricci curvature and the almost rigidity of warped products. *Ann. of Math*, 144:189–237, 1996.

[29] J. Cheeger and T. Colding. On the structure of spaces with curvature bounded below I. *Journal of Diff. Geom.*, 46:406–480, 1997.

[30] J. Cheeger and T. Colding. On the structure of spaces with curvature bounded below II. *Journal of Diff. Geom.*, 54:13–35, 2000.

[31] J. Cheeger and T. Colding. On the structure of spaces with curvature bounded below III. *Journal of Diff. Geom.*, 54:37–74, 2000.

[32] T. Colding. Large manifolds with positive Ricci curvature. *Invent. Math.*, 124:193–214, 1996.

[33] T. Colding. Shape of manifolds with positive Ricci curvature. *Invent. Math.*, 124:175–191, 1996.

[34] T. Colding. Ricci curvature and volume convergence. *Annals of Math*, 145:477–501, 1997.

[35] D. Cordero-Erausquin, B. Nazaret, and C. Villani. A Mass-Transportation approach to sharp Sobolev and Gaglardo-Nirenberg inequalities. *Advances in Math.*, 2004. to appear.

[36] R. Courant. Beweis des satzes daßvon allen homogenen Membranen gegebenen Umfanges und gegebener Spannung die Kreisförmige den tiefsten Grundton besitzt. *Math. Zeitschrift*, 1:321–328, 1928.

[37] R. Courant and D. Hilbert. *Methods of mathematical physics*, volume I and II. Wiley-Interscience, I 1953, II 1962.

[38] G. Faber. Beweis, dass unter allen homogenen Membranen von gleicher Fläche und gleicher Spannung die Kreisförmige des tiefsten Grundton gibt. *Sitzungsber. Bayer. Akad. Wiss., Math. Phys. München*, pages 169–172, 1923.

[39] H. Federer. *Geometric Measure Theory*, volume 153 of *Grundlehren der mathematischen Wissenschaften*. Springer-Verlag, 1969.

[40] S. Gallot. Inégalités isopérimétriques et analytiques sur les variétés riemanniennes. In Ida Cattaneo Gasparini, editor, *On the geometry of differentiable manifolds, Roma, june 23-27, 1986*, volume 163-164 of *Astérisque*, pages 31–91. Société Mathématique de France, 1988.

[41] S. Gallot. Isoperimetric inequalities based on integral norms of the Ricci curvature. In *Colloque Paul Lévy sur les processus stochastiques*, volume 157-158 of *Astérisque*, pages 191–216. Société Mathématique de France, 1988.

[42] S. GALLOT, J. LAFONTAINE, and D. HULIN. *Riemannian Geometry*. Universitext. Springer-Verlag, 2nd edition, 1990.

[43] R. J. GARDNER. The Brunn-Minkowski inequality. *Bull. Amer. Math. Soc.*, 39(3):355–405, 2002.

[44] P.B. GILKEY. *Invariance theory, heat equation and the Atiyah-Singer index theory*. Studies in Advanced Mathematics. CRC Press, 2nd edition, 1995.

[45] M. GROMOV. *Metric structures for Riemannian and non-Riemannian spaces. With appendices by M. Katz, P. Pansu and S. Semmes*, volume 152 of *Progress in Mathematics*. Birkhäuser, 2nd edition, 1999.

[46] G. HARDY, J. LITTLEWOOD, and G. POLYA. *Inequalities*. Cambridge University Press, Cambridge, 1967.

[47] E. HEINTZE and H. KARCHER. A general comparison theorem with applications to volume estimates for submanifolds. *Ann. Sci. Ec. Norm. Sup.*, 11:451–470, 1978.

[48] H. HESS, R. SCHRADER, and D. UHLENBROCK. Kato's inequality and the spectral distribution of Laplacians on compact Riemannian manifolds. *Journal of Diff. Geom.*, 15:27–37, 1980.

[49] A. HURWITZ. Sur le problème des isopérimètres. *C. R. Acad. Sci. Paris Sér. I Math.*, 132:401–403, 1901.

[50] A. HURWITZ. Sur quelques applications géométriques des séries de Fourier. *Ann. Sci. École Norm. Sup. (3)*, 19:357–408, 1902.

[51] B.H. LAWSON J. and M-L. MICHELSON. *Spin Geometry*. Number 38 in Princeton mathematical series. Princeton University Press, 1989.

[52] A. KASUE. Convergence of Riemannian manifolds and Laplace operators I. *Ann. Inst. Fourier, Grenoble*, 52(4):1219–1257, 2002.

[53] A. KASUE and H. KUMURA. Spectral convergence of Riemannian manifolds. *Tôhoku Math. Journal*, 46:147–179, 1994.

[54] A. KASUE and H. KUMURA. Spectral convergence of Riemannian manifolds II. *Tôhoku Math. Journal*, 48:71–120, 1996.

[55] E. KRAHN. Über eine von rayleigh formulierte Minimaleigenshaft des Kreises. *Math. Annalen*, 94:97–100, 1925.

[56] P. KRÖGER. On the spectral gap for compact manifolds. *J. Diff. Geom.*, 36:315–330, 1992.

[57] H. LEBESGUE. *Leçons sur les séries trigonométriques*. Gauthier-Villars, Paris, 1906.

[58] P. LÉVY. *Problèmes concrets d'analyse fonctionnelle*. Gauthier-Villars, Paris, 1951.

[59] P. LI and S.T. YAU. Estimates of eigenvalues of a compact riemannian manifold. volume 36 of *Proc. Sympos. Pure Math.*, pages 205–239. American Math. Soc., Providence, R.I., 1980.

[60] A. LICHNÉROWICZ. *Géométrie des groupes de transformations*, volume 3 of *Travaux et recherches mathématiques*. Dunod, 1958.

[61] R. MCCANN. Existence and uniqueness of monotone measure-preserving maps. *Duke Math. Journal*, 80:309–323, 1995.

[62] V.D. MILMAN and G. SCHECHTMAN. *Asymptotic Theory of Finite Dimensional Normed Spaces*, volume 1200 of *Lecture Notes in Math.* Springer-Verlag, 1980.

[63] F. MORGAN and D.L. JOHNSON. Some sharp isoperimetric theorems for Riemannian manifolds. *Indiana Univ. Math. J.*, 49(3):1017–1041, 2000.

[64] M. OBATA. Certain conditions for a Riemannian manifold to be isometric with a sphere. *J. Math. Soc. Japan*, 14:333–340, 1962.

[65] G. PEREL'MAN. The entropy formula for the Ricci flow and its geometric applications. ArXiv: Math. DG/ 0211159v1, Nov. 11 2002.

[66] P. PETERSEN. Convergence theorems in Riemannian geometry. In Karsten Grove and Peter Petersen, editors, *Comparison geometry*, volume 30 of *M.R.S.I. Publications*, pages 167–202. Cambridge University Press, 1997.

[67] P. PETERSEN. On eigenvalue pinching in positive Ricci curvature. *Invent. Math.*, 138:1–21, 1999.

[68] P. PETERSEN, S. SHTEINGOLD, and G. WEI. Comparison geometry with integral curvature bounds. *Geom. and Funct. Anal.*, 7:1011–1030, 1997.

[69] P. PETERSEN and C. SPROUSE. Integral curvature bounds, distance estimates and applications. *J. Diff. Geom.*, 50:269–298, 1998.

[70] P. PETERSEN and G. WEI. Relative volume comparison with integral curvature bounds. *Geom. and Funct. Anal.*, 7:1031–1045, 1997.

[71] G. Pólya and G. Szegö. *Isoperimetric inequalities in mathematical physics.* Number 27 in Annals of Mathematics Studies. Princeton University Press, 2nd edition, 1966.
[72] M.H. Protter and H.F. Weinberger. *Maximum principles in differential equations.* Prentice-Hall, Englewood Cliffs, N.J., 1967.
[73] M. Reed and B. Simon. *Methods of modern mathematical physics*, volume I-IV. Academic Press, 1975.
[74] E. Schmidt. Beweis der isoperimetrischen Eigenshaft der Kugel im hyperbolischen und sphärischen Raum jeder Dimensionenzahl. *Math. Zeitschrift*, 49:1–109, 1943.
[75] Z.I. Szabo. The Lichnérowicz conjecture on harmonic manifolds. *Journal of Diff. Geom.*, 31:1–28, 1990.
[76] G. Talenti. Elliptic equations and rearrangements. *Ann. Scuola Norm. Sup. Pisa*, 3:697–718, 1976.
[77] J. W. Strutt (Lord Rayleigh). *The Theory of Sound.* reprinted by Dover in 1945, 1894.
[78] C. Villani. *Topics in Optimal Transportation*, volume 58 of *Graduate Studies in Mathematics.* American Mathematical Society, Providence R.I., 2003.
[79] J.Q. Zhong and H.C. Yang. On the estimate of the first eigenvalue of a compact Riemannian manifold. *Scientia Sinica Ser. A*, 27(12):1265–1273, 1984.

Institut Fourier, U.M.R. 5582 du C.N.R.S., Université de Grenoble I, B.P. 74, 38402 Saint Martin d'Hères Cedex, France
E-mail address: G.Besson@ujf-grenoble.fr

Surveys in Differential Geometry
IX, International Press

Discrete Isoperimetric Inequalities

Fan Chung

CONTENTS

1. Introduction

One of the earliest problems in geometry is the isoperimetric problem, which was considered by the ancient Greeks. The problem is to find, among all closed curves of a given length, the one which encloses the maximum area. Isoperimetric problems for the discrete domain are in the same spirit but with different complexity. A basic model for communication and computational networks is a graph $G = (V, E)$ consisting of a set V of vertices and a prescribed set E of unordered pairs of vertices. For a subset X of vertices, there are two types of boundaries:

- The *edge boundary* $\partial(X) = \{\{u, v\} \in E \ : \ u \in X, v \in V \setminus X\}$.
- The *vertex boundary* $\delta(X) = \{v \in V \setminus X \ : \ \{v, u\} \in E$ for some $u \in X\}$.

Numerous questions arise in examining the relations between $\partial(X), \delta(X)$ and the sizes of X. Here the size of a subset of vertices may mean the number of vertices, the number of incident edges, or some other appropriate measure defined on graphs.

In this paper, we will survey spectral techniques for studying discrete isoperimetric inequalities and the like. In addition, a number of applications in extremal graph theory and random walks will be included. This paper is organized as follows:

Section 1: A brief introduction and an outline are given.

Section 2: Definitions and notations of the combinatorial Laplacian and normalized Laplacian are introduced as well as the historical matrix-tree theorem.

Section 3: Enumeration of spanning trees using eigenvalues of the combinatorial and normalized Laplacians

Section 4: Complimentary to the isoperimetric inequalities is the discrepancy inequalities. Instead of tracking the number of edges leaving X, discrepancy inequalities provide estimates of the edges remaining in X. Here we give several versions of discrepancy inequalities for general graphs.

Section 5: The usual Cheeger inequality is for bounding the edge boundary. Here we also consider vertex boundaries and their generalizations.

Section 6: Eigenvalues are related to the diameter of a graph, i.e., the maximum distance between any two vertices. We will give several eigenvalue inequalities for bounding the diameter and the distance between two or more subsets both for graphs and manifolds.

Section 7: We consider discrete Sobolev inequalities which include Cheeger inequalities as a special case.

Section 8: Harnack inequalities, which provide pointwise estimates for eigenfunctions, can be established for certain convex subgraphs of homogeneous graphs.

Section 9: The heat kernel of a graph contains all the spectral information about the graph. For example, the heat kernel can be used to deduce bounds for eigenvalues of certain induced subgraphs.

Section 10-14: The containment or avoidance of specified subgraphs, such as paths, cycles and cliques of given sizes are central themes in classical Ramsey and Turán theory. Our focus here is to derive such extremal graph properties as direct consequences of spectral bounds. The applications include the forcing of long paths and cycles, universal graphs for trees, chromatic numbers and list chromatic numbers, and the Turán numbers, as well as random walks involving the enumeration of contingency tables.

In the applicational sections, there are some overlaps with the study of the so-called (n, d, λ)-graphs (i.e., regular graphs on n vertices having degree d with all but one eigenvalue of the adjacency matrix bounded above by λ). Such graphs are extensively examined in many papers by Alon [**4**] and others [**46, 60**]. There is a recent comprehensive survey by Krivelevich and Sudakov [**46**] on (n, d, λ)-graphs. Here we deal with general graphs with no degree constraints.

Throughout the paper, we consider only finite graphs. The reader is referred to the book by Woess [**63**], which explores isoperimetric properties and random walks on infinite graphs. There is a close connection between discrete isoperimetric inequalities and their continuous counterpart, as evidenced in Section 7 to 9. Isoperimetric inequalities for Riemannian geometry have been long studied and well developed (see [**12, 64**]). As a result, one of the earlier approaches on discrete isoperimetric inequalities focuses on discretizations of manifolds [**35, 42**]. Another approach is to study graphs with group symmetry [**58**] or random walks on finite groups [**36**]. In this paper, we consider general graphs and our approach here is from a graph-theoretic point of view.

2. Combinatorial and normalized Laplacian

One of the classical results in graph theory is the matrix-tree theorem by Kirchhoff [**43**], which states that the number of spanning trees in a graph is determined by the determinant of a principle minor of the combinatorial Laplacian. For a graph G with vertex set V and edge set E, the combinatorial Laplacian L is a matrix with rows and columns indexed by vertices in V and can be written as

$$L = D - A$$

where D is the diagonal matrix with $D(v, v)$ equal to the degree d_v of v and A is the adjacency matrix of G. If G is a simple graph with no loops or multiple edges, $A(u, v) = 1$ if u and v are adjacent or else $A(u, v) = 0$. In this paper, we restrict ourselves to simple graphs although most of the statements and results can be easily carried out to general graphs or weighted graphs.

The combinatorial Laplacian has its root in homological algebra and spectral geometry. We can write L as

$$L = BB^* \tag{1}$$

where B is the incidence matrix whose rows are indexed by the vertices and whose columns are indexed by the edges of G such that each column corresponding to an edge $e = \{u, v\}$ has an entry 1 in the row corresponding to u, an entry -1 in the row corresponding to v, and has zero entries elsewhere. (As it turns out, the choice of signs can be arbitrary as long as one is positive and the other is negative.) Also, B^* denotes the transpose of B. Here B can be viewed as a "boundary operator" mapping "1-chains" defined on edges (denoted by C_1) of a graph to "0-chains" defined on vertices (denoted by C_0). Then, B^* is the corresponding "coboundary operator" and we have

$$C_1 \underset{B^*}{\overset{B}{\rightleftarrows}} C_0$$

This fact can be used to give a short proof of the matrix-tree theorem as follows: For a fixed vertex v, let L' denote the submatrix obtained by deleting the vth row and vth column of L. Since $L = BB^*$, we have $L' = B_0 B_0^*$ where B_0 denotes the submatrix of B without the vth column. By the Binet-Cauchy Theorem [**53**] we have

$$\det B_0 B_0^* = \sum_X \det B_X \det B_X^*$$

where the sum ranges over all possible choices of size $n - 1$ subsets X of $E(G)$ and B_X denotes the square submatrix of B_0 whose $n - 1$ columns correspond to the edges in X and whose rows are indexed by all the vertices except for v. Standard graph-theoretical arguments can be used to show that $|\det B_X| = 1$ if edges in X form a tree and 0 otherwise. Thus, $\det B_0 B_0^*$ is exactly the number of spanning trees in G, as asserted by the matrix-tree theorem.

Geometrically, the combinatorial Laplacian L can be viewed as the discrete analog of of the Laplace-Beltrami operator, especially for graphs that are the Cartesian products of paths, for example. For a path $v_0, v_1, \ldots, v_n$ and a function f that assigns a real value to each v_i, the combinatorial Laplacian can be written as :

$$\begin{aligned} Lf(v) &= \sum_{u \sim v} (f(v) - f(u)) \\ Lf(v_i) &= (f(v_i) - f(v_{i-1})) - (f(v_{i+1}) - f(v_i)) \\ &= \nabla f(v_i, v_{i-1}) - \nabla f(v_{i+1}, v_i) \\ &= \Delta f(v_i) \end{aligned}$$

for $0 < i < n$. Here we use the following notation:

$$\begin{aligned} \text{For an edge } \{u, v\},\ \nabla f(u, v) &= f(u) - f(v). \\ \text{For a vertex } v,\ \Delta f(v) &= \sum_{u \sim v} (f(v) - f(u)). \\ \|\nabla f\|^2 = \sum_{u \sim v} (f(u) - f(v))^2 &= \langle f, Lf \rangle. \end{aligned}$$

where $u \sim v$ means $\{u, v\} \in E(G)$. Note that $\nabla f(u, v)$ can be viewed as the first derivative in the direction of the edge $\{u, v\}$ and $\Delta f(v)$ can be regarded as the sum of second derivatives over all directions along the edges incident to v. The so-called Dirichlet sum is just $\|\nabla f\|^2$ as indicated above.

Since L is self-adjoint as seen in (1), its eigenvalues are non-negative. We denote the eigenvalues of L of a graph G on n vertices by

$$0 = \sigma_0 \le \sigma_1 \le \ldots \le \sigma_{n-1}.$$

One of the main approaches in spectral graph theory is to deduce various graph properties from eigenvalue distributions. In order to so so, it is sometimes appropriate to consider the normalized Laplacian $\mathcal{L}$, especially for diffusion-type problems such as random walks.

$$\mathcal{L} = D^{-1/2} L D^{-1/2}$$

Here we preclude isolated vertices in order to guarantee that D is invertible, an inconsequential constraint in practice. For regular graphs L and $\mathcal{L}$ are basically the same (up to a scale factor). However, for general graphs, it is often advantageous to utilize the normalized Laplacian. We denote the eigenvalues of $\mathcal{L}$ of G by

$$0 = \lambda_0 \le \lambda_1 \le \ldots \le \lambda_{n-1}.$$

To compare $\mathcal{L}$ and L, we note that

$$\lambda_1 = \inf_{f \perp \phi_0} \frac{\langle f, \mathcal{L} f \rangle}{\langle f, f \rangle} = \inf_{\sum g(x) d_x = 0} \frac{\langle g, Lg \rangle}{\langle g, Dg \rangle} = \inf_{\sum g(x) d_x = 0} \frac{\|\nabla g\|^2}{\langle g, Dg \rangle}$$

where ϕ_0 denotes the eigenfunction associated with eigenvalue 0 of $\mathcal{L}$ and $g = D^{-1/2} f$. In contrast,

$$\sigma_1 = \inf_{\sum_x f(x) = 0} \frac{\langle f, Lf \rangle}{\langle f, f \rangle} = \inf_{\sum_x f(x) = 0} \frac{\|\nabla f\|^2}{\|f\|^2}.$$

From the above equations, we see that $\mathcal{L}$ is the Laplace operator with the weight/measure of a vertex taken to be the degree of the vertex while L is the corresponding operator having all vertices with weights equal. The combinatorial

Laplacian is simpler, but the normalized Laplacian is sometimes better for capturing graph properties that are sensitive to degree distributions.

3. Eigenvalues and spanning trees

We consider separately the cases for the combinatorial Laplacian and normalized Laplacian.

3.1. Eigenvalues of the combinatorial Laplacian and spanning trees. The number of spanning trees, denoted by $\tau(G)$, can be related to the eigenvalues of L in the following folklore theorem: For completeness, we briefly describe the proof here.

THEOREM 1. *For a graph G on n vertices , the number of spanning trees $\tau(G)$ is :*

$$\tau(G) = \frac{1}{n} \prod_{i \neq 0} \sigma_i$$

where σ_i are eigenvalues of the combinatorial Laplacian L.

PROOF. Consider the characteristic polynomial $p(x)$ of the combinatorial Laplacian L.

$$p(x) = det(L - xI).$$

The coefficient of the linear term is exactly

$$-\prod_{i \neq 0} \sigma_i.$$

On the other hand, the coefficient of the linear term of $p(x)$ is -1 times the sum of the determinant of n principle submatrices of L obtained by deleting the ith row and ith column. By the matrix-tree theorem, the product $\prod_{i \neq 0} \sigma_i$ is exactly n times the number of spanning trees of G. □

3.2. Eigenvalues of the normalized Laplacian and spanning trees.

THEOREM 2. *For a graph G with degree sequences (d_v), the number of spanning trees $\tau(G)$ is*

$$\tau(G) = \frac{\Pi_v d_v}{\sum_v d_v} \; \Pi_{i \neq 0} \lambda_i$$

where λ_i are eigenvalues of the normalized Laplacian $\mathcal{L}$.

PROOF. We consider the coefficient of the linear term in

$$\mathcal{P}(x) = \det(\mathcal{L} - xI).$$

We have

$$-\prod_{i \neq 0} \lambda_i = -\sum_v \det \mathcal{L}_v$$

where $\mathcal{L}_v$ is the submatrix obtained by deleting the row and column corresponding to v. From the matrix-tree theorem, we have

$$\det \mathcal{L}_v = \frac{\det L_v}{\Pi_{u \neq v} d_u} \det \mathcal{L}_v = \frac{\tau(G)}{\Pi_{u \neq v} d_u}.$$

Thus we have

$$\tau(G) = \frac{\Pi_v d_v}{\sum_v d_v} \Pi_{i \neq 0} \lambda_i$$

as desired. □

4. A discrepancy inequality

For a graph G and a subset X of vertices in G, the volume $\mathrm{vol}(X)$ is defined by

$$\mathrm{vol}(X) = \sum_{v \in X} d_v$$

where d_v is the degree of v. We note that for a simple graph G, the degree d_v of v is the number of neighbors of v in G. We will denote the volume of G by $\mathrm{vol}(G) = \sum_v d_v$.

For two subsets X and Y of vertices in G, we write

$$e(X,Y) = \{(x,y) : x \in X, y \in Y, \{x,y\} \in E(G)\}.$$

Eigenvalues can be used to estimate $e(X,Y)$, as summarized in the discrepancy inequalities in this section.

4.1. A general discrepancy inequality.

THEOREM 3. *For a graph G and two subsets of vertices X and Y, we have*

$$|e(X,Y) - \frac{\mathrm{vol}(X)\mathrm{vol}(Y)}{\mathrm{vol}(G)}| \le \bar{\lambda}\frac{\sqrt{\mathrm{vol}(X)\mathrm{vol}(\bar{X})\mathrm{vol}(Y)\mathrm{vol}(\bar{Y})}}{\mathrm{vol}(G)}$$

where the normalized Laplacian has eigenvalues λ_i, and $\bar{\lambda} = \max_{i \neq 0} |1 - \lambda_i|$.

PROOF. We consider the characteristic function of X.

$$\psi_X(u) = \begin{cases} 1 & \text{if } u \in X, \\ 0 & \text{otherwise.} \end{cases}$$

Then we can write

$$\begin{aligned} e(X,Y) &= \langle \psi_X, A\psi_Y \rangle \\ &= \langle \psi_X, D^{1/2}(I - \mathcal{L})D^{1/2}\psi_Y \rangle \\ &= a_0 b_0 + \sum_{i \neq 0} a_i b_i (1 - \lambda_i) \end{aligned}$$

where $D^{1/2}\psi_X = \sum_i a_i \phi_i$, $D^{1/2}\psi_Y = \sum_i b_i \phi_i$, and the ϕ_i's are orthonormal eigenfunctions of $\mathcal{L}$ associated with λ_i. In particular, $\phi_0(v) = \sqrt{d_v/\mathrm{vol}(G)}$ and $a_0 b_0 = \mathrm{vol}(X)\mathrm{vol}(Y)/\mathrm{vol}(G)$. Thus

$$\begin{aligned} |e(X,Y) - \frac{\mathrm{vol}(X)\mathrm{vol}(Y)}{\mathrm{vol}(G)}| &= |e(X,Y) - a_0 b_0| \\ &= \sum_{i \neq 0} a_i b_i (1 - \lambda_i) \\ &\le \bar{\lambda} \sum_{i \neq 0} |a_i b_i| \\ &\le \bar{\lambda} \sqrt{\sum_{i \neq 0} a_i^2 \sum_{j \neq 0} b_j^2} \\ &= \bar{\lambda} \frac{\sqrt{\mathrm{vol}(X)\mathrm{vol}(\bar{X})\mathrm{vol}(Y)\mathrm{vol}(\bar{Y})}}{\mathrm{vol}(G)} \end{aligned}$$

as desired. □

4.2. Discrepancy inequalities using combinatorial Laplacians. The discrepancy inequalities using eigenvalues of the combinatorial Laplacian are more complicated. We will give several versions of the discrepancy inequalities depending on the intersection of subsets X and Y. We will use the fact that an eigenfunction of L associated with eigenvalue 0 is the all 1's function (but it is in general not an eigenfunction for A).

THEOREM 4. *Suppose that a graph G with n vertices has average degree d and the eigenvalues σ_i of the combinatorial Laplacian satisfy $|d - \sigma_i| \leq \theta$ for $i \neq 0$. Then for any two subsets X and Y of vertices in G, the number $e(X,Y)$ of edges with one endpoint in X and the other in Y satisfies*

$$\left| e(X,Y) - \frac{d}{n}|X|\ |Y| + d \mid X \cap Y \mid - \mathrm{vol}(X \cap Y) \right| \leq \frac{\theta}{n}\sqrt{|X|\ (n - |X|)|Y|(n - |Y|)}$$

The proof of Theorem 4 will be given later. We will first state several immediate consequences of Theorem 4.

THEOREM 5. *Suppose that a graph G with n vertices has average degree d and the eigenvalues σ_i of the combinatorial Laplacian satisfy $|d - \sigma_i| \leq \theta$ for $i \neq 0$. Then for any two disjoint subsets X and Y of vertices in G, the number $e(X,Y)$ of (ordered) edges with first endpoint in X and the second endpoint in Y satisfies*

$$\left| e(X,Y) - \frac{d}{n}|X|\ |Y| \right| \leq \frac{\theta}{n}\sqrt{|X|\ (n - |X|)|Y|(n - |Y|)}. \tag{2}$$

COROLLARY 1. *Suppose that a graph G with n vertices has average degree d and the eigenvalues σ_i of the combinatorial Laplacian satisfy $|d - \sigma_i| \leq \theta$ for $i \neq 0$. For two disjoint subsets X and Y each with at least $\theta n/d$ vertices in G, there is at least one edge joining a vertex in X to a vertex in Y.*

PROOF. If $e(X,Y) = 0$, we have, by substituting into (2),

$$\frac{d}{n}\sqrt{|X|\ |Y|} \leq \frac{\theta}{n}\sqrt{|\bar{X}|\ |\bar{Y}|}$$

which is impossible for $|X|, |Y| \geq \theta n/d$. □

By setting Y to be the complement of X in (2), we have the following:

COROLLARY 2. *Suppose that a graph G with n vertices has average degree d and the eigenvalues σ_i of the combinatorial Laplacian satisfy $|d - \sigma_i| \leq \theta$ for $i \neq 0$. Then for a subset X of vertices and its complement $\bar{X}$, the number $e(X, \bar{X})$ of (ordered) edges with first endpoint in X and the second endpoint in $\bar{X}$ satisfies*

$$\frac{d-\theta}{n}|X|\ |\bar{X}| \leq e(X, \bar{X}) \leq \frac{d+\theta}{n}|X|\ |\bar{X}|.$$

By setting $Y = \{v\}$ in (2), we have the following:

COROLLARY 3. *Suppose that a graph G with n vertices has average degree d and the eigenvalues σ_i of the combinatorial Laplacian satisfy $|d - \sigma_i| \leq \theta$ for $i \neq 0$. Then for all vertices v, we have*

$$\frac{n}{n-1}(d - \theta) \leq d_v \leq \frac{n}{n-1}(d + \theta) < d + \theta.$$

Using Theorem 5, we will prove the following:

THEOREM 6. *Suppose that a graph G with n vertices has average degree d and the eigenvalues σ_i of the combinatorial Laplacian satisfy $|d - \sigma_i| \le \theta$ for $i \neq 0$. Suppose that vertex v has l_v loops. Then for a subset X of vertices in G, the number $e(X,X)$ of ordered pairs corresponding to edges with both endpoints in X satisfies*

$$\left| e(X,X) - \frac{d|X|^2}{n} + (\frac{d|X|}{n} - \sum_{v \in X} l_v) \right| \le \frac{2\theta}{n}|X| \; (n - |X|/2).$$

PROOF. Let $x = |X|$ and X' denote a subset of X of size $x' = \lfloor x/2 \rfloor$. We apply Theorem 5 on X' and $X \setminus X'$ and we have

$$\left| e(X', X \setminus X') - \frac{d}{n}\lfloor\frac{x}{2}\rfloor \lceil\frac{x}{2}\rceil \right| \le \frac{\theta}{n}\sqrt{x'(n-x')(x-x')(n-x+x')}.$$

Since

$$\binom{x-2}{x'-1}(e(X,X) - \sum_{v \in X} l_v) = \sum_{X' \subseteq X, |X'| = x'} e(X', X \setminus X'),$$

we have

$$\begin{aligned}
&\binom{x-2}{x'-1}\left| e(X,X) - \frac{d}{n}x^2 + (\frac{dx}{n} - \sum_{v \in X} l_v) \right| \\
&= \left| \left(\sum_{X' \subseteq X, |X'| = x'} e(X', X \setminus X') - \binom{x-2}{x'-1}\frac{d}{n}x(x-1) \right) \right| \\
&= \left| \sum_{X' \subseteq X, |X'| = x'} (e(X', X \setminus X') - \frac{d}{n}\lfloor\frac{x}{2}\rfloor\lceil\frac{x}{2}\rceil) \right| \\
&\le \binom{x}{x'}\frac{\theta}{n}\sqrt{x'(n-x')(x-x')(n-x+x')}.
\end{aligned}$$

Therefore we have

$$\begin{aligned}
\left| e(X,X) - \frac{d}{n}x^2 + (\frac{dx}{n} - \sum_{v \in X} l_v) \right| &\le \frac{\binom{x}{x'}}{\binom{x-2}{x'-1}}\frac{\theta}{n}\sqrt{x'(n-x')(x-x')(n-x+x')} \\
&\le \frac{2\theta}{n}x(n - x/2)
\end{aligned}$$

as desired. □

COROLLARY 4. *Suppose that a graph G with n vertices has average degree d and the eigenvalues σ_i of the combinatorial Laplacian satisfy $|d - \sigma_i| \le \theta$ for $i \neq 0$. Then for a subset X of vertices in G, the number $e(X)$ of edges with both endpoints in X satisfy*

$$\left| 2e(X) - \frac{d|X|(|X|-1)}{n} \right| \le \frac{2\theta}{n}|X| \; (n - |X|/2).$$

COROLLARY 5. *Suppose that a graph G with n vertices has average degree d and the eigenvalues σ_i of the combinatorial Laplacian satisfy $|d - \sigma_i| \le \theta$ for $i \neq 0$. Then an independent subset in G can have no more than $\theta n/d + 1$ vertices.*

It remains to prove Theorem 4.

Proof of Theorem 4: Let φ_i, $i = 0, \ldots, n-1$, denote orthonormal eigenvectors of the combinatorial Laplacian. The eigenvalue associated with φ_i is σ_i. Then for $i = 0$, $\varphi_0 = \mathbf{1}/\sqrt{n}$ where $\mathbf{1}$ is the all 1's vector. Suppose we write

$$\begin{aligned} \psi_X &= \sum_{i=0}^{n-1} a_i \varphi_i, \\ \psi_Y &= \sum_{i=0}^{n-1} b_i \varphi_i. \end{aligned}$$

Here, let I denote the identity matrix. We consider

$$\begin{aligned} e(X,Y) + d|X \cap Y| &= \langle \psi_X, (A + dI)\psi_Y \rangle \\ &= \langle \psi_X, (D - L)\psi_Y \rangle + d\langle \psi_X, \psi_Y \rangle \\ &= \mathrm{vol}(X \cap Y) - \sum_{i \neq 0} \sigma_i a_i b_i + d \sum_i a_i b_i \\ &= \mathrm{vol}(X \cap Y) - \sum_{i \neq 0} (d - \sigma_i) a_i b_i + d a_0 b_0. \end{aligned}$$

We note that

$$\begin{aligned} a_0 &= \langle \psi_X, \phi_0 \rangle = \frac{|X|}{\sqrt{n}}, \\ b_0 &= \langle \psi_Y, \phi_0 \rangle = \frac{|Y|}{\sqrt{n}}. \end{aligned}$$

Thus, we have

$$e(X,Y) + d|X \cap Y| = \mathrm{vol}(X \cap Y) - \sum_{i \neq 0} (d - \sigma_i) a_i b_i + \frac{d}{n}|X|\,|Y|.$$

Hence,

$$\begin{aligned} &| \, e(X,Y) - \frac{d}{n}|X|\,|Y| + d|X \cap Y| - \mathrm{vol}(X \cap Y) \, | \\ \leq \; & \sum_{i \neq 0} (d - \sigma_i) a_i b_i \\ \leq \; & \theta \sum_{i \neq 0} |a_i b_i| \\ \leq \; & \theta \sqrt{\sum_{i \neq 0} a_i^2 \sum_{j \neq 0} b_j^2} \\ \leq \; & \theta \sqrt{(\|\psi_X\|^2 - a_0^2)(\|\psi_Y\|^2 - b_0^2)} \\ \leq \; & \theta \frac{\sqrt{(|X|(n - |X|)|Y|(n - |Y|)}}{n} \end{aligned}$$

as claimed. □

In the preceding proof, if we use $d' = (\sigma_1 + \sigma_{n-1})/2$ in place of d, then we have the following:

THEOREM 7. *Suppose that in a graph G with n vertices, the combinatorial Laplacian has eigenvalues $0 = \sigma_0 \le \sigma_1 \ldots \le \sigma_{n-1}$. Then for any two subsets X and Y of vertices in G, the number $e(X,Y)$ of edges with one endpoint in X and the other in Y satisfies*

$$\left| e(X,Y) - \frac{d'}{n}|X|\,|Y| + d' \mid X \cap Y \mid - \mathrm{vol}(X \cap Y) \right|$$
$$\le \frac{(\sigma_{n-1} - \sigma_1)}{2} \sqrt{|X|(n-|X|)|Y|(n-|Y|)}$$

where $d' = (\sigma_{n-1} + \sigma_1)/2$.

5. Cheeger inequalities

In a graph $G = (V, E)$ and a subset X of vertices, we define several versions of neighborhoods and boundaries.

- The *neighborhood* $N(X)$ is defined to be

$$N(X) = \{v \ : \ v \in X \text{ or } v \sim u \in X\}$$

where $v \sim u$ means $\{u, v\}$ is an edge.

- The *exact neighborhood* $\Gamma(X)$ is just

$$\Gamma(X) = \{v \ : \ v \sim u \in X\}.$$

- The *vertex boundary* $\delta(X)$ of a subset X of vertices is $\delta(X) = \{v \notin X \ : \ v \sim u \in X\}$ In general, for an integer $k \ge 1$, the k-neighborhood $N_k(X)$ of X is defined to be $N_k(X) = N(N_{k-1}(X))$ and $N_0(X) = X$. The k-boundary $\delta_k(X)$ is just $N_k(X) \setminus N_{k-1}(X)$.
- The *edge boundary* $\partial(X)$ is $\partial(X) = e(X, \bar{X})$. where $\bar{X}$ is the complement of X.

5.1. Isoperimetric inequalities for edge boundaries. The edge boundary is closely related to the discrete Cheeger's constant, which is defined as follows (see [**15, 13**]).

$$h_G = \inf_X \frac{|\partial(X)|}{\min\{\mathrm{vol}(X), \mathrm{vol}(\bar{X})\}}.$$

The eigenvalues of the normalized Laplacian are related to Cheeger's constant by the discrete Cheeger inequality:

$$2h_G \ge \lambda_1 \ge \frac{h_G^2}{2}.$$

Clearly, this implies

$$\frac{|\partial(X)|}{\mathrm{vol}(X)} \ge h_G \ge \frac{\lambda_1}{2}$$

if $\mathrm{vol}(X) \le \mathrm{vol}(\bar{X})$. For subsets with given cardinality, a slightly stronger lower bound is as follows:

LEMMA 1. *For a graph G and a subset of vertices X with* $\mathrm{vol}(X) \le \mathrm{vol}(G)/2$, *we have*

$$\frac{|\partial(X)|}{\mathrm{vol}(X)} \ge \lambda_1\left(1 - \frac{\mathrm{vol}(X)}{\mathrm{vol}(G)}\right) \ge \frac{\lambda_1}{2}$$

where λ_1 is the least non-trivial eigenvalue of the normalized Laplacian of G.

The proof follows from the definition of λ_1 and can be found in [**13**]. Along the same line, the following holds for the combinatorial Laplacian.

LEMMA 2. *For a graph G on n vertices and a subset of vertices X with $|X| \leq n/2$, we have*

$$\frac{|\partial(X)|}{|X|} \geq \sigma_1(1 - \frac{|X|}{n}) \geq \frac{\sigma_1}{2}$$

where σ_1 is the least non-trivial eigenvalue of the combinatorial Laplacian of G.

We note that the Cheeger inequality used in Lemma 1 can be slightly improved (see [**13**]).

$$2h_G \geq \lambda_1 > 1 - \sqrt{1 - h_G^2}.$$

Another characterization of the Cheeger constant h_G of a graph G can be described as follows (see [**13**]):

$$h_G = \inf_f \sup_{c \in \mathbb{R}} \frac{\sum_{x \sim y} |f(x) - f(y)|}{\sum_{x \in V} |f(x) - c| d_x} \tag{3}$$

where f ranges over all functions $f : V \to \mathbb{R}$ which are not identically zero.

A variation of (3) seems to be particularly useful, e.g., for deriving isoperimetric relationships between graphs and their Cartesian products [**24**].

$$h_G \geq \inf_{f \neq 0} \frac{\sum_{x \sim y} |f(x) - f(y)|}{\sum_{x \in V} |f(x)| d_x} \geq \frac{1}{2} h_G$$

where $f : V(G) \to \mathbb{R}$ satisfies $\sum_{x \in V} f(x) d_x = 0$.

5.2. Isoperimetric inequalities for vertex boundaries. We will prove the following basic isoperimetric inequality:

THEOREM 8. *Suppose that a graph G with n vertices has average degree d and the eigenvalues σ_i of the combinatorial Laplacian satisfy $|d - \sigma_i| \leq \theta$ for $i \neq 0$. Then for any nonempty subset X, the boundary $\delta(X)$ of X satisfies*

$$\frac{|\delta(X)|}{|X|} \geq \frac{d^2 - \theta^2}{\theta^2 + d^2|X|/(n - |X|)}.$$

PROOF. We use Theorem 4 and choose $Y = V(G) \backslash X \backslash \delta(X)$. Since $e(X, Y) = \emptyset$, we have

$$\frac{d}{n}|X|\,|Y| \leq \theta \frac{\sqrt{|X|(n - |X|)|Y|(n - |Y|)}}{n}.$$

Thus,

$$d^2|X|\,|Y| \leq \theta^2(|X| + |\delta(X)|)(|Y| + |\delta(X)|).$$

This implies

$$\begin{aligned} \frac{d^2 - \theta^2}{\theta^2}|X|\,|Y| &\leq |\delta(X)|(|X| + |Y| + |\delta(X)|) \\ &\leq |\delta(X)|\, n. \end{aligned}$$

Therefore,

$$\frac{d^2-\theta^2}{\theta^2 n}|X|\,(n-|X|-|\delta(X)|) \;\le\; |\delta(X)|$$

which is equivalent to

$$\frac{d^2-\theta^2}{\theta^2 n}|X|\,(n-|X|) \;\le\; |\delta(X)|(1+\frac{d^2-\theta^2}{\theta^2 n}|X|).$$

Finally, we have

$$\begin{aligned}\frac{|\delta(X)|}{|X|} &\ge \frac{(d^2-\theta^2)(n-|X|)}{\theta^2 n+(d^2-\theta^2)|X|}\\ &= \frac{(d^2-\theta^2)(n-|X|)}{\theta^2(n-|X|)+d^2|X|}\\ &= \frac{d^2-\theta^2}{\theta^2+d^2|X|/(n-|X|)}\end{aligned}$$

as claimed. □

As an immediate consequence of Theorem 8, we have the following three corollaries:

COROLLARY 6. *Suppose that a graph G with n vertices has average degree d and the eigenvalues σ_i of the combinatorial Laplacian satisfy $|d-\sigma_i| \le \theta$ for $i \ne 0$. For any subset X with $|X| \ll n\theta^2/d^2$, the neighborhood $N(X) = X \cup \delta(X)$ of X satisfies*

$$\frac{|N(X)|}{|X|} \ge (1+o(1))\frac{d^2}{\theta^2}.$$

COROLLARY 7. *Suppose that a graph G with n vertices has average degree d and the eigenvalues σ_i of the combinatorial Laplacian satisfy $|d-\sigma_i| \le \theta$ for $i \ne 0$. For any subset X with $|X| \ll n\theta^{2t}/d^{2t}$, the t-neighborhood $N_t(X) = N(N_{t-1}(X))$ of X satisfies*

$$\frac{|N_t(X)|}{|X|} \ge (1+o(1))\frac{d^{2t}}{\theta^{2t}}.$$

COROLLARY 8. *Suppose that a graph G with n vertices has average degree d and the eigenvalues σ_i of the combinatorial Laplacian satisfy $|d-\sigma_i| \le \theta$ for $i \ne 0$. For any vertex v, we have*

$$\delta_t(v) \;\ge\; \frac{(d-\theta)(d^{2t-2}-\theta^{2t-2})}{\theta^{2t-2}}(1-\frac{d^2}{n(d^2-\theta^2)}(\frac{d^2}{\theta^2})^{t-1}). \tag{4}$$

In particular, for $\theta = o(d)$, we have

$$\delta_t(v) \;\ge\; (1+o(1))\frac{d^{2t-1}}{\theta^{2t-2}}(1-\frac{1}{n}(\frac{d^2}{\theta^2})^{t-1}).$$

PROOF. For the case of $t = 1$, inequality (4) follows from Corollary 3. For $t > 1$, inequality (4) is proved by using Corollary 6 and 7.

Remark 1: The isoperimetric inequality for regular graphs by Tanner [**61**] and Alon [**4**] is a special case of Theorem 8.

6. Diameter-eigenvalue inequalities

In a graph G, the distance between two vertices u and v, denoted by $d(u,v)$, is defined to be the length of a shortest path joining u and v in G. The diameter of G, denoted by $D(G)$, is the maximum distance over all pairs of vertices in G. Although the diameter is a combinatorial invariant, it is closely related to eigenvalues. In [**14**], the following relation between the diameter and eigenvalues holds for regular graphs (except for complete graphs).

$$D(G) \leq \left\lceil \frac{\log(n-1)}{\log(1/(1-\lambda))} \right\rceil . \tag{5}$$

Here, λ is $\lambda = \lambda_1$ if $1 - \lambda_1 \geq \lambda_{n-1} - 1$.

6.1. Eigenvalues and diameters. Inequality (5) can be generalized to all graphs by using the combinatorial Laplacian [**19**].

THEOREM 9. *Suppose a graph G on n vertices has eigenvalues $0 \leq \sigma_1 \leq \ldots \leq \sigma_{n-1}$. The diameter of G satisfies*

$$D(G) \leq \left\lceil \frac{\log(n-1)}{\log \dfrac{\sigma_{n-1}+\sigma_1}{\sigma_{n-1}-\sigma_1}} \right\rceil .$$

We note that for some graphs the above bound gives a pretty good upper bound for the diameter. For example, for k-regular Ramanujan graphs, we have $1 - \lambda_1 = \lambda_{n-1} - 1 = 1/(2\sqrt{k-1})$, so we get $D \leq \log(n-1)/(2\log(k-1))$, which is within a factor of 2 of the best possible bound.

The bound in (5) can be further improved by using the Chebyshev polynomial of degree t. We can then replace the logarithmic function by $cosh^{-1}$ (see [**19**]) :

$$D(G) \leq \left\lceil \frac{\cosh^{-1}(n-1)}{\cosh^{-1} \frac{\sigma_{n-1}+\sigma_1}{\sigma_{n-1}-\sigma_1}} \right\rceil .$$

6.2. Distances between two subsets. Instead of considering distances between two vertices, we can relate the eigenvalue λ_1 to distances between two subsets of vertices. For two subsets X, Y of vertices in G, the distance between X and Y, denoted by $d(X,Y)$, is the minimum distance between a vertex in X and a vertex in Y; i.e., $d(X,Y) = \min\{d(x,y) : x \in X, y \in Y\}$. In [**31**], the distance between two sets can be related to eigenvalues as follows:

THEOREM 10. *Suppose G is not a complete graph. For $X, Y \subset V(G)$,*

$$d(X,Y) \leq \left\lceil \frac{\log \sqrt{\frac{\mathrm{vol}\bar{X}\,\mathrm{vol}\bar{Y}}{\mathrm{vol}X\ \mathrm{vol}Y}}}{\log \frac{\lambda_{n-1}+\lambda_1}{\lambda_{n-1}-\lambda_1}} \right\rceil . \tag{6}$$

6.3. Higher order eigenvalues and distances among many subsets. For any $k > 1$, we can relate the eigenvalue λ_k to distances among $k+1$ distinct subsets of vertices [**31**].

THEOREM 11. *Suppose G is not a complete graph. For $X_i \subset V(G)$, $i = 0, 1, \cdots, k$, we have*

$$\min_{i \neq j} d(X_i, X_j) \leq \max_{i \neq j} \left\lceil \frac{\log \sqrt{\frac{\mathrm{vol}\bar{X}_i \, \mathrm{vol}\bar{X}_j}{\mathrm{vol}X_i \, \mathrm{vol}X_j}}}{\log \frac{1}{1-\lambda_k}} \right\rceil$$

if $1 - \lambda_k \geq \lambda_{n-1} - 1$.

We note that the condition $1 - \lambda_k \geq \lambda_{n-1} - 1$ can be eliminated by modifying λ_k: For $X_i \subset V(G)$, $i = 0, 1, \cdots, k$, we have

$$\min_{i \neq j} d(X_i, X_j) \leq \max_{i \neq j} \left\lceil \frac{\log \sqrt{\frac{\mathrm{vol}\bar{X}_i \, \mathrm{vol}\bar{X}_j}{\mathrm{vol}X_i \, \mathrm{vol}X_j}}}{\log \frac{\lambda_{n-1}+\lambda_k}{\lambda_{n-1}-\lambda_k}} \right\rceil$$

if $\lambda_k \neq \lambda_{n-1}$.

Another useful generalization is the following: For $X_i \subset V(G)$, $i = 0, 1, \cdots, k$, we have

$$\min_{i \neq j} d(X_i, X_j) \leq \min_{0 \leq j < k} \max_{i \neq j} \left\lceil \frac{\log \sqrt{\frac{\mathrm{vol}\bar{X}_i \, \mathrm{vol}\bar{X}_j}{\mathrm{vol}X_i \, \mathrm{vol}X_j}}}{\log \frac{\lambda_{n-j-1}+\lambda_{k-j}}{\lambda_{n-j-1}-\lambda_{k-j}}} \right\rceil$$

where j satisfies $\lambda_{k-j} \neq \lambda_{n-j-1}$.

6.4. Eigenvalue upper bounds for manifolds. The above discrete methods can be used to derive new eigenvalue upper bounds for compact smooth Riemannian manifolds [**21, 22**]. Let M be a complete Riemannian manifold with finite volume and let $\mathcal{L}$ be the self-adjoint operator $-\Delta$, where Δ is the Laplace operator associated with the Riemannian metric on M. Or, we could consider a compact Riemannian manifold M with boundary and let $\mathcal{L}$ be a self-adjoint operator $-\Delta$ subject to the Neumann or Dirichlet boundary conditions.

The operator $\mathcal{L} = -\Delta$ is self-adjoint and has a discrete spectrum in $L^2(M, \mu)$, where μ denotes the Riemannian measure. Let the eigenvalues be denoted by $0 = \lambda_0 < \lambda_1 \leq \lambda_2 \leq \cdots$. Let $\mathrm{dist}(x, y)$ be a distance function on $M \times M$ which is Lipschitz and satisfies

$$|\nabla \mathrm{dist}(x, y)| \leq 1$$

for all $x, y \in M$. For example, $\mathrm{dist}(x, y)$ may be taken to be the geodesic distance, but we don't necessarily assume this is the case.

Using very similar methods as in the discrete case, it can be shown that (see [**21**]):

For two arbitrary measurable disjoint sets X and Y on M, we have

$$\lambda_1 \leq \frac{1}{\mathrm{dist}(X, Y)^2} \left(1 + \log \frac{(\mu M)^2}{\mu X \, \mu Y}\right)^2.$$

Moreover, if we have $k+1$ disjoint subsets $X_0, X_1, \cdots, X_k$ such that the distance between any pair of them is greater than or equal to $D > 0$, then we have for any $k \geq 1$,

$$\lambda_k \leq \frac{1}{D^2}(1 + \sup_{i \neq j} \log \frac{(\mu M)^2}{\mu X_i \, \mu X_i})^2.$$

Although differential geometry and spectral graph theory have a great deal in common, there is no question that significant differences exist. Obviously, a graph is not "differentiable", and many geometrical techniques involving high-order derivatives could be very difficult, if not impossible, to utilize for graphs. There are substantial obstacles for deriving the discrete analogs of many of the known results in the continuous case. Nevertheless, the above result is an example of mutual fertilization that shed insight to both the continuous and discrete cases.

7. Sobolev constants and Sobolev inequalities

The characterization of the Cheeger constant is basically the Rayleigh quotient using the L_1-norm both in the numerator and denominator. In general, we can consider the so-called Sobolev constant for all $p, q > 0$:

$$s_{p,q} = \inf_f \frac{\sum_{u \sim v} |f(u) - f(v)|^p}{\sum_v |f(v)|^q d_v} = \inf_f \frac{\int |\nabla f|^p}{\int |f|^q}$$

where f ranges over functions satisfying

$$\sum_x |f(x) - c|^q d_x \geq \sum_x |f(x)|^q d_x$$

for any c, or, equivalently,

$$\int |f - c|^q \geq \int |f|^q.$$

The eigenvalue λ_1 corresponds to the case of $p = q = 2$, while the Cheeger constant is associated with the case of $p = q = 1$.

There are many common concepts that provide connections and interactions between spectral graph theory and Riemannian geometry. For example, the Sobolev inequalities for graphs can be proved almost entirely by classical techniques which can be traced back to Nash [**64**]. We will describe Sobolev inequalities which hold for all general graphs. However, such inequalities depend on a graph invariant, the so-called isoperimetric dimension.

We say that a graph G has isoperimetric dimension δ with an isoperimetric constant c_δ if for all subsets X of $V(G)$, the number of edges between X and the complement $\bar{X}$ of X, denoted by $e(X, \bar{X})$, satisfies

$$e(X, \bar{X}) \geq c_\delta (\mathrm{vol} X)^{\frac{\delta - 1}{\delta}} \tag{7}$$

where vol $X \leq$ vol $\bar{X}$ and c_δ is a constant depending only on δ. Let f denote an arbitrary function $f : V(G) \to \mathbb{R}$. The following Sobolev inequalities hold [**25**].

(i) For $\delta > 1$, $\sum_{u \sim v} |f(u) - f(v)| \geq c_1 \min_m (\sum_v |f(v) - m|^{\frac{\delta}{\delta-1}} d_v)^{\frac{\delta-1}{\delta}}$

(ii) For $\delta > 2$, $(\sum_{u \sim v} |f(u) - f(v)|^2)^{\frac{1}{2}} \geq c_2 \min_m (\sum_v |(f(v) - m)^\alpha d_v)^{\frac{1}{\alpha}}$

where $\alpha = \frac{2\delta}{\delta - 2}$ and c_1, c_2 are constants depending only on c_δ.

The above two inequalities can be used to derive the following eigenvalue inequalities for a graph G (see [**25**]):

$$\sum_{i>0} e^{-\lambda_i t} \leq c\frac{\text{vol}(G)}{t^{\frac{\delta}{2}}}$$

$$\lambda_k \geq c'(\frac{k}{\text{vol}(G)})^{\frac{2}{\delta}}$$

for suitable constants c and c' which depend only on δ and c_δ. In a way, a graph can be viewed as a discretization of a Riemannian manifold in $\mathbb{R}^n$ where n is roughly equal to δ. The eigenvalue bounds above are analogs of the Polyá conjecture for Dirichlet eigenvalues of a regular domain M.

$$\lambda_k \geq \frac{2\pi}{w_n}(\frac{k}{\text{vol } M})^{\frac{2}{n}}$$

where w_n is the volume of the unit disc in $\mathbb{R}^n$.

In a later paper [**23**], the condition in (7) is further relaxed. It was shown that if in a graph $G = (V, E)$, any subset $X \subseteq V$ satisfies

$$e(X, \bar{X}) \geq c(\text{vol}(X))^{(\delta-1)/\delta}$$

for $\text{vol}(X) \leq c_1$, then the Dirichlet eigenvalue $\lambda_k(S)$ for the induced subgraph S satisfies

$$\lambda_k(S) \geq c'(\frac{k}{\text{vol}(S)})^{2/\delta}$$

where c' depends on δ and c, provided $|S| \geq k \geq \frac{(2+\delta)e}{c_1}\text{vol}(S)$.

An interesting question is to deduce the isoperimetric dimension or inequalities such as (7) from an arbitrarily given graph. In [**23**], we examine certain sufficient conditions on graph distance functions and their modifications for deriving (7).

8. Harnack inequalities

A crucial part of spectral graph theory concerns understanding the behavior of eigenfunctions. Intuitively, an eigenfunction maps the vertices of a graph to the real line in such a way that edges serve as "elastic bands" with the effect of pulling adjacent vertices closer together. To be specific, let f denote an eigenfunction with eigenvalue λ in a graph G (or for an induced subgraph S with nonempty boundary). Locally, at each vertex, the eigenfunction stretches the incident edges in a balanced way. That is, for each vertex x, f satisfies

$$\sum_{\substack{y \\ y\sim x}} (f(x) - f(y)) = \lambda f(x) d_x.$$

Globally, we would like to have some notion that adjacent vertices are close to one another. In spectral geometry, Harnack inequalities are exactly the tools for capturing the essence of eigenfunctions. There are many different versions of Harnack inequalities (involving constants depending on the dimension of the manifold, for example). We consider the following inequality for graphs.

A Harnack inequality:
For every vertex x in a graph G and some absolute constant c, any eigenfunction

f with eigenvalue λ satisfies

$$\frac{1}{d_x}\sum_{\substack{y\\ y\sim x}}(f(x)-f(y))^2 \le c\lambda \max_z f^2(z).$$

However, the above inequality does not hold for all graphs in general. An easy counterexample is the graph formed by joining two complete graphs of the same size by a single edge. Nevertheless, we can establish a Harnack inequality for certain homogeneous graphs and their "strongly convex subgraphs".

A homogeneous graph is a graph Γ together with a group H acting on the vertices satisfying:

1. For any $g \in H, u \sim v$ if and only if $gu \sim gv$.
2. For any $u, v \in V(\Gamma)$ there exists $g \in H$ such that $gu = v$.

In other words, Γ is vertex transitive under the action of H, and the vertices of Γ can be labeled by cosets H/I where $I = \{g : gv = v\}$ for a fixed v. Also, there is an edge-generating set $K \subset H$ such that for all vertices $v \in V(\Gamma)$ and $g \in K$, we have $\{v, gv\} \in E(\Gamma)$. A homogeneous graph is said to be invariant if K is invariant as a set under conjugation by elements of K, i.e., for all $a \in K$, $aKa^{-1} = K$.

Let f denote an eigenfunction in an invariant homogeneous graph with edge-generating set K consisting of k generators. Then it can be shown [**26**] that

$$\frac{1}{k}\sum_{a\in K}(f(x)-f(ax))^2 \le 8\lambda \sup_y f^2(y).$$

An induced subgraph S of a graph Γ is said to be *strongly convex* if for all pairs of vertices u and v in S, all shortest paths joining u and v in Γ are contained in S. The main theorem in [**26**] asserts that the following Harnack inequality holds.

Suppose S is a strongly convex subgraph in an abelian homogeneous graph with edge-generating set K consisting of k generators. Let f denote an eigenfunction of S associated with the Neumann or Dirichlet eigenvalue λ. Then for all $x \in S$, $x \sim y$,

$$|f(x)-f(y)|^2 \le 8k\lambda \sup_{z\in S} f^2(z).$$

(The detailed definition of Neumann or Dirichlet eigenvalues of an induced subgraph will be given in the next section.)

A direct consequence of the Harnack inequalities is the following lower bound for the Neumann or Dirichlet eigenvalue λ_1 of S:

$$\lambda_1 \ge \frac{1}{8kD^2}$$

where k is the maximum degree and D is the diameter of S. Such eigenvalue bounds are particularly useful for deriving polynomial approximation algorithms when enumeration problems of combinatorial structures can be represented as random walk problems on "convex" subgraphs of appropriate homogeneous graphs. However, the condition of being a strongly convex subgraph poses quite severe constraints, which will be relaxed in the next section.

9. Heat kernel eigenvalue inequalities

In a graph G, for a subset S of the vertex set $V = V(G)$, the induced subgraph determined by S has edge set consisting of all edges of G with both endpoints in S. Although an induced subgraph can also be viewed as a graph in its own right, it is natural to consider an induced subgraph S as having a boundary. For an induced subgraph S with non-empty boundary, there are, in general, two kinds of eigenvalues — the Neumann eigenvalues and the Dirichlet eigenvalues – subject to different boundary conditions.

For the Neumann eigenvalues, the Laplacian $\mathcal{L}$ acts on functions $f : S \cup \delta S \to \mathbb{R}$ with the Neumann boundary condition, i.e., for every $x \in \delta S$, $\sum_{y \in S, y \sim x}(f(x) - f(y)) = 0$.

For the Dirichlet eigenvalues, the Laplacian $\mathcal{L}$ acts on functions with the Dirichlet boundary condition. In other words, we consider the space of functions $\{f : V \to \mathbb{R}\}$ which satisfy the Dirichlet condition $f(x) = 0$ for any vertex x in the vertex boundary δS of S.

The Neumann boundary condition corresponds to the Neumann boundary condition for Riemannian manifolds: $\frac{\partial f(x)}{\partial \nu} = 0$ for x on the boundary where ν is the normal direction orthogonal to the tangent hyperplane at x. Neumann eigenvalues are closely associated with random walk problems, whereas the Dirichlet eigenvalues are related to many boundary-value problems (Details can be found in [**13**]). In the remainder of this section, we will focus on Neumann eigenvalues and the heat kernel techniques to obtain eigenvalue lower bounds. In the literature, there are many general formulations for discrete heat kernel in connection with the continuous heat kernels [**34, 40**]. Here we define a natural heat kernel for general graphs.

Let ϕ_i denote the eigenfunction for the Laplacian corresponding to eigenvalue λ_i. We now define the heat kernel of S as a $n \times n$ matrix

$$\begin{aligned} H_t &= \sum e^{\lambda_i t} P_i \\ &= e^{-t\mathcal{L}} \\ &= I - t\mathcal{L} + \frac{t^2}{2}\mathcal{L} + \cdots \end{aligned}$$

where $\mathcal{L} = \sum \lambda_i P_i$ is the decomposition of the Laplacian $\mathcal{L}$ into projections on its eigenspaces. In particular, we have $H_0 = I$, and for $F(x,t) = \sum_{y \in S \cup \delta S} H_t(x,y) f(y) = (H_t f)(x)$ with $F(x, 0) = f(x)$, F satisfies the heat equation

$$\frac{\partial F}{\partial t} = -\mathcal{L}F.$$

By using the heat kernel, the following eigenvalue inequality can be derived, for all $t > 0$:

$$\lambda_S \geq \frac{1}{2t} \sum_{x \in S} \inf_{y \in S} H_t(x,y) \frac{\sqrt{d_x}}{\sqrt{d_y}}. \tag{8}$$

One way to use the above theorem is to bound the heat kernel of a graph by the (continuous) heat kernel of Riemannian manifolds. We say Γ is a lattice graph if Γ is embedded into a d-dimensional Riemannian manifold $\mathcal{M}$ with a metric μ such that $\epsilon = \mu(x, gx) = \mu(y, g'y)$ for all $g, g' \in K$ and $x, y \in V(\Gamma)$. An induced subgraph on a subset S of a lattice graph Γ is said to be *convex* if there is a submanifold $M \subset \mathcal{M}$ with a convex boundary $\partial M \neq \emptyset$ such that S consists of all vertices of Γ in the interior of M. Furthermore, we require that for any vertex x, the *Voronoi* region $R_x = \{y : \ \mu(y, x) < \mu(y, z)$ for all $z \in \Gamma \cap \mathcal{M}\}$ is contained in M.

One additional condition is needed here. Basically, ϵ has to be "small" enough so that the count of vertices in S can be used to approximate the volume of the manifold M. Namely, let us define

$$r = \frac{U\ |S|}{\mathrm{vol}(M)} \tag{9}$$

where U denotes the maximaum volume of a *Voronoi* region. Then the main result in [**28**] states that the Neumann eigenvalue λ_1 of S satisfies the following inequality:

$$\lambda_1 \geq \frac{c\ r\ \epsilon^2}{d^2 D^2(M)}$$

for some absolute constant c, which depends only on Γ, and $D(M)$ denotes the diameter of the manifold M. We note that r in (9) can be bounded below by a constant if the diameter of M measured in L_1 norm is at least ϵd. The applications on random walks in Section 14 will use the above eigenvalue inequality.

10. Paths and cycles

One of the major theorems in studying the paths of a graph is a result of Pósa [**57**] (see [**54**], Problem 10.20, for an elegant solution).

Pósa's Theorem
In a graph H if every subset X of vertices with $|X| \leq k$ satisfies

$$|\delta(X)| \geq 2|X| - 1,$$

then H contains a path with $3k - 2$ vertices.

THEOREM 12. *Suppose that a graph G with n vertices has average degree d and the eigenvalues σ_i of the combinatorial Laplacian satisfy $|d - \sigma_i| \leq \theta$ for $i \neq 0$. Then G contains a path of at least*

$$(1 - \frac{2\theta^2}{d^2 - \theta^2})n - 2 \textit{ vertices.}$$

PROOF. To deduce the existence of a path of $\frac{d^2-3\theta^2}{d^2-\theta^2}n - 2$ vertices, it suffices to show that we have $\delta(X) \geq 2|X| - 1$ for any subset X with cardinality at most

$$|X| \leq \frac{d^2 - 3\theta^2}{3(d^2 - \theta^2)}n = x_0$$

by using Pósa's Theorem. From Theorem 8, we know that

$$\frac{|\delta(X)|}{|X|} \geq \frac{d^2 - \theta^2}{\theta^2 + d^2 x_0/(n - x_0)}.$$

After substituting for x_0, we have

$$\frac{|\delta(X)|}{|X|} \geq 2.$$

Theorem 12 is proved. □

THEOREM 13. *Suppose that a graph G with n vertices has average degree d and the eigenvalues σ_i of the combinatorial Laplacian satisfy $|d - \sigma_i| \leq \theta$ for $i \neq 0$. If $d^{2k} \gg n\theta^{2k-1}$, then G contains a cycle of length $2k + 1$, if n is sufficiently large.*

PROOF. We consider $\delta_i(v) = \{u : d(u,v) = i\}$. In a paper [**38**] by Erdős et al., it was shown that if a graph G contains no cycle of length $2k+1$, then for any $1 \le i \le k$ the induced subgraph on $\delta_i(v)$ contains an independent set S of size at least $|\delta_i(v)|/(2k-1)$. From Theorem 6, we have

$$d|S|(|S|-1) < 2\theta|S|(n - |S|/2).$$

This implies that

$$|S| \le \frac{2\theta n}{d}.$$

(Here we use the fact that $\theta \ge \sqrt{d}$.) Hence, we have $|\delta_i(v)|/(2k-1) \le 2\theta n/d$.

Since $\theta = o(d)$, by Corollary 8, we have

$$|\delta_k(v)| \ge (1+o(1))\frac{d^{2k-1}}{\theta^{2k-2}}(1 - \frac{1}{n}(\frac{d^2}{\theta^2})^{t-1}).$$

Thus, we have

$$\frac{2(2k-1)\theta n}{d} \ge (1+o(1))\frac{d^{2k-1}}{\theta^{2k-2}}(1 - \frac{1}{n}(\frac{d^2}{\theta^2})^{t-1}).$$

This implies that

$$n \ge (1+o(1))\frac{d^{2k}}{2(2k-1)\theta^{2k-1}}$$

which is a contradiction to the assumption that $n \ll d^{2k}/\theta^{2k-1}$. □

In [**45**] Krivelevich and Sudakov showed that a d-regualar graph on n vertices is Hamiltonian if the eigenvalues of the combinatorial Laplacian satisfy

$$|d - \sigma_i| \le c\frac{(\log\log n)^2}{\log n(\log\log\log n)}$$

for $i \ne 0$ and for some constant c. The method is a modified version of Posa's technique developed by Komlós and Szemerédi [**44**] for examining Hamiltonian cycles in random graphs. By using the discrepancy inequalities and the isoperimetric inequalities in previous sections, the above result can be extended to general graphs as well.

11. Universal graphs for trees of bounded degrees

There is quite a literature on so-called "*universal graphs*" that contain all trees on n vertices or other families of graphs such as trees with bounded degree [**5, 8, 16, 20, 39**]. One of the main avenues in the study of universal graphs is the connection with expanding properties of the graph. Friedman and Pippenger [**39**] proved the following beautiful result:

Theorem [**39**] *Suppose that H is a graph such that for every subset X of vertices with $|X| \le 2n-2$, X has exact neighborhood $\Gamma(X) = \{u : u \sim v \in X\}$ satisfying*

$$|\Gamma(X)| \ge (k+1)|X|.$$

Then H contains every tree with n vertices and maximum degree at most k.

Here we will prove the following slightly stronger result the proof of which will be given later.

THEOREM 14. *Suppose that H is a graph such that for every subset X of vertices with $|X| \leq 2n-2$ has boundary $\delta(X)$ satisfying*

$$|\delta(X)| \geq k|X|,$$

then H contains every tree with n vertices and maximum degree at most k.

As an immediate consequence of Theorem 14, we have the following:

THEOREM 15. *Suppose that a graph G with n vertices has average degree d and the eigenvalues σ_i of the combinatorial Laplacian satisfy $|d-\sigma_i| \leq \theta$ for $i \neq 0$. Then G contains all trees with maximum degree k and having at least*

$$\frac{d^2-(k+1)\theta^2}{2(k+1)(d^2-\theta^2)}n+1 \text{ vertices.}$$

PROOF. To deduce the existence of a tree having at least

$$t = \frac{d^2-(k+1)\theta^2}{2(k+1)(d^2-\theta^2)}n+1$$

vertices and degree bounded above by k it suffices to show that we have $\delta(X) \geq k|X|$ for any subset X with size at most

$$|X| \leq \frac{d^2-(k+1)\theta^2}{(k+1)(d^2-\theta^2)}n = z_0 \leq 2t-2$$

by using Theorem 14. From Theorem 8, we know that

$$\frac{|\delta(X)|}{|X|} \geq \frac{d^2-\theta^2}{\theta^2+d^2z_0/(n-z_0)}.$$

After substituting for z_0, we have

$$\frac{|\delta(X)|}{|X|} \geq k.$$

Theorem 15 is proved. □

COROLLARY 9. *Suppose that a graph G with n vertices has average degree d and the eigenvalues σ_i of the combinatorial Laplacian satisfy $|d-\sigma_i| \leq \theta$ for $i \neq 0$. If $\theta = o(d)$, then G contains all trees with maximum degree k and having at most*

$$(1+o(1))\frac{n}{2(k+1)}$$

vertices, if n is sufficiently large.

It remains to prove Theorem 14. The proof is quite similar to that in [**39**]. For completeness, we sketch the proof here.

Proof of Theorem 14: Suppose that T is a tree on m vertices with maximum degree at most k. For an embedding $f : V(T) \to V(H)$, we define the excess $C(f,X)$ for $X \subseteq V(H)$ by

$$\begin{aligned} C(f,X) &:= A(f,X) - B(f,X) \\ \text{where } A(f,X) &= |\, X \cup \delta(X) \setminus f(V(T))\,| \\ \text{and } B(f,X) &= \sum_{x \in X} (k - deg_T f^{-1}(x)) \end{aligned}$$

where $deg_T f^{-1}(x)$ denotes the degree in T of the vertex u that is mapped to x under f, or $deg_T f^{-1}(x)$ is zero if no such u exists. Since the maximum degree in the tree is k, $B(f, X) \geq 0$. An embedding f is said to be *good* if for every $X \subseteq V(H)$ with $|X| \leq 2m - 2$, we have

$$C(f, X) \geq 0. \tag{10}$$

Instead of proving Theorem 14, We will show a stronger statement:

(*) *Suppose that H is a graph such that for every subset X of vertices with $|X| \leq 2m - 2$ has neighborhood $\Gamma(X)$ satisfying $|\delta(X)| \geq k|X|$.*
Suppose that S is a tree on m vertices and maximum degree at most k. Furthermore, we assume that S has a good embedding in H. Then a tree T formed by adding any leaf to T also has a good embedding.

We will prove (*) by induction on m. Clearly, (*) holds for the case of S consisting of one single vertex. Suppose (*) holds for any tree on m vertices having maximum degree k. We now consider a tree T obtained from S by adding a leaf v and the incident edge $\{v, w\}$. Suppose f is a good embedding for S. By the definition of being good, for each $g \in \delta(f(w)) \setminus f(V(S))$, we can extend f to be an embedding f_g for T by mapping v to g. We want to show that there is some good extended embedding f_g.

Suppose to the contrary that no extended embedding f_g is good. For every $g \in \delta(f(w)) \setminus f(V(S))$, there is a subset X_g of $V(H)$ such that $C(f_g, X_g) < 0$. We have

$$C(f_g, X_q) < 0 \leq C(f, X_g \cap V(S)).$$

Note that

$$C(f_g, X_q) = C(f, X_q) - \epsilon(g, X \cup \delta X) + \epsilon(f(w), X) + \epsilon(q, X),$$

where $\epsilon(x, Y)$ is 1 if $x \in Y$ and 0 otherwise. Therefore, we have $g \notin X_g$, $f(w) \notin X_g$, $g \in \delta(X_g)$ and $C(f, X_g) = 0$.

We say that X is critical under f if $C(f, X) = 0$. We need the following facts, some of which have the same proofs as those in [**39**].

Fact 1: Suppose that X is critical under f and $|S| \leq m - 1$. If $|X| \leq 2m - 2$, then $|X| \leq m - 1$.

Proof: We have

$$\begin{aligned} 0 &= C(f, X) \geq |X \cup \delta X| - |V(S)| - k|X| \\ &\geq (k+1)|X| - (m-1) - k|X| \\ &\geq |X| - (m-1). \end{aligned}$$

Fact 2: The excess $C(f, \cdot)$ is submodular:

$$C(f, X \cup Y) + C(f, X \cap Y) \leq C(f, X) + C(f, Y).$$

Fact 3: If $X, Y \subseteq V(H)$ are critical under f so that $|X|, |Y| \leq m - 1$, then $X \cup Y$ is critical under f and $|X \cup Y| \leq m - 1$.

Now we return to the proof of Theorem 14. For every $g \in \delta(f(w)) \setminus f(V(S))$, X_g is critical under f and $|X_g| \leq 2m - 2$. By Fact 1, $|X_g| \leq m - 1$. Furthermore, we consider $X^* = \cup_g X_g$, which by Fact 3 is critical under f and $|X^*| \leq m - 1$. Now consider $X' = X^* \cup f(w)$ and $C(f, X')$. Since f is a good embedding of S and $|X'| \leq m$, we have $C(f, X') \geq 0$.

For every $g \in \delta(f(w)) \setminus f(V(S))$, we have $g \in X^* \cup \delta(X^*)$ which implies $A(f, X') = A(f, X^*)$. However, $B(f, X') = B(f, X^*) + B(f, f(w))$ since $f(w) \notin X_g$ for every g. Thus, $B(f, X') < B(f, X^*)$, and we have $C(f, X^*) < 0$. This contradicts the assumption that f is good. We conclude that there is an extended embedding f_g which is good, and the proof of Theorem 14 is complete. □

12. Chromatic numbers and list chromatic numbers

One of the basic topics in graph theory is graph coloring. For a graph G, the chromatic number $\chi(G)$ of G is the least integer m so that each vertex can be assigned one of the k colors such that adjacent vertices have different colors. In the past ten years, there has been a great deal of work on extensions of chromatic numbers. In a graph G, suppose each vertex is associated with a list of k colors. A proper coloring assigns to each vertex a color from its list so that no two adjacent vertices have the same color. The *list chromatic number* $\chi_l(G)$ is the least integer k so that there is a proper coloring for any given color lists of length k. Clearly, we have

$$\chi(G) \leq \chi_l(G).$$

We will use the following theorem by Vu [**62**]:

Vu's Theorem : *Suppose a graph G has maximum degree d. If the neighborhood of each vertex in $V(G)$ contains at most d^2/f edges, for some $d^2 > f > 1$, then*

$$\chi_l(G) \leq K\frac{d}{\log f}$$

for some constant K.

THEOREM 16. *Suppose that a graph G with n vertices has average degree d and the eigenvalues σ_i of the combinatorial Laplacian satisfy $|d - \sigma_i| \leq \theta$ for $i \neq 0$. Then the chromatic number $\chi(G)$ and the list chromatic number $\chi_l(G)$ satisfy*

$$\frac{d}{\theta} < \chi(G) \leq \chi_l(G) \leq O(\frac{d}{\log(\min\{\frac{n}{d}, \frac{d}{\theta}\})}).$$

PROOF. From Corollary 3, we know that the maximum degree is at most $d+\theta$. Using Corollary 4, any subgraph on $d + \theta$ vertices can have at most

$$\frac{d(d+\theta)^2}{n} + 2\theta(d+\theta)$$

edges. So, we can use Vu's Theorem by choosing f satisfying

$$\frac{1}{f} = \frac{d}{n} + \frac{2\theta}{d},$$

and therefore

$$\begin{aligned}\chi_l(G) &= O(\frac{d+\theta}{\log f}) \\ &= O(\frac{d}{\log(\min\{\frac{n}{d}, \frac{d}{\theta}\})}).\end{aligned}$$

In the other direction, we can establish a lower bound for $\chi(G)$ as well as $\chi_l(G)$, by using Corollary 5 again. Namely, an independent set in G can have at most $\theta n/d$ vertices. Therefore

$$\frac{d}{\theta} \leq \chi(G) \leq \chi_l(G) = O(\frac{d}{\log(\min\{\frac{n}{d}, \frac{d}{\theta}\})}).$$

□

13. Turán numbers

A celebrated result in extremal graph theory is Turán's Theorem which states that a graph on n vertices containing no K_{t+1} can have at most $(1-1/t+o(1))\binom{n}{2}$ edges. Sudakov, Szabó and Vu [**60**] consider a generalization of Turán's Theorem. A graph G with $e(G)$ edges is said to be t-Turán if any subgraph of G containing no K_{t+1} has at most $(1-1/t+o(1))e(G)$ edges. In [**60**], it is shown that a regular graph on n vertices with degree d is t-Turán if the second largest eigenvalue of its adjacency matrix λ is sufficiently small. Their result can be extended to general graphs by using the isoperimetric and discrepancy inequalities [**17**].

THEOREM 17. *Suppose a graph G on n vertices has eigenvalues of the normalized Laplacian*
$0=\lambda_0 \le \lambda_1 \le \ldots \le \lambda_{n-1}$ *with* $\bar{\lambda} = \max_{i\neq 0}|1-\lambda_i|$ *satisfying*

$$\bar{\lambda} = o(\frac{1}{\text{vol}_{-2t+1}(G)\text{vol}(G)^{t-1}}) \tag{11}$$

where $\text{vol}_i(S) = \sum_{x\in S} d_x^i$.

Then, G is t-Turán for $t \ge 2$, i.e., any subgraph of G containing no K_{t+1} has at most $(1+1/t+o(1))e(G)$ edges where $e(G)$ is the number of edges in G.

14. Random walks and contingency tables

A *walk* is a sequence of vertices $w = (v_0, v_1, \cdots, v_s)$ such that $\{v_{i-1}, v_i\}$ is an edge. A Markov chain can be viewed as a random walk, defined by its transition probability matrix P, where the probability of moving from u to v is $P(u,v)$. Clearly, $P(u,v) > 0$ only if (u,v) is an edge. Also, $\sum_v P(u,v) = 1$. For a weighted graph with edge weights $w_{u,v} \ge 0$, a typical transition probability matrix P can be defined as

$$P_{uv} = \frac{w_{uv}}{\sum\limits_z w_{uz}}.$$

For any initial distribution $f : V \to \mathbf{R}$, the distribution after one step of a random walk is just

$$f\ P\ (v) \ = \ \sum_u \frac{w_{uv}}{d_u} f(u). \tag{12}$$

Here we treat f as a row vector. The distribution after k steps is $f\ P^k(v)$.

In the terminology of Markov chains, a random walk is said to be *ergodic* if there is a unique stationary distribution π satisfying $\pi P = \pi$. Necessary conditions for the ergodicity of a random walk on a graph with n vertices are *irreducibility* (i.e., no 0 submatrix of P of size $k \times (n-k)$ for any k) and *aperiodicity*, (i.e., g.c.d. $\{s : P^s(u,v) > 0\} = 1$). As it turns out, these necessary conditions are also sufficient. An ergodic Markov chain is said to be *reversable* if for any two vertices u and v, we have

$$\pi(u)P(u,v) = \pi(v)P(v,u).$$

A reversible Markov chain can be studied as a weighted graph as follows [**13**]:

THEOREM 18. *The following three statements are equivalent:*

(a): *A Markov chain with transition probability matrix P is ergodic and reversible.*

(b): *The weighted graph defined by edge weight $w_{u,v} = \pi(u)P(u,v)$ is connected and non-bipartite.*

(c): *The weighted graph defined by edge weight $w_{u,v} = \pi(u)P(u,v)$ has one eigenvalue 0 and all other eigenvalues greater than 0 and strictly less than 2.*

It can be easily shown that the stationary distribution π satisfies

$$\pi(v) = \frac{d_v}{\text{vol}(G)}.$$

A natural question of interest is what is the rate of convergence to the stationary distribution. The answer to this question again lies in the eigenvalues of the associated graph.

In the study of rapidly mixing Markov chains, the convergence in the L_2 distance is rather weak since it it does not require convergence to the stationary distribution at every vertex. A strong notion of convergence that is often used is measured by the *relative pointwise distance* (r.p.d). After s steps the relative pointwise distance of P to the stationary distribution $\phi(x)$ is given by

$$\Delta(s) = \max_{x,y} \frac{|P^s(y,x) - \pi(x)|}{\pi(x)}$$

Another notion of distance for measuring convergence is the so-called it total variation distance:

$$\begin{aligned}\Delta_{TV}(s) &= \max_{A\subset V(G)} \max_{y\in V(G)} \|\sum_{x\in A}(P^s(y,x) - \pi(x))\| \\ &= \frac{1}{2} \max_{y\in V(G)} \sum_{x\in V(G)} \|(P^s(y,x) - \pi(x))\|.\end{aligned}$$

It is easy to see that $\Delta_{TV}(s) \le \frac{1}{2}\Delta(s)$. Thus a convergence upper bound for $\Delta(s)$ implies one for $\Delta_{TV}(s)$.

The rate of convergence of a random walk on a graph on n vertices depends on the spectral gaps, λ_1 and $2 - \lambda_{n-1}$. However, the gap $2 - \lambda_{n-1}$ can often be circumvented by considering a modified random walk, so- called *lazy* walk. For a transition probability matrix P, a C-lazy walk is defined by the transition probablity matrix P_C, for some $C < 1$:

$$P_C(u,v) = \begin{cases} C + \frac{P(u,u)}{1+C} & \text{if } u = v, \\ \frac{P(u,v)}{1+C} & \text{otherwise.} \end{cases}$$

The value for C is often chosen to be $1/2$. Suppose we choose

$$C = \begin{cases} 0 & \text{if } \lambda_1 \ge 2 - \lambda_{n-1} \\ \frac{\lambda_1 + \lambda_{n-1} - 2}{2} & \text{otherwise.} \end{cases}$$

Then we have the following:

THEOREM 19. *Suppose that a graph G on n vertices has Laplacian eigenvalues $0 = \lambda_0 \le \lambda_1 \le \ldots \le \lambda_{n-1}$. Then G has a lazy random walk with the rate of convergence of order*

$$\lambda^{-1} \log(\frac{\text{vol}(G)}{\min_x d_x})$$

where $\lambda = \lambda_1$ if $2 \ge \lambda_1 + \lambda_{n-1}$ and $\lambda = 2\lambda_1/(\lambda_1 + \lambda_{n-1})$ otherwise.

Namely, after at most

$$t \geq \lambda^{-1}(\log(\frac{\text{vol}(G)}{\min_x d_x}) + c)$$

steps, we have

$$\Delta(t) \leq e^{-c}.$$

For random walks on groups or, equivalently, on graphs defined by groups, the eigenvalues of the Laplacian can often be evaluated exactly using group representation theory. For example, various applications in card shuffling, are associated with graphs with vertex set the symmetric group S_n and edge set defined by permutations corresponding to the allowable shuffling moves. There are extensive surveys and books [**36, 58**] on this subject. Here, instead, we consider two applications for general graphs.

Logarithmic Sobolev inequalities The upper bound for the rate of convergence in Theorem 19 can sometimes be further improved by using the *log-Sobolev* constant α defined as follows [**37**]:

$$\alpha(G) \quad = \quad \alpha = \inf_{f \neq 0} \frac{\sum_{x \sim y}(f(x) - f(y))^2 w_{x,y}}{\sum_x f^2(x) d_x \log \frac{f^2(x)\text{vol}(G)}{\sum_z f^2(z) d_z}}$$

where f ranges over all nontrivial vectors $f : V \to \mathbb{R}$.

Then we have the following [**18**]:

THEOREM 20. *For a weighted graph G with log-Sobolev constant α, there is a lazy walk satisfying*

$$\Delta(t) < e^{2-c}$$

provided that

$$t \geq \frac{1}{\alpha} \log \log \frac{\text{vol}(G)}{\min_x d_x} + \frac{c}{\lambda}$$

where λ is as defined in Theorem 19.

Enumerating contingency tables As an application of the eigenvalue inequalities in Section 9, we consider the classical problem of sampling and enumerating the family S of $n \times n$ matrices with nonnegative integral entries and given row and column sums. Although the problem is presumed to be computationally intractable (in the so-called $\#P$-complete class), the eigenvalue bounds in the previous section can be used to obtain a polynomial approximation algorithm. To see this, we consider the homogeneous graph Γ with the vertex set consisting of all $n \times n$ matrices with integral entries (possibly negative) with given row and column sums. Two vertices u and v are adjacent if u and v differ at just the four entries of a 2×2 submatrix with entries

$$u_{ik} = v_{ik} + 1, u_{jk} = v_{jk} - 1, u_{im} = v_{im} - 1, u_{jm} = v_{jm} + 1.$$

The family S of matrices with all nonnegative entries is then a convex subgraph of Γ.

On the vertices of S, we consider the following random walk. The transition probability $P(u, v)$ of moving from a vertex u in S to a neighboring vertex v is $\frac{1}{k}$ if v is in S, where k is the degree of Γ. If a neighbor v of u (in Γ) is not in S, then we move from u to each neighbor z of v, z in S, with the (additional) probability $\frac{1}{d'_v}$ where $d'_v = |\{z \in S : z \sim v \text{ in } \Gamma\}|$ for $v \notin S$. In other words, for $u, v \in S$,

$$P(u,v) = \frac{w_{uv}}{d_u} + \sum_{\substack{z \notin S \\ u \sim z, v \sim z}} \frac{w_{uz}}{d_v d'_z} w_{zv}$$

where w_{uv} denotes the weight of the edge $\{u, v\}$ ($w_{uv} = 1$ or 0 for simple graphs) and $d_u = \sum_{u \sim v} d_{uv}$. The stationary distribution for this walk is uniform. Let ρ denote the second largest eigenvalue of P. It can be easily checked that $1 - \rho \geq \lambda_1(S)$, where $\lambda_1(S)$ denotes the first Neumann eigenvalue of the induced subgraph S of Γ. In particular, if the total row sum (minus the maximum row sum) is $\geq c'\, n^2$, it can be shown (see [**27**]) that $\lambda_1(S) \geq \frac{c}{kD^2}$, where D is the diameter of S. This implies that a random walk converges to the uniform distribution in $O(k^2 D^2 (\log n))$ steps. (In some cases, the factor $\log n$ can be further reduced by using logarithmic Sobolev inequalities and logarithmic Harnack inequalities (see [**37, 13**]).)

It is reasonable to expect that the above techniques can be useful for developing approximation algorithms for many other difficult enumeration problems by considering random walk problems in appropriate convex subgraphs. Further applications using the eigenvalue bounds in previous sections can be found in [**31**].

15. Concluding remarks

1. In this paper, we mainly deal with simple graphs. For a weighted graph G with edge weight $w_{u,v}$, we define $d_v = \sum_u w_{u,v}$ and $e(X,Y) = \sum_{u \in X, v \in Y} w_{u,v}$. Then the isoperimetric inequalities in Sections 4 and 5 still hold.

2. We consider three families of graphs on n vertices:

$\mathcal{F}_1 = \{d - \text{regular graphs that are } (n, d, \lambda) - \text{graphs}\}$,

$\mathcal{F}_2 = \{\text{graphs satisfying } |\text{average degree } - \sigma_i| \leq \theta \text{ for } i \neq 0\}$,

$\mathcal{F}_3 = \{\text{graphs satisfying } |1 - \lambda_i| \leq \bar{\lambda} \text{ for } i \neq 0\}$,

where the σ_i's are eigenvalues of the combinatorial Laplacian and λ_i's are eigenvalues of the normalized Laplacian.

Clearly, we have

$$\mathcal{F}_1 \subset \mathcal{F}_2$$

if $\lambda = \theta$. Also

$$\mathcal{F}_2 \subset \mathcal{F}_3$$

if θ is $\bar{\lambda}$ times the average degree . Hence, the isoperimetric inequalities involving eigenvalues of the normalized Laplacian have stronger implications than that of the combinatorial Laplacian. For applications using eigenvalues of the combinatorial Laplacian, it is natural to ask if the same results hold for the normalized Laplacian. For example, is it true that graphs in $\mathcal{F}_3$ are Hamiltonian provided that $\bar{\lambda}$ is small enough?

3. Graph theory has 250 years of history. In the very early days, graphs were used to study the structure of molecules and in particular, the eigenvalues of graphs are associated with stability of chemicals [**9**]. In recent years, many realistic graphs that arise in Internet and biological networks can be modeled as graphs with certain "power law" degree distribution [**1, 2, 3**]. Again, eigenvalues come into play since random graphs with given expected degrees are shown to have eigenvalue distribution as predicted [**32, 33**]. In this paper, we discuss only a few

applications of isoperimetric inequalities. It would be of interest to find further applications especially for power law graphs.

Acknowledgement: The author wishes to thank Robert Ellis and Josh Cooper for careful reading and making many valuable suggestions for an earlier draft of this paper.

References

[1] W. Aiello, F. Chung and L. Lu, A random graph model for massive graphs, *STOC* 2001, 171-180. The paper version appeared in *Experimental Math.* **10**, (2001), 53-66.

[2] W. Aiello, F. Chung and L. Lu, Random evolution in massive graphs, *Handbook of Massive Data Sets*, Volume 2, (Eds. J. Abello et al.), Kluwer Academic Publishers, (2002), 97-122. Extended abstract appeared in FOCS 2001, 510-519.

[3] R. Albert, H. Jeong and A. Barabási, Diameter of the World Wide Web, *Nature* **401** (1999), 130-131.

[4] N. Alon, Eigenvalues and expanders, *Combinatorica* **6**, (1986) 86-96.

[5] N. Alon and F. R. K. Chung, Explicit constructions of linear-sized tolerant networks, *Discrete Math.* 72 (1988), 15-20.

[6] N. Alon, F. R. K. Chung and R. L. Graham, Routing permutations on graphs via matchings, *SIAM J. Discrete Math.* **7** (1994), 513-530.

[7] N. Alon and V. D. Milman, λ_1 isoperimetric inequalities for graphs and superconcentrators, *J. Comb. Theory* B 38 (1985), 73-88.

[8] J. Beck, On size Ramsey number of paths, trees, and circuits. I, *J. Graph Theory*, **7**, (19xx), 115-129.

[9] N.L. Biggs, E.K. Lloyd and R.J. Wilson, *Graph Theory 1736-1936*, Clarendon Press, Oxford, 1976.

[10] B. Bollobás, *Extremal Graph Theory*, Academic Press, London (1978).

[11] J. Cheeger, A lower bound for the smallest eigenvalue of the Laplacian, *Problems in Analysis*, (R. C. Gunning, ed.) Princeton Univ. Press (1970), 195-199.

[12] I. Chavel, *Isoperimetric inequalities,* Differential geometric and analytic perspectives, Cambridge Tracts in Mathematics, 145, *Cambridge University Press*, Cambridge, 2001.

[13] F. R. K. Chung, *Spectral Graph Theory,* CBMS Lecture Notes, AMS Publications, 1997.

[14] F. R. K. Chung, Diameters and eigenvalues, *J. of Amer. Math. Soc.* **2** (1989), 187-196.

[15] F. R. K. Chung, Eigenvalues of graphs and Cheeger inequalities, in *Combinatorics, Paul Erdős is Eighty,* Volume 2, edited by D. Miklós, V. T. Sós, and T. Szőnyi, János Bolyai Mathematical Society, Budapest (1996), 157-172.

[16] F. R. K. Chung, Universal graphs and induced-universal graphs, *J. Graph Theory* 14 (1990), 443-454

[17] Fan Chung, A spectral Turán theorem, preprint.

[18] F. R. K. Chung, Logarithmic Sobolev techniques for random walks on graphs, *Emerging Applications of Number Theory, IMA Volumes in Math. and its Applications*, **109**, (eds. D. A. Hejhal et. al.), 175-186, Springer, 1999.

[19] F. R. K. Chung, V. Faber and T. A. Manteuffel, An upper bound on the diameter of a graph from eigenvalues associated with its Laplacian, *SIAM J. Discrete Math.* **7** (1994), 443-457.

[20] F. R. K. Chung and R. L. Graham, On universal graphs for spanning trees, *Journal of London Math. Soc.* 27 (1983), 203-211.

[21] F. R. K. Chung, A. Grigor'yan, and S.-T. Yau, Eigenvalues and diameters for manifolds and graphs, Tsing Hua Lectures on Geometry and Analysis, (ed. S.-T. Yau), International Press, Cambridge, Massachusetts, (1997), 79-106.

[22] F. R. K. Chung, A. Grigor'yan, and S.-T. Yau, Upper bounds for eigenvalues of the discrete and continuous Laplace operator, *Advances in Mathematics*, 117 (1996), 165-178.

[23] F. R. K. Chung, A. Grigor'yan, and S.-T. Yau, Higher eigenvalues and isoperimetric inequalities on Riemannian manifolds and graphs, *Communications on Analysis and Geometry,* **8** (2000), 969-1026.

[24] F. R. K. Chung and Prasad Tetali, Isoperimetric inequalities for cartesian products of graphs, **7** (1998), 141-148.

[25] F. R. K. Chung and S.-T. Yau, Eigenvalues of graphs and Sobolev inequalities, *Combinatorics, Probability and Computing* **4** (1995), 11-26.

[26] F. R. K. Chung and S.-T. Yau, A Harnack inequality for homogeneous graphs and subgraphs, *Communications in Analysis and Geometry* **2** (1994), 628-639.
[27] F. R. K. Chung and S.-T. Yau, Eigenvalue inequalities of graphs and convex subgraphs, *Communications in Analysis and Geometry,* **5** (1998), 575-624.
[28] F. R. K. Chung and S.-T. Yau, Logarithmic Harnack inequalities, *Mathematical Research Letters* **3** (1996), 793-812.
[29] F. R. K. Chung and R. L. Graham, Quasi-random set systems *J. Amer. Math. Soc.***4** (1991), 151-196
[30] F. R. K. Chung, R. L. Graham and R. M. Wilson, Quasi-random graphs, *Combinatorica* **9** (1989), 345-362.
[31] F. R. K. Chung, R. L. Graham and S.-T. Yau, On sampling with Markov chains, *Random Structures and Algorithms* **9** (1996), 55-78.
[32] F. Chung, L. Lu and Van Vu, Eigenvalues of random power law graphs, *Annals of Combinatorics,* **7** (2003), 21-33.
[33] F. Chung, L. Lu and Van Vu, The spectra of random graphs with given expected degrees, *Proceedings of National Academy of Sciences,* **100**, no. 11, (2003), 6313-6318.
[34] T. Coulhon, A. Grigor'yan and C. Pittet, A geometric approach to on-diagonal heat kernel lower bounds on groups, *Ann. Inst. Fourier(Grenoble)* **51** (2001), 1763-1827.
[35] T. Coulhon and L. Saloff-Coste, Variétés riemanniennes isométriques à l'infini, *Rev. Mat. Iberoamericana,* **11** (1995), 687-726.
[36] P. Diaconis, *Group representations in probability and statistics, Institute of Mathematical Statistics Lecture Notes — Monograph Series, 11,* Institute of Mathematical Statistics, Hayward, CA, 1988.
[37] P. Diaconis and L. Saloff-Coste, Logarithmic Sobolev inequalities for finite Markov chains, *Ann. Appl. Probab.* **6** (1996), 695-750.
[38] P. Erdős, R. Faudree, C. Rousseau and R. Schelp, On cycle-complete graph Ramsey numbers, *J. Graph Theory* **2** (1978), 53-64.
[39] J. Friedman and N. Pippenger, Expanding graphs contain all small trees, *Combinatorica,* **7** (1987), 71-76.
[40] A. Grigor'yan, Heata kernels on manifolds, graphs and fractals, *European Congress of Mathematics,* vol. 1 (Barcelona, 2000), 393-406, *Progr. Math.*, 201, Birkh auser, Basel, 2001.
[41] M. Jerrum and A. Sinclair, Approximating the permanent, *SIAM J. Computing* **18** (1989), 1149-1178.
[42] M. Kanai, Rough isometries and combinatorial approximations of geometries of noncompact Riemannian manifolds, *J. Math. Soc. Japan* **37** (1985), 391-413.
[43] F. Kirchhoff, Über die Auflösung der Gleichungen, auf welche man bei der Untersuchung der linearen Verteilung galvanischer Ströme geführt wird, *Ann. Phys. Chem.* 72 (1847), 497-508.
[44] J. Komlós amd E. Szemerédi, Limit distributions for the existence of Hamilton circuits in a random graph, *Discrete Math.* **43**, (1983), 55-63.
[45] M. Krivelevich and B. Sudakov, Sparse pseudo-random graphs are Hamiltonian, *J. Graph Theory* **42** (2003), 17-33.
[46] M. Krivelevich and B. Sudakov, Pseudo-random graphs, preprint.
[47] F. T. Leighton, *Introduction to Parallel Algorithms and Architectures: Arrays, Trees, Hypercubes*, Morgan-Kauffman, San Mateo, CA, 1992.
[48] F.T. Leighton and Satish Rao, An approximate max-flow min-cut theorem for uniform multicommodity flow problem with applications to approximation algorithms, *29nd Symposium on Foundations of Computer Science*, IEEE Computer Society Press, (1988), 422-431.
[49] P. Li and S. T. Yau, On the parabolic kernel of the Schrödinger operator, Acta Mathematica 156, (1986) 153-201
[50] A. Lubotsky, R. Phillips and P. Sarnak, Ramanujan graphs, *Combinatorica* **8** (1988),261-278.
[51] R. J. Lipton and R.E. Tarjan, A separator theorem for planar graphs, *SIAM J. Appl. Math.* **36** (1979), 177-189.
[52] G. A. Margulis, Arithmetic groups and graphs without short cycles, *6th Internat. Symp. on Information Theory, Tashkent* (1984) *Abstracts* **1**, 123-125 (in Russian).
[53] P. Lancaster, Theory of matrices, Academic Press, 1969.
[54] L. Lovász, *Combinatorial Problems and Exercises*, North-Holland, 1979.
[55] B. Mohar, Isoperimetric number of graphs, *J. of Comb. Theory (B)* **47** (1989), 274-291.

[56] G. Polyá and S. Szegö, *Isoperimetric Inequalities in Mathematical Physics*, Annals of Math. Studies, no. 27, Princeton University Press, (1951).
[57] L. Pósa, Hamiltonian circuits in random graphs, *Discrete Math.*, **14** (1976), 359-364.
[58] L. Saloff-Coste, Lectures on finite Markov chains, *Lectures on Probability Theory and Statistics* (Saint-Flour, 1996), *Lecture Notes in Math.,1665* Springer, Berlin, (1997), 301-413.
[59] R. M. Tanner, Explicit construction of concentrators from generalized N-gons, *SIAM J. Algebraic Discrete Methods* **5** (1984), 287-294.
[60] B. Sudakov, T. Szabo and V. Vu, A generalization of Turán's theorem, preprint.
[61] R.M. Tanner, Explicit construction of concentrators from generalized N-gons, *SIAM J. Algebraic Discrete Methods* **5** (1984), 287-294.
[62] V. Vu, A general upper bound on the list chromatic number of locally sparse graphs, *Combinatorics, Probability and Computing* **11** (2002), 103-111.
[63] W. Woess, Random walks on infinite graphs and groups, *Cambridge Tracts in Mathematics*, 138, Cambridge University Press, Cambridge, 2000.
[64] S. T. Yau and R. M. Schoen, *Differential Geometry*, (1994), International Press, Cambridge, Massachusetts.

University of California at San Diego, La Jolla, CA 92093-0112, USA
E-mail address: fan@euclid.ucsd.edu

Surveys in Differential Geometry
IX, International Press

An excursion into geometric analysis

Tobias H. Colding and William P. Minicozzi II

CONTENTS

The authors were partially supported by NSF Grants DMS 0104453 and DMS 0104187.

1. Introduction

This is a guided tour through some selected topics in geometric analysis. Most of the results here can be found in the literature but some are new and do not appear elsewhere. We have chosen to illustrate many of the basic ideas as they apply to the theory of minimal surfaces. This is, in part, because minimal surfaces is, if not the oldest, then certainly one of the oldest areas of geometric analysis dating back to Euler's work in the 1740's and in fact many of the basic ideas in geometric analysis originated in the study of minimal surfaces. In any case, the ideas apply to a variety of different fields and we will mention some of these as we go along.

Part 1 reviews some of the classical ideas and results in geometric analysis. We begin with the definition and basic results for minimal surfaces, including the first variation formula and maximum principle in Section 2. Section 3 gives some applications of the Bochner formula to comparison theorems, vanishing theorems (such as the famous Bochner theorem), and harmonic functions. We turn next to the monotonicity formula and mean value inequality in Section 4. These play a fundamental role in many areas of geometric analysis, however, we have chosen to focus on the special cases of minimal surfaces and manifolds with non–negative Ricci curvature. Section 5 recalls the Bernstein theorem (for entire solutions) and Bers' theorem (for exterior solutions) of the minimal surface equation. These results illustrate an interesting rigidity for solutions of the minimal surface equation which comes from the nonlinearity (and does not occur for solutions of linear equations). In Sections 6 and 7, we briefly review the basic facts for mean curvature flow and Ricci flow. The next two sections discuss some fundamental a priori estimates in pde. First, Section 8 gives various gradient estimates for linear/nonlinear, elliptic/parabolic equations, all based on the maximum principle. The importance of this fundamental estimate has been well-understood since the work of Bernstein in the early 1900's. Section 9 recalls a much more recent tool for a priori estimates, namely, Simons' inequality, and illustrates its usefulness for proving a priori estimates. The original inequality of Simons was for the Laplacian of the norm squared of the second fundamental form of a minimal hypersurface, but variations of this inequality appear in a surprising number of fields (Einstein manifolds, harmonic maps, Yang-Mills connections, various parabolic equations, etc.). Finally, in Section 10 we derive the basic estimates for minimal annuli with small total curvature, including a quantitative form of Bers' theorem. This last section also sets the stage for some estimates in Part 2 for multi–valued graphs.

In Part 2, we turn our attention to embedded multi–valued minimal graphs (the basic example is half of the helicoid). These are graphs of multi–valued functions and should be thought of as "spiral staircases." The analysis of these has played a major role in recent developments in minimal surface theory. The first two sections, 11 and 12, prove the fundamental estimates on the separation and curvature. Section 13 extends the Bers' theorem from Part 1 to this setting. Finally, Section 14 proves some original results, including a representation formula showing that an embedded multi–valued minimal graph can be written as a sum of a helicoid, a catenoid, and a small perturbation.

In Part 3, we survey some of the key ideas in classical regularity theory, recent developments on embedded minimal disks, and some global results for minimal surfaces in $\mathbf{R}^3$. Sections 15 and 16 focus on Reifenberg type conditions, where a set is assumed to be close to a plane at all points and at all scales ("close" is in the Hausdorff or Gromov–Hausdorff sense and is defined in Section 15). This condition automatically gives Hölder regularity (and hence higher regularity if the set is also a weak solution to a natural equation). Section 17 surveys the role of monotonicity and scaling in regularity theory, including ϵ-regularity theorems (such as Allard's theorem) and tangent cone analysis (such as Almgren's refinement of Federer's dimension reducing). Section 18 briefly reviews recent results of the authors for embedded minimal disks, developing a regularity theory in a setting where the classical methods cannot be applied and in particular where there is no monotonicity. The estimates and ideas discussed in Section 18 have applications to the global theory of minimal surfaces in $\mathbf{R}^3$. In Section 19, we give a quick tour of some recent results in this classical, but rapidly developing, area.

Thus far, we have mainly dealt with regularity and a priori estimates but have ignored questions of existence. Part 4 surveys some of the most useful existence results for minimal surfaces and gives an application to Ricci flow. Section 20 recalls the classical Weierstrass representation, including a few modern applications, and the Kapouleas desingularization method. Section 21 deals with area minimizing surfaces (whether for fixed boundary, fixed homotopy class, etc.) and questions of embeddedness. The next section discusses unstable (hence *not* minimizing) surfaces and the corresponding questions for geodesics, concentrating on whether the Morse index can be bounded uniformly. Section 23 recalls the min–max construction for producing unstable minimal surfaces and, in particular, doing so while controlling the topology and guaranteeing embeddedness. Finally, Section 24 discusses a recent application of min–max surfaces to bound the extinction time for Ricci flow, answering a question of Perelman.

Finally, in Part 5, we discuss some global results for harmonic functions and a few applications of function theory. We begin by reviewing the basic theory of harmonic functions on Euclidean space. This starts with the Liouville theorem and the relationship between polynomial growth harmonic functions and eigenfunctions on the sphere; see Section 25. In Section 26 we sketch the proof that the spaces of harmonic functions of polynomial growth are finite dimensional on manifolds with non–negative Ricci curvature. Section 27 gives a version of this for minimal submanifolds and a geometric application of this. Finally, Section 28 discusses two estimates related to nodal sets of eigenfunctions.

Part 1. Classical and almost classical results

2. Minimal surfaces

Let $\Sigma \subset \mathbf{R}^3$ be a smooth orientable surface (possibly with boundary) with unit normal n_Σ. Given a function ϕ in the space $C_0^\infty(\Sigma)$ of infinitely differentiable (i.e., smooth), compactly supported functions on Σ, consider the one–parameter variation

$$\Sigma_{t,\phi} = \{x + t\,\phi(x)\,\mathrm{n}_\Sigma(x) | x \in \Sigma\}\,. \tag{2.1}$$

The so called first variation formula of area is the equation (integration is with respect to $d$area)

$$\left.\frac{d}{dt}\right|_{t=0} \mathrm{Area}(\Sigma_{t,\phi}) = \int_\Sigma \phi\, H\,, \tag{2.2}$$

where H is the mean curvature of Σ. (When Σ is noncompact, then $\Sigma_{t,\phi}$ in (2.2) is replaced by $\Gamma_{t,\phi}$, where Γ is any compact set containing the support of ϕ.) The surface Σ is said to be a *minimal* surface (or just minimal) if

$$\left.\frac{d}{dt}\right|_{t=0} \mathrm{Area}(\Sigma_{t,\phi}) = 0 \qquad \text{for all } \phi \in C_0^\infty(\Sigma) \tag{2.3}$$

or, equivalently by (2.2), if the mean curvature H is identically zero. Thus Σ is minimal if and only if it is a critical point for the area functional. (Since a critical point is not necessarily a minimum the term "minimal" is misleading, but it is time honored. The equation for a critical point is also sometimes called the Euler–Lagrange equation.) Moreover, a computation shows that if Σ is minimal, then

$$\left.\frac{d^2}{dt^2}\right|_{t=0} \mathrm{Area}(\Sigma_{t,\phi}) = -\int_\Sigma \phi\, L_\Sigma \phi\,, \tag{2.4}$$

where

$$L_\Sigma \phi = \Delta_\Sigma \phi + |A|^2 \phi \tag{2.5}$$

is the second variational (or Jacobi) operator. Here Δ_Σ is the Laplacian on Σ and A is the second fundamental form. So $|A|^2 = \kappa_1^2 + \kappa_2^2$, where κ_1, κ_2 are the principal curvatures of Σ and $H = \kappa_1 + \kappa_2$. A minimal surface Σ is said to be stable if

$$\left.\frac{d^2}{dt^2}\right|_{t=0} \mathrm{Area}(\Sigma_{t,\phi}) \geq 0 \qquad \text{for all } \phi \in C_0^\infty(\Sigma)\,. \tag{2.6}$$

A graph (i.e., the set $\{(x_1, x_2, u(x_1, x_2)) \,|\, (x_1, x_2) \in \Omega\}$) of a real valued function u on a domain Ω in $\mathbf{R}^2$ is minimal iff the function satisfies the minimal surface equation

$$\mathrm{div}\left(\frac{du}{\sqrt{1 + |du|^2}}\right) = 0\,, \tag{2.7}$$

where du is the $\mathbf{R}^2$ gradient of the function u and div is the divergence in $\mathbf{R}^2$. One can show that a minimal graph is stable and, more generally, so is a multi–valued minimal graph (see below for the precise definition).

We will next derive the weak form of the minimal surface equation, i.e., the so–called first variation formula, which is the basic tool for working with "weak solutions" (typically, stationary varifolds). Let X be a vector field on $\mathbf{R}^3$. We can write the divergence $\operatorname{div}_\Sigma X$ of X on Σ as

$$\operatorname{div}_\Sigma X = \operatorname{div}_\Sigma X^T + X \cdot H\,, \tag{2.8}$$

where X^T and X^N are the tangential and normal projections of X. From this and Stokes' theorem, we see that Σ is minimal if and only if for all vector fields X with compact support and vanishing on the boundary of Σ,

$$\int_\Sigma \operatorname{div}_\Sigma X = 0\,. \tag{2.9}$$

This equation is known as the first variation formula. It has the benefit that (2.9) makes sense as long as we can define the divergence on Σ. As a consequence of (2.9), we will show the following proposition:

PROPOSITION 2.1. *$\Sigma^k \subset \mathbf{R}^n$ is minimal if and only if the restrictions of the coordinate functions of $\mathbf{R}^n$ to Σ are harmonic functions.*

PROOF. Let η be a smooth function on Σ with compact support and $\eta|\partial\Sigma = 0$, then

$$\int_\Sigma \langle \nabla_\Sigma \eta, \nabla_\Sigma x_i\rangle = \int_\Sigma \langle \nabla_\Sigma \eta\,, e_i\rangle = \int_\Sigma \operatorname{div}_\Sigma(\eta\, e_i)\,. \tag{2.10}$$

From this, the claim follows easily. □

Recall that if $\Xi \subset \mathbf{R}^n$ is a compact subset, then the smallest convex set containing Ξ (the convex hull, $\operatorname{Conv}(\Xi)$) is the intersection of all half–spaces containing Ξ. The maximum principle forces a minimal submanifold to lie in the convex hull of its boundary (this is the "convex hull property"):

PROPOSITION 2.2. *If $\Sigma^k \subset \mathbf{R}^n$ is a compact minimal submanifold, then $\Sigma \subset \operatorname{Conv}(\partial\Sigma)$.*

PROOF. A half–space $H \subset \mathbf{R}^n$ can be written as

$$H = \{x \in \mathbf{R}^n \,|\, \langle x, e\rangle \leq a\}\,, \tag{2.11}$$

for a vector $e \in \mathbf{S}^{n-1}$ and constant $a \in \mathbf{R}$. By Proposition 2.1, the function $u(x) = \langle e, x\rangle$ is harmonic on Σ and hence attains its maximum on $\partial\Sigma$ by the maximum principle. □

The argument in the proof of the convex hull property can be rephrased as saying that as we translate a hyperplane towards a minimal surface, the first point of contact must be on the boundary. When Σ is a hypersurface, this is a special case of the strong maximum principle for minimal surfaces:

LEMMA 2.3. *Let $\Omega \subset \mathbf{R}^{n-1}$ be an open connected neighborhood of the origin. If $u_1,\ u_2 : \Omega \to \mathbf{R}$ are solutions of the minimal surface equation with $u_1 \leq u_2$ and $u_1(0) = u_2(0)$, then $u_1 \equiv u_2$.*

See [**CM1**] for a proof of Lemma 2.3 and further discussion.

3. The Bochner formula

On a Riemannian manifold M a very useful formula of S. Bochner asserts that for any function u on M

$$\frac{1}{2}\Delta|\nabla u|^2 = |\mathrm{Hess}_u|^2 + \langle \nabla \Delta u, \nabla u\rangle + \mathrm{Ric}_M(\nabla u, \nabla u)\,. \tag{3.1}$$

Two special cases of this formula are particularly useful. When u is a distance function, that is, when $|\nabla u| = 1$, then the above formula reduces to

$$0 = |U|^2 + \mathrm{Tr}(U') + \mathrm{Ric}_M\,, \tag{3.2}$$

where U is the Hessian of u and the Ricci curvature and the derivative U' is taken in the direction of the unit vector ∇u. This is the so–called Ricatti equation. The other useful special case of the Bochner formula is when u is a harmonic function. In this case, the Bochner formula reduces to

$$\frac{1}{2}\Delta|\nabla u|^2 = |\mathrm{Hess}_u|^2 + \mathrm{Ric}_M(\nabla u, \nabla u)\,. \tag{3.3}$$

So when M has non–negative Ricci curvature, this formula implies that the energy density of u is subharmonic.

The Laplacian and Hessian comparison theorems relate the distance function on M to a space of constant curvature. These comparisons are essentially integrated forms of the Ricatti equation (3.2). For simplicity, we will not state the most general forms of these theorems, but rather only state the comparisons with $\mathbf{R}^n$.

The Laplacian comparison theorem compares Δr, where r is the distance to a point, on M with $\Delta|x| = (n-1)/|x|$ on Euclidean space:

THEOREM 3.1. *If M has non–negative Ricci curvature and r is the distance function to a fixed point p, then*

$$\Delta r \leq \frac{n-1}{r}\,. \tag{3.4}$$

Moreover, (3.4) holds weakly even where r is not smooth.

PROOF. We will prove (3.4) assuming that r is smooth so that $|\nabla r| = 1$ (see [**Ca**] for the extension to the general case). Let γ be a geodesic from p parametrized by arclength and set

$$L(t) = \Delta r \circ \gamma(t)\,. \tag{3.5}$$

Note that $L'(t) = \langle \nabla \Delta r, \nabla r\rangle$ by the chain rule so that (3.2) gives

$$L' = \mathrm{Tr}(U') \leq -|U|^2 \leq -L^2/(n-1)\,. \tag{3.6}$$

Here the second inequality used the Cauchy-Schwarz inequality

$$\left(\sum_{i=1}^{n-1} \lambda_i\right)^2 \leq (n-1)\sum_{i=1}^{n-1} \lambda_i^2 \tag{3.7}$$

for the eigenvalues λ_i of the matrix U (there are at most $(n-1)$ non–zero eigenvalues since $U(\nabla r, \nabla r) = 0$). We can rewrite (3.6) as

$$(1/L)' \geq 1/(n-1)\,. \tag{3.8}$$

(Notice that we get equality in (3.8) for $L(t) = (n-1)/t$.) Since any manifold is "almost Euclidean" for r small, it is easy to see that

$$\lim_{r\to 0} r\,\Delta r = (n-1)\,. \tag{3.9}$$

Integrating the differential equality (3.8) and substituting the "boundary condition" (3.9) gives

$$\Delta r \circ \gamma(t) = L(t) \leq \frac{n-1}{t}\,. \tag{3.10}$$

Since γ was arbitrary, the theorem follows. □

We note two immediate consequences of Theorem 3.1:

- Since ∇r is the unit normal to the geodesic spheres, the mean curvature of these spheres is at most $(n-1)/r$.
- The square of the distance function satisfies

$$\Delta r^2 = 2|\nabla r|^2 + 2r\,\Delta r \leq 2n\,. \tag{3.11}$$

The Hessian comparison theorem is somewhat more restrictive since it requires bounds on the sectional curvatures of M; of course, the conclusion is correspondingly stronger. The following theorem is a useful special case of the Hessian comparison theorem:

THEOREM 3.2. *If M is simply connected with non–positive sectional curvature and r is the distance function to a fixed point p, then*

$$Hess_r(X,X) \geq \frac{|X - \langle X, \nabla r\rangle \nabla r|^2}{r}\,, \tag{3.12}$$

for any vector X.

An important application of (3.1) (and similar formulas) is to prove vanishing theorems relating a pointwise curvature condition to global properties of M. The prototype is the Bochner theorem (see also [**C1**], [**C2**] for an extension of this famous theorem of Bochner that had been conjectured by M. Gromov):

THEOREM 3.3. [**Bc**] *If M^n is closed with $Ric_M \geq 0$, then each harmonic 1–form is parallel. In particular, the space of harmonic 1–forms is at most n-dimensional.*

PROOF. (Sketch) A harmonic 1–form α can be written locally as du where u is a (locally defined) harmonic function. In particular, (3.1) implies that

$$\Delta|\alpha|^2 = \Delta|\nabla u|^2 \geq 2\,|\mathrm{Hess}_u|^2 = 2|\nabla\alpha|^2\,. \tag{3.13}$$

Since M is closed (in particular, $\partial M = \emptyset$), Stokes' theorem gives

$$0 = \int \Delta|\alpha|^2 \geq 2\int |\nabla\alpha|^2\,. \tag{3.14}$$

Therefore, $|\nabla\alpha|^2$ vanishes identically. □

Therefore, by the Hodge theorem, the first betti number of a closed manifold M with non–negative Ricci curvature is at most n with equality only if the universal cover of M is $\mathbf{R}^n$. There have been many geometric applications of this method, where analytic methods (like the Hodge theorem) use topology to produce solutions of a pde and then a curvature condition (like the Bochner formula) places restrictions on these solutions.

Finally, we note that (3.1) can be used to prove an eigenvalue comparison theorem when M has positive Ricci curvature. Namely, A. Lichnerowicz showed that if $\mathrm{Ric}_M \geq \mathrm{Ric}_{\mathbf{S}^n}$, then the first (non–zero) eigenvalue $\lambda_1(M) \geq \lambda_1(\mathbf{S}^n)$:

THEOREM 3.4. [**Lc**] *If M^n is closed with $Ric_M \geq (n-1)$, then $\lambda_1(M) \geq n$.*

PROOF. Let u be a (non–constant) eigenfunction on M with $\Delta u = -\lambda u$. We will show that $\lambda \geq n$. After normalizing so $\int u^2 = 1$, Stokes' theorem gives

$$\int |\nabla u|^2 = -\int u \, \Delta u = \lambda \int u^2 = \lambda \, . \tag{3.15}$$

Substituting the equation for u into the Bochner formula (3.1) gives

$$\frac{1}{2} \Delta |\nabla u|^2 \geq |\text{Hess}_u|^2 + (n-1-\lambda) \, |\nabla u|^2 \geq \frac{\lambda^2}{n} \, u^2 + (n-1-\lambda) \, |\nabla u|^2 \, , \tag{3.16}$$

where the last inequality used the Cauchy-Schwarz inequality as in (3.7). Integrating (3.16) over M gives

$$0 = \frac{1}{2} \int \Delta |\nabla u|^2 \geq \frac{\lambda^2}{n} \int u^2 + (n-1-\lambda) \int |\nabla u|^2 = \lambda \, \frac{n-1}{n} \, (n - \lambda) \, . \tag{3.17}$$

□

Remark 3.5. These comparison theorems are sharp in the sense that equality is achieved on the model spaces. The converse of this is also true, i.e., equality is achieved only for the model spaces, and is known as "rigidity." For example, if M^n is closed with $\text{Ric}_M \geq (n-1)$ and $\lambda_1(M) = n$, then M. Obata, [**Ob**], proved that $M = \mathbf{S}^n$. It is then natural to ask how stable is this rigidity – i.e., what happens if equality is almost achieved? These questions, known as "almost rigidity," were answered by Colding and J. Cheeger–Colding; see [**C2**], [**ChC4**] and references therein. Moreover, almost rigidity theorems have played a key role in regularity theory; see [**ChC1**], [**ChC2**], [**ChC3**].

4. Monotonicity and the mean value inequality

Monotonicity formulas and mean value inequalities play a fundamental role in many areas of geometric analysis. In this section, we focus on the specific cases of minimal surfaces and manifolds with non–negative Ricci curvature.

Before we state and prove the monotonicity formula of volume for minimal submanifolds, we will need to recall the coarea formula. This formula asserts (see, for instance, [**Fe**] for a proof) that if Σ is a manifold and $h : \Sigma \to \mathbf{R}$ is a proper (i.e., $h^{-1}((-\infty, t])$ is compact for all $t \in \mathbf{R}$) Lipschitz function on Σ, then for all locally integrable functions f on Σ and $t \in \mathbf{R}$

$$\int_{\{h \leq t\}} f \, |\nabla h| = \int_{-\infty}^{t} \int_{h = \tau} f \, d\tau \, . \tag{4.1}$$

PROPOSITION 4.1. *Suppose that $\Sigma^k \subset \mathbf{R}^n$ is a minimal submanifold and $x_0 \in \mathbf{R}^n$; then for all $0 < s < t$*

$$t^{-k} \, Vol(B_t(x_0) \cap \Sigma) - s^{-k} \, Vol(B_s \cap \Sigma) = \int_{(B_t(x_0) \setminus B_s(x_0)) \cap \Sigma} \frac{|(x - x_0)^N|^2}{|x - x_0|^{k+2}} \, . \tag{4.2}$$

PROOF. Within this proof, we set $B_t = B_t(x_0)$. Since Σ is minimal,

$$\Delta_\Sigma |x - x_0|^2 = 2 \, \text{div}_{\,\Sigma} (x - x_0) = 2k \, . \tag{4.3}$$

By Stokes' theorem integrating this gives

$$2 \, k \, \text{Vol}(B_s \cap \Sigma) = \int_{B_s \cap \Sigma} \Delta_\Sigma |x - x_0|^2 = 2 \int_{\partial B_s \cap \Sigma} |(x - x_0)^T| \, . \tag{4.4}$$

Using this and the coarea formula (i.e., (4.1)), an easy calculation gives

$$\begin{aligned}\frac{d}{ds}\left(s^{-k}\,\mathrm{Vol}(B_s\cap\Sigma)\right) &= -k\,s^{-k-1}\,\mathrm{Vol}(B_s\cap\Sigma)+s^{-k}\int_{\partial B_s\cap\Sigma}\frac{|x-x_0|}{|(x-x_0)^T|}\\ &= s^{-k-1}\int_{\partial B_s\cap\Sigma}\left(\frac{|x-x_0|^2}{|(x-x_0)^T|}-|(x-x_0)^T|\right)\\ &= s^{-k-1}\int_{\partial B_s\cap\Sigma}\frac{|(x-x_0)^N|^2}{|(x-x_0)^T|}\,.\end{aligned}\tag{4.5}$$

Integrating and applying the coarea formula once more gives the claim. □

Notice that $(x-x_0)^N$ vanishes precisely when Σ is conical about x_0, i.e., when Σ is invariant under dilations about x_0. As a corollary, we get the following:

COROLLARY 4.2. *Suppose that $\Sigma^k\subset\mathbf{R}^n$ is a minimal submanifold and $x_0\in\mathbf{R}^n$; then the function*

$$\Theta_{x_0}(s)=\frac{\mathit{Vol}(B_s(x_0)\cap\Sigma)}{\mathit{Vol}(B_s\subset\mathbf{R}^k)}\tag{4.6}$$

is a nondecreasing function of s. Moreover, $\Theta_{x_0}(s)$ is constant in s if and only if Σ is conical about x_0.

Of course, if x_0 is a smooth point of Σ, then $\lim_{s\to 0}\Theta_{x_0}(s)=1$; the Allard regularity theorem gives the converse of this.

The monotonicity of area is an very useful tool in the regularity theory for minimal surfaces — at least when there is some *a priori* area bound. For instance, this monotonicity and a compactness argument allow one to reduce many regularity questions to questions about minimal cones (this was a key observation of W. Fleming in his work on the Bernstein problem; see Section 5). Similar monotonicity formulas have played key roles in other geometric problems, including harmonic maps, Yang–Mills connections, J–holomorphic curves, and regularity of limit spaces with a lower Ricci curvature bound.

Arguing as in Proposition 4.1, we get a weighted monotonicity:

PROPOSITION 4.3. *If $\Sigma^k\subset\mathbf{R}^n$ is a minimal submanifold, $x_0\in\mathbf{R}^n$, and f is a function on Σ, then*

$$t^{-k}\int_{B_t(x_0)\cap\Sigma}f-s^{-k}\int_{B_s(x_0)\cap\Sigma}f\tag{4.7}$$

$$=\int_{(B_t(x_0)\setminus B_s(x_0))\cap\Sigma}f\,\frac{|(x-x_0)^N|^2}{|x-x_0|^{k+2}}+\frac{1}{2}\int_s^t\tau^{-k-1}\int_{B_\tau(x_0)\cap\Sigma}(\tau^2-|x-x_0|^2)\,\Delta_\Sigma f\,d\tau\,.$$

We get immediately the following mean value inequality for the special case of non–negative subharmonic functions:

COROLLARY 4.4. *Suppose that $\Sigma^k\subset\mathbf{R}^n$ is a minimal submanifold, $x_0\in\mathbf{R}^n$, and f is a non–negative subharmonic function on Σ; then*

$$s^{-k}\int_{B_s(x_0)\cap\Sigma}f\tag{4.8}$$

is a nondecreasing function of s. In particular, if $x_0\in\Sigma$, then for all $s>0$

$$f(x_0)\le\frac{\int_{B_s(x_0)\cap\Sigma}f}{\mathit{Vol}(B_s\subset\mathbf{R}^k)}\,.\tag{4.9}$$

Another interesting (and crucial) appearance of monotonicity is the volume comparison theorem of Bishop–Gromov for manifolds with non–negative Ricci curvature, [**GLPa**]; see also [**Pe1**] for a generalization to Ricci flow. In the case of non–negative Ricci curvature, the monotonicity goes the opposite direction:

THEOREM 4.5. *If a k–dimensional manifold M has non–negative Ricci curvature, then*

$$\frac{Vol(B_s(x_0) \subset M)}{Vol(B_s \subset \mathbf{R}^k)} \tag{4.10}$$

is a non–increasing function of s.

PROOF. (Sketch) By the Laplacian comparison theorem, we have

$$\Delta_M r^2 \leq 2k \tag{4.11}$$

where r is the distance function to x_0. Integrating this by parts gives

$$2k \, \mathrm{Vol}(B_s(x_0)) \geq \int_{B_s(x_0)} \Delta r^2 = 2s \int_{\partial B_s(x_0)} |\nabla r| = 2s \, \frac{d}{ds} \, \mathrm{Vol}(B_s(x_0)) \, , \tag{4.12}$$

where the last equality used the co–area formula (since $|\nabla r| = 1$ almost everywhere). This differential inequality gives (4.10). □

Remark 4.6. Equation (4.10) immediately implies a volume doubling property for manifolds with non–negative Ricci curvature:

$$\mathrm{Vol}(B_{2s}(x_0) \subset M) \leq 2^n \, \mathrm{Vol}(B_s(x_0) \subset M) \, . \tag{4.13}$$

This property is very useful for covering arguments, cf. [**G**].

We conclude this section with a well–known intrinsic mean value inequality which is often useful but difficult to find in the literature (it is often stated only for subharmonic functions).

PROPOSITION 4.7. *There exists $C = C(k)$ so that if M is k–dimensional, $Ric_M \geq -(k-1)\, s^{-2}$, and $u \geq 0$ satisfies $\Delta_M u \geq -s^{-2}\, u$, then*

$$u^2(x) \leq \frac{C}{Vol(B_s(x))} \int_{B_s(x)} u^2 \, . \tag{4.14}$$

PROOF. After rescaling the metric by s, it suffices to prove the case $s = 1$. Let $N = M \times [-1, 1]$ have the product metric, so that $\mathrm{Ric}_N \geq -(k-1)$. Define an auxiliary function w on N by

$$w(x,t) = u(x) \, \mathrm{e}^t \, . \tag{4.15}$$

An easy calculation shows that

$$\Delta_N w = \mathrm{e}^t \, \Delta_M u + \mathrm{e}^t \, u \geq 0 \, , \tag{4.16}$$

so that w is subharmonic. The mean value inequality for subharmonic functions (see theorem 6.2 on page 77 of [**ScYa1**]) then gives

$$\begin{aligned} w^2(x,0) &\leq \frac{C}{\mathrm{Vol}(B_1(x,0) \subset N)} \int_{B_1(x,0) \subset N} w^2 \\ &\leq 2\mathrm{e}^2 \, \frac{C}{\mathrm{Vol}(B_{1/2}(x) \subset M)} \int_{B_1(x) \subset M} u^2 \, . \end{aligned} \tag{4.17}$$

The proposition follows from this after we use the Bishop–Gromov volume comparison (cf. Theorem 4.5) to bound

$$\frac{\mathrm{Vol}(B_1(x))}{\mathrm{Vol}(B_{1/2}(x))}\,. \tag{4.18}$$

□

5. The theorems of Bernstein and Bers

A classical theorem of S. Bernstein from 1916 says that entire (i.e., defined over all of $\mathbf{R}^2$) minimal graphs are planes. This remarkable theorem of Bernstein was one of the first illustrations of the fact that the solutions to a nonlinear PDE, like the minimal surface equation, can behave quite differently from the solutions to a linear equation. Rather surprisingly, this result very much depended on the dimension. The combined efforts of S. Bernstein [**Be**], E. De Giorgi [**DG**], F. J. Almgren, Jr. [**Am1**], and J. Simons [**Sim**] finally gave:

THEOREM 5.1. *If $u : \mathbf{R}^{n-1} \to \mathbf{R}$ is an entire solution to the minimal surface equation and $n \leq 8$, then u is an affine function.*

However, in 1969 E. Bombieri, De Giorgi, and E. Giusti [**BDGG**] constructed entire non–affine solutions to the minimal surface equation on $\mathbf{R}^8$ and an area–minimizing singular cone in $\mathbf{R}^8$. In fact, they showed that for $m \geq 4$ the cones

$$C_m = \{(x_1, \dots, x_{2m}) \mid x_1^2 + \dots + x_m^2 = x_{m+1}^2 + \dots + x_{2m}^2\} \subset \mathbf{R}^{2m} \tag{5.1}$$

are area–minimizing (and obviously singular at the origin).

One way to prove the Bernstein theorem is to prove a curvature estimate for minimal graphs. The basic example is the estimate of E. Heinz for surfaces:

THEOREM 5.2. [**He**] *If $D_{r_0} \subset \mathbf{R}^2$ and $u : D_{r_0} \to \mathbf{R}$ satisfies the minimal surface equation, then for $\Sigma = \mathrm{Graph}_u$ and $0 < \sigma \leq r_0$*

$$\sigma^2 \sup_{D_{r_0-\sigma}} |A|^2 \leq C\,. \tag{5.2}$$

The original Bernstein Theorem follows from Theorem 5.2 by taking $r_0 \to \infty$. By the same reasoning, the examples of [**BDGG**] show that (5.2) cannot hold for all dimensions. However, curvature estimates for graphs over $B_{r_0} \subset \mathbf{R}^{n-1}$ were proven in [**ScSiYa**] for $n \leq 6$ (the remaining cases, i.e., $n = 7$ and 8, were proven in [**Si2**]).

In contrast to the entire case, exterior solutions of the minimal graph equation, i.e., solutions on $\mathbf{R}^2 \setminus B_1$, are much more plentiful. In this case, Theorem 5.2 only gives quadratic curvature decay $|A|^2 \leq C\,|x|^{-2}$. In particular, it is not even clear that $|\nabla u|$ is bounded since

$$|\nabla |\nabla u|| \leq C\,|x|^{-1} \tag{5.3}$$

is not integrable along rays. However, L. Bers proved that ∇u actually has an asymptotic limit:

THEOREM 5.3. [**Ber**] *If u is a C^2 solution to the minimal surface equation on $\mathbf{R}^2 \setminus B_1$, then ∇u has a limit at infinity (i.e., there is an asymptotic tangent plane).*

Bers' theorem was extended to higher dimensions by L. Simon:

THEOREM 5.4. [**Si1**] *If u is a C^2 solution to the minimal surface equation on $\mathbf{R}^n \setminus B_1$, then either*

- *$|\nabla u|$ is bounded and ∇u has a limit at infinity.*
- *All tangent cones at infinity are of the form $\Sigma \times \mathbf{R}$ where Σ is singular.*

Bernstein's theorem has had many other interesting generalizations, including, e.g., curvature estimates of R. Schoen for stable surfaces and Schoen–Simon–Yau for stable hypersurfaces with bounded density. In the early nineteen–eighties Schoen and Simon extended the theorem of Bernstein to complete simply connected embedded minimal surfaces in $\mathbf{R}^3$ with quadratic area growth. A surface Σ is said to have quadratic area growth if for all $r > 0$, the intersection of the surface with the ball in $\mathbf{R}^3$ of radius r and center at the origin is bounded by $C\,r^2$ for a fixed constant C independent of r. In corollary 1.18 in [**CM4**], this was generalized to quadratic area growth for *intrinsic* balls:

COROLLARY 5.5. [**CM4**] *Given a constant C_I, there exists C_P so that if $\mathcal{B}_{2s} \subset \Sigma \subset \mathbf{R}^3$ is an embedded minimal disk with*

$$\int_{\mathcal{B}_{2s}} |A|^2 \leq C_I \,, \tag{5.4}$$

then

$$\sup_{\mathcal{B}_s} |A|^2 \leq C_P\, s^{-2} \,.$$

6. Mean curvature flow

Just as the Laplace equation has the heat equation as a parabolic analog, mean curvature flow is the parabolic analog of the minimal surface equation. A one–parameter family of smooth hypersurfaces $\{M_t\} \subset \mathbf{R}^{n+1}$ *flows by mean curvature* if

$$z_t = \mathbf{H}(z) = \Delta_{M_t} z \,, \tag{6.1}$$

where z are coordinates on $\mathbf{R}^{n+1}$ and $\mathbf{H} = -H\mathrm{n}$ is the mean curvature vector.

Example 6.1. Let M_t be the family of concentric shrinking n–spheres of radius

$$\sqrt{R^2 - 2nt} \,. \tag{6.2}$$

It is easy to see that M_t flows by mean curvature and is smooth up to $t = R^2/(2n)$ when it shrinks to a point.

Suppose now that each M_t is the graph of a function $u(\cdot, t)$. So, if $z = (x, y)$ with $x \in \mathbf{R}^n$, then M_t is given by $y = u(x, t)$ which satisfies

$$u_t = (1 + |du|^2)^{1/2} \operatorname{div} \left(\frac{du}{(1 + |du|^2)^{1/2}} \right) , \tag{6.3}$$

where du is the $\mathbf{R}^n$ gradient of the function u and div is divergence in $\mathbf{R}^n$.

The monotonicity formula and mean value inequality of Section 4 have analogs in mean curvature flow. The monotonicity formula, proven by G. Huisken (and extended to more general weak solutions by T. Ilmanen and B. White), is:

THEOREM 6.2. [**H**] *If a smooth one–parameter family of hypersurfaces M_t flows by mean curvature in $\mathbf{R}^{n+1} \times [-T, 0]$, then*

$$\frac{d}{dt} \int_{M_t} (-4\pi t)^{-n/2}\, e^{\frac{|x|^2}{4t}} = - \int_{M_t} \left| \mathbf{H} - \frac{x^N}{2t} \right|^2 (-4\pi t)^{-n/2}\, e^{\frac{|x|^2}{4t}} \,. \tag{6.4}$$

In particular, the "density ratio"

$$\int_{M_t} (-4\pi t)^{-n/2} \, e^{\frac{|x|^2}{4t}} \tag{6.5}$$

is non–increasing.

The restrictions of the coordinate functions to M_t satisfy the heat equation

$$\partial_t x_i = \Delta_M x_i \,. \tag{6.6}$$

From this, we see that the restriction of $|x|^2$ satisfies

$$(\partial_t - \Delta_M)\, |x|^2 = -2n \,. \tag{6.7}$$

As in the stationary case (i.e., for minimal surfaces), this is the key to the proof of Theorem 6.2.

The mean value inequality in this case applies to non–negative solutions of the heat equation on M_t; we refer to [**E1**] for more detail on this as well as discussion of the local monotonicity formula for mean curvature flow proven in [**E2**].

The parabolic maximum principle has been very useful in mean curvature flow (somewhat similarly to the convex hull property for minimal surfaces). Two immediate, but useful, consequences are:

1. Disjoint surfaces stay disjoint.
2. An embedded surface stays embedded (as long as it evolves smoothly).

The reason for (1) and (2) is quite simple. Suppose that two initially disjoint surfaces touch at a first time t at a point x. Clearly, they will be tangent at (x, t) so nearby we see two graphs, one above the other. Hence, at x the mean curvature of the upper graph is larger (or equal to). Therefore, the upper graph crossed over the lower at a slightly early time, contradicting that t is first time of contact.

Combining (1) with the shrinking spheres of Example 6.1, we see that any compact hypersurface flowing by mean curvature has a finite extinction time.

7. Ricci flow

The Ricci flow is the parabolic analog of the Einstein equation $\mathrm{Ric}_M = \ \text{Constant}\, g$, where g is the metric. Namely, let M^3 be a fixed smooth manifold and let $g(t)$ be a one–parameter family of metrics on M evolving by the Ricci flow, so

$$\partial_t g = -2\, \mathrm{Ric}_{M_t} \,. \tag{7.1}$$

Short–time existence for the Ricci flow was established by Hamilton:

THEOREM 7.1. [**Ha2**] *Given any smooth compact Riemannian manifold (M, g_0), there exists a unique smooth solution $g(t)$ to (7.1) with initial condition $g(0) = g_0$ on some time interval $[0, \epsilon)$.*

Long–time existence is quite a bit more subtle, see [**Ha2**] and [**Pe1**].

There are many formal similarities between the Ricci flow and the mean curvature flow, including similarities between the evolution equations for various geometric quantities. One interesting distinction is the evolution equation for the scalar curvature $R = R(t)$ under the Ricci flow (see, for instance, page 16 of [**Ha3**])

$$\partial_t R = \Delta R + 2|\mathrm{Ric}|^2 \geq \Delta R + \frac{2}{n}\, R^2 \,, \tag{7.2}$$

where the inequality used the Cauchy–Schwarz inequality (M is n–dimensional). This differential inequality has an interesting consequence: After flowing for any positive amount of time, there is a lower bound for the scalar curvature (the mean curvature flow has no such analog). To make this precise, a straightforward maximum principle argument gives that at time $t > 0$

$$R(t) \geq \frac{1}{1/[\min R(0)] - 2t/n} = -\frac{n}{2(t+C)} \,. \tag{7.3}$$

In the derivation of (7.3) we implicitly assumed that $\min R(0) < 0$. If this was not the case, then (7.3) trivially holds with $C = 0$, since, by (7.2), $\min R(t)$ is always non–decreasing.

8. Gradient estimates

Gradient estimates have played a key role in geometry and pde since at least the early work of Bernstein. These are probably the most fundamental a priori estimates for elliptic and parabolic equations, leading to Harnack inequalities, Liouville theorems, and compactness theorems for both linear and nonlinear pde.

A typical example for linear equations is the well–known gradient estimate of S.Y. Cheng and S.T. Yau for harmonic functions:

THEOREM 8.1. [**CgYa**] *If $\Delta u = 0$ on $B_r(0)$ with non–negative Ricci curvature, then*

$$|\nabla u|(0) \leq C\, r^{-1}\, \|u\|_\infty \,, \tag{8.1}$$

where $\|u\|_\infty$ is the sup norm of the function u on $B_r(0)$.

To give something of the flavor, we will use the maximum principle to prove 8.1 on the Euclidean unit ball $B_1(0) \subset \mathbf{R}^n$.

PROOF. (of Theorem 8.1 for $B_1(0) \subset \mathbf{R}^n$.) Define the cutoff function

$$\eta(x) = 1 - |x|^2 \,, \tag{8.2}$$

so that $|\nabla \eta| \leq 2$ and $\Delta \eta = -2n$. We compute that

$$\begin{aligned} \Delta(\eta^2\, |\nabla u|^2) &\geq -2n\, |\nabla u|^2 - 16\, \eta\, |\nabla u|\, |\text{Hess}_u| + 2\, \eta^2\, |\text{Hess}_u|^2 \\ &\geq -(2n+32)\, |\nabla u|^2 \,, \end{aligned} \tag{8.3}$$

where the last inequality used the absorbing inequality

$$16ab \leq 2a^2 + 32b^2 \,. \tag{8.4}$$

In particular, the function

$$w = (n+16)\, u^2 + \eta^2\, |\nabla u|^2 \tag{8.5}$$

is subharmonic on $B_1(0)$ (i.e., $\Delta w \geq 0$). By the maximum principle, the maximum of w occurs on the boundary so that

$$|\nabla u|^2(0) \leq w(0) \leq \max_{\partial B_1(0)} w = (n+16) \max_{\partial B_1(0)} u^2 \,. \tag{8.6}$$

□

In fact, Cheng and Yau prove a stronger estimate: If in addition u is positive on $B_r(0)$, then

$$|\nabla \log u|(0) \leq C\, r^{-1}\,. \tag{8.7}$$

An important consequence is the Harnack inequality for positive harmonic functions **Elliptic Harnack inequality**:

$$\sup_{B_{r/2}(0)} u \leq C' \inf_{B_{r/2}(0)} u\,. \tag{8.8}$$

PROOF. Suppose the sup and inf are achieved at $p, q \in \partial B_{r/2}(0)$. Fix a curve $\gamma_{p,q} \subset B_{r/2}(0)$ from p to q of length at most r (e.g., connect each point to 0 by a ray). Integrating the bound

$$\sup_{B_{r/2}(0)} |\nabla \log u| \leq 2C\, r^{-1} \tag{8.9}$$

over $\gamma_{p,q}$ gives

$$\log \frac{u(p)}{u(q)} \leq \int_{\gamma_{p,q}} |\nabla \log u| \leq 2\,C\,. \tag{8.10}$$

□

This gradient estimate also gives the global Liouville theorem of Yau, [**Ya3**]: **Liouville theorem**: If u is a positive harmonic function on a complete manifold with non–negative Ricci curvature, then u is constant.

PROOF. We can take $r \to \infty$ in (8.7) to get that $|\nabla u| = 0$. □

The parabolic analog of Theorem 8.1 is the gradient estimate for the heat equation of P. Li and Yau (we will state the version for M complete):

THEOREM 8.2. [**LiYa**] *If u is a positive solution of $\partial_t u = \Delta u$ for $0 \leq t$ on a complete M with non–negative Ricci curvature, then*

$$2t\left(|\nabla \log u|^2 - \partial_t \log u\right) \leq n\,, \tag{8.11}$$

or, equivalently, $2t\,\Delta \log u \geq -n$.

PROOF. (Sketch) Set

$$w = -t\,\Delta \log u = t\left(|\nabla \log u|^2 - \partial_t \log u\right)\,. \tag{8.12}$$

The key calculation is (cf. lemma 1 on page 155 of [**ScYa1**])

$$t\,(\Delta - \partial_t)\,w \geq \frac{2}{n}\,w^2 - w - 2t\,\nabla w \cdot \nabla \log u\,. \tag{8.13}$$

Suppose that w achieves its maximum on $M \times [0, t]$ at (x, t) (for example, when M is compact; otherwise we use a cutoff). The parabolic maximum principle then gives $\nabla w(x, t) = 0$ and $(\Delta - \partial_t)\,w(x, t) \leq 0$. Substituting this into (8.13) gives

$$0 \geq \frac{2}{n}\,w^2(x,t) - w(x,t)\,, \tag{8.14}$$

so that $w(x, t) \leq n/2$ as desired. □

Integrating this along curves as in the elliptic case gives for $t_1 < t_2$ that **Parabolic Harnack inequality**:

$$u(x_1, t_1) \leq u(x_2, t_2) \left(\frac{t_2}{t_1}\right)^{\frac{n}{2}} \mathrm{e}^{\frac{\mathrm{dist}^2(x_1, x_2)}{4(t_2 - t_1)}} . \tag{8.15}$$

In [**Ha1**], R. Hamilton gave an extension of (8.11) to a full matrix estimate whose trace was (8.11). For example, if u is positive solution of the heat equation on $\mathbf{R}^n \times [0, T]$, then [**Ha1**] implies that

$$2t \,\mathrm{Hess}_{\log u} + \delta_{ij} \geq 0 . \tag{8.16}$$

Taking the trace of (8.16) gives $|\nabla \log u|^2 - \partial_t \log u \leq n/(2t)$.

8.1. Gradient estimates for nonlinear equations. For the (nonlinear) minimal surface equation, the situation is somewhat different. In this case, i.e., when the graph of u is minimal on $B_r(0)$, then Bombieri, De Giorgi, and M. Miranda proved in [**BDM**] that

$$\log |du|(0) \leq C \left(1 + r^{-1} \, \|u\|_\infty\right), \tag{8.17}$$

where du is the $\mathbf{R}^n$ gradient of the function u (the case of surfaces was done by R. Finn in [**Fi**]). By an earlier example of Finn, this exponential dependence cannot be improved. In [**K**], N. Korevaar gave a maximum principle proof of a weaker form of [**BDM**]; this weaker form had $\|u\|_\infty^2$ in place of $\|u\|_\infty$.

In [**CM12**], we proved a sharp gradient estimate for graphs flowing by mean curvature:

THEOREM 8.3. [**CM12**] *There exists $C = C(n)$ so if the graph of*

$$u : B_{\sqrt{2n+1} r} \times [0, r^2] \to \mathbf{R} \tag{8.18}$$

flows by mean curvature, then

$$\log |du|(0, r^2/[4n]) \leq C \left(1 + r^{-1} \, \|u(\cdot, 0)\|_\infty\right)^2 . \tag{8.19}$$

The quadratic dependence on $\|u(\cdot, 0)\|_\infty$ in (8.19) should be compared with the linear dependence which holds when the graph of u is minimal (i.e., $u_t = 0$). Somewhat surprisingly, examples in [**CM12**] show that this quadratic dependence on $\|u(\cdot, 0)\|_\infty$ is sharp.

The first gradient estimate for mean curvature flow was proven by Ecker and Huisken who adapted Korevaar's argument to mean curvature flow in theorem 2.3 of [**EH2**] to get

$$\log |du|(0, r^2/[4n]) \leq 1/2 \, \log \left(1 + \|du(\cdot, 0)\|_\infty^2\right) + C \left(1 + r^{-1} \, \|u(\cdot, 0)\|_\infty\right)^2 . \tag{8.20}$$

Note that, unlike (8.19), the gradient bound (8.20) depends also on the initial bound for the gradient.

8.2. Generalizations. The Harnack inequality actually holds for much more general spaces. For instance, L. Saloff–Coste and A. Grigor'yan (see [**SC**] and [**Gr**]) have shown that the following two properties suffice

Volume doubling: There exists C_D so that

$$\mathrm{Vol}(B_{2r}(x)) \leq C_D \, \mathrm{Vol}(B_r(x)) , \tag{8.21}$$

for all $r > 0$ and points x.

Neumann Poincaré inequality: There exists C_N so that if $\int_{B_r(x)} f = 0$, then

$$\int_{B_r(x)} f^2 \leq C_N\, r^2 \int_{B_r(x)} |\nabla f|^2 \,, \tag{8.22}$$

for all $r > 0$ and points x.

These properties, however, do not imply the gradient estimate. Note that manifolds with non–negative Ricci curvature satisfy both conditions (the Poincaré inequality essentially follows from [**Bu**], cf. also [**Je**]).

The De Giorgi, Nash, Moser theory (see chapter 8 in [**GiTr**] or section 4.4 in [**HnLn**]) gives a Harnack inequality as long as we have a volume doubling and a Sobolev inequality. The difference between a Sobolev and Poincaré inequality is that a Sobolev controls an L^p–norm of f for some $p > 2$. Surprisingly, in [**HzKo**], P. Hajlasz and P. Koskela showed that the volume doubling and Neumann Poincaré inequality together imply a Sobolev inequality, thereby recovering the above result of Saloff–Coste and Grigor'yan. See [**ChC2**], [**ChC3**], [**Hj**] for more such "low regularity" analysis (including analysis on singular spaces).

9. Simons type inequalities

In this section, we recall a very useful differential inequality for the Laplacian of the norm squared of the second fundamental form of a minimal hypersurface Σ in $\mathbf{R}^n$ and illustrate its role in a priori estimates. This inequality, originally due to J. Simons (see [**CM1**] for a proof and further discussion), is:

LEMMA 9.1. [**Sim**] *If* $\Sigma^{n-1} \subset \mathbf{R}^n$ *is a minimal hypersurface, then*

$$\Delta_\Sigma |A|^2 = -2\,|A|^4 + 2|\nabla_\Sigma A|^2 \geq -2\,|A|^4 \,. \tag{9.1}$$

An inequality of the type (9.1) on its own does not lead to pointwise bounds on $|A|^2$ because of the nonlinearity. However, it does lead to estimates if a "scale–invariant energy" is small. For example, H. Choi and Schoen used (9.1) to prove:

THEOREM 9.2. [**CiSc**] *If* $0 \in \Sigma \subset B_r(0)$ *with* $\partial\Sigma \subset \partial B_r(0)$ *is a minimal surface with sufficiently small total curvature* $\int |A|^2$, *then*

$$|A|^2(0) \leq r^{-2} \,. \tag{9.2}$$

Analogs of (9.1) occur in a surprising number of geometric problems. For example, when $u : M^m \to N^n$ is a harmonic map, the energy density $|du|^2$ satisfies this type of inequality, leading to an *a priori* estimate when u has small scale–invariant energy

$$r^{2-m} \int_{B_r} |du|^2 \tag{9.3}$$

(see [**SaUh**], [**Sc2**]). Similar inequalities hold for the curvature of a Yang–Mills connection or the curvature of an Einstein manifold. When M is an Einstein manifold, its curvature tensor R satisfies (see [**Ha2**] or equation (2.6) in [**An1**]; [**Ha2**] also establishes a parabolic analog for the Ricci flow)

$$\Delta_M |R| \geq -C\,|R|^2 \,. \tag{9.4}$$

We next use (9.4) to prove an estimate for Einstein manifolds (cf. lemma 2.1 in [**An1**]). For simplicity, we restrict to the case $\mathrm{Ric}_M = 0$.

THEOREM 9.3. *There exist $\epsilon = \epsilon(n) > 0$, such that if M^n is an n–dimensional Ricci–flat (Einstein) manifold and for some $x \in M$ either*

$$Vol(B_r(x)) \geq [1 - \epsilon]\; Vol(B_r \subset \mathbf{R}^n) \tag{9.5}$$

or

$$\int_{B_r(x)} |R|^{n/2} < \epsilon\, r^{-n}\; Vol(B_r(x))\,, \tag{9.6}$$

then $|R|(x) \leq C\, r^{-2}$.

PROOF. We will prove that (9.6) gives the pointwise curvature bound; the other case is similar. Set

$$F(z) = (r - 4\text{dist}_M(x,z))\, |R|^{1/2}(z)\,, \tag{9.7}$$

so that

$$F(x) = r\, |R|^{1/2}(x) \geq 0\,, \text{ and } F\, |_{B_r(x) \setminus B_{r/4}(x)} \leq 0\,. \tag{9.8}$$

Therefore, it suffices to prove that $F \leq C$ for some fixed constant C.

We will assume that $\max_{B_r(x)} F > 32$ and deduce a contradiction if $\epsilon > 0$ is sufficiently small. Let y be a point where the maximum of F is achieved and set $s = |R|^{-1/2}(y)$. Since $F(y) > 32$, we have $32 s < |r - 4\text{dist}_M(x,y)|$ so that for $z \in B_s(y)$

$$1/2 \leq \frac{|r - 4\text{dist}_M(x,y)|}{|r - 4\text{dist}_M(x,z)|} \leq 2\,. \tag{9.9}$$

Since $F(z) \leq F(y)$, it follows that $B_s(y)$ satisfies

$$\sup_{B_s(y)} |R|^{1/2} \leq 2 |R|^{1/2}(y) = 2/s\,, \tag{9.10}$$

so that (9.4) gives on $B_s(y)$ that

$$\Delta_M\, |R| \geq -C\, |R|^2 \geq -4\, C\, s^{-2}\, |R|\,. \tag{9.11}$$

Furthermore, the Bishop–Gromov volume comparison , i.e., Theorem 4.5, gives

$$\frac{\text{Vol}(B_s(y))}{s^n} \geq \frac{\text{Vol}(B_{r/2}(y))}{(r/2)^n} \geq 2^{-n}\, \frac{\text{Vol}(B_r(x))}{r^n}\,. \tag{9.12}$$

It follows from this and (9.6) that $B_s(y)$ also satisfies

$$\int_{B_s(y)} |R|^{n/2} < 2^n\, \epsilon\, \frac{\text{Vol}(B_s(y))}{s^n}\,. \tag{9.13}$$

Using (9.10), (9.11), and (9.13), the mean value inequality (Proposition 4.7) gives

$$s^{-n} = |R|^{n/2}(y) \leq \frac{C}{\text{Vol}(B_s(y))} \int_{B_s(y)} |R|^{n/2} < C\, 2^n\, \epsilon\, s^{-n}\,. \tag{9.14}$$

This gives a contradiction for ϵ sufficiently small. □

Remark 9.4. We could alternatively have proven Theorem 9.3 by integral methods, i.e., using Moser iteration. However, the above proof by scaling is both shorter and more elementary.

Finally, we mention that (9.1) has a parabolic version as well (see proposition 2.15 in [**E1**]): If M_t flows by mean curvature, then

$$\left(\frac{\partial}{\partial t} - \Delta_{M_t}\right) |A|^2 = 2\,|A|^4 - 2|\nabla_{M_t} A|^2 \,. \tag{9.15}$$

As in the elliptic case, this Simons' type inequality is a crucial ingredient for establishing curvature estimates.

10. Minimal annuli with small total curvature are graphs

It is easy to see that a minimal disk with small total curvature must be a graph away from its boundary:

If $\int_{B_R \cap \Sigma} |A|^2 < \epsilon$, then Theorem 9.2 gives

$$\sup_{B_{R/2} \cap \Sigma} |A|^2 < C\,\epsilon / R^2 \,. \tag{10.1}$$

Integrating this (since $|\nabla \mathrm{n}| \leq |A|$) implies that each component of $B_{R/2} \cap \Sigma$ is a graph if $\epsilon > 0$ is small enough.

However, the corresponding question for minimal annuli is more subtle. We shall discuss this and some related problems in this section.

In [**CM10**], we gave three proofs that a minimal annulus with small total curvature is a graph. The first used a singular integral formula which had previously been useful for estimating nodal and singular sets; see Proposition 10.1 below and compare [**Do**]. The second, and easiest, applies more generally to surfaces with quasi–conformal Gauss maps; see Proposition 10.2. The third, which is outlined in Lemma 10.3, was the one which could be extended to "annuli with slits" — i.e., embedded double–valued minimal graphs.

In this section, $\Sigma \subset \mathbf{R}^3$ is a compact connected oriented immersed surface. If $a \in \mathbf{S}^2$, $a^\perp$ denotes $\{x \in \mathbf{R}^3 \,|\, \langle x, a \rangle = 0\}$. For $a, b \in \mathbf{S}^2$, $\mathrm{Angle}(a,b)$ is the angle between $a^\perp, b^\perp$; i.e.,

$$\mathrm{Angle}(a,b) = \mathrm{dist}_{\mathbf{S}^2}(a, \{b, -b\}) \,. \tag{10.2}$$

Let f be harmonic on Σ^2 with critical points $\{y_i\}$ with multiplicities $\{m_i\}$. Suppose that none of the y_i's lie on $\partial\Sigma$. The Bochner formula on $\Sigma \setminus \{y_i\}$ gives

$$\Delta_\Sigma \log |\nabla_\Sigma f|^2 = 2\frac{|\mathrm{Hess}_f|^2}{|\nabla_\Sigma f|^2} + 2\,K - \frac{|\nabla_\Sigma |\nabla_\Sigma f|^2|^2}{|\nabla_\Sigma f|^4} = 2\,K \,. \tag{10.3}$$

Here we used that since $\Delta_\Sigma f = 0$ and Σ is 2–dimensional, then

$$2\,|\mathrm{Hess}_f|^2\, |\nabla_\Sigma f|^2 = |\nabla_\Sigma |\nabla_\Sigma f|^2|^2 \,. \tag{10.4}$$

Hence, by Stokes' theorem

$$\begin{aligned} \int_{\partial \Sigma} \frac{d \log |\nabla_\Sigma f|^2}{dn} &= \int_{\Sigma \setminus \{y_i\}} \Delta_\Sigma \log |\nabla_\Sigma f|^2 + 4\,\pi \sum_i m_i \\ &= 2 \int_\Sigma K + 4\,\pi \sum_i m_i \,. \end{aligned} \tag{10.5}$$

PROPOSITION 10.1. [**CM10**] *If Σ is connected and minimal with boundaries σ_1 and σ_2, $\int_{\sigma_1 \cup \sigma_2} |A| < \pi/8$, and $\int_\Sigma K \geq -\pi$, then Σ is graphical.*

PROOF. Fix $q_i \in \sigma_i$. Since

$$|\nabla_\Sigma \text{dist}_{\mathbf{S}^2}(\text{n}(q_i), \text{n}(\cdot))| \leq |A| , \tag{10.6}$$

the assumption on $\partial \Sigma$ gives

$$\sum_i \sup_{z_i \in \sigma_i} \text{dist}_{\mathbf{S}^2}(\text{n}(q_i), \text{n}(z_i)) \leq \sum_i \int_{\sigma_i} |A| < \pi/8 . \tag{10.7}$$

Choose $b \in \mathbf{S}^2$ with $\text{Angle}(\text{n}(q_i), b) \leq \pi/4$ for $i = 1, 2$. We will show that Σ is graphical over the plane $b^\perp$. By the triangle inequality and (10.7), for $i = 1, 2$,

$$\sup_{z_i \in \sigma_i} \text{Angle}(b, \text{n}(z_i)) \leq \pi/4 + \sup_{z_i \in \sigma_i} \text{dist}_{\mathbf{S}^2}(\text{n}(q_i), \text{n}(z_i)) < 3\,\pi/8 . \tag{10.8}$$

Rotate coordinates so that $b = (0, 0, 1)$ and $b^\perp$ is the x_1–x_2–plane. Fix θ and set

$$f = x_1 \cos\theta + x_2 \sin\theta . \tag{10.9}$$

Given $x \in \Sigma$,

$$|\nabla_\Sigma f|^2(x) = 1 - \langle (\cos\theta, \sin\theta, 0), \text{n}(x)\rangle^2 \geq \langle b, \text{n}(x)\rangle^2 . \tag{10.10}$$

On $\partial\Sigma = \sigma_1 \cup \sigma_2$, (10.8) and (10.10) imply that

$$\inf_{\partial\Sigma} |\nabla_\Sigma f| \geq \inf_{\partial\Sigma} |\langle b, \text{n}(x)\rangle| > \cos(3\,\pi/8) > 1/3 . \tag{10.11}$$

Since $|\nabla_\Sigma |\nabla_\Sigma f|| \leq |\text{Hess}_f| \leq |A|$, (10.11) gives on $\partial\Sigma$

$$|\nabla_\Sigma \log |\nabla_\Sigma f|^2| = \frac{2\,|\nabla_\Sigma |\nabla_\Sigma f||}{|\nabla_\Sigma f|} \leq 6\,|A| . \tag{10.12}$$

Integrating (10.12), we get

$$\int_{\partial\Sigma} \left| \frac{d \log |\nabla_\Sigma f|^2}{dn} \right| \leq 6 \int_{\partial\Sigma} |A| < 3\,\pi/4 . \tag{10.13}$$

Since Σ is minimal, $\Delta_\Sigma f = 0$. Substituting (10.13) into (10.5),

$$4\,\pi \sum_i m_i = \int_{\partial\Sigma} \frac{d \log |\nabla_\Sigma f|^2}{dn} - 2 \int_\Sigma K < 3\,\pi/4 + 2\pi < 4\,\pi ; \tag{10.14}$$

hence, f has no critical points. Since this is true for any θ, Σ is graphical over $b^\perp$. □

In fact, Proposition 10.1 holds for Σ whose Gauss map n is quasi–conformal:

PROPOSITION 10.2. [**CM10**] *If $\Sigma \subset \mathbf{R}^3$ is connected, $\int_\Sigma |K| \leq \pi$, $|A|^2 \leq C\,|K|$, and $\partial\Sigma$ has components $\{\sigma_i\}_{1 \leq i \leq n}$ with*

$$\sum_{i=1}^n \inf_{a \in \mathbf{S}^2} \sup_{x \in \sigma_i} \{dist_{\mathbf{S}^2}(n(x), a)\} < \epsilon < \pi/8 , \tag{10.15}$$

then $n(\Sigma) \subset \mathcal{B}_{2\,\epsilon}(a)$ for some $a \in \mathbf{S}^2$ and Σ is the graph of u over $a^\perp$ with $|\nabla u| \leq 4\,\epsilon$.

10.1. Holomorphic functions on annuli. We next estimate the oscillation of a holomorphic function with small gradient on annuli. Since the Gauss map of a minimal annulus is conformal and its derivative is the second fundamental form A, this serves as a good model (in fact, the proof can easily be adapted to that case).

Suppose that f is holomorphic on an annulus $D_R \setminus D_\delta$ and satisfies

$$|\nabla f| \leq \epsilon/(2\pi\,|z|)\,. \tag{10.16}$$

Integrating this around each circle, we see that the oscillation of f on each circle is at most ϵ. However, if we simply integrate this bound radially to compare f on the different circles, we get the log of the ratio of the radii. The next lemma improves on this to get a bound independent of this ratio; the key is to keep track of cancellation rather than integrating the bound on $|\nabla f|$.

LEMMA 10.3. *If $f : D_R \setminus D_\delta \to \mathbf{C}$ is holomorphic and*

$$\int_{\partial D_R \cup \partial D_\delta} |\nabla f| \leq \epsilon\,, \tag{10.17}$$

then

$$\min_{c \in \mathbf{C}} \max_z |f(z) - c| \leq \epsilon\,. \tag{10.18}$$

PROOF. For $\delta \leq s \leq R$, define the circular average of f by

$$I(s) = (2\,\pi\,s)^{-1} \int_{\partial D_s} f = (2\,\pi)^{-1} \int_0^{2\,\pi} f(s\,e^{i\,\theta})\,d\theta\,, \tag{10.19}$$

and $c = I(\delta)$. Differentiating (10.19), we have

$$\begin{aligned} 2\,\pi\,s\,I'(s) &= \int_{\partial D_s} \frac{\partial f}{\partial r} = -i\,s^{-1} \int_{\partial D_s} \frac{\partial f}{\partial \theta} = -i \int_0^{2\,\pi} \frac{\partial f}{\partial \theta}\,d\theta \\ &= -i\,[f(s\,e^{i\,2\,\pi}) - f(s\,e^0)] = 0\,, \end{aligned} \tag{10.20}$$

where we used that

$$\frac{\partial f}{\partial r} = -i\,r^{-1}\,\frac{\partial f}{\partial \theta} \tag{10.21}$$

since f is holomorphic. In particular, $I(R) = c$. Since $I(s)$ is the average of f over ∂D_s, there exist $y_1, y_2 \in \partial D_R$ with

$$c = \mathrm{Re}\,(f(y_1)) + i\,\mathrm{Im}\,(f(y_2))\,. \tag{10.22}$$

Combining (10.22) with $\int_{\partial D_R \cup \partial D_\delta} |\nabla f| \leq \epsilon$,

$$\max_{y \in \partial D_R} |\mathrm{Re}\,(f(y) - c)| \leq \epsilon/2 \text{ and } \max_{y \in \partial D_R} |\mathrm{Im}\,(f(y) - c)| \leq \epsilon/2\,, \tag{10.23}$$

so that $|f - c| \leq \epsilon$ on $\partial(D_R \setminus D_\delta)$. The maximum principle then gives (10.18). □

Note that Lemma 10.3 does not hold for harmonic functions (in particular, (10.20) does not hold); e.g., take $\epsilon\,\log r/(4\,\pi)$ and $R > e^{8\,\pi}\,\delta$.

10.2. Bers' Theorem revisited. We have seen that an annulus with small total curvature must be graphical, even as the outer radius goes to infinity. However, Bers' Theorem indicates that much more should be true: the unit normal actually goes to a limit at infinity. This follows from a sharper decay estimate for the curvature. To keep things simple, we will give the argument for a holomorphic function with small gradient as in the previous lemma.

LEMMA 10.4. *If* $f : D_R \setminus D_1 \to \mathbf{C}$ *is holomorphic and* $|\nabla f(z)| \leq 1/|z|$, *then*

$$\int_{D_{R^{1/2}e^t} \setminus D_{R^{1/2}e^{-t}}} |\nabla f|^2 \leq 2\pi \, e^{2t}/R \,. \tag{10.24}$$

PROOF. By (10.20), we can subtract a constant so that $\int_{\partial D_r} f = 0$ for all r. Set

$$E(t) = \int_{D_{R^{1/2}e^t} \setminus D_{R^{1/2}e^{-t}}} |\nabla f|^2 \,. \tag{10.25}$$

Since f is holomorphic, the Cauchy–Riemann equations give

$$d\bar{f} \wedge df = i \, |\nabla f|^2 \, dx \wedge dy \,. \tag{10.26}$$

Applying Stokes' theorem, we get

$$E(t) = -i \int_{D_{R^{1/2}e^t} \setminus D_{R^{1/2}e^{-t}}} d\bar{f} \wedge df = -i \int_{\partial(D_{R^{1/2}e^t} \setminus D_{R^{1/2}e^{-t}})} \bar{f} \, df \,. \tag{10.27}$$

Differentiating (10.25), the chain rule and $|\nabla f| \leq 1/|z|$ give

$$E'(t) = R^{1/2} e^{-t} \int_{\partial D_{R^{1/2}e^{-t}}} |\nabla f|^2 + R^{1/2} e^t \int_{\partial D_{R^{1/2}e^t}} |\nabla f|^2 \leq 4\pi \,. \tag{10.28}$$

The Cauchy–Schwarz and Wirtinger inequalities (recall that $\int_{\partial D_r} f = 0$) give

$$\begin{aligned} \left| \int_{\partial D_r} \bar{f} \, df \right| &= \left| \int_{\partial D_r} \bar{f} \, f_\theta \, d\theta \right| \leq \left(\int_{\partial D_r} |\bar{f}|^2 \, d\theta \right)^{1/2} \left(\int_{\partial D_r} |f_\theta|^2 \, d\theta \right)^{1/2} \\ &\leq \int_{\partial D_r} |f_\theta|^2 \, d\theta = r/2 \int_{\partial D_r} |\nabla f|^2 \,, \end{aligned} \tag{10.29}$$

where the last equality used the Cauchy–Riemann equations to relate f_θ and f_r. By (10.27)–(10.29), we get

$$2\, E(t) \leq E'(t) \tag{10.30}$$

and $E(1/2 \log R) \leq 2\pi$. Integrating the differential inequality (10.30) yields (10.24). □

Combining Lemma 10.4 (with $t = \log 2$) and the mean value inequality, we get

$$\max_{\partial D_{R^{1/2}}} |\nabla f|^2 \leq (\pi \, R/4)^{-1} \int_{D_{2R^{1/2}} \setminus D_{R^{1/2}/2}} |\nabla f|^2 \leq 32 \, R^{-2} \,. \tag{10.31}$$

If we now take $R \to \infty$ as in Bers' theorem, then we see that

$$|\nabla f(z)| \leq C \, |z|^{-2} \,. \tag{10.32}$$

The bound (10.32) is integrable radially (as $|z| \to \infty$), so we see that f has an asymptotic limit.

Finally, note that $f(z) = 1/z$ shows that (10.31) is sharp (up to the constant).

Part 2. The role of multi–valued graphs in minimal surfaces

There are two local models for embedded minimal disks (by an *embedded disk* we mean a smooth injective map from the closed unit ball in $\mathbf{R}^2$ into $\mathbf{R}^3$). One model is the plane (or, more generally, a minimal graph) and the other is a piece of a helicoid.

The second model comes from the helicoid which was discovered by Meusnier in 1776. Meusnier had been a student of Monge. He also discovered that the surface now known as the catenoid is minimal in the sense of Lagrange, and he was the first to characterize a minimal surface as a surface with vanishing mean curvature. Unlike the helicoid, the catenoid is not topologically a plane but rather a cylinder.

The helicoid is a "double spiral staircase" (see [**CM17**]):

Example 2: (Helicoid; see fig. 1). The helicoid is the minimal surface in $\mathbf{R}^3$ given by the parametrization

$$(s\cos t, s\sin t, t)\,, \quad \text{where } s,\, t \in \mathbf{R}\,. \tag{10.33}$$

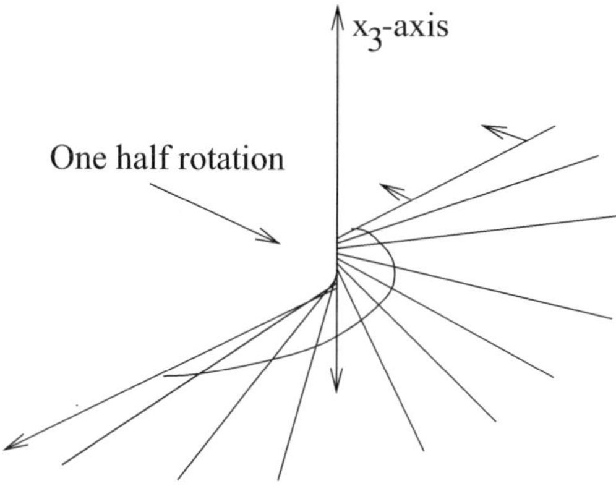

FIGURE 1. Multi–valued graphs. The helicoid is obtained by gluing together two ∞–valued graphs along a line.

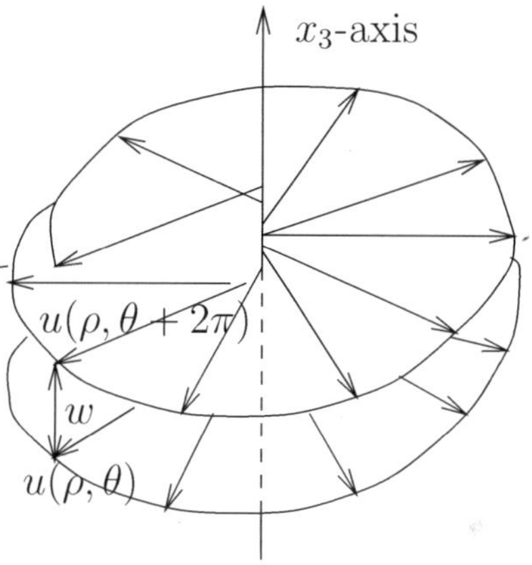

FIGURE 2. The separation w grows/decays in ρ at most sublinearly for a multi–valued minimal graph; see (11.6).

11. Basic properties of multi–valued graphs

We will need the notion of a multi–valued graph, each staircase will be a multi–valued graph. Intuitively, an (embedded) multi–valued graph is a surface such that over each point of the annulus, the surface consists of N graphs. To make this notion precise, let D_r be the disk in the plane centered at the origin and of radius r and let $\mathcal{P}$ be the universal cover of the punctured plane $\mathbf{C} \setminus \{0\}$ with global polar coordinates (ρ, θ) so $\rho > 0$ and $\theta \in \mathbf{R}$. An *N–valued graph* on the annulus $D_s \setminus D_r$ is a single valued graph of a function u over

$$\{(\rho, \theta)\,|\, r < \rho \leq s\,,\, |\theta| \leq N\,\pi\}\,. \tag{11.1}$$

For working purposes, we generally think of the intuitive picture of a multi–sheeted surface in $\mathbf{R}^3$, and we identify the single–valued graph over the universal cover with its multi–valued image in $\mathbf{R}^3$.

The multi–valued graphs that we will consider will all be embedded, which corresponds to a nonvanishing separation between the sheets (or the floors). Here the *separation* is the function (see fig. 2)

$$w(\rho,\theta) = u(\rho,\theta+2\pi) - u(\rho,\theta)\,. \tag{11.2}$$

If Σ is the helicoid, then $\Sigma \setminus \{x_3 - \text{axis}\} = \Sigma_1 \cup \Sigma_2$, where Σ_1, Σ_2 are ∞–valued graphs on $\mathbf{C}\setminus\{0\}$. Σ_1 is the graph of the function $u_1(\rho,\theta) = \theta$ and Σ_2 is the graph of the function $u_2(\rho,\theta) = \theta + \pi$. ($\Sigma_1$ is the subset where $s > 0$ in (10.33) and Σ_2 the subset where $s < 0$.) In either case the separation $w = 2\,\pi$. A *multi–valued minimal graph* is a multi–valued graph of a function u satisfying the minimal surface equation.

Note that for an embedded multi–valued graph, the sign of w determines whether the multi–valued graph spirals in a left–handed or right–handed manner, in other words, whether upwards motion corresponds to turning in a clockwise direction or in a counterclockwise direction.

11.1. The sublinear growth of the separation. As we have seen, the separation is constant for the multi–valued graphs coming from each half of the helicoid. This can be viewed as a type of Liouville Theorem reflecting the conformal properties of an infinite–valued graph. In Proposition II.2.12 of [**CM3**], we proved a corresponding gradient estimate:

PROPOSITION 11.1. [**CM3**] *Given $\alpha > 0$, there exists N_g so that if u satisfies the minimal surface equation on*

$$\{e^{-N_g}\,R \leq \rho \leq e^{N_g}\,R\,,\, -N_g \leq \theta \leq 2\pi + N_g\}\,, \tag{11.3}$$

$|\nabla u| \leq 1$, *and has separation $w \neq 0$, then*

$$|Hess_u|(R,0) + |\nabla \log |w||(R,0) \leq \alpha/R\,. \tag{11.4}$$

One important consequence of (11.4) is that, for $\alpha < 1$, the separation grows sublinearly:

COROLLARY 11.2. *Given $\alpha > 0$, there exists N_g so that if u satisfies the minimal surface equation on*

$$\{e^{-N_g}\,r_1 \leq \rho \leq e^{N_g}\,r_2\,,\, -N_g \leq \theta \leq 2\pi + N_g\}\,, \tag{11.5}$$

$|\nabla u| \leq 1$, *and has separation $w \neq 0$, then*

$$|w|(r_2,0) \leq |w|(r_1,0)\left(\frac{r_2}{r_1}\right)^{\alpha}\,. \tag{11.6}$$

PROOF. Integrate $|\nabla \log |w|| \leq \alpha/\rho$ along the ray $\theta = 0$ to get

$$\log \frac{w(r_2,0)}{w(r_1,0)} \leq \alpha \int_{r_1}^{r_2} \rho^{-1}\, d\rho = \log \left(\frac{r_2}{r_1}\right)^{\alpha}\,. \tag{11.7}$$

□

Since $u(\cdot,\cdot)$ and its 2π–rotation $u(\cdot,\cdot+2\pi)$ are both solutions of the minimal surface equation, the difference w is almost a solution of the linearized equation (which is the Jacobi equation in this case). Since the graphs have bounded gradient, this equation is not too far from the Laplace equation. To give some indication of why (11.4) holds, we will give an elementary proof when u and w are harmonic.

PROOF. (of Proposition 11.1 when u is harmonic.) After rescaling, we can assume that $R = 1$. By making the conformal change of coordinates

$$(\rho, \theta) \to (\log \rho, \theta) \tag{11.8}$$

we get a positive harmonic function

$$\tilde{w}(x, y) = w(\mathrm{e}^x, y) \tag{11.9}$$

defined on the square $[-N_g, N_g] \times [-N_g, N_g]$. Since the chain rule gives

$$\nabla \log w(1, 0) = \nabla \log \tilde{w}(0, 0)\,, \tag{11.10}$$

applying the Euclidean gradient estimate to $\tilde{w}$ yields

$$|\nabla \log w(1, 0)| = |\nabla \log \tilde{w}(0, 0)| \leq C/N_g\,. \tag{11.11}$$

This gives the sublinear gradient estimate for w if N_g is sufficiently large. The bound on Hess_u follows similarly. □

Proposition 11.1 allows us to assume (after rotating so $\nabla u(1, 0) = 0$) that

$$|\nabla u| + \rho\, |\text{Hess}_u| + 4\, \rho\, |\nabla w|/|w| + \rho^2\, |\text{Hess}_w|/|w| \leq \epsilon < 1/(2\pi)\,. \tag{11.12}$$

The bound on $|\text{Hess}_w|$ follows from the other bounds and standard elliptic theory.

11.2. Curvature decay. In corollary 1.14 of **[CM8]**, we proved faster than quadratic curvature decay for double–valued minimal graphs whose separation grows sublinearly:

PROPOSITION 11.3. *There exists C so if u satisfies the minimal surface equation and (11.12) on*

$$\{1 \leq \rho \leq R\,,\ -\pi \leq \theta \leq 3\pi\}\,, \tag{11.13}$$

then on $\{1 \leq \rho \leq R^{1/2}\,,\ 0 \leq \theta \leq 2\pi\}$ we have

$$\rho\, |Hess_u| \leq C\, \epsilon\, \rho^{-5/12}\,, \tag{11.14}$$

and therefore, after possibly rotating $\mathbf{R}^3$ (and replacing u), we get

$$|\nabla u| \leq C\, \epsilon\, \rho^{-5/12}\,. \tag{11.15}$$

Of course, Proposition 11.3 is a generalization of Lemma 10.4 which proved curvature decay for annuli. As in the annuli case, the proof uses the quasi–conformality of the Gauss map to deduce a differential inequality. However, the "slit" (i.e., where the double–valued graph does not close up) contributes new terms which are controlled using the estimate for the separation. Notice that the second conclusion (11.15) of the proposition proves a generalization of Bers theorem: Embedded multi–valued minimal graphs have an asymptotic tangent plane.

12. Sharp estimates on the separation for multi–valued graphs

We will describe in this section two sharp estimates on the separation of a multi–valued minimal graph; see Propositions 12.1 and 12.2 below. These estimates will, unlike the earlier estimates (11.6) and (11.14), require a rapidly growing number of sheets (growing in ρ).

Suppose for a moment that we are looking at an embedded surface which is the ∞–valued graph of a harmonic function so that in particular the separation w is a harmonic function which after reflection we may assume is positive. By making the conformal change of coordinates $(\rho, \theta) \to (\log \rho, \theta)$ we get a positive harmonic

$$\tilde{w}(x, y) = w(\mathrm{e}^x, y) \tag{12.1}$$

defined on the half–plane $\{x \geq 0,\ y \in \mathbf{R}\}$. By the mean value inequality and the Harnack inequality (since $\tilde{w}$ is positive)

$$\tilde{w}(y, 0) = \frac{1}{2\pi\, y} \int_{\partial D_y(y,0)} \tilde{w} \geq \frac{1}{2\pi\, y} \int_{\partial D_y(y,0) \cap D_1} \tilde{w} \geq \frac{C}{2\pi\,(y+1)}\, \tilde{w}(0, 0)\,. \tag{12.2}$$

Similarly, by an inversion formula one may show that

$$\tilde{w}(y, 0) \leq \frac{C}{2\pi\,(y+1)}\, \tilde{w}(0, 0)\,. \tag{12.3}$$

For the original function w, (12.2) combined with (12.3) gives for some constant C independent of w

$$\frac{1}{C\, \log \rho} \leq \frac{w(\rho, 0)}{w(1, 0)} \leq C\, \log \rho\,. \tag{12.4}$$

In the case of embedded multi–valued minimal graphs we get similarly:

PROPOSITION 12.1. **[CM8]**. *Let Σ be an embedded multi–valued minimal graph of a function u and with a rapidly growing number of sheets, then for the separation w we have for some constant C*

$$\frac{1}{C\, \log \rho} \leq \frac{w(\rho, 0)}{w(1, 0)} \leq C\, \log \rho\,. \tag{12.5}$$

Suppose again for a moment that u and hence w is harmonic. Similarly to (12.2) we get that

$$\tilde{w}(0, y) \leq C\, 2\, \pi\, (y+1)\, \tilde{w}(y, y)\,. \tag{12.6}$$

By the Harnack inequality, we have

$$\tilde{w}(y, y) \leq C\, \tilde{w}(y, 0)\,. \tag{12.7}$$

Combining this with (12.6) and (12.3), we get

$$\tilde{w}(0, y) \leq C\, (y^2 + 1)\, \tilde{w}(0, 0)\,. \tag{12.8}$$

For the original function w, this gives

$$\frac{1}{C\, (\theta^2 + 1)} \leq \frac{w(\rho, \theta)}{w(\rho, 0)} \leq C\, (\theta^2 + 1)\,. \tag{12.9}$$

Again in the case of embedded multi–valued minimal graphs we get similarly:

PROPOSITION 12.2. *Let Σ be an embedded multi–valued minimal graph of a function u and with a rapidly growing number of sheets, then for the separation w we have for some constant C*

$$\frac{1}{C\, (\theta^2 + 1)} \leq \frac{w(\rho, \theta)}{w(\rho, 0)} \leq C\, (\theta^2 + 1)\,. \tag{12.10}$$

The lower bound in (12.6) for the decay of the separation is sharp. It is achieved for the ∞–valued graph of the harmonic function (graphs of multi–valued harmonic functions are good models for multi–valued minimal graphs)

$$u(\rho, \theta) = \arctan \frac{\theta}{\log \rho}\,. \tag{12.11}$$

Note that the graph of u is embedded and lies in a slab in $\mathbf{R}^3$, i.e., $|u| \leq \pi/2$, and hence in particular is not proper. On the top it spirals into the plane $\{x_3 = \pi/2\}$ and on the bottom into $\{x_3 = -\pi/2\}$, yet it never reaches either of these planes.

QUESTION 1. *It would be interesting to construct an infinite–valued exterior solution of the minimal graph equation with the same properties as* $\arctan \frac{\theta}{\log \rho}$*; i.e., one which spirals infinitely in a slab (see* [**CM18**] *for a local example).*

13. Double–valued minimal graphs

We will now describe how to bound the oscillation of the Gauss map of a double–valued minimal graph. This bound was proven in [**CM10**]. Rather than give the precise statement here, we will instead illustrate a few of the key ideas by considering the analogous situation for a holomorphic function f. Here we are of course thinking of f as being the stereographic projection of the Gauss map. The sublinear growth of the separation, i.e., (11.12) and its integrated form, correspond to

$$|f| + \rho\,|\nabla f| \leq \epsilon < 1/(2\pi)\,, \tag{13.1}$$

$$|f(\rho, 2\pi) - f(\rho, 0)| \leq 2\pi\,\epsilon \left(\frac{\delta}{\rho}\right)^{1-\epsilon}. \tag{13.2}$$

The bound on the oscillation of f (Lemma 13.1 below) now follows by modifying the argument for annuli (Lemma 10.3). Since f does not match up at $\theta = 0$ and 2π, we get an additional term which is estimated using the sublinear growth (13.2).

LEMMA 13.1. *If* $f : \{\delta \leq \rho \leq R\,,\, 0 \leq \theta \leq 2\pi\} \to \mathbf{C}$ *is holomorphic and satisfies (13.1) and (13.2), then*

$$\min_{c \in \mathbf{C}} \max |f - c| \leq \frac{\epsilon}{1 - \epsilon} + 2\pi\,\epsilon\,. \tag{13.3}$$

PROOF. (Following Lemma 10.3.) For $\delta \leq s \leq R$, define the circular average

$$I(s) = (2\,\pi)^{-1} \int_0^{2\,\pi} f(s, \theta)\, d\theta\,. \tag{13.4}$$

Note that integrating (13.1) gives

$$|I(s) - f(s, \theta)| \leq 2\pi\,\epsilon\,. \tag{13.5}$$

Differentiating (13.4) and using $\frac{\partial f}{\partial \rho} = -i\,\rho^{-1}\,\frac{\partial f}{\partial \theta}$ since f is holomorphic, we have

$$\begin{aligned} 2\,\pi\,s\,I'(s) &= s \int_0^{2\pi} \frac{\partial f(s, \theta)}{\partial \rho}\, d\theta = -i \int_0^{2\,\pi} \frac{\partial f(s, \theta)}{\partial \theta}\, d\theta \\ &= -i\,[f(s, 2\,\pi) - f(s, 0)]\,. \end{aligned} \tag{13.6}$$

Using the bound (13.2) along the slit, (13.6) gives

$$s\,|I'(s)| \leq \epsilon\,\delta^{1-\epsilon}\,s^{\epsilon - 1}\,. \tag{13.7}$$

In particular, integrating this gives for $\delta \leq \rho \leq R$ that

$$|I(\rho) - I(\delta)| \leq \epsilon\, \delta^{1-\epsilon} \int_{\delta}^{\rho} s^{\epsilon-2}\, ds \leq \epsilon/(1-\epsilon)\,. \tag{13.8}$$

The bound (13.3) follows from (13.5) and (13.8). □

Modifying this argument to apply to double–valued minimal graphs introduces new difficulties. In that case, one works directly on the graph (where the Gauss map is holomorphic) and uses averages over "geodesic sectors" rather than circles in the plane. One difficulty is a new term in the analog of (13.6) which results from differentiating the measures of the level sets.

14. Approximation by standard pieces

In this section we show that any embedded multi–valued minimal graph has a sub–graph which is close to the sum of a piece of a catenoid and a piece of a helicoid. This generalizes a similar representation for minimal graphs over an annulus given in proposition 1.5 in [**CM9**]; of course, there was no helicoid term in that case. These results are new and have not appeared in the literature elsewhere.

Recall that half of a catenoid, i.e.,

$$\{(x_1/s)^2 + (x_2/s)^2 = \cosh^2(x_3/s)\,,\ \pm x_3 > 0\}\,, \tag{14.1}$$

is a minimal graph of

$$u(z) = \pm s \cosh^{-1}(|z|/s) \tag{14.2}$$

over $\mathbf{C} \setminus D_s$. Note that $s \cosh^{-1}(|z|/s)$ is asymptotic to $s \log[2\,|z|/s]$. Recall also that half of the helicoid is the multi–valued graph of the function u given in polar coordinates by $u(\rho, \theta) = \theta$. Our approximation result (see Corollary 14.3 below) is therefore that any embedded multi–valued minimal graph has a sub–graph which is close to the graph of a multi–valued function v given by (a, b, $c \in \mathbf{R}$ are constants)

$$v(\rho, \theta) = a + b\, \log(\rho/r) + c\, \theta/(2\pi)\,. \tag{14.3}$$

As in the two previous sections, u will be a multi-valued function and w will be its separation. For convenience, we will write $S_{r_1,r_2}^{\theta_1,\theta_2}$ to denote the "rectangle"

$$\{(\rho,\theta)\,|\, r_1 \leq \rho \leq r_2,\ \theta_1 \leq \theta \leq \theta_2\}\,. \tag{14.4}$$

We begin with a representation formula for the gradient of an "almost harmonic" function. Recall that if $\Delta u = 0$ on an annulus, then the function $f = u_x - i\, u_y$ is holomorphic. In particular, f has a Laurent expansion which can be recovered using the Cauchy integral formula. The next lemma uses a variation on this for multi–valued "almost harmonic" functions:

LEMMA 14.1. *Given a function u on $S_{1,\sqrt{R}}^{-\pi,3\pi}$, set $f = u_x - i\, u_y$. If*

$$|f(\zeta)| \leq C\, |\zeta|^{-5/12}\,, \tag{14.5}$$

$$|\Delta u(\zeta)| \leq C\, |\zeta|^{-9/4}\,, \tag{14.6}$$

$$\rho\, |\nabla w|/|w| \leq \epsilon < 1/(2\pi)\,, \tag{14.7}$$

then for $r_1 \geq 1$ and $\zeta \in S_{2r_1,\sqrt{R}/2}^{0,2\pi}$

$$f(\zeta) = (b + i\, c/(2\pi))\, \zeta^{-1} + g(\zeta) \tag{14.8}$$

where b, $c \in \mathbf{R}$ *and*

$$|g(\zeta)| \leq C_1 \, R^{-5/24} + C_1 \, r_1^{-1/4} \, |\zeta|^{-1} + C_1 \, \epsilon \, r_1^{-1} \, |w(r_1, 0)| \, . \tag{14.9}$$

PROOF. We will first use the Cauchy integral formula (this is just Stokes' theorem applied to the one–form $f(z)/(z-\zeta)\,dz$) on the domain $S^{0,2\pi}_{1,\sqrt{R}}$ to get a representation formula on $S^{\pi/2,3\pi/2}_{2\,r_1,\sqrt{R}/2}$. Assume that $\zeta \in S^{\pi/2,3\pi/2}_{2\,r_1,\sqrt{R}/2}$. The Cauchy integral formula gives

$$\begin{aligned} 2\,\pi\,i\,f(\zeta) = & \int_{S^{0,2\pi}_{\sqrt{R},\sqrt{R}}} \frac{f(z)}{z-\zeta}\,dz - \int_{S^{0,2\pi}_{1,1}} \frac{f(z)}{z-\zeta}\,dz - i \int_{S^{0,2\pi}_{1,\sqrt{R}}} \frac{\Delta u}{z-\zeta} \\ & + \left(\int_{\sqrt{R}}^{1} \frac{f(\rho, 2\,\pi)}{\rho - \zeta}\,d\rho + \int_1^{\sqrt{R}} \frac{f(\rho,0)}{\rho-\zeta}\,d\rho \right) . \end{aligned} \tag{14.10}$$

The first two terms correspond to the usual formula for annuli, the third term vanishes when f is holomorphic, and the last term arises since f is multi–valued. (The first three are almost identical to the corresponding ones in lemma 1.7 in [**CM9**].) To prove the lemma, we will show that the first term is small and the other three are small after subtracting a multiple of $1/\zeta$.

First, $|f(z)| \leq C\,|z|^{-5/12}$ and $|\zeta| \leq \sqrt{R}/2$ give that

$$\left| \int_{S^{0,2\pi}_{\sqrt{R},\sqrt{R}}} \frac{f(z)}{z-\zeta}\,dz \right| \leq 4\,\pi \sup_{S^{0,2\pi}_{\sqrt{R},\sqrt{R}}} |f| \leq 4\pi\,C\,R^{-5/24} \, . \tag{14.11}$$

For the remaining terms, we will use the identity

$$\frac{1}{z-\zeta} = \frac{-1}{\zeta} + \frac{z}{\zeta(z-\zeta)} \, . \tag{14.12}$$

Second, using (14.12), $2 \leq |\zeta|$, and $|f| \leq C$,

$$\left| \int_{S^{0,2\pi}_{1,1}} \frac{f(z)}{z-\zeta}\,dz + \frac{1}{\zeta} \int_{S^{0,2\pi}_{1,1}} f(z)\,dz \right| \leq \frac{2}{|\zeta|^2} \int_{S^{0,2\pi}_{1,1}} |z\,f(z)| \leq \frac{4\pi\,C}{|\zeta|^2} \, . \tag{14.13}$$

To bound the third integral, we separate out the disk of radius $|\zeta|/2$ about ζ and divide the remainder into two regions using the circle of radius r_1. Using (14.12), $2\,r_1 \leq |\zeta|$, and $|\Delta u(z)| \leq C\,|z|^{-9/4}$, we get

$$\begin{aligned} \left| \int_{S^{0,2\pi}_{1,\sqrt{R}}} \frac{\Delta u}{z-\zeta} + \frac{1}{\zeta} \int_{S^{0,2\pi}_{1,r_1}} \Delta u \right| & \leq 4\,\pi\,C\,|\zeta|^{-2} \int_1^{r_1} r^{-1/4}\,dr + 2^{9/4}\,C \int_{D_{\frac{|\zeta|}{2}}(\zeta)} \frac{|\zeta|^{-9/4}}{|z-\zeta|} \\ & \quad + 4\,\pi C\,|\zeta|^{-1} \int_{r_1}^{\sqrt{R}} r^{-5/4}\,dr \\ & \leq 12\,\pi\,C\,|\zeta|^{-5/4} + 16\,\pi\,C\,r_1^{-1/4}\,|\zeta|^{-1} \, . \end{aligned} \tag{14.14}$$

Finally, this leaves the slit (i.e., the terms which arise because u is not well-defined over the annulus), where we divide the integral into three parts

$$\begin{aligned} \int_1^{\sqrt{R}} \frac{f(\rho,2\pi) - f(\rho,0)}{\rho-\zeta}\,d\rho = & \int_{r_1}^{\sqrt{R}} \frac{f(\rho,2\pi) - f(\rho,0)}{\rho-\zeta}\,d\rho \\ & - \zeta^{-1} \int_1^{r_1} (f(\rho,2\pi) - f(\rho,0))\,d\rho + \int_1^{r_1} \frac{\rho(f(\rho,2\pi) - f(\rho,0))}{\zeta(\rho-\zeta)}\,d\rho \, . \end{aligned} \tag{14.15}$$

Using $\rho\,|\nabla w|/|w| \leq \epsilon$, we get for $\rho \geq r_1$ that

$$|f(\rho,2\pi) - f(\rho,0)| = |\nabla w(\rho,0)| \leq \epsilon\, \frac{|w|(\rho,0)}{\rho} \leq \epsilon \left(\frac{\rho}{r_1}\right)^{\epsilon} \frac{|w|(r_1,0)}{\rho}\,. \tag{14.16}$$

Note that the real part of ζ is negative since $\zeta \in S^{\pi/2,3\pi/2}_{2r_1,\sqrt{R}/2}$ and hence $|\rho - \zeta| \geq \rho$. To bound the first term in (14.15), we use this and (14.16) to get

$$\left| \int_{r_1}^{\sqrt{R}} \frac{f(\rho,2\pi) - f(\rho,0)}{\rho - \zeta}\, d\rho \right| \leq \frac{\epsilon}{r_1^{\epsilon}}\, |w|(r_1,0) \int_{r_1}^{\sqrt{R}} \rho^{\epsilon-2}\, d\rho \leq 2\epsilon\, \frac{|w|(r_1,0)}{r_1}\,. \tag{14.17}$$

Similarly, to bound the third, use $\rho\,|\nabla w|/|w| \leq \epsilon$ to get

$$\begin{aligned} \left| \int_1^{r_1} \frac{\rho\,(f(\rho,2\pi) - f(\rho,0))}{\zeta(\rho - \zeta)}\, d\rho \right| &\leq \epsilon\, r_1^{-2} \int_1^{r_1} |w|(\rho,0)\, d\rho \\ &\leq \epsilon\, r_1^{-2} \int_1^{r_1} |w|(r_1,0)\, (r_1/\rho)^{\epsilon}\, d\rho \leq 2\epsilon\, \frac{|w|(r_1,0)}{r_1}\,. \end{aligned} \tag{14.18}$$

Putting all of this together, we now get the desired representation formula (14.8) and remainder estimate (14.9) for $\zeta \in S^{\pi/2,3\pi/2}_{2r_1,\sqrt{R}/2}$. Namely, we set $g = g_1 + g_2 + g_3$ and

$$\begin{aligned} 2\pi i\,(b + i\,\frac{c}{2\pi}) &= \int_{S^{0,2\pi}_{1,1}} f(z)\, dz + \int_1^{r_1} (f(\rho,2\pi) - f(\rho,0))\, d\rho + i \int_{S^{0,2\pi}_{1,r_1}} \Delta u\,, \\ 2\pi i\, g_1(\zeta) &= \int_{S^{0,2\pi}_{\sqrt{R},\sqrt{R}}} \frac{f(z)}{z - \zeta}\, dz\,, \\ 2\pi i\, g_2(\zeta) &= -\int_{r_1}^{\sqrt{R}} \frac{f(\rho,2\pi) - f(\rho,0)}{\rho - \zeta}\, d\rho - \int_1^{r_1} \frac{f(\rho,2\pi) - f(\rho,0)}{\zeta(\rho - \zeta)}\, \rho\, d\rho\,, \\ 2\pi i\, g_3(\zeta) &= -\int_{S^{0,2\pi}_{1,1}} \frac{f(z)}{z - \zeta}\, dz - \frac{1}{\zeta} \int_{S^{0,2\pi}_{1,1}} f(z)\, dz - i \int_{S^{0,2\pi}_{1,\sqrt{R}}} \frac{\Delta u}{z - \zeta} - \frac{i}{\zeta} \int_{S^{0,2\pi}_{1,r_1}} \Delta u\,. \end{aligned} \tag{14.19}$$

We can repeat this by integrating over $S^{-\pi/2,3\pi/2}_{1,\sqrt{R}}$ and $S^{\pi/2,5\pi/2}_{1,\sqrt{R}}$ to get similar representations on $S^{0,\pi}_{2r_1,\sqrt{R}/2}$ and $S^{\pi,2\pi}_{2r_1,\sqrt{R}/2}$ (with different values of b and c). By continuity, the same representation holds on all three regions (since they overlap), giving the lemma. □

Definition 14.2. Fix $\mu > 0$. A standard piece Σ_v on the scale r is a graph of v over $S^{-\pi,3\pi}_{r,\mu r}(y)$ given in polar coordinates (ρ_p, θ_p) centered at p by

$$v(\rho_p, \theta_p) = a + b\, \log(\rho_p/r) + c\,\theta_p/(2\pi)\,. \tag{14.20}$$

The constant a is the "plane coefficient", b is the "catenoid coefficient" and c is the "helicoid coefficient" (this gives the separation w). Note that v is harmonic.

COROLLARY 14.3. *Let u be a solution of the minimal graph equation on $S^{-3\pi,5\pi}_{1/2,R}$ with $w < 0$ satisfying (11.12) and $|u| \leq \epsilon\,\rho$. There is a rotation of $\mathbf{R}^3$ so given $r_1 > 2$, we get a standard piece*

$$v = a + b\, \log\rho + c\,\theta/(2\pi)\,, \tag{14.21}$$

for $a, b, c \in \mathbf{R}$ *with*

$$\sup_{S^{0,2\pi}_{r_1,\mu r_1}} |u(\rho,\theta) - v(\rho,\theta)| \leq C_2\, \mu\, \epsilon\, |w|(r_1,0) + C_2\, \mu\, r_1\, R^{-5/24} + C_2\, \mu\, r_1^{-1/4}\,. \tag{14.22}$$

PROOF. Corollary 1.14 in [**CM8**] give a rotation of $\mathbf{R}^3$ so that for $z \in S^{-2\pi,4\pi}_{1,\sqrt{R}}$

$$|\nabla u(z)| + |z|\, |\text{Hess}_u(z)| \leq C\, \epsilon\, |z|^{-5/12}\,. \tag{14.23}$$

Since, by (1.6) of [**CM8**], $|\Delta\, u| \leq |\nabla u|^2\, |\text{Hess}_u|$, (14.23) gives on $S^{-2\pi,4\pi}_{1,\sqrt{R}}$ that

$$|\Delta u(z)| \leq C\, \epsilon^3\, |z|^{-9/4}\,. \tag{14.24}$$

Note that

$$(\log \rho)_x - i\,(\log \rho)_y = 1/(x+iy)\,, \tag{14.25}$$

$$\theta_x - i\,\theta_y = i/(x+iy)\,. \tag{14.26}$$

Using this and Lemma 14.1, we get that on $S^{0,2\pi}_{r_1,\mu r_1}$

$$|\nabla(u - b\, \log \rho - c\,\theta/(2\pi))| \leq C_1 \left(R^{-5/24} + 2\, r_1^{-1/4}\, \rho^{-1} + 2\,\epsilon\, \frac{|w(r_1,0)|}{r_1} \right). \tag{14.27}$$

Integrating (14.27) gives (14.22). □

QUESTION 2. *Corollary 14.3 shows that embedded multi-valued minimal graphs are closely approximated by helicoids (plus a catenoid term) on each scale. However, a priori, the helicoid coefficient can change from scale to scale (the point of* [**CM10**] *is that the axis of the helicoid does not change so that all are vertical). It would be interesting to estimate how quickly this helicoid coefficient can change and construct examples demonstrating this.*

Part 3. Regularity theory

In this part, we survey some of the key ideas in classical regularity theory, recent developments on embedded minimal disks, and some global results for minimal surfaces in $\mathbf{R}^3$. Sections 15 and 16 focus on Reifenberg type conditions, where a set is assumed to be close to a plane at all points and at all scales ("close" is in the Hausdorff or Gromov–Hausdorff sense and is defined in Section 15). This condition automatically gives Hölder regularity (and hence higher regularity if the set is also a weak solution to a natural equation). Section 17 surveys the role of monotonicity and scaling in regularity theory, including ϵ-regularity theorems (such as Allard's theorem) and tangent cone analysis (such as Almgren's refinement of Federer's dimension reducing). Section 18 briefly reviews recent results of the authors for embedded minimal disks, developing a regularity theory in a setting where the classical methods cannot be applied and in particular where there is no monotonicity. The estimates and ideas discussed in Section 18 have applications to the global theory of minimal surfaces in $\mathbf{R}^3$. In Section 19, we give a quick tour of some recent results in this classical, but rapidly developing, area.

15. Hausdorff and Gromov–Hausdorff distances

Recall that the Hausdorff distance between two subsets A and B of a Euclidean space is no greater than $\delta > 0$ provided that each is contained in a δ-neighborhood of the other.

There is a natural generalization of this classical Hausdorff distance between subsets of Euclidean space to a distance function on all metric spaces. This is the Gromov-Hausdorff distance and it gives a good tool to study metric spaces.

Suppose that (X, d_X) and (Y, d_Y) are two compact metric spaces. We say that the Gromov-Hausdorff distance between them is at most $\epsilon > 0$ if there exist maps $f : X \to Y$ and $g : Y \to X$ such that

$$\forall x_1, x_2 \in X : \; |d_X(x_1, x_2) - d_Y(f(x_1), f(x_2))| < \epsilon \,, \tag{15.1}$$

$$\forall x \in X : \; d_X(x, g \circ f(x)) < \epsilon \,, \tag{15.2}$$

and the two symmetric properties in Y hold. The Gromov-Hausdorff distance between X and Y, denoted by $d_{GH}(X, Y)$, is then the infimum of all such ϵ.

Using this topology we can then say that a sequence of metric space converges to another metric space. For noncompact metric spaces there is a more useful notion of convergence which is essentially convergence on compact subsets. Namely, if (X_i, x_i, d_{X_i}) is a pointed sequence of metric spaces then we say that (X_i, x_i, d_{X_i}) converges to some pointed metric space (X, x, d_X) in the pointed Gromov-Hausdorff topology if for all $r_0 > 0$ fixed the compact metric spaces $B_{r_0}(x_i)$ converges to $B_{r_0}(x)$.

Gromov's compactness theorem is the statement that any pointed sequence, (M_i^n, m_i), of n-dimensional manifolds, with

$$\mathrm{Ric}_{M_i^n} \geq (n-1)\,\Lambda \,, \tag{15.3}$$

has a subsequence, (M_j^n, m_j), which converges in the pointed Gromov-Hausdorff topology to some length space (M_∞, m_∞).

The proof of this compactness theorem relies only on the volume comparison theorem. In fact it only uses the volume doubling which is implied by the volume comparison.

16. Reifenberg type conditions

One concept or idea that often plays a central role in regularity theory is that of being Reifenberg flat, a notion introduced by E. R. Reifenberg [**Re**]. In the case of Reifenberg it was introduced to measure the closeness of a subset of Euclidean space to being an affine space. The deviation was measured on all scales and a number was assigned which was the maximal scale invariant Hausdorff distance to a affine plane, see [**To**], [**Se**], [**DaTo**], [**DaKeTo**] for interesting results in this classical direction.

To explain this point about the importance of this condition further we recall the following definition from [**ChC1**] which was inspired by the classical work of Reifenberg.

Let (Z, d_Z) be a complete metric space. We will say that Z satisfies the (ϵ, ρ, n)-$\mathcal{G}_{\mathcal{R}}$ (or *Generalized Reifenberg*) condition at $z \in Z$ if for all $0 < \sigma < \rho$ and all $y \in B_{\rho-\sigma}(z)$,

$$d_{GH}(B_\sigma(y), B_\sigma(0)) < \epsilon\, \sigma \,, \tag{16.1}$$

where $B_{r_0}(0) \subset \mathbf{R}^n$.

THEOREM 16.1. *(Appendix of* [**ChC2**].*) Given $\epsilon > 0$ and $n \geq 2$, there exists $\delta > 0$ such that if (Z, d_Z) is a complete metric space and Z satisfies the (δ, r_0, n)-$\mathcal{G}_{\mathcal{R}}$ condition at $z \in Z$ then there exists a bi-Hölder homeomorphism $\Phi : B_{\frac{r_0}{2}}(z) \to B_{\frac{r_0}{2}}(0)$ such that for all z_1, $z_2 \in Z$,*

$$r_0^{-\epsilon} |\Phi(z_1) - \Phi(z_2)|^{1+\epsilon} \leq d_Z(z_1, z_2) \leq r_0^{\epsilon} |\Phi(z_1) - \Phi(z_2)|^{1-\epsilon} . \tag{16.2}$$

Here is one example where such a Reifenberg type condition naturally come up; see [**C2**], [**ChC1**] for more on this.

THEOREM 16.2. *(*[**C1**], [**ChC1**]*). Given $\epsilon > 0$ and $n \geq 2$, there exist $\delta = \delta(\epsilon, n) > 0$ and $\rho = \rho(\epsilon, n) > 0$, such that if M^n has $Ric_{M^n} \geq -(n-1)$ and $0 < r_0 \leq \rho$ with*

$$d_{GH}(B_{2r_0}(x), B_{2r_0}(0)) < \delta\, r_0 , \tag{16.3}$$

or

$$Vol(B_{2r_0}(x)) \geq (1 - \delta)\, V_0^n(2r_0) , \tag{16.4}$$

then M^n satisfies the (ϵ, r_0, n)-$\mathcal{G}_{\mathcal{R}}$ condition at x.

17. Monotonicity and regularity theory

As we have already seen, the monotonicity of a "scale–invariant energy" has played a key role in the regularity theory for geometric variational problems. In many cases, this monotonicity is useful for establishing two key tools:

1. An ϵ–regularity theorem which guarantees that a weak solution is actually smooth when the scale–invariant energy is small.
2. The existence of tangent cones which are dilation invariant.

In this section, we briefly review the role of monotonicity in regularity theory, emphasizing these two tools, and give some examples.

To keep things concrete, it may be useful to mention some examples of variational problems and their scale–invariant energies, some of which we have already encountered. The primary example is for a minimal k–dimensional submanifold; in this case, the appropriate scale–invariant energy is the density

$$\Theta_{x_0}(s) = \frac{\text{Vol}(B_s(x_0) \cap \Sigma)}{\text{Vol}(B_s \subset \mathbf{R}^k)} . \tag{17.1}$$

There are many other examples, including

- If $\Sigma \subset \mathbf{R}^n$ is a (2–dimensional) minimal surface, then another possible scale–invariant energy is $\int_{B_r} |A|^2$.
- If $u : \mathbf{R}^k \to M$ is a harmonic map, then the scale–invariant energy is $r^{2-k} \int_{B_r} |\nabla u|^2$, see [**ScUh**].
- If M^n has non–negative Ricci curvature, then the scale–invariant energy is the volume density $r^{-n}\, \text{Vol}(B_r(x_0) \subset M)$ – here, the monotonicity goes the opposite direction.
- If A is a Yang–Mills connection on M^k with curvature F_A, then the scale–invariant energy is $r^{4-k} \int_{B_r} |F_A|^2$, see [**Pr**].

Interestingly, there are also parabolic analogs for all of these examples (cf. Huisken's monotonicity formula for the mean curvature flow, i.e., Theorem 6.2 or M. Struwe's monotonicity for the harmonic map heat flow, theorem 1.10 in [**St**]).

Using this monotonicity, we can define the density Θ_{x_0} at the point x_0 to be the limit as $r \to 0$ of the scale–invariant energy in $B_r(x_0)$. It follows easily from the monotonicity that the density is semi–continuous.

Remark 17.1. These energies are all scale–invariant in the sense that if a solution is rescaled by a factor λ, then the energy of the new solution on a ball of radius λr is the same as the original solution on a ball of radius r. It is not difficult to find scale–invariant quantities – the difficulty lies in finding one that is monotone and can be estimated. For example, if $u : \mathbf{R}^k \to M$ is a harmonic map, then $\int |\nabla u|^k$ is scale–invariant and monotone; however, it is not at all clear that there should be an a priori bound on this when $k > 2$.

17.1. ϵ–regularity and the singular set. This monotonicity is then important in proving an ϵ–regularity theorem. Recall that an ϵ–regularity theorem gives that a weak (or generalized) solution is actually smooth at a point if the scale–invariant energy is small enough there. The standard example is the Allard regularity theorem:

THEOREM 17.2. [**Al**] *There exists $\delta(k,n) > 0$ such that if $\Sigma \subset \mathbf{R}^n$ is a k–rectifiable stationary varifold (with density at least one a.e.), $x_0 \in \Sigma$, and*

$$\Theta_{x_0} = \lim_{r \to 0} \frac{Vol(B_r(x_0) \cap \Sigma)}{Vol(B_r \subset \mathbf{R}^k)} < 1 + \delta\,, \tag{17.2}$$

then Σ is smooth in a neighborhood of x_0.

Another well–known example is the ϵ–regularity theorem of Schoen and K. Uhlenbeck for harmonic maps:

THEOREM 17.3. [**ScUh**] *There exists $\epsilon(k,N) > 0$ such that if $u : B_r \subset \mathbf{R}^k \to N$ is an energy minimizing map and*

$$r^{2-k} \int_{B_r} |\nabla u|^2 < \epsilon\,, \tag{17.3}$$

then u is smooth in a neighborhood of 0 and $|\nabla u|(0) \leq C\, r^{-1}$.

See [**ChCTi**] for ϵ-regularity results for limits of Kähler–Einstein metrics and their applications to regularity of such limit spaces and see F.H. Lin, [**Ln**], for recent developments on the regularity of harmonic maps which are not minimizing.

Remark 17.4. The ϵ–regularity theorem is closely related to a priori estimates when the energy is small (cf. Section 9). In one direction, combining the regularity with a compactness argument usually directly gives an a priori estimate.

The singular set $\mathcal{S}$ is defined to be the set where the scale–invariant energy is not small. The first application of these ϵ–regularity theorems is some control on the singular set. For example, the semi–continuity of the density immediately gives that $\mathcal{S}$ is closed. In order to bound the size of the singular set (e.g., the Hausdorff measure), we combine the ϵ–regularity with simple covering arguments:

LEMMA 17.5. *If $u : B_1 \subset \mathbf{R}^k \to M$ is an energy minimizing (harmonic) map, then the $(k-2)$–dimensional Hausdorff measure of $\mathcal{S}$ is zero.*

PROOF. Given $\delta > 0$ and $x \in \mathcal{S}$, then the ϵ–regularity theorem (Theorem 17.3) yields a ball $B_{r_x}(x)$ so that

$$0 < \epsilon \leq r_x^{2-k} \int_{B_{r_x}(x)} |\nabla u|^2 . \tag{17.4}$$

Using the 5–times covering lemma, we can find a disjoint collection of balls $B_{r_i}(x_i)$ so that

$$\epsilon \leq r_i^{2-k} \int_{B_{r_i}(x_i)} |\nabla u|^2 , \tag{17.5}$$

$$\mathcal{S} \subset \cup_i B_{5r_i}(x_i) . \tag{17.6}$$

Since the balls $B_{r_i}(x_i)$ are disjoint, (17.5) implies that

$$\epsilon \sum_i r_i^{k-2} \leq \sum_i \int_{B_{r_i}(x_i)} |\nabla u|^2 \leq \int_{\cup_i B_{r_i}(x_i)} |\nabla u|^2 . \tag{17.7}$$

Combining this with (17.6), we get a uniform bound for the $(k-2)$–dimensional Hausdorff measure of $\mathcal{S}$. In particular, $\mathcal{S}$ has Lebesgue measure zero.

Finally, we show that the $(k-2)$–dimensional Hausdorff measure of $\mathcal{S}$ is zero (and not just finite). First, notice that as $\delta \to 0$, the Lebesgue measure of $\cup_i B_{r_i}(x_i)$ goes to zero. Since $|\nabla u|^2$ is an L^1 function, the dominated convergence theorem implies that

$$\lim_{\delta \to 0} \int_{\cup_i B_{r_i}(x_i)} |\nabla u|^2 = 0 . \tag{17.8}$$

Substituting this back into (17.7), gives the claim. □

This preliminary analysis of the singular set can be refined by doing a so–called tangent cone analysis.

17.2. Tangent cone analysis. Each of these variational problems comes with a natural scaling which preserves the space of solutions. For example, if $\Sigma \subset \mathbf{R}^n$ is a minimal submanifold, then so is

$$\Sigma_{y,\lambda} = \{ y + \lambda^{-1} (x - y) \, | \, x \in \Sigma \} . \tag{17.9}$$

(To see this, simply note that this scaling multiplies the principal curvatures by λ.) Similarly, if $u : \mathbf{R}^n \to M$ is an energy minimizing map, then so is the map $u_{y,\lambda}$ defined by

$$u_{y,\lambda}(x) = u(y + \lambda(x - y)) . \tag{17.10}$$

Suppose now that we fix the point y and take a sequence $\lambda_j \to 0$. The monotonicity formula (for either area or energy) bounds the density of the rescaled solution, allowing us to extract a convergent subsequence and limit. This limit, which is called a *tangent cone* at y, achieves equality in the monotonicity formula and, hence, must be homogeneous (i.e., invariant under dilations about y).

The usefulness of tangent cone analysis in regularity theory is based on two key facts. For simplicity, we illustrate these when $\Sigma \subset \mathbf{R}^n$ is an area minimizing hypersurface. (See [**ChC1**] for similar results for singular limit spaces of manifolds with lower Ricci curvature bounds.) First, if any tangent cone at y is a hyperplane $\mathbf{R}^{n-1}$, then Σ is smooth in a neighborhood of y. This follows easily from the Allard regularity theorem since the density at y of the tangent cone is the same as the density at y of Σ. The second key fact, known as "dimension reducing," is due to Almgren, [**Am2**], and is a refinement of an argument of Federer. To state this, we first stratify the singular set $\mathcal{S}$ of Σ into subsets

$$\mathcal{S}_0 \subset \mathcal{S}_1 \subset \cdots \subset \mathcal{S}_{n-2}\,, \tag{17.11}$$

where we define $\mathcal{S}_i$ to be the set of points $y \in \mathcal{S}$ so that any linear space contained in any tangent cone at y has dimension at most i. (Note that $\mathcal{S}_{n-1} = \emptyset$ by the previous fact.) The dimension reducing argument then gives that

$$\dim(\mathcal{S}_i) \leq i\,, \tag{17.12}$$

where dimension means the Hausdorff dimension. In particular, the solution of the Bernstein problem then gives codimension 7 regularity of Σ, i.e., $\dim(\mathcal{S}) \leq n-8$. See lecture 2 in [**Si3**] for a proof of (17.12).

Remark 17.6. Using that the $(k-2)$–dimensional Hausdorff measure of $\mathcal{S}$ is zero for an energy minimizing map $u : \mathbf{R}^k \to M$ by Lemma 17.5, it is not hard to see that $\mathcal{S}_{k-2} = \emptyset$. In particular, we get that $\mathcal{S}$ is at most $(k-3)$-dimensional

This approach has been applied fruitfully to many problems since it requires only a natural scaling, a monotonicity formula, and a compactness theorem. A variation of this, giving tangent flows rather than tangent cones, has also been useful in parabolic problems (see, e.g., [**W**] for an application to mean curvature flow).

Note that the tangent cones produced in this way may very well depend on the particular convergent subsequence. In some cases, one can prove uniqueness of the tangent cone and this is often quite useful (see, for instance, section 3.4 in [**Si3**] for one such application). However, in many settings tangent cones are not unique; see, for instance, [**Pe4**], [**ChC1**].

18. Embedded minimal disks

We next survey recent results of the authors for embedded minimal disks. The main result is a compactness and singular convergence theorem (Theorem 18.1 below) in a setting where the classical methods cannot be applied. In particular, there is no useful monotonicity formula or natural a priori bound. The main tools for overcoming these difficulties are a "classification of singularities" which describes a neighborhood of points of large curvature (Theorem 18.2) and our one-sided curvature estimate (Theorem 18.3 below). We will keep things brief here, attempting to highlight a few key points. We refer to [**CM22**] for a more detailed survey.

As we will see, a fundamental theorem about embedded minimal disks is that such a disk is either a minimal graph or can be approximated by a piece of a rescaled helicoid depending on whether the curvature is small or not; see Theorem 18.1 below. To avoid tedious dependence of various quantities we state this, our main result, not for a single embedded minimal disk with sufficiently large curvature at a given point but instead for a sequence of such disks where the curvatures are blowing up. Theorem 18.1 says that a sequence of embedded minimal disks mimics the following behavior of a sequence of rescaled helicoids:

> Consider the sequence $\Sigma_i = a_i\,\Sigma$ of rescaled helicoids where $a_i \to 0$. (That is, rescale $\mathbf{R}^3$ by a_i, so points that used to be distance d apart will in the rescaled $\mathbf{R}^3$ be distance $a_i\,d$ apart.) The curvatures of

this sequence of rescaled helicoids are blowing up along the vertical axis. The sequence converges (away from the vertical axis) to a foliation by flat parallel planes. The singular set $\mathcal{S}$ (the axis) then consists of removable singularities.

Let now $\Sigma_i \subset B_{2R}$ be a sequence of embedded minimal disks with $\partial \Sigma_i \subset \partial B_{2R}$. Clearly (after possibly going to a subsequence) either (A) or (B) occur:

(A) $\sup_{B_R \cap \Sigma_i} |A|^2 \leq C < \infty$ for some constant C.
(B) $\sup_{B_R \cap \Sigma_i} |A|^2 \to \infty$.

In (A) (by a standard argument) the intrinsic ball $\mathcal{B}_s(y_i)$ is a graph for all $y_i \in B_R \cap \Sigma_i$, where s depends only on C. Thus the main case is (B) which is the subject of the next theorem.

Using the notion of multi–valued graphs, the lamination theorem (the main theorem of [**CM6**]), can now be stated:

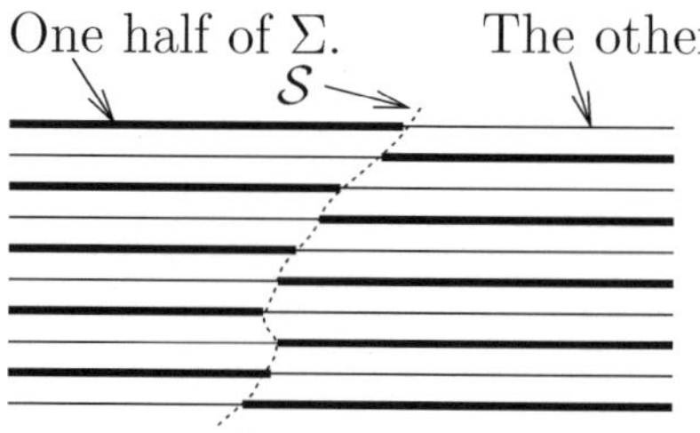

FIGURE 3. Theorem 18.1 – the singular set, $\mathcal{S}$, and the two multi–valued graphs.

THEOREM 18.1. *(Theorem 0.1 in* [**CM6**]*). See fig. 3. Let* $\Sigma_i \subset B_{R_i} = B_{R_i}(0) \subset \mathbf{R}^3$ *be a sequence of embedded minimal disks with* $\partial \Sigma_i \subset \partial B_{R_i}$ *where* $R_i \to \infty$. *If*

$$\sup_{B_1 \cap \Sigma_i} |A|^2 \to \infty \,, \tag{18.1}$$

then there exists a subsequence, Σ_j, *and a Lipschitz curve* $\mathcal{S} : \mathbf{R} \to \mathbf{R}^3$ *such that after a rotation of* $\mathbf{R}^3$:

1. $x_3(\mathcal{S}(t)) = t$. *(That is,* $\mathcal{S}$ *is a graph over the* x_3*–axis.)*
2. *Each* Σ_j *consists of exactly two multi–valued graphs away from* $\mathcal{S}$ *(which spiral together).*
3. *For each* $1 > \alpha > 0$, $\Sigma_j \setminus \mathcal{S}$ *converges in the* C^α*–topology to the foliation,* $\mathcal{F} = \{x_3 = t\}_t$, *of* $\mathbf{R}^3$.
4. $\sup_{B_r(\mathcal{S}(t)) \cap \Sigma_j} |A|^2 \to \infty$ *for all* $r > 0$, $t \in \mathbf{R}$. *(The curvatures blow up along* $\mathcal{S}$.*)*

In (2), (3) that $\Sigma_j \setminus \mathcal{S}$ are multi–valued graphs and converges to $\mathcal{F}$ means that for each compact subset $K \subset \mathbf{R}^3 \setminus \mathcal{S}$ and j sufficiently large $K \cap \Sigma_j$ consists of multi–valued graphs over (part of) $\{x_3 = 0\}$ and $K \cap \Sigma_j \to K \cap \mathcal{F}$ in the sense of graphs.

Theorem 18.1 (as many of the other results discussed below) is modelled by the helicoid and its rescalings. Take a sequence $\Sigma_i = a_i \Sigma$ of rescaled helicoids where $a_i \to 0$. The curvatures of this sequence are blowing up along the vertical axis. The sequence converges (away from the vertical axis) to a foliation by flat parallel planes. The singular set $\mathcal{S}$ (the axis) then consists of removable singularities.

We now come to our key results for embedded minimal disks. These are some of the main ingredients in the proof of Theorem 18.1. The first says that if the curvature of such a disk Σ is large at some point $x \in \Sigma$, then nearby x a multi–valued graph forms (in Σ) and this extends (in Σ) almost all the way to the boundary. Precisely this is:

THEOREM 18.2. *(Theorem 0.2 in* [**CM4**]*). See fig. 4 and fig. 5. Given $N \in \mathbf{Z}_+$, $\epsilon > 0$, there exist $C_1, C_2 > 0$ so that the following holds:*
Let $0 \in \Sigma^2 \subset B_R \subset \mathbf{R}^3$ be an embedded minimal disk with $\partial \Sigma \subset \partial B_R$. If for some $R > r_0 > 0$ we have

$$\max_{B_{r_0} \cap \Sigma} |A|^2 \geq 4\, C_1^2\, r_0^{-2}\,,$$

then there exists (after a rotation of $\mathbf{R}^3$) an N-valued graph Σ_g over $D_{R/C_2} \setminus D_{2r_0}$ with gradient $\leq \epsilon$ and contained in $\Sigma \cap \{x_3^2 \leq \epsilon^2\,(x_1^2 + x_2^2)\}$.

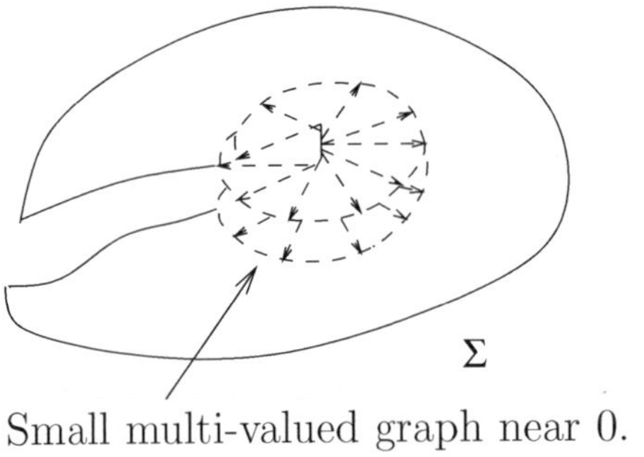

FIGURE 4. Part 1 of the proof of Theorem 18.2; finding a small multi–valued graph in a disk near a point of large curvature.

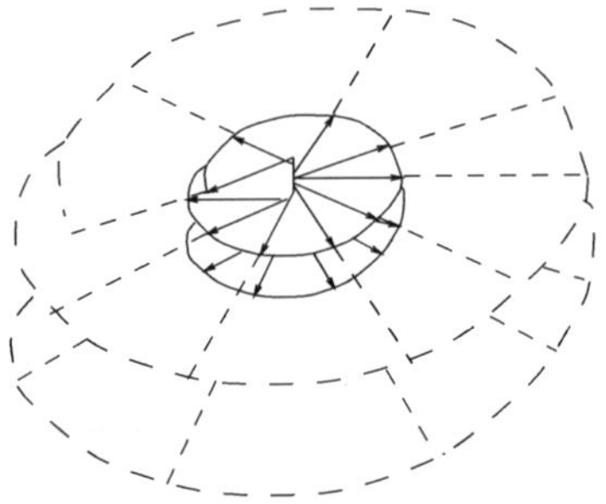

FIGURE 5. Part 2 of the proof of Theorem 18.2; extending a small multi–valued graph in a disk.

As a consequence of Theorem 18.2, one easily gets that if $|A|^2$ is blowing up near 0 for a sequence of embedded minimal disks Σ_i, then there is a sequence of 2–valued graphs $\Sigma_{i,d} \subset \Sigma_i$, where the 2–valued graphs start off on a smaller and smaller scale (namely, r_0 in Theorem 18.2 can be taken to be smaller as the curvature gets larger). Consequently, by the sublinear separation growth, such 2–valued graphs collapse and, hence, a subsequence converges to a smooth minimal graph through 0. To be precise, given any fixed $\rho > 0$, (11.4) bounds the separation w at $(\rho, 0)$ by

$$|w(\rho,0)| \leq \left(\frac{\rho}{r_0}\right)^{\alpha} |w(r_0,0)| \leq 2\pi\, \epsilon\, \rho^{\alpha}\, r_0^{1-\alpha}\,, \tag{18.2}$$

and this goes to 0 as $r_0 \to 0$ since $\alpha < 1$. The bound

$$|w(r_0,0)| \leq 2\pi\, \epsilon\, r_0 \tag{18.3}$$

in (18.2) came from integrating the gradient bound on the graph around the circle of radius r_0. (Here 0 is a removable singularity for the limit.) Moreover, if the sequence of such disks is as in Theorem 18.1, i.e., if $R_i \to \infty$, then the minimal graph in the limit is entire and hence, by Bernstein's theorem (theorem 1.16 in [**CM1**]), is a plane.

The second key result is the curvature estimate for embedded minimal disks in a half–space. This theorem says roughly that if an embedded minimal disk lies in a half–space above a plane and comes close to the plane, then it is a graph over the plane. Precisely, this is the following theorem:

THEOREM 18.3. *(Theorem 0.2 in [**CM6**]). See fig. 6. There exists $\epsilon > 0$, so that if*

$$\Sigma \subset B_{2r_0} \cap \{x_3 > 0\} \subset \mathbf{R}^3$$

is an embedded minimal disk with $\partial\Sigma \subset \partial B_{2r_0}$, then for all components Σ' of $B_{r_0} \cap \Sigma$ which intersect $B_{\epsilon r_0}$ we have

$$\sup_{\Sigma'} |A_\Sigma|^2 \leq r_0^{-2}\,. \tag{18.4}$$

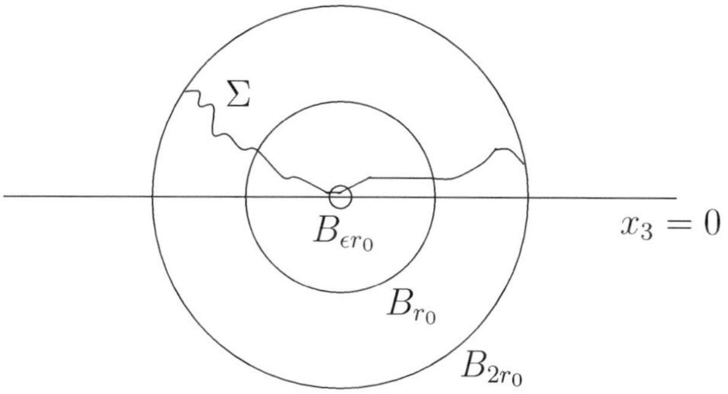

FIGURE 6. Theorem 18.3 – the one–sided curvature estimate for an embedded minimal disk Σ in a half–space with $\partial\Sigma \subset \partial B_{2r_0}$: The components of $B_{r_0} \cap \Sigma$ intersecting $B_{\epsilon r_0}$ are graphs.

Using the minimal surface equation and that Σ' has points close to a plane, it is not hard to see that, for $\epsilon > 0$ sufficiently small, (18.4) is equivalent to the statement that Σ' is a graph over the plane $\{x_3 = 0\}$.

An embedded minimal surface Σ which is as in Theorem 18.3 is said to satisfy the (ϵ, r_0)*–effective one–sided Reifenberg condition*; cf. appendix A of [**CM6**] and the appendix of [**ChC1**]. We will often refer to Theorem 18.3 as *the one–sided curvature estimate.*

Note that the assumption in Theorem 18.3 that Σ is simply connected is crucial as can be seen from the example of a rescaled catenoid. The catenoid is the minimal surface in $\mathbf{R}^3$ given by

$$(\cosh s \,\cos t, \cosh s \,\sin t, s) \tag{18.5}$$

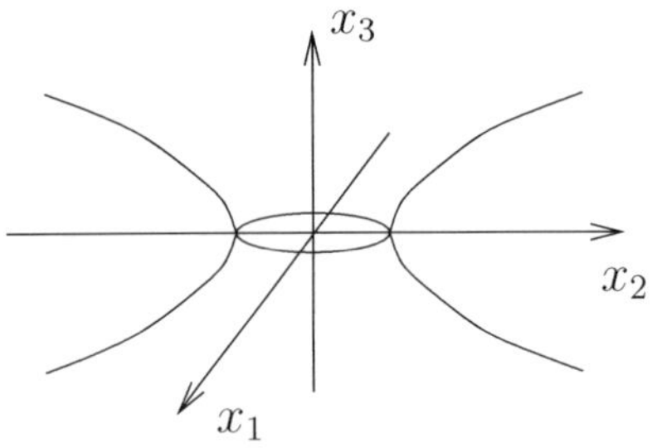

FIGURE 7. The catenoid given by revolving $x_1 = \cosh x_3$ around the x_3–axis.

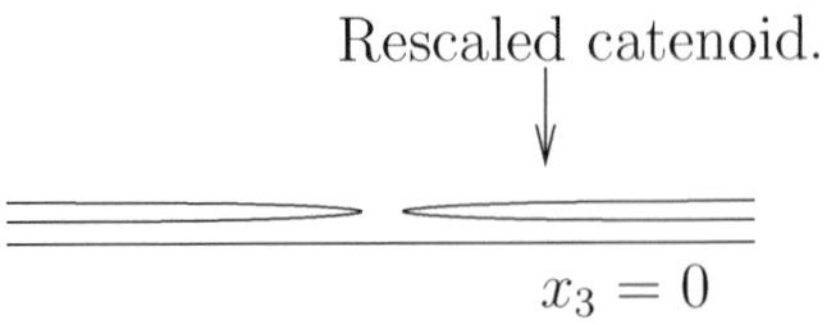

FIGURE 8. Rescaling the catenoid shows that simply connected (and embedded) is needed in the one–sided curvature estimate.

where $s, t \in \mathbf{R}$; see fig. 7. Under rescalings this converges (with multiplicity two) to the flat plane; see fig. 8. Likewise, by considering the universal cover of the catenoid, one sees that embedded, and not just immersed, is needed in Theorem 18.3.

As an almost immediate consequence of Theorem 18.3 and a simple barrier argument we get that if in a ball two embedded minimal disks come close to each other near the center of the ball then each of the disks are graphs. Precisely, this is the following:

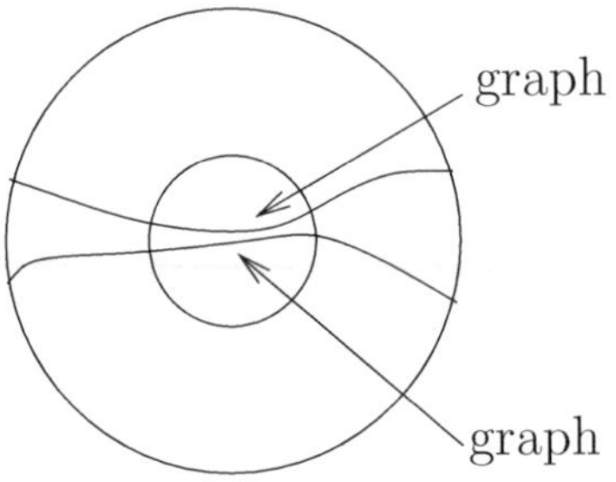

FIGURE 9. Corollary 18.4: Two sufficiently close components of an embedded minimal disk must each be a graph.

COROLLARY 18.4. *(Corollary 0.4 in [**CM6**]). See fig. 9. There exist $c > 1$, $\epsilon > 0$ so that the following holds:*
Let Σ_1 and $\Sigma_2 \subset B_{cr_0} \subset \mathbf{R}^3$ be disjoint embedded minimal surfaces with $\partial \Sigma_i \subset \partial B_{cr_0}$ and $B_{\epsilon r_0} \cap \Sigma_i \neq \emptyset$. If Σ_1 is a disk, then for all components Σ_1' of $B_{r_0} \cap \Sigma_1$ which intersect $B_{\epsilon r_0}$

$$\sup_{\Sigma_1'} |A|^2 \leq r_0^{-2} \,. \tag{18.6}$$

This estimate has also been useful in the global theory of minimal surfaces, cf. [**CM9**], [**CM22**], and [**MeRo**]. It would be very interesting to find an intrinsic version of it (i.e., for intrinsic balls on one side of a plane):

QUESTION 3. *It would be very useful to prove an intrinsic version of the one–sided curvature estimate. Namely, does Theorem 18.3 hold when Σ is an intrinsic ball (in an embedded minimal disk)? If true, then this would likely have important consequences for proving properness of embedded minimal disks.*

One of the topics that we have suppressed is what we call "properness" of the limit. Basically, this is proving that we get a foliation in the limit or, equivalently, that the points of $\mathcal{S}$ cannot be isolated. This is the one place where the assumption $R_i \to \infty$ is used in an essential way; see [**CM18**] for a nonproper limit when R_i does not go to ∞.

QUESTION 4. *Suppose that a sequence of embedded minimal planar domains $\Sigma_i \subset B_1 \subset \mathbf{R}^3$ with $\partial \Sigma_i \subset \partial B_1$ converges away from 0 to a minimal lamination $\mathcal{L}'$ of $B_1 \setminus \{0\}$. Does $\mathcal{L}'$ extend to a smooth lamination of B_1? In other words, is 0 a removable singularity? An example constructed in* [**CM18**] *shows that this need not be the case when the Σ_i's are disks. It would be interesting to find non–disk examples (cf.* [**Ka**], [**Tr**]*).*

19. Global theory of minimal surfaces in $\mathbf{R}^3$

Recent years have seen breakthroughs on many long–standing problems in the global theory of minimal surfaces in $\mathbf{R}^3$. This is an enormous subject and, rather than give a comprehensive treatment, we will mention a few important results which fit well with the theme of this survey. Throughout this section, Σ will be a complete properly embedded minimal surface in $\mathbf{R}^3$ (recall that properness here means that the intersection of Σ with any compact subset of $\mathbf{R}^3$ is compact).

We say that Σ has finite topology if it is homeomorphic to a closed Riemann surface with a finite number of punctures; the genus of Σ is then the genus of this Riemann surface and the number of punctures is the number of ends. It follows that a neighborhood of each puncture corresponds to a properly embedded annular end of Σ. Perhaps surprisingly at first, the more restrictive case is when Σ has more than one end. The reason for this is that a barrier argument gives a stable minimal surface between any pair of ends. This stable surface is then asymptotic to a plane (or catenoid), essentially forcing each end to live in a half–space. Using this restriction, P. Collin proved:

THEOREM 19.1. [**Co**] *Each end of a complete properly embedded minimal surface with finite topology and at least two ends is asymptotic to a plane or catenoid.*

In particular, outside some compact set, Σ is given by a finite collection of disjoint graphs over a common plane (and has finite total curvature). See [**CM21**] for a proof of Theorem 19.1 using the one–sided curvature estimate.

When Σ has only one end (e.g., for the helicoid), it need not have finite total curvature so the situation is more delicate. However, the regularity results of the previous section can be applied. For example, if Σ is a (non–planar) embedded minimal disk, then we get a multi–valued graph structure away from a "one–dimensional singular set." Using Theorems 18.1 and 18.3, W. Meeks and H. Rosenberg proved the uniqueness of the helicoid:

THEOREM 19.2. [**MeRo**] *The plane and helicoid are the only complete properly embedded simply–connected minimal surfaces in $\mathbf{R}^3$.*

This uniqueness should have many applications. Recall that if we take a sequence of rescalings of the helicoid, then the singular set $\mathcal{S}$ for the convergence is the vertical axis perpendicular to the leaves of the foliation. In [**Me1**], W. Meeks used this fact together with the uniqueness of the helicoid to prove that the singular set $\mathcal{S}$ in Theorem 18.1 is always a straight line perpendicular to the foliation. Recently, W. Meeks and M. Weber have constructed a *local* example (i.e., a sequence of embedded minimal disks in a unit ball) where $\mathcal{S}$ is a circle.

Properness is an important ingredient in many of these results and it is not known to what extent this assumption can be relaxed. In [**Na**], N. Nadirashvili constructed complete non–proper minimal *immersions* (in fact, contained in a ball). It would be interesting to know whether this is possible for embeddings:

QUESTION 5. *Suppose that $\Sigma \subset \mathbf{R}^3$ is a complete embedded minimal surface with finite topology. Does Σ have to be proper?*

We have not even touched on the case where Σ has infinite topology (e.g., when Σ is one of the Riemann examples). This is an area of much current research, see [**CM5**], the work of Meeks, J. Perez and A. Ros, [**MePRs1**], [**MePRs2**], and [**MePRs3**], the survey [**MeP**] and references therein.

We close this section with a local analog of the two–ended case. Namely, in [**CM9**], we proved that any embedded minimal annulus in a ball (with boundary in the boundary of the ball and) with a small neck can be decomposed by a simple closed geodesic into two graphical sub–annuli. Moreover, we gave a sharp bound for the length of this closed geodesic in terms of the separation (or height) between the graphical sub–annuli. This serves to illustrate our "pair of pants" decomposition from [**CM5**] in the special case where the embedded minimal planar domain is an annulus (we will not touch on this further here). The catenoid

$$\{x_1^2 + x_2^2 = \cosh^2 x_3\} \tag{19.1}$$

is the prime example of an embedded minimal annulus.

The precise statement of this decomposition for annuli is:

THEOREM 19.3. [**CM9**] *There exist $\epsilon > 0$, C_1, C_2, $C_3 > 1$ so: If $\Sigma \subset B_R \subset \mathbf{R}^3$ is an embedded minimal annulus with $\partial \Sigma \subset \partial B_R$ and $\pi_1(B_{\epsilon R} \cap \Sigma) \neq 0$, then there exists a simple closed geodesic $\gamma \subset \Sigma$ of length ℓ so that:*

- *The curve γ splits the connected component of $B_{R/C_1} \cap \Sigma$ containing it into two annuli Σ^+, Σ^- each with $\int |A|^2 \leq 5\,\pi$.*
- *Furthermore, $\Sigma^\pm \setminus \mathcal{T}_{C_2\,\ell}(\gamma)$ are graphs with gradient ≤ 1.*
- *Finally, $\ell \log(R/\ell) \leq C_3\, h$ where the separation h is given by*

$$h = \min_{x_\pm \in \partial B_{R/C_1} \cap \Sigma^\pm} |x_+ - x_-| \, . \tag{19.2}$$

Here $\mathcal{T}_s(S) \subset \Sigma$ denotes the intrinsic s–tubular neighborhood of a subset $S \subset \Sigma$.

Part 4. Constructing minimal surfaces and applications

Thus far, we have mainly dealt with regularity and a priori estimates but have ignored questions of existence. In this part we surveys some of the most useful existence results for minimal surfaces and gives an application to Ricci flow. Section 20 recalls the classical Weierstrass representation, including a few modern applications, and the Kapouleas desingularization method. Section 21 deals with producing area minimizing surfaces (whether for fixed boundary, fixed homotopy class, etc.) and questions of embeddedness. The next section discusses unstable (hence *not* minimizing) surfaces and the corresponding questions for geodesics, concentrating on whether the Morse index can be bounded uniformly. Section 23 recalls the min–max construction for producing unstable minimal surfaces and, in particular, doing so while controlling the topology and guaranteeing embeddedness. Finally, Section 24 discusses a recent application of min–max surfaces to bound the extinction time for Ricci flow, answering a question of Perelman.

20. The Weierstrass representation

The classical Weierstrass representation (see [**HoK**] or [**Os**]) takes holomorphic data (a Riemann surface, a meromorphic function, and a holomorphic one–form) and associates a minimal surface in $\mathbf{R}^3$. To be precise, given a Riemann surface Ω, a meromorphic function g on Ω, and a holomorphic one–form ϕ on Ω, then we get a (branched) conformal minimal immersion $F : \Omega \to \mathbf{R}^3$ by

$$F(z) = \text{Re} \int_{\zeta \in \gamma_{z_0,z}} \left(\frac{1}{2}\,(g^{-1}(\zeta) - g(\zeta)), \frac{i}{2}\,(g^{-1}(\zeta) + g(\zeta)), 1 \right) \phi(\zeta)\,. \tag{20.1}$$

Here $z_0 \in \Omega$ is a fixed base point and the integration is along a path $\gamma_{z_0,z}$ from z_0 to z. The choice of z_0 changes F by adding a constant. In general, the map F may depend on the choice of path (and hence may not be well–defined); this is known as "the period problem" (see M. Weber and M. Wolf, [**WeWo**], for the latest developments). However, when g has no zeros or poles and Ω is simply connected, then $F(z)$ does not depend on the choice of path $\gamma_{z_0,z}$.

Two standard constructions of minimal surfaces from Weierstrass data are

$$g(z) = z,\ \phi(z) = dz/z,\ \Omega = \mathbf{C} \setminus \{0\} \text{ giving a catenoid}\,, \tag{20.2}$$

$$g(z) = \mathrm{e}^{iz},\ \phi(z) = dz,\ \Omega = \mathbf{C} \text{ giving a helicoid}\,. \tag{20.3}$$

The Weierstrass representation is particularly useful for constructing immersed minimal surfaces. For example, in [**Na**], Nadirashvili used it to construct a complete immersed minimal surface in the unit ball in $\mathbf{R}^3$ (see also [**JXa**] for the case of a slab). In particular, Nadirashvili's surface is not proper, i.e., the intersections with compact sets are not necessarily compact.

Typically, it is rather difficult to prove that the resulting immersion is an embedding (i.e., is 1–1), although there are some interesting cases where this can be done. The first modern example was [**HoMe**] where D. Hoffman and Meeks proved that the surface constructed by Costa was embedded; this was the first new complete finite topology properly embedded minimal surface discovered since the classical catenoid, helicoid, and plane. This led to the discovery of many more such surfaces (see [**HoK**] and [**Ro**] for more discussion).

Very recently in [**HoWeWo**], Hoffman, Weber, and Wolf have used the Weierstrass representation to construct a genus one properly embedded minimal surface asymptotic to the helicoid. They construct this "genus one helicoid" as the limit of a continuous one-parameter family of screw-motion invariant minimal surfaces–also asymptotic to the helicoid–that have genus equal to one in the quotient.

In [**CM18**], we used the Weierstrass representation to construct a sequence of embedded minimal disks

$$\Sigma_i \subset B_1 = B_1(0) \subset \mathbf{R}^3 \tag{20.4}$$

with $\partial \Sigma_i \subset \partial B_1$ where the curvatures blow up only at 0 and $\Sigma_i \setminus \{x_3\text{–axis}\}$ consists of two multi–valued graphs for each i. Furthermore, $\Sigma_i \setminus \{x_3 = 0\}$ converges to two embedded minimal disks $\Sigma^- \subset \{x_3 < 0\}$ and $\Sigma^+ \subset \{x_3 > 0\}$ each of which spirals into $\{x_3 = 0\}$ and thus is not proper. (This should be contrasted with Theorem 18.1 where the *complete* limits are planes and hence proper.)

N. Kapouleas has developed another method to construct complete embedded minimal surfaces with finite total curvature. For instance, in [**Ka**], he shows that (most) collections of coaxial catenoids and planes can be desingularized to get complete embedded minimal surfaces with finite total curvature. The Costa surface above had genus one and three ends (that is to say, it is homeomorphic to a torus with three punctures). In the Kapouleas construction, one could start with a plane and catenoid intersecting in a circle and then desingularize this circle using suitably scaled and bent Scherk surfaces to get a finite genus embedded surface with three ends. (This desingularization process adds handles, i.e., increases the genus.) In this manner, Kapouleas gets an enormous number of new examples; see also the gluing construction of S.D. Yang, [**Y**], which uses catenoid necks to glue together nearby minimal surfaces.

21. Area–minimizing surfaces

Perhaps the most natural way to construct minimal surfaces is to look for ones which minimize area, e.g., with fixed boundary, or in a homotopy class, etc. This has the advantage that often it is possible to show that the resulting surface is embedded. We mention a few results along these lines.

The first embeddedness result, due to Meeks and Yau, shows that if the boundary curve is embedded and lies on the boundary of a smooth mean convex set (and it is null–homotopic in this set), then it bounds an embedded least area disk.

THEOREM 21.1. [**MeYa1**] *Let M^3 be a compact Riemannian three–manifold whose boundary is mean convex and let γ be a simple closed curve in ∂M which is null–homotopic in M; then γ is bounded by a least area disk and any such least area disk is properly embedded.*

Note that some restriction on the boundary curve γ is certainly necessary. For instance, if the boundary curve was knotted (e.g., the trefoil), then it could not be spanned by any embedded disk (minimal or otherwise). Prior to the work of Meeks and Yau, embeddedness was known for extremal boundary curves in $\mathbf{R}^3$ with small total curvature by the work of R. Gulliver and J. Spruck [**GuSp**]; see chapter 4 in [**CM1**] for other results and further discussion.

If we instead fix a homotopy class of maps, then the two fundamental existence results are due to Sacks–Uhlenbeck and Schoen–Yau (with embeddedness proven by Meeks–Yau and Freedman–Hass–Scott, respectively):

THEOREM 21.2. [**SaUh**], [**MeYa2**] *Given M^3, there exist conformal (stable) minimal immersions $u_1, \dots, u_m : \mathbf{S}^2 \to M$ which generate $\pi_2(M)$ as a $\mathbf{Z}[\pi_1(M)]$ module. Furthermore,*

- *If $u : \mathbf{S}^2 \to M$ and $[u]_{\pi_2} \neq 0$, then Area$(u) \geq \min_i$ Area(u_i).*

- *Each u_i is either an embedding or a 2–1 map onto an embedded 2–sided* $\mathbf{R}P^2$.

THEOREM 21.3. [**ScYa2**], [**FHS**] *If Σ^2 is a closed surface with genus $g > 0$ and $i_0 : \Sigma \to M^3$ is an embedding which induces an injective map on π_1, then there is a least area embedding with the same action on π_1.*

In [**MeSiYa**], Meeks, Simon, and Yau find an embedded sphere minimizing area in an isotopy class in a closed 3–manifold.

We end this section by mentioning two applications of Theorem 21.3. First, in [**CM20**], we showed that any topological 3–manifold M had an open set of metrics so that, for each such metric, there was a sequence of embedded minimal tori whose area went to infinity. In [**De**], B. Dean showed that this was true for every genus $g \geq 1$. This leaves an obvious interesting question:

QUESTION 6. *Given a topological 3–manifold M, does there exist an open set of metrics which have embedded minimal spheres with arbitrarily large area?*

It would be interesting to answer this question even when the minimal spheres are stable (the examples constructed in [**CM20**] and [**De**] were all locally minimizing and hence also stable)

22. Index bounds for geodesics and minimal surfaces

The minimal surfaces discussed in the previous section were all stable and in fact locally area minimizing. This is a very special and strong property of a minimal surface. In general, like, for instance the helicoid and the catenoid, most minimal surfaces are not stable but have non–zero index. In this section we will discuss the Morse index of simple closed geodesics on surfaces and of embedded minimal surfaces in 3–manifolds. First let us discuss the situation of simple closed geodesics in surfaces.

Let M^2 be a closed orientable surface with curvature K and $\gamma \subset M$ a closed geodesic. The *Morse index* of γ is the index of the critical point γ for the length functional, i.e., the number of negative eigenvalues (counted with multiplicity) of the second derivative of length (throughout curves will always be in H^1). Since the second derivative of length at γ in the direction of a normal variation $u\,\mathrm{n}$ is

$$-\int_\gamma u\, L_\gamma\, u\,, \tag{22.1}$$

where

$$L_\gamma\, u = u'' + K\, u\,, \tag{22.2}$$

the Morse index is the number of negative eigenvalues of L_γ. (By convention, an eigenfunction ϕ with eigenvalue λ of L_γ is a solution of $L_\gamma\, \phi + \lambda\, \phi = 0$.) Note that if $\lambda = 0$, then ϕ (or $\phi\, \mathrm{n}$) is a (normal) Jacobi field. γ is *stable* if the index is zero. The *index* of a noncompact geodesic is the dimension of a maximal vector space of compactly supported variations for which the second derivative of length is negative definite. We also say that such a geodesic is *stable* if the index is 0.

As the following result shows then it turns out that in general there are no Morse index bounds for simple closed geodesics on surfaces.

THEOREM 22.1. [**CH1**]. *On any M^2, there exists a metric with a geodesic lamination with infinitely many unstable leaves. Moreover, there is such a metric with simple closed geodesics of arbitrary high Morse index.*

A codimension one *lamination* on a surface M^2 is a collection $\mathcal{L}$ of smooth disjoint curves (called leaves) such that $\cup_{\ell \in \mathcal{L}} \ell$ is closed. Moreover, for each $x \in M$ there exists an open neighborhood U of x and a C^0 coordinate chart, (U, Φ), with $\Phi(U) \subset \mathbf{R}^2$ so that in these coordinates the leaves in $\mathcal{L}$ pass through $\Phi(U)$ in slices of the form $(\mathbf{R} \times \{t\}) \cap \Phi(U)$. A *foliation* is a lamination for which the union of the leaves is all of M and a *geodesic lamination* is a lamination whose leaves are geodesics.

Similarly, to the geodesic case, for an immersed minimal surface Σ in a 3–manifold M, we set

$$L_\Sigma \, \phi = \Delta_\Sigma \, \phi + |A|^2 \, \phi + \operatorname{Ric}_M(\mathrm{n}, \mathrm{n}) \, \phi \, . \tag{22.3}$$

(Note that by the second variational formula (see, for instance, section 1.7 of [**CM1**]), then

$$\frac{\partial^2}{\partial r^2}_{r=0} \operatorname{Area}(\Sigma_r) = - \int_\Sigma \phi \, L_\Sigma \, \phi \, , \tag{22.4}$$

where $\Sigma_r = \{x + r\, \phi(x) \, \mathrm{n}_\Sigma(x) \, | \, x \in \Sigma\}$.) Recall also that by definition the index of a minimal surface Σ is the number of negative eigenvalues (counted with multiplicity) of L_Σ. (A function η is an eigenfunction of L_Σ with eigenvalue λ if $L_\Sigma \, \eta + \lambda \, \eta = 0$.) Thus in particular, since Σ is assumed to be closed, the index is always finite.

Theorem 22.1 was proven by first constructing a metric on the disk with convex boundary having no Morse index bounds and then completing the metric to a metric on the given M^2. By taking the product of this metric on the disk with a circle we get, on a solid torus, a metric with convex boundary and without Morse index bounds for embedded minimal tori, and with a minimal lamination with infinitely many unstable leaves. By completing this metric we get:

THEOREM 22.2. [**CH2**] *On any M^3, there exists a metric with a minimal lamination with infinitely many unstable leaves. Moreover, there is such a metric with embedded minimal tori of arbitrary high Morse index.*

By construction the embedded minimal tori in Theorem 22.2 and the leaves of the lamination can be taken to be totally geodesic.

We will equip the space of metrics on a given manifold with the C^∞-topology. A subset of the set of metrics on the manifold is said to be *residual* if it is a countable intersection of open dense subsets. A metric on a surface is *bumpy* if each closed geodesic is a nondegenerate critical point, i.e., $L_\gamma u = 0$ implies $u \equiv 0$. It follows from results of Abraham and Anosov that bumpy metrics are generic; that is the set of bumpy metrics contain a residual set.

To check that any given metric is bumpy is virtually impossible; however it seems that the metric in Theorem 22.1 can be chosen to be bumpy. Thus it seems unlikely that a bumpy metric is enough to ensure a bound for the Morse index of simple closed geodesics on M^2. What is needed is a nondegeneracy condition for noncompact simple geodesics, rather than one for closed geodesics; cf. [**CH2**], [**CH3**].

In [**HaNoRu**] examples were given of metrics on any M^3 that have embedded minimal spheres without bounds and in [**CD2**] the following was shown: For any

3-manifold M^3 and any nonnegative integer $\mathbf{g}$, there are examples of metrics on M each of which has a sequence of embedded minimal surfaces of genus $\mathbf{g}$ and without Morse index bounds. On any spherical space form $\mathbf{S}^3/\Gamma$ [**CD2**] constructed such a metric with positive scalar curvature. More generally [**CD2**] constructed such a metric with Scal > 0 (and such surfaces) on any 3-manifold which carries a metric with Scal > 0. In all but one of the examples in [**CD2**] the Hausdorff limit is a singular minimal lamination. The singularities being in each case exactly two points lying on a closed leaf (the leaf is a strictly stable sphere).

[**CD2**] used in part ideas of Hass-Norbury-Rubinstein [**HaNoRu**]. As in [**HaNoRu**], but unlike the examples in [**CH1**], the surfaces in [**CD2**] have no uniform curvature bounds. In fact, it follows easily (see appendix B of [**CM4**]) that if $\Sigma_i \subset M^3$ is a sequence of embedded minimal surfaces with uniformly bounded curvatures, then a subsequence converges to a smooth lamination. Moreover, with the right notion of being generic, the following seems likely (by [**CH1**] bumpy is not the right generic notion):

QUESTION 7. *Let M^3 be a closed 3-manifold with a generic metric and $\Sigma_i \subset M$ a sequence of embedded minimal surfaces of a given genus. If any limit of the Σ_i's is a smooth (minimal) lamination, then the sequence Σ_i has a uniform Morse index bound.*

A codimension one *lamination* of M^3 is a collection $\mathcal{L}$ of smooth disjoint connected surfaces (called leaves) such that $\cup_{\Lambda \in \mathcal{L}} \Lambda$ is closed. Moreover, for each $x \in M$ there exists an open neighborhood U of x and a local coordinate chart, (U, Φ), with $\Phi(U) \subset \mathbf{R}^3$ such that in these coordinates the leaves in $\mathcal{L}$ pass through the chart in slices of the form $(\mathbf{R}^2 \times \{t\}) \cap \Phi(U)$.

A lamination is said to be minimal if the leaves are (smooth) minimal surfaces. If the union of the leaves is all of M, then it is a foliation.

There are two results that support this question. The first concerns the corresponding question in one dimension less (that is for geodesics on surfaces); see [**CH2**], [**CH3**]. The second concerns the question for 3-manifolds with positive scalar curvature. However, there are examples where the limit is not smooth; see [**CD2**].

Finally, we refer to [**CH2**] and [**CH3**] for further discussion of Morse index bounds for geodesics including some positive results about when one has such bounds.

23. The min–max construction of minimal surfaces

Variational arguments can also be used to construct higher index (i.e., non–minimizing) minimal surfaces using the topology of the space of surfaces. There are two basic approaches:

- Applying Morse theory to the energy functional on the space of maps from a fixed surface Σ to M.
- Doing a min–max argument over families of (topologically non–trivial) sweep–outs of M.

The first approach has the advantage that the topological type of the minimal surface is easily fixed; however, the second approach has been more successful at producing embedded minimal surfaces. We will highlight a few key results below but refer to [**CD1**] for a thorough treatment.

Unfortunately, one cannot directly apply Morse theory to the energy functional on the space of maps from a fixed surface because of a lack of compactness (the Palais–Smale Condition C does not hold). To get around this difficulty, [**SaUh**] introduce a family of perturbed energy functionals which do satisfy Condition C and then obtain minimal surfaces as limits of critical points for the perturbed problems:

THEOREM 23.1. [**SaUh**] *If $\pi_k(M) \neq 0$ for some $k > 1$, then there exists a branched immersed minimal 2–sphere in M (for any metric).*

This was sharpened somewhat by [**MiMo**] (showing that the index of the minimal sphere was at most $k-2$), who used it to prove a generalization of the sphere theorem. See [**Jo**] and [**St**] for approaches which avoid using the perturbed functionals and [**Fr**] for a generalization to a free boundary problem.

The basic idea of constructing minimal surfaces via min–max arguments and sweep–outs goes back to Birkhoff, who developed it to construct simple closed geodesics on spheres. In particular, when M is a topological 2–sphere, we can find a 1–parameter family of curves starting and ending at point curves so that the induced map $F : \mathbf{S}^2 \to \mathbf{S}^2$ (see fig. 10) has nonzero degree. The min–max argument produces a nontrivial closed geodesic of length less than or equal to the longest curve in the initial one–parameter family. A curve shortening argument gives that the geodesic obtained in this way is simple.

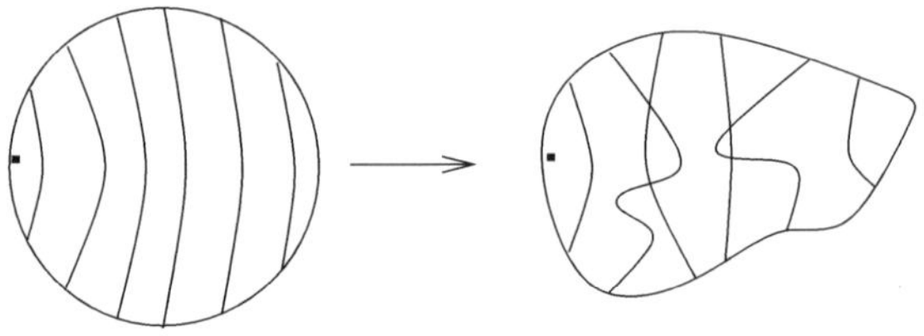

FIGURE 10. A 1–parameter family of curves on a 2–sphere which induces a map $F : \mathbf{S}^2 \to \mathbf{S}^2$ of degree 1.

In [**Pi**], J. Pitts applied a similar argument and geometric measure theory to get that every closed Riemannian three manifold has an embedded minimal surface (his argument was for dimensions up to seven), but he did not estimate the genus of the resulting surface. Finally, F. Smith (under the direction of L. Simon) proved (see [**CD1**]):

THEOREM 23.2. [**Sm**] *Every metric on a topological 3–sphere M admits an embedded minimal 2–sphere.*

The main new contribution of Smith was to control the topological type of the resulting minimal surface while keeping it embedded; see also Pitts and Rubinstein, [**PiRu**], for some generalizations.

24. An application of min–max surfaces to Ricci flow

We will in this section give an application of the min–max construction of minimal surfaces to the Ricci flow. This application is that on $\mathbf{S}^3$ starting at any given metric the Ricci flow becomes extinct in finite time. The treatment here follows [**CM19**] and was inspired by a question of Perelman; see [**CM19**], [**Pe3**]. (See also paragraph 4.4 of [**Pe2**] for the precise definition of extinction time in the case that surgery occurs.)

Throughout this section we let M^3 be the 3–sphere and let $g(t)$ be a one–parameter family of metrics on M evolving by the Ricci flow, so

$$\partial_t g = -2\,\mathrm{Ric}_{M_t}\,. \tag{24.1}$$

Since $\pi_3(M) = \mathbf{Z}$, it follows from suspension, as in lemma 3 of [**MiMo**], that the space of maps from $\mathbf{S}^2$ to M is not simply connected.

Fix a continuous map

$$\beta : [0,1] \to C^0 \cap L_1^2(\mathbf{S}^2, M) \tag{24.2}$$

where $\beta(0)$ and $\beta(1)$ are constant maps so that β is in the nontrivial homotopy class $[\beta]$. We define the width $W = W(g, [\beta])$ by

$$W(g) = \min_{\gamma\in[\beta]} \max_{s\in[0,1]} \mathrm{Energy}(\gamma(s))\,. \tag{24.3}$$

One could equivalently define the width using the area rather than the energy, but the energy is somewhat easier to work with. As for the Plateau problem, this equivalence follows using the uniformization theorem and the inequality $\mathrm{Area}(u) \leq \mathrm{Energy}(u)$ (with equality when u is a branched conformal map); cf. lemma 4.12 in [**CM1**].

The next theorem gives an upper bound for the derivative of $W(g(t))$ under the Ricci flow which forces the solution $g(t)$ to become extinct in finite time.

THEOREM 24.1. *([**CM19**] and cf. [**Pe3**]). Let M^3 be the 3–sphere equipped with a Riemannian metric $g = g(0)$. Under the Ricci flow, the width $W(g(t))$ satisfies*

$$\frac{d}{dt} W(g(t)) \leq -4\pi + \frac{3}{4(t+C)}\, W(g(t))\,, \tag{24.4}$$

in the sense of the limsup of forward difference quotients. Hence, $g(t)$ must become extinct in finite time.

Suppose that $\Sigma \subset M$ is a closed immersed surface (not necessarily minimal), then using (24.1) an easy calculation gives (cf. page 38–41 of [**Ha3**])

$$\frac{d}{dt}_{t=0} \mathrm{Area}_{g(t)}(\Sigma) = -\int_\Sigma [R - \mathrm{Ric}_M(\mathrm{n},\mathrm{n})]\,. \tag{24.5}$$

If Σ is also minimal, then

$$\begin{aligned} \frac{d}{dt}_{t=0} \mathrm{Area}_{g(t)}(\Sigma) &= -2\int_\Sigma \mathrm{K}_\Sigma - \int_\Sigma [|A|^2 + \mathrm{Ric}_M(\mathrm{n},\mathrm{n})] \\ &= -\int_\Sigma K_\Sigma - \frac{1}{2}\int_\Sigma [|A|^2 + R]\,. \end{aligned} \tag{24.6}$$

Here K_Σ is the (intrinsic) curvature of Σ, n is a unit normal for Σ (our Σ's below will be $\mathbf{S}^2$'s and hence have a well–defined unit normal), A is the second fundamental form of Σ so that $|A|^2$ is the sum of the squares of the principal curvatures, Ric_M is the Ricci curvature of M, and R is the scalar curvature of M. (The curvature is normalized so that on the unit $\mathbf{S}^3$ the Ricci curvature is 2 and the scalar curvature is 6.) To get (24.6), we used that by the Gauss equations and minimality of Σ

$$\mathrm{K}_\Sigma = \mathrm{K}_M - \frac{1}{2}|A|^2 \,, \tag{24.7}$$

where K_M is the sectional curvature of M on the two–plane tangent to Σ.

Our first lemma gives an upper bound for the rate of change of area of minimal 2–spheres.

LEMMA 24.2. *If $\Sigma \subset M^3$ is a branched minimal immersion of the 2–sphere, then*

$$\frac{d}{dt}_{t=0} Area_{g(t)}(\Sigma) \leq -4\pi - \frac{Area_{g(0)}(\Sigma)}{2} \min_M R(0) \,. \tag{24.8}$$

PROOF. Let $\{p_i\}$ be the set of branch points of Σ and $b_i > 0$ the order of branching at p_i. By (24.6)

$$\frac{d}{dt}_{t=0} \mathrm{Area}_{g(t)}(\Sigma) \leq - \int_\Sigma K_\Sigma - \frac{1}{2}\int_\Sigma R = -4\pi - 2\pi \sum b_i - \frac{1}{2}\int_\Sigma R \,, \tag{24.9}$$

where the equality used the Gauss–Bonnet theorem with branch points. □

We need to recall a result on harmonic maps which gives the existence of minimal spheres realizing the width $W(g)$. The results of Sacks and Uhlenbeck give the harmonic maps but potentially allow some loss of energy. This energy loss was ruled out by Siu and Yau (using also arguments of Meeks and Yau), see Chapter VIII in [**ScYa3**]. For our purposes, the most convenient statement of this is given in theorem 4.2.1 of [**Jo**]. (The bound for the index is not stated explicitly in [**Jo**] but follows immediately as in [**MiMo**].)

PROPOSITION 24.3. *Given a metric g on M and a nontrivial $[\beta] \in \pi_1(C^0 \cap L_1^2(\mathbf{S}^2, M))$, there exists a sequence of sweep–outs $\gamma^j : [0,1] \to C^0 \cap L_1^2(\mathbf{S}^2, M)$ with $\gamma^j \in [\beta]$ so that*

$$W(g) = \lim_{j\to\infty} \max_{s\in[0,1]} Energy(\gamma_s^j) \,. \tag{24.10}$$

Furthermore, there exist $s_j \in [0,1]$ and branched conformal minimal immersions $u_0, \dots, u_m : \mathbf{S}^2 \to M$ with index at most one so that, as $j \to \infty$, the maps $\gamma^j_{s_j}$ converge to u_0 weakly in L_1^2 and uniformly on compact subsets of $\mathbf{S}^2 \setminus \{x_1, \dots, x_k\}$, and

$$W(g) = \sum_{i=0}^{m} Energy(u_i) = \lim_{j\to\infty} Energy(\gamma^j_{s_j}) \,. \tag{24.11}$$

Finally, for each $i > 0$, there exists a point x_{k_i} and a sequence of conformal dilations $D_{i,j} : \mathbf{S}^2 \to \mathbf{S}^2$ about x_{k_i} so that the maps $\gamma^j_{s_j} \circ D_{i,j}$ converge to u_i.

We will also need a standard additional property for the min–max sequence of sweep–outs γ^j of Proposition 24.3 which can be achieved by modifying the sequence as in section 4 of [**CD1**] (cf. proposition 4.1 on page 85 in [**CD1**]). Loosely speaking this is the property that any subsequence $\gamma^k_{s_k}$ with energy converging to $W(g)$ converges (after possibly going to a further subsequence) to the union of branched immersed minimal 2–spheres each with index at most one. Precisely this is that we can choose γ^j so that: Given $\epsilon > 0$, there exist J and $\delta > 0$ (both depending on g and γ^j) so that if $j > J$ and

$$\text{Energy}(\gamma_s^j) > W(g) - \delta\,, \tag{24.12}$$

then there is a collection of branched minimal 2–spheres $\{\Sigma_i\}$ each of index at most one and with

$$\text{dist}\,(\gamma_s^j, \cup_i \Sigma_i) < \epsilon\,. \tag{24.13}$$

Here, the distance means varifold distance (see, for instance, section 4 of [**CD1**]). Below we will use that, as an immediate consequence of (24.13), if F is a quadratic form on M and Γ denotes γ_s^j, then

(24.14)
$$\left| \int_\Gamma [\text{Tr}(F) - F(\mathbf{n}_\Gamma, \mathbf{n}_\Gamma)] - \sum_i \int_{\Sigma_i} [\text{Tr}(F) - F(\mathbf{n}_{\Sigma_i}, \mathbf{n}_{\Sigma_i})] \right| < C\,\epsilon\, \|F\|_{C^1}\, \text{Area}(\Gamma)\,.$$

Proof. (of Theorem 24.1) Fix a time τ. Below $\tilde{C}$ denotes a constant depending only on τ but will be allowed to change from line to line. Let $\gamma^j(\tau)$ be the sequence of sweep–outs for the metric $g(\tau)$ given by Proposition 24.3. We will use the sweep–out at time τ as a comparison to get an upper bound for the width at times $t > \tau$. The key for this is the following claim:
Given $\epsilon > 0$, there exist J and $\bar{h} > 0$ so that if $j > J$ and $0 < h < \bar{h}$, then

$$\begin{aligned} &\text{Area}_{g(\tau+h)}(\gamma_s^j(\tau)) - \max_s \text{Energy}_{g(\tau)}(\gamma_s^j(\tau)) \\ &\leq [-4\pi + \tilde{C}\,\epsilon + \frac{3}{4(\tau + C)} \max_s \text{Energy}_{g(\tau)}(\gamma_s^j(\tau))]\, h + \tilde{C}\, h^2\,. \end{aligned} \tag{24.15}$$

To see why (24.15) implies (24.4), we use the definition of the width to get

$$W(g(\tau + h)) \leq \max_{s\in[0,1]} \text{Area}_{g(\tau+h)}(\gamma_s^j(\tau))\,, \tag{24.16}$$

and then take the limit as $j \to \infty$ (so that $\max_s \text{Energy}_{g(\tau)}(\gamma_s^j(\tau)) \to W(g(\tau))$) in (24.15) to get

$$\frac{W(g(\tau + h)) - W(g(\tau))}{h} \leq -4\pi + \tilde{C}\,\epsilon + \frac{3}{4(\tau + C)}\, W(g(\tau)) + \tilde{C}\, h\,. \tag{24.17}$$

Taking $\epsilon \to 0$ in (24.17) gives (24.4).

It remains to prove (24.15). First, let $\delta > 0$ and J, depending on ϵ (and on τ), be given by (24.12)–(24.14). If $j > J$ and

$$\text{Energy}_{g(\tau)}(\gamma_s^j(\tau)) > W(g) - \delta\,, \tag{24.18}$$

then let $\cup_i \Sigma_{s,i}^j(\tau)$ be the collection of minimal spheres in (24.14). Combining (24.5), (24.14) with $F = \text{Ric}_M$, and Lemma 24.2 gives

$$\begin{aligned} \frac{d}{dt}_{t=\tau} \text{Area}_{g(t)}(\gamma_s^j(\tau)) &\leq \frac{d}{dt}_{t=\tau} \text{Area}_{g(t)}(\cup_i \Sigma_{s,i}^j(\tau)) + \tilde{C}\,\epsilon\, \|\text{Ric}_M\|_{C^1}\, \text{Area}_{g(t)}(\gamma_s^j(\tau)) \\ &\leq -4\pi - \frac{\text{Energy}_{g(\tau)}(\gamma_s^j(\tau))}{2} \min_M R(\tau) + \tilde{C}\,\epsilon \\ &\leq -4\pi + \frac{3}{4(\tau + C)} \max_s \text{Energy}_{g(\tau)}(\gamma_s^j(\tau)) + \tilde{C}\,\epsilon\,, \end{aligned} \tag{24.19}$$

where the last inequality used the lower bound (7.3) for $R(\tau)$. Since the metrics $g(t)$ vary smoothly and every sweep–out γ^j has uniformly bounded energy, it is easy to see that

$$\text{Energy}_{g(\tau+h)}(\gamma^j_s(\tau)) \tag{24.20}$$

is a smooth function of h with a uniform C^2 bound independent of both j and s near $h = 0$ (cf. (24.5)). In particular, (24.19) and Taylor expansion gives $\bar{h} > 0$ (independent of j) so that (24.15) holds for s with

$$\text{Energy}_{g(\tau)}(\gamma^j_s(\tau)) > W(g) - \delta\,. \tag{24.21}$$

In the remaining case, we have

$$\text{Energy}(\gamma^j_s(\tau)) \leq W(g) - \delta \tag{24.22}$$

so the continuity of $g(t)$ implies that (24.15) automatically holds after possibly shrinking $\bar{h} > 0$.

Finally, we claim that (24.4) implies finite extinction time. Namely, rewriting (24.4) as $\frac{d}{dt}\left(W(g(t))\,(t+C)^{-3/4}\right) \leq -4\pi\,(t+C)^{-3/4}$ and integrating gives

$$(T+C)^{-3/4}\,W(g(T)) \leq C^{-3/4}\,W(g(0)) - 16\,\pi\,\left[(T+C)^{1/4} - C^{1/4}\right]\,. \tag{24.23}$$

Since $W \geq 0$ by definition and the right hand side of (24.23) would become negative for T sufficiently large, the theorem follows. □

Part 5. Growth of harmonic functions

We next discuss some global results for harmonic functions and a few applications of function theory. We will focus on the function theory of manifolds with non–negative Ricci curvature, with the exception of Section 28 where we discuss two estimates related to nodal sets of eigenfunctions.

Recall that the classical Liouville theorem states that any bounded (or even just positive) harmonic function is constant on Euclidean space. In fact, the Euclidean gradient estimate shows that a nonconstant harmonic function must grow at least linearly. Since the partial derivatives of a Euclidean harmonic function are again harmonic, iterating this gives that on Euclidean space any harmonic function of polynomial growth is a harmonic polynomial. In particular, the dimensions of these spaces are finite on Euclidean space.

The picture gets quite a bit more complicated when we look at more general manifolds. For example, one can prescribe asymptotic values at infinity on hyperbolic space (cf. [**An3**]), so that even the space of bounded harmonic functions is infinite dimensional in this case.

However, [**CM14**] proved that each space of harmonic functions of polynomial growth is finite dimensional for manifolds with non–negative Ricci curvature. (This had been conjectured by S.T. Yau; see [**Ya1**], [**Ya2**]. The case of surfaces was settled in [**LiTa2**].). An interesting feature of [**CM14**] was that only two properties were used: a volume doubling and a Neumann Poincaré inequality, cf. 8.2.

Given an open manifold M and $d > 0$, we define the spaces of harmonic functions of polynomial growth of order at most d, $\mathcal{H}_d(M)$, using the distance function from a fixed point p:

Definition 24.4. A function u is in $\mathcal{H}_d(M)$ if u is harmonic on M and

$$|u(x)| \leq C(1 + \mathrm{dist}_M(p, x)^d)\,, \tag{24.24}$$

for some constant C and point $p \in M$.

25. Harmonic functions and spherical harmonics

It is worthwhile to recall the Euclidean case using polar coordinates (ρ, θ), where $\theta \in \mathbf{S}^{k-1}$. In these coordinates, the Laplacian is

$$\Delta_{\mathbf{R}^k} = \rho^{-2}\Delta_\theta + (k-1)\,\rho^{-1}\frac{\partial}{\partial\rho} + \frac{\partial^2}{\partial\rho^2}\,. \tag{25.1}$$

In particular, the restriction of a homogeneous harmonic polynomial of degree d to $\mathbf{S}^{k-1}$ gives an eigenfunction with eigenvalue

$$d^2 + (k-2)d\,. \tag{25.2}$$

It is then not hard to see that understanding $\mathcal{H}_d(\mathbf{R}^k)$ is a spectral problem on the compact manifold $\mathbf{S}^{k-1}$.

A similar "cone construction" holds more generally. Given a manifold N^{k-1}, the *cone* over N is the manifold $C(N) = N \times [0, \infty)$ with the metric

$$ds^2_{C(N)} = dr^2 + r^2\, ds^2_N\,. \tag{25.3}$$

Usually we identify $N \times \{0\}$ and refer to this point as the vertex. A direct computation shows that the Laplacians of N and $C(N)$ are related by the following simple formula at $(x, r) \in N \times (0, \infty)$:

$$\Delta_{C(N)} u = r^{-2}\Delta_N u + (k-1)\,r^{-1}\frac{\partial}{\partial r}u + \frac{\partial^2}{\partial r^2}u\,. \tag{25.4}$$

Using (25.4), we can now reinterpret the spaces $\mathcal{H}_d(C(N))$:

LEMMA 25.1. *If $\Delta_N g = -\lambda\, g$ on N^{k-1}, then $r^p\, g \in \mathcal{H}_p(C(N))$ where*

$$p^2 + (k-2)p = \lambda\,. \tag{25.5}$$

As a consequence of Lemma 25.1, the spectral properties of N are equivalent to properties of harmonic functions of polynomial growth on $C(N)$.

When N^{k-1} is a submanifold of $\mathbf{S}^{n-1} \subset \mathbf{R}^n$, the cone over N can be isometrically embedded in $\mathbf{R}^n$ as

$$C(N) = \{x \in \mathbf{R}^n \mid x/|x| \in N\}\,. \tag{25.6}$$

Note that $C(N)$ is then invariant under dilations about the origin. We get the following simple lemma whose proof is left for the reader:

LEMMA 25.2. *Suppose that $N^{k-1} \subset \mathbf{S}^{n-1}$. The following are equivalent:*

- *N is minimal.*
- *The Euclidean mean curvature of $N \subset \mathbf{R}^n$ is normal to $\mathbf{S}^{n-1} \subset \mathbf{R}^n$.*
- *The coordinate functions are eigenfunctions on N with eigenvalue $k-1$.*
- *The cone $C(N)$ is minimal.*

26. Manifolds with non–negative Ricci curvature

In [**Ya3**], Yau extended the classical Liouville theorem to complete manifolds M with non–negative Ricci curvature:

A positive (or bounded) harmonic function on M must be constant.

In fact, by the gradient estimate of Cheng and Yau (Theorem 8.1), any harmonic function with sublinear growth must be constant:

COROLLARY 26.1. [**CgYa**] *If M is complete with $Ric_M \geq 0$ and $d < 1$, then $\mathcal{H}_d(M) = \{$Constant functions$\}$.*

Since $\mathbf{R}^n$ has non–negative Ricci curvature and the coordinate functions are harmonic, this is obviously sharp. Therefore, when $d \geq 1$ a different approach is needed. Namely, instead of showing a Liouville theorem, the point is to control the size of the space of solutions. Over the years, there were many interesting partial results (including two proofs when M is a surface with non–negative sectional curvature, [**LiTa2**] and [**DF**]). For example, in [**LiTa1**], Li and L.F. Tam obtained the borderline case $d = 1$, showing that

$$\dim(\mathcal{H}_1(M)) \leq n + 1\,, \tag{26.1}$$

for an n-dimensional manifold with $\mathrm{Ric}_M \geq 0$. This is similar in spirit to the classical comparison theorems since $\dim(\mathcal{H}_1(\mathbf{R}^n)) = n + 1$ (the n coordinate functions plus the constant functions). This corresponding rigidity theorem was proven in [**ChCM**] (see [**Li**] for the special case where M is Kähler):

THEOREM 26.2. [**ChCM**] *If M is complete with $Ric_M \geq 0$, then every tangent cone at infinity M_∞ splits isometrically as*

$$M_\infty = N \times \mathbf{R}^{\dim(\mathcal{H}_1(M))-1}\,. \tag{26.2}$$

*Hence, if $\dim(\mathcal{H}_1(M)) = n + 1$, then [**C1**] implies that $M = \mathbf{R}^n$.*

Finally, in [**CM14**], the spaces of polynomial growth harmonic functions were shown to be finite dimensional:

THEOREM 26.3. [**CM14**] *If M is complete with $Ric_M \geq 0$, then $\mathcal{H}_d(M)$ is finite dimensional for each d.*

The proof of Theorem 26.3 consists of two independent steps (the first does not use harmonicity):

- Given a $2k$-dimensional subspace $H \subset \mathcal{H}_d(M)$ and $h \in (0, 1]$, there exists a k-dimensional subspace $K \subset H$ and $R > 0$ so that

$$\sup_{v \in K \setminus \{0\}} \frac{\int_{B_{(1+h)^2 R}} v^2}{\int_{B_R} v^2} \leq C_1\,(1 + h)^{8d}\,. \tag{26.3}$$

- We bound the dimension of a subspace K of harmonic functions satisfying (26.3) in terms of h and d.

To give some feel for the argument, we will sketch a proof of the second step.

PROOF. (Sketch of second step) For simplicity, suppose that $R = 1$ and $h = 1$. Fix a scale $r \in (0, 1)$ to be chosen small. We will use two properties of manifolds with $\mathrm{Ric}_M \geq 0$:

First, we can find $N \leq C_n \, r^{-n}$ balls $B_r(x_i)$ with

$$\chi_{B_1} \leq \sum_i \chi_{B_r(x_i)} \leq C_n \, \chi_{B_2} \, , \tag{26.4}$$

where χ_E is the characteristic function of a set E. (To do this, choose a maximal disjoint collection of balls of radius $r/2$ and then use the volume comparison to get the second inequality in (26.4) and bound N.)

Second, there is a uniform Neumann Poincaré inequality: If $\int_{B_s(x)} f = 0$, then

$$\int_{B_s(x)} f^2 \leq C_N \, s^2 \int_{B_s(x)} |\nabla f|^2 \, . \tag{26.5}$$

To bound the dimension of K, we will construct a linear map $\mathcal{M} : K \to \mathbf{R}^N$ and show that $\mathcal{M}$ is injective for $r > 0$ sufficiently small. We define $\mathcal{M}$ by

$$\mathcal{M}(v) = \left(\int_{B_r(x_1)} v \, , \cdots , \int_{B_r(x_N)} v \right) \, . \tag{26.6}$$

We will deduce a contradiction if $v \in K \setminus \{0\}$ is in the kernel of $\mathcal{M}$. In particular, (26.5) gives that for each i

$$\int_{B_r(x_i)} v^2 \leq C_N \, r^2 \int_{B_r(x_i)} |\nabla v|^2 \, . \tag{26.7}$$

Combining this with (26.4) gives

$$\int_{B_1} v^2 \leq \sum_{i=1}^N \int_{B_r(x_i)} v^2 \leq C_N \, r^2 \sum_{i=1}^N \int_{B_r(x_i)} |\nabla v|^2 \leq C_n \, C_N \, r^2 \int_{B_2} |\nabla v|^2 \, . \tag{26.8}$$

We now (for the only time) use that v is harmonic. Namely, the Caccioppoli inequality (or reverse Poincaré inequality) for harmonic functions gives

$$\int_{B_2} |\nabla v|^2 \leq \int_{B_4} v^2 \, . \tag{26.9}$$

Combining (26.8) and (26.9), we get

$$\int_{B_1} v^2 \leq C_n \, C_N \, r^2 \int_{B_4} v^2 \, . \tag{26.10}$$

This contradicts (26.3) if r is sufficiently small, completing the proof. □

On Euclidean space $\mathbf{R}^n$, the spaces $\mathcal{H}_d$ are given by harmonic polynomials of degree at most d. In particular, it is not hard to see that

$$\dim(\mathcal{H}_d(\mathbf{R}^n)) \approx C \, d^{n-1} \, . \tag{26.11}$$

Using the correspondence between harmonic polynomials and eigenfunctions on $\mathbf{S}^{n-1}$ (see Lemma 25.1), this is closely related to Weyl's asymptotic formula on $\mathbf{S}^{n-1}$. In [**CM15**], the authors proved a similar sharp polynomial bound for manifolds with non–negative Ricci curvature:

THEOREM 26.4. [**CM15**] *If M^n is complete with $Ric_M \geq 0$ and $d \geq 1$, then*

$$\dim(\mathcal{H}_d(M)) \leq C \, d^{n-1} \, . \tag{26.12}$$

Taking $M = \mathbf{R}^n$, (26.11) illustrates that the exponent $n-1$ is sharp in (26.12). However, as in Weyl's asymptotic formula, the constant in front of d^{n-1} can be related to the volume. Namely, we actually showed the stronger statement

$$\dim(\mathcal{H}_d(M)) \leq C_n \, \mathrm{V}_M \, d^{n-1} + o(d^{n-1}) \, , \tag{26.13}$$

where

- C_n depends only on the dimension n.
- V_M is the "asymptotic volume ratio" $\lim_{r\to\infty} \mathrm{Vol}(B_r)/r^n$.
- $o(d^{n-1})$ is a function of d with $\lim_{d\to\infty} o(d^{n-1})/d^{n-1} = 0$.

As noted above, Theorem 26.4 also gives lower bounds for eigenvalues on a manifold N^{n-1} with $\mathrm{Ric}_N \geq (n-2) = \mathrm{Ric}_{\mathbf{S}^{n-1}}$. Using the sharper estimate (26.13) introduces the volume of N into these eigenvalue estimates (as predicted by Weyl's asymptotic formula).

An interesting feature of these dimension estimates is that they follow from "rough" properties of M and are therefore surprisingly stable under perturbation. For instance, in [**CM14**], we actually proved Theorem 26.3 for manifolds with a volume doubling and a Neumann Poincaré inequality; unlike a Ricci curvature bound, these properties are stable under bi–Lipschitz transformations.

This finite dimensionality was not previously known even for manifolds bi–Lipschitz to $\mathbf{R}^n$ (except under additional hypotheses, cf. results of Avellenada–Lin, [**AvLn**], and Moser–Struwe, [**MrSt**]).

There are two particularly interesting directions which have not been adequately explored. The first is to develop machinery to produce harmonic functions of polynomial growth.

QUESTION 8. *Suppose that M^n has non–negative Ricci curvature and Euclidean volume growth. When can we produce harmonic functions of polynomial growth on M? The most interesting would be to solve a "Dirichlet problem at infinity," where polynomially growing harmonic functions on a tangent cone at infinity give rise to harmonic functions on M. In complete generality, this is likely to be rather delicate since these tangent cones need not be unique.*

The second direction is to get sharper dimension estimates for holomorphic functions of polynomial growth.

QUESTION 9. *If M^n is Kähler, then each holomorphic function is harmonic so Theorem 26.4 bounds the dimension of the space of polynomial growth holomorphic functions by $C\, d^{n-1}$. However, on $\mathbf{C}^{n/2}$, one get only $C\, d^{n/2}$ holomorphic functions of degree d. Does the sharper bound $C\, d^{n/2}$ hold? A stronger curvature condition may be necessary (cf.* [**Ni**] *for one such result).*

This is just a very brief overview (omitting many interesting results), but we hope that it gives something of the flavor of the subject; the interested reader may consult [**CM13**] and references therein for more information.

27. Minimal surfaces and a generalized Bernstein theorem

We next describe a similar finite dimensionality result for minimal submanifolds and an application of this – a "generalized Bernstein theorem" – proven in [**CM16**]. Recall that the Bernstein theorem implies that, through dimension seven, area–minimizing hypersurfaces are affine. A weaker form of this is true in all dimensions by the Allard regularity theorem [**Al**]:

There exists $\delta = \delta(k, n) > 0$ such that if $\Sigma^k \subset \mathbf{R}^n$ is a complete immersed minimal submanifold with

$$\frac{\text{Vol}(B_r \cap \Sigma)}{\text{Vol}(B_r \subset \mathbf{R}^k)} \leq (1 + \delta) \tag{27.1}$$

for all r, then Σ is affine.

The generalized Bernstein theorem, which should perhaps be called a generalized Allard theorem instead, shows that *any* upper bound on the density gives a corresponding upper bound for the dimension of the smallest affine subspace containing the minimal surface.

The results of this section apply to a large class of generalized minimal submanifolds $\Sigma^k \subset \mathbf{R}^n$: stationary rectifiable k–varifolds with density at least 1 a.e. on the support. This includes the case of embedded minimal submanifolds and, for simplicity, we will focus on this case below.

THEOREM 27.1. [**CM16**] *If $\Sigma^k \subset \mathbf{R}^n$ has density bounded by V_Σ, then Σ must be contained in an affine subspace of dimension at most $C_k\, V_\Sigma$.*

Another way to think of Theorem 27.1 is that it bounds the number of linearly independent coordinate functions on Σ in terms of its volume. The linear dependence on V_Σ in Theorem 27.1 is sharp; namely, any bound of the form $C_k V_\Sigma^\alpha$ must have $\alpha \geq 1$.

For a submanifold $\Sigma \subset \mathbf{R}^n$, we will define the spaces of harmonic functions of polynomial growth using the extrinsic distance; it will be clear from the context which definition we are using. Namely, given $\Sigma \subset \mathbf{R}^n$ and $d > 0$, we define the vector spaces $\mathcal{H}_d(\Sigma)$ of harmonic functions of polynomial growth by:

Definition 27.2. A function u is in $\mathcal{H}_d(\Sigma)$ if u is harmonic on Σ and

$$|u(x)| \leq C(1 + |x|^d)\,, \tag{27.2}$$

for some C. Thus, the coordinate functions x_i are in $\mathcal{H}_1(\Sigma)$.

Since the coordinate functions are harmonic on Σ (cf. Proposition 2.1), Theorem 27.1 follows from a bound for the dimensions of the spaces of harmonic functions on Σ of polynomial growth:

THEOREM 27.3. [**CM16**] *Let $\Sigma^k \subset \mathbf{R}^n$ have density bounded by V_Σ. For any $d \geq 1$,*

$$\dim \mathcal{H}_d(\Sigma) \leq C_k\, V_\Sigma\, d^{k-1}\,. \tag{27.3}$$

(The spectral properties of spherical minimal submanifolds have been studied in their own right; see, for instance, Cheng–Li–Yau [**CgLiYa**] or Choi–Wang [**CiWa**].)

27.1. Other applications of function theory. We have just seen an application of function theory to describe the geometry of the underlying space (in this case, a bound on the dimension of the space of linear growth harmonic functions controlled the complexity of the minimal submanifold). Another example is the Bochner technique and resulting topological restrictions of curvature. There are many other examples and, indeed, often these sorts of applications motivate developments in function theory. This is perhaps most evident in (one variable) complex analysis, where function theory has played a major role.

Function theory has also played an important role in the theory of quasi–regular maps, see [**G**]. Recall that a map $F : \mathbf{R}^n \to \mathbf{R}^n$ is K–quasi–regular if F and dF are in L^n and at almost every point we have

$$|dF| \leq K \det(dF)\,. \tag{27.4}$$

For instance, M. Bonk and J. Heinonen used function theoretic arguments to prove:

THEOREM 27.4. [**BoHj**] *If M is a compact n–dimensional manifold and $F : \mathbf{R}^n \to M$ is a (non–trivial) K–quasi–regular map, then the dimension of the de Rham cohomology ring of M is at most $C = C(n,K)$.*

Finally, we note that in the theory of quasi–regular maps, the most natural functions to study are no longer the harmonic ones. Rather, one is interested in $\mathcal{A}$–harmonic functions, i.e., functions u satisfying

$$\text{div}\,(\mathcal{A}(\nabla u)) = 0\,, \tag{27.5}$$

where $\mathcal{A}$ is a *nonlinear* map on the tangent space satisfying several natural conditions (e.g., taking $\mathcal{A}(x) = |x|^{p-2}\,x$ gives the so–called p–Laplacian).

28. Volumes for eigensections

Let M^n be a closed n-dimensional Riemannian manifold and V a vector bundle over M. Suppose that $\{f_i\}$ is an L^2-orthonormal set of eigensections of V of the Laplacian with eigenvalues λ_i (where $0 = \lambda_0 \leq \lambda_1 \leq \lambda_2 \leq \cdots$), that is

$$\int_M f_i\, f_j = \delta_{i,j} \text{ and } \Delta f_i + \lambda_i\, f_i = 0\,. \tag{28.1}$$

Given $a = (a_1, \cdots, a_i) \in \mathbf{S}^{i-1}$, we define a function $F_i^a(\epsilon)$ by

$$F_i^a(\epsilon) = \text{Vol}(\{x \in M \,|\, |\sum_{j=1}^{i} a_j\, f_j(x)| < \epsilon\})\,. \tag{28.2}$$

Set $F_i(\epsilon) = F_i^{e_i}(\epsilon)$ where $e_i = (\delta_{i,j})_j$ and define $F_i^A(\epsilon)$ by averaging $F_i^a(\epsilon)$ over $a \in \mathbf{S}^{i-1}$

$$F_i^A(\epsilon) = ⨍_{a \in \mathbf{S}^{i-1}} F_i^a(\epsilon)\,. \tag{28.3}$$

In this section we will discuss the answer to the following question of S.T. Yau:

> Let M^n be closed and $V = \Omega^1(M)$ the bundle of one forms on M. Then $\limsup_{i\to\infty} F_i(\epsilon)$ and $\liminf_{i\to\infty} F_i(\epsilon)$ are interesting functions of ϵ. Are they positive? Can one estimate the behavior of $\epsilon^{-n} \liminf_{i\to\infty} F_i(\epsilon)$ as $\epsilon \to 0$? The problem may be easier if we replace $F_i(\epsilon)$ by $F_i^A(\epsilon)$. We can of course consider problems for p-forms with $p > 1$.

It turns out that on a flat square torus $\liminf_{i\to\infty} F_i = 0$ on Ω^1. However as the next theorem shows then for eigenfunctions on any manifold F_i is positive. We will also see below in Theorem 28.2 that the average F_i^A for p-forms is positive and we give a sharp lower bound.

THEOREM 28.1. [**CM23**] *Let M^n be closed with*

$$Ric_M \geq -(n-1)\,. \tag{28.4}$$

There exists $C = C(n) > 0$ and $\Lambda = \Lambda(n) > 0$ such that if f is a eigenfunction of Δ on M with eigenvalue $\lambda \geq \Lambda$ and $0 < \epsilon \leq 1$, then

$$\text{(28.5)} \qquad Vol\left(\left\{x \in M \,|\, |f|^2 < \epsilon^2 \fint_M |f|^2\right\}\right) \geq C\,\epsilon^n\, Vol(M)\,.$$

It is possible to generalize Theorem 28.1 to the case where M is assumed to have the doubling property and satisfy the Neumann Poincaré inequality for $r \leq 1$. In this case, however, the exponent in ϵ may not be n but rather will depend on the doubling constant C_D and the constant C_p in the Poincaré inequality.

In contrast to F_i for eigenforms, then the next theorem shows that the average F_i^A has always a positive lower bound.

THEOREM 28.2. [**CM23**] *There exists $C = C(q)$ so that if M^n is closed and V^q is a rank q vector bundle over M with Laplace-type operator Δ_V, then for $0 < \epsilon \leq 1$ and $i > q$*

$$\text{(28.6)} \qquad F_i^A(\epsilon) \geq C\,\epsilon^q\,[Vol(M)]^{1+q/2}\,.$$

In contrast to eigenfunctions, $F_i(\epsilon)$ need not be positive for eigenforms:

Let T^2 be a flat square torus with side lengths $2\,\pi$ and define one forms by

$$\text{(28.7)} \qquad \sigma_m = \cos(m\,x_1)\,dx_1 + \sin(m\,x_1)\,dx_2\,,$$

then for all m

$$\text{(28.8)} \qquad \{x \in T^2 \,|\, |\sigma_m| < 1\} = \emptyset \text{ and hence } \liminf_{i \to \infty} F_i(\epsilon) = 0 \text{ for } \epsilon < 1/(4\pi^2)\,.$$

References

[Al] W. Allard, *On the first variation of a varifold*, Ann. of Math. (2) 95 (1972) 417–491.

[Am1] F. J. Almgren, Jr., *Some interior regularity theorems for minimal surfaces and an extension of Bernstein's theorem*, Ann. of Math. (2) 84 (1966) 277–292.

[Am2] F. J. Almgren, Jr., *Q-valued functions minimizing Dirichlet's integral and the regularity of area minimizing rectifiable currents up to codimension two*, preprint.

[An1] M. T. Anderson, *Ricci curvature bounds and Einstein metrics on compact manifolds*, J. Amer. Math. Soc. 2 (1989) 455–490.

[An2] M. T. Anderson, *Convergence and rigidity of manifolds under Ricci curvature bounds*, Invent. Math. 102 (1990) 429–445.

[An3] M. T. Anderson, *The Dirichlet problem at infinity for manifolds of negative curvature*, J. Diff. Geom. 18 (1983) 701–721.

[AvLn] M. Avellanada and F.H. Lin, *Un theoreme de Liouville pour des equations elliptiques a coefficients periodiques*, C. R. Acad. Sci. Paris 309 (1989) 245–250.

[Be] S. Bernstein, *Über ein geometrisches Theorem und seine Anwendung auf die partiellen Differentialgleichungen vom ellipschen* Typos. Math. Zeit. 26 (1927) 551–558 (translation of the original version in Comm. Soc. Math. Kharkov 2-ème sér. 15 (1915–17) 38–45).

[Ber] L. Bers, *Isolated singularities of minimal surfaces*, Ann. of Math. (2) 53 (1951) 364–386.

[BDGG] E. Bombieri, E. De Giorgi, and E. Giusti, *Minimal cones and the Bernstein problem*, Invent. Math. 7 (1969) 243–268.

[BDM] E. Bombieri, E. De Giorgi, and M. Miranda, *Una maggiorazione a priori relativa alla ipersuperfici minimali non parametriche*, Arch. Rational Mech. Anal., 32 (1969) 255–267.

[Bc] S. Bochner, *Curvature and Betti numbers, I*, Ann. of Math. (2) 49 (1948) 379–390.

[BoHj] M. Bonk and J. Heinonen, *Quasi-regular maps and cohomology*, Acta Math. 86 (2001) 219–238.

[Bu] P. Buser, *A note on the isoperimetric constant*, Ann. Scient. Ec. Norm. 15 (1982) 213–230.

[Ca] E. Calabi, *An extension of E. Hopf's maximum principle with an application to Riemannian geometry*, Duke Math. Jour. 25 (1958) 45–56.

[ChC1] J. Cheeger and T.H. Colding, *On the Structure of Spaces with Ricci Curvature Bounded Below; I*, J. Diff. Geom. 46 (1997) 406–480.

[ChC2] J. Cheeger and T.H. Colding, *On the structure of spaces with Ricci curvature bounded below. II*, J. Differential Geom. 54 (2000), no. 1, 13–35.

[ChC3] J. Cheeger and T.H. Colding, *On the structure of spaces with Ricci curvature bounded below. III*, J. Differential Geom. 54 (2000), no. 1, 37–74.

[ChC4] J. Cheeger and T.H. Colding, *Lower bounds on Ricci curvature and the almost rigidity of warped products.* Ann. of Math. (2) 144 (1996), no. 1, 189–237.

[ChCM] J. Cheeger, T.H. Colding, and W.P. Minicozzi II, *Linear growth harmonic functions on complete manifolds with nonnegative Ricci curvature*, Geom. Funct. Anal. 5 (1995), no. 6, 948–954.

[ChCTi] J. Cheeger, T.H. Colding, and G. Tian, *On the singularities of spaces with bounded Ricci curvature*, Geom. Funct. Anal. 12 (2002), no. 5, 873–914.

[CgLiYa] S.Y. Cheng, P. Li, and S.T. Yau, *Heat equations on minimal submanifolds and their applications*, Amer. J. Math. 106 (1984) 1033–1065.

[CgYa] S.Y. Cheng and S.T. Yau, *Differential equations on Riemannian manifolds and their geometric applications*, Comm. Pure Appl. Math. 28 (1975) 333–354.

[CiSc] H.I. Choi and R. Schoen, *The space of minimal embeddings of a surface into a three-dimensional manifold of positive Ricci curvature*, Invent. Math. 81 (1985) 387–394.

[CiWa] H.I. Choi and A.N. Wang, *A first eigenvalue estimate for minimal hypersurfaces*, J. Differential Geom. 18 (1983) 559–562.

[C1] T.H. Colding, *Ricci curvature and volume convergence.* Ann. of Math. (2) 145 (1997), no. 3, 477–501.

[C2] T.H. Colding, *Spaces with Ricci curvature bounds.* Proceedings of the International Congress of Mathematicians, Vol. II (Berlin, 1998). Doc. Math. 1998, Extra Vol. II, 299–308.

[CD1] T.H. Colding and C. De Lellis, *The min–max construction of minimal surfaces*, Surveys in differential geometry, Vol. 8, Lectures on Geometry and Topology held in honor of Calabi, Lawson, Siu, and Uhlenbeck at Harvard University, May 3–5, 2002, Sponsored by the Journal of Differential Geometry, (2003) 75–107, math.AP/0303305.

[CD2] T.H. Colding and C. De Lellis, *Singular limit laminations, Morse index, and positive scalar curvature*, math.DG/0208100.

[CH1] T.H. Colding and N. Hingston, *Metrics without Morse index bounds*, Duke Math. Jour., vol. 119, issue 2. (2003) 345–365.

[CH2] T.H. Colding and N. Hingston, *Geodesic laminations with closed ends on surfaces and Morse index; Kupka–Smale metrics*, math.DG/0208133.

[CH3] T.H. Colding and N. Hingston, in preparation.

[CM1] T.H. Colding and W.P. Minicozzi II, *Minimal surfaces, Courant Lecture Notes in Math.*, v. 4, 1999.

[CM2] T.H. Colding and W.P. Minicozzi II, *Estimates for parametric elliptic integrands*, International Mathematics Research Notices, no. 6 (2002) 291-297.

[CM3] T.H. Colding and W.P. Minicozzi II, *The space of embedded minimal surfaces of fixed genus in a 3-manifold I; Estimates off the axis for disks*, Ann. of Math., to appear, math.AP/0210106.

[CM4] T.H. Colding and W.P. Minicozzi II, *The space of embedded minimal surfaces of fixed genus in a 3-manifold II; Multi-valued graphs in disks*, Ann. of Math., to appear, math.AP/0210086.

[CM5] T.H. Colding and W.P. Minicozzi II, *The space of embedded minimal surfaces of fixed genus in a 3-manifold III; Planar domains*, Ann. of Math., to appear, math.AP/0210141.

[CM6] T.H. Colding and W.P. Minicozzi II, *The space of embedded minimal surfaces of fixed genus in a 3-manifold IV; Locally simply connected*, Ann. of Math., to appear, math.AP/0210119.

[CM7] T.H. Colding and W.P. Minicozzi II, *The space of embedded minimal surfaces of fixed genus in a 3-manifold V; Fixed genus*, in preparation.

[CM8] T.H. Colding and W.P. Minicozzi II, *Multi-valued minimal graphs and properness of disks*, International Mathematics Research Notices, no. 21 (2002) 1111-1127.

[CM9] T.H. Colding and W.P. Minicozzi II, *On the structure of embedded minimal annuli*, International Mathematics Research Notices, no. 29 (2002) 1539–1552.

[CM10] T.H. Colding and W.P. Minicozzi II, *Minimal annuli with and without slits*, Jour. of Symplectic Geometry, vol. 1, issue 1 (2001) 47–61.

[CM11] T.H. Colding and W.P. Minicozzi II, *Harmonic functions with polynomial growth*, J. Diff. Geom., v. 46, no. 1 (1997) 1–77.

[CM12] T.H. Colding and W.P. Minicozzi II, *Sharp estimates for mean curvature flow of graphs*, Crelle's Journal, to appear, math.AP/0305099.

[CM13] T.H. Colding and W.P. Minicozzi II, *Generalized Liouville properties of manifolds*, Math. Res. Lett., 3 (1996) 723–729.

[CM14] T.H. Colding and W.P. Minicozzi II, *Harmonic functions on manifolds*, Ann. of Math. (2), 146, no. 3 (1997) 725–747.

[CM15] T.H. Colding and W.P. Minicozzi II, *Weyl type bounds for harmonic functions*, Inventiones Math., 131 (1998) 257–298.

[CM16] T.H. Colding and W.P. Minicozzi II, *Liouville theorems for harmonic sections and applications*, Comm. Pure Appl. Math., 52 (1998) 113–138.

[CM17] T.H. Colding and W.P. Minicozzi II, *Disks that are double spiral staircases*, Notices of the AMS, Vol. 50, no. 3, March (2003) 327–339.

[CM18] T.H. Colding and W.P. Minicozzi II, *Embedded minimal disks: Proper versus nonproper - global versus local*, Transactions of the AMS, to appear. math.DG/0210328.

[CM19] T.H. Colding and W.P. Minicozzi II, *Estimates for the extinction time for the Ricci flow on certain 3–manifolds and a question of Perelman*, math.AP/0308090.

[CM20] T.H. Colding and W.P. Minicozzi II, *Examples of embedded minimal tori without area bounds*, International Mathematics Research Notices, 99 no. 20 (1999) 1097–1100.

[CM21] T.H. Colding and W.P. Minicozzi II, *Complete properly embedded minimal surfaces in* $\mathbf{R}^3$, Duke Math. J. 107 (2001) 421–426.

[CM22] T.H. Colding and W.P. Minicozzi II, *Embedded minimal disks*, Minimal surfaces (MSRI, 2001), ed. D. Hoffman, Clay Mathematics Proceedings, AMS, Providence (2004), 405–438, math.DG/0206146.

[CM23] T.H. Colding and W.P. Minicozzi II, *Volumes for eigensections*, Geom. Ded. 102 (2003) 19–24.

[Co] P. Collin, *Topologie et courbure des surfaces minimales proprement plonges de* $\mathbf{R}^3$, Ann. of Math. (2) 145 (1997) 1–31.

[DaKeTo] G. David, C. Kenig, and T. Toro, *Asymptotically optimally doubling measures and Reifenberg flat sets with vanishing constant.* Comm. Pure Appl. Math. 54 (2001), no. 4, 385–449.

[DaSe] G. David and S. Semmes, *Uniform rectifiability and quasiminimizing sets of arbitrary codimension*, Mem. Amer. Math. Soc. 144 (2000), no. 687.

[DaTo] G. David, and T. Toro, *Reifenberg flat metric spaces, snowballs, and embeddings.* Math. Ann. 315 (1999), no. 4, 641–710.

[DG] E. De Giorgi, *Frontiere orientate di misura minima*, Sem. Mat. Scuola Norm. Sup. Pisa (1961) 1–56.

[De] B. Dean, *Compact Embedded Minimal Surfaces of Positive Genus Without Area Bounds*, Geom. Ded. 102 (2003), 45–52.

[DF] H. Donnelly and C. Fefferman, *Nodal domains and growth of harmonic functions on noncompact manifolds*, J. Geom. Anal. 2 (1992) 79-93.

[Do] R.T. Dong, *Nodal sets of eigenfunctions on Riemann surfaces*, J. Diff. Geom., 36 (1992) 493–506.

[E1] K. Ecker, *Lectures on regularity for mean curvature flow*, preprint.

[E2] K. Ecker, *A local monotonicity formula for mean curvature flow*, Ann. of Math. (2), 154 (2001) 503–525.

[EH1] K. Ecker and G. Huisken, *Mean curvature evolution of entire graphs*, Ann. of Math. (2), 130 (1989) 453–471.

[EH2] K. Ecker and G. Huisken, *Interior estimates for hypersurfaces moving by mean curvature*, Invent. Math., 105 (1991) 547–569.

[EvSp] L.C. Evans and J. Spruck, *Motion of level sets by mean curvature I*, J. Diff. Geom., 33 (1991) 635–681.

[Fe] H. Federer, *Geometric measure theory*, Springer-Verlag, Berlin–Heidelberg–New York, 1969.

[Fi] R. Finn, *On equations of minimal surface type*, Ann. of Math. (2), 60 (1954) 397–416.

[FHS] M.H. Freedman, J. Hass, and P. Scott, *Least area incompressible surfaces in 3-manifolds*, Invent. Math. 71 (1983), no. 3, 609–642.

[Fr] A. Fraser, *On the free boundary variational problem for minimal disks*, Comm. Pure Appl. Math. 53 (2000) 931–971.

[Ga] S. Gallot, *Volumes, courbure de Ricci et convergence des varits (d'aprs T. H. Colding et Cheeger-Colding)*. Sminaire Bourbaki. Vol. 1997/98. Astrisque No. 252, (1998), Exp. No. 835, 3, 7–32.

[GiTr] D. Gilbarg and N. Trudinger, *Elliptic Partial Differential Equations of Second Order*, Springer Verlag, 2nd Ed., (1983).

[Gr] A. Grigor'yan, *The heat equation on noncompact Riemannian manifolds*, English translation in Math. USSR Sb. 72 (1) (1992) 47–77.

[G] M. Gromov, *Metric structures for Riemannian and non-Riemannian spaces* (With appendices by M. Katz, P. Pansu and S. Semmes) Progress in Mathematics, 152. Birkhuser Boston, Inc., Boston, MA (1999).

[GLPa] M. Gromov, J. Lafontaine, and P. Pansu, *Structures métriques pour les variétés riemanniennes*, Cedid/Fernand Nathan, Paris (1981).

[GuSp] R. Gulliver and J. Spruck, *On embedded minimal surfaces*, Ann. of Math. (2) 103 (1976) 331–347; Ann. of Math. (2) 109 (1979) 407–412.

[HzKo] P. Hajlasz and P. Koskela, *Sobolev meets Poincare*, C. R. Acad. Sci. Paris S. I Math. 320 (1995), no. 10, 1211–1215.

[HnLn] Q. Han and F.H. Lin, *Elliptic Partial Differential Equations*, Courant Lecture Notes in Math., v. 1, 1997.

[Ha1] R. Hamilton, *A matrix Harnack estimate for the heat equation*, Comm. Anal. Geom., 1 (1993) 113–126.

[Ha2] R. Hamilton, *Three–manifolds with positive Ricci curvature*, J. Diff. Geom. 17 (1982) 255–306.

[Ha3] R. Hamilton, *The formation of singularities in the Ricci flow*, Surveys in differential geometry, Vol. II (Cambridge, MA, 1993), International Press, Cambridge, MA, 1995, 7–136.

[HaNoRu] J. Hass, P. Norbury, and J.H. Rubinstein, *Minimal spheres of arbitrary high Morse index*, Comm. Anal. Geometry, to appear.

[Hj] J. Heinonen, *Lectures on analysis on metric spaces*, Springer-Verlag (2001).

[He] E. Heinz, *Über die Lösungen der Minimalflächengleichung*, Nachr. Akad. Wiss. Göttingen Math.-Phys. Kl, II (1952) 51–56.

[HoK] D. Hoffman and H. Karcher, *Complete embedded minimal surfaces with finite total curvature*, Geometry V (R. Osserman, ed.) Encyclopaedia Math. Sci. 90, Springer-Verlag, New York (1997) 5–93.

[HoMe] D. Hoffman and W. Meeks III, *A complete embedded minimal surface in* $\mathbf{R}^3$ *with genus one and three ends*, J. Diff. Geom. 21 (1985) 109–127.

[HoWeWo] D. Hoffman, M. Weber, and M. Wolf, *An embedded genus-one helicoid*, math.DG/0401080.

[H] G. Huisken, *Asymptotic behavior for singularities of the mean curvature flow*, J. Diff. Geom., 31 (1990) 285–299.

[I] T. Ilmanen, *Elliptic regularization and partial regularity for motion by mean curvature*, Mem. Amer. Math. Soc. 108 (1994), no. 520.

[Je] D. Jerison, *The Poincaré inequality for vector fields satisfying Hörmander's condition*, Duke Math. J. 53 (1986) 503–523.

[JXa] L. Jorge and F. Xavier, *A complete minimal surface in* $\mathbf{R}^3$ *between two parallel planes*, Ann. of Math. (2) 112 (1980) 203–206.

[Jo] J. Jost, *Two-dimensional geometric variational problems*, J. Wiley and Sons, Chichester, N.Y. (1991).

[Ka] N. Kapouleas, *Complete embedded minimal surfaces of finite total curvature*, J. Diff. Geom., 47 (1997) 95–169.

[K] N. Korevaar, *An easy proof of the interior gradient bound for solutions to the prescribed mean curvature equation*, Nonlinear functional analysis and its applications, Part 2, Proc. of Symposia in Pure Math. 45 (1986) 81–89.

[Li] P. Li, *Linear growth harmonic functions on Kähler manifolds with non–negative Ricci curvature*, Math. Res. Lett. 2 (1995) 79–94.

[LiTa1] P. Li and L.F. Tam, *Linear growth harmonic functions on a complete manifold*, J. Diff. Geom. 29 (1989) 421–425.

[LiTa2] P. Li and L.F. Tam, *Complete surfaces with finite total curvature*, J. Diff. Geom. 33 (1991) 139–168.

[LiYa] P. Li and S.T. Yau, *On the parabolic kernel of the Schrödinger operator*, Acta Math. 156 (1986) 153–201.

[Lc] A. Lichnerowicz, *Geometrie des Groupes des Transformationes*, Dunod, Paris (1958).

[Ln] F.H. Lin, *Gradient estimates and blow-up analysis for stationary harmonic maps*, Ann. of Math. (2) 149 (1999), no. 3, 785–829.

[Me1] W. Meeks III, *The regularity of the singular set in the Colding and Minicozzi lamination theorem*, Duke Math. Jour., to appear.

[Me2] W. Meeks III, *The lamination metric for the Colding-Minicozzi minimal lamination*, preprint.

[MeP] W. Meeks III and J. Perez, *Conformal properties in classical minimal surface theory*, preprint.

[MePRs1] W. Meeks III, J. Perez, and A. Ros, *The geometry of minimal surfaces of finite genus I; Curvature estimates and quasiperiodicity*, preprint.

[MePRs2] W. Meeks III, J. Perez, and A. Ros, *The geometry of minimal surfaces of finite genus II; Nonexistence of one limit end examples*, preprint.

[MePRs3] W. Meeks III, J. Perez, and A. Ros, *The geometry of minimal surfaces of finite genus III; bounds on the topology and index of classical minimal surfaces*, preprint.

[MeRo] W. Meeks III and H. Rosenberg, *The uniqueness of the helicoid and the asymptotic geometry of properly embedded minimal surfaces with finite topology*, preprint.

[MeSiYa] W. Meeks III, L. Simon, and S.T. Yau, *Embedded minimal surfaces, exotic spheres and manifolds with positive Ricci curvature*, Ann. of Math. (2) 116 (1982) 621–659.

[MeWe] W. Meeks III and M. Weber, *Existence of bent helicoids and the regularity of the singular set in the Colding-Minicozzi lamination theorem*, in preparation.

[MeYa1] W. Meeks III and S.T. Yau, *The classical Plateau problem and the topology of three dimensional manifolds*, Topology 21 (1982) 409–442.

[MeYa2] W.H. Meeks and S.T. Yau, *Topology of three-dimensional manifolds and the embedding problems in minimal surface theory*, Ann. of Math. (2) 112 (1980), no. 3, 441–484.

[MiMo] M.J. Micallef and J.D. Moore, *Minimal two–spheres and the topology of manifolds with positive curvature on totally isotropic two–planes*, Ann. of Math. (2) 127 (1988) no. 1 199–227.

[MrSt] J. Moser and M. Struwe, *On a Liouville–type theorem for linear and nonlinear elliptic differential equations on a torus*, Bol. Soc. Bra. Mat. 23 (1992) 1–20.

[Na] N. Nadirashvili, *Hadamard's and Calabi-Yau's conjectures on negatively curved and minimal surfaces*, Invent. Math. 126 (1996) 457–465.

[Ni] L. Ni, *A monotonicity formula on complete Kähler manifolds with nonnegative bisectional curvature math*, DG/0307275.

[Ob] M. Obata, *Certain conditions for a Riemannian manifold to be isometric to a sphere*, J. Math. Soc. Japan 14 (1960) 333–340.

[Os] R. Osserman, *A survey of minimal surfaces*, Dover, 2nd. edition (1986).

[Pe1] G. Perelman, *The entropy formula for the Ricci flow and its geometric applications*, math.DG/0211159.

[Pe2] G. Perelman, *Ricci flow with surgery on three–manifolds*, math.DG/0303109.

[Pe3] G. Perelman, *Finite extinction time for the solutions to the Ricci flow on certain three–manifolds*, math.DG/0307245.

[Pe4] G. Perelman, *A complete Riemannian manifold of positive Ricci curvature with Euclidean volume growth and nonunique asymptotic cone.* Comparison geometry (Berkeley, CA, 1993–94), 165–166, Math. Sci. Res. Inst. Publ., 30, Cambridge Univ. Press, Cambridge, 1997.

[Pi] J.T. Pitts, *Existence and regularity of minimal surfaces on Riemannian manifolds*, Princeton University Press, Princeton, N.J.; University of Tokyo Press, Tokyo (1981).

[PiRu] J.T. Pitts and J.H. Rubinstein, *Applications of minmax to minimal surfaces and the topology of 3–manifolds*, Miniconference on geometry and partial differential equations, Proceedings of the CMA, Australia National University (1986).

[Pr] P. Price, *A monotonicity formula for Yang-Mills fields. Manuscripta Math*, 43 (1983), no. 2-3, 131–166.

[Re] E.R. Reifenberg, *Solution of the Plateau Problem for m-dimensional surfaces of varying topological type.* Acta Math. 104 1960 1–92.

[Ro] H. Rosenberg, *Some recent developments in the theory of properly embedded minimal surfaces in* $\mathbf{R}^3$, Seminare Bourbaki 1991/92, Asterisque No. 206 (1992) 463–535.

[SaUh] J. Sacks and K. Uhlenbeck, *The existence of minimal immersions of 2–spheres*, Ann. of Math. (2) 113 (1981) no. 1, 1–24.

[SC] L. Saloff-Coste, *Uniformly elliptic operators on Riemannian manifolds*, J. Diff. Geom. 36 (1992) 417–450.

[Sc1] R. Schoen, *Estimates for stable minimal surfaces in three–dimensional manifolds*, In Seminar on Minimal Submanifolds, Ann. of Math. Studies, vol. 103, Princeton University Press, Princeton, N.J., (1983) 111–126.

[Sc2] R. Schoen, *Analytic aspects of the harmonic map problem*, MSRI publ. 2, Springer, New York-Berlin, (1984) 321–358.

[ScSi] R. Schoen and L. Simon, *Regularity of simply connected surfaces with quasi-conformal Gauss map*, In Seminar on Minimal Submanifolds, Annals of Math. Studies, vol. 103, Princeton University Press, Princeton, N.J., (1983) 127–145.

[ScSiYa] R. Schoen, L. Simon, and S.T. Yau, *Curvature estimates for minimal hypersurfaces*, Acta Math. 134 (1975) 275–288.

[ScUh] R. Schoen and K. Uhlenbeck, *A regularity theory for harmonic maps*, J. Diff. Geom. 17 (1982) 307–335.

[ScYa1] R. Schoen and S.T. Yau, *Lectures on differential geometry*, International Press 1994.

[ScYa2] R. Schoen and S.T. Yau, *Existence of incompressible minimal surfaces and the topology of three dimensional manifolds with nonnegative scalar curvature*, Ann. of Math. (2) 110 (1979) 127–142.

[ScYa3] R. Schoen and S.T. Yau, *Lectures on harmonic maps*, International Press 1997.

[Se] S. Semmes, *Chord-arc surfaces with small constant II: Good parameterizations*, Adv. Math. 88 (1991) 170–199.

[Si1] L. Simon, *Asymptotic behaviour of minimal graphs over exterior domains*, Ann. Inst. H. Poincar Anal. Non Linaire 4 (1987) 231–242.

[Si2] L. Simon, *Remarks on curvature estimates for minimal hypersurfaces*, Duke Math. J. 43 (1976) 545–553.

[Si3] L. Simon, *Singularities of Geometric Variational Problems*, In Nonlinear Partial Differential Equations in Differential Geometry (R. Hardt and M. Wolf, Ed.), American Mathematical Society, Providence (1996) 185–223.

[Sim] J. Simons, *Minimal varieties in Riemannian manifolds*, Ann. of Math. (2) 88 (1968) 62–105.

[Sm] F. Smith, *On the existence of embedded minimal 2–spheres in the 3–sphere, endowed with an arbitrary Riemannian metric*, supervisor L. Simon, University of Melbourne (1982).

[St] M. Struwe, *Geometric evolution problems*, In Nonlinear Partial Differential Equations in Differential Geometry (R. Hardt and M. Wolf, Ed.), American Mathematical Society, Providence (1996) 259–339.

[To] T. Toro, *Geometric conditions and existence of bi-Lipschitz parameterizations.* Duke Math. J. 77 (1995), no. 1, 193–227.

[Tr] M. Traizet, *Adding handles to Riemann's minimal surfaces*, J. Inst. Math. Jussieu 1 (2002) 145–174.

[WeWo] M. Weber and M. Wolf, *Teichmüller theory and handle addition for minimal surfaces*, Ann. of Math. (2), 156 (2002) 713–795.

[W] B. White, *Evolution of curves and surfaces by mean curvature*, Proceedings of the ICM (2002).

[Y] S.D. Yang, *A connected sum construction for complete minimal surfaces of finite total curvature*, Comm. Anal. Geom. 9 (2001), no. 1, 115–167.

[Ya1] S.T. Yau, *Nonlinear analysis in geometry*, L'Eseignement Mathematique (2) 33 (1987) 109–158.

[Ya2] S.T. Yau, *Open problems in geometry*, Proc. Sympos. Pure Math., 54, Part 1, American Mathematical Society, Providence, 1993.

[Ya3] S.T. Yau, *Harmonic functions on complete Riemannian manifolds*, Comm. Pure Appl. Math. 28 (1975) 201–228.

Courant Institute of Mathematical Sciences, 251 Mercer Street, New York, NY 10012

Department of Mathematics, Johns Hopkins University, 3400 N. Charles St., Baltimore, MD 21218

E-mail address: colding@cims.nyu.edu and minicozz@math.jhu.edu

Surveys in Differential Geometry
IX, International Press

Eigenvalues of elliptic operators and geometric applications

Alexander Grigor'yan, Yuri Netrusov, and Shing-Tung Yau

Contents

1. Introduction

Eigenvalues and capacitors. Let X be a Riemannian manifold and Δ be the Laplace operator on X. It is well-known that if X is compact then the spectrum of $-\Delta$ is discrete and consists of an increasing sequence $\{\lambda_k\}_{k=1}^{\infty}$ of the eigenvalues (counted with the multiplicities) where $\lambda_1 = 0$ and $\lambda_k \to \infty$ as $k \to \infty$. Moreover, if $n = \dim X$ then Weyl's asymptotic formula says that

$$\lambda_k \sim c_n \left(\frac{k}{\mu(X)}\right)^{2/n}, \quad k \to \infty, \tag{1.1}$$

where μ is the Riemannian measure on X and $c_n > 0$ is a constant depending only on n.

In this paper we develop a method of obtaining *upper* bounds for eigenvalues via *capacities.* A capacitor on X is a couple (F, G) of Borel sets in X such that $F \subset G$. The *capacity* of the capacitor (F, G) is defined by

$$\operatorname{cap}(F, G) = \inf_{\varphi \in \mathcal{T}} \int_X |\nabla \varphi|^2 \, d\mu \tag{1.2}$$

where $\mathcal{T} = \mathcal{T}(F, G)$ is the class of test function, which consists of all functions $\varphi \in C_0^\infty(X)$ such that $\operatorname{supp} \varphi \subset \overset{o}{G}$ and $\varphi \equiv 1$ in a neighborhood of F.

The second author was supported by an EPSRC Advanced Research Fellowship AF/97/1657.
The third author was supported in part by NSF Grant No. DMS 95-04834.

Assume that there exist k *disjoint* capacitors (F_i, G_i) in X (that is, the sets G_i are disjoint) satisfying the following properties, for all $i = 1, 2, ..., k$, and for some positive constants υ and $\varkappa$:

(a) $\mu(F_i) \geq \upsilon$;
(b) $\operatorname{cap}(F_i, G_i) \leq \varkappa$.

Then, for any $\varepsilon > 0$, there exists a test function φ_i of the capacitor (F_i, G_i) such that

$$\int_X |\nabla \varphi_i|^2 \, d\mu \leq \varkappa + \varepsilon.$$

The condition (a) implies

$$\int_X \varphi_i^2 d\mu \geq \mu(F_i) \geq \upsilon,$$

whence

$$\int_X |\nabla \varphi_i|^2 \, d\mu \leq \Lambda \int_X \varphi_i^2 d\mu,$$

where $\Lambda = \frac{\varkappa + \varepsilon}{\upsilon}$. Since the supports of the functions φ_i are disjoint, it follows that

$$\int_X |\nabla \varphi|^2 \, d\mu \leq \Lambda \int_X \varphi^2 d\mu \tag{1.3}$$

for any $\varphi \in \mathcal{V} := \operatorname{span}(\varphi_1, ..., \varphi_k)$. Note that $\mathcal{V}$ is a k-dimensional subspace of $C_0^\infty(X)$. The fact that any function $\varphi \in \mathcal{V}$ satisfies (1.3) implies that $\lambda_k \leq \Lambda$. Since this is true for any $\varepsilon > 0$, we obtain that the hypotheses (a) and (b) imply that

$$\lambda_k \leq \frac{\varkappa}{\upsilon}.$$

Clearly, (a) implies $\upsilon \leq \frac{\mu(X)}{k}$. Under certain conditions, one can hope to get k disjoint capacitors satisfying (a) and (b) with $\upsilon = c\frac{\mu(X)}{k}$, where c is a small enough positive constant. In this case one obtains the following upper bound for λ_k:

$$\lambda_k \leq \frac{\varkappa}{c} \frac{k}{\mu(X)}. \tag{1.4}$$

Comparison with (1.1) shows that if $\varkappa$ is independent of k and the dimension n is equal to 2 then the estimate (1.4) is sharp up to a constant factor.

This approach was successfully used by Korevaar [**37**] in the proof of the following result. Let $X = (\Sigma_\gamma, g)$ where Σ_γ is the oriented compact Riemann surface of genus γ and g is an arbitrary Riemannian metric on Σ_γ. Then the eigenvalues of the Laplace operator on X admit for all $k \geq 1$ the estimate[1]

$$\lambda_k \leq C(\gamma + 1) \frac{k}{\mu(X)}, \tag{1.5}$$

where C is an *absolute* constant. Note that the metric g is involved in the estimate (1.5) only through the total volume $\mu(X)$.

[1]Since $\lambda_1 = 0$, by changing the constant C in (1.5) one obtains

$$\lambda_k \leq C(\gamma + 1) \frac{k-1}{\mu(X)}.$$

The same applies to estimates (1.14 and (1.16) below. However, the distinction between k and $k-1$ is marginal for our purpose.

For λ_2, Yang and the third author [**56**] proved earlier a sharper estimate[2]

$$\lambda_2 \leq \frac{8\pi (\gamma + 1)}{\mu (X)}. \tag{1.6}$$

The estimate (1.5) was stated in [**57**] as a conjecture, which was eventually settled by Korevaar.

Both works [**37**] and [**56**] have used the fact that there is a *conformal* mapping $T : X \to \mathbb{S}^2$ of degree at most $\gamma + 1$. Let μ^* be the measure on $\mathbb{S}^2$, which is the T-pullback of the measure μ on X, and let cap* be the capacity on $\mathbb{S}^2$ associated with the *standard* Riemannian metric of $\mathbb{S}^2$. Assume that there exist k disjoint capacitors (F_i^*, G_i^*) on $\mathbb{S}^2$ satisfying the following conditions, for all $i = 1, 2, ..., k$:

(a') $\mu^* (F_i^*) \geq \upsilon := c\frac{\mu^*(\mathbb{S}^2)}{k}$;
(b') $\text{cap}^*(F_i^*, G_i^*) \leq C$,

where c and C are positive absolute constants. Taking then $F_i = T^{-1}(F_i^*)$ and $G_i = T^{-1}(G_i^*)$, we obtain k disjoint capacitors (F_i, G_i) on X satisfying (a) with $\upsilon = c\frac{\mu(X)}{k}$. In fact, they also satisfy (b) with $\varkappa = C(\gamma + 1)$, which follows from the fact that the Dirichlet integral is locally preserved by the mapping T, and the degree of T is at most $\gamma + 1$. Substituting these values of υ and $\varkappa$ into (1.4), we obtain (1.5).

Since the measure μ^* on $\mathbb{S}^2$ is a pull-back of a measure on X, one may not have enough control over how μ^* is distributed on $\mathbb{S}^2$, contrary to the capacity cap*, which is related to the standard metric on $\mathbb{S}^2$. Nevertheless, suppose for a moment, that one can find k geodesic balls B_i on $\mathbb{S}^2$ such that

$$\mu^* (B_i) \geq \upsilon, \tag{1.7}$$

and the balls $2B_i$ are disjoint (here $2B$ denotes the ball with the same center as B and with the radius equal to twice the radius of B). It is easy to show that the capacity $\text{cap}^*(B, 2B)$ admits an upper bound by an absolute constant C (see inequality (5.7) in the proof of Theorem 5.3). Hence, in this case, both conditions (a') and (b') are satisfied for the capacitors $(B_i, 2B_i)$.

However, in general one may not find k disjoint balls on $\mathbb{S}^2$ with the property (1.7). Korevaar introduced a very ingenious and extremely elaborate method for choosing more complicated sets to be used for the capacitors in questions. His argument was designed to spot the places of concentration of measure μ^* on $\mathbb{S}^2$, while still having control over the corresponding capacities. This method was further developed by two of the authors [**27**] in the setting of abstract measure metric spaces.[3]

Decomposition of a metric measure space by annuli. In this paper we present a new, significantly simpler method of constructing the capacitors satisfying the above properties (a) and (b). Let (X, d) be a metric space. By an *annulus* in X we mean any set $A \subset X$ of the following form

$$A = \{x \in X : r \leq d(x, a) < R\}, \tag{1.8}$$

[2]The constant 8π in (1.6) is sharp for the case $\gamma = 0$, that is for a sphere , but not in general. The sharp value for a torus was obtained by Nadirashvili [**46**]. For non-oriented surfaces, the sharp estimate for λ_2 of a projective plane was obtained by Li and Yau [**40**], and for a Klein bottle – by Jakobson, Nadirashvili, and Polterovich [**34**].

[3]For another general method of obtaining upper bounds for higher eigenvalues see [**8**], [**9**]. For lower bounds of eigenvalues via capacities see [**23**], [**42**, Theorem 2.3.2/1].

where $a \in X$ and $0 \leq r < R < \infty$ (in particular, if $r = 0$ then A is the ball $B(x, R)$). Also, denote by $2A$ the following annulus:

$$2A = \{x \in X : \frac{1}{2}r \leq d(x, a) < 2R\}.$$

The following theorem is a key result that underpins the above approach for constructing disjoint capacitors and estimating of the eigenvalues.

Theorem 1.1. (=Corollary 3.12) *Let (X, d) be a metric space satisfying the following covering property: there exists a constant N such that any metric ball of radius r in X can be covered by at most N balls of radii $r/2$. Let all metric balls in X be precompact sets, and let ν be a non-atomic Radon measure on X. Then, for any positive integer k, there exists a sequence $\{A_i\}_{i=1}^k$ of k annuli in X such that, for any $i = 1, 2, ..., k$,*

$$\nu(A_i) \geq c\frac{\nu(X)}{k},$$

and the annuli $2A_i$ are disjoint. Here c is a positive constant depending only on N.

If in addition one has a properly defined capacity cap on X such that $\text{cap}(B, 2B) \leq Q$ for any ball B in X then one can show that $\text{cap}(A, 2A) \leq 4Q$ for any annulus A in X. Thus, applying Theorem 1.1 to $X = \mathbb{S}^2$ with the measure $\nu = \mu^*$ we obtain k capacitors $(A_i, 2A_i)$ satisfying (a') and (b'), thus giving a new proof of the theorem of Korevaar (see Corollary 4.11 and Theorem 5.4 for more details).

A modification of the above method can be used to estimate the eigenvalues of a Schrödinger type operator. Let X be as above a Riemannian manifold, and fix a function $q \in L^1_{loc}(X, \mu)$. Consider the operator

$$L = -\Delta - q$$

and the associated energy form

$$\mathcal{E}_q[f] := \int_X f\, Lf\, d\mu = \int_X \left(|\nabla f|^2 - qf^2\right) d\mu,$$

defined for all $f \in \mathcal{F} := C_0^\infty(X)$. For any real λ, define the *counting function* $\mathcal{N}_\lambda(L)$ as the supremum of the dimensions of all vector spaces $\mathcal{V} \subset \mathcal{F}$ such that

$$\mathcal{E}_q[f] < \lambda \|f\|^2_{L^2(X,\mu)} \quad \text{for all } f \in \mathcal{V},\ f \not\equiv 0.$$

If the operator L with the domain $\mathcal{F}$ admits the Friedrichs extension to a self-adjoint operator in $L^2(X, \mu)$ (also denoted by L), then

$$\mathcal{N}_\lambda(L) = \dim \operatorname{Im} \mathbf{1}_{(-\infty,\lambda)}(L).$$

In particular, if the spectrum of L below λ is discrete then $\mathcal{N}_\lambda(L)$ is just the number of the eigenvalues of L, which are smaller than λ, counted with the multiplicity. In the case $\lambda = 0$ we will also use the notation

$$Neg(L) = \mathcal{N}_0(L).$$

Denote by $B(x, r)$ the geodesic ball on X of radius r centered at $x \in X$. The following result is a particular case of Theorem 5.3.

Theorem 1.2. *Let X be a complete Riemannian manifold. Assume that, for some constants N and M, the following is true:*

(*i*) *any ball $B(x, r)$ in X can be covered by at most N balls of radii $r/2$;*

(ii) for all $x \in X$ and $r > 0$,

$$\mu(B(x,r)) \leq Mr^2. \tag{1.9}$$

Then, for any function $q \in L^1_{loc}(X,\mu)$,

$$Neg(-\Delta - q) \geq c \int_X q_\delta d\mu, \tag{1.10}$$

where

$$q_\delta := \delta q_+ - q_-,$$

the constant $\delta \in (0,1)$ depends only on N, and the constant $c > 0$ depends on N and M.

The right hand side of (1.10) is undefined if $\int_X q_+ d\mu = \int_X q_- d\mu = \infty$. In this case, let us set it to be $-\infty$ so that the statement is still trivially valid.

The constant δ comes from the proof for technical reasons. We conjecture that in fact one can take $\delta = 1$ so that $q_\delta = q$ and

$$Neg(-\Delta - q) \geq c \int_X q d\mu. \tag{1.11}$$

If $q \geq 0$ then certainly (1.11) follows from (1.10) just by renaming $c\delta$ by c. It would be interesting to obtain (1.11) also for a signed function q.

Note also that the hypothesis (1.9) is essential for the result: without it the estimate (1.11) may fail even for positive q (see Example 4.20).

Let us mention for comparison a theorem of Cwickel–Lieb–Rozenblum saying that, for any non-negative function q in $\mathbb{R}^n$, $n > 2$,

$$Neg(-\Delta - q) \leq C \int_{\mathbb{R}^n} q^{n/2} d\mu, \tag{1.12}$$

where C depends only on n (see, for example, [**28**], [**38**], [**51**]). It is known that (1.12) is not true in $\mathbb{R}^2$ whereas, by Theorem 1.2, the *opposite* inequality (1.11) holds in $\mathbb{R}^2$.

Applying Theorem 1.2 to the potential $q + \lambda$ instead of q and using elementary estimates for $(q+\lambda)_\delta$ we obtain that, for any $\lambda \in \mathbb{R}$,

$$\mathcal{N}_\lambda(L) \geq c \int_X q_{\delta^2} d\mu + c'\lambda\mu(X), \tag{1.13}$$

where $c' = c\delta$. If the spectrum of L is discrete and hence consists of an increasing sequence $\{\lambda_k(L)\}_{k=1}^\infty$ of eigenvalues counted with the multiplicities, then (1.13) implies

$$\lambda_k(L) \leq \frac{Ck - \int_X (\delta q_+ - \delta^{-1} q_-)\, d\mu}{\mu(X)}, \tag{1.14}$$

where C depends on N and M. Again, in would be interesting to prove (1.14) with $\delta = 1$, that is

$$\lambda_k(L) \leq \frac{Ck - \int_X q d\mu}{\mu(X)}.$$

In the case when the operator L is *positive definite*, we have proved that

$$\lambda_k(L) \leq \frac{Ck - \int_X q d\mu}{\varepsilon\mu(X)},$$

where $\varepsilon \in (0,1)$ depends on N (see Theorem 5.15). Note that the function q here may be signed, and the positivity of L implies $\int_X q d\mu \leq 0$.

We deduce Theorem 5.3 (and hence Theorem 1.2) from more general Theorems 4.1, 4.17, which estimate the counting function of an abstract energy form on a (pseudo)metric space. The idea of the proof is as follows. Assume for simplicity that $q \geq 0$. In order to estimate from below $Neg(-\Delta - q)$ we will construct k disjoint capacitors (F_i, G_i) on X such that $\text{cap}(F_i, G_i)$ is controlled from above while $\nu(F_i)$ is controlled from below, where the measure ν is defined by $d\nu = q d\mu$. Indeed, applying Theorem 1.1 to the space X with this measure ν, we obtain k annuli A_i such that

$$\nu(A_i) \geq c\frac{\nu(X)}{k}$$

and $2A_i$ are disjoint. The hypothesis (1.9) implies that

$$\text{cap}(A_i, 2A_i) \leq CM.$$

Assuming that k is taken so that

$$CM < c\frac{\nu(X)}{k}, \tag{1.15}$$

we choose nearly optimal test functions for the capacitors $(A_i, 2A_i)$ and consider the linear space $\mathcal{V}$ spanned by them. Then $\mathcal{V}$ is a k-dimensional subspace of $C_0^\infty(X)$ such that for any $f \in \mathcal{V} \setminus \{0\}$

$$\int_X |\nabla f|^2 d\mu < \int_X f^2 d\nu,$$

that is $\mathcal{E}_q[f] < 0$, whence it follows that $Neg(-\Delta - q) \geq k$. Taking the largest k satisfying (1.15), we obtain

$$Neg(-\Delta - q) \geq \lfloor \frac{c}{CM}\nu(X) \rfloor = \lfloor c' \int_X q d\mu \rfloor.$$

An additional argument allows to get rid of the floor function here and to obtain (1.11).

If as above $X = (\Sigma_\gamma, g)$ then, using the conformal mapping T between X and $\mathbb{S}^2$, one obtains from Theorem 1.2 the following extension of the Korevaar estimate (1.5): for any function $q \in L^1_{loc}(X)$ and for all $k \geq 1$,

$$\lambda_k(-\Delta - q) \leq \frac{C(\gamma+1)k - \int_X (\delta q_+ - \delta^{-1} q_-) d\mu}{\mu(X)}, \tag{1.16}$$

where $C > 0$ and $\delta > 0$ are absolute constants (see Theorem 5.4).

Stability index of minimal surfaces. Another application of Theorem 1.2 occurs for minimal surfaces[4] in $\mathbb{R}^3$. Let X be a complete oriented immersed minimal surface in $\mathbb{R}^3$. The *Jacobi operator* (or the *stability operator*) of X is the operator $L = -\Delta - 2K$, where Δ is the Laplace operator on X associated with the induced Riemannian metric, and K is the Gauss curvature of X. By definition, the *stability index* $\text{ind}(X)$ of the minimal surface X is $Neg(L)$. It is well known that if $ind(X) < \infty$ then the total curvature

$$K_{total}(X) := \int_X |K| d\mu$$

[4]For a detailed account of minimal surfaces see the surveys **[13]** and **[43]** in the same volume.

is finite, and X has a finite number of ends and a finite genus. Theorem 1.2 allows to prove the following result (see also Theorem 6.6).

Theorem 1.3. *Let X be a connected complete oriented minimal surface immersed in $\mathbb{R}^3$ and let* $\operatorname{ind}(X) < \infty$. *If X has k ends and all the ends are embedded then*

$$\operatorname{ind}(X) \geq \frac{c}{k} K_{total}(X), \tag{1.17}$$

where c is an absolute positive constant.

The factor $\frac{1}{k}$ in (1.17) comes from an estimate of the constant M in (1.9). We conjecture that in fact

$$\operatorname{ind}(X) \geq c K_{total}(X). \tag{1.18}$$

Note that the inequality in the opposite direction is true for any minimal surface (not necessarily complete) – see [**28**] and the discussion in Section 6.2 below. As a step towards (1.18) we prove the following result.

Theorem 1.4. (=Corollary 6.8) *Let X be a connected complete oriented minimal surface embedded in $\mathbb{R}^3$. Then*

$$\operatorname{ind}(X) \geq c\sqrt{K_{total}(X)}, \tag{1.19}$$

where c is an absolute positive constant.

The inequality (1.19) is obtained from (1.17) and the following estimate:

$$\operatorname{ind}(X) \geq k - 1,$$

where k is the number of ends of X (it suffices to assume that $\operatorname{ind}(X) < \infty$). The latter estimate is proved in Theorem 6.7 using the techniques specific to minimal surfaces.

Manifolds of higher dimension. Let (X, g) be a compact Riemannian manifold of dimension $n \geq 2$, μ be its Riemannian measure, and $q \geq 0$ be a L^1_{loc}-function on X. We claim that

$$Neg(-\Delta - q) \geq \frac{c}{\mu(X)^{n/2-1}} \left(\int_X q d\mu \right)^{n/2}, \tag{1.20}$$

where c is a positive constant depending only on the conformal class of the Riemannian metric g (see Theorem 5.9 and Example 5.12).

Assuming for simplicity that q is positive and continuous, we obtain from (1.20) that, for all $k = 1, 2, \ldots$,

$$\lambda_k \left(\frac{-\Delta}{q} \right) \leq C \frac{\mu(X)}{\int_X q d\mu} \left(\frac{k}{\mu(X)} \right)^{2/n},$$

where $C = c^{-2/n}$. In the case $q \equiv 1$ this estimate was proved by Korevaar [**37**].

Eigenvalues of a boundary surface. Let us mention another amusing application of the estimate (1.5). Let M be a 3-dimensional Cartan-Hadamard manifold, and let $\Omega \subset M$ be a bounded open set with a smooth boundary Γ that is diffeomorphic to Σ_γ. Let $\lambda_k(\Omega)$ denote the k-th smallest eigenvalue of the Laplace operator in Ω with the Dirichlet boundary condition, and $\lambda_k(\Gamma)$ be the k-th smallest eigenvalue of the Laplace operator on Γ (in both cases, k starts with 1). Then, for any $k = 1, 2, ...$, we have

$$\lambda_k(\Omega) \geq \frac{c}{\gamma+1} \frac{\lambda_{k+1}(\Gamma)}{k^{1/3}},$$

where c is an absolute positive constant (see Theorem 5.14).

Higher order operators. Let m be a positive integer, and consider in $\mathbb{R}^n$ the operator

$$L = (-\Delta)^m - q$$

where $q \in L^1_{loc}$. One defines the counting functions $\mathcal{N}_\lambda(L)$ and $Neg(L)$ in the same way as above using the associated energy form. We claim that if $n = 2m$ and $q \geq 0$ then

$$Neg((-\Delta)^m - q) \geq c \int_{\mathbb{R}^n} q d\mu,$$

where $c = c(n) > 0$ (see Example 4.19).

Fractals sets. As we have already mentioned, the main estimates of counting functions in Theorems 4.1, 4.17 are obtained in the general setting of energy forms on metric spaces. This makes it possible to apply the present results to *fractal sets*. Without going into details of the theory of fractals[5], let us just say that a fractal set can typically be regarded as a metric space (X, d) endowed with a Radon measure μ and an energy functional $\mathcal{E}$. The properties of these spaces resemble many properties of $\mathbb{R}^n$ but with fractional dimensions.

Let $B(x, r)$ denote a ball of the metric d. Then, normally, there exists a positive exponent α such that

$$\mu(B(x,r)) \simeq r^\alpha,$$

for all $x \in X$ and $r > 0$ (or for a bounded range of r if X is bounded). With the energy functional $\mathcal{E}$ one associates the capacity $\operatorname{cap}_{\mathcal{E}}$ defined similarly to (1.2) (see Section 2.2 for details), which normally admits the following estimate:

$$\operatorname{cap}_{\mathcal{E}}(B(x,r), B(x,2r)) \simeq r^{\alpha-\beta},$$

where $\beta > 0$. The next result (which is a particular case of Corollary 4.14) is obtained using the techniques based on Theorem 1.1.

Theorem 1.5. *Let (X, d) be a metric space, μ be a non-atomic Radon measure on X such that $0 < \mu(X) < \infty$, and $\mathcal{E}$ be a local, positive definite, closable energy form on (X, μ), whose generator H has a discrete spectrum. Assume that, for some positive constants N, C_1, C_2,*

(*i*) *any ball $B(x, r)$ in X can be covered by at most N balls of radii $r/2$;*
(*ii*) *for some $\alpha > 0$ and $\beta \geq \alpha$ and for any ball $B(x, r)$ in X, the following estimates hold:*

$$\mu(B(x,r)) \leq C_1 r^\alpha \quad \text{and} \quad \operatorname{cap}_{\mathcal{E}}(B(x,r), B(x,2r)) \leq C_2 r^{\alpha-\beta}.$$

[5] For a detailed account of fractals, we refer the reader to lecture notes [**2**] by Barlow as well as to his article [**4**] in this volume. See also [**25**] for function theory on fractal spaces.

Then, for all $k = 1, 2, ...,$

$$\lambda_k (H) \leq C \left(\frac{k}{\mu (X)} \right)^{\beta/a},$$

where the constant C *depends on* $N, C_1, C_2, \alpha, \beta$.

The structure of the paper. In Section 2 we introduce the abstract notion of an energy form and the associated capacity, prove their general properties, and give examples.

In Section 3 we present the proof of Theorem 1.1.

In Section 4 we prove general estimates for the counting function of an abstract energy form, using decomposition of a metric space by capacitors.

In Section 5 we estimate the counting function and the eigenvalues of Schrödinger type operators on Riemannian manifolds and Riemann surfaces (most of these results were surveyed above).

In Section 6 we apply these estimates to the Jacobi operator and deduce lower bounds of the stability index of minimal surfaces in $\mathbb{R}^3$.

Acknowledgment. The first author is thankful to David Hoffman for the useful discussions about minimal surfaces.

2. Energy forms on measure spaces

2.1. Energy form. Let X be a topological space and let $C_0(X)$ be the space of all continuous functions on X with compact supports, endowed with the sup-norm.

Definition 2.1. A C_0*-energy form* $(\mathcal{E}, \mathcal{F})$ on a topological space X is a symmetric bilinear form $\mathcal{E}(f, g)$, defined on a dense subspace $\mathcal{F} \subset C_0(X)$.

Below, we will introduce also L^2-energy forms. By default, by an energy form we will mean a C_0-energy form.

An energy form $(\mathcal{E}, \mathcal{F})$ is called *positive definite* if $\mathcal{E}[f] \geq 0$ for all $f \in \mathcal{F}$. An energy form $(\mathcal{E}, \mathcal{F})$ is called *local* if $\mathcal{E}(f, g) = 0$ whenever function $f, g \in \mathcal{F}$ have disjoint supports. The form $(\mathcal{E}, \mathcal{F})$ is called *strongly local* if, for all $f, g \in \mathcal{F}$,

$$f \equiv \text{const in a neighborhood of } \operatorname{supp} g \implies \mathcal{E}(f, g) = 0.$$

Clearly, a strongly local energy form is local.

A measure μ on X is called a *Radon measure* if μ is defined on all Borel sets of X and μ is finite on all compact sets. A couple (X, μ) is called a *measure space* if X is a topological space and μ is a Radon measure on X. In the presence of measure, we can consider more general energy forms.

Definition 2.2. An L^2*-energy form* $(\mathcal{E}, \mathcal{F})$ in a measure space (X, μ) is a symmetric bilinear form $\mathcal{E}(f, g)$, defined on a dense subspace $\mathcal{F} \subset L^2(X, \mu)$.

Clearly, any C_0-energy form is also an L^2-energy form. Note also that any signed Radon measure σ can be considered as an energy form with domain $C_0(X)$, as follows:

$$\sigma(f, g) := \int_X f g \, d\sigma. \tag{2.1}$$

Moreover, (2.1) is defined for all $f, g \in L^2(X, |\sigma|)$ (where $|\sigma|$ is the total variation of σ) so that (2.1) defines an L^2-energy form in $L^2(X, |\sigma|)$.

An L^2-energy form $(\mathcal{E},\mathcal{F})$ is called *semi-bounded below* if there exists a constant $K\geq 0$ such that for all $f\in\mathcal{F}$

$$\mathcal{E}\left[f\right]:=\mathcal{E}\left(f,f\right)\geq -K\|f\|_2, \tag{2.2}$$

where $\|f\|_2$ is the $L^2\left(X,\mu\right)$-norm of f. In particular, a positive definite form is semi-bounded below with the constant $K=0$.

An L^2-energy form $(\mathcal{E},\mathcal{F})$ is called *closed* if it is semi-bounded below and the domain $\mathcal{F}$ is a Hilbert space with respect to the inner product

$$(f,g)_{\mathcal{E}}:=\mathcal{E}(f,g)+(K+1)\,\mu\left(f,g\right), \tag{2.3}$$

where K is the constant from (2.2).

An L^2-energy form $(\mathcal{E},\mathcal{F})$ is called *closable* if it is semi-bounded below and, for any sequence $\{f_n\}\subset\mathcal{F}$,

$$\|f_n\|_2\to 0 \quad\text{and}\quad \mathcal{E}\left[f_n-f_m\right]\to 0 \quad\Longrightarrow\quad \mathcal{E}\left[f_n\right]\to 0.$$

It is well-known that a closable L^2-form $(\mathcal{E},\mathcal{F})$ has a unique extension to a subspace $\widetilde{\mathcal{F}}$ of $L^2\left(X,\mu\right)$ so that $(\mathcal{E},\widetilde{\mathcal{F}})$ is closed and $\mathcal{F}$ is dense in $\widetilde{\mathcal{F}}$ with respect to the inner product (2.3). The extension of $\mathcal{E}$ to $\widetilde{\mathcal{F}}$ is also denoted by $\mathcal{E}$, and the form $(\mathcal{E},\widetilde{\mathcal{F}})$ is called the *closure* of $(\mathcal{E},\mathcal{F})$.

2.2. Capacity. Let X be a topological space. For any Borel set $G\subset X$ denote by $C_0\left(G\right)$ the set of all continuous functions f on X such that $\operatorname{supp} f$ is compact and is contained in the interior of G. Any couple (F,G) of Borel subsets of X such that $F\subset G$, will be referred to as a *capacitor.*

Let $(\mathcal{E},\mathcal{F})$ be a positive definite energy form on X. For any capacitor (F,G), define the class $\mathcal{T}\left(F,G\right)$ of test functions as follows:

$$\mathcal{T}(F,G):=\left\{f\in\mathcal{F}\cap C_0\left(G\right):0\leq f\leq 1,\ f=1\text{ in a neighborhood of }\overline{F}\right\}, \tag{2.4}$$

and define the *capacity* $\operatorname{cap}_{\mathcal{E}}(F,G)$ by

$$\operatorname{cap}_{\mathcal{E}}(F,G):=\inf_{f\in\mathcal{T}(F,G)}\mathcal{E}\left[f\right]. \tag{2.5}$$

If $\mathcal{T}(F,G)$ is empty then $\operatorname{cap}_{\mathcal{E}}(F,G)=+\infty$.

For a general theory of capacities see [**20**] or [**42**]. Here we will need only two elementary facts.

Lemma 2.3. *Let $(\mathcal{E},\mathcal{F})$ be a positive definite energy form. If $F\subset G\subset F'\subset G'$ are Borel sets then*

$$\operatorname{cap}_{\mathcal{E}}(F'\setminus G,G'\setminus F)^{1/2}\leq\operatorname{cap}_{\mathcal{E}}(F,G)^{1/2}+\operatorname{cap}_{\mathcal{E}}(F',G')^{1/2} \tag{2.6}$$

Proof. If $\mathcal{T}\left(F,G\right)$ or $\mathcal{T}\left(F',G'\right)$ is empty then (2.7) trivially holds. Otherwise, observe that if $f\in\mathcal{T}\left(F,G\right)$ and $g\in\mathcal{T}\left(F',G'\right)$ then the function $g-f$ is in $\mathcal{T}\left(F'\setminus G,G'\setminus F\right)$ (see Fig. 1).

Since the form $\mathcal{E}$ is positive definite, it satisfies the Cauchy-Schwarz inequality inequality. Hence, we obtain

$$\begin{aligned}\operatorname{cap}_{\mathcal{E}}(F'\setminus G,G'\setminus F) &\leq \mathcal{E}\left[g-f\right]\\ &= \mathcal{E}\left[f\right]+\mathcal{E}\left[g\right]-2\mathcal{E}\left(f,g\right)\\ &\leq \mathcal{E}\left[f\right]+\mathcal{E}\left[g\right]+2\sqrt{\mathcal{E}\left[f\right]\mathcal{E}\left[g\right]}\\ &= \left(\sqrt{\mathcal{E}\left[f\right]}+\sqrt{\mathcal{E}\left[g\right]}\right)^2.\end{aligned}$$

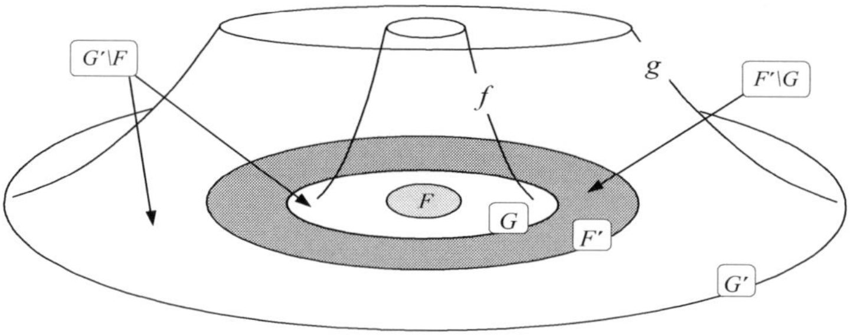

FIGURE 1. Capacitor $(F' \setminus G, G' \setminus F)$

Taking infimum over f and g, we obtain (2.6). ∎

Remark 2.4. If the form $(\mathcal{E}, \mathcal{F})$ is in addition strongly local then (2.6) improves as follows:

$$\mathrm{cap}_{\mathcal{E}}(F' \setminus G, G' \setminus F) \leq \mathrm{cap}_{\mathcal{E}}(F, G) + \mathrm{cap}_{\mathcal{E}}(F', G'). \tag{2.7}$$

Indeed, in the above notation, $f \in C_0(G)$ and $g|_{F'} \equiv 1$ imply that $g \equiv 1$ in a neighborhood of $\mathrm{supp}\, f$, whence $\mathcal{E}(f, g) = 0$. Hence, we obtain

$$\mathrm{cap}_{\mathcal{E}}(F' \setminus G, G' \setminus F) \leq \mathcal{E}[g - f] = \mathcal{E}[f] + \mathcal{E}[g],$$

whence (2.7) follows.

Lemma 2.5. *Let $(\mathcal{E}, \mathcal{F})$ be a strongly local positive definite energy form. If $E \subset F \subset G$ are Borel sets then*

$$\mathrm{cap}_{\mathcal{E}}(E, F)^{-1} + \mathrm{cap}_{\mathcal{E}}(F, G)^{-1} \leq \mathrm{cap}_{\mathcal{E}}(E, G)^{-1}. \tag{2.8}$$

PROOF. Note that by the monotonicity property of the capacity,

$$\mathrm{cap}_{\mathcal{E}}(E, F) \geq \mathrm{cap}_{\mathcal{E}}(E, G) \quad \text{and} \quad \mathrm{cap}_{\mathcal{E}}(F, G) \geq \mathrm{cap}_{\mathcal{E}}(E, G). \tag{2.9}$$

Therefore, if one of the capacities $\mathrm{cap}_{\mathcal{E}}(E, F)$, $\mathrm{cap}_{\mathcal{E}}(F, G)$ is equal to ∞ then (2.8) follows from (2.9). Otherwise the classes $\mathcal{T}(E, F)$ and $\mathcal{T}(F, G)$ are non-empty. Observe that, for any test functions $f \in \mathcal{T}(E, F)$ and $g \in \mathcal{T}(F, G)$, the function $h = tf + (1 - t)g$ belongs to $\mathcal{T}(E, G)$, for any $t \in [0, 1]$ (see Fig. 2).

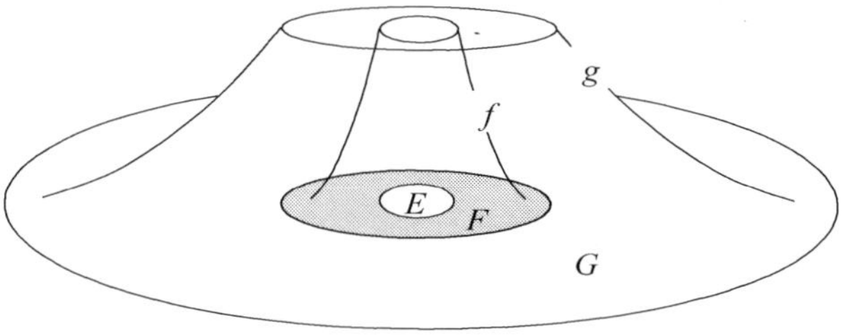

FIGURE 2. Capacitors (E, F) and (F, G)

By the strong locality, $f \in C_0(F)$ and $g|_F \equiv 1$ imply $\mathcal{E}(f, g) = 0$, whence

$$\mathrm{cap}_{\mathcal{E}}(E, G) \leq \mathcal{E}[h] = t^2 \mathcal{E}[f] + (1 - t)^2 \mathcal{E}[g]. \tag{2.10}$$

If $\mathcal{E}[f] = \mathcal{E}[g] = 0$ then (2.10) yields $\operatorname{cap}_{\mathcal{E}}(E, G) = 0$, and (2.8) is trivially satisfied. Otherwise, taking in (2.10) $t = \frac{\mathcal{E}[g]}{\mathcal{E}[f]+\mathcal{E}[g]}$, we obtain

$$\operatorname{cap}_{\mathcal{E}}(E, G) \leq \frac{\mathcal{E}[f]\,\mathcal{E}[g]}{\mathcal{E}[f] + \mathcal{E}[g]},$$

whence (2.8) follows. ∎

2.3. Generator. Let (X, μ) be a measure space and $(\mathcal{E}, \mathcal{F})$ be a L^2-energy form on X.

Definition 2.6. A densely defined linear operator H in $L^2(X, \mu)$ is called a *generator* of the form $(\mathcal{E}, \mathcal{F})$ if $\operatorname{dom}(H) \subset \mathcal{F}$ and

$$\mathcal{E}(f, g) = \mu(Hf, g) \quad \text{for all } f \in \operatorname{dom}(H) \text{ and } g \in \mathcal{F}. \tag{2.11}$$

It follows from (2.11) that a generator is a symmetric operator, in the sense that

$$\mu(Hf, g) = \mu(f, Hg) \quad \text{for all } f, g \in \operatorname{dom}(H).$$

Conversely, any densely defined symmetric operator H in $L^2(X, \mu)$ determines the L^2-energy form $(\mathcal{E}_H, \operatorname{dom}(H))$ by

$$\mathcal{E}_H(f, g) := \mu(Hf, g). \tag{2.12}$$

Clearly, H is a generator of $\mathcal{E}_H$.

If the form $(\mathcal{E}, \mathcal{F})$ is closable then its closure $(\mathcal{E}, \widetilde{\mathcal{F}})$ has a unique generator H, which is a *self-adjoint* operator on $L^2(X, \mu)$ (see for example [**16**, Theorem 4.4.2]). We will refer to H as the *self-adjoint generator* of $(\mathcal{E}, \widetilde{\mathcal{F}})$, and also of $(\mathcal{E}, \mathcal{F})$ (although H may be not *a* generator of $(\mathcal{E}, \mathcal{F})$). If in addition the operator H is non-negative definite then $H^{1/2}$ is defined. In it known that $\operatorname{dom}(H^{1/2}) = \widetilde{\mathcal{F}}$ and

$$\mathcal{E}(f, g) = \mu\left(H^{1/2} f, H^{1/2} g\right) \quad \text{for all } f, g \in \widetilde{\mathcal{F}} \tag{2.13}$$

(see [**15**]). If H is a self-adjoint, semi-bounded below operator in $L^2(X, \mu)$ then the form $\mathcal{E}_H$ defined by (2.12) is closable, and its self-adjoint generator is H.

2.4. Counting function. Let $(\mathcal{E}, \mathcal{F})$ be an L^2-energy form on a measure space (X, μ). Define the *counting function* $\mathcal{N}_\lambda(\mathcal{E}, \mu)$ as the supremum of the dimensions of all linear spaces $\mathcal{V} \subset \mathcal{F}$ such that

$$\mathcal{E}[f] < \lambda \mu[f] \text{ for any } f \in \mathcal{V} \setminus \{0\}, \tag{2.14}$$

where λ is a real parameter; that is,

$$\mathcal{N}_\lambda(\mathcal{E}, \mu) := \sup \{\dim \mathcal{V} : \mathcal{V} \prec \mathcal{F} \text{ and } \mathcal{E}[f] < \lambda \mu[f] \ \forall f \in \mathcal{V} \setminus \{0\}\}, \tag{2.15}$$

where the relation $\mathcal{V} \prec \mathcal{F}$ means that $\mathcal{V}$ is a linear subspace of $\mathcal{F}$. If the family of spaces $\mathcal{V}$ in (2.15) is empty then we assign to the sup the value 0.

If $(\mathcal{E}, \mathcal{F})$ is a C_0-energy form on X then $\mathcal{N}_0(\mathcal{E}, \mu)$ does not depend on μ so we will use in this case notation $Neg(\mathcal{E})$; that is

$$Neg(\mathcal{E}) := \sup \{\dim \mathcal{V} : \mathcal{V} \prec \mathcal{F} \text{ and } \mathcal{E}[f] < 0 \, \forall f \in \mathcal{V} \setminus \{0\}\}. \tag{2.16}$$

For a *closable* L^2-energy form $(\mathcal{E}, \mathcal{F})$, consider also the *modified* counting function

$$\mathcal{N}_\lambda^*(\mathcal{E}, \mu) = \sup \left\{\dim \mathcal{V} : \mathcal{V} \prec \widetilde{\mathcal{F}} \text{ and } \mathcal{E}[f] \leq \lambda \mu[f] \ \forall f \in \mathcal{V}\right\}, \tag{2.17}$$

where $\widetilde{\mathcal{F}}$ is the domain of the closure of $(\mathcal{E}, \mathcal{F})$. Note that there are two differences between $\mathcal{N}^*_\lambda(\mathcal{E}, \mu)$ and $\mathcal{N}_\lambda(\mathcal{E}, \mu)$: using $\widetilde{\mathcal{F}}$ instead of $\mathcal{F}$ and using a non-strict inequality.

Lemma 2.7. *Let $(\mathcal{E}, \mathcal{F})$ be a closable L^2-energy form on a measure space (X, μ) and let H be its self-adjoint generator. Then, for any real λ, we have*

$$\mathcal{N}_\lambda(\mathcal{E}, \mu) = \dim \operatorname{Im} \mathbf{1}_{(-\infty,\lambda)}(H) \tag{2.18}$$

and

$$\mathcal{N}^*_\lambda(\mathcal{E}, \mu) = \dim \operatorname{Im} \mathbf{1}_{(-\infty,\lambda]}(H). \tag{2.19}$$

Here $\mathbf{1}_A$ is the indicator function of the set $A \subset \mathbb{R}$, and the operators $\mathbf{1}_{(-\infty,\lambda)}(H)$ are $\mathbf{1}_{(-\infty,\lambda]}(H)$ are understood in the sense of spectral theory. In particular, if the spectrum of H is discrete then $\dim \operatorname{Im} \mathbf{1}_{(-\infty,\lambda)}(H)$ is the number of the eigenvalues of H below λ, counted with the multiplicities, and $\dim \operatorname{Im} \mathbf{1}_{(-\infty,\lambda]}(H)$ is the number of the eigenvalues of H, which are at most λ, also counted with the multiplicities.

The two quantities $\dim \operatorname{Im} \mathbf{1}_{(-\infty,\lambda)}(H)$ and $\dim \operatorname{Im} \mathbf{1}_{(-\infty,\lambda]}(H)$ are normally referred to as the counting functions of the operator H. Here we prefer to give a more general definition of the counting functions as follows. For any densely defined symmetric operator H in $L^2(X, \mu)$ let us define the *counting function* $\mathcal{N}_\lambda(H)$ by

$$\mathcal{N}_\lambda(H) := \mathcal{N}_\lambda(\mathcal{E}_H, \mu),$$

where the form $\mathcal{E}_H$ is defined by (2.12). Also, set $Neg(H) = \mathcal{N}_0(H)$.

If the operator H is self-adjoint and semi-bounded below then the form $\mathcal{E}_H$ is closable and its self-adjoint generator is H. Therefore, by (2.18) we obtain that

$$\mathcal{N}_\lambda(H) = \dim \operatorname{Im} \mathbf{1}_{(-\infty,\lambda)}(H). \tag{2.20}$$

For such an operator H, define also the *modified counting function* by

$$\mathcal{N}^*_\lambda(H) := \mathcal{N}^*_\lambda(\mathcal{E}_H, \mu). \tag{2.21}$$

Then by (2.19) we obtain that

$$\mathcal{N}^*_\lambda(H) = \dim \operatorname{Im} \mathbf{1}_{(-\infty,\lambda]}(H). \tag{2.22}$$

Hence, the identities (2.20) and (2.22) justify the above definitions of $\mathcal{N}_\lambda(H)$ and $\mathcal{N}^*_\lambda(H)$.

Proof of Lemma 2.7. Without loss of generality, we can assume that $\mathcal{E}$ is positive definite, and hence, the generator H is also positive definite. For any $\lambda < 0$ all terms in (2.18) and (2.19) vanish, so we can assume in the sequel $\lambda \geq 0$.

Let $\{E_t\}_{t\in\mathbb{R}}$ be the spectral resolution of the operator H in $L^2 = L^2(X, \mu)$, that is

$$E_t = \mathbf{1}_{(-\infty,t)}(H) = \mathbf{1}_{[0,t)}(H).$$

By spectral theory, for any $f \in L^2(X, \mu)$,

$$\|f\|^2 = \int_{[0,+\infty)} d\|E_t f\|^2 \tag{2.23}$$

(where $\|\cdot\|$ is the norm in L^2), and, for any $f \in \widetilde{\mathcal{F}} = \operatorname{dom}(H^{1/2})$,

$$\mathcal{E}[f] = \|H^{1/2} f\|^2 = \int_{[0,+\infty)} t d\|E_t f\|^2. \tag{2.24}$$

Let us first prove (2.19). To prove the inequality

$$\mathcal{N}_\lambda^*(\mathcal{E},\mu) \le \dim \operatorname{Im} \mathbf{1}_{[0,\lambda]}(H), \tag{2.25}$$

it suffices to show that, for any finitely dimensional subspace $\mathcal{V} \subset \widetilde{\mathcal{F}}$ satisfying (2.17), we have

$$\dim \mathcal{V} \le \dim \operatorname{Im} \mathbf{1}_{[0,\lambda]}(H).$$

Assuming the contrary, we find a function $v \in \mathcal{V} \setminus \{0\}$ such that v is orthogonal to $\operatorname{Im} \mathbf{1}_{[0,\lambda]}(H)$. In particular, $E_t v = 0$ for all $t \le \lambda$ so that from (2.23) and (2.24) imply

$$\|v\|^2 = \int_{[\lambda,+\infty)} d\|E_t v\|^2 \quad \text{and} \quad \mathcal{E}[v] = \int_{[\lambda,+\infty)} t d\|E_t v\|^2 \tag{2.26}$$

whence $\mathcal{E}[v] \ge \lambda \|v\|^2$. By $v \in \mathcal{V}$ the opposite inequality also holds, that is $\mathcal{E}[v] = \lambda\|v\|^2$. In the view of (2.26) it is only possible when, for any $\varepsilon > 0$,

$$\int_{[\lambda+\varepsilon,+\infty)} d\|E_t v\|^2 = 0,$$

which implies that $v \in \operatorname{Im} \mathbf{1}_{\{\lambda\}}(H)$, while v is orthogonal to this space.

To prove the opposite inequality, that is

$$\mathcal{N}_\lambda^*(\mathcal{E},\mu) \ge \dim \operatorname{Im} \mathbf{1}_{[0,\lambda]}(H), \tag{2.27}$$

let us observe that, for any $w \in \operatorname{Im} \mathbf{1}_{[0,\lambda]}(H)$,

$$\|w\|^2 = \int_{[0,\lambda]} d\|E_t v\|^2 \quad \text{and} \quad \mathcal{E}[v] = \int_{[0,\lambda]} t d\|E_t v\|^2, \tag{2.28}$$

whence $\mathcal{E}[w] \le \lambda \|w\|^2$. Therefore, the space $\mathcal{V} := \operatorname{Im} \mathbf{1}_{(-\infty,\lambda]}(H) \subset \widetilde{\mathcal{F}}$ satisfies (2.17), which implies (2.27).

Let us now prove (2.18), that is

$$\mathcal{N}_\lambda(\mathcal{E},\mu) = \dim \operatorname{Im} E_\lambda.$$

The inequality

$$\mathcal{N}_\lambda(\mathcal{E},\mu) \le \dim \operatorname{Im} E_\lambda \tag{2.29}$$

is proved similarly to (2.25). Assume from the contrary that there exists a finitely dimensional subspace $\mathcal{V} \subset \mathcal{F}$ satisfying (2.14) and such that $\dim \mathcal{V} > \dim \operatorname{Im} E_\lambda$. Then there exists a vector $v \in \mathcal{V} \setminus \{0\}$ such that v is orthogonal to $\operatorname{Im} E_\lambda$. Therefore, $E_t v = 0$ for all $t \le \lambda$, which implies (2.26) and hence $\mathcal{E}[v] \ge \lambda \|v\|^2$, which contradicts $v \in \mathcal{V}$.

Before we prove the opposite inequality, that is

$$\dim \operatorname{Im} E_\lambda \le \mathcal{N}_\lambda(\mathcal{E},\mu), \tag{2.30}$$

let us verify that

$$\mathcal{E}[w] < \lambda \|w\|^2 \quad \text{for any } w \in \operatorname{Im} E_\lambda \setminus \{0\}. \tag{2.31}$$

Indeed, for all $w \in \operatorname{Im} E_\lambda$ we have (2.28), whence $\mathcal{E}[w] \le \lambda\|w\|^2$. In the view of (2.28) the equality is possible only if $w \in \operatorname{Im} \mathbf{1}_{\{\lambda\}}(H)$ which is impossible because the spaces $\operatorname{Im} \mathbf{1}_{\{\lambda\}}(H)$ and $\operatorname{Im} E_\lambda$ are orthogonal.

Let us now prove (2.30). If $\operatorname{Im} E_\lambda \subset \mathcal{F}$ then taking $\mathcal{V} = \operatorname{Im} E_\lambda$ we conclude the proof. However, in general we can only ensure that $\operatorname{Im} E_\lambda \subset \widetilde{\mathcal{F}}$. In order to prove

(2.30) in full generality, we will show that, for any positive integer $n \leq \dim \operatorname{Im} E_\lambda$, there exists a subspace $\mathcal{V} \subset \mathcal{F}$ of the dimension n satisfying (2.14).

Consider the following inner product in $\widetilde{\mathcal{F}}$

$$(f, g)_{\mathcal{E}} := (f, g)_{L^2} + \mathcal{E}(f, g).$$

Since $\operatorname{Im} E_\lambda \subset \widetilde{\mathcal{F}}$, there exists a sequence $\{w_k\}_{k=1}^n$ in $\operatorname{Im} E_\lambda$, which is orthonormal with respect to the inner product $(\cdot, \cdot)_{\mathcal{E}}$. Consider the following subspace of $\widetilde{\mathcal{F}}$

$$\mathcal{W} := \operatorname{span}\{w_1, w_2, ... w_n\}$$

and show that there exists $\beta < \lambda$ such that

$$\mathcal{E}[w] \leq \beta \|w\|^2 \quad \text{for any } w \in \mathcal{W}. \tag{2.32}$$

Indeed, set

$$S := \{f \in \mathcal{W} : \|f\| = 1\} \quad \text{and} \quad \beta := \sup_{f \in S} \mathcal{E}[f],$$

so that (2.32) holds by linearity. Let us verify that $\beta < \lambda$. The sphere S is a compact subset of the finite dimensional space $\mathcal{W}$, and $\mathcal{E}[f]$ is a continuous functional on $\mathcal{W}$. Hence, there is a point $f \in S$ such that $\mathcal{E}[f] = \beta$. On the other hand, by (2.31) we have $\mathcal{E}[f] < \lambda$ for any $f \in S$, whence $\beta < \lambda$.

Since $\mathcal{F}$ is dense in $\widetilde{\mathcal{F}}$ in the norm

$$\|f\|_{\mathcal{E}} := (f, f)_{\mathcal{E}}^{1/2} = \left(\|f\|^2 + \mathcal{E}[f]\right)^{1/2}, \tag{2.33}$$

for any $\varepsilon > 0$ there exists a sequence $\{v_k\}_{k=1}^n$ in $\mathcal{F}$ such that

$$\|v_k - w_k\|_{\mathcal{E}} < \varepsilon \quad \text{for all } k = 1, 2, ..., n.$$

Set

$$\mathcal{V} = \operatorname{span}(v_1, v_2, ..., v_n)$$

and observe that $\mathcal{V} \subset \mathcal{F}$ and $\dim \mathcal{V} = n$ provided ε is small enough. Furthermore, let us show that ε can be chosen so small that

$$\mathcal{E}[v] < \lambda \|v\|^2 \quad \text{for any } v \in \mathcal{V} \setminus \{0\}, \tag{2.34}$$

which would finish the proof of (2.30). For any $v \in \mathcal{V} \setminus \{0\}$ there exists a vector $\xi = (\xi_1, \xi_2, ..., \xi_n) \in \mathbb{R}^n \setminus \{0\}$ such that

$$v = \sum_k \xi_k v_k.$$

Set

$$w = \sum_k \xi_k w_k$$

and observe that $w \in \mathcal{W}$ and

$$\|w\|_{\mathcal{E}} = \left(\sum_k \xi_k^2\right)^{1/2} = |\xi|.$$

Then we have

$$\begin{aligned}\|v-w\|_{\mathcal{E}} &= \|\sum_k \xi_k (v_k - w_k)\|_{\mathcal{E}} \\ &\le \sum_k |\xi_k| \, \|v_k - w_k\|_{\mathcal{E}} \\ &\le \left(\sum_k |\xi_k|^2\right)^{1/2} \left(\sum_k \|v_k - w_k\|_{\mathcal{E}}^2\right)^{1/2} \\ &\le |\xi| \sqrt{n\varepsilon},\end{aligned}$$

whence

$$\|v-w\|_{\mathcal{E}} \le \sqrt{n\varepsilon}\|w\|_{\mathcal{E}}. \tag{2.35}$$

In particular, we obtain from (2.35)

$$\|v\| \ge \|w\| - \sqrt{n\varepsilon}\|w\|_{\mathcal{E}} \tag{2.36}$$

and

$$\mathcal{E}[v]^{1/2} \le \mathcal{E}[w]^{1/2} + \sqrt{n\varepsilon}\|w\|_{\mathcal{E}}. \tag{2.37}$$

Setting

$$R[w] := \frac{\mathcal{E}[w]^{1/2}}{\|w\|} \le \sqrt{\beta}$$

and using

$$\|w\|_{\mathcal{E}} \le \|w\| + \mathcal{E}[w]^{1/2} = (1 + R[w])\|w\| \le (1+\sqrt{\beta})\|w\| =: c\|w\|,$$

we obtain

$$\frac{\mathcal{E}[v]^{1/2}}{\|v\|} \le \frac{\mathcal{E}[w]^{1/2} + \sqrt{n\varepsilon}c\|w\|}{\|w\| - \sqrt{n\varepsilon}c\|w\|} = \frac{R[w] + \sqrt{n\varepsilon}c}{1 - \sqrt{n\varepsilon}c} \le \frac{\sqrt{\beta} + \sqrt{n\varepsilon}c}{1-\sqrt{n\varepsilon}c}.$$

Since $\beta < \lambda$, the right hand side here is smaller than $\sqrt{\lambda}$ provided ε is small enough, whence (2.34) follows. ∎

For the rest of this section we fix an L^2-energy form $(\mathcal{E}, \mathcal{F})$ on (X, μ) and use the short notation $\mathcal{N}_\lambda = \mathcal{N}_\lambda(\mathcal{E}, \mu)$. The function $\lambda \mapsto \mathcal{N}_\lambda$ is monotone increasing and takes only values $0, 1, 2, ..., +\infty$. It is useful to observe that this function is left continuous, that is

$$\mathcal{N}_\lambda = \lim_{t \to \lambda -} \mathcal{N}_t. \tag{2.38}$$

Indeed, let $\mathcal{V}$ be a finite dimensional space satisfying (2.14). As it follows from the compactness argument (see the above proof), there exists $t < \lambda$ such that

$$\mathcal{E}[v] < t\|v\|^2 \quad \text{for any } v \in \mathcal{V} \setminus \{0\},$$

which implies $\mathcal{N}_t \ge \dim \mathcal{V}$ and hence (2.38).

It is natural to interpret the jumps of the function $\mathcal{N}_\lambda$ as the *eigenvalues* of the form $\mathcal{E}$. Namely, for any positive integer $k = 1, 2, 3, ...$, set

$$\lambda_k = \lambda_k(\mathcal{E}, \mu) := \inf\{\lambda \in \mathbb{R} : \mathcal{N}_\lambda \ge k\}. \tag{2.39}$$

Here we allow λ_k to take also values $\pm\infty$. It is straightforward to see that, for any real λ,

$$\mathcal{N}_\lambda = \sup\{k : \lambda_k < \lambda\}, \tag{2.40}$$

where the supremum is taken over all positive integers k (see Fig. 3).

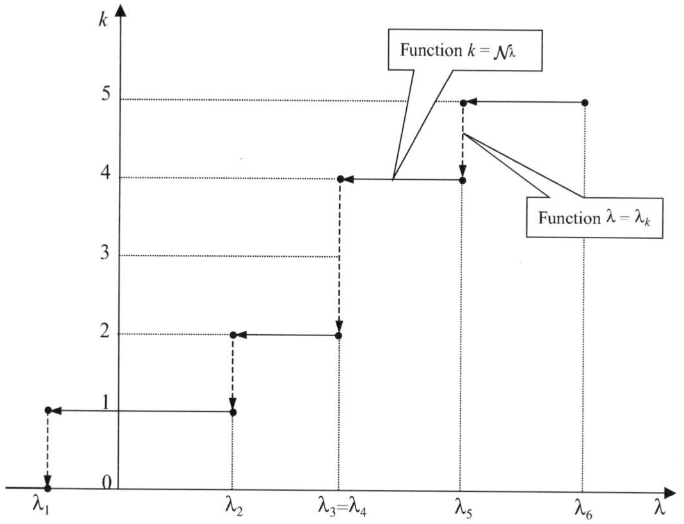

FIGURE 3. Sample graphs of functions $\lambda \mapsto \mathcal{N}_\lambda$ (horizontal) and $k \mapsto \lambda_k$ (vertical)

The following statement will allow us to switch between the estimates of $\mathcal{N}_\lambda$ and those of λ_k.

Lemma 2.8. *Let a, b be reals, and $a > 0$.*
(a) If, for all $\lambda > b$,

$$\mathcal{N}_\lambda \geq \lfloor \frac{\lambda - b}{a} \rfloor,$$

then, for all $k = 1, 2, ...$

$$\lambda_k \leq ak + b.$$

(b) If, for all $\lambda > b$,

$$\mathcal{N}_\lambda \geq \lceil \frac{\lambda - b}{a} \rceil,$$

then for all $k = 1, 2, ...$

$$\lambda_k \leq a(k-1) + b.$$

Here $\lfloor \cdot \rfloor$ is the *floor* function, that is $\lfloor x \rfloor$ is the maximal integer, which is at most x, and $\lceil \cdot \rceil$ is the *ceiling* function, that is $\lceil x \rceil$ is the minimal integer, which is at least x.

PROOF. (*a*) Set $\lambda = ak + b$ and observe that for this λ, we have $\mathcal{N}_\lambda \geq \lfloor k \rfloor = k$. Therefore, by (2.39) we obtain $\lambda_k \leq \lambda$, which was to be proved.

(*b*) Choose $\lambda = a(k-1) + b + \varepsilon$ for some $\varepsilon > 0$. Then by the hypothesis

$$\mathcal{N}_\lambda \geq \lceil k - 1 + \frac{\varepsilon}{a} \rceil = k,$$

whence by (2.39) $\lambda_k \leq \lambda$. The claim follows by letting $\varepsilon \to 0$. ∎

Remark 2.9. Assume that the form $(\mathcal{E},\mathcal{F})$ is closable and H is its self-adjoint generator. The spectrum spec (H) is bounded below, and let λ_{ess} be the bottom of the *essential* spectrum of H. In particular, the spectrum of H below λ_{ess} is discrete.

By (2.18), for any $\lambda < \lambda_{\text{ess}}$, $\mathcal{N}_\lambda$ is the sum of the multiplicities of all the eigenvalues of H below λ, and, for any $\lambda > \lambda_{\text{ess}}$, $\mathcal{N}_\lambda = +\infty$. If $n := \mathcal{N}_{\lambda_{\text{ess}}} < \infty$ then there are exactly n eigenvalues of H below λ_{ess}, and they coincide with $\lambda_1, \lambda_2, ..., \lambda_n$, while for all $k > n$ we have $\lambda_k = \lambda_{ess}$. If $n = \infty$ then all the eigenvalues of H below λ_{ess} are given by the sequence $\{\lambda_k\}_{k=1}^{\infty}$.

2.5. Perturbation of an energy form. As already was mentioned above, any signed Radon measure σ on a topological space X defines an energy form on $C_0(X)$ by

$$\sigma(f,g) := \int_X fg\,d\sigma.$$

For any C_0-energy form $(\mathcal{E},\mathcal{F})$ on X, consider a new energy form $(\mathcal{E}-\sigma,\mathcal{F})$ defined by

$$(\mathcal{E}-\sigma)(f,g) := \mathcal{E}(f,g) - \sigma(f,g),$$

which is called the *perturbation* of the form $(\mathcal{E},\mathcal{F})$ by the signed measure σ. If σ is a measure then we obviously have the following identity:

$$\mathcal{N}_\lambda(\mathcal{E},\sigma) = Neg(\mathcal{E}-\lambda\sigma). \tag{2.41}$$

Lemma 2.10. *Let an L^2-energy form $(\mathcal{E},\mathcal{F})$ be non-negative definite and closable in $L^2(X,\mu)$, and let σ be a signed Radon measure on X, absolutely continuous with respect to μ. If there exist constants $0 < c < 1$ and $C > 0$ such that, for all $f \in \mathcal{F}$,*

$$\sigma_+[f] \le c\mathcal{E}[f] + C\mu[f], \tag{2.42}$$

then the form $(\mathcal{E}-\sigma,\mathcal{F})$ is closable.

PROOF. Consider first the case when $\sigma_+ = 0$. Then the form $\mathcal{E}-\sigma = \mathcal{E}+\sigma_-$ is non-negative definite, and if $\{f_n\}_{n=1}^{\infty}$ is a sequence in $\mathcal{F}$ such that

$$\mu[f_n] \to 0 \quad \text{and} \quad (\mathcal{E}+\sigma_-)[f_n - f_m] \to 0 \quad \text{as } n,m \to \infty, \tag{2.43}$$

then also $\mathcal{E}[f_n - f_m] \to 0$. By the closability of the form $(\mathcal{E},\mathcal{F})$ we conclude that $\mathcal{E}[f_n] \to 0$. By (2.43) we have also $\sigma_-[f_n - f_m] \to 0$ whence it follows that the sequence $\{f_n\}$ converges in $L^2(X,\sigma_-)$. Since $\{f_n\}$ converges to 0 in $L^2(X,\mu)$ and σ_- is absolutely continuous with respect to μ then the limit of $\{f_n\}$ in $L^2(X,\sigma_-)$ is also 0. Therefore, $(\mathcal{E}+\sigma_-)[f_n] \to 0$, which means that the form $(\mathcal{E}+\sigma_-,\mathcal{F})$ is closable.

In general, note that $\mathcal{E}-\sigma = (\mathcal{E}+\sigma_-) - \sigma_+$. By the above argument, the form $(\mathcal{E}+\sigma_-,\mathcal{F})$ is closable, so rename it to $\mathcal{E}$ and rename σ_+ to σ. Hence, we are left to consider the particular case when $\sigma \ge 0$.

Rewrite (2.42) in the form

$$(1-c)\mathcal{E}[f] \le C\mu[f] + (\mathcal{E}-\sigma)[f]. \tag{2.44}$$

In particular, it implies that $\mathcal{E}-\sigma$ is semi-bounded below. Let $\{f_n\}_{n=1}^{\infty}$ be a sequence in $\mathcal{F}$ such that

$$\mu[f_n] \to 0 \quad \text{and} \quad (\mathcal{E}-\sigma)[f_n - f_m] \to 0 \quad \text{as } n,m \to \infty. \tag{2.45}$$

Then (2.44) and (2.45) imply that $\mathcal{E}[f_n - f_m] \to 0$, and the closability of $(\mathcal{E}, \mathcal{F})$ yields that $\mathcal{E}[f_n] \to 0$. Using again (2.45) we obtain $\sigma[f_n - f_m] \to 0$, and as in part (a) we conclude that $(\mathcal{E} - \sigma)[f_n] \to 0$, which was to be proved. ∎

Consider two topological spaces X and X'. A continuous mapping $T : X \to X'$ is said to be *proper* if, for any compact set $K \subset X'$, the preimage $T^{-1}(K)$ is also compact in X. If T is a proper mapping then any signed Radon measure σ on X can be lifted to a signed Radon measure σ' on X' by

$$\sigma'(\cdot) = \sigma(T^{-1}(\cdot)).$$

Obviously, we have then

$$f \in C_0(X') \implies f \circ T \in C_0(X) \text{ and } \sigma[f \circ T] = \sigma'[f].$$

Definition 2.11. Let $(\mathcal{E}, \mathcal{F})$ and $(\mathcal{E}', \mathcal{F}')$ be positive definite C_0-energy forms on topological spaces X and X', respectively. We say that a proper mapping $T : X \to X'$ has the *energy degree* at most D, where $0 < D < \infty$, if the form $\mathcal{E}$ is dominated by $\mathcal{E}'$ in the following sense:

$$f \in \mathcal{F}' \implies f \circ T \in \mathcal{F} \text{ and } \mathcal{E}[f \circ T] \le D\mathcal{E}'[f]. \tag{2.46}$$

Lemma 2.12. *If a proper mapping $T : X \to X'$ has the energy degree at most D then*

$$Neg\,(\mathcal{E} - \sigma) \ge Neg\,(D\mathcal{E}' - \sigma'). \tag{2.47}$$

PROOF. Let $\mathcal{V}'$ be a linear subspace of $\mathcal{F}'$ such that

$$D\mathcal{E}'[f] - \sigma'[f] < 0 \quad \text{for any } f \in \mathcal{V}' \setminus \{0\}.$$

Consider the space

$$\mathcal{V} = \{f \circ T : f \in \mathcal{V}'\}.$$

By (2.46), $\mathcal{V}$ is a linear subspace of $\mathcal{F}$ and, for any $\varphi = f \circ T \in \mathcal{V} \setminus \{0\}$, we have

$$\mathcal{E}[\varphi] - \sigma[\varphi] \le D\mathcal{E}'[f] - \sigma'[f] < 0.$$

Finally, (2.47) follows from $\dim \mathcal{V} \ge \dim \mathcal{V}'$. ∎

2.6. Weighted Riemannian manifolds. Let X be a Riemannian manifold and μ_0 be the Riemannian measure on X. For any Radon measure μ on X (which may be equal to μ_0 or not), the couple (X, μ) is called a *weighted manifold.* On any weighted manifold (X, μ), there is a natural energy form $\mathcal{E}_\mu$ defined on $C_0^1(X)$ by

$$\mathcal{E}_\mu(f, g) = \int_X \nabla f \cdot \nabla g \, d\mu, \tag{2.48}$$

where ∇ is the Riemannian gradient. Clearly, $\mathcal{E}_\mu$ is a strongly local positive definite energy form on X.

Frequently it is more convenient to define the domain of $\mathcal{E}_\mu$ by $\mathcal{F} = Lip_0(X)$ where $Lip_0(X)$ is the space all locally Lipschitz functions on X with compact support (note that for any locally Lipschitz function f on X, the gradient ∇f exists μ_0-almost everywhere). This will be our default choice of the domain of $\mathcal{E}_\mu$.

Let μ be absolutely continuous with respect to μ_0, and the density $\psi = d\mu/d\mu_0$ be a smooth positive function. Then the form $(\mathcal{E}_\mu, \mathcal{F})$ is closable in $L^2(X, \mu)$, and its generator is $-\Delta_\mu$, where Δ_μ is the *Laplace operator* of (X, μ), defined by

$$\Delta_\mu f = \psi^{-1} \text{div}\,(\psi \nabla f),$$

where div is the Riemannian divergence. If $\mu = \mu_0$ is the Riemannian measure then $\mathcal{E}_\mu$ is the Riemannian energy form and $\Delta_\mu = \Delta = \operatorname{div} \circ \nabla$ is the Laplace-Beltrami operator of X.

For any positive integer m, consider also the energy form of the order m:

$$\mathcal{E}_\mu^{(m)}(f,g) = \int_X \nabla^m f \cdot \nabla^m g \, d\mu\,, \tag{2.49}$$

where m is a positive integer and

$$\nabla^m = \begin{cases} \Delta_\mu^{\frac{m}{2}}, & m \text{ even}, \\ \nabla \Delta_\mu^{\frac{m-1}{2}}, & m \text{ odd} \end{cases}$$

(in the case of even m, the dot "$\cdot$" in (2.49) denotes the product of scalars, whereas for odd m it is the inner product of vectors). The form $\mathcal{E}_\mu^{(m)}$ with the domain $C_0^\infty(X)$ is closable, and its generator is $(-\Delta_\mu)^m$.

Given a signed Radon measure σ on a weighted manifold (X,μ), consider a perturbed form $\mathcal{E}_\mu - \sigma$ with the domain $\mathcal{F}$. Its generator is a Schrödinger type operator $-\Delta_\mu - \sigma$. In Section 5.2 we will need the following fact.

Lemma 2.13. *Let (X,μ) be a weighted manifold, $\mathcal{E} = \mathcal{E}_\mu$ be the energy form of (X,μ), and σ be a signed Radon measure on X. Let K be a compact subset of X with empty interior, such that $|\sigma|(K) = 0$ and $\operatorname{cap}_{\mathcal{E}}(K,U) = 0$ for any open set $U \subset X$ containing K. Set $X' = X \setminus K$, $\mu' = \mu|_{X'}$, $\sigma' = \sigma|_{X'}$, and let $\mathcal{E}'$ be the energy form of (X',μ'). Then*

$$Neg(\mathcal{E}' - \sigma') = Neg(\mathcal{E} - \sigma)\,. \tag{2.50}$$

If in addition $\mu(K) = 0$ then, for all real λ,

$$\mathcal{N}_\lambda(\mathcal{E}' - \sigma', \mu') = \mathcal{N}_\lambda(\mathcal{E} - \sigma, \mu)\,. \tag{2.51}$$

Proof. We have $\mathcal{F} = Lip_0(X)$ and $\mathcal{F}' = Lip_0(X')$. From $\mathcal{F}' \subset \mathcal{F}$ it follows

$$\mathcal{N}_\lambda(\mathcal{E}' - \sigma', \mu) \le \mathcal{N}_\lambda(\mathcal{E} - \sigma, \mu)\,,$$

so that we need to prove the opposite inequality. To that end, it suffices to show that, for any finite dimensional subspace $\mathcal{V}$ of $\mathcal{F}$ such that

$$\mathcal{E}[f] < \sigma[f] + \lambda\mu[f] \quad \text{for all } f \in \mathcal{V} \setminus \{0\}\,, \tag{2.52}$$

there exists a subspace $\mathcal{V}'$ of $\mathcal{F}'$ of the same dimension as $\mathcal{V}$ and such that

$$\mathcal{E}[f] < \sigma[f] + \lambda\mu[f] \quad \text{for all } f \in \mathcal{V}' \setminus \{0\}\,. \tag{2.53}$$

Set $\nu := \sigma + \lambda\mu$ and observe that by (2.52) the bilinear form $\nu(f,g)$ is an inner product in $\mathcal{V}$. We will regard $\mathcal{V}$ as a finite dimensional Euclidean space with this inner product. It follows from (2.52) by a compactness argument that there exists $c < 1$ such that

$$\mathcal{E}[f] \le c\nu[f] \quad \text{for any } f \in \mathcal{V}. \tag{2.54}$$

Let U be a precompact open neighborhood of K in X to be specified below. Since $\sup_U |f|$ is a semi-norm in $\mathcal{V}$ and any semi-norm in a finite dimensional space is dominated by any norm, there exists a constant $C = C_U$ such that

$$\sup_{x \in U} |f(x)| \le C\sqrt{\nu[f]} \quad \text{for any } f \in \mathcal{V}. \tag{2.55}$$

Set $\psi = 1 - \varphi$ where $\varphi \in \mathcal{T}(K,U)$, and define $\mathcal{V}'$ by

$$\mathcal{V}' = \{\psi f : f \in \mathcal{V}\}.$$

Clearly, $\mathcal{V}' \subset \mathcal{F}'$ and $\dim \mathcal{V}' \leq \dim \mathcal{V}$. Let us show that if U is small enough then $\dim \mathcal{V}' = \dim \mathcal{V}$. Indeed, let $\{U_n\}_{n=1}^{\infty}$ be a shrinking sequence of precompact open neighborhoods of K such that the intersection of all U_n is K. The expression

$$S_n(f) := \sup_{X \setminus U_n} |f|$$

defines an increasing sequence $\{S_n\}$ of semi-norms in $\mathcal{V}$. Since K has no interior, we obtain

$$\lim_{n\to\infty} S_n(f) = \sup_{X\setminus K} |f| = \sup_X |f|,$$

that is $\lim_{n\to\infty} S_n$ is a norm in $\mathcal{V}$. By a compactness argument, we conclude that S_n is a norm in $\mathcal{V}$ for some finite n. So, choose $U = U_n$ for this n. Since $\psi = 1$ on $X \setminus U$ we obtain, for any $f \in \mathcal{V}$,

$$\sup_X |\psi f| \geq \sup_{X\setminus U} |f| = S_n(f),$$

which implies that $\psi f \equiv 0$ if and only if $f \equiv 0$, and hence $\dim \mathcal{V}' = \dim \mathcal{V}$.

We are left to prove (2.53), which is equivalent to

$$\mathcal{E}[\psi f] < \nu[\psi f] \quad \text{for any } f \in \mathcal{V} \setminus \{0\}. \tag{2.56}$$

Indeed, using $\nabla\psi = -\nabla\varphi$, (2.55), and (2.54), we obtain

$$\begin{aligned}
\mathcal{E}[\psi f] &= \int_X |\nabla(\psi f)|^2 d\mu \\
&= \int_X |\nabla f|^2 \psi^2 d\mu + \int_U \left(-2f\psi\nabla f \cdot \nabla\varphi + f^2 |\nabla\varphi|^2\right) d\mu \\
&\leq \mathcal{E}[f] + 2C\sqrt{\nu[f]\,\mathcal{E}[f]\,\mathcal{E}[\varphi]} + C^2\nu[f]\,\mathcal{E}[\varphi] \\
&= \left(\sqrt{\mathcal{E}[f]} + C\sqrt{\nu[f]\,\mathcal{E}[\varphi]}\right)^2 \\
&\leq \left(\sqrt{c} + C\sqrt{\mathcal{E}[\varphi]}\right)^2 \nu[f]
\end{aligned}$$

and

$$\begin{aligned}
\nu[\psi f] &= \int_X f^2\psi^2 d\nu \\
&= \int_X f^2 d\nu + \int_U f^2\left(\psi^2 - 1\right) d\nu \\
&\geq \nu[f] - \sup_U |f|^2 |\nu|(U) \\
&\geq \left(1 - C^2|\nu|(U)\right)\nu[f],
\end{aligned}$$

whence

$$\frac{\mathcal{E}[\psi f]}{\nu[\psi f]} \leq \frac{\left(\sqrt{c} + C\sqrt{\mathcal{E}[\varphi]}\right)^2}{1 - C^2|\nu|(U)}. \tag{2.57}$$

Note that the best constant $C = C_U$ in (2.55) is decreasing when U is shrinking. Since $|\nu|(U) \leq |\sigma|(U) + \lambda\mu(U)$, $|\sigma|(K) = 0$, and $\mu(K) = 0$ (the latter is needed only in the case $\lambda \neq 0$), by choosing U small enough we can make $C_U^2|\nu|(U)$

arbitrarily close to 0. Since $\operatorname{cap}_{\mathcal{E}}(K,U) = 0$ we can choose φ to make $\mathcal{E}[\varphi]$ also arbitrarily close to 0. Finally, using $c < 1$, we see that the right hand side of (2.57) can be made smaller than 1, whence (2.56) follows. ∎

2.7. Fractal spaces. By fractal spaces we mean the fractal sets[6] in $\mathbb{R}^n$, obtained by certain self-similar constructions, like the celebrated Sierpinski gasket. Recall that the Sierpinski gasket SG is constructed from a unit equilateral triangle T in $\mathbb{R}^2$ by, firstly, removing the triangle with the vertices in the middles of the sides of T, then removing similar middle triangles from the three remaining equilateral triangles with the sides $\frac{1}{2}$, and so on (see Fig. 4).

FIGURE 4. Construction of the Sierpinski gasket: after 5 steps.

One can define a distance function d on SG as the induced Euclidean distance from $\mathbb{R}^2$, and a measure μ on SG as the Hausdorff measure $\mathcal{H}_\alpha$ where α is the Hausdorff dimension of SG (in fact, $\alpha = \log_2 3$). Furthermore, approximating SG by a sequence of graphs and considering a scaling limit of discrete energy forms on the approximating graphs, one defines a positive definite local energy form $(\mathcal{E},\mathcal{F})$ on SG, which is closable in $L^2(SG,\mu)$. Moreover, its closure $(\mathcal{E},\widetilde{\mathcal{F}})$ is a regular Dirichlet form (see for example [**2**]).

Similar structures can be introduced on most other fractals sets. Let (X,d) be a metric space, μ be a Radon measure on X, and $(\mathcal{E},\mathcal{F})$ be a positive definite energy form on X. If X is obtained by a self-similar construction like above then normally the metric, measure, and the energy structures on X exhibit certain homogeneity. Denote by $B(x,r)$ the metric ball of the radius r centered $x \in X$. Typically the following estimates hold on fractal spaces, for some positive parameters α and β:

$$\mu(B(x,r)) \simeq r^\alpha \tag{2.58}$$

and

$$\operatorname{cap}_{\mathcal{E}}(B(x,r), B(x,2r)) \simeq r^{\alpha-\beta}. \tag{2.59}$$

Here the sign $\simeq$ means that the ratio of the left hand side and the right hand side is bounded from above and below by positive constants, for the specified range of the arguments. The relations (2.58) and (2.59) are supposed to be true for all $x \in X$ and for some range $0 < r < R$ of the radius.

[6]For detailed account of fractals we refer the reader to [**2**], [**3**], [**4**], [**36**].

The parameter α from (2.58) is equal to the Hausdorff dimension of (X, d). The nature of the parameter β, which is called the *walk dimension*, is more complicated (see [**26**] or [**25**]). For example, for SG we have $\beta = \log_2 5$. The Euclidean space $\mathbb{R}^n$ with the Lebesgue measure μ satisfies (2.58) with $\alpha = n$, and the standard energy form $\mathcal{E}_\mu$ in $\mathbb{R}^n$ satisfies (2.59) with $\beta = 2$. Moreover, the capacity of the form $\mathcal{E}_\mu^{(m)}$ in $\mathbb{R}^n$ (defined by (2.49)) satisfies

$$\operatorname{cap}_{\mathcal{E}_\mu^{(m)}}(B(x,r), B(x,2r)) = c_{n,m}\, r^{n-2m}, \tag{2.60}$$

where $c_{n,m} > 0$, that is (2.59) with $\beta = 2m$ (see [**42**]). Hence, in this case β is equal to the order of the generator $(-\Delta)^m$.

Also in the general case β can be regarded as the order of the generator of the energy form. For most fractal spaces one has $\beta > 2$ although the generator is always a Markov operator (that is, satisfies the maximum principle) unlike the operators of order > 2 in $\mathbb{R}^n$.

We will come back to fractal spaces in Section 4.3.

3. Decomposition of a pseudometric space by annuli

In this section we prove Theorem 3.5, which is the main technical tool of this paper.

Definition 3.1. Given a set X, we say that a function $d : X \times X \to [0, +\infty)$ is a *pseudometric* (or a pseudodistance function) if d is symmetric, that is $d(x,y) = d(y,x)$, and if d satisfies the triangle inequality, that is

$$d(x,y) \le d(x,z) + d(z,y)$$

for all $x, y, z \in X$.

So, unlike the notion of a metric, we allow $d(x,y) = 0$ for distinct x, y. For any $x \in X$ and $r \ge 0$, define a ball $B(x,r)$ associated with a pseudometric d as follows:

$$B(x,r) = \{y \in X : d(x,y) < r\}.$$

Definition 3.2. A couple (X, d) is called a *pseudometric space* if X is a topological space, d is a pseudometric on X, and the function $y \mapsto d(x,y)$ is continuous for any $x \in X$ (consequently, all balls in a pseudometric space are open sets).

In particular, any metric space is a pseudometric space. Given any set X and a pseudometric d on it, one can define a topology on X using the balls of d as a base (although this topology is not necessarily Hausdorff). With this topology, (X, d) is a pseudometric space. However, in applications the set X may be a priori endowed with a different topology, for example, if X is a manifold, in which case we require that this topology is richer that the one induced by d.

Typically, a pseudometric space arises as follows. Let (X', d') be a metric space and let $T : X \to X'$ be a continuous mapping from a topological space X to X'. Then the identity

$$d(x,y) = d'(T(x), T(y))$$

defines a continuous pseudometric on X so that (X, d) is a pseudometric space.

Definition 3.3. Given $\kappa > 1$ and a positive integer N, we say that a pseudometric space (X, d) satisfies (κ, N)-*covering property* if, for any ball $B(x,r)$ in X, there exists a family of at most N balls of radii r/κ, which cover $B(x,r)$.

Lemma 3.4. *If a pseudometric space (X,d) satisfies (κ,N)-covering property then it satisfies (λ,M)-covering property for any $\lambda>1$ and some $M=M(\lambda,\kappa,N)$.*

Proof. Indeed, let n be a positive integer such that $\kappa^{n-1}<\lambda\leq\kappa^n$. Since the (λ,M)-covering property is monotone in λ, it suffices to assume that $\lambda=\kappa^n$. If $n=1$ then the the claim is trivial. Let us make the inductive step from n to $n+1$. Indeed, by the inductive hypothesis, any ball $B(x,r)$ can be covered by at most M_n balls $B(x_i,r/\kappa^n)$. By the assumption, each ball $B(x_i,r/\kappa^n)$ can be covered by at most N balls of radius r/κ^{n+1} each. Hence, $B(x,r)$ can be covered by at most $M_{n+1}:=NM_n$ balls of radius r/κ^{n+1} each, which settles the claim. ∎

It is useful to note that if (X,d) admits a *doubling* measure μ, that is, a Borel measure μ such that, for all $x\in X$ and $r>0$ and for some constant C,

$$0<\mu(B(x,2r))\leq C\mu(B(x,r))<\infty,$$

then (X,d) satisfies $(2,N)$ covering property with $N\leq C^3$.

In a pseudometric space (X,d), for any $x\in X$ and $0\leq r\leq R$ define the annulus

$$A(x,r,R):=B(x,R)-B(x,r)=\{y\in X:r\leq d(x,y)<R\}.$$

Note that $A(x,0,R)=B(x,R)$.

For any annulus $A=A(x,r,R)$ and $\lambda\geq1$ denote by λA the following annulus:

$$\lambda A=A\left(x,\lambda^{-1}r,\lambda R\right).$$

Similarly, for $B=B(x,r)$ and $\lambda>0$, set $\lambda B=B(x,\lambda r)$.

Theorem 3.5. *Let a pseudometric space (X,d) satisfy $(2,N)$-covering property. Let ν be a Borel measure on X, and assume that there exist positive reals υ and ρ such that*

$$\forall x\in X\quad \nu(B(x,\rho/2))\leq\upsilon\quad \text{and}\quad \exists x_0\in X\quad \nu(B(x_0,\rho))>\upsilon. \tag{3.1}$$

Then, for any $\lambda>1$, there exists a family $\mathcal{A}$ of $\lceil c\frac{\nu(X)}{\upsilon}\rceil$ annuli in X satisfying the following properties:

(a) *$\nu(A)\geq\upsilon$ for any $A\in\mathcal{A}$;*
(b) *the annuli $\{\lambda A\}_{A\in\mathcal{A}}$ are disjoint.*

Here c is a positive constant depending only on λ and N (for example, one can define it by $c^{-1}=2+4M\left(200\lambda^3,2,N\right)$).

Remark 3.6. *If $\nu(X)=\infty$ then we interpret $\lceil c\frac{\nu(X)}{\upsilon}\rceil$ as ∞. In this case we claim the existence of an infinite (countable) family $\mathcal{A}$ of annuli with the properties (a) and (b).*

If $\nu(X)<\infty$ then one cannot have more than $\frac{\nu(X)}{\upsilon}$ disjoint sets each with measure at least υ. Hence, the number $\lceil c\frac{\nu(X)}{\upsilon}\rceil$ of disjoint annuli guarantied by Theorem 3.5 is optimal up to a constant factor.

The hypothesis (3.1) excludes, in particular, a situation when measure ν is concentrated in a few atoms, in which case the conclusion of Theorem 3.5 is no longer true. Lemma 3.10 will provide a simple sufficient condition for (3.1).

Remark 3.7. For some applications it is useful to know that each annulus in the family $\mathcal{A}$, which is constructed in the proof, either has the internal radius at least $\rho/2$ or is a ball of the radius at least $\rho/2$.

We precede the proof of Theorem 3.5 by an elementary lemma.

Lemma 3.8. *Let $B(x,r)$ and $B(y,s)$ be two balls in (X,d) and $\lambda>1$ be a real such that*

$$\lambda B(x,r)\cap\frac{1}{2}B(y,s)\neq\emptyset \quad \text{and} \quad B(x,r)\setminus B(y,s)\neq\emptyset\,. \tag{3.2}$$

Then the following inclusions take place:

$$B(x,r) \subset B(y,(2\lambda+2)r)\,, \tag{3.3}$$

$$B(y,s) \subset B(x,(4\lambda+3)r)\,, \tag{3.4}$$

$$B(y,\eta r) \subset B(x,(\eta+2\lambda+1)r) \quad \text{for any } \eta>0. \tag{3.5}$$

PROOF. The hypotheses (3.2) imply that

$$\begin{aligned} d(x,y) &\leq \lambda r+\frac{s}{2} \\ s &\leq d(x,y)+r\,. \end{aligned}$$

(see Fig. 5).

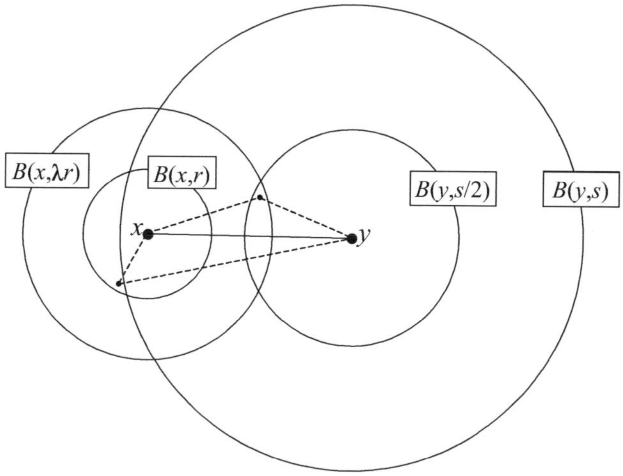

FIGURE 5. Illustration to Lemma 3.8

It follows that

$$d(x,y)\leq\lambda r+\frac{d(x,y)+r}{2},$$

and

$$d(x,y)\leq(2\lambda+1)\,r,$$

whence

$$s\leq(2\lambda+1)\,r+r=(2\lambda+2)\,r. \tag{3.6}$$

Using the above two inequalities, we obtain

$$B(x,r)\subset B(y,r+d(x,y))\subset B(y,(2\lambda+2)\,r)\,.$$

$$B(y,s)\subset B(x,s+d(x,y))\subset B(x,(4\lambda+3)\,r)\,,$$

$$B(y,\eta r)\subset B(x,\eta r+d(x,y))\subset B(x,(\eta+2\lambda+1)\,r)\,,$$

which was to be proved. ∎

PROOF OF THEOREM 3.5. Set for simplicity

$$m(\kappa) = M(\kappa, 2, N),$$

where $M(\kappa, 2, N)$ is the constant from Lemma 3.4. Hence, any ball of radius r can be covered by at most $m(\kappa)$ balls of radii r/κ.

Let n be a positive integer such that

$$n \leq \lceil c\frac{\nu(X)}{v}\rceil. \tag{3.7}$$

We will construct two sequences $\{\mathcal{A}_i\}$ and $\{\mathcal{B}_i\}$ where $i = 1, 2, ..., 2n$, $\mathcal{A}_i$ is a family of annuli in X, and $\mathcal{B}_i$ is a family of balls in X. These families will satisfy the following properties, for all $i = 1, 2, ..., 2n$:

(i) for any $a \in \mathcal{A}_i$,

$$\nu(a) \geq v;$$

(ii) the annuli $\{\lambda a\}_{a\in\mathcal{A}_i}$ are disjoint;

(iii) the following inclusion takes place:

$$\bigcup_{a\in\mathcal{A}_i} \lambda a \subset \bigcup_{b\in\mathcal{B}_i} \frac{1}{2\lambda} b; \tag{3.8}$$

(iv) the following inequality takes place:

$$\nu\left(\bigcup_{b\in\mathcal{B}_i} b\right) \leq Cvi, \tag{3.9}$$

where $C = m\left(200\lambda^3\right)$;

(v) $|\mathcal{A}_1| = |\mathcal{B}_1| = 1$, and if $i > 1$ then
- either $|\mathcal{A}_i| = |\mathcal{A}_{i-1}| + 1$ and $|\mathcal{B}_i| \leq |\mathcal{B}_{i-1}| + 1$,
- or $|\mathcal{A}_i| = |\mathcal{A}_{i-1}|$ and $|\mathcal{B}_i| \leq |\mathcal{B}_{i-1}| - 1$;

(vi) if $i > 1$ then $\mathcal{A}_i \supset \mathcal{A}_{i-1}$;

(vii) each annulus in $\mathcal{A}_i$ either has the internal radius at least $\rho/2$ or is a ball of the radius at least $\rho/2$, and each ball in $\mathcal{B}_i$ has the radius at least ρ.

Before we actually construct these sequences, let us show how the existence of them proves the theorem. Indeed, each family $\mathcal{A}_i$ obviously satisfies (a) and (b). Let us show that one of the families $\mathcal{A}_i$ contains at least n annuli. If $n = 1$ then we have already $|\mathcal{A}_1| = 1$. For an arbitrary n, let us verify that $|\mathcal{A}_{2n}| \geq n$. To that end, observe that by (v) the numerical sequence $\{2|\mathcal{A}_i| - |\mathcal{B}_i|\}_{i=1}^{2n}$ is strictly increasing whereas $2|\mathcal{A}_1| - |\mathcal{B}_1| = 1$. This implies

$$2|\mathcal{A}_{2n}| - |\mathcal{B}_{2n}| \geq 2n$$

and hence $|\mathcal{A}_{2n}| \geq n$. If $\nu(X) < \infty$ then this finishes the proof because we can take $n = \lceil c\frac{\nu(X)}{v}\rceil$. If $\nu(X) = \infty$ then we can do the above construction for any n. The union $\mathcal{A} = \bigcup_{i=1}^{\infty} \mathcal{A}_i$ is infinite and, since the sequence $\{\mathcal{A}_i\}$ is increasing, $\mathcal{A}$ satisfies the conditions (a) and (b), which finishes the proof in this case. The condition (vii) is not needed for the proof of Theorem 3.5 but settles the claim of Remark 3.7.

Now we construct the families $\mathcal{A}_i$ and $\mathcal{B}_i$ by induction in i. For the inductive basis $i = 1$, fix $x_0 \in X$ and $\rho > 0$ satisfying (3.1). Define the set $\mathcal{A}_1$ to consist of a single annulus $A(x_0, 0, \rho)$, and the set $\mathcal{B}_1$ to consist of a single ball $B\left(x_0, 2\lambda^2\rho\right)$. Then the properties ($i$), ($ii$), ($iii$), ($v$), ($vi$), ($vii$) are trivially satisfied. To prove

(iv), let us observe that $B\left(x_0, 2\lambda^2\rho\right)$ is covered by at most $m\left(4\lambda^2\right) \le C$ balls of radii $\rho/2$, and by (3.1) the measure of each ball of radius $\rho/2$ is at most v. Hence,

$$\nu\left(B\left(x_0, 2\lambda^2\rho\right)\right) \le Cv,$$

which is exactly (3.9). If $n = 1$ then the construction of $\mathcal{A}_1$ finishes the proof. Hence, in the sequel we assume $n > 1$.

Assuming that $i \le 2n - 1$ and that $\mathcal{A}_i$ and $\mathcal{B}_i$ are already defined, let us construct $\mathcal{A}_{i+1}$ and $\mathcal{B}_{i+1}$. Set

$$\overline{r} := \sup\left\{r : \nu\left(B(x,r) \setminus \bigcup_{b \in \mathcal{B}_i} b\right) \le v \quad \text{for all } x \in X\right\} \tag{3.10}$$

and observe that $\overline{r} \ge \rho/2$ and hence $\overline{r} > 0$. It follows from (3.7) and $n > 1$ that

$$n < c\frac{\nu(X)}{v} + 1 < 2c\frac{\nu(X)}{v}. \tag{3.11}$$

Hence, we obtain from (3.9), (3.11), and $i < 2n$ that

$$\nu\left(X \setminus \bigcup_{b \in \mathcal{B}_i} b\right) > \nu(X) - 2nCv > \left(\frac{1}{2c} - 2C\right) nv > v$$

because $c^{-1} = 2 + 4C$ by the definitions of c and C. We conclude that there exists a ball $B(x,r)$ such that

$$\nu\left(B(x,r) \setminus \bigcup_{b \in \mathcal{B}_i} b\right) > v, \tag{3.12}$$

and, furthermore, we can assume that $\overline{r} \le r < 2\overline{r}$ (see Fig. 6). In particular, we have $r \ge \rho/2$.

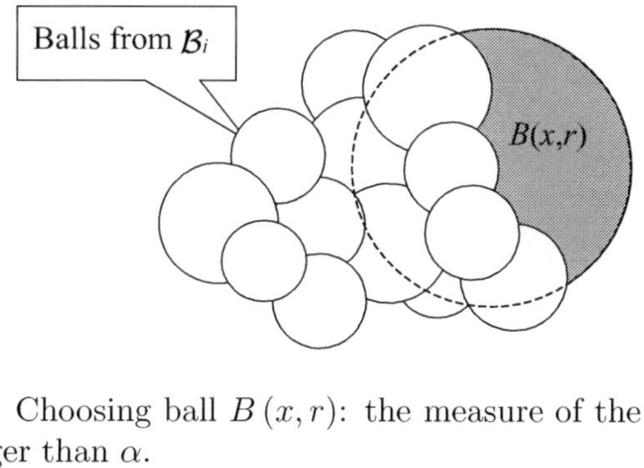

FIGURE 6. Choosing ball $B(x,r)$: the measure of the shaded region is larger than α.

Set

$$k := \operatorname{card}\left\{b \in \mathcal{B}_i : B\left(x, \lambda' r\right) \cap \frac{1}{2} b \ne \emptyset\right\} \tag{3.13}$$

where

$$\lambda' := 7\lambda^2,$$

and consider the following three cases.

Case $k = 0$. Since $1 < \lambda < \lambda'$, in this case we have

$$B(x, \lambda r) \cap \frac{1}{2\lambda} b = \emptyset \quad \text{for any } b \in \mathcal{B}_i. \tag{3.14}$$

Define $\mathcal{A}_{i+1}$ and $\mathcal{B}_{i+1}$ by

$$\mathcal{A}_{i+1} = \mathcal{A}_i \cup \{A(x, 0, r)\} \quad \text{and} \quad \mathcal{B}_{i+1} = \mathcal{B}_i \cup \{B(x, 2\lambda^2 r)\}.$$

Then condition (i) holds by (3.12), and conditions (iii), (v), (vi) are trivially satisfied. Condition (vii) is satisfied because $r \geq \rho/2$.

Let us prove (ii). By the inductive hypothesis, we have

$$\bigcup_{a \in \mathcal{A}_i} \lambda a \subset \bigcup_{b \in \mathcal{B}_i} \frac{1}{2\lambda} b$$

(see Fig. 7)

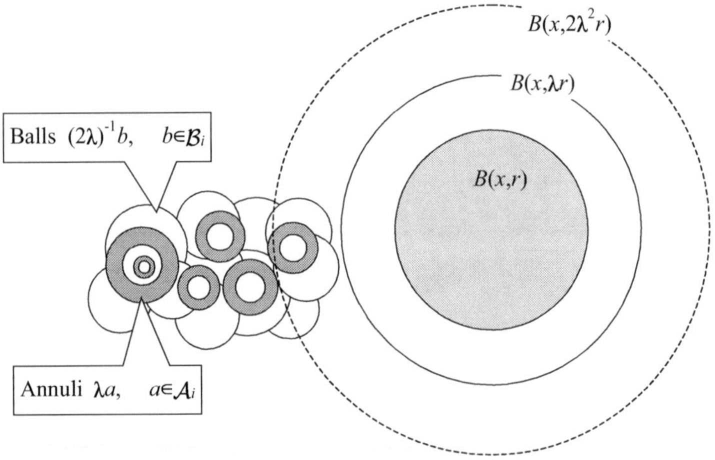

FIGURE 7. Ball $B(x, \lambda r)$ does not intersect $\frac{1}{2\lambda} b$, whereas the union of all λa is covered by the union of all $\frac{1}{2\lambda} b$.

Together with (3.14), this implies that $\lambda A(x, 0, r) = B(x, \lambda r)$ does not intersect any annulus λa, $a \in \mathcal{A}_i$, whence (ii) follows.

Let us prove (iv). Since $r/2 < \overline{r}$, it follows from (3.10) that, for any $z \in X$,

$$\nu\left(B(z, r/2) \setminus \bigcup_{b \in \mathcal{B}_i} b\right) \leq v. \tag{3.15}$$

The ball $B(x, 2\lambda^2 r)$ can be covered by at most $m(4\lambda^2) \leq C$ balls of radii $r/2$ whence it follows that

$$\nu\left(B(x, 2\lambda^2 r) \setminus \bigcup_{b \in \mathcal{B}_i} b\right) \leq Cv.$$

By the inductive hypothesis, we have

$$\nu\left(\bigcup_{b \in \mathcal{B}_i} b\right) \leq Cvi,$$

whence it follows that

$$\nu\left(\bigcup_{b\in\mathcal{B}_{i+1}} b\right) \le Cv\,(i+1)\,.$$

Case $k \ge 2$. Let us say that a ball $b \in \mathcal{B}_i$ is *selected* if

$$B\left(x,\lambda' r\right) \cap \frac{1}{2}b \neq \emptyset,$$

so that the number of selected balls is exactly k. In this case, let us set

$$\mathcal{A}_{i+1} = \mathcal{A}_i$$

and

$$\mathcal{B}_{i+1} = \mathcal{B}_i \setminus \{\text{all selected balls}\} \cup \left\{B\left(x,\lambda'' r\right)\right\}$$

where

$$\lambda'' := 14\lambda\lambda' = 98\lambda^3.$$

Conditions (i), (ii), (vi), (vii) are trivially satisfied. Condition (v) is satisfied because the number k of the selected balls removed from $\mathcal{B}_i$ is at least 2, whereas only a single ball $B\left(x,\lambda'' r\right)$ is added.

Let us prove (iii). Let $b \in \mathcal{B}_i$ be a selected ball. By definition, $B\left(x,\lambda' r\right) \cap \frac{1}{2}b$ is non-empty and, by (3.12), $B(x,r)\setminus b$ is non-empty, too. Then Lemma 3.8 (inclusion (3.4)) yields

$$b \subset B\left(x,\left(4\lambda'+3\right)r\right) \subset B\left(x,7\lambda' r\right) = \frac{1}{2\lambda}B\left(x,\lambda'' r\right)$$

(see Fig. 8).

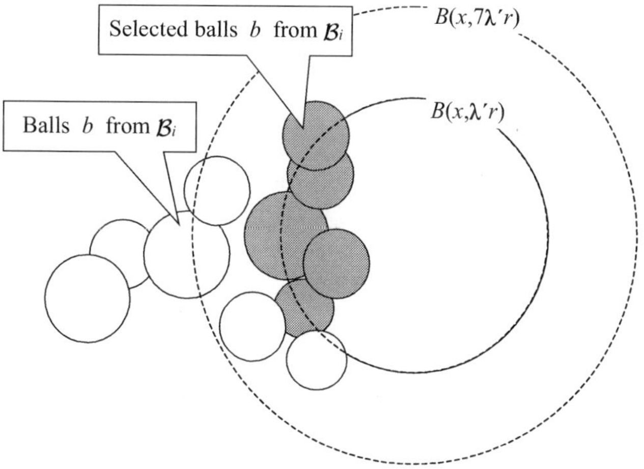

FIGURE 8. The selected balls (shaded) are inside $B\left(x,7\lambda' r\right)$.

In particular, the union of all balls $\frac{1}{2\lambda}b$, where b is selected, is covered by a single ball $\frac{1}{2\lambda}B\left(x,\lambda'' r\right)$, whence we obtain

$$\bigcup_{b\in\mathcal{B}_i}\frac{1}{2\lambda}b \subset \bigcup_{b\in\mathcal{B}_{i+1}}\frac{1}{2\lambda}b$$

The claim of (iii) follows then by the inductive hypothesis.

Let us prove (iv). The ball $B(x,\lambda''r)$ can be covered by at most $m(2\lambda'') \leq C$ balls of radii $r/2$ whence it follows from (3.15) that

$$\nu\left(B(x,\lambda''r)\setminus\bigcup_{b\in\mathcal{B}_i} b\right)\leq Cv.$$

By definition of $\mathcal{B}_{i+1}$ and by the inductive hypothesis, we obtain

$$\bigcup_{b\in\mathcal{B}_{i+1}} b\subset\left(\bigcup_{b\in\mathcal{B}_i} b\right)\cup\left(B(x,\lambda''r)\setminus\bigcup_{b\in\mathcal{B}_i} b\right)$$

and

$$\nu\left(\bigcup_{b\in\mathcal{B}_{i+1}} b\right)\leq Cvi+Cv=Cv(i+1).$$

Case $k=1$. Set

$$k_0 := \operatorname{card}\left\{b\in\mathcal{B}_i : B(x,\lambda r)\cap\frac{1}{2}b\neq\emptyset\right\} \tag{3.16}$$

and observe that $k_0\leq k$. Hence, either $k_0=0$ or $k_0=1$. If $k_0=0$ then the condition (3.14) is satisfied, and we can argue exactly as in the case $k=0$.

Let us consider the main case $k_0=1$. Let $B(y,s)\in\mathcal{B}_i$ be the unique ball such that

$$B(x,\lambda r)\cap\frac{1}{2}B(y,s)\neq\emptyset. \tag{3.17}$$

Set

$$\begin{aligned}\widetilde{a} &= A(y,\frac{1}{2}s,4\lambda r)\\ \widetilde{b} &= B(x,2\lambda\lambda' r)\end{aligned}$$

(see Fig. 9) and define the families $\mathcal{A}_{i+1}$, $\mathcal{B}_{i+1}$ by

$$\mathcal{A}_{i+1}=\mathcal{A}_i\cap\{\widetilde{a}\}\quad\text{and}\quad\mathcal{B}_{i+1}=\mathcal{B}_i\cup\{\widetilde{b}\}.$$

By (3.12) the difference $B(x,r)\setminus B(y,s)$ is non-empty, which together with (3.17) shows that the hypotheses of Lemma 3.8 are satisfied. By inequality (3.6) obtained in the proof of this lemma, we see that $s/2<4\lambda r$ so that the annulus $\widetilde{a}$ is well-defined.

Conditions (v), (vi) are obviously true. Condition (vii) is satisfied because by the inductive hypothesis the ball $B(y,s)$ (being an element of $\mathcal{B}_i$) has the radius $s\geq\rho$ and hence the annulus $\widetilde{a}$ has the internal radius at least $\rho/2$.

Let us verify (i) – (iv). By Lemma 3.8 we have

$$B(x,r)\subset B(y,(2\lambda+2)r)\subset B(y,4\lambda r).$$

Therefore,

$$\widetilde{a}=B(y,4\lambda r)\setminus B(y,\frac{1}{2}s)\supset B(x,r)\setminus B(y,s),$$

which together with (3.12) implies

$$\nu(\widetilde{a})\geq\nu\left(B(x,r)\setminus\bigcup_{b\in\mathcal{B}_i} b\right)\geq v,$$

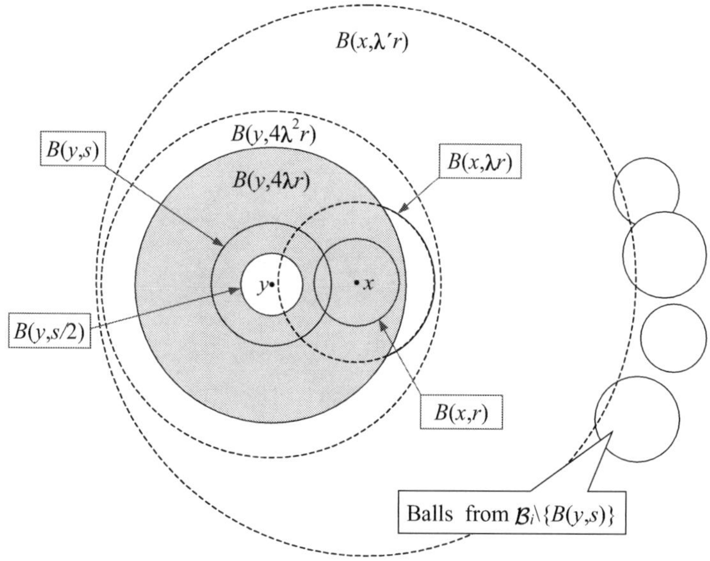

FIGURE 9. The annulus $\widetilde{a} = A\left(y, \frac{1}{2}s, 4\lambda r\right)$ is shaded.

whence (i) follows.

To verify (ii) we need to prove that

$$\lambda\widetilde{a} \cap \lambda a = \emptyset \quad \text{for all } a \in \mathcal{A}_i\,.$$

It it suffices to prove that

$$\lambda\widetilde{a} \cap \bigcup_{b \in \mathcal{B}_i} \frac{1}{2\lambda} b = \emptyset, \tag{3.18}$$

because by the inductive hypothesis,

$$\lambda a \subset \bigcup_{b \in \mathcal{B}_i} \frac{1}{2\lambda} b \quad \text{for all } a \in \mathcal{A}_i\,.$$

To prove (3.18), observe that

$$\lambda\widetilde{a} = A(y, \frac{1}{2\lambda}s, 4\lambda^2 r) = B\left(y, 4\lambda^2 r\right) \setminus \frac{1}{2\lambda} B\left(y, s\right).$$

Applying again Lemma 3.8 to the balls $B\left(x, r\right)$ and $B\left(y, s\right)$, we obtain by (3.5)

$$B\left(y, 4\lambda^2 r\right) \subset B\left(x, (4\lambda^2 + 2\lambda + 1)r\right) \subset B\left(x, \lambda' r\right),$$

whence

$$\lambda\widetilde{a} \subset B\left(x, \lambda' r\right) \setminus \frac{1}{2\lambda} B\left(y, s\right). \tag{3.19}$$

In particular, we immediately see that

$$\lambda\widetilde{a} \cap \frac{1}{2\lambda} B\left(y, s\right) = \emptyset. \tag{3.20}$$

By the hypothesis $k = 1$, there is a unique ball $b \in \mathcal{B}_i$ such that $\frac{1}{2}b$ intersects $B\left(x, \lambda' r\right)$, and this ball must be $B\left(y, s\right)$. Hence,

$$B\left(x, \lambda' r\right) \cap \frac{1}{2}b = \emptyset \quad \text{for any } b \in \mathcal{B}_i \setminus \{B\left(y, s\right)\},$$

whence it follows that

$$\lambda \widetilde{a} \cap \frac{1}{2\lambda}b = \emptyset \quad \text{for any } b \in \mathcal{B}_i \setminus \{B\left(y, s\right)\}. \tag{3.21}$$

Clearly, (3.18) follows from (3.20) and (3.21).

Let us verify (iii). Indeed, by (3.19) we have

$$\lambda \widetilde{a} \subset B\left(x, \lambda' r\right) = \frac{1}{2\lambda}\widetilde{b},$$

which together with the inductive hypothesis settles the claim.

Finally, let us verify (iv). To that end, it suffices to show that

$$\nu\left(\widetilde{b} \setminus \bigcup_{b \in \mathcal{B}_i} b\right) \leq Cv.$$

The ball $\widetilde{b} = B\left(x, 2\lambda\lambda' r\right)$ can be covered by at most $m\left(4\lambda\lambda'\right) \leq C$ balls of radii $r/2$, whence the claim follows from (3.15). ∎

Definition 3.9. A measure ν on a pseudometric space (X, d) is called *d-non-atomic* if, for any $x \in X$,

$$\lim_{r \to 0} \nu\left(B\left(x, r\right)\right) = 0.$$

In other words, this means that

$$\nu\left(B\left(x, 0+\right)\right) = 0$$

where

$$B\left(x, 0+\right) := \bigcap_{r > 0} B\left(x, r\right) = \{y \in X : d\left(x, y\right) = 0\}.$$

If d is a metric then $B\left(x, 0+\right) = \{x\}$ and, hence, ν is d-non-atomic if and only if ν is non-atomic in the usual sense, that is $\nu\left(\{x\}\right) = 0$ for any point x.

The following lemma provides a sufficient condition for hypothesis (3.1).

Lemma 3.10. *Let (X, d) be a pseudometric space and ν be a Borel measure on X. Assume that*

(i) *all balls in X are precompact;*
(ii) *measure ν is d-non-atomic and $0 < \nu\left(X\right) < \infty$.*

Then, for any $0 < v < \nu(X)$, there exists $\rho > 0$ satisfying (3.1), that is

$$\forall x \in X \quad \nu\left(B(x, \rho/2)\right) \leq v \quad \text{and} \quad \exists x_0 \in X \quad \nu\left(B(x_0, \rho)\right) > v. \tag{3.22}$$

PROOF. Define on the interval $(0, \infty)$ a function

$$V\left(r\right) := \sup_{x \in X} \nu\left(B\left(x, r\right)\right). \tag{3.23}$$

It is obvious that $V\left(r\right) \to \nu\left(X\right)$ as $r \to \infty$. Let us verify that $V\left(r\right) \to 0$ as $r \to 0$. Assuming the contrary, we obtain that there exists a number $\varepsilon > 0$ and sequences $\{x_k\}_{k=1}^{\infty} \subset X$ and $\{r_k\}_{k=1}^{\infty} \to 0$ such that $\nu\left(B(x_k, r_k)\right) \geq \varepsilon$ for all k. If there is a convergent subsequence $\{x_{k_i}\} \to x$ then this implies $\nu(B\left(x, r\right)) \geq \varepsilon$ for any $r > 0$, which contradicts the hypothesis that ν is d-non-atomic. If there is no convergent subsequence of $\{x_k\}$ then the compactness of balls implies that the sequence $\{x_k\}$ eventually leaves every ball. Therefore, for any fixed $r > 0$, there exists a subsequence $\{x_{k_i}\}$ such that the balls $B(x_{k_i}, r)$ are disjoint. By the finiteness of $\nu(X)$ we obtain that $\nu(B(x_{k_i}, r)) \to 0$, whence it follows that $\nu(B(x_{k_i}, r_{k_i})) \to 0$.

Fix $r > 0$ and set $r_k = r/2^k$, $k = 0, 1, 2, ...$ Then $V(r_k) \to 0$ as $k \to \infty$ and hence the union of the intervals $[V(r_{k+1}), V(r_k))$ is $(0, V(r))$. Therefore, for any $v \in (0, V(r))$, there exists an index k such that

$$V(r_{k+1}) \le v < V(r_k).$$

Letting $r \to \infty$, we obtain that for any $v \in (0, \nu(X))$ there exists $\rho > 0$ such that

$$V(\rho/2) \le v < V(\rho),$$

whence (3.22) follows.

Let us mention for the record that the radius ρ from (3.22) satisfies

$$\rho \ge V^{-1}(v), \tag{3.24}$$

where V^{-1} is the generalized inverse to $V(r)$. ∎

Corollary 3.11. *Let (X, d) be a pseudometric space and ν be a Radon measure on X. Assume that*

(i) X satisfies $(2, N)$-covering property:
(ii) all balls in X are precompact;
(iii) measure ν is d-non-atomic.

Then, for any $0 < v < \nu(X)$ and any positive integer $n \le \lceil c\frac{\nu(X)}{v}\rceil$ there exists a family $\mathcal{A}$ of n annuli in X satisfying the following properties:

(a) $\nu(A) \ge v$ for any $A \in \mathcal{A}$;
(b) the annuli $\{2A\}_{A\in\mathcal{A}}$ are disjoint.

Here c is a positive constant depending only on N (for example, one can define it by $c^{-1} = 2 + 4M(1600, 2, N)$).

Proof. If $\nu(X) = 0$ then the statement is void. If $0 < \nu(X) < \infty$ then, by Lemma 3.10, the hypothesis (3.1) of Theorem 3.5 is satisfied, whence the claim follows.

If $\nu(X) = \infty$ then, for given v and n, choose a ball $B \subset X$ so big that

$$v < \nu(B) \quad \text{and} \quad n \le \lceil c\frac{\nu(B)}{v}\rceil.$$

Note that $\nu(B) < \infty$ because ν is a Radon measure and B is precompact. Applying the previous case to the measure $\nu' = \mathbf{1}_B\nu$, we obtain a family $\mathcal{A}$ of

$$\lceil c\frac{\nu'(X)}{v}\rceil = \lceil c\frac{\nu(B)}{v}\rceil \ge n$$

annuli satisfying (a) and (b), which was to be proved. ∎

Corollary 3.12. *Under the hypotheses of* Corollary 3.11, *if in addition $0 < \nu(X) < \infty$ then, for any positive integer n, there exists a family $\mathcal{A}$ of n annuli in X such that*

(a) $\nu(A) \ge c\frac{\nu(X)}{n}$ for any $A \in \mathcal{A}$;
(b) the annuli $\{2A\}_{A\in\mathcal{A}}$ are disjoint.

Here c is the same as in Corollary 3.11.

PROOF. Indeed, setting $v = c\frac{\nu(X)}{n}$ we immediately obtain the claim from Corollary 3.11. ∎

Remark 3.13. In the case $0 < \nu(X) < \infty$ the radius ρ from (3.1) satisfies (3.24). Hence, by Remark 3.7, each annulus $A \in \mathcal{A}$ either has the internal radius $\geq \frac{1}{2}V^{-1}(v)$ or is a ball of the radius $\geq \frac{1}{2}V^{-1}(v)$.

4. Estimating the counting function of an energy form

In this section we will prove the main estimates of the counting function of an energy form on a pseudometric measure space.

4.1. Main estimate of the counting function. Let σ be a signed measure on a topological space X, and let σ_+ and σ_- be the positive and negative parts of σ, respectively, so that $\sigma = \sigma_+ - \sigma_-$. The latter means that for any Borel set $A \subset X$ we have $\sigma(A) = \sigma_+(A) - \sigma_-(A)$, so that $\sigma(A)$ makes sense and takes value in $[-\infty, +\infty]$ provided at least one of the values $\sigma_+(A)$ and $\sigma_-(A)$ is finite (note that both $\sigma_+(A)$ and $\sigma_-(A)$ are finite if σ is Radon and A is compact). If $\sigma_+(A) = \sigma_-(A) = +\infty$ then, strictly speaking, $\sigma(A)$ is not defined, but for the sake of simplification of the statements we use here the convention that $\sigma(A) = -\infty$. The same convention we will use for the difference of any two measures.

For any signed measure and for a constant $0 \leq \delta \leq 1$, denote by σ_δ the following signed measure:

$$\sigma_\delta := \delta\sigma_+ - \sigma_- . \tag{4.1}$$

Clearly, we have $(\sigma_\delta)_+ = \delta\sigma_+$ and $(\sigma_\delta)_- = \sigma_-$.

Now we can state the main result of this paper.

Theorem 4.1. *Let (X, d) be a pseudometric space and $(\mathcal{E}, \mathcal{F})$ be an energy form on X. Assume that, for some positive constants N, Q, the following conditions hold:*

(i) (X, d) satisfies $(2, N)$-covering property;
(ii) the energy form $(\mathcal{E}, \mathcal{F})$ is local and positive definite, and for any ball B in X, we have

$$\mathrm{cap}_{\mathcal{E}}(B, 2B) \leq Q. \tag{4.2}$$

Then, for any signed Radon measure σ on X such that σ_+ is d-non-atomic, we have

$$Neg\,(\mathcal{E} - \sigma) \geq \lfloor \frac{\sigma_\delta(X)}{5Q} \rfloor, \tag{4.3}$$

where $\delta \in (0, 1)$ depends only on N.

We do not claim that the constants 5 and δ in (4.3) are sharp.

Example 4.2. Let X be the Euclidean space $\mathbb{R}^n$ and $\mathcal{E} = \mathcal{E}_\mu^{(m)}$ be the m-th order energy form defined by (2.49) (where μ is the Lebesgue measure on $\mathbb{R}^n$). By (2.60), the hypothesis (4.2) holds provided $n = 2m$. Hence, for any non-atomic Radon measure σ on $\mathbb{R}^n$, we obtain by Theorem 4.1

$$Neg\,(\mathcal{E} - \sigma) \geq \lfloor c\sigma(\mathbb{R}^n) \rfloor$$

where $c > 0$ is a constant depending on n and m. In fact, one has here a better inequality

$$Neg\,(\mathcal{E} - \sigma) \geq c\sigma(\mathbb{R}^n),$$

as will follow from Theorem 4.17. Since the generator of the form $\mathcal{E}-\sigma$ is the operator $(-\Delta)^m - \sigma$, the above inequalities give also lower bounds for $Neg\left((-\Delta)^m - \sigma\right)$.

The proof of Theorem 4.1 will be given in Section 4.4 after a preparatory result in Section 4.2. Here we show what consequences can be deduced from (4.3).

Lemma 4.3. *Let σ and ν be two signed measures and $0 < \delta \leq 1$. Then the following is true.*

(a) $(\sigma+\nu)_\delta \geq \sigma_\delta + \nu_\sigma$;
(b) $\delta\sigma_\delta \geq \sigma_{\delta^2}$;
(c) If a measurable set A is such that either $\nu_+(A) = 0$ or $\nu_-(A) = 0$ then

$$(\sigma+\nu)_\delta (A) \geq (\sigma_{\delta^2} + \delta\nu)(A), \tag{4.4}$$

provided the right hand side of (4.4) is non-negative.

PROOF. (a) Let us first show that if μ_1 and μ_2 are two measures then

$$(\mu_1 - \mu_2)_\delta \geq \delta\mu_1 - \mu_2. \tag{4.5}$$

Indeed, set $\mu := \mu_1 - \mu_2$ and observe that there exists a measure α such that

$$\mu_1 = \mu_+ + \alpha \quad \text{and} \quad \mu_2 = \mu_- + \alpha.$$

Then

$$\delta\mu_1 - \mu_2 = \delta\left(\mu_+ + \alpha\right) - \left(\mu_- + \alpha\right) = \mu_\delta + (\delta - 1)\,\alpha \leq \mu_\delta,$$

that is (4.5).

Applying (4.5) to $\mu_1 = \sigma_+ + \nu_+$ and $\mu_2 = \sigma_- + \nu_-$, we obtain

$$(\sigma+\nu)_\delta \geq \delta\left(\sigma_+ + \nu_+\right) - \left(\sigma_- + \nu_-\right) = \sigma_\delta + \nu_\delta,$$

which was to be proved.

(b) We have

$$\delta\sigma_\delta = \delta^2\sigma_+ - \delta\sigma_- \geq \delta^2\sigma_+ - \sigma_- = \sigma_{\delta^2}.$$

(c) By part (a) we have

$$(\sigma+\nu)_\delta (A) \geq \sigma_\delta (A) + \delta\nu_+ (A) - \nu_- (A).$$

If $\nu_-(A) = 0$ then

$$(\sigma+\nu)_\delta (A) \geq \sigma_\delta (A) + \delta\nu (A) \geq (\sigma_{\delta^2} + \delta\nu)(A).$$

If $\nu_+(A) = 0$ then

$$(\sigma+\nu)_\delta (A) \geq \sigma_\delta (A) - \nu_- (A) = \sigma_\delta (A) + \nu (A). \tag{4.6}$$

By hypothesis, we have

$$\sigma_{\delta^2} (A) + \delta\nu (A) \geq 0,$$

which implies by part (b) than

$$\delta\sigma_\delta (A) + \delta\nu (A) \geq 0.$$

Hence, $\sigma_\delta (A) + \nu (A) \geq 0$ whence, by $\delta \leq 1$,

$$\sigma_\delta (A) + \nu (A) \geq \delta\sigma_\delta (A) + \delta\nu (A) \geq \sigma_{\delta^2} (A) + \delta\nu (A).$$

Substituting into (4.6) we finish the proof. ∎

Lemma 4.4. *Let $(\mathcal{E},\mathcal{F})$ be an energy form on a pseudometric space (X,d), and assume that, for any signed Radon measure σ on X such that σ_+ is d-non-atomic,*

$$Neg\,(\mathcal{E}-\sigma)\geq\lfloor\frac{\sigma_\delta(X)}{C}\rfloor, \tag{4.7}$$

where $0<\delta\leq 1$ and $C>0$ are some constant. If μ is a d-non-atomic measure on X then, for all real λ,

$$\mathcal{N}_\lambda\,(\mathcal{E}-\sigma,\mu)\geq\lfloor\frac{\sigma_{\delta^2}\,(X)+\delta\lambda\mu(X)}{C}\rfloor. \tag{4.8}$$

If $0<\mu\,(X)<\infty$ then, for any $k=1,2,...,$

$$\lambda_k(\mathcal{E}-\sigma,\mu)\leq\frac{Ck-\sigma_{\delta^2}(X)}{\delta\mu(X)}. \tag{4.9}$$

PROOF. Set $\widetilde{\sigma}=\sigma+\lambda\mu$. Then $\widetilde{\sigma}$ is a signed Radon measure on X, and $\widetilde{\sigma}_+$ is d-non-atomic. By (2.41) we have

$$\mathcal{N}_\lambda\,(\mathcal{E}-\sigma,\mu)=Neg\,(\mathcal{E}-\widetilde{\sigma})\,.$$

Applying (4.3) to $\widetilde{\sigma}$ instead of σ, we obtain

$$\mathcal{N}_\lambda\,(\mathcal{E}-\sigma,\mu)\geq\lfloor\frac{(\sigma+\lambda\mu)_\delta\,(X)}{C}\rfloor. \tag{4.10}$$

If $\sigma_{\delta^2}\,(X)+\delta\lambda\mu(X)\leq 0$ then (4.8) holds trivially. Otherwise, we have by Lemma 4.3,

$$(\sigma+\lambda\mu)_\delta\geq\sigma_{\delta^2}\,(X)+\delta\lambda\mu(X),$$

whence (4.8) follows. Finally, (4.9) follows from (4.8) by Lemma 2.8. ∎

Remark 4.5. Hence, the estimates (4.8) and (4.9) are true in the setting of Theorem 4.1 with $C=5Q$. It is not known yet whether the constant δ in Theorem 4.1 can be taken $\delta=1$. If so then $\sigma_\delta\,(X)$ and $\sigma_{\delta^2}\,(X)$ could be replaced by $\sigma\,(X)$, which would be most convenient for applications. If $\sigma\geq 0$ or $\sigma\leq 0$ then this can be achieved by changing the other constants. For example, if $\sigma\geq 0$ then $\sigma_{\delta^2}=\delta^2\sigma$ and (4.8) becomes

$$\mathcal{N}_\lambda\,(\mathcal{E}-\sigma,\mu)\geq\lfloor\frac{\delta\sigma\,(X)+\lambda\mu(X)}{C'}\rfloor, \tag{4.11}$$

where $C'=C/\delta$. Similarly, (4.9) becomes

$$\lambda_k(\mathcal{E}-\sigma,\mu)\leq\frac{C'k-\delta\sigma(X)}{\mu(X)}. \tag{4.12}$$

It would be very useful to prove (4.11) and/or (4.12) also for a signed measure σ. In Section 5.5 we prove (4.12) for a signed measure σ assuming in addition that $\mathcal{E}$ is a Riemannian energy form and the perturbed form $\mathcal{E}-\sigma$ is positive definite. We conjecture that (4.11) and (4.12) are true in general, without these assumptions.

Remark 4.6. If the floor function in (4.7) is replaced by the ceiling function (as will be in Theorem 4.17 below) then the floor function in (4.8) is also replaced by the ceiling function, and the term Ck in the right hand side of (4.9) is replaced by $C\,(k-1)$.

4.2. Decomposition of a pseudometric space by capacitors. We say that a sequence $\{(F_i, G_i)\}_{i=1}^n$ of capacitors is *disjoint* if the sequence of sets $\{G_i\}_{i=1}^n$ is disjoint. The following result plays the crucial role in the proof of Theorem 4.1.

Theorem 4.7. *Let (X, d) be a pseudometric space, ν be a Radon measure on X, and $(\mathcal{E}, \mathcal{F})$ be an energy form on X. Assume that the following hypotheses are satisfied, for some positive constants N, Q:*

(i) (X, d) satisfies $(2, N)$-covering property;
(ii) measure ν is d-non-atomic;
(iii) the energy form $(\mathcal{E}, \mathcal{F})$ is positive definite and, for any ball B in X,

$$\operatorname{cap}_{\mathcal{E}}(B, 2B) \leq Q. \tag{4.13}$$

Then, for any $0 < \upsilon < \nu(X)$ and any positive integer n such that

$$n \leq \lceil c\frac{\nu(X)}{\upsilon} \rceil, \tag{4.14}$$

there exist n disjoint capacitors (F_i, G_i) in X such that, for all $i = 1, 2, ..., n$,

$$\nu(F_i) \geq \upsilon \tag{4.15}$$

and

$$\operatorname{cap}_{\mathcal{E}}(F_i, G_i) \leq 4Q. \tag{4.16}$$

Here c is a positive constant depending only on N.

A version of this Theorem was proved in [**27**, Theorem 1] using an abstract version of Korevaar's argument [**37**]. The present Theorem 4.7 needs fewer assumptions about the energy form, unlike [**27**, Theorem 1], which required the energy form to satisfy the Markov property.

PROOF. The finiteness of $\operatorname{cap}_{\mathcal{E}}(B, 2B)$ implies that the ball B is precompact. Indeed, the class $\mathcal{T}(B, 2B)$ is non-empty, let $f \in \mathcal{T}(B, 2B)$. Since $\operatorname{supp} f$ is compact and $f|_B = 1$, it follows that $B \subset \operatorname{supp} f$ and B is precompact.

Hence, all the hypotheses of Corollary 3.11 are satisfied. By Corollary 3.11, there exists a sequence $\{F_i\}_{i=1}^n$ of n annuli such that $\nu(F_i) \geq \upsilon$ and the annuli $G_i := 2F_i$ are disjoint.

We are left to verify (4.16). Indeed, let $F_i = A(x, r, R)$ and, hence, $G_i = A(x, r/2, 2R)$. Let us write for simplicity $B_s = B(x, s)$. If $r = 0$ then by (4.13) we have

$$\operatorname{cap}_{\mathcal{E}}(F_i, G_i) = \operatorname{cap}_{\mathcal{E}}(B_R, B_{2R}) \leq Q.$$

If $r > 0$ then, by Lemma 2.3 and (4.13), we obtain

$$\begin{aligned} \operatorname{cap}_{\mathcal{E}}(F_i, G_i)^{1/2} &= \operatorname{cap}_{\mathcal{E}}(B_R \setminus B_r, B_{2R} \setminus B_{r/2})^{1/2} \\ &\leq \operatorname{cap}_{\mathcal{E}}(B_{r/2}, B_r)^{1/2} + \operatorname{cap}_{\mathcal{E}}(B_R, B_{2R})^{1/2} \\ &\leq 2Q^{1/2}, \end{aligned}$$

whence (4.16) follows. ∎

Remark 4.8. Let $0 < \nu(X) < \infty$. As follows from Remark 3.13, for each annulus $F_i = A(x, r, R)$ we have either $r \geq \frac{1}{2}V^{-1}(\upsilon)$ or $r = 0$ and $R \geq \frac{1}{2}V^{-1}(\upsilon)$, where V^{-1} is the generalized inverse to the function $V(r)$ defined by (3.23). Therefore, as we see from the above proof, the hypothesis (4.13) can be restricted to the balls of radii $\geq \frac{1}{4}V^{-1}(\upsilon)$.

Corollary 4.9. *Under the hypotheses of* Theorem 4.7, *if* $0 < \nu(X) < \infty$ *then, for any positive integer* n, *there exist* n *disjoint capacitors* (F_i, G_i) *such that, for all* $i = 1, 2, ..., n$,

$$\nu(F_i) \geq c\frac{\nu(X)}{n}, \tag{4.17}$$

and

$$\operatorname{cap}_{\mathcal{E}}(F_i, G_i) \leq 4Q, \tag{4.18}$$

where c *is the same as in* Theorem 4.7.

PROOF. Indeed, given a positive integer n, define υ by

$$\upsilon = c\frac{\nu(X)}{n},$$

where c is the constant from Theorem 4.7. Then $0 < \upsilon < \nu(X)$, and the conclusion follows from Theorem 4.7. ∎

Remark 4.10. As follows from Remark 4.8, the hypothesis (4.13) can be restricted to the balls of radii $\geq \frac{1}{4}V^{-1}\left(\frac{c\nu(X)}{n}\right)$, where $V(r)$ defined by (3.23).

To show an example of application of Corollary 4.9, let us give a direct proof of the estimate (4.9) in the case $\sigma = 0$ (without using Theorem 4.1), which contains the main idea of the proof of Theorem 4.1.

Corollary 4.11. *Let* (X, d) *be a pseudometric space,* ν *be a Radon measure on* X, *and* $(\mathcal{E}, \mathcal{F})$ *be an energy form on* X. *Assume that, for some positive constants* N, Q, *the following conditions hold:*

(*i*) (X, d) *satisfies* $(2, N)$*-covering property;*
(*ii*) *measure* ν *is a* d*-non-atomic and* $0 < \nu(X) < \infty$;
(*iii*) *the energy form* $(\mathcal{E}, \mathcal{F})$ *is local and positive definite, and for any ball* B *in* X, *we have*

$$\operatorname{cap}_{\mathcal{E}}(B, 2B) \leq Q\ ; \tag{4.19}$$

Then, for any $k = 1, 2, ...,$

$$\lambda_k(\mathcal{E}, \nu) \leq CQ\frac{k}{\nu(X)}, \tag{4.20}$$

where the positive constant C *depends only on* N.

PROOF. By Corollary 4.9, there exists k disjoint capacitors (F_i, G_i) satisfying (4.17) and (4.18). For any $\varepsilon > 0$ there exists a test function $f_i \in \mathcal{T}(F_i, G_i)$ such that

$$\mathcal{E}[f_i] < \operatorname{cap}_{\mathcal{E}}(F_i, G_i) + \varepsilon \leq 4Q + \varepsilon.$$

Take $\varepsilon = Q$ and fix such a function f_i. Since $f_i|_{F_i} = 1$, we have

$$\nu[f_i] \geq \nu(F_i) \geq c\frac{\nu(X)}{k},$$

whence

$$\frac{\mathcal{E}[f_i]}{\nu[f_i]} < \frac{5Q}{c\frac{\nu(X)}{k}} = CQ\frac{k}{\nu(X)} =: \lambda,$$

where $C := 5/c$.

Hence, for any $f = f_i$ we have

$$\mathcal{E}[f] < \lambda\nu[f]. \tag{4.21}$$

Since the supports of the functions f_i are disjoint and the form $\mathcal{E}$ is local, we conclude that (4.21) holds for any function $f \in \operatorname{span}(f_1, ..., f_n)$ except for $f = 0$. By the definition of the counting function, we obtain $\mathcal{N}_\lambda(\mathcal{E}, \nu) \geq k$, whereas

$$\mathcal{N}_{\lambda_k}(\mathcal{E}, \nu) = \sup\{n : \lambda_n < \lambda_k\} \leq k - 1.$$

Therefore, $\lambda_k \leq \lambda$ which was to be proved. ∎

Remark 4.12. As follows from Remark 4.10, in order to obtain (4.20) for a *fixed* index k, it suffices to assume the hypothesis (4.19) only for the balls of radii at least $\frac{1}{4}V^{-1}\left(\frac{c\nu(X)}{k}\right)$.

Example 4.13. Let X be the Euclidean space $\mathbb{R}^n$, μ be the Lebesgue measure on $\mathbb{R}^n$, and $\mathcal{E} = \mathcal{E}_\mu^{(m)}$ be the m-th order energy form on $\mathbb{R}^n$ defined by (2.49). If $n = 2m$ then the capacity of the form $\mathcal{E}$ satisfies (4.19) as follows from (2.60). Let q be a smooth positive function on $\mathbb{R}^n$, and consider the measure ν defined by $d\nu = qd\mu$. A generator of the form $\mathcal{E}$ is the operator $\frac{1}{q}(-\Delta)^m$. We obtain from Corollary 4.11 that if $2m = n$ and

$$\nu(\mathbb{R}^n) = \int_{\mathbb{R}^n} qd\mu < \infty$$

then

$$\lambda_k\left(\frac{1}{q}(-\Delta)^m\right) \leq C\frac{k}{\nu(\mathbb{R}^n)}.$$

4.3. Eigenvalues on fractal spaces. Let us show how Corollary 4.11 can be used in conjunction with Remark 4.12 to handle the case when the capacity uniform bound (4.19) is not available but, instead, one has (2.58) and (2.59). The next statement applies to most fractal spaces.

Corollary 4.14. *Let (X, d) be a pseudometric space, ν be a Radon measure on X, and $(\mathcal{E}, \mathcal{F})$ be an energy form on X. Assume that, for some positive constants C_1, C_2, N the following conditions hold:*

(i) *(X, d) satisfies $(2, N)$-covering property;*

(ii) *measure ν is a d-non-atomic, $0 < \nu(X) < \infty$, and, for any ball $B(x, r)$ in X,*

$$\nu(B(x, r)) \leq C_1 r^\alpha, \tag{4.22}$$

where $\alpha > 0$;

(iii) *the energy form $(\mathcal{E}, \mathcal{F})$ is local and positive definite, and, for any ball $B(x, r)$ in X, we have*

$$\operatorname{cap}_{\mathcal{E}}(B(x, r), B(x, 2r)) \leq C_2 r^{\alpha - \beta}, \tag{4.23}$$

where $\beta \geq \alpha$;

Then, for all $k = 1, 2, ...,$

$$\lambda_k(\mathcal{E}, \nu) \leq C\left(\frac{k}{\nu(X)}\right)^{\beta/a}, \tag{4.24}$$

where the constant C depends on $N, C_1, C_2, \alpha, \beta$.

Remark 4.15. In the case $\alpha = \beta$ the hypothesis (4.23) becomes (4.19) and hence, by Corollary 4.11, the hypothesis (4.22) can be omitted. In the case $\alpha \neq \beta$ the hypothesis (4.22) is essential because the value of $\alpha - \beta$ from (4.23) does not allow to recover the value of β/α necessary for (4.24). It is not clear whether the condition $\beta \geq \alpha$ can be dropped here.

PROOF. By (4.22) and by definition (3.23) of $V(r)$, we have $V(r) \leq C_1 r^{\alpha}$ whence

$$V^{-1}(v) \geq (v/C_1)^{1/\alpha}.$$

Fix an index k. If $r \geq \frac{1}{4}V^{-1}\left(\frac{c\nu(X)}{k}\right)$ (where $c > 0$ is the constant from Corollary 4.11) then

$$r \geq \frac{1}{4}\left(\frac{c\nu(X)}{C_1 k}\right)^{1/\alpha}.$$

Therefore, using $\alpha - \beta \leq 0$ and (4.23), we obtain, for any ball B of such a radius r,

$$\operatorname{cap}_{\mathcal{E}}(B, 2B) \leq C_2 r^{\alpha-\beta} \leq C_2\left(\frac{1}{4}\left(\frac{c\nu(X)}{C_1 k}\right)^{1/\alpha}\right)^{\alpha-\beta} = C'\left(\frac{k}{\nu(X)}\right)^{\beta/\alpha-1} =: Q.$$

By Remark 4.12, we can apply Corollary 4.11 and, hence, obtain by (4.20)

$$\lambda_k(\mathcal{E}, \nu) \leq CQ\frac{k}{\nu(X)} = CC'\left(\frac{k}{\nu(X)}\right)^{\beta/\alpha},$$

which was to be proved. ∎

4.4. Proof of the main estimate. Here we prove Theorem 4.1, that is the following estimate

$$Neg(\mathcal{E} - \sigma) \geq \lfloor\frac{\sigma_\delta(X)}{5Q}\rfloor =: k. \tag{4.25}$$

If $k \leq 0$ then there is nothing to do, so we assume in the sequel that $k > 0$. In particular, this implies $\sigma_\delta(X) > 0$ and hence $\sigma_-(X) < \infty$.

Without loss of generality we can assume that also $\sigma_+(X) < \infty$. Indeed, if $\sigma_+(X) = \infty$ then consider a signed measure

$$\sigma^{(r)} := \mathbf{1}_{B(x,r)}\sigma_+ - \sigma_-,$$

where $x \in X$ is fixed and $r > 0$. Hypothesis (4.2) implies that the ball $B(x,r)$ is precompact (see the proof of Theorem 4.7). Since the measure σ_+ is Radon, we see that

$$|\sigma^{(r)}|(X) = \sigma_+(B(x,r)) + \sigma_-(X) < \infty.$$

If we can prove that

$$Neg\left(\mathcal{E} - \sigma^{(r)}\right) \geq \lfloor\frac{\sigma_\delta^{(r)}(X)}{5Q}\rfloor$$

then passing to the limit as $r \to \infty$ and using

$$Neg(\mathcal{E} - \sigma) \geq Neg\left(\mathcal{E} - \sigma^{(r)}\right),$$

we will obtain (4.25). Hence, we can assume in the sequel that $|\sigma|(X) < \infty$.

In order to prove (4.25) it suffices to construct k linearly independent functions $f_1, f_2, ..., f_k$ in $\mathcal{F}$ such that

$$\mathcal{E}[f] - \sigma[f] < 0 \tag{4.26}$$

for any $f \in \operatorname{span}\{f_i\} \setminus \{0\}$. The functions f_i in our construction will have disjoint supports. Hence, by the locality of $\mathcal{E} - \sigma$, we have $(\mathcal{E} - \sigma)(f_i, f_j) = 0$ for all $i \neq j$, so it suffices to establish (4.26) for $f = f_i$, $i = 1, 2, ..., k$.

Suppose that we have k disjoint capacitors (F_i, G_i) in X of finite capacity. Then we choose $f_i \in \mathcal{T}(F_i, G_i)$ to be a nearly optimal test function in the sense that

$$\mathcal{E}[f_i] < \operatorname{cap}_{\mathcal{E}}(F_i, G_i) + \varepsilon,$$

where $\varepsilon > 0$ will be specified later on (see Fig. 10).

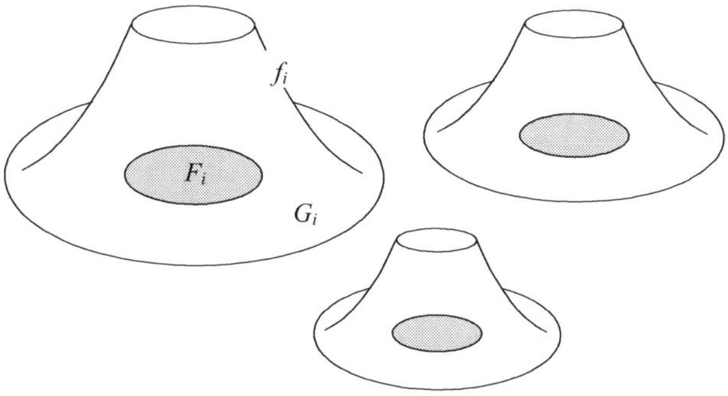

FIGURE 10. Capacitors (F_i, G_i) and their test functions.

By the definition of $\mathcal{T}(F_i, G_i)$, we have $f_i \in \mathcal{F} \cap C_0(G_i)$, $f_i = 1$ on F_i, and $0 \leq f_i \leq 1$. Therefore,

$$\sigma_-[f_i] \leq \sigma_-(G_i) \quad \text{and} \quad \sigma_+[f_i] \geq \sigma_+(F_i).$$

Then (4.26) will follow if we know that

$$\operatorname{cap}_{\mathcal{E}}(F_i, G_i) + \varepsilon + \sigma_-(G_i) \leq \sigma_+(F_i). \tag{4.27}$$

The assumption $\sigma_\delta(X) > 0$ implies $\sigma_+(X) > 0$. Hence, the metric-measure space (X, d, σ_+) with the form $(\mathcal{E}, \mathcal{F})$ satisfies all the hypotheses of Corollary 4.9. By Corollary 4.9, for any $k = 1, 2, ...,$ there exists $n = 2k$ disjoint capacitors (F_i, G_i) on X such that

$$\sigma_+(F_i) \geq c\frac{\sigma_+(X)}{2k} \quad \text{and} \quad \operatorname{cap}_{\mathcal{E}}(F_i, G_i) \leq 4Q. \tag{4.28}$$

Since

$$\sum_{i=1}^{2k} \sigma_-(G_i) \leq \sigma_-(X),$$

there are at most k sets G_i for which

$$\sigma_-(G_i) > \frac{\sigma_-(X)}{k},$$

and, hence, there are at least $n - k = k$ sets G_i, for which

$$\sigma_-(G_i) \leq \frac{\sigma_-(X)}{k}. \tag{4.29}$$

We can assume that these sets are G_i, $i=1,2,...,k$. Substituting (4.28) and (4.29) into (4.27) and choosing $\varepsilon = Q$, we see that (4.27) will be satisfied for capacitors (F_i, G_i), $i=1,2,...,k$, provided

$$5Q + \frac{\sigma_-(X)}{k} \leq c\frac{\sigma_+(X)}{2k}. \tag{4.30}$$

Set $\delta := c/2$ and observe that (4.30) is equivalent to

$$k \leq \frac{\delta\sigma_+(X) - \sigma_-(X)}{5Q} = \frac{\sigma_\delta(X)}{5Q},$$

which is true by the choice of k.

Remark 4.16. In the case $|\sigma|(X) < \infty$, by Remark 4.10, the hypothesis (4.2) of Theorem 4.1 can be assumed only for the balls of radii $\geq \frac{1}{4}V^{-1}\left(c\frac{\sigma_+(X)}{2k}\right)$ where $k = \lfloor \frac{\sigma_\delta(X)}{5Q} \rfloor$ and V^{-1} is the generalized inverse to the function

$$V(r) = \sup_{x\in X} \sigma_+(B(x,r)). \tag{4.31}$$

4.5. Strongly local energy forms. In this section, we slightly improve the estimate of Theorem 4.1 for *strongly* local energy forms.

Theorem 4.17. *Suppose that, under the hypotheses of* Theorem 4.1 *the energy form* $(\mathcal{E},\mathcal{F})$ *is strongly local. Then the estimate* (4.3) *in the conclusion of* Theorem 4.1 *can be replaced by*

$$Neg(\mathcal{E} - \sigma) \geq \lceil \frac{\sigma_\delta(X)}{10Q} \rceil. \tag{4.32}$$

The point of this theorem is that the *floor* function in the estimate (4.3) can replaced by the *ceiling* function as in (4.32) This improvement is only noticeable when, for example, $0 < \frac{\sigma_\delta(X)}{10Q} < 1$. In this case, the estimate (4.3) becomes trivial whereas (4.32) still gives $Neg(\mathcal{E} - \sigma) \geq 1$. The strong locality hypothesis is essential for that, as it will be shown in the example at the end of this section.

Similarly to Lemma 4.4 (cf. Remark 4.6), the estimate (4.32) implies the following. If μ is a d-non-atomic Radon measure on X then, for all real λ,

$$\mathcal{N}_\lambda(\mathcal{E} - \sigma, \mu) \geq \lceil \frac{\sigma_{\delta^2}(X) + \delta\lambda\mu(X)}{10Q} \rceil. \tag{4.33}$$

If in addition $0 < \mu(X) < \infty$ then, for any $k = 1,2,...,$

$$\lambda_k(\mathcal{E} - \sigma, \mu) \leq \frac{10Q(k-1) - \sigma_{\delta^2}(X)}{\delta\mu(X)}. \tag{4.34}$$

The next lemma will be used in the proof of Theorem 4.17.

Lemma 4.18. *Let* $(\mathcal{E},\mathcal{F})$ *be a strongly local, positive definite energy form on a pseudometric space* (X,d). *Assume that, for any ball* B *in* X,

$$\mathrm{cap}_{\mathcal{E}}(B, 2B) \leq Q, \tag{4.35}$$

where Q *is a positive constant. Then, for any signed Radon measure* σ *on* X *such that* $\sigma(X) > 0$, *we have*

$$Neg(\mathcal{E} - \sigma) \geq 1. \tag{4.36}$$

PROOF. The hypothesis $\sigma(X) > 0$ implies, in particular, that $\sigma_-(X) < \infty$. By the same argument as in the proof of Theorem 4.1, we can assume that also $\sigma_+(X) < \infty$.

Consider a sequence of balls $B_n = B(x, 2^n)$ where x is a fixed reference point in X. By hypothesis (4.35), we have

$$\mathrm{cap}_{\mathcal{E}}(B_n, B_{n+1}) \le Q.$$

By Lemma 2.5, for all indices $n < m$,

$$\mathrm{cap}_{\mathcal{E}}(B_n, B_m)^{-1} \ge \sum_{i=n}^{m-1} \mathrm{cap}_{\mathcal{E}}(B_i, B_{i+1})^{-1} \ge (m-n)\, Q^{-1},$$

whence

$$\mathrm{cap}_{\mathcal{E}}(B_n, B_m) \le \frac{Q}{m-n}. \tag{4.37}$$

Let $f \in \mathcal{T}(B_n, B_m)$ be such that

$$\mathcal{E}[f] < \frac{2Q}{m-n}.$$

Using the fact that $0 \le f \le 1$ and $f = 1$ on B_n, we obtain

$$\begin{aligned}\mathcal{E}[f] - \sigma[f] &< \frac{2Q}{m-n} - \sigma(B_n) - \int_{X \setminus B_n} f^2 d\sigma \\ &\le \frac{2Q}{m-n} - \sigma(B_n) + |\sigma|(X \setminus B_n).\end{aligned}$$

Since $|\sigma|(X) < \infty$, for large enough n and m we obtain

$$(\mathcal{E} - \sigma)[f] < -\sigma(X) + \varepsilon, \tag{4.38}$$

where $\varepsilon > 0$ is prescribed. Choosing $\varepsilon < \sigma(X)/2$, we conclude $(\mathcal{E} - \sigma)[f] < 0$ whence (4.36) follows. ∎

PROOF OF THEOREM 4.17. We need to prove (4.32), that is

$$Neg(\mathcal{E} - \sigma) \ge \lceil \frac{1}{2}\kappa \rceil, \tag{4.39}$$

where $\kappa = \frac{\sigma_\delta(X)}{5Q}$. If $\kappa \le 0$ then there is nothing to do, so assume $\kappa > 0$. In this case $\sigma(X) > 0$ and hence, by Lemma 4.18,

$$Neg(\mathcal{E} - \sigma) \ge 1.$$

On the other hand, by Theorem 4.1,

$$Neg(\mathcal{E} - \sigma) \ge \lfloor \kappa \rfloor.$$

Then (4.39) follows by the elementary inequality

$$\max(\lfloor \kappa \rfloor, 1) \ge \lceil \frac{1}{2}\kappa \rceil.$$

∎

Example 4.19. As in Example 4.13, let $X = \mathbb{R}^n$, μ be the Lebesgue measure on $\mathbb{R}^n$, and $\mathcal{E} = \mathcal{E}_\mu^{(m)}$ be the m-th order energy form on $\mathbb{R}^n$ defined by (2.49). Assume that $n = 2m$ so that the form $\mathcal{E}$ satisfies (4.35). Let q be a non-negative L^1_{loc}-function in $\mathbb{R}^n$. Considering the measure σ defined by $d\sigma = qd\mu$, we obtain by (4.32) that

$$Neg\left(\left(-\Delta\right)^m - q\right) \geq c \int_{\mathbb{R}^n} qd\mu$$

where $c = c\left(n\right) > 0$.

Example 4.20. Let us show that if the form $\mathcal{E}$ is local but not strongly local it can happen that $\sigma \geq 0$, $\sigma\left(X\right) > 0$ but $Neg\left(\mathcal{E} - \sigma\right) = 0$. Indeed, take $X = \mathbb{R}^2$ with the Euclidean distance d, and consider the form $\mathcal{E} = \mathcal{E}_\mu + \nu$ with domain $Lip_0(\mathbb{R}^2)$, that is

$$\mathcal{E}\left(f, g\right) = \int_{\mathbb{R}^2} \nabla f \cdot \nabla g d\mu + \int_{\mathbb{R}^2} fg\, d\nu$$

where μ is the Lebesgue measure on $\mathbb{R}^2$ and ν is a measure on $\mathbb{R}^2$ such that

$$0 < \nu(\mathbb{R}^2) < \infty.$$

It is easy to see that, for any capacitor (F, G) in $\mathbb{R}^2$,

$$\operatorname{cap}_{\mathcal{E}}(F, G) \leq \operatorname{cap}_{\mathcal{E}_\mu}(F, G) + \nu\left(G\right)$$

whence it follows that, for any ball B,

$$\operatorname{cap}_{\mathcal{E}}(B, 2B) \leq \operatorname{cap}_{\mathcal{E}_\mu}(B, 2B) + \nu(\mathbb{R}^2) = \text{const}.$$

Therefore, all the hypotheses of Theorem 4.1 are satisfied for the form $\mathcal{E}$. However, the claim of Lemma 4.18 (and that of Theorem 4.17) is not true in this case. Indeed, just take $\sigma = \nu$ so that $\mathcal{E} - \sigma = \mathcal{E}_\mu$. Then $\sigma(\mathbb{R}^2) > 0$ but $Neg\left(\mathcal{E} - \sigma\right) = Neg\left(\mathcal{E}_\mu\right) = 0$.

Let us show that the hypothesis (4.35) is also essential for Lemma 4.18. For that, consider $X = \mathbb{R}^3$ with the standard form $\mathcal{E}_\mu$, for which (4.35) does not hold. The form $\mathcal{E}_\mu$ is strongly local, but nevertheless there exists a positive measure σ in $\mathbb{R}^3$ such that $Neg\left(\mathcal{E}_\mu - \sigma\right) = 0$. For example, this is the case whenever σ satisfies the estimate

$$\frac{d\sigma}{d\mu}\left(x\right) \leq \frac{1}{4\left|x\right|^2},$$

because of the Hardy inequality

$$\int_{\mathbb{R}^3} \frac{1}{4\left|x\right|^2} f^2\left(x\right) d\mu\left(x\right) \leq \int_{\mathbb{R}^3} \left|\nabla f\right|^2 d\mu,$$

which is true for any $f \in C_0^\infty(\mathbb{R}^3)$ (see, for example, [**50**, Section X.2]).

5. Eigenvalues on Riemannian manifolds

Let X be a Riemannian manifold and d_0 be the geodesic distance on X (note that d_0 may take value ∞ if X is disconnected).

Definition 5.1. A pseudometric d on X is called Riemannian if d is dominated by d_0, that is

$$d\left(x, y\right) \leq d_0\left(x, y\right) \quad \text{for all } x, y \in X. \tag{5.1}$$

The condition (5.1) and the triangle inequality imply that, for any $x \in X$, the function $y \mapsto d\left(x, y\right)$ is locally Lipschitz and $\left|\nabla d\left(x, \cdot\right)\right| \leq 1$.

Definition 5.2. A pseudometric space (X, d) is called *Riemannian* if X is a Riemannian manifold and d is a Riemannian pseudometric on X.

For example, if X is a connected Riemannian manifold then (X, d_0) is a Riemannian (pseudo)metric space. Let $T : X \to X'$ be an isometric immersion of a Riemannian manifold X into a connected Riemannian manifold X', and let d' be the geodesic distance on X'. Then the identity

$$d(x, y) = d'(T(x), T(y)) \tag{5.2}$$

defines the *extrinsic metric* on X, which obviously is a Riemannian pseudometric. Hence, (X, d) is a Riemannian pseudometric space.

In this section, we adapt the results of the previous sections to a Riemannian pseudometric space (X, d). As before, we denote by $B(x, r)$ the balls of the pseudometric d. Set $\mathcal{F} = Lip_0(X)$ and recall that any Radon measure μ on X induces a strongly local positive definite energy form $(\mathcal{E}_\mu, \mathcal{F})$ on the weighted manifold (X, μ) as follows:

$$\mathcal{E}_\mu(f, g) = \int_X \nabla f \cdot \nabla g \, d\mu. \tag{5.3}$$

5.1. Quadratic volume growth.

Theorem 5.3. *Let (X, d) be a Riemannian pseudometric space, μ be a Radon measure on X, and $\mathcal{E} = \mathcal{E}_\mu$. Assume that the following properties are satisfied, for some positive constants M, N:*

(a) *space (X, d) satisfies $(2, N)$-covering property;*
(b) *all balls in (X, d) are precompact;*
(c) *for all $x \in X$ and $r > 0$*

$$\mu(B(x, r)) \le Mr^2. \tag{5.4}$$

Then, for any signed Radon measure σ on X such that σ_+ is d-non-atomic,

$$Neg(\mathcal{E} - \sigma) \ge \frac{\sigma_\delta(X)}{100M}, \tag{5.5}$$

where $\delta \in (0, 1)$ depends only on N.

The estimate (5.5) implies, by Lemma 4.4, that for any d-non-atomic Radon measure ν on X and for any real λ,

$$\mathcal{N}_\lambda(\mathcal{E} - \sigma, \nu) \ge \frac{\sigma_{\delta^2}(X) + \delta\lambda\nu(X)}{100M}. \tag{5.6}$$

If $0 < \nu(X) < \infty$ then for all $k = 1, 2, \ldots$,

$$\lambda_k(\mathcal{E} - \sigma, \nu) \le \frac{C(k-1) - \sigma_{\delta^2}(X)}{\nu(X)}$$

where $C = C(N)$. In particular, these estimate hold for $\nu = \mu$ because μ is d-non-atomic by (5.4).

PROOF. Let us show that (5.4) implies

$$\mathrm{cap}_{\mathcal{E}}(B, 2B) \le 11M. \tag{5.7}$$

Fix a point $x \in X$ and denote for simplicity $B_r = B(x,r)$ and $V_r = \mu(B(x,r))$. For all $0 < r < R$, the following inequality is always true:

$$\operatorname{cap}_{\mathcal{E}}(B_r, B_R) \leq 2\left(\int_r^R \frac{(s-r)ds}{V_s - V_r}\right)^{-1} \tag{5.8}$$

(see [**54**] or [**24**, Theorem 7.1] – note that the proof of (5.8) uses the fact that $|\nabla d| \leq 1$, which is the case by (5.1)). By (5.8) and (5.4), we obtain

$$\operatorname{cap}_{\mathcal{E}}(B_r, B_{2r}) \leq 2\left(\int_r^{2r} \frac{(s-r)ds}{Ms^2}\right)^{-1} = 2M\left(\log 2 - \frac{1}{2}\right)^{-1} < 11M, \tag{5.9}$$

which was claimed.

Given (5.7), we see that all hypotheses of Theorem 4.17 are satisfied, so that (5.5) follows from (4.32). We need only to mention that (4.32) and (5.7) yield the coefficient 110 in (5.5), while we prefer 100, for the obvious esthetic reason. However, as one can see from the proof of Theorems 4.1 and 4.17, the constant factor 10 in (4.32) can be replaced by any number > 8, for example, by 9, which is enough to achieve the factor 100 in (5.5). ∎

5.2. Riemann surfaces. Denote by Σ_γ a closed orientable Riemann surface of genus γ.

Theorem 5.4. *Let g be a Riemannian metric on Σ_γ. Let μ be the Riemannian measure on the Riemannian manifold $X = (\Sigma_\gamma, g)$, and $\mathcal{E} = \mathcal{E}_\mu$ be the Riemannian energy form on X. Then, for any signed Radon measure σ on X such that σ_+ is non-atomic,*

$$Neg\,(\mathcal{E} - \sigma) \geq \frac{\sigma_\delta(X)}{C(\gamma+1)}, \tag{5.10}$$

where $C > 0$ and $0 < \delta < 1$ are absolute constants.

Consequently, we obtain from (5.10) by Lemma 4.4 that for any non-atomic Radon measure ν on X such that $0 < \nu(X) < \infty$, and for any real λ, we have

$$\mathcal{N}_\lambda(\mathcal{E} - \sigma, \nu) \geq \frac{\sigma_{\delta^2}(X) + \delta\lambda\nu(X)}{C(\gamma+1)}, \tag{5.11}$$

and, for any $k = 1, 2, \ldots$,

$$\lambda_k(\mathcal{E} - \sigma, \nu) \leq \frac{C(\gamma+1)(k-1) - \sigma_{\delta^2}(X)}{\delta\nu(X)}. \tag{5.12}$$

Remark 5.5. Already the case $\sigma = 0$, $\nu = \mu$ of (5.12) is highly non-trivial. In this case (5.12) becomes

$$\lambda_k(-\Delta) \leq C'(\gamma+1)\frac{k-1}{\mu(X)}, \tag{5.13}$$

where Δ is the Laplace-Beltrami operator of X (which is a generator of $\mathcal{E}$) and $C' = C/\delta$. It is not difficult to prove that if X is a connected compact n-dimensional Riemannian manifold then

$$\lambda_k(-\Delta) \leq C_X\left(\frac{k-1}{\mu(X)}\right)^{2/n}. \tag{5.14}$$

However, the constant C_X in (5.14), as it is suggested by the notation, depends on various geometric properties of X (cf. Theorem 5.9 below) whereas the constant

C' in (5.13) is universal, and only the genus γ reflects the geometry (or rather the topology) of X in (5.13). The estimate (5.13) for $k = 2$ was first proved by Hersch [**29**] in the case $\gamma = 0$ and by Yang and Yau [**56**] for any γ. For $k > 2$ it was conjectured by Yau [**57**] and was eventually proved by Korevaar [**37**]. It was shown by Colbois and Dodziuk [**11**] that in the case $n > 2$ one cannot have (5.14) with a universal constant C instead of C_X.

PROOF. The Riemannian metric g determines a conformal class of Σ_γ. A well-known consequence of the Riemann-Roch theorem says that the Riemann surface Σ_γ (with a fixed conformal class) admits a non-constant meromorphic function of the topological degree at most $D := \gamma+1$. Hence, there exists a conformal mapping $T : X \to \mathbb{S}^2$ of the topological degree $\leq D$ (see for example [**56**]).

Here we consider $\mathbb{S}^2$ as a Riemannian manifold with the canonical Riemannian metric. Let d' be the geodesic distance on $\mathbb{S}^2$, μ' the Riemannian measure on $\mathbb{S}^2$, and $\mathcal{E}' = \mathcal{E}_{\mu'}$ be the Riemannian energy form on $\mathbb{S}^2$. Since the conformal mapping of two-dimensional Riemannian manifolds locally preserves the Riemannian energy form and the mapping T has topological degree $\leq D$, we see that T has the energy degree at most D, in the sense of Definition 2.11. Hence, by Lemma 2.12, we have

$$Neg\,(\mathcal{E} - \sigma) \geq Neg\,(D\mathcal{E}' - \sigma') = Neg\left(\mathcal{E}' - D^{-1}\sigma'\right), \tag{5.15}$$

where $\sigma'(\cdot) := \sigma(T^{-1}(\cdot))$.

Obviously, $(\mathbb{S}^2, d')$ admits $(2, N)$-covering property with an absolute constant N, all balls on $\mathbb{S}^2$ are precompact, and, for any ball $B(x, r)$ on $\mathbb{S}^2$,

$$\mu'(B(x,r)) \leq \pi r^2.$$

Applying Theorem 5.3 to the Riemannian metric space $(\mathbb{S}^2, d')$ and a signed measure $D^{-1}\sigma'$ (clearly, σ'_+ is non-atomic) we conclude

$$Neg(\mathcal{E}' - D^{-1}\sigma') \geq \frac{D^{-1}\sigma'_\delta(\mathbb{S}^2)}{100\pi} = \frac{\sigma_\delta(X)}{CD},$$

where $C = 100\pi$ and $\delta \in (0, 1)$ is an absolute constant. Combining with (5.15) we obtain (5.10). ∎

The estimate (5.10) admits the following extension.

Corollary 5.6. *Let g be a Riemannian metric on Σ_γ and let X be a Riemannian manifold conformal to $(\Sigma_\gamma \setminus P, g)$ where P is a finite subset of Σ_γ. Let $\mathcal{E}$ be the Riemannian energy form on X. Then, for any signed Radon measure σ on X such that σ_+ is non-atomic, we have*

$$Neg\,(\mathcal{E} - \sigma) \geq \frac{\sigma_\delta(X)}{C(\gamma+1)}, \tag{5.16}$$

where δ and C are the same as in Theorem 5.4.

PROOF. Consider the manifold $X' = (\Sigma_\gamma, g)$ and let $\mathcal{E}'$ be the Riemannian energy form on X' with the domain $\mathcal{F}' = Lip_0(X')$. The conformal mapping identifies X with $X' \setminus P$. Set $\mathcal{F} = Lip_0(X)$ and observe that $\mathcal{F} \subset \mathcal{F}'$ and $\mathcal{E}[f] = \mathcal{E}'[f]$ for any $f \in \mathcal{F}$, because the Riemannian metric of X and the metric g are conformal.

Let σ' be the extension of the measure σ to X' by setting $\sigma'|_P = 0$. Then σ' is a signed Radon measure on X' such that σ'_+ is non-atomic. Since $|\sigma'|(P) = 0$ and

$\operatorname{cap}_{\mathcal{E}'}(P,U)=0$ for any open set $U\subset X'$ containing P, Lemma 2.13 and Theorem 5.4 yield

$$Neg\left(\mathcal{E}-\sigma\right)=Neg\left(\mathcal{E}'-\sigma'\right)\geq\frac{\sigma_{\delta}\left(X'\right)}{C\left(\gamma+1\right)}=\frac{\sigma_{\delta}\left(X\right)}{C\left(\gamma+1\right)}.$$

∎

Example 5.7. Let $X=(\Sigma_{\gamma},g)$ and let $K=K(x)$ be the Gauss curvature of the metric g on X. Fix a real constant α and define a signed measure σ on X by $d\sigma=-\alpha Kd\mu$ where μ is the the Riemannian measure on X. The energy form $(\mathcal{E}-\sigma,\mathcal{F})$ (where $\mathcal{F}=Lip_0(X)$) is closable in $L^2(X,\mu)$ and its generator

$$H=-\Delta+\alpha K$$

has a discrete spectrum that can be estimated by (5.12) as follows. Observe that, by the Gauss-Bonnet formula,

$$\sigma\left(X\right)=-\alpha\int_{X}Kd\mu=-2\pi\chi\alpha, \tag{5.17}$$

where $\chi=2-2\gamma$. Hence, Theorem 5.4 yields the following estimates, for all $k=1,2,...$:

If $\alpha K(x)\geq0$ for all $x\in X$ (and hence $\sigma\leq0$ and $\sigma_{\delta^2}=\sigma$) then

$$\lambda_k\left(H\right)\leq\frac{C\left(\gamma+1\right)\left(k-1\right)+2\pi\chi\alpha}{\delta\mu\left(X\right)}.$$

If $\alpha K(x)\leq0$ all $x\in X$ (and hence $\sigma\geq0$ and $\sigma_{\delta^2}=\delta^2\sigma$) then

$$\lambda_k\left(H\right)\leq\frac{C\left(\gamma+1\right)\left(k-1\right)+2\delta^2\pi\chi\alpha}{\delta\mu\left(X\right)}.$$

Example 5.8. Let $X=(\Sigma_{\gamma}\setminus P,g)$ where $\operatorname{card}P=\kappa$, and H be as above a generator of the energy form $(\mathcal{E}-\sigma,\mathcal{F})$ in $L^2(X,\mu)$. In this case (5.17) still holds but with the Euler characteristic $\chi=2-2\gamma-\kappa$. Let $K(x)\leq0$ on X, and $K\not\equiv0$. Then, for any $\alpha>0$, we have $\sigma\geq0$, and Corollary 5.6 yields

$$Neg\left(H\right)\geq\frac{\delta\sigma\left(X\right)}{C\left(\gamma+1\right)}=\frac{2\delta\pi\left(\kappa+2\gamma-2\right)\alpha}{C\left(\gamma+1\right)}. \tag{5.18}$$

By hypotheses we have $\chi<0$ and hence $\kappa+2\gamma\geq3$, which implies

$$\kappa+2\gamma-2\geq\frac{1}{2}\left(\gamma+1\right). \tag{5.19}$$

Indeed, if (5.19) fails then $2\kappa+3\gamma\leq4$, which is not compatible with $\kappa+2\gamma\geq3$. Substituting (5.19) into (5.18) we obtain that

$$Neg\left(H\right)\geq\frac{\delta\pi}{C}\alpha=c\alpha, \tag{5.20}$$

where c is an absolute positive constant.

5.3. Manifolds of higher dimension.

Theorem 5.9. *Let (X,g_0) be a Riemannian manifold of dimension $n\geq2$, μ_0 be its Riemannian measure, and d be a Riemannian pseudometric on X. Assume that that the following properties are satisfied, for some positive constants M,N:*

(a) *space (X,d) satisfies $(2,N)$-covering property;*
(b) *all balls in (X,d) are precompact;*

(c) for any $r > 0$ and for any d-ball B_r of radius r on X,

$$\mu_0(B_r) \le Mr^n. \tag{5.21}$$

Let g be another metric on X, which is conformal to g_0, μ be the Riemannian measure of g, and $\mathcal{E} = \mathcal{E}_\mu$ be the Riemannian energy form of g. If $\mu(X) < \infty$ then, for any non-atomic Radon measure σ on X,

$$Neg(\mathcal{E} - \sigma) \ge \lfloor c \frac{\sigma(X)^{n/2}}{\mu(X)^{n/2-1}} \rfloor, \tag{5.22}$$

where $c = c(n, N, M) > 0$.

Remark 5.10. Applying (5.22) to $\lambda\sigma$ instead of σ, we obtain, for any $\lambda \ge 0$,

$$\mathcal{N}_\lambda(\mathcal{E}, \sigma) = Neg(\mathcal{E} - \lambda\sigma) \ge \lfloor c \frac{\sigma(X)^{n/2}}{\mu(X)^{n/2-1}} \lambda^{n/2} \rfloor. \tag{5.23}$$

Similarly to Lemma 2.8, one obtains from (5.23) that, for all $k = 1, 2, ...,$

$$\lambda_k(\mathcal{E}, \sigma) \le C \frac{\mu(X)^{1-2/n}}{\sigma(X)} k^{2/n}, \tag{5.24}$$

where $C = C(n, N, M)$.

Remark 5.11. Theorem 5.9 is to some extent a higher order generalization of Theorem 5.3. Indeed, assuming that in Theorem 5.9 $n = 2$ and $g = g_0$, and that in Theorem 5.3 μ is the Riemannian measure, we obtain the same statements. However, in general Theorem 5.3 is not reduced to Theorem 5.9 because in the former the measure μ does not have to be Riemannian, and the measure σ can be signed.

Example 5.12. Let (X, g_0) be a compact connected n-manifold and d be the geodesic distance on X. Then the hypotheses (a), (b), (c) are automatically satisfied with the constants N, M depending on the metric g_0. The estimate (5.22) of $Neg(\mathcal{E}_\mu - \sigma)$ depends on the measures μ and σ only via their total mass, provided μ is the Riemannian measure of a metric g that is conformal to g_0. The constant c in (5.22) depends on the metric g only via its conformal class.

In the compact case the floor function in (5.22) can be dropped, that is the following is true:

$$Neg(\mathcal{E} - \sigma) \ge c \frac{\sigma(X)^{n/2}}{\mu(X)^{n/2-1}}. \tag{5.25}$$

Indeed, if $\sigma(X) > 0$ then the function $\varphi \equiv 1 \in Lip_0(X)$ satisfies $\mathcal{E}[\varphi] < \sigma[\varphi]$ so that $Neg(\mathcal{E} - \sigma) \ge 1$. Combining with (5.22) we obtain (5.25).

Note that in the case of a compact manifold and $\sigma = \mu$, the estimate (5.24) was first proved by Korevaar [**37**].

Example 5.13. Let $X = \mathbb{R}^n$, g_0 be the standard Euclidean metric, and d be the Euclidean distance. Then all the hypotheses (a), (b), (c) are satisfied. Let $a(x)$ be a smooth positive function on $\mathbb{R}^n$, $n > 2$, and set $g = ag_0$ so that $d\mu = a^{n/2} d\mu_0$. Let measure σ be defined by $d\sigma = b d\mu_0$, where $b(x)$ is a continuous positive function on $\mathbb{R}^n$. Then the following operator

$$L = \frac{1}{b} \operatorname{div}\left(a^{n/2-1} \nabla\right)$$

is a generator of the form $\mathcal{E}_\mu$ in $L^2(X,\sigma)$ (where ∇ and div are related to g_0). Hence, (5.24) yields

$$\lambda_k(L) \le C \frac{\left(\int_{\mathbb{R}^n} a^{n/2} d\mu_0\right)^{1-2/n}}{\int_{\mathbb{R}^n} b d\mu_0} k^{2/n},$$

provided the both integrals are finite.

PROOF OF THEOREM 5.9. Recall that for any capacitor (F,G) on (X,g), the capacity associated with the energy form $\mathcal{E}$ is defined by

$$\operatorname{cap}(F,G) = \inf_{\mathcal{T}} \int_X |\nabla_g \varphi|^2 \, d\mu,$$

where $\mathcal{T} = \mathcal{T}(F,G)$ is the class of test functions, and ∇_g is the gradient associated with the metric g. Consider also the *n-capacity* defined by

$$\operatorname{cap}^{(n)}(F,G) = \inf_{\mathcal{T}} \int_X |\nabla_g \varphi|^n \, d\mu.$$

Since n is the dimension of X, the n-capacity is preserved by a conformal change of the metric, that is

$$\operatorname{cap}^{(n)}(F,G) = \inf_{\mathcal{T}} \int_X |\nabla_{g_0} \varphi|^n \, d\mu_0.$$

In the metric g_0, the n-capacity of the capacitor (B_r, R_R) (where $0 < r < R$ and the balls B_r, B_R are concentric) admits the following estimate

$$\operatorname{cap}^{(n)}(B_r, R_R) \le C \left(\int_r^R \left(\frac{s}{\mu_0(B_s)} \right)^{\frac{1}{n-1}} ds \right)^{1-n},$$

(see [**14**], [**32**]), whence by (5.21)

$$\operatorname{cap}^{(n)}(B_r, B_{2r}) \le CM \tag{5.26}$$

(here C denotes any positive constant depending only on n, N, and the value of C may be different at different occurrences).

By the Hölder inequality, we obtain, for any $\varphi \in \mathcal{T}(F,G)$,

$$\int_X |\nabla_g \varphi|^2 \, d\mu \le \left(\int_G |\nabla_g \varphi|^n \, d\mu \right)^{2/n} \mu(G)^{1-2/n},$$

whence it follows that

$$\operatorname{cap}(F,G) \le \left(\operatorname{cap}^{(n)}(F,G)\right)^{2/n} \mu(G)^{1-2/n}. \tag{5.27}$$

Similarly to the proof of Theorem 4.1, we can assume in the sequel that $0 < \sigma(X) < \infty$. By Corollary 3.12, for any positive integer k, there exists a family $\{A_i\}_{i=1}^{2k}$ of annuli in (X,d) such that

$$\sigma(A_i) \ge c \frac{\sigma(X)}{k} \quad \text{for all } i = 1, 2, ..., 2k, \tag{5.28}$$

and the annuli $\{2A_i\}_{i=1}^{2k}$ are disjoint. It follows from (5.26) that

$$\operatorname{cap}^{(n)}(A_i, 2A_i) \le CM$$

(cf. the proof of Theorem 4.7), whence (5.27) implies

$$\operatorname{cap}(A_i, 2A_i) \le CM^{2/n} \mu(2A_i)^{1-2/n}.$$

Since

$$\sum_{i=1}^{2k} \mu(2A_i) \leq \mu(X),$$

there exists at least k sets $2A_i$ such that

$$\mu(2A_i) \leq \frac{\mu(X)}{k}.$$

Without loss of generality, we can assume that this is the case for $i = 1, 2, ..., k$, whence it follows that

$$\operatorname{cap}(A_i, 2A_i) \leq CM^{2/n} \left(\frac{\mu(X)}{k}\right)^{1-2/n} \quad \text{for } i = 1, 2, ..., k. \tag{5.29}$$

Assume for a moment that the following inequality is true:

$$CM^{2/n} \left(\frac{\mu(X)}{k}\right)^{1-2/n} \leq \frac{c}{2} \frac{\sigma(X)}{k}, \tag{5.30}$$

which implies by (5.29) and (5.28)

$$\operatorname{cap}(A_i, 2A_i) \leq \frac{1}{2}\sigma(A_i) \quad \text{for } i = 1, 2, ..., k.$$

Then choosing nearly optimal test functions for the capacitors $(A_i, 2A_i)$, $i = 1, 2, .., k$, we obtain a k-dimensional subspace $\mathcal{V} \subset Lip_0(X)$ such that $\mathcal{E}[\varphi] < \sigma[\varphi]$ for any $\varphi \in \mathcal{V} \setminus \{0\}$, whence $Neg(\mathcal{E} - \sigma) \geq k$.

Finally, noticing that the inequality (5.30) holds for any k such that

$$k \leq \frac{c'}{M} \frac{\sigma(X)^{n/2}}{\mu(X)^{n/2-1}},$$

we obtain (5.22). ∎

5.4. Boundary surfaces. Let X be a Riemannian manifold. In this section, we use the notation

$$\lambda_k(X) = \lambda_k(\mathcal{E}_\mu, \mu)$$

where μ is the Riemannian measure on X and $\mathcal{E}_\mu$ is the energy form given by (5.3) with the domain $\mathcal{F} = Lip_0(X)$.

Recall that a Riemannian M is called a *Cartan-Hadamard manifold* if M is complete, non-compact, simply connected manifold of non-positive sectional curvature. In particular, $\mathbb{R}^n$ and $\mathbb{H}^n$ are Cartan-Hadamard manifolds.

Theorem 5.14. *Let Ω be a bounded open set in a 3-dimensional Cartan-Hadamard manifold M, and let the boundary Γ of Ω be smooth so that Γ is a compact oriented Riemannian 2-manifold. Let γ be the genus of Γ. Then, for all positive integers k, m,*

$$\lambda_k(\Omega)^{3/2} \geq \frac{c}{(\gamma+1)^{3/2}} \lambda_{k+1}(\Gamma) \sqrt{\frac{\lambda_{m+1}(\Gamma)}{m}}, \tag{5.31}$$

where $c > 0$ is an absolute constant.

In particular, for all $k = 1, 2, ...,$

$$\lambda_k(\Omega) \geq \frac{c}{\gamma+1} \frac{\lambda_{k+1}(\Gamma)}{k^{1/3}}. \tag{5.32}$$

Proof. By a theorem of Hoffman and Spruck [**31**], a Cartan-Hadamard 3-manifold admits the following isoperimetric inequality:

$$V(\Omega) \le CA(\Gamma)^{3/2}, \tag{5.33}$$

where V stands for the volume in M, A is the area on Γ, and C is an absolute constant. It follows from (5.33) that $\lambda_k(\Omega)$ admits the lower bound

$$\lambda_k(\Omega) \ge c\left(\frac{k}{V(\Omega)}\right)^{2/3}, \tag{5.34}$$

(see for example [**10**], [**41**], [**22**]), where $c > 0$ is an absolute constant.

On the other hand, by (5.13) we have the following upper bound for $\lambda_{k+1}(\Gamma)$:

$$\lambda_{k+1}(\Gamma) \le C(\gamma+1)\frac{k}{A(\Gamma)}, \tag{5.35}$$

with an absolute constant C. Combining (5.35) and (5.33) we obtain

$$\lambda_{k+1}(\Gamma)\sqrt{\frac{\lambda_{m+1}(\Gamma)}{m}} \le \left(\frac{C(\gamma+1)}{A(\Gamma)}\right)^{3/2} k \le \frac{C^{5/2}(\gamma+1)^{3/2}k}{V(\Omega)},$$

which together with (5.34) implies (5.31). Clearly, (5.31) implies (5.32) for $m = k$. ∎

5.5. Positive definite perturbations. The purpose of this section is to present a partial result towards the conjecture that the constant δ in Theorem 4.1 can be taken to be 1.

Theorem 5.15. *Let X be a Riemannian manifold, d be a pseudometric on X, μ be a Radon measure on X, and $\mathcal{E} = \mathcal{E}_\mu$. Assume that the following conditions hold, for some positive constants N, Q:*

(i) (X, d) satisfies $(2, N)$-covering property;
(ii) measure μ is d-non-atomic and $0 < \mu(X) < \infty$;
(iii) for any d-ball B in X,

$$\operatorname{cap}_{\mathcal{E}}(B, 2B) \le Q.$$

Let σ be a finite signed Radon measure on X such that the form $(\mathcal{E}-\sigma, \mathcal{F})$ is positive definite (where $\mathcal{F} = Lip_0(X)$). Then, for any $\lambda \ge 0$,

$$\mathcal{N}_\lambda(\mathcal{E}-\sigma, \mu) \ge \frac{\sigma(X) + \varepsilon\lambda\mu(X)}{10Q}, \tag{5.36}$$

where $0 < \varepsilon < 1$ is a constant depending only on N. Also, for any $k = 1, 2, \ldots$,

$$\lambda_k(\mathcal{E}-\sigma, \mu) \le \frac{10Q(k-1) - \sigma(X)}{\varepsilon\mu(X)}. \tag{5.37}$$

Proof. By Lemma 4.18, if $\sigma(X) > 0$ then $Neg(\mathcal{E}-\sigma) \ge 1$ which contradicts the hypothesis that $\mathcal{E}-\sigma$ is positive definite. Therefore, $\sigma(X) \le 0$. Assuming that $\varepsilon^{-1}\sigma(X) + \lambda\mu(X) > 0$ (otherwise, (5.36) is trivial), we obtain $\sigma(X) + \lambda\mu(X) > 0$, whence by Lemma 4.18,

$$\mathcal{N}_\lambda(\mathcal{E}-\sigma, \mu) = Neg(\mathcal{E} - (\sigma + \lambda\mu)) \ge 1. \tag{5.38}$$

We will show that there exist k functions $f_i \in \mathcal{F}$ with disjoint supports such that

$$(\mathcal{E}-\sigma)[f_i] < \lambda\mu[f_i], \tag{5.39}$$

where

$$k := \lfloor \frac{\sigma(X) + \varepsilon\lambda\mu(X)}{5Q} \rfloor \tag{5.40}$$

This will imply

$$\mathcal{N}_\lambda(\mathcal{E} - \sigma, \mu) \geq k$$

which together with (5.38) yields

$$\mathcal{N}_\lambda(\mathcal{E} - \sigma, \mu) \geq \max(1, k) \geq \frac{\sigma(X) + \varepsilon\lambda\mu(X)}{10Q},$$

thus finishing the proof of (5.36). Clearly, the estimate (5.37) follows from (5.36) by Lemma 2.8.

To prove the above claim observe that, by Corollary 4.9, there exist $2k$ disjoint capacitors (F_i, G_i) on X such that

$$\mu(F_i) \geq c\frac{\mu(X)}{2k} \quad \text{and} \quad \operatorname{cap}_{\mathcal{E}}(F_i, G_i) \leq 4Q,$$

where $c \in (0,1)$ depends only on N. Choose a test function $2f_i \in \mathcal{T}(F_i, G_i)$ such that $\mathcal{E}[2f_i] < 5Q$. Recall that $2f_i \in \mathcal{T}(F_i, G_i)$ implies

$$f \in C_0(G_i), \ f|_{F_i} = \frac{1}{2}, \ 0 \leq f_i \leq \frac{1}{2}.$$

Hence, we have, for $\varepsilon := c/8$,

$$\mu[f_i] \geq \varepsilon\frac{\mu(X)}{k} \quad \text{and} \quad \mathcal{E}[f_i] < \frac{5}{4}Q, \tag{5.41}$$

which, in particular, implies

$$\sum_{i=1}^{2k} \mathcal{E}[f_i] < \frac{5}{2}Qk. \tag{5.42}$$

Let us prove that

$$\sigma(X) - \sum_{i=1}^{2k} \sigma[f_i] \leq \sum_{i=1}^{2k} \mathcal{E}[f_i]. \tag{5.43}$$

Assume for the moment that (5.43) has been proved. Then (5.42) and (5.43) imply

$$\sigma(X) + \sum_{i=1}^{2k} (\mathcal{E} - \sigma)[f_i] < \frac{5}{2}Qk + \sum_{i=1}^{2k} \mathcal{E}[f_i] \leq 5Qk$$

and, hence,

$$\sum_{i=1}^{2k} (\mathcal{E} - \sigma)[f_i] \leq 5Qk - \sigma(X).$$

Since $(\mathcal{E} - \sigma)[f_i] \geq 0$, there exists at least k functions f_i such that

$$(\mathcal{E} - \sigma)[f_i] < \frac{5Qk - \sigma(X)}{k} \tag{5.44}$$

By (5.40) we have

$$5Qk \leq \sigma(X) + \varepsilon\lambda\mu(X),$$

whence by (5.41)

$$\frac{5Qk - \sigma(X)}{k} \leq \lambda\mu[f_i].$$

Together with (5.44), this yields (5.39).

We are left to prove (5.43). Define a function $h \geq 0$ on X by the identity

$$h^2 + \sum_i f_i^2 = 1. \tag{5.45}$$

Since the supports of f_i are disjoint and $0 \leq f_i \leq 1/2$, we obtain $h > 1/2$. Integrating (5.45) against σ, we obtain

$$\sigma(X) - \sum_i \sigma[f_i] = \sigma[h]. \tag{5.46}$$

Since the form $\mathcal{E} - \sigma$ is positive definite and $h > 1/2$, we obtain

$$\sigma[h] \leq \mathcal{E}[h] \leq 2 \inf |h| \, \mathcal{E}[h] \leq \mathcal{E}\left[h^2\right]. \tag{5.47}$$

Next, it follows from (5.45) that

$$\nabla\left(h^2\right) = -\sum_i \nabla\left(f_i^2\right)$$

whence

$$\mathcal{E}\left[h^2\right] = \sum_i \mathcal{E}[f_i^2]. \tag{5.48}$$

Using

$$\mathcal{E}[f_i^2] \leq 2 \sup |f_i| \, \mathcal{E}[f_i] \leq \mathcal{E}[f_i]$$

we obtain

$$\sum_i \mathcal{E}[f_i^2] \leq \sum_i \mathcal{E}[f_i]. \tag{5.49}$$

Combining (5.46), (5.47), (5.48), and (5.49) we obtain (5.43). ∎

6. Eigenvalues of the Jacobi operator

Throughout this section, except for Subsection 6.3, X will be an oriented two-dimensional manifold immersed into a three dimensional Riemannian manifold M. For simplicity of notation, we will not distinguish between the points of X and their images in M (although some points in X may merge in M). We assume that X is endowed with the induced Riemannian metric, and denote by μ the Riemannian measure on X. Let K be the Gauss curvature of X, R_M be the scalar curvature of M, and Ric_M be the Ricci curvature of M. Let $\mathbf{n}$ be an orthonormal vector field on X in M.

Let A be the operator of the second fundamental form of X, that is, at any point $x \in X$, $A = A(x)$ is a linear operator in T_xX acting by $A\xi = -\nabla_\xi \mathbf{n}$. Denote $\|A\|^2 := \operatorname{trace}(AA^*)$, set

$$q := Ric_M(\mathbf{n}, \mathbf{n}) + \|A\|^2, \tag{6.1}$$

and consider the energy form

$$\mathcal{A}[f] := \int_X \left(|\nabla f|^2 - qf^2\right) d\mu$$

with the domain $\mathcal{F} = Lip_0(X)$. In other words, $\mathcal{A} = \mathcal{E} - \sigma$ where $\mathcal{E}$ is the Riemannian energy form on X and σ is a signed measure defined by $d\sigma = q d\mu$.

It is known that the energy form$\mathcal{A}$ determines the second variation of the area functional under the normal deformation of X (see for example [**12**], [**39**,

Section 6], [**47**]), while the first variation is determined by the *mean curvature*. We will be concerned with estimates of the counting function of the energy form $\mathcal{A}$, in particular, $Neg(\mathcal{A})$. If in addition X is a minimal surface (that is, the mean curvature of X vanishes everywhere) then the number $Neg(\mathcal{A})$ is called the *stability index* of X and is denoted by $\operatorname{ind}(X)$. The minimal surface X is called *stable* if $\operatorname{ind}(X) = 0$.

A generator of the form $(\mathcal{A}, \mathcal{F})$ in $L^2(X, \mu)$ is the following operator

$$L := -\Delta - q = -\Delta - (Ric_M(\mathbf{n}, \mathbf{n}) + \|A\|^2),$$

which is called the *stability operator* or the *Jacobi operator*.

6.1. Riemann surfaces. Let X and M be as above.

Theorem 6.1. *Assume that $Ric_M \geq 0$ and let X be conformally equivalent to $(\Sigma_\gamma \setminus P, g)$, where g is a Riemannian metric on Σ_γ and P is a finite subset of Σ_γ. Then*

$$Neg(\mathcal{A}) \geq \frac{c_0}{\gamma+1}\left(\int_X R_M d\mu - \int_X K d\mu\right), \tag{6.2}$$

where c_0 is an absolute positive constant.

If, in addition, $\mu(X) < \infty$ then, for any $k = 1, 2, \ldots$,

$$\lambda_k(\mathcal{A}, \mu) \leq \frac{C(\gamma+1)k}{\mu(X)} - \frac{c}{\mu(X)}\int_X R_M d\mu, \tag{6.3}$$

where C and c are absolute positive constants.

Remark 6.2. Recall that by the Gauss-Bonnet formula

$$\int_X K d\mu = 2\pi\chi, \tag{6.4}$$

where χ is the Euler characteristic of X. In the present setting we have

$$\chi = 2 - 2\gamma - \kappa,$$

where $\kappa := \operatorname{card} P$.

Remark 6.3. In the case when X is compact we have by Theorem 5.4 that

$$\lambda_k(\mathcal{E}, \mu) \leq \frac{C(\gamma+1)k}{\mu(X)}.$$

The additional non-negative term $\int_X R_M d\mu$ in (6.3) reflects the distinction between the Jacobi operator and the Laplace operator.

Proof. We use the following identity on X:

$$R_M - Ric_M(\mathbf{n}, \mathbf{n}) = K + \frac{\|A\|^2}{2} - \frac{|H|^2}{2},$$

where H is the mean curvature vector of X. It implies

$$q = Ric_M(\mathbf{n}, \mathbf{n}) + \|A\|^2 = R_M - K + \frac{\|A\|^2}{2} + \frac{|H|^2}{2} \geq R_M - K, \tag{6.5}$$

whence

$$\sigma(X) = \int_X q d\mu \geq \int_X R_M d\mu - \int_X K d\mu = \int_X R_M d\mu - 2\pi\chi. \tag{6.6}$$

By $Ric_M \geq 0$ and (6.1) we have $q \geq 0$ and hence $\sigma \geq 0$. In particular, $\sigma_\delta = \delta\sigma$, and by Corollary 5.6 we conclude

$$Neg(\mathcal{A}) = Neg(\mathcal{E} - \sigma) \geq \frac{\delta\sigma(X)}{C(\gamma+1)},$$

whence (6.2) follows with $c_0 = \delta/C$.

The second claim follows from (5.12) and (6.6) using also $\chi \leq 2 - 2\gamma$:

$$\begin{aligned} \lambda_k(\mathcal{A}, \mu) &\leq \frac{C(\gamma+1)(k-1) - \delta^2\sigma(X)}{\delta\mu(X)} \\ &\leq \frac{C(\gamma+1)k - \delta^2 \int_X R_M d\mu - \left[C(\gamma+1) + \delta^2 2\pi(2\gamma-2)\right]}{\delta\mu(X)} \\ &\leq \frac{C}{\delta}\frac{(\gamma+1)k}{\mu(X)} - \frac{\delta}{\mu(X)}\int_X R_M d\mu, \end{aligned}$$

provided $C \geq 4\pi\delta^2$, which can be assumed to be true. Renaming the constants we obtain (6.3). ∎

Remark 6.4. The hypothesis $Ric_M \geq 0$ is only needed to conclude that $q \geq 0$. One can also obtain $q \geq 0$ using different assumptions. For example, it is true provided $R_M \geq 0$ and $K \leq 0$, as one can see from (6.5).

Theorem 6.1 may have many applications. For example, (6.2) and (6.4) imply the following statement.

Corollary 6.5. *Under the hypotheses of* Theorem 6.1, *if in addition* $R_{\min} := \inf_X R_M > 0$ *and* X *is immersed in* M *as a minimal surface then*

$$\mu(X) \leq \frac{C_0(\gamma+1)\operatorname{ind}(X) + 2\pi\chi}{R_{\min}},$$

where C_0 *is an absolute positive constant.*

6.2. Minimal surfaces in $\mathbb{R}^3$ with finite total curvature. In this section, we assume by default that X is an oriented immersed minimal surface in $\mathbb{R}^3$. Then we have $\|A\|^2 = -2K$ and, hence, the second variation form $\mathcal{A}$ is given by $\mathcal{A} = \mathcal{E} - \sigma$ where σ is defined by

$$d\sigma = -2K d\mu.$$

In particular, we have $\sigma \geq 0$ and

$$\sigma(X) = 2K_{total}(X)$$

where

$$K_{total}(X) := \int_X |K| \, d\mu.$$

The first result related ind (X) to the total curvature is due to Barbosa and do Carmo [**1**] who proved that

$$K_{total}(X) < 2\pi \quad \Longrightarrow \quad \operatorname{ind}(X) = 0.$$

A number of authors [**5**], [**18**], [**49**] independently proved the following extension of Bernstein's theorem: the only complete stable minimal surface is a plane. In other words, if X is complete then

$$K_{total}(X) \equiv 0 \quad \Longleftrightarrow \quad \operatorname{ind}(X) = 0.$$

Fischer-Colbrie [**17**] proved that if X is complete then

$$K_{total}(X) < \infty \iff \operatorname{ind}(X) < \infty.$$

These results suggests that there may exist inequalities relating $\operatorname{ind}(X)$ and $K_{total}(X)$. Indeed, it was proved by Tysk [**55**] that if X is compete then

$$\operatorname{ind}(X) \leq c_0 K_{total}(X) \tag{6.7}$$

where $c_0 \approx 0.6133$. Two of the authors proved in [**28**] that (6.7) holds for *any* (not necessarily complete) minimal surface X, although with a very large constant c_0.

For a complete X the estimate (6.7) was improved by Micallef [**44**]: if X is complete and not a plane then

$$\operatorname{ind}(X) \leq \frac{1}{\pi} K_{total}(X) + 2\gamma - 3,$$

where γ is the genus of X. If in addition $K_{total} < \infty$ and all branching values of the extended Gauss map of X lie on an equator of $\mathbb{S}^2$, then by a theorem of Montiel and Ros [**45**],

$$\operatorname{ind}(X) = \frac{1}{2\pi} K_{total}(X) - 1.$$

It was conjectured in [**7**] and [**30**] that if X is complete and non-planar then

$$\operatorname{ind}(X) \leq \frac{1}{2\pi} K_{total}(X) - 1.$$

The known examples of complete oriented minimal surfaces suggest that $\operatorname{ind}(X)$ may admit also a *lower* bound via $K_{total}(X)$. We prove here some weak versions of this conjecture. Before we do so, let us briefly recall some results about the structure of minimal surfaces in $\mathbb{R}^3$. We refer a reader to the surveys by Hoffman and Karcher [**30**] and by Meeks and Perez [**43**] for more details.

Let X be a complete minimal surface. Since we are interested in lower estimates of $\operatorname{ind}(X)$, we can assume that $\operatorname{ind}(X) < \infty$ and hence $K_{total}(X) < \infty$. Then, by a theorem of Huber [**33**] (see also [**48**]) X is conformally equivalent to $\Sigma_\gamma \setminus P$ where $P = \{p_1, ..., p_k\}$ is a finite subset of Σ_γ. Moreover, a punctured neighborhood in Σ_γ of each point p_i corresponds to an end E_i of X.

Let $n(x)$ be a normal unit vector field on X in $\mathbb{R}^3$. When a point $x \in E_i$ escapes to ∞ along E_i then $n(x)$ has a limit, say n_i. Denote by C_i the large circle on the unit sphere $\mathbb{S}^2$ such that the plane through C_i has the normal n_i.

For any $r > 0$, consider the set

$$C(r) := \mathbb{S}^2 \cap \frac{1}{r} X,$$

where $\frac{1}{r}X$ is the scaling transformation of X in $\mathbb{R}^3$. By a theorem of Jorge and Meeks [**35**], for large enough r, the set $C(r)$ consists of k immersed closed curves on $\mathbb{S}^2$, say $C_1(r), ..., C_k(r)$ (assuming that the ordering of γ_r^i matches that of E_i), and when $r \to \infty$, the curve $C_i(r)$ converges in C^∞-sense to the circle C_i, with a multiplicity m_i, where m_i is a positive integer (see Fig. 11).

In particular, the length of the circle C_i is $2\pi m_i$. We say that the end E_i has multiplicity m_i. It is known that $m_i = 1$ if and only if the end E_i is embedded.

Theorem 6.6. *Let X be a complete oriented minimal surface immersed in $\mathbb{R}^3$. If $\operatorname{ind}(X) < \infty$ then*

$$\operatorname{ind}(X) \geq \frac{c}{m} K_{total}(X), \tag{6.8}$$

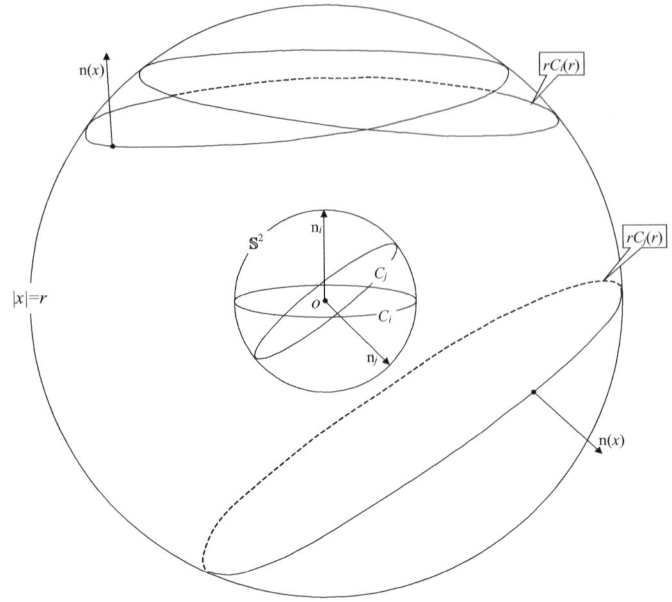

FIGURE 11. Circle C_i with multiplicity 2 and circle C_j with multiplicity 1.

where $m = m_1 + ... + m_k$ *is the total multiplicity of the ends of* X, *and* c *is an absolute positive constant.*

If in addition the ends of X *are embedded then*

$$\operatorname{ind}(X) \geq \frac{c}{k} K_{total}(X), \tag{6.9}$$

where k *is the number of ends of* X.

PROOF. Let d_0 be the geodesic distance on X with respect to the induced metric. Denote by d the extrinsic distance on X, that is the restriction to X of the Euclidean distance in $\mathbb{R}^3$. Then (X, d) is a Riemannian pseudometric space (see Section 5). Let μ be the Riemannian measure on X and $\mathcal{E}$ be the Riemannian energy form on X. Let us show that the hypotheses (a)-(c) of Theorem 5.3 are satisfied.

Let $B(x,r)$ be a d-ball on X, that is $B(x,r)$ is the intersection of the Euclidean ball $\widetilde{B}(x,r)$ in $\mathbb{R}^3$ with X. The ball $\widetilde{B}(x,r)$ can be covered by at most N euclidean balls in $\mathbb{R}^3$ of radii $r/4$, where N is an absolute constant. Select out of them those balls that have non-empty intersection with X, and let their centers be $y_1, y_2, ..., y_k$, where $k \leq N$. Let x_i be a point in the intersection of $\widetilde{B}(y_i, r/4)$ with X. Then $\widetilde{B}(x_i, r/2)$ covers $\widetilde{B}(y_i, r/4)$ whence it follows that all balls $B(x_i, r/2)$ cover $B(x,r)$. Hence, (X, d) satisfies $(2, N)$-covering property, that is the hypothesis (a) holds.

Since X is complete and $K_{total}(X) < \infty$, the immersion of X into $\mathbb{R}^3$ is *proper*, that is the intersection of any compact set in $\mathbb{R}^3$ with X is compact in the topology of X (see [**43**, Section 2.3]). This immediately implies that d-balls in X are precompact, that is the hypothesis (b).

Let us prove that, for any d-ball $B(x,r)$ in X,

$$\mu(B(x,r)) \leq \pi m r^2, \tag{6.10}$$

which will settle the hypothesis (c). It is a consequence of the minimality of X that the function

$$r \mapsto \frac{\mu(B(x,r))}{r^2}$$

is increasing (see [**53**, p.84]). Therefore, it suffices to prove (6.10) asymptotically, that is

$$\mu(B(x,r)) \sim \pi m r^2 \quad \text{as } r \to \infty, \tag{6.11}$$

for any fixed $x \in X$. Without loss of generality, we will prove this for $x = o$, where o is the origin of $\mathbb{R}^3$.

Set $S(r) = \partial B(o,r)$, $\rho(x) = |x|$ (where $x \in \mathbb{R}^3$) and observe that by the coarea formula,

$$\mu(B(o,R)) = \int_0^R \left(\int_{S(r)} |\nabla \rho|^{-1} \, dl \right) dr, \tag{6.12}$$

where ∇ is the Riemannian gradient on X and dl is the length element on $S(r)$. Let $\widetilde{\nabla}$ be the Euclidean gradient in $\mathbb{R}^3$. Then $\nabla\rho(x)$ is the projection of $\widetilde{\nabla}\rho(x)$ onto T_xX (see Fig. (12)) and since $n(x)$ is a normal to T_xX, we obtain

$$|\nabla \rho|^2 = |\widetilde{\nabla}\rho|^2 - (\widetilde{\nabla}\rho \cdot n)^2 = 1 - (\frac{x}{\rho} \cdot n)^2.$$

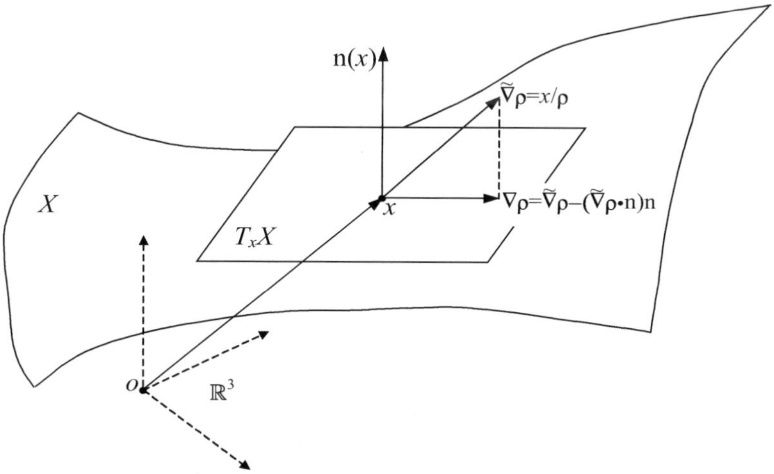

FIGURE 12. Gradients $\nabla\rho$ and $\widetilde{\nabla}\rho$

If $x \to \infty$ along the end E_i, then $x/\rho \in C_i(\rho)$ and hence x/ρ tends on $\mathbb{S}^2$ to the circle C_i whereas $n(x)$ tends to n_i. Since n_i is orthogonal to C_i, we obtain

$$\frac{x}{\rho} \cdot n \longrightarrow 0$$

and hence $|\nabla\rho| \to 1$. For large enough r, $S(r)$ is the union of the curves $rC_i(r)$. Therefore, we obtain that, for $r \to \infty$,

$$\int_{S(r)} |\nabla\rho|^{-1} dl \sim l(S(r)) = \sum_{i=1}^{k} l(rC_i(r)) \sim \sum_{i=1}^{k} l(C_i)\, r = 2\pi m r,$$

whence by (6.12)

$$\mu(B(o,R)) \sim \int_0^R 2\pi m r\, dr = \pi m R^2.$$

This finishes the proof of (6.11) and hence (6.10).

Finally, we claim that measure σ on X (given by $d\sigma = -2K\, d\mu$) is d-non-atomic. Let I be the immersion in question of the manifold X into $\mathbb{R}^3$. It follows from the definition of the extrinsic distance d that, for any $x \in X$,

$$\{y \in X : d(x,y) = 0\} = I^{-1}(x).$$

By the definition of an immersion, for any point $y \in X$ there is an open neighbourhood U of y in X such that $I|_U$ is an injection. Therefore, $I^{-1}(x)$ consists of isolated points and hence $\sigma(I^{-1}(x)) = 0$, that is σ is d-non-atomic.

Applying Theorem 5.3 we obtain

$$\operatorname{ind}(X) = Neg(\mathcal{E} - \sigma) \geq \frac{\delta\sigma(X)}{100\pi m} = \frac{c}{m} K_{total}(X),$$

where $c = \frac{\delta}{50\pi}$ is an absolute positive constant. In the case when the ends of X are embedded, we have $m = k$, whence (6.9) follows. ∎

Note that by Corollary 5.6 we have also in the above setting that

$$\operatorname{ind}(X) = Neg(\mathcal{E} - \sigma) \geq \frac{\sigma_\delta(X)}{C(\gamma+1)} = \frac{c'}{\gamma+1} K_{total}(X), \tag{6.13}$$

where $c' = 2\sigma/C$. However, in most applications (6.9) gives a better lower bound for $\operatorname{ind}(X)$ than (6.13).

Theorem 6.7. *Let X be a connected complete oriented minimal surface embedded in $\mathbb{R}^3$. If $\operatorname{ind}(X) < \infty$ then*

$$\operatorname{ind}(X) \geq k - 1, \tag{6.14}$$

where k is the number of ends of X.

This theorem will be proved Section 6.4 after introducing the necessary techniques.

Corollary 6.8. *For any connected complete oriented minimal surface X embedded in $\mathbb{R}^3$, we have*

$$\operatorname{ind}(X) \geq c' \sqrt{K_{total}(X)} \tag{6.15}$$

and

$$\operatorname{ind}(X) \geq c'' \sqrt{genus(X)}, \tag{6.16}$$

where c', c'' are absolute positive constants.

PROOF. If $\operatorname{ind}(X) = \infty$ then there is nothing to prove, so assume $\operatorname{ind}(X) < \infty$ and hence $K_{total}(X) < \infty$. Let k be the number of ends of X. If $k = 1$ then (6.15) follows from (6.9) and the fact that $\operatorname{ind}(X)$ is an integer. If $k \geq 2$ then (6.9) and (6.14) imply

$$\operatorname{ind}(X) \geq \frac{1}{2}\left(\frac{c}{k}K_{total} + \frac{k}{2}\right) \geq c'\sqrt{K_{total}}$$

where $c' = \frac{1}{2}\sqrt{c/2}$.

To prove (6.16) observe that, by a theorem of Osserman (see also [**35**]), we have

$$K_{total}(X) = 4\pi(\gamma + k - 1) \geq 4\pi\gamma,$$

where $\gamma = genus(X)$. Hence, (6.16) follows from (6.15) with $c'' = c'\sqrt{4\pi}$. ∎

Let us mention for comparison the following result of Jorge and Meeks [**35**]: there exists a function $F : [0, +\infty) \to [0, +\infty)$ such that if M is a properly embedded minimal surface in $\mathbb{R}^3$then

$$\operatorname{ind}(X) \leq F(genus(X)).$$

Here no assumption is made about finiteness of the total curvature.

6.3. Counting functions of subsets. In this section, we assume that X is a Riemannian manifold, d is a Riemannian pseudometric on X, and μ is a Radon measure on X having a continuous positive density with respect to the Riemannian measure. Let $\mathcal{E} = \mathcal{E}_\mu$ be the associated energy form with the domain $\mathcal{F} = Lip_0(X)$. As was already mentioned, the form $(\mathcal{E}, \mathcal{F})$ is closable in $L^2(X, \mu)$, and its generator is $-\Delta_\mu$.

Let σ be another Radon measure on X defined by

$$d\sigma = q d\mu,$$

where q is a positive continuous function on X. The operator $-\frac{1}{q}\Delta_\mu$ is a generator of the form $(\mathcal{E}, \mathcal{F})$ in $L^2(X, \sigma)$. For any open set $\Omega \subset X$, consider the form $\mathcal{E}$ with the domain $\mathcal{F}(\Omega) := \mathcal{F} \cap C_0(\Omega) = Lip_0(\Omega)$. The form $(\mathcal{E}, \mathcal{F}(\Omega))$ is closable in $L^2(\Omega, \sigma)$. Let H_Ω be its self-adjoint generator, and $\widetilde{\mathcal{F}}(\Omega)$ be the domain of the closure. Set

$$N(\Omega) = \sup\{\dim \mathcal{V} : \mathcal{V} \prec \mathcal{F}(\Omega) \quad \text{and} \quad \mathcal{E}[f] < \sigma[f] \ \ \forall f \in \mathcal{V} \setminus \{0\}\}$$

and

$$N^*(\Omega) = \sup\left\{\dim \mathcal{V} : \mathcal{V} \prec \widetilde{\mathcal{F}}(\Omega) \quad \text{and} \quad \mathcal{E}[f] \leq \sigma[f] \ \ \forall f \in \mathcal{V}\right\}.$$

By Lemma 2.7, we have

$$N(\Omega) = \dim \operatorname{Im} \mathbf{1}_{(-\infty,1)}(H_\Omega),$$

and

$$N^*(\Omega) = \dim \operatorname{Im} \mathbf{1}_{(-\infty,1]}(H_\Omega).$$

Lemma 6.9. *Let (X, μ) be a connected weighted manifold, and let $\Omega_0, \Omega_1, ..., \Omega_n$ be non-empty disjoint open sets in X. Then*

$$N(X) \geq N(\Omega_0) + \sum_{i=1}^{n} N^*(\Omega_i). \tag{6.17}$$

PROOF. The proof follows an argument of Montiel and Ros [**45**]. Set $\mathcal{V} = \operatorname{Im} \mathbf{1}_{(-\infty,1)}(H)$ where H is the self-adjoint generator of the form $(\mathcal{E}, \mathcal{F})$ in $L^2(X,\sigma)$. Let also $\mathcal{V}_i$ be finite dimensional linear spaces such that

$$\mathcal{V}_0 \subset \operatorname{Im} \mathbf{1}_{(-\infty,1)}(H_{\Omega_0})$$

and

$$\mathcal{V}_i \subset \operatorname{Im} \mathbf{1}_{(-\infty,1]}(H_{\Omega_i}) \quad \text{for } i = 1, 2, ..., n.$$

Since Ω_i are disjoint sets, the spaces $\mathcal{V}_i$ are all mutually orthogonal in $L^2(X,\sigma)$.

To prove (6.17), it suffices to show that

$$\dim \mathcal{V} \geq \dim \mathcal{V}_0 + \sum_{i=1}^{n} \dim \mathcal{V}_i =: m.$$

Assume from the contrary that $\dim \mathcal{V} < m$. Then there exists a non-zero function $v \in \bigoplus_{i=0}^{n} \mathcal{V}_i$ such that v is orthogonal to $\mathcal{V}$ in $L^2(X,\sigma)$. Therefore,

$$v = \int_{[1,+\infty)} dE_t(v),$$

where $\{E_t\}_{t\in\mathbb{R}}$ is the spectral resolution of the operator H. Similarly to (2.26) we have

$$\sigma[v] = \int_{[1,+\infty)} d\|E_t v\|^2 \quad \text{and} \quad \mathcal{E}[v] = \int_{[1,+\infty)} t d\|E_t v\|^2 \tag{6.18}$$

whence

$$\mathcal{E}[v] \geq \sigma[v].$$

On the other hand, for any $f \in \mathcal{V}_i$ we have $\mathcal{E}[f] \leq \sigma[f]$. Since v is a linear combination of functions from $\mathcal{V}_i \subset \mathcal{F}(\Omega_i)$ and the sets Ω_i are disjoint, we obtain

$$\mathcal{E}[v] \leq \sigma[v].$$

Hence, $\mathcal{E}[v] = \sigma[v]$ which is only possible if the measure $d\|E_t v\|^2$ does not charge $(1,+\infty)$, that is

$$v = \int_{\{1\}} dE_t(v),$$

so that v is an eigenfunction of H with the eigenvalue 1. In particular, v satisfies on X the elliptic equation

$$\Delta_\mu v + qv = 0. \tag{6.19}$$

On the other hand, since for any $f \in \mathcal{V}_0 \setminus \{0\}$ we have

$$\mathcal{E}[f] < \sigma[f],$$

the projection of v onto $\mathcal{V}_0$ must vanish (otherwise, we would get $\mathcal{E}[v] < \sigma[v]$). This means that $v \in \bigoplus_{i=1}^{n} \mathcal{V}_i$ and hence $v \equiv 0$ in Ω_0. Since X is connected, the well-known property of solutions to the elliptic equations yields that $v \equiv 0$ in X, which contradicts the construction of v. ∎

Definition 6.10. A weighted manifold (X, μ) is called *parabolic* if $\operatorname{cap}_{\mathcal{E}}(K, X) = 0$ for any compact set $K \subset X$.

Lemma 6.11. *A weighted manifold (X,μ) is parabolic provided anyone of the following conditions is satisfied:*

(a) There exists a constant Q such that for any ball B in X,

$$\operatorname{cap}_{\mathcal{E}}(B,2B)\leq Q. \tag{6.20}$$

(b) All balls of pseudometric d are precompact and there exists a constant C such that, for any ball $B(x,r)$,

$$\mu(B(x,r))\leq Cr^2. \tag{6.21}$$

PROOF. Assume that (a) holds. Any compact set in X is bounded and hence is covered by a ball. Therefore, it suffices to show that $\operatorname{cap}_{\mathcal{E}}(B,X)=0$ for any ball B. It was shown in the proof of Lemma 4.18 that for balls $B_n=B(x,2^n)$ the following inequality holds:

$$\operatorname{cap}_{\mathcal{E}}(B_n,B_m)\leq\frac{Q}{m-n},$$

where $m>n$ are positive integers (see (4.37). Letting $m\to\infty$ we obtain $\operatorname{cap}_{\mathcal{E}}(B_n,X)=0$, which settles the claim.

The fact that (b) implies the parabolicity of X was essentially proved in [**6**] (see also [**21**], [**24**], [**54**]). Alternatively, one can use that $(b)\Longrightarrow(a)$, which was shown in the proof of Theorem 5.3 (cf. inequality (5.9)). ∎

Definition 6.12. A non-empty open set $\Omega\subset X$ is called *q-massive* if there exists a function $u\in C^2(\Omega)\cap C(\overline{\Omega})$ such that

$$0\leq u\leq 1,\quad(\Delta_\mu+q)\,u\geq 0\text{ in }\Omega,\quad u|_{\partial\Omega}=0,\quad u\not\equiv 0. \tag{6.22}$$

Lemma 6.13. *Let the weighted manifold (X,μ) be geodesically complete and parabolic. If $\Omega\subset X$ is a q-massive open set and $\sigma(\Omega)<\infty$ then $N^*(\Omega)\geq 1$.*

PROOF. We shall prove that $u\in\widetilde{\mathcal{F}}(\Omega)$ and $\mathcal{E}[u]\leq\sigma[u]$, which will imply $N^*(\Omega)\geq 1$. Fix a smooth non-negative function $\eta:\mathbb{R}\to\mathbb{R}$ such that $0\leq\eta'\leq 1$ and $\eta|_{(-\infty,\varepsilon]}\equiv 0$, for some $\varepsilon>0$ (see Fig. 13).

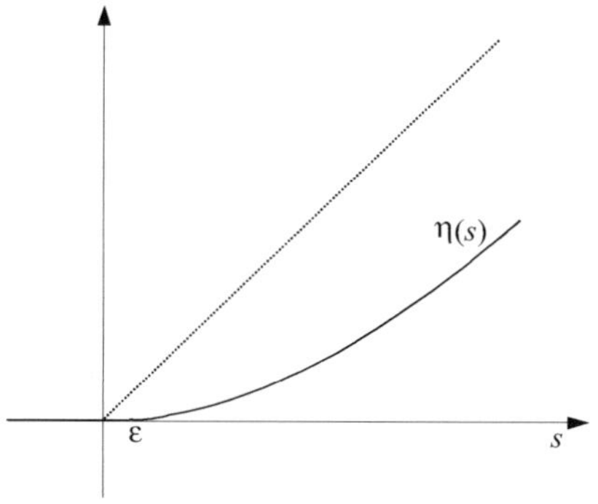

FIGURE 13. A function $\eta(s)$

Set $v=\eta(u)$ and observe that the function v vanishes in a neighborhood of $\partial\Omega$. For any non-negative function $\varphi\in C_0^\infty(X)$, multiplying the inequality

$$\Delta_\mu u+qu\geq 0$$

by $v\varphi^2$ and integrating by parts in Ω, we obtain

$$\int_\Omega \nabla u \cdot \nabla v\, \varphi^2 d\mu + 2\int_\Omega \nabla u \cdot \nabla \varphi\, v\varphi\, d\mu \leq \int_\Omega quv\varphi^2 d\mu. \tag{6.23}$$

Let us allow the parameter ε in the definition of function η to vary and to tend to 0 so that we have a functional family $\{\eta_\varepsilon\}_{\varepsilon>0}$ and respectively the family $v_\varepsilon = \eta_\varepsilon(u)$. The function $v_\varepsilon\varphi$ is clearly in the class $\mathcal{F}(\Omega) = Lip_0(\Omega)$, for any $\varepsilon > 0$. Choose the family $\{\eta_\varepsilon\}$ so that

$$\eta'_\varepsilon(s) \to 1 \quad \text{as } \varepsilon \to 0, \text{ for all } s > 0.$$

In particular, $\eta_\varepsilon(s) \leq s$ and $\eta_\varepsilon(s) \to s$ as $\varepsilon \to 0$. This implies, by the dominated convergence theorem, that the following convergences take place, both in $L^2(\Omega, \mu)$ and $L^2(\Omega, \sigma)$:

$$v_\varepsilon\varphi \to u\varphi, \quad \varphi\nabla v_\varepsilon \to \varphi\nabla u, \quad \nabla(v_\varepsilon\varphi) \to \nabla(u\varphi).$$

In particular, it follows that $u\varphi \in \widetilde{\mathcal{F}}(\Omega)$. Setting $v = v_\varepsilon$ in (6.23) and letting $\varepsilon \to 0$ we obtain

$$\int_\Omega |\nabla u|^2 \varphi^2 d\mu + 2\int_\Omega \nabla u \cdot \nabla\varphi\, u\varphi\, d\mu \leq \int_\Omega qu^2\varphi^2 d\mu.$$

Adding $\int_\Omega u^2 |\nabla\varphi|^2 d\mu$ to the both sides, we obtain in the left hand side a complete square, that is

$$\int_\Omega |\nabla(u\varphi)|^2 d\mu \leq \int_\Omega qu^2\varphi^2 d\mu + \int_\Omega u^2 |\nabla\varphi|^2 d\mu.$$

Finally, using $|u| \leq 1$, we obtain

$$\mathcal{E}[u\varphi] \leq \sigma[u\varphi] + \int_\Omega |\nabla\varphi|^2 d\mu. \tag{6.24}$$

Next, let us construct by induction a sequence of functions $\{\varphi_n\} \subset \mathcal{F}$ such that

$$0 \leq \varphi_n \leq 1, \quad \varphi_n \leq \varphi_{n+1}, \quad \varphi_n \not\equiv 0, \quad \text{and } \mathcal{E}[\varphi_n] < 1/n. \tag{6.25}$$

Indeed, fix a point $x \in X$ and set $B_r = B(x,r)$. Since $\operatorname{cap}_\mathcal{E}(B_1, X) = 0$, we can choose a test function $\varphi_1 \in \mathcal{T}(B_1, X)$ so that $\mathcal{E}[\varphi_1] < 1$. Assuming that φ_n is already constructed, find such a large number r that $\operatorname{supp}\varphi_n \subset B_r$. Since $\operatorname{cap}_\mathcal{E}(B_r, X) = 0$, we can choose a test function $\varphi_{n+1} \in \mathcal{T}(B_n, X)$ so that $\mathcal{E}[\varphi_{n+1}] < 1/(n+1)$. Finally, the monotonicity condition $\varphi_n \leq \varphi_{n+1}$ is satisfied because $\varphi_n \leq 1$ while $\varphi_{n+1} = 1$ on $\operatorname{supp}\varphi_n$.

Setting $u_n = u\varphi_n$ and using (6.24), (6.25), we obtain

$$\mathcal{E}[u_n] \leq \sigma[u_n] + \frac{1}{n}. \tag{6.26}$$

By construction, the sequence $\{u_n\}$ is monotone increasing and converges to u pointwise. By the dominated convergence theorem, we obtain that $u_n \to u$ in $L^2(\Omega, \sigma)$. Let us prove that also $\mathcal{E}[u_n - u] \to 0$. Indeed, by the construction of the functions φ_n, we have

$$\nabla u_n = \varphi_n\nabla u + u\nabla\varphi_n \longrightarrow \nabla u \quad \text{pointwise as } n \to \infty.$$

By Fatou's lemma and (6.26), we obtain

$$\mathcal{E}[u] \leq \liminf_{n\to\infty} \mathcal{E}[u_n] \leq \liminf_{n\to\infty} \sigma[u_n] = \sigma[u] \leq \sigma(\Omega). \tag{6.27}$$

On the other hand, we have

$$\begin{aligned} \mathcal{E}[u-u_n] &= \int_\Omega |\nabla((1-\varphi_n)u)|^2 d\mu \\ &\le 2\int_\Omega |\nabla\varphi_n|^2 u^2 d\mu + 2\int_\Omega (1-\varphi_n)^2 |\nabla u|^2 d\mu \quad (6.28) \\ &\le \frac{2}{n} + 2\int_\Omega (1-\varphi_n)^2 |\nabla u|^2 d\mu. \quad (6.29) \end{aligned}$$

Since the sequence $\{1-\varphi_n\}_{n=1}^\infty$ is bounded and goes to 0 pointwise, while by (6.27) the measure $|\nabla u|^2 d\mu$ is finite, we obtain by the dominated convergence theorem that $\mathcal{E}[u-u_n] \to 0$. Since $u_n \in \widetilde{\mathcal{F}}(\Omega)$, we conclude that also $u \in \widetilde{\mathcal{F}}(\Omega)$. By (6.27) we have $\mathcal{E}[u] \le \sigma[u]$, which finishes the proof. ∎

Corollary 6.14. *Let the weighted manifold (X,μ) be geodesically complete, connected, and parabolic, and let $\sigma(X) < \infty$. Assume that, for some positive integer n, there exist disjoint non-empty open sets $\Omega_0, \Omega_1, ..., \Omega_n$ in X such that Ω_i are q-massive for all $i = 1, 2, ..., n$. Then*

$$Neg(\mathcal{E}-\sigma) \ge n.$$

PROOF. Indeed, by Lemma 6.13, we have, for any $i = 1, 2, ..., n$,

$$N^*(\Omega_i) \ge 1,$$

and, by Lemma 6.9,

$$Neg(\mathcal{E}-\sigma) = \mathcal{N}_1(\mathcal{E},\sigma) = N(X) \ge \sum_{i=1}^n N^*(\Omega_i) \ge n.$$

∎

6.4. Lower bound of the stability index via the number of ends. Here we prove Theorem 6.7. We assume throughout that X is a connected complete oriented minimal surface embedded into $\mathbb{R}^3$such that $\operatorname{ind}(X) < \infty$ and hence $K_{total}(X) < \infty$. As before, let μ be the Riemannian measure on X and $\mathcal{E}$ be the Riemannian energy form on X. Set $q := -2K \ge 0$ and define a measure σ on X by $d\sigma = q d\mu$. We need to prove (6.14), that is

$$Neg(\mathcal{E}-\sigma) \ge k-1, \quad (6.30)$$

where k is the number of ends of X.

Let d be the extrinsic distance on X. It was shown in the proof of Theorem 6.6 that, for any d-ball $B(x,r)$ on X,

$$\mu(B(x,r)) \le \text{const}\, r^2.$$

Therefore, by Lemma 6.11, X is parabolic. Hence, all the hypotheses of the first sentence of Corollary 6.14 are satisfied. Therefore, (6.30) will be proved if we construct k disjoint non-empty open sets $\Omega_1, ..., \Omega_k$ on X such that each Ω_i is q-massive[7].

[7]For application of Corollary 6.14, it suffices to show that $k-1$ sets out of the family $\{\Omega_i\}_{i=1}^k$ are q-massive. In our construction all k sets Ω_i happen to be q-massive. However, this does not imply $\operatorname{ind}(X) \ge k$ because the closures $\overline{\Omega_i}$ may cover all X so that there may be no place for one more non-empty open set as it is required by Corollary 6.14.

Fix a normal unit vector field $n(x)$ on X, a unit vector $v \in \mathbb{R}^3$, and consider the following function on X

$$u(x) := n(x) \cdot v, \tag{6.31}$$

which is known to satisfy on X the Jacobi equation

$$\Delta u + qu = 0.$$

Consider the open set

$$\Omega := \{x \in X : u(x) \neq 0\},$$

and let Ω' be a connected component of Ω. Then either u or $-u$ satisfies (6.22) in Ω' so that Ω' is q-massive. Hence, it suffices to show that, for an appropriate choice of the vector v, the set Ω has at least k connected components.

For that we will use the additional information about the structure of the ends of X, which comes from the fact that X is embedded. By a result of Schoen [**52**], after a rigid rotation of X in $\mathbb{R}^3$, each end E of X can be represented (far enough from the origin) as the graph in $\mathbb{R}^3$ of the following function

$$x_3 = a + b \log r + \frac{c' x_1 + c'' x_2}{r^2} + O\left(r^{-2}\right),$$

where $r = \sqrt{x_1^2 + x_2^2}$ and a, b, c', c'' are real constants. If $b = 0$ then the end E is asymptotic to the horizontal plane $x_3 = a$, whereas in the case $b \neq 0$ the end E is asymptotic to the catenoid

$$r = 2 \cosh\left(\frac{x_3 - a}{b}\right).$$

In the former case, we refer to E as a *planar* end, and in the latter case - as a *catenoidal* end (see Fig. 14).

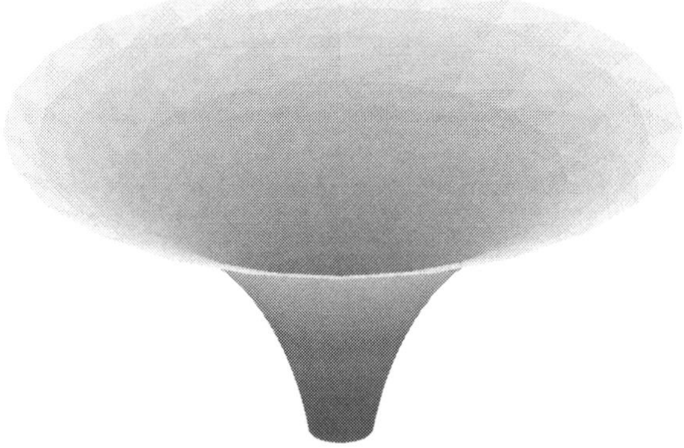

FIGURE 14. A catenoidal end.

All ends of X are naturally ordered by the way they intersect a remote vertical line l. Namely, let $h(E)$ be the x_3-coordinate of the point where E meets l (see Fig. 15).

Definition 6.15. We say that the end E is *below* the end E' if $h(E) < h(E')$.

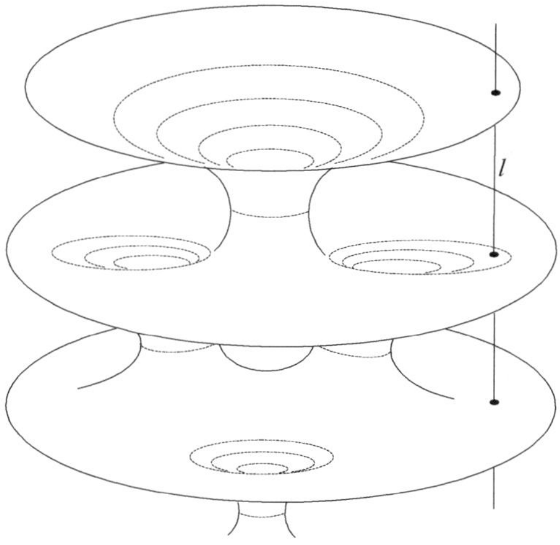

FIGURE 15. Ordering the ends of an embedded minimal surface according to their intersections with a vertical line l.

Clearly, this definition does not depend on the choice of l provided the distance from l to the origin o is large enough (for a more general result on ordering of the ends of embedded minimal surfaces see [**19**]). We say that the ends E and E' are *neighbors* if there is no end E'' between E and E' in the sense of the order "below".

For any end E, the normal vector field $n(x)$ has the limit as x goes to ∞ along E, so let $n(E)$ denote this limit. Clearly, $n(E)$ is vertical, that is $n(E) = (0,0,1)$ or $n(E) = (0,0,-1)$.

Lemma 6.16. *If E and E' are two ends of X, which are neighbors, then* $n(E) = -n(E')$.

PROOF. Let x and y be the points of intersection of respectively E and E' with a remote vertical line l. Choose l far enough so that $n(x)$ and $n(y)$ are "nearly" vertical and that the segment $[x,y]$ of l does not intersect X except for the points x, y. Since X is connected, there is a path $\gamma : [0,1] \to X$ connecting x and y on X. Fix $\varepsilon > 0$ and consider the deformed path in $\mathbb{R}^3$

$$\gamma_\varepsilon(t) = \gamma(t) + \varepsilon n(\gamma(t)), \quad t \in [0,1].$$

The path γ_ε connects in $\mathbb{R}^3$ the points x_ε and y_ε where

$$x_\varepsilon = x + \varepsilon n(x) \quad \text{and} \quad y_\varepsilon = y + \varepsilon n(y).$$

If ε is small enough then $\gamma_\varepsilon(t)$ does not intersect X (see Fig. 16). Therefore, any other path from x_ε to y_ε must have even number of intersections with X.

Contrary to what we need to prove, assume that

$$n(E) = n(E'). \tag{6.32}$$

Then there is a path from x_ε to y_ε that crosses X exactly *once*, at the point y: this path is obtained by slightly modifying the path $[x_\varepsilon, x] \# [x,y] \# [y, y_\varepsilon]$ near the point x so that it does not meet X in a neighborhood of x. This contradiction finishes the proof. ∎

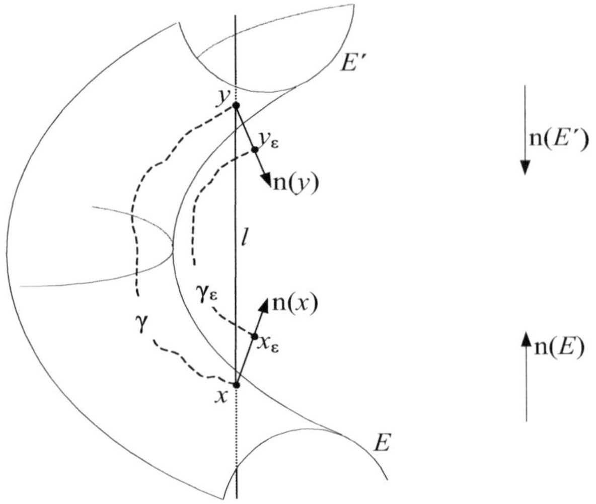

FIGURE 16. Paths γ and γ_ε.

Choose the vector v in (6.31) as follows: $v = (0,0,1)$. If $x \to \infty$ along an end E then $u(x) \to n(E) \cdot v = \pm 1$. Let us say that an end E is *positive* if $u(x) \to 1$ on E and E is *negative* of $u(x) \to -1$ on E. It follows from Lemma 6.16 that positive and negative ends alternate relative to the order "below". Let $E_1, E_2, ..., E_k$ be all ends of X and let Ω_i be the connected component of the set $\Omega = \{u \neq 0\}$ containing a neighborhood of ∞ in E_i. Clearly, if the end E_i is positive then $u > 0$ in Ω_i, and if E_i is negative then $u < 0$ in Ω_i. This implies that the components Ω_i and Ω_j, which correspond to neighboring ends E_i and E_j, are disjoint. Therefore, all components Ω_i, $i = 1, 2, ..., k$, are disjoint, which finishes the proof of Theorem 6.7.

Remark 6.17. The main point of the above proof was to show that the function $u = n(x) \cdot v$ has at least k components of constant sign. Having proved that, we could have used instead of Corollary 6.14 a result of Choe [**7**] about a *vision number*, which says that if this particular function u has k components of constant sign then $\operatorname{ind}(X) \geq k - 1$. We have preferred a more general approach based on q-massive sets, because this approach does not need a function u to be defined on the *entire* manifold X. This approach might work for *immersed* minimal surfaces where one can expect to be able to construct a function u satisfying (6.22) *separately* for each end.

References

[1] **Barbosa J.L., do Carmo M.**, On the size of a stable minimal surface in $\mathbb{R}^3$, *Amer. J. Math.*, **98** (1976) no.2, 515–528.

[2] **Barlow M.T.**, Diffusions on fractals, *in:* "Lectures on Probability Theory and Statistics, Ecole d'été de Probabilités de Saint-Flour XXV - 1995", Lecture Notes Math. 1690, Springer, 1998. 1-121.

[3] **Barlow M.T.**, Heat kernels and sets with fractal structure, *in:* "Heat kernels and analysis on manifolds, graphs, and metric spaces", *Contemporary Mathematics*, **338** (2003) 11-40.

[4] **Barlow M.T.**, Anomalous diffusion and stability of Harnack inequalities, *in:* "Surveys in Differential Geometry", 2004.

[5] **do Carmo M., Peng C.K.**, Stable minimal surfaces in $\mathbb{R}^3$ are planes, *Bulletin of the AMS*, **1** (1979) 903-906.
[6] **Cheng S.Y., Yau S.-T.**, Differential equations on Riemannian manifolds and their geometric applications, *Comm. Pure Appl. Math.*, **28** (1975) 333-354.
[7] **Choe J.**, Index, vision number and stability of complete minimal surfaces, *Arch. Rat. Mech. Anal.*, **109** (1990) no.3, 195–212.
[8] **Chung F.R.K., Grigor'yan A., Yau S.-T.**, Upper bounds for eigenvalues of the discrete and continuous Laplace operators, *Advances in Math.*, **117** (1996) 165-178.
[9] **Chung F.R.K., Grigor'yan A., Yau S.-T.**, Eigenvalues and diameters for manifolds and graphs, *in:* "Tsing Hua Lectures on Geometry and Analysis", Ed. S.-T.Yau, International Press, 1997. 79-105.
[10] **Chung F.R.K., Grigor'yan A., Yau S.-T.**, Higher eigenvalues and isoperimetric inequalities on Riemannian manifolds and graphs, *Comm. Anal. Geom*, **8** (2000) no.5, 969-1026.
[11] **Colbois B., Dodziuk J.**, Riemannian metrics with large λ_1, *Proceedings of AMS*, **122** (1994) no.3, 905–906.
[12] **Colding T.H., Minicozzi W.P. II**, "Minimal surfaces", Courant Lecture Notes in Math.4, 1999.
[13] **Colding T.H., Minicozzi W.P. II**, An excursion into geometric analysis, *in:* "Surveys in Differential Geometry", 2004.
[14] **Coulhon T., Holopainen I., Saloff-Coste L.**, Harnack inequality and hyperbolicity for subelliptic p-Laplacians with applications to Picard type theorems, *Geom. Funct. Anal.*, **11** (2001)
[15] **Davies E.B.**, "Heat kernels and spectral theory", Cambridge University Press, 1989.
[16] **Davies E.B.**, "Spectral theory and differential operators", Cambridge University Press, 1995.
[17] **Fischer-Colbrie D.**, On complete minimal surfaces with finite Morse index in three manifolds, *Invent. Math.*, **82** (1985) 121-132.
[18] **Fischer-Colbrie D., Schoen R.**, The structure of complete stable minimal surfaces in 3-manifolds of non-negative scalar curvature, *Comm. Pure Appl. Math.*, **33** (1980) 199-211.
[19] **Frohman C., Meeks W.H. III**, The ordering theorem for the ends of properly embedded minimal surfaces, *Topology*, **36** (1997) no.3, 605-617.
[20] **Fukushima M., Oshima Y., Takeda M.**, "Dirichlet forms and symmetric Markov processes", Studies in Mathematics 19, De Gruyter, 1994.
[21] **Grigor'yan A.**, On the existence of positive fundamental solution of the Laplace equation on Riemannian manifolds, (in Russian) *Matem. Sbornik*, **128** (1985) no.3, 354-363. Engl. transl. *Math. USSR Sb.*, **56** (1987) 349-358.
[22] **Grigor'yan A.**, Heat kernel upper bounds on a complete non-compact manifold, *Revista Matemática Iberoamericana*, **10** (1994) no.2, 395-452.
[23] **Grigor'yan A.**, Isoperimetric inequalities and capacities on Riemannian manifolds, *Operator Theory: Advances and Applications*, **109** (1999) 139-153.
[24] **Grigor'yan A.**, Analytic and geometric background of recurrence and non-explosion of the Brownian motion on Riemannian manifolds, *Bull. Amer. Math. Soc.*, **36** (1999) 135-249.
[25] **Grigor'yan A.**, Heat kernels and function theory on metric measure spaces, *in:* "Heat kernels and analysis on manifolds, graphs, and metric spaces", *Contemporary Mathematics*, **338** (2003) 143-172.
[26] **Grigor'yan A., Hu J., Lau K.S.**, Heat kernels on metric-measure spaces and an application to semi-linear elliptic equations, *Trans. Amer. Math. Soc.*, **355** (2003) no.5, 2065-2095.
[27] **Grigor'yan A., Yau S.-T.**, Decomposition of a metric space by capacitors, *in:* "Differential Equations: La Pietra 1996", Ed. Giaquinta et. al., *Proceeding of Symposia in Pure Mathematics*, **65** 1999. 39-75.
[28] **Grigor'yan A., Yau S.-T.**, Isoperimetric properties of higher eigenvalues of elliptic operator, *Amer. J. Math*, **125** (2003) 893-940.
[29] **Hersch J.**, Quatre properiétés isopérimétriques de membranes sphériques homogènes, *C.R. Acad. Sci. Paris*, **270** (1970) 1645-1648.
[30] **Hoffman D., Karcher H.**, Complete embedded minimal surfaces of finite total curvature, *in:* "Geometry, V", Encyclopaedia Math. Sci. 90, Springer, Berlin, 1997. 5–93, 267–272.
[31] **Hoffman D., Spruck J.**, Sobolev and isoperimetric inequalities for Riemannian submanifolds, *Comm. Pure Appl. Math.*, **27** (1974) 715–727. See also "A correction to: Sobolev

and isoperimetric inequalities for Riemannian submanifolds", *Comm. Pure Appl. Math.*, **28** (1975) no.6, 765–766.

[32] **Holopainen I.**, Volume growth, Green's functions and parabolicity of ends, *Duke Math. J.*, **97** (1999) no.2, 319-346.

[33] **Huber A.**, On subharmonic functions and differential geometry in the large, *Comment. Math. Helvetici*, **32** (1957) 181–206.

[34] **Jakobson D., Nadirashvili N., Polterovich I.**, Extremal metric for the first eigenvalue on a Klein bottle, to appear in *Canad. J. Math*

[35] **Jorge L., Meeks W.H. III**, The topology of complete minimal surfaces of finite total Gaussian curvature, *Topology*, **22** (1983) no.2, 203-221.

[36] **Kigami J.**, "Analysis on fractals", Cambridge University Press, 2001.

[37] **Korevaar N.**, Upper bounds for eigenvalues of conformal metric, *J. Diff. Geom.*, **37** (1993) 73-93.

[38] **Levin D., Solomyak M.**, The Rozenblum-Lieb-Cwikel inequality for Markov generators, *J. d'Analyse Math.*, **71** (1997) 173-193.

[39] **Li P., Treibergs A.**, Applications of eigenvalue techniques to geometry, *in:* "Contemporary Geometry", Univ. Ser. Math., Plenum, New York, 1991. 21-52.

[40] **Li P., Yau S.-T.**, A new conformal invariant and its applications to the Willmore conjecture and the first eigenvalue of compact surfaces, *Invent. Math.*, **69** (1982) 269–291.

[41] **Li P., Yau S.-T.**, On the Schrödinger equation and the eigenvalue problem, *Comm. Math. Phys.*, **88** (1983) 309–318.

[42] **Maz'ya V.G.**, "Sobolev spaces", (in Russian) Izdat. Leningrad Gos. Univ. Leningrad, 1985. Engl. transl. Springer, 1985.

[43] **Meeks W. H. III, Pérez J.**, Conformal properties in classical minimal surface theory, *in:* "Surveys in Differential Geometry", 2004.

[44] **Micallef M.**, Comparison of index of energy with index of area of minimal surfaces, in preparation

[45] **Montiel S., Ros A.**, Schrödinger operators associated to a holomorphic map, *in:* "Global Differental Geometry and Global Analysis", Lecture Notes Math. 1481, Springer, 1990. 147-174.

[46] **Nadirashvili N.**, Berger's isoperimetric problem and minimal immersions of surfaces, *Geom. Funct. Anal.*, **6** (1996) 877-897.

[47] **Nitsche J.C.C.**, "Lectures on minimal surfaces, vol.1", Cambridge University Press, 1989.

[48] **Osserman R.**, "A survey of minimal surfaces", Dover, New York, 1986.

[49] **Pogorelov A.V.**, On the stability of minimal surfaces, *Soviet Math. Dokl.*, **24** (1981) 274-276.

[50] **Reed M., Simon B.**, "Methods of modern mathematical physics. II: Fourier analysis, self-adjointness", Academic Press, 1975.

[51] **Saloff-Coste L.**, "Aspects of Sobolev inequalities", LMS Lecture Notes Series 289, Cambridge Univ. Press, 2002.

[52] **Schoen R.**, Uniqueness, symmetry, and embeddedness of minimal surfaces, *J. Diff. Geom.*, **18** (1983) 791–809.

[53] **Simon L.M.**, "Lectures on geometric measure theory", Proceedings of the Centre for Mathematical Analysis, Australian National University, 3, Australian National University, Centre for Mathematical Analysis, Canberra, 1983.

[54] **Sturm K-Th.**, Sharp estimates for capacities and applications to symmetrical diffusions, *Probability theory and related fields*, **103** (1995) no.1, 73-89.

[55] **Tysk J.**, Eigenvalue estimates with applications to minimal surfaces, *Pacific J. Math.*, **128** (1987) 361-366.

[56] **Yang P., Yau S.-T.**, Eigenvalues of the Laplacian of compact Riemann surfaces and minimal submanifolds, *Ann. Scuola Norm. Sup. Pisa Cl. Sci. (4)*, **7** (1980) 55-63.

[57] **Yau S.-T.**, Survey on partial differential equations in differential geometry, *Ann. Math. Studies*, **102** (1982) 3-70.

Imperial College London, London SW7 2AZ, United Kingdom
E-mail address: a.grigoryan@imperial.ac.uk

University of Bristol, University Walk, Bristol, BS8 1TW, United Kingdom
E-mail address: y.netrusov@bristol.ac.uk

Harvard University, Cambridge MA 02138, USA
E-mail address: yau@math.harvard.edu

Surveys in Differential Geometry
IX, International Press

Spectral gap, logarithmic Sobolev constant, and geometric bounds

Michel Ledoux

ABSTRACT. We survey recent works on the connection between spectral gap and logarithmic Sobolev constants, and exponential integrability of Lipschitz functions. In particular, tools from measure concentration are used to describe bounds on the diameter of a (compact) Riemannian manifold and of Markov chains in terms of the first eigenvalue of the Laplacian and the logarithmic Sobolev constant. We examine similarly dimension free isoperimetric bounds using these parameters.

CONTENTS

1. Introduction

In the recent years, it has been realized that simple measure theoretic arguments may be used to produce (sharp) bounds on geometric objects, such as the diameter of manifolds or graphs, in terms of spectral or logarithmic Sobolev constants. This question has of course a long run in Riemannian geometry and this work focuses more on the (elementary) methods than on the conclusions themselves. The key argument is described through the measure concentration and exponential integrability properties of distance functions under Poincaré or logarithmic Sobolev inequalities going back to the work of M. Gromov and V. Milman [G-M] and I. Herbst (cf. [Le5]). In particular, the approach avoids any type of purely geometric arguments and delicate heat kernel bounds, and produces bounds of the correct order of magnitude in the dimension. The investigation at the level of logarithmic Sobolev constants turns out to be of crucial interest in the study of rates of convergence to equilibrium, especially for Markov chains as developed by P. Diaconis and L. Saloff-Coste (cf. [D-SC], [SC3]).

To describe the connection between spectral and logarithmic Sobolev inequalities, and integrability properties of distance functions, it will be convenient to use the language of metric measure spaces and of measure concentration (cf. [Le5]). Let thus (X, d, μ) be a metric measure space in the sense of [Grom2], that is (X, d) is a metric space and μ is a finite non-negative Borel measure on (X, d), normalized to be a probability measure ($\mu(X) = 1$). Define, for $\lambda \in \mathbb{R}$, the Laplace functional of μ on (X, d) as

$$\mathrm{E}_{(X,d,\mu)}(\lambda) = \sup \int_X \mathrm{e}^{\lambda F} d\mu$$

where the supremum runs over all (bounded) 1-Lipschitz functions F on (X, d) such that $\int_X F d\mu = 0$. By 1-Lipschitz, we understand that $|F(x) - F(y)| \leq d(x, y)$ for all $x, y \in X$. We often write more simply $\mathrm{E}_\mu = \mathrm{E}_{(X,d,\mu)}$. Note that E_μ is an even function, non-decreasing on $[0, \infty)$. We will say, following [Grom2] (cf. [Le5]) that (X, d, μ) has exponential concentration whenever $\mathrm{E}_{(X,d,\mu)}(\lambda_0) < \infty$ for some $\lambda_0 > 0$, and that (X, d, μ) has normal concentration if for some constant $C > 0$,

$$\mathrm{E}_{(X,d,\mu)}(\lambda) \leq \mathrm{e}^{C\lambda^2/2}, \quad \lambda \in \mathbb{R}.$$

This terminology is motivated by the following elementary lemma that describes the geometric aspects, in terms of measure concentration, of these properties. For any two sets A, B in (X, d), set

$$d(A, B) = \inf \big\{ d(x, y); x \in A, y \in B \big\}.$$

LEMMA 1.1. *For any two sets A, B in (X, d), and any $\lambda \geq 0$,*

$$\mu(A)\mu(B) \leq \mathrm{e}^{-\lambda d(A,B)}\, \mathrm{E}_\mu(\lambda)^2.$$

In particular, if (X, d, μ) has normal concentration (with constant C)

$$\mu(A)\mu(B) \leq \mathrm{e}^{-d(A,B)^2/4C}.$$

PROOF. By definition of the Laplace functional E_μ of (X, d, μ), for any (bounded) 1-Lipschitz function F on X and any $\lambda \geq 0$,

$$\int_A \int_B \mathrm{e}^{\lambda[F(x) - F(y)]} d\mu(x) d\mu(y) \leq \int_X \mathrm{e}^{\lambda(F-M)} d\mu \int_X \mathrm{e}^{\lambda(M-F)} d\mu \leq \mathrm{E}_\mu(\lambda)^2$$

where $M = \int_X F d\mu$. Choose now $F(x) = \min(d(x, B), r)$, $x \in X$, so that

$$\int_A \int_B \mathrm{e}^{\lambda[F(x) - F(y)]} d\mu(x) d\mu(y) \geq \mathrm{e}^{\lambda \min(d(A,B), r)} \mu(A)\mu(B).$$

Let then $r \to \infty$. If (X, d, μ) has normal concentration, optimize in λ. Lemma 1.1 is proved. □

If A is a subset of (X, d), denote by

$$A_r = \{x \in X; d(x, A) < r\}$$

its (open) neighborhood of order $r > 0$.

COROLLARY 1.2. *If (X, d, μ) has exponential concentration (respectively normal concentration), for every Borel set A in (X, d) such that $\mu(A) \geq \frac{1}{2}$,*

$$\mu(A_r) \geq 1 - 2\, \mathrm{E}_\mu(\lambda_0)^2\, \mathrm{e}^{-\lambda_0 r}$$

(respectively

$$\mu(A_r) \geq 1 - 2\, \mathrm{e}^{-r^2/4C})$$

for all $r > 0$.

It is worthwhile mentioning for further comparison that whenever the diameter D of (X, d) is finite, (X, d, μ) has normal concentration with constant D^2 for any probability measure μ on the Borel sets of (X, d). Indeed, let F be a mean zero 1-Lipschitz function on (X, d). By Jensen's inequality, for every $\lambda \in \mathbb{R}$,

$$\int_X e^{\lambda F} d\mu \leq \int_X \int_X e^{\lambda [F(x) - F(y)]} d\mu(x) d\mu(y) \leq \sum_{i=0}^{\infty} \frac{(D\lambda)^{2i}}{(2i)!} \leq e^{D^2 \lambda^2 / 2}.$$

The claim follows.

In these notes, we survey recent developments on spectral and logarithmic Sobolev bounds by the preceding measure concentration tools, mainly taken from the references [SC2], [D-SC], [Le3]. In Section 2, we show how the existence of a spectral gap, or Poincaré inequality, implies exponential concentration. This observation turns out to have rather useful consequences to bounds on the diameter to which we turn next in Section 3. All these results have analogous counterparts on graphs and discrete structures. In Section 4, we develop the corresponding investigation at the level of logarithmic Sobolev inequalities that provide more precise bounds. Namely, together with the Herbst argument, we show how the logarithmic Sobolev constant entails normal concentration of Lipschitz functions and deduce then sharp bounds on the diameter. Again, the discrete case leads to a number of considerations of interest in connection with geometric bounds. We briefly discuss the analogous conclusions under the entropic constant. In the last part, we investigate isoperimetric bounds under spectral and logarithmic Sobolev constants using some simple semigroup tools, and describe, as a main feature, dimension free inequalities of isoperimetric type.

2. Spectrum and exponential integrability

Assume first we are given a smooth complete connected Riemannian manifold (X, g) (without boundary) with Riemannian metric g and finite volume V. Denote by d the distance function induced by g on X and by $d\mu = \frac{dv}{V}$ the normalized Riemannian volume element on (X, g). Let furthermore $\lambda_1 = \lambda_1(X)$ be the first non-trivial eigenvalue of the Laplacian Δ_g on X. By the Raleigh-Ritz variational principle (cf. [Cha1], [G-H-L]...), λ_1 is characterized by the spectral gap, or Poincaré, inequality

$$\lambda_1 \int_X f^2 d\mu \leq \int_X f(-\Delta_g f) d\mu = \int_X |\nabla f|^2 d\mu \tag{2.1}$$

for all smooth real-valued functions f on (X, g) such that $\int_X f d\mu = 0$, where $|\nabla f|$ denotes the Riemannian length of the gradient of f.

The following result goes back independently to M. Gromov and V. Milman [G-M] (in a geometric context) and A. Borovkov and S. Utev [B-U] (in a probabilistic context). (See also [Br].) It has been investigated in [A-M-S] and [A-S] using moment bounds, and in [Sc] using a differential inequality on Laplace transforms (similar to the Herbst argument presented in Section 4). We follow here the approach by S. Aida and D. Stroock [A-S].

THEOREM 2.1. *Let (X, g) be a smooth complete connected Riemannian manifold with finite volume and normalized Riemannian measure μ. Denote by $\lambda_1 = \lambda_1(X)$*

the first non-trivial eigenvalue of the Laplacian Δ_g *on* (X,g). *Then,*

$$\mathrm{E}_{(X,d,\mu)}\big(\sqrt{\lambda_1}\big) \le 3.$$

In particular (X,d,μ) *has exponential concentration whenever* $\lambda_1 > 0$.

PROOF. Set $u(\lambda) = \int_X e^{\lambda F} d\mu$, $\lambda \ge 0$, where F is bounded and such that $\int_X F d\mu = 0$. Since F may be assumed smooth enough, we can have that $|\nabla F| \le 1$ everywhere. Apply (2.1) to $f = e^{\lambda F/2}$, $\lambda \ge 0$. Since

$$\int_X |\nabla f|^2 d\mu = \frac{\lambda^2}{4} \int_X |\nabla F|^2 e^{\lambda F} d\mu \le \frac{\lambda^2}{4} \int_X e^{\lambda F} d\mu,$$

we get that

$$u(\lambda) - u\Big(\frac{\lambda}{2}\Big)^2 \le \frac{\lambda^2}{4\lambda_1}\, u(\lambda).$$

Hence, for every $\lambda < 2\sqrt{\lambda_1}$,

$$u(\lambda) \le \frac{1}{1 - \lambda^2/4\lambda_1}\, u\Big(\frac{\lambda}{2}\Big)^2.$$

Applying the same inequality for $\lambda/2$ and iterating, yields, after n steps,

$$u(\lambda) \le \prod_{k=0}^{n-1} \Big(\frac{1}{1 - \lambda^2/4^{k+1}\lambda_1}\Big)^{2^k} u\Big(\frac{\lambda}{2^n}\Big)^{2^n}.$$

Since $u(\lambda) = 1 + o(\lambda)$, we have that $u(\lambda/2^n)^{2^n} \to 1$ as $n \to 0$. Therefore,

$$u(\lambda) \le \prod_{k=0}^{\infty} \Big(\frac{1}{1 - \lambda^2/4^{k+1}\lambda_1}\Big)^{2^k}$$

where the infinite product converges whenever $\lambda < 2\sqrt{\lambda_1}$. Setting for example $\lambda = \sqrt{\lambda_1}$ yields that

$$\int_X e^{\sqrt{\lambda_1} F} d\mu = u\big(\sqrt{\lambda_1}\big) \le 3.$$

The proof of Theorem 2.1 is complete. □

It is a simple yet non-trivial observation that $\lambda_1(X \times Y) = \min(\lambda_1(X), \lambda_1(Y))$ for Riemannian manifolds X and Y. Theorem 2.1 therefore provides a useful tool to concentration in product spaces (cf. [Le5]).

Theorem 2.1 has an analogue on graphs to which we turn now. It is convenient to deal with finite state Markov chains.

Let X be a finite (or countable) set. Let $\Pi(x,y) \ge 0$, $x, y \in X$, satisfy

$$\sum_{y \in X} \Pi(x,y) = 1$$

for every $x \in X$. Assume furthermore that there is a symmetric invariant probability measure μ on X for Π, that is $\Pi(x,y)\mu(\{x\})$ is symmetric in x and y and $\sum_x \Pi(x,y)\mu(\{x\}) = \mu(\{y\})$ for every $y \in X$. In other words, (Π, μ) is a reversible Markov chain (cf. e.g. [SC3] and the references therein). Define, for $f, g : X \to \mathbb{R}$ say finitely supported, the Dirichlet form

$$\mathcal{Q}(f,g) = \sum_{x,y \in X} \big[f(x) - f(y)\big]\big[g(x) - g(y)\big]\Pi(x,y)\mu(\{x\}).$$

We may speak of the spectral gap, or the Poincaré constant, of the chain (Π, μ) as the largest λ_1 such that for all f's (with finite support) such that $\int_X f d\mu = 0$,

$$\lambda_1 \int_X f^2 d\mu \leq \mathcal{Q}(f, f). \tag{2.2}$$

Set also

$$|||f|||_\infty^2 = \sup_{x \in X} \sum_{y \in X} \big|f(x) - f(y)\big|^2 \Pi(x, y).$$

The triple norm $|||\cdot|||_\infty$ may be thought of as a discrete version of the Lipschitz norm in the continuous setting. Although it may not be well adapted to all discrete structures, it behaves similarly for what concerns spectrum and exponential concentration. Equip X with the distance associated with $|||\cdot|||_\infty$ defined as

$$d_\mathcal{Q}(x, y) = \sup_{|||f|||_\infty \leq 1} \big[f(x) - f(y)\big], \quad x, y \in X.$$

Theorem 2.2 below is the analogue of Theorem 2.1 in this discrete setting (cf. [A-S]). The proof is essentially the same.

THEOREM 2.2. *Let (Π, μ) be a reversible Markov chain on X as before with spectral gap λ_1. Then*

$$\mathrm{E}_{(X, d_\mathcal{Q}, \mu)}\big(\sqrt{\lambda_1/2}\big) \leq 3.$$

In particular $(X, d_\mathcal{Q}, \mu)$ has exponential concentration whenever $\lambda_1 > 0$.

PROOF. We proceed as for Theorem 2.1. The main observation is that, for every F on X and every $\lambda \geq 0$,

$$\mathcal{Q}\big(\mathrm{e}^{\lambda F/2}, \mathrm{e}^{\lambda F/2}\big) \leq \frac{1}{2} \, |||F|||_\infty^2 \, \lambda^2 \int_X \mathrm{e}^{\lambda F} d\mu. \tag{2.3}$$

Indeed, by symmetry,

$$\begin{aligned}
\mathcal{Q}\big(\mathrm{e}^{\lambda F/2}, \mathrm{e}^{\lambda F/2}\big) &= \sum_{x,y \in X} \big[\mathrm{e}^{\lambda F(x)/2} - \mathrm{e}^{\lambda F(y)/2}\big]^2 \Pi(x, y) \mu(\{x\}) \\
&= 2 \sum_{F(y) < F(x)} \big[\mathrm{e}^{\lambda F(x)/2} - \mathrm{e}^{\lambda F(y)/2}\big]^2 \Pi(x, y) \mu(\{x\}) \\
&\leq \frac{\lambda^2}{2} \sum_{x,y \in X} \big[F(x) - F(y)\big]^2 \mathrm{e}^{\lambda F(x)} \Pi(x, y) \mu(\{x\})
\end{aligned}$$

from which (2.3) follows by definition of $|||F|||_\infty$.

Assume now that F is bounded with mean zero and $|||F|||_\infty \leq 1$. Set $u(\lambda) = \int_X \mathrm{e}^{\lambda F} d\mu$, $\lambda \geq 0$. Applying (2.2) to $\mathrm{e}^{\lambda F/2}$ yields, together with (2.3),

$$u(\lambda) - u\Big(\frac{\lambda}{2}\Big)^2 \leq \frac{\lambda^2}{2\lambda_1} \, u(\lambda),$$

that is, for every $0 \leq \lambda < \sqrt{2\lambda_1}$,

$$u(\lambda) \leq \frac{1}{1 - \lambda^2/2\lambda_1} \, u\Big(\frac{\lambda}{2}\Big)^2.$$

We then conclude as for Theorem 2.1, with λ_1 replaced by $\lambda_1/2$. □

The distance most often used in such contexts is however not d_Q but the combinatoric distance d_c associated with the graph with vertex-set X and edge-set $\{(x,y) : \Pi(x,y) > 0\}$. This distance can be defined as the minimal number of edges one has to cross to go from x to y. Equivalently,

$$d_c(x,y) = \sup_{\|\nabla f\|_\infty \le 1} \big[f(x) - f(y)\big]$$

where

$$\|\nabla f\|_\infty = \sup\big\{\big|f(x) - f(y)\big|; \Pi(x,y) > 0\big\}.$$

Now since $\sum_y \Pi(x,y) = 1$,

$$|||f|||_\infty^2 \le \|\nabla f\|_\infty^2.$$

In particular $d_c \le d_Q$ so that Theorem 2.2 also holds for the metric measure space (X, d_c, μ).

As an example, let $X = (V, \mathcal{E})$ be a finite connected graph with set of vertices V and symmetric set of edges $\mathcal{E}$. Equip V with the normalized uniform measure μ and the graph distance d_c. We may consider $\Pi(x,y) = \frac{1}{k(x)}$ whenever x and y are adjacent in V and 0 otherwise where $k(x)$ is the number of neighbors of x. Consider the quadratic form

$$\widetilde{Q}(f,f) = \sum_{x \sim y} \big[f(x) - f(y)\big]^2$$

where the sum runs over all neighbors $x \sim y$ in X. Since X is connected, $\widetilde{Q}(f,f) \ge 0$ for all f's and is zero whenever f is constant. Let $\lambda_1 > 0$ be the first non-trivial eigenvalue of $\widetilde{Q}$ (that is of the Laplace operator on X). As a consequence of Theorem 2.2, we have the following result.

Corollary 2.3. *Let $k_0 = \max\{k(x); x \in V\} < \infty$. Then*

$$\mathrm{E}_{(X,d,\mu)}\left(\sqrt{\frac{\lambda_1}{2k_0}}\right) \le 3.$$

The most important examples of applications of the preceding corollary are the Cayley graphs. If V is a finite group and $S \subset V$ a symmetric set of generators of V, we may join x and y in V by an edge if $x = s^{-1}y$ for some $s \in S$. The path distance on $X = (V, \mathcal{E})$ is the word distance in V induced by S and $k_0 = \mathrm{Card}(S)$. For oriented graphs, see [Al], [A-M].

To conclude this section, we show, on the basis of Lemma 1.1, how the preceding exponential integrability results may be applied to bounds on the spectral gap λ_1 in terms of distances between disjoints sets. Letting indeed $\lambda_0 = \sqrt{\lambda_1}$ in Lemma 1.1, it follows from Theorem 2.1 that, for every sets A, B in X,

$$\lambda_1 \le \frac{1}{d(A,B)^2} \log^2\left(\frac{C}{\mu(A)\mu(B)}\right) \tag{2.4}$$

with $C = 9$. (In the context of Theorem 2.2, replace λ_1 by $\lambda_1/2$.) As discussed in [Bo-L], the preceding arguments may easily be improved to reach $C = 1$ in (2.4). Inequalities such as (2.4) have been considered by F. R. K. Chung, A. Grigory'an and S.-T. Yau [C-G-Y1] who showed (2.4) (with $C = 4$) using heat kernel expansions, and then using the wave equation [C-G-Y2] (with $C = \mathrm{e}$). They actually establish similar inequalities for the all sequence of eigenvalues, something not considered here. They also establish similar results on graphs.

3. Spectral and diameter bounds

On the basis of the results of the preceding section, we investigate here some relationships between spectral and diameter bounds. The following result is taken from [Le5], and may be traced back in the work by R. Brooks [Br].

Assume that we are given a smooth complete connected Riemannian manifold (X, g), (without boundary), not necessarily compact but with finite volume V. Denote by $d\mu = \frac{dv}{V}$ the normalized Riemannian volume element. Let as before $\lambda_1 = \lambda_1(X)$ be the first non-trivial eigenvalue of the Laplace operator Δ_g on (X, g).

If $B(x, r)$ is the (open) ball with center x and radius $r > 0$ in X, it follows from Theorem 2.1 together with Corollary 1.2 that $\lambda_1 = \lambda_1(X) = 0$ as soon as

$$\limsup_{r\to\infty} \frac{1}{r} \log\Big(1 - \mu\big(B(x,r)\big)\Big) = 0 \tag{3.1}$$

for some (all) x in X (cf. [Br]). The following is a kind of converse.

THEOREM 3.1. *Let (X, g) be a smooth complete connected Riemannian manifold with dimension n and finite volume. Let μ be the normalized Riemannian volume on (X, g). Assume that the Ricci curvature of (X, g) is bounded below. Then (X, g) is compact as soon as*

$$\liminf_{r\to\infty} \frac{1}{r} \log\Big(1 - \mu\big(B(x,r)\big)\Big) = -\infty$$

for some (or all) $x \in X$. In particular $\lambda_1 = \lambda_1(X) > 0$ under this condition. Furthermore, if (X, g) has non-negative Ricci curvature and if D is the diameter of X, then

$$\lambda_1 \le \frac{C_n}{D^2} \tag{3.2}$$

where $C_n > 0$ only depends on the dimension n of X.

The upper bound (3.2) goes back to the work by S.-Y. Cheng [Chen] in Riemannian geometry (see also [Cha1], [L-Y1] and below). In the opposite direction, it has been shown by P. Li [Li] and H. C. Yang and J. Q. Zhong [Y-Z] that when (X, g) has non-negative Ricci curvature,

$$\lambda_1 \ge \frac{\pi^2}{D^2}\,. \tag{3.3}$$

This lower bound is optimal since achieved on the one-dimensional torus.

PROOF OF THEOREM 3.1. We proceed by contradiction and assume that X is not compact. Choose $B(x, r_0)$ a geodesic ball in X with center x and radius $r_0 > 0$ such that $\mu(B(x, r_0)) \ge \frac{1}{2}$. By non-compactness (and completeness), for every $r > 0$, we can take z at distance $r_0 + 2r$ from x. In particular, $B(x, r_0) \subset B(z, 2(r_0 + r))$. By the Riemannian volume comparison theorem [C-E], [Cha2], for every $y \in X$ and $0 < s < t$,

$$\frac{\mu(B(y,t))}{\mu(B(y,s))} \le \Big(\frac{t}{s}\Big)^n e^{t\sqrt{(n-1)K}} \tag{3.4}$$

where $-K$, $K \ge 0$, is the lower bound on the Ricci curvature of (X, g). Therefore,

$$\begin{aligned}\mu\big(B(z,r)\big) &\ge \Big(\frac{r}{2(r_0+r)}\Big)^n e^{-2(r+r_0)\sqrt{(n-1)K}}\,\mu\big(B((z, 2(r_0+r))\big)\\ &\ge \frac{1}{2}\Big(\frac{r}{2(r_0+r)}\Big)^n e^{-2(r_0+r)\sqrt{(n-1)K}}\end{aligned}$$

where we used that $\mu(B(z, 2(r_0 + r))) \geq \mu(B(x, r_0)) \geq \frac{1}{2}$. Since $B(z, r)$ is included into the complement of $B(x, r_0 + r)$,

$$1 - \mu\big(B(x, r + r_0)\big) \geq \frac{1}{2}\Big(\frac{r}{2(r_0 + r)}\Big)^n \mathrm{e}^{-2(r_0+r)\sqrt{(n-1)K}} \tag{3.5}$$

which is impossible as $r \to \infty$ by the assumption. The first part of the theorem is established.

Thus (X, g) is compact. Denote by D its diameter. Assume that (X, g) has non-negative Ricci curvature. That is, we may take $K = 0$ in (3.4) and (3.5). By Theorem 2.1 together with Corollary 1.2, for every measurable subset A in X such that $\mu(A) \geq \frac{1}{2}$, and every $r > 0$,

$$1 - \mu(A_r) \leq 18\,\mathrm{e}^{-\sqrt{\lambda_1}\, r}. \tag{3.6}$$

We distinguish between two cases. If $\mu(B(x, \frac{D}{8})) \geq \frac{1}{2}$, apply (3.6) to $A = B(x, \frac{D}{8})$. By definition of D, we may choose $r = r_0 = \frac{D}{8}$ in (3.5) to get

$$\frac{1}{2 \cdot 4^n} \leq 1 - \mu(A_{D/8}) \leq 18\,\mathrm{e}^{-\sqrt{\lambda_1}\, D/8}.$$

If $\mu(B(x, \frac{D}{8})) < \frac{1}{2}$, apply (3.6) to A the complement of $B(x, \frac{D}{8})$. Since the ball $B(x, \frac{D}{16})$ is included into the complement of $A_{D/16}$ and since by (3.4) with $t = D$,

$$\mu\Big(B\Big(x, \frac{D}{16}\Big)\Big) \geq \frac{1}{16^n},$$

we get from (3.6) with $r = \frac{D}{16}$ that

$$\frac{1}{16^n} \leq 1 - \mu(A_{D/16}) \leq 18\,\mathrm{e}^{-\sqrt{\lambda_1}\, D/16}.$$

The conclusion easily follows from either case, with a constant C_n of the order of n^2 as n is large. Theorem 3.1 is established. □

Analogous conclusions may be obtained in the discrete case. Let as before $\Pi(x, y)$ be a Markov chain on a finite state space X with symmetric invariant probability measure μ. Recall λ_1 the spectral gap of (Π, μ) and d_Q the distance defined from the norm

$$|||f|||_\infty^2 = \sup_{x \in X} \sum_{y \in X} \big|f(x) - f(y)\big|^2 \Pi(x, y).$$

Denote by D_Q the diameter of X for d_Q.

PROPOSITION 3.2. *If μ is nearly constant, that is if there exists $C > 0$ such that, for every x, $\mu(\{x\}) \leq C \min_{y \in X} \mu(\{y\})$, then*

$$D_Q^2 \leq \frac{\big(4 \log(3C|X|)\big)^2}{\lambda_1}$$

where $|X|$ is the cardinal of X.

PROOF. Consider two points $x, y \in X$ such that $d(x, y) = D_Q$. By Lemma 1.1 and Theorem 2.2,

$$\mu(\{x\})\mu(\{y\}) \leq 9\,\mathrm{e}^{-D_Q\sqrt{\lambda_1}/2}.$$

Since, by the hypothesis on μ, $\min_{z \in X} \mu(\{z\}) \geq (C|X|)^{-1}$, the conclusion follows. □

Recall that the combinatoric diameter D_c is less than or equal to D_Q. As such, Proposition 3.2 goes back to [A-M] (see also [Al], [Chu]), where it is observed that the bound on D_c of Proposition 3.2 is optimal on the class of regular graphs.

4. Logarithmic Sobolev constant and diameter bounds

In this section, we turn to the corresponding investigation for logarithmic Sobolev inequalities. Logarithmic Sobolev constants actually provide sharper bounds than spectral gaps, and are of importance in the study of rates of convergence to equilibrium.

To start with, let as before (X, g) be a complete connected Riemannian manifold (without boundary) with finite volume V. Let d be the distance function associated to g and $d\mu = \frac{dv}{V}$ the normalized volume element. In analogy with the first non-trivial eigenvalue $\lambda_1 = \lambda_1(X)$ of Δ_g on X, define the logarithmic Sobolev constant $\rho_0 = \rho_0(X)$ of Δ_g as the largest constant ρ such that for every smooth function f on X with $\int_X f^2 d\mu = 1$,

$$\rho \int_X f^2 \log f^2 d\mu \leq 2 \int_X f(-\Delta_g f) d\mu = 2 \int_X |\nabla f|^2 d\mu. \tag{4.1}$$

Applying (4.1) to $1 + \varepsilon f$, a simple Taylor expansion as $\varepsilon \to 0$ shows that $\lambda_1 \geq \rho_0$.

Note that, as for $\lambda_1 = \lambda_1(X)$, one may show (cf. [Gros], [Le3]) that $\rho_0(X \times Y) = \min(\rho_0(X), \rho_0(Y))$ for Riemannian manifolds X and Y.

It is a non-trivial result, due to O. Rothaus [Ro1], that whenever X is compact,

$$\lambda_1 \geq \rho_0 > 0. \tag{4.2}$$

When the Ricci curvature of (X, g) is uniformly bounded below by a strictly positive constant R, it goes back to A. Lichnerowicz (cf. [Cha1], [G-H-L]) that $\lambda_1 \geq R_n$ where $R_n = \frac{R}{1-\frac{1}{n}}$, with equality if and only if X is a sphere (Obata's theorem). This lower bound has been shown to hold similarly for the logarithmic Sobolev constant by D. Bakry and M. Emery [B-E] (cf. [Ba]) so that

$$\lambda_1 \geq \rho_0 \geq R_n. \tag{4.3}$$

The case of equality for ρ_0 is a consequence of Obata's theorem due to an improvement of the preceding by O. Rothaus [Ro2] who showed that when (X, g) is compact and $\mathrm{Ric}_g \geq R$ $(R \in \mathbb{R})$,

$$\rho_0 \geq \alpha_n \lambda_1 + (1 - \alpha_n) R_n \tag{4.4}$$

where $\alpha_n = 4n/(n+1)^2$. In particular, λ_1 and ρ_0 are of the same order if (X, g) has non-negative Ricci curvature. As examples, $\rho_0 = \lambda_1 = n$ on the n-sphere [M-W]. On the n-dimensional torus, $\lambda_1 = \rho_0 = 1$. The question whether $\rho_0 < \lambda_1$ in this setting has been open for some time until the geometric investigation by L. Saloff-Coste [SC2]. He actually proved, using heat kernel bounds and equilibrium rates, that the existence of a logarithmic Sobolev inequality in a Riemannian manifold with finite volume and Ricci curvature bounded from below forces the manifold to be compact. It is known that there exist non-compact manifolds of finite volume with $\lambda_1 > 0$. In particular, there exist compact manifolds of constant negative sectional curvature with spectral gaps uniformly bounded away from zero, and arbitrarily large diameters (cf. [SC2]. This yield examples for which the ratio ρ_0/λ_1 can be made arbitrarily small.

We present below a simplified argument of this result based on Theorem 3.1 and normal concentration. To this task, we develop, for the logarithmic Sobolev inequality, the connection with exponential integrability and measure concentration as for the spectral inequalities in Sections 2 and 3. This will be achieved by the Herbst argument from a logarithmic Sobolev inequality to exponential integrability. It goes back to an unpublished argument by I. Herbst [Da-S], revived in the past years by S. Aida, T. Masuda and I. Shigekawa [A-M-S]. Relevance to measure concentration was emphasized in [Le2], and further developed in [Le3] (cf. [Le5] for the historical developments).

The principle is similar to the application of spectral properties to concentration presented in Section 2.1, but logarithmic Sobolev inequalities allow us to reach normal concentration. Let F be a smooth bounded 1-Lipschitz function on X such that $\int_X F d\mu = 0$. In particular, since F is assumed to be regular enough, we can have that $|\nabla F| \leq 1$ at every point. We apply (4.1) to $f^2 = e^{\lambda F}$ for every $\lambda \in \mathbb{R}$. We have

$$\int_X |\nabla f|^2 d\mu = \frac{\lambda^2}{4} \int_X |\nabla F|^2 e^{\lambda F} d\mu \leq \frac{\lambda^2}{4} \int_X e^{\lambda F} d\mu.$$

Setting $u(\lambda) = \int_X e^{\lambda F} d\mu$, $\lambda \in \mathbb{R}$, by the definition of entropy,

$$\lambda u'(\lambda) - u(\lambda) \log u(\lambda) \leq \frac{1}{2\rho_0} \lambda^2 u(\lambda).$$

In other words, if $U(\lambda) = \frac{1}{\lambda} \log u(\lambda)$, $U(0) = \int_X F d\mu = 0$, then

$$U'(\lambda) \leq \frac{1}{2\rho_0}, \quad \lambda \in \mathbb{R}.$$

Therefore $U(\lambda) \leq \frac{\lambda}{2\rho_0}$ from which we immediately conclude that

$$u(\lambda) = \int_X e^{\lambda F} d\mu \leq e^{\lambda^2/2\rho_0}$$

for every $\lambda \in \mathbb{R}$.

We summarize the preceding argument in the following statement. Recall the Laplace functional $E_{(X,d,\mu)}$ of μ on X.

THEOREM 4.1. *Let (X, g) be a smooth complete connected Riemannian manifold with finite volume and normalized Riemannian measure μ. Denote by ρ_0 the logarithmic Sobolev constant of Δ_g on (X, g). Then,*

$$E_{(X,d,\mu)}(\lambda) \leq e^{\lambda^2/2\rho_0}, \quad \lambda \in \mathbb{R}.$$

In particular, (X, d, μ) has normal concentration whenever $\rho_0 > 0$.

As announced, it was shown by L. Saloff-Coste [SC2] that the existence of $\rho_0 = \rho_0(X) > 0$ forces a Riemannian manifold with finite volume and Ricci curvature bounded from below to be compact. Together with Theorem 4.1, we may present, along the lines of the proof of Theorem 3.1, a sharp improvement of the quantitative bound on the diameter of X in terms of the logarithmic Sobolev constant ρ_0.

THEOREM 4.2. *Let (X, g) be a smooth complete connected Riemannian manifold with dimension n and finite volume. Let μ be the normalized Riemannian volume element on (X, g) and denote by $\rho_0 = \rho_0(X)$ the logarithmic Sobolev constant of Δ_g on (X, g). Assume that $\mathrm{Ric}_g \geq -K$, $K \geq 0$. If $\rho_0 > 0$, then (X, g) is compact.*

Furthermore, if D is the diameter of X, there exists a numerical constant $C > 0$ such that

$$D \le C\sqrt{n}\,\max\left(\frac{\sqrt{K}}{\rho_0}, \frac{1}{\sqrt{\rho_0}}\right).$$

It is known from the theory of hypercontractive semigroups (cf. [De-S]) that conversely there exists $C(n, K, \varepsilon)$ such that

$$\rho_0 \ge \frac{C(n, K, \varepsilon)}{D}$$

whenever $\lambda_1 \ge \varepsilon > 0$.

PROOF. By Theorem 4.1 and Corollary 1.2,

$$1 - \mu(A_r) \le 2\,\mathrm{e}^{-\rho_0 r^2/4} \tag{4.5}$$

for every $r > 0$ and $A \subset X$ such that $\mu(A) \ge \frac{1}{2}$. It is thus clear that

$$\liminf_{r\to\infty} \frac{1}{r} \log\Big(1 - \mu\big(B(x, r)\big)\Big) = -\infty$$

so that (X, g) is compact by Theorem 3.1. To establish the bound on the diameter D, we repeat the proof of Theorem 3.1, replacing (3.6) by (4.5). Let $B(x, \frac{D}{8})$ be the ball with center x and radius $\frac{D}{8}$. We distinguish between two cases. If $\mu(B(x, \frac{D}{8})) \ge \frac{1}{2}$, apply (4.5) to $A = B(x, \frac{D}{8})$. By definition of D, we may choose $r = r_0 = \frac{D}{8}$ in (3.5) to get

$$\frac{1}{2} \cdot \frac{1}{4^n}\,\mathrm{e}^{-\sqrt{(n-1)K}D/2} \le 1 - \mu(A_{D/8}) \le 2\,\mathrm{e}^{-\rho_0 D^2/256}.$$

If $\mu(B(x, \frac{D}{8})) < \frac{1}{2}$, apply (4.5) to A the complement of $B(x, \frac{D}{8})$. Since the ball $B(x, \frac{D}{16})$ is included into the complement of $A_{D/16}$ and since by (3.4)

$$\mu\Big(x_0, \frac{D}{16}\Big) \ge \frac{1}{16^n}\,\mathrm{e}^{-\sqrt{(n-1)K}D},$$

it follows from (4.5) with $r = \frac{D}{16}$ that

$$\frac{1}{16^n}\,\mathrm{e}^{-\sqrt{(n-1)K}D} \le 1 - \mu(A_{D/16}) \le 2\,\mathrm{e}^{-\rho_0 D^2/1024}.$$

In both cases,

$$\rho_0 D^2 - C\sqrt{(n-1)K}\,D - Cn \le 0$$

for some numerical constant $C > 0$. Hence

$$D \le \frac{C\sqrt{(n-1)K} + \sqrt{C^2(n-1)K + 4C\rho_0 n}}{2\rho_0}$$

and thus

$$D \le \frac{C\sqrt{(n-1)K} + \sqrt{C\rho_0 n}}{\rho_0}$$

which yields the conclusion. The theorem is established. □

COROLLARY 4.3. *Let X be a compact Riemannian manifold with dimension n and non-negative Ricci curvature. Then*

$$\rho_0 \le \frac{Cn}{D^2}$$

for some numerical constant $C > 0$.

Corollary 4.3 has to be compared to Cheng's upper bound [Chen] on the spectral gap of compact manifolds with non-negative Ricci curvature

$$\lambda_1 \le \frac{2n(n+4)}{D^2}\,. \tag{4.6}$$

Hence, generically, the difference between the upper bound on λ_1 and ρ_0 seems to be of the order of n. Moreover, it is mentioned in [Chen] that there exist examples with $\lambda_1 \approx n^2/D^2$. They indicate that both Rothaus' lower bound (4.4) and Corollary 4.3 could be sharp. Note also that (4.4) together with Corollary 4.3 allows us to recover Cheng's upper bound on λ_1 of the same order in n.

Corollary 4.3 is stated for (compact) manifolds without boundary but it also holds for compact manifolds of non-negative Ricci curvature with convex boundary (and Neuman's conditions). In particular, this result applies to convex bounded domains in $\mathbb{R}^n$ equipped with normalized Lebesgue measure. If we indeed closely inspect the proof of Theorem 4.2 in the latter case for example, we see that what is only required is (4.5), that holds similarly, and the volume comparisons. These are however well-known and easy to establish for bounded convex domains in $\mathbb{R}^n$. In this direction, it might be worthwhile mentioning moreover that the first non-zero Neumann eigenvalue λ_1 of the Laplacian on radial functions on the Euclidean ball B in $\mathbb{R}^n$ behaves as n^2. It may be identified indeed as the square of the first positive zero κ_n of the Bessel function $J_{n/2}$ of order $n/2$ (cf. [Cha1] e.g.). (On a sphere of radius r, there will be a factor r^{-2} by homogeneity.) In particular, standard methods or references [Wat] show that $\kappa_n \approx n$ as n is large. Denoting by ρ_0 the logarithmic Sobolev constant on radial functions on B, a simple adaption of the proof of Theorem 4.2 shows that $\rho_0 \le Cn$ for some numerical constant $C > 0$. Actually, ρ_0 is of the order of n and this may be shown directly in dimension one by a simple analysis of the measure with density nx^{n-1} on the interval $[0,1]$. We are indebted to S. Bobkov for this observation. One can further measure on this example the difference between the spectral gap and the logarithmic Sobolev constant as the dimension n is large. (On general functions, λ_1 and ρ_0 are both of the order of n, see [Bo].)

As another application, assume $\mathrm{Ric}_g \ge R > 0$. As we have seen, by the Bakry-Emery inequality [B-E], $\rho_0 \ge R_n$ where $R_n = \frac{R}{1-\frac{1}{n}}$. Therefore, by Corollary 4.3,

$$D \le C\sqrt{\frac{n-1}{R}}\,.$$

Up to the numerical constant, this is just Myers' theorem on the diameter of a compact manifold $D \le \pi\sqrt{\frac{n-1}{R}}$ (cf. [Cha2]). This could suggest that the best numerical constant in Corollary 4.3 is π^2.

Dimension free lower bounds on the logarithmic Sobolev constant in manifolds with non-negative Ricci curvature, similar to the lower bound (3.3) on the spectral gap, are also available. It has been shown by F.-Y. Wang [Wan] (see also [B-L-Q] and [Le3] for slightly improved quantitative estimates) that, if $\mathrm{Ric}_g \ge 0$,

$$\rho_0 \ge \frac{\lambda_1}{1 + 2D\sqrt{\lambda_1}}\,.$$

In particular, together with (3.1),

$$\rho_0 \ge \frac{\pi^2}{(1+2\pi)D^2}\,.$$

The preceding lower bound holds more generally for the logarithmic Sobolev constants of Laplace operators with drift $\mathrm{L} = \Delta_g - \nabla U \cdot \nabla$ for a smooth function U (with finite, time reversible measure $d\mu = \mathrm{e}^{-U} dv$) of non-negative curvature in the sense that, as symmetric tensors,

$$\mathrm{Ric}_g - \nabla\nabla U \geq 0.$$

Under this condition, it is actually shown in [Wan] that if for some $c > 0$ and some (all) x in X,

$$\int_X \mathrm{e}^{cd(x,\cdot)^2} d\mu \leq C < \infty,$$

then $\rho_0 > 0$, with a lower bound depending on c, C. In $\mathbb{R}^n$ with U convex, S. Bobkov [Bo] showed that $\lambda_1 > 0$ with a lower bound depending on n. It would be a challenging question in this context to establish a lower bound on λ_1 only depending on $c, C > 0$ such that

$$\int_X \mathrm{e}^{cd(x,\cdot)} d\mu \leq C < \infty.$$

We refer to the previous references for further details.

Next, we describe analagous results in the discrete case. As in Section 2.1, let $\Pi(x,y)$ be a Markov chain on a finite state space X with symmetric invariant probability measure μ. Let ρ_0 be the logarithmic Sobolev constant of (Π, μ) defined as the largest ρ such that

$$\rho \int_X f^2 \log f^2 d\mu \leq 2\mathcal{Q}(f,f)$$

for every f on X with $\int_X f^2 d\mu = 1$. Recall that here

$$\mathcal{Q}(f,f) = \sum_{x,y \in X} \big[f(x) - f(y)\big]^2 \Pi(x,y)\mu\big(\{x\}\big).$$

Recall also we set

$$|||f|||_\infty^2 = \sup_{x \in X} \sum_{y \in X} \big|f(x) - f(y)\big|^2 \Pi(x,y)$$

and denote by $d_{\mathcal{Q}}$ the associated metric.

Arguing as for Theorem 4.1, and using (2.3), we may obtain similarly normal concentration from the logarithmic Sobolev constant ρ_0.

THEOREM 4.4. *Let (Π, μ) be a reversible Markov chain on X as before with logarithmic Sobolev constant ρ_0. Then*

$$\mathrm{E}_{(X, d_{\mathcal{Q}}, \mu)}(\lambda) \leq \mathrm{e}^{\lambda^2/\rho_0}, \quad \lambda \in \mathbb{R}.$$

In particular $(X, d_{\mathcal{Q}}, \mu)$ has normal concentration whenever $\rho_0 > 0$.

In the context of Corollary 2.3, we have similarly that whenever $k_0 = \max\{k(x); x \in V\} < \infty$,

$$\mathrm{E}_{(X, d_c, \mu)}(\lambda) \leq \mathrm{e}^{k_0 \lambda^2/\rho_0}, \quad \lambda \in \mathbb{R}.$$

The next statement is analagous to Proposition 3.2 for the logarithmic Sobolev constant. Denote by $D_{\mathcal{Q}}$ the diameter of X for $d_{\mathcal{Q}}$. The proof is an immediate consequence of Lemma 1.1 and Theorem 4.4. The numerical constant is not sharp.

PROPOSITION 4.5. *If μ is nearly constant, that is if there exists C such that, for every x, $\mu(\{x\}) \leq C \min_{y \in X} \mu(\{y\})$, then*

$$D_{\mathcal{Q}}^2 \leq \frac{16 \log(C|X|)}{\rho_0}$$

where $|X|$ is the cardinal of X.

As already discussed, the distance most often used in the present setting is not d but the combinatoric distance d_c associated with the graph with vertex-set X and edge-set $\{(x,y) : \Pi(x,y) > 0\}$. Recall that $d_c \leq d$ (so that in particular, the combinatoric diameter D_c satisfies $D_c \leq D_{\mathcal{Q}}$.) Hence, Proposition 4.5 also holds with D_c.

It is worthwhile mentioning that a small improvement of the concentration bound may be obtained with the graph distance d_c. Indeed, reproducing the argument leading to (2.3) actually shows that when

$$\|\nabla f\|_\infty = \sup\big\{|f(x) - f(y)|; \Pi(x,y) > 0\big\} \leq 1,$$

for every $\lambda \geq 0$,

$$\begin{aligned}
\mathcal{Q}\big(e^{\lambda F/2}, e^{\lambda F/2}\big) &= 2 \sum_{F(y)<F(x)} \Big(1 - e^{-\lambda[F(x)-F(y)]/2}\Big)^2 e^{\lambda F(x)} \Pi(x,y)\mu(\{x\}) \\
&\leq 2\big(1 - e^{-\lambda/2}\big)^2 \int_X e^{\lambda F} d\mu \\
&\leq 2 \min\left(1, \frac{\lambda^2}{4}\right) \int_X e^{\lambda F} d\mu. \qquad (4.7)
\end{aligned}$$

The proof of Theorem 4.1 then yields that

$$E_{(X,d_c,\mu)}(\lambda) \leq e^{2\phi(\lambda)/\rho_0}, \quad \lambda \geq 0,$$

where $\phi(\lambda) = \lambda^2$ if $\lambda \leq 2$ and $\phi(\lambda) = 4(\lambda - 1)$ if $\lambda \geq 2$. Together with Lemma 1.1, we thus draw, under the assumption of Proposition 4.5, an upper bound on ρ_0 in term of the graph diameter D_c as

$$\rho_0 \leq \min\left(\frac{8}{D_c}, \frac{16\log(C|X|)}{D_c^2}\right). \qquad (4.8)$$

Results such as Proposition 4.5 and (4.8) may be used as efficient upper bounds on the logarithmic Sobolev constant ρ_0 in terms of simple geometric objects such as the graph diameter D_c. These are of interest in the study of rates of convergence to equilibrium for finite Markov chains. While it is classical that the spectral gap λ_1 governs the asymptotic exponential rate of convergence to equilibrium, it has been shown by P. Diaconis and L. Saloff-Coste [SC1], [SC2], [SC3], [D-SC], both in the continuous and discrete cases actually, that the logarithmic Sobolev constant ρ_0 is more closely related to convergence to stationarity than λ_1 is. Let us now survey a few of examples of interest, kindly communicated to us by L. Saloff-Coste (cf. [D-SC], [SC3] for the necessary background).

Consider first the hypercube $\{0,1\}^n$ with $\Pi(x,y) = 1/n$ if x, y differ by exactly one coordinate and $\Pi(x,y) = 0$ otherwise. The reversible measure is the uniform distribution and it is classical (see [D-SC]) that $\rho_0 = 4/n$. The bound (4.8) tells us that $\rho_0 \leq 8/n$.

Consider the Bernoulli-Laplace model of diffusion. This is a Markov chain on the n-sets of an N-set with $n \le N/2$. If the current state is an n-set A, we pick an element x at random in A, an element y at random in the complement A^c of A and change A to $B = (A \setminus \{x\}) \cup \{y\}$. The kernel Π is given by $\Pi(A,B) = 1/[n(N-n)]$ if $|A \cap B| = n-2$ and $\Pi(A,B) = 0$ otherwise. The uniform distribution $\pi(A) = \binom{N}{n}^{-1}$ is the reversible measure. Clearly, $D_c = n$. Hence, by (4.8), $\rho_0 \le 8/n$ which is the right order of magnitude [L-Y].

Let now the random transposition chain on the symmetric group S_n, $n \ge 2$. Here, $\Pi(\sigma,\theta) = 2/[n(n-1)]$ if $\theta = \sigma\tau$ for some transposition τ and $\Pi(\sigma,\theta) = 0$ otherwise and $\pi \equiv (n!)^{-1}$, The diameter is $D_c = n-1$ and one knows that ρ_0 is of order $1/n \log n$ [D-SC], [L-Y]. Here (4.8) only yields $\rho_0 \le 16/n$. Since we know that $\rho_0 \ge (2n \log n)^{-1}$ [D-SC], we can also conclude from Proposition 4.5 that

$$D_Q \le \sqrt{\frac{32n \log n}{\rho_0}} \le 8n \log n. \tag{4.9}$$

It is not clear whether or not this bound can be obtained more easily. Note that the upper bound

$$d_Q(\sigma,\theta) \le \Big(\min_{\Pi(\sigma,\theta)>0} \Pi(\sigma,\theta)\Big)^{-1/2} d_c(x,y)$$

only yields $D_Q \le n^2$, up to a multiplicative constant. It might be worthwhile observing that in this example, ρ_0 is of order $1/n \log n$ while it has been shown by B. Maurey [Ma] that concentration (with respect to the combinatoric metric) is satisfied at a rate of the order of $1/n$ (see below).

Consider a N-regular graph with N fixed. Let $\Pi(x,y) = 1/N$ if they are neighbors and $\Pi(x,y) = 0$ otherwise. Then $\mu(\{x\}) = 1/|X|$. Assume that for some constant $C > 0$, and all $x \in X$ and $t > 0$,

$$\big|B(x,2t)\big| \le C\big|B(x,t)\big| \tag{4.10}$$

where $B(x,t)$ is the ball with center x and radius t in the graph distance d_c, and $|B(x,t)|$ its number of elements. Fix $x, y \in X$ such that $d_c(x,y) = D_c$. Set $A = B(x, \frac{D_c}{2})$ and $B = B(y, \frac{D_c}{4})$. By (4.10), $|B(x, \frac{D_c}{2})| \ge C^{-1}|X|$ and $|B(y, \frac{D_c}{4})| \ge C^{-2}|X|$ so that

$$\mu(A) \ge \frac{1}{C} \quad \text{and} \quad \mu(B) \ge \frac{1}{C^2}\,.$$

Since $d_c(A,B) \ge \frac{D_c}{4}$, by Theorem 4.4 and Lemma 1.1,

$$\rho_0 \le \frac{192 \log C}{D_c^2}\,.$$

For N and C fixed, this is the right order of magnitude in the class of Cayley graphs of finite groups satisfying the volume doubling condition (4.10). See [D-SC, Theorem 4.1].

As a last example, consider any N-regular graph on a finite set X. Let $\Pi(x,y) = 1/N$ if they are neighbors and $\Pi(x,y) = 0$ otherwise. Then $\mu(\{x\}) = 1/|X|$ and $|X| \le N^{D_c}$ (at least if $D_c \ge 2$). From (4.8), $\rho_0 \le 8/D_c$. Compare with the results of [D-SC] and the Riemannian case. This is, in a sense, optimal generically. Indeed, if $|X| \ge 4$, one also have the lower bound [D-SC]

$$\rho_0 \ge \frac{\lambda}{D_c \log N}$$

where $1-\lambda$ is the second largest eigenvalue of Π. There are many known families of N-regular graphs (N fixed) such that $|X| \to \infty$ whereas $\lambda \geq \epsilon > 0$ stays bounded away from zero (the so-called expanders graphs). Moreover graphs with this property are "generic" amongst N-regular graphs [Al].

In the study of birth and death Markov chains, and especially Poisson point processes, some modified versions of the logarithmic Sobolev inequalities have been recently considered [B-T]. One of them is the entropic inequality that gives rise to the entropic constant ρ_1 defined as the largest ρ such that

$$2\rho \int f \log f d\mu \leq \mathcal{Q}(f, \log f)$$

when f runs over all finitely supported functions $f: X \to \mathbb{R}^+$ such that $\int_X f d\mu = 1$. The entropic inequality and constant have been considered in [Wu] for Poisson measure on $\mathbb{N}$, and in the present context in the recent contributions [B-T], [G-Q], [Go]. It is pointed out there that, in general,

$$\lambda_1 \geq \rho_1 \geq \rho_0$$

and that ρ_1 is also suited to control convergence to equilibrium in the total variation distance. In some typical examples, the entropic constant ρ_1 however turns out to be a much better rate than the logarithmic Sobolev constant ρ_0. For example, while on the symmetric discrete cube $\{0,1\}^n$, $\rho_0 = \rho_1 = \lambda_1 = 4/n$, on the cube with weight p and q ($p+q=1$), $\rho_1 \sim \lambda_1$ but $\rho_0 \ll \lambda_1$ as $pq \to 0$. Similarly, on the symmetric group S_n with the random transposition chain, $\rho_1 \sim \lambda_1 \sim 1/n$ (as $n \to \infty$) (cf. the previous references), while, as mentioned above, $\rho_0 \sim 1/n \log n$. With respect to the logarithmic Sobolev constant, the entropic constant seems however inadequate to control convergence in ℓ^2 (cf. [Go]).

Arguing as for the proof of Theorem 4.4, we may get a concentration bound from the entropic inequality. Indeed, as for (2.3), we see that for every $\lambda \geq 0$,

$$\mathcal{Q}(\lambda F, e^{\lambda F}) \leq 2\lambda^2 |||F|||^2 \int_X e^{\lambda F} d\mu,$$

yielding thus Theorem 4.4, and similarly Proposition 4.5, but with the improved entropic constant $\rho_1 \geq \rho_0$. In particular, since $\rho_1 \geq 1/2(n-1)$ on S_n, we recover Maurey's concentration result [Ma]. Furthermore,

$$D_{\mathcal{Q}} \leq 8n\sqrt{\log n}$$

that improves upon (4.9).

Again, the graph distance d_c yields some improved bounds (cf. [B-T], [Go]). Namely, arguing as for (4.7), we have that, for every $\lambda \geq 0$ and every F with $\|\nabla F\|_\infty \leq 1$,

$$\mathcal{Q}(\lambda F, e^{\lambda F}) \leq 2\lambda(1 - e^{-\lambda}) \int_X e^{\lambda F} d\mu \leq 2\min(\lambda, \lambda^2) \int_X e^{\lambda F} d\mu.$$

The Herbst argument then applies to yield the following consequence.

THEOREM 4.6. *Let (Π, μ) be a reversible Markov chain on X as before with entropic constant ρ_1. Then*

$$E_{(X,d_c,\mu)}(\lambda) \leq e^{\min(\lambda^2, \lambda(1+\log^+\lambda))/\rho_1}, \quad \lambda \geq 0.$$

By Lemma 1.1, we deduce from Theorem 4.6 a bound, to be compared to (4.8), on the entropic constant ρ_1 in terms of the graph diameter D_c, namely

$$\rho_1 \leq \min\left(\frac{2}{D_c}\log\left(\frac{2\mathrm{e}^2}{D_c}\log\left(C|X|\right)\vee \mathrm{e}\right), \frac{16\log(C|X|)}{D_c^2}\right). \tag{4.11}$$

The previous Markov chains examples may then be analyzed via the entropic constant ρ_1 to yield, depending on the cases, possibly sharper bounds.

We refer to [B-T], [G-Q], [Go] for further results and details on the entropic constant.

5. Dimension free isoperimetric bounds

In this last section, we investigate some inequalities of isoperimetric type that may be produced from spectral and logarithmic Sobolev informations. While sharper dimensional bounds are known and classical, we emphasize here dimension free estimates of interest in the study of diffusion operators with drifts (cf. [Ba], [Le4]). The result improve upon [Le1] (see also [Ba-L]).

Let (X, g) be a smooth complete connected Riemanian manifold, and let Δ_g be the Laplace operator on (X, g). Let $(P_t)_{t\geq 0}$ be the heat semigroup (cf. [Da]). It is worthwhile mentioning that, whenever (X, g) is of finite volume V, both the spectral gap λ_1 and logarithmic Sobolev constants admit equivalent description in terms of smoothing properties of $(P_t)_{t\geq 0}$. Denote by $d\mu = \frac{dv}{V}$ the normalized volume element. By the spectral theorem

$$\|P_t f\|_2 \leq \mathrm{e}^{-\lambda_1 t}\|f\|_2, \quad t \geq 0, \tag{5.1}$$

for every f with $\int_X f\,d\mu = 0$, where $\|\cdot\|_p$ is the L^p-norm $(1 \leq p \leq \infty)$ with respect to μ. A fundamental theorem of L. Gross [Gros] shows that ρ_0 may be characterized by the hypercontractivity property

$$\|P_t f\|_q \leq \|f\|_p \tag{5.2}$$

for every f whenever $1 < p < q < \infty$ and $\mathrm{e}^{\rho_0 t} \geq [(q-1)/(p-1)]^{1/2}$.

The next lemma is a reversed Poincaré inequality for heat kernel measures (cf. [Le4]). We use it below as a weak, dimension free, form of the Li-Yau parabolic gradient inequality [L-Y2].

LEMMA 5.1. *Assume that* $\mathrm{Ric}_g \geq -K$, $K \geq 0$. *Then, for every* $t \geq 0$ *and every smooth function* f *on* (X, g), *at every point,*

$$c(t)\,|\nabla P_t f|^2 \leq P_t(f^2) - (P_t f)^2$$

where

$$c(t) = \frac{1 - \mathrm{e}^{-2Kt}}{K} \qquad (= 2t \text{ if } K = 0).$$

PROOF. For a smooth function f on (X, g), and $t > 0$ fixed, set

$$\varphi(s) = \mathrm{e}^{2Ks}P_s(|\nabla P_{t-s}f|^2),\ 0 \leq s \leq t$$

(evaluated at some point in X). By the chain rule for differentiation,

$$\varphi'(s) = 2\,\mathrm{e}^{2Ks}\Big[KP_s\quad(|\nabla P_{t-s}f|^2) + P_s\big(\tfrac{1}{2}\,\Delta_g P_{t-s}f - \nabla P_{t-s}f\cdot\nabla\Delta_g P_{t-s}f\big)\Big].$$

By the Bochner formula,

$$\frac{1}{2}\Delta_g P_{t-s}f - \nabla P_{t-s}f\cdot\nabla\Delta_g P_{t-s}f \geq -K|\nabla P_{t-s}f|^2.$$

Hence φ is non-decreasing so that, for every $t \geq 0$, and at every point,

$$|\nabla P_t f|^2 \leq e^{2Kt} P_t\big(|\nabla f|^2\big). \tag{5.3}$$

Write then

$$P_t(f^2) - (P_t f)^2 = \int_0^t \frac{d}{ds} P_s\big((P_{t-s} f)^2\big) ds = 2 \int_0^t P_s\big(|\nabla P_{t-s} f|^2\big) ds.$$

By (5.3), $P_s(|\nabla P_{t-s} f|^2) \geq e^{-2Ks} |\nabla P_t f|^2$ so that the claim follows. The proof is complete. □

As a consequence of this lemma, and since $1 - e^{-u} \geq \frac{u}{2}$ for every $0 \leq u \leq 1$, note that for every $0 < t \leq \frac{1}{2K}$, and every smooth bounded function f on (X, g),

$$\|\nabla P_t f\|_\infty \leq \frac{1}{\sqrt{t}} \|f\|_\infty. \tag{5.4}$$

In particular, integrating (5.4) yields, by duality, that for every smooth function f and every $0 < t \leq \frac{1}{2K}$,

$$\|f - P_t f\|_1 \leq 2\sqrt{t}\, \|\nabla f\|_1. \tag{5.5}$$

Indeed, for every g smooth with $\|g\|_\infty \leq 1$,

$$\begin{aligned}
\int_X g\,(f - P_t f)\, d\mu &= -\int_0^t \Big(\int_X g\, \Delta_g P_s f\, d\mu\Big) ds \\
&= \int_0^t \Big(\int_X \nabla P_s g \cdot \nabla f\, d\mu\Big) ds \\
&\leq \|\nabla f\|_1 \int_0^t \|\nabla P_s g\|_\infty\, ds \\
&\leq 2\sqrt{t}\, \|\nabla f\|_1
\end{aligned}$$

and the claim follows.

Provided with this result, the next theorems describe isoperimetric type bounds under spectral and logarithmic Sobolev constants. If A is an open subset of X with smooth boundary ∂A, we denote by $\mu_s(\partial A)$ the surface measure of ∂A.

THEOREM 5.2. *Let (X, g) be a smooth complete connected Riemannian manifold without boundary with finite volume V, and denote by $d\mu = \frac{dv}{V}$ the normalized volume element. Assume that $\mathrm{Ric}_g \geq -K$, $K \geq 0$. Then, if λ_1 denotes the first non-trivial eigenvalue of Δ_g on (X, g), for any open subset A of X with smooth boundary ∂A,*

$$\mu_s(\partial A) \geq \frac{1}{3} \min\Big(\frac{\lambda_1}{\sqrt{K}}, \sqrt{\lambda_1}\Big) \mu(A)\big(1 - \mu(A)\big).$$

Theorem 5.2 produces equivalently a lower bound on the Cheeger constant defined as the largest h such that

$$\mu_s(\partial A) \geq h \min\big(\mu(A), 1 - \mu(A)\big) \tag{5.6}$$

for all open subsets A of X with smooth boundary ∂A. Recall from [Chee] that $h \leq 2\sqrt{\lambda_1}$. In this form, Theorem 5.2 goes back to the work by P. Buser [Bu]. It is a remarkable fact however that Theorem 5.2 yields constants independent of the dimension of the manifold.

PROOF. We apply (5.5) to smooth functions approximating the characteristic function χ_A of an open set A in X with smooth boundary ∂A. It yields, for every $0 < t \leq \frac{1}{2K}$,

$$\begin{aligned} 2\sqrt{t}\,\mu_s(\partial A) &\geq \int_A [1 - P_t(\chi_A)]\,d\mu + \int_{A^c} P_t(\chi_A)\,d\mu \\ &= 2\Big(\mu(A) - \int_A P_t(\chi_A)\,d\mu\Big) \\ &= 2\Big(\mu(A) - \big\|P_{t/2}(\chi_A)\big\|_2^2\Big) \end{aligned}$$

where we used reversibility of the heat semigroup $(P_t)_{t\geq 0}$ with respect to the Riemannian volume element (cf. [Da]). Now, by (5.1),

$$\begin{aligned} \big\|P_{t/2}(\chi_A)\big\|_2^2 &= \mu(A)^2 + \big\|P_{t/2}\big(\chi_A - \mu(A)\big)\big\|_2^2 \\ &\leq \mu(A)^2 + \mathrm{e}^{-\lambda_1 t}\big\|\chi_A - \mu(A)\big\|_2^2 \end{aligned}$$

so that, with the preceding,

$$\sqrt{t}\,\mu_s(\partial A) \geq \mu(A)\big(1 - \mu(A)\big)\left(1 - \mathrm{e}^{-\lambda_1 t}\right)$$

for every $0 < t \leq \frac{1}{2K}$. We need simply optimize in $0 < t \leq \frac{1}{2K}$ to conclude: if $\lambda_1 \geq 2K$, we can choose $t = \frac{1}{\lambda_1}$ while if $\lambda_1 \leq 2K$, we simply take $t = \frac{1}{2K}$. The result follows. □

The next statement is the corresponding result for the logarithmic Sobolev constant. The numerical constant is not sharp.

THEOREM 5.3. *Let (X, g) be a smooth complete connected Riemannian manifold without boundary with finite volume V, and denote by $d\mu = \frac{dv}{V}$ the normalized volume element. Assume that $\mathrm{Ric}_g \geq -K$, $K \geq 0$. Then, if ρ_0 denotes the logarithmic Sobolev constant of Δ_g on (X, g), for any open subset A of X with smooth boundary ∂A such that $\mu(A) \leq \frac{1}{2}$,*

$$\mu_s(\partial A) \geq \frac{1}{34}\min\Big(\frac{\rho_0}{\sqrt{K}}, \sqrt{\rho_0}\Big)\mu(A)\Big(\log\frac{1}{\mu(A)}\Big)^{1/2}.$$

PROOF. We follow the proof of Theorem 5.2 using now that ρ_0 is characterized by the hypercontractivity property (5.2). As we have seen in the proof of Theorem 5.2, for every open set A in X with smooth boundary ∂A, and every $0 < t \leq \frac{1}{2K}$,

$$\sqrt{t}\,\mu_s(\partial A) \geq \mu(A) - \big\|P_{t/2}(\chi_A)\big\|_2^2.$$

Now, by (5.2) with $p = 2$ and $q = 1 + \mathrm{e}^{-\rho_0 t}$,

$$\big\|P_{t/2}(\chi_A)\big\|_2^2 \leq \mu(A)^{2/(1+\mathrm{e}^{-\rho_0 t})}.$$

Since $1 - \mathrm{e}^{-u} \geq \frac{u}{2}$, $0 \leq u \leq 1$, it follows that

$$\sqrt{t}\,\mu_s(\partial A) \geq \mu(A)\bigg[1 - \exp\Big(-\frac{\rho_0 t}{4}\log\frac{1}{\mu(A)}\Big)\bigg]. \tag{5.7}$$

Set $t_0 = \min(\frac{1}{2K}, \frac{1}{\rho_0})$. Choose then $0 < t \leq t_0$ such that

$$t = 4t_0\Big(\log\frac{1}{\mu(A)}\Big)^{-1}$$

provided $\mu(A)$ is small enough so that $\mu(A) \leq \mathrm{e}^{-4}$. For this value of t, (5.7) yields

$$\begin{aligned}\mu_s(\partial A) &\geq \frac{1}{2\sqrt{t_0}}\left(1-\mathrm{e}^{-\rho_0 t_0}\right)\mu(A)\left(\log\frac{1}{\mu(A)}\right)^{1/2}\\ &\geq \frac{1}{4}\,\rho_0\sqrt{t_0}\,\mu(A)\left(\log\frac{1}{\mu(A)}\right)^{1/2}\end{aligned}$$

since $\rho_0 t_0 \leq 1$. This inequality holds for $\mu(A) \leq \mathrm{e}^{-4}$. In general however, when $0 \leq \mu(A) \leq \frac{1}{2}$, we can always apply (5.7) with $t = t_0$ to get

$$\mu_s(\partial A) \geq \frac{1}{\sqrt{t_0}}\,\mu(A)\Big[1-\exp\Big(-\frac{\rho_0 t_0}{4}\,\log 2\Big)\Big] \geq \frac{1}{12}\,\rho_0\sqrt{t_0}\,\mu(A).$$

Combined with the preceding, Theorem 5.3 is established. □

The preceding results hold, with the same proofs, in the context of diffusion operators $\Delta_g - \nabla U\cdot\nabla$ with non-negative curvature in the sense that

$$\mathrm{Ric}_g - \nabla\nabla U \geq 0$$

(cf. [Ba-L]). In particular, if μ is a log-concave probability measure on $\mathbb{R}^n$, its Cheeger constant h (of (5.6)) and Poincaré constant λ_1 satisfy

$$\frac{1}{6}\sqrt{\lambda_1} \leq h \leq 2\sqrt{\lambda_1}\,. \tag{5.8}$$

A deep conjecture of R. Kannan, L. Lovàsz and M. Simonovits [K-L-S] asserts that the Cheeger constant h should be bounded below by a universal strictly positive constant in the class of all log-concave probability measures μ under the isotropic condition

$$\int_{\mathbb{R}^n}\langle x,\theta\rangle^2 d\mu(x) = |\theta|^2 \quad \text{for all} \quad \theta\in\mathbb{R}^n.$$

By (5.8), the question is thus reduced to the corresponding one for the easier Poincaré constant.

Discrete versions of Theorems 5.2 and 5.3 are studied in [B-H-T] and [H-T].

While only of logarithmic type with respect to the (power type) isoperimetric comparison theorems of M. Gromov [Grom1] (cf. [Grom2]) and [B-B-G], the isoperimetric bound of Theorem 5.3, on the other hand, involves ρ_0 rather than the diameter of the manifold, and is independent of the dimension of the manifold (dimension is actually hidden in Theorem 4.2). In the context of diffusion operators of the preceding type, this information is a weaker one since (in contrast with the Sobolev constants) the hypercontractivity constant does not usually control the diameter of the manifold, as is shown by the example of $\Delta - x\cdot\nabla$ on $\mathbb{R}^n$ (with the standard Gaussian measure as invariant measure). Actually, the isoperimetric function in Theorem 5.3 is a form of the isoperimetric function in Gauss space (cf. [Le4], [Le5]) for which the "infinite dimensional" extension of the Lévy-Gromov isoperimetric inequality of [Grom1 (cf. [Grom2]) is studied in [Ba-L].

Acknowledgement. Thanks are due to Professors A. Grigor'yan and S.T. Yau for their invitation to write this paper. We also thank L. Saloff-Coste for several comments, years ago, about the subject of these notes, and S. Bobkov for several precious observations.

References

[A-M-S] S. Aida, T. Masuda, I. Shigekawa. Logarithmic Sobolev inequalities and exponential integrability. J. Funct. Anal. 126, 83–101 (1994).

[A-S] S. Aida, D. Stroock. Moment estimates derived from Poincaré and logarithmic Sobolev inequalities. Math. Res. Lett. 1, 75–86 (1994).

[Al] N. Alon. Eigenvalues and expanders. J. Combin. Theory, Ser. B, 38, 78–88 (1987).

[A-M] N. Alon, V. Milman. λ_1, isoperimetric inequalities for graphs ans superconcentrators. J. Combin. Theory, Ser. B, 38, 78–88 (1985).

[Ba] D. Bakry. L'hypercontractivité et son utilisation en théorie des semigroupes. Ecole d'Eté de Probabilités de St-Flour. Lecture Notes in Math. 1581, 1–114 (1994). Springer.

[B-E] D. Bakry, M. Emery. Diffusions hypercontractives. Séminaire de Probabilités XIX. Lecture Notes in Math. 1123, 177–206 (1985). Springer.

[Ba-L] D. Bakry, M. Ledoux. Lévy-Gromov's isoperimetric inequality for an infinite dimensional diffusion generator. Invent. math. 123, 259–281 (1996).

[B-L-Q] D. Bakry, M. Ledoux, Z. Qian. Unpublished manuscript (1997).

[B-B-G] P. Bérard, G. Besson, S. Gallot. Sur une inégalité isopérimétrique qui généralise celle de Paul Lévy-Gromov. Invent. math. 80, 295–308 (1985).

[Bo] S. Bobkov. Isoperimetric and analytic inequalities for log-concave probability measures. Ann. Probab., 27, 1903–1921 (1999).

[B-H-T] S. Bobkov, C. Houdré, P. Tetali. λ_∞, vertex isoperimetry and concentration. Combinatorica 20, 153–172 (2000).

[Bo-L] S. Bobkov, M. Ledoux. Poincaré's inequalities and Talagrand's concentration phenomenon for the exponential measure. Probab. Theory Relat. Fields 107, 383–400 (1997).

[B-T] S. Bobkov, P. Tetali. Modified logarithmic Sobolev inequalities in discrete settings (2003).

[B-U] A. Borovkov, S. Utev. On a inequality and a related characterization of the normal distribution. Theor. Probab. Appl. 28, 209–218 (1983).

[Br] R. Brooks. On the spectrum of non-compact manifolds with finite volume. Math. Z. 187, 425–437 (1984).

[Bu] P. Buser. A note on the isoperimetric constant. Ann. scient. Éc. Norm. Sup. 15, 213–230 (1982).

[Cha1] I. Chavel. Eigenvalues in Riemannian geometry. Academic Press (1984).

[Cha2] I. Chavel. Riemannian geometry - A modern introduction. Cambridge Univ. Press (1993).

[Chee] J. Cheeger. A lower bound for the smallest eigenvalue of the Laplacian. Problems in Analysis, Symposium in honor of S. Bochner. Princeton Univ. Press., 195–199. Princeton (1970).

[C-E] J. Cheeger, D. Ebin. Comparison theorems in Riemannian geometry. North-Holland (1975).

[Chen] S.-Y. Cheng. Eigenvalue comparison theorems and its geometric applications. Math. Z. 143, 289–297 (1975).

[Chu] F. R. K. Chung. Diameters and eigenvalues. J. Amer. Math. Soc. 2, 187–196 (1989).

[C-G-Y1] F. R. K. Chung, A. Grigor'yan, S.-T. Yau. Upper bounds for eigenvalues of tha discrete and continuous Laplace operators. Advances in Math. 117, 165–178 (1996).

[C-G-Y2] F. R. K. Chung, A. Grigor'yan, S.-T. Yau. Eigenvalues and diameters for manifolds and graphs. Tsing Hua lectures on geometry and & analysis (1990/91) 79–105. Internat. Press Cambridge MA (1997).

[Da] E. B. Davies. Heat kernel and spectral theory. Cambridge Univ. Press (1989).

[Da-S] E. B. Davies, B. Simon. Ultracontractivity and the heat kernel for Schrödinger operators and Dirichlet Laplacians. J. Funct. Anal. 59, 335–395 (1984).

[De-S] J.-D. Deuschel, D. Stroock. Large deviations. Academic Press (1989).

[D-SC] P. Diaconis, L. Saloff-Coste. Logarithmic Sobolev inequalities for finite Markov chains. Ann. Appl. Probab. 6, 695–750 (1996).

[G-H-L] S. Gallot, D. Hulin, J. Lafontaine. Riemannian Geometry. Second Edition. Springer (1990).

[G-Q] F. Gao, J. Quastel. Exponential decay of entropy in random transposition and Bernoulli-Laplace models (2002).

[Go] S. GOEL. Modified logarithmic Sobolev inequalities for some models of rnadom walk (2003).

[Grom1] M. GROMOV. Paul Lévy's isoperimetric inequality. Preprint I.H.E.S. (1980).

[Grom2] M. GROMOV. Metric structures for Riemannian and non-Riemannian spaces. Birkhäuser (1998).

[G-M] M. GROMOV, V. D. MILMAN. A topological application of the isoperimetric inequality. Amer. J. Math. 105, 843–854 (1983).

[Gros] L. GROSS. Logarithmic Sobolev inequalities. Amer. J. Math. 97, 1061–1083 (1975).

[H-T] C. HOUDRÉ, P. TETALI. Concentration of measure for products of Markov kernels via functional inequalities. Combin. Probab. Comput. 10 1–28 (2001).

[Le1] M. LEDOUX. A simple analytic proof of an inequality by P. Buser. Proc. Amer. Math. Soc. 121, 951–959 (1994).

[Le2] M. LEDOUX. Remarks on logarithmic Sobolev constants, exponential integrability and bounds on the diameter. J. Math. Kyoto Univ. 35, 211–220 (1995).

[Le3] M. LEDOUX. Concentration of measure and logarithmic Sobolev inequalities. Séminaire de Probabilités XXXIII. Lecture Notes in Math. 1709, 120–216 (1999). Springer.

[Le4] M. LEDOUX. The geometry of Markov diffusion generators. Ann. Fac. Sci. Toulouse IX, 305–366 (2000).

[Le5] M. LEDOUX. The concentration of measure phenomenon. Math. Surveys and Monographs 89. Amer. Math. Soc. (2001).

[L-Y] T.Y. LEE, H.-T. YAU. Logarithmic Sobolev inequality fo some models of random walks. Ann. Probab. 26, 1855–1873 (1998).

[Li] P. LI. A lower bound for the first eigenvalue of the Laplacian on a compact manifold. Indiana Univ. Math. J. 28, 1013–1019 (1979).

[L-Y1] P. LI, S.-T. YAU. Estimates of eigenvalues of a compact Riemannian manifold. Proc. Symp. Pure Math. 36, 205–239. Amer. Math. Soc. (1980).

[L-Y2] P. LI, S.-T. YAU. On the parabolic kernel of the Schrödinger operator. Acta Math. 156, 153–201 (1986).

[Ma] B. MAUREY. Constructions de suites symétriques. C. R. Acad. Sci. Paris 288, 679–681 (1979).

[Ro1] O. ROTHAUS. Diffusion on compact Riemannian manifolds and logarithmic Sobolev inequalities. J. Funct. Anal. 42, 358–367 (1981).

[Ro2] O. ROTHAUS. Hypercontractivity and the Bakry-Emery criterion for compact Lie groups. J. Funct. Anal. 65, 358–367 (1986).

[SC1] L. SALOFF-COSTE. Precise estimates on the rate at which certain diffusions tend to equilibrium. Math. Z. 217, 641–677 (1994).

[SC2] L. SALOFF-COSTE. Convergence to equilibrium and logarithmic Sobolev constant on manifolds with Ricci curvature bounded below. Colloquium Math. 67, 109–121 (1994).

[SC3] L. SALOFF-COSTE. Lectures on finite Markov chains. Ecole d'Eté de Probabilités de St-Flour 1996. Lecture Notes in Math. 1665, 301–413 (1997). Springer-Verlag.

[Sc] M. SCHMUCKENSCHLÄGER. Martingales, Poincaré type inequalities and deviations inequalities. J. Funct. Anal. 155, 303–323 (1998).

[Wan] F.-Y. WANG. Logarithmic Sobolev inequalities on noncompact Riemannian manifolds. Probab. Theory Relat. Fields 109, 417–424 (1997).

[Wat] G. N. WATSON. A treatise on the theory of Bessel functions. Cambridge Univ. Press (1944).

[Ya] S.-T. YAU. Isoperimetric constants and the first eigenvalue of a compact Riemannian manifold. Ann. scient. Éc. Norm. Sup. 8, 487–507 (1975).

[Z-Y] J. Q. ZHONG, H. C. YANG On the estimate of the first eigenvalue of a compact Riemanian manifold. Sci. Sinica Ser. A 27 (12), 1265–1273 (1984).

INSTITUT DE MATHÉMATIQUES, UNIVERSITÉ PAUL-SABATIER, 31062 TOULOUSE, FRANCE

Surveys in Differential Geometry
IX, International Press

Discrete Analytic Functions: An Exposition

László Lovász

Abstract. Harmonic and analytic functions have natural discrete analogues. Harmonic functions can be defined on every graph, while analytic functions (or, more precisely, holomorphic forms) can be defined on graphs embedded in orientable surfaces. Many important properties of the "true" harmonic and analytic functions can be carried over to the discrete setting.

Contents

1. Introduction

Discrete and continuous mathematics study very different structures, by very different methods. But they have a lot in common if we consider which *phenomena* they study: Symmetry, dispersion, expansion, and other general phenomena have interesting formulations both in the discrete and continuous setting, and the influence of ideas from one to the other can be most fruitful. One such notion we should more explicitly mention here are *discrete harmonic functions*, which can be defined on every graph, and have been studied quite extensively. See [**23**] for a lot of information on harmonic functions on (infinite) graphs and their connections with electrical networks and random walks. In this paper we show that *analycity* (most notably the uniqueness of analytic continuation and the long-range dependence it implies) is an important phenomenon in discrete mathematics as well.

Discrete analytic functions were introduced for the case of the square grid in the 40's by Ferrand [**11**] and studied quite extensively in the 50's by Duffin [**8**]. For the case of a general map, the notion of discrete analytic functions is implicit in a paper of Brooks, Smith, Stone and Tutte [**5**] (cf. section 7.2) and more recent work by Benjamini and Schramm [**4**]. They were formally introduced recently by Mercat [**18**].

Discrete analytic functions and holomorphic forms can be defined on orientable maps, i.e., graphs embedded in orientable surfaces. (Much of this could be extended to non-orientable surfaces, but we don't go into this in this paper.) In graph-theoretic terms, they can be defined as rotation-free circulations (which is the same as requiring that the circulation is also a circulation on the dual graph).

Many important properties of the "true" harmonic and analytic functions can be carried over to the discrete setting: maximum principles, Cauchy integrals etc. Some of these translations are straightforward, sometimes it is not so easy to find the right formulation. But discreteness brings in several new aspects as well, like connections with network flows, matroid theory, various embeddings of graphs, tiling the plane by squares, circle representations etc.

Other aspects of analytic functions are worse off. Integration can be defined on the grid [**8**], but we run into trouble if we want to extend it to more general maps. Mercat [**18**] introduced a (rather restrictive) condition called "criticality", under which integrals can be defined. Multiplication is problematic even on the grid. Analogues of polynomials and exponential functions can be defined on the grid [**8**], and can be extended to to critical maps [**19, 20**].

In this paper we start with briefly surveying two related topics: harmonic functions on graphs and discrete analytic functions on grids. This is not our main topic, and we concentrate on some aspects only that we need later. In particular, we show the connection of harmonic functions with random walks, electrical networks and rubber band structures.

We discuss in detail zero-sets of discrete analytic functions, in particular how to extend to discrete analytic functions the fact that a nonzero analytic function can vanish only on a very small connected piece [**2, 3**]. As an application, we describe a simple local random process on maps, which has the property that observing it in a small neighborhood of a node through a polynomial time, we can infer the genus of the surface.

2. Notation

We recall some terminology from graph theory. Let $G = (V, E)$ be a graph, where V is the set of its nodes and E is the set of its edges. An edge of G is a *loop*, if both endpoints are the same. Two edges are called *parallel*, if they connect the same pair of nodes. A graph G is called *simple*, if it has no loops or parallel edges. The set of nodes connected to a given node $v \in V$ (called its neighbors) is denoted by $N(v)$.

A graph is k-connected, if deleting fewer than k nodes always leaves a connected graph.

A *directed graph* is a graph in which every edges has an orientation. So each edge $e \in E$ has a *tail* $t_e \in V$ and a *head* $h_e \in V$. Our main concern will be undirected graphs, but we will need to orient the edges for reference purposes.

Let G be a directed graph. For each node v, let $\delta v \in \mathbb{R}^E$ denote the coboundary of v:

$$(\delta v)_e = \begin{cases} 1 & \text{if } t_e = v, \\ -1 & \text{if } h_e = v, \\ 0 & \text{otherwise.} \end{cases}$$

Thus $|\delta v|^2 = d_v$ is the degree of v. We say that a node $v \in V$ is a *source* [*sink*] if all edges incident with it are directed away from [toward] the node. Every function $\pi \in \mathbb{R}^V$ gives rise to a vector $\delta\pi \in \mathbb{R}^E$, where

$$(\delta\pi)(uv) = \pi(v) - \pi(u). \tag{1}$$

In other words,

$$\delta\pi = \sum_v \pi(v)\delta v. \tag{2}$$

For an edge e, let $\partial e \in \mathbb{R}^V$ be the boundary of e:

$$(\partial e)_i = \begin{cases} 1 & \text{in } i = h(e), \\ -1 & \text{in } i = t(e), \\ 0 & \text{otherwise.} \end{cases}$$

For $\phi : E \to \mathbb{R}$, we define

$$\partial\phi(v) = (\delta v)^T\phi = \sum_{e:\ t(e)=v} \phi(e) - \sum_{e:\ h(e)=v} \phi(e)$$

In other words,

$$\partial\phi = \sum_e \phi(e)\partial e.$$

We say that ϕ *satisfies the flow condition at* v if $\partial\phi(v) = 0$. We say that ϕ is a *circulation* if it satisfies the flow condition at every node v. Note that this depends on the orientation of the edges, but if we reverse an edge, we can compensate for it by switching the sign of $\phi(e)$.

3. Discrete harmonic functions

3.1. Definition. Let $G = (V, E)$ be a connected graph. A function $f : V \to \mathbf{C}$ is called *harmonic* at node i if

$$\frac{1}{d_i} \sum_{j \in N(i)} f(j) = f(i), \tag{3}$$

and is said to have a *pole* at i otherwise. Note that the condition can be re-written as

$$\sum_{j \in N(i)} (f(j) - f(i)) = 0. \tag{4}$$

More generally, if we also have a "length" $\ell_{ij} > 0$ assigned to each edge ij, then we say that f is harmonic on the weighted graph $G = (V, E, \ell)$ at node i if

$$\sum_{j \in N(i)} \frac{f(j) - f(i)}{\ell_{ij}} = 0. \tag{5}$$

If S is the set of poles of a function f, we call f a *harmonic function with poles S*.

In the definition we allowed complex values, but since the condition applies separately to the real and imaginary parts of f, it is usually enough to consider real valued harmonic functions.

PROPOSITION 3.1. *Every non-constant function has at least two poles.*

This follows simply by looking at the minimum and maximum of the function. In fact, the maximum of a function cannot be attained at a node where it is harmonic, unless the same value is attained at all of its neighbors. This argument can be though of as a (very simple) discrete version of the Maximum Principle.

For any two nodes $a, b \in V$ there is a harmonic function with exactly these poles. More generally, we have the following fact.

PROPOSITION 3.2. *For every set $S \subseteq V$, $S \neq \emptyset$, every function $f_0 : S \to \mathbf{C}$ has a unique extension to a function $f : V \to \mathbf{C}$ that is harmonic at each node in $V \setminus S$.*

The proof of uniqueness is easy (consider the maximum or minimum of the difference of any two extensions). The existence of the extension follows from any of several constructions, some of which will be given in the next section. Note that the case $|S| = 1$ does not contradict Proposition 3.1: the unique extension is a constant function.

If $S = \{a, b\}$, then a harmonic function with poles S is uniquely determined up to scaling by a real number and translating by a constant. There are various natural ways to normalize; we'll somewhat arbitrarily decide on the following one:

$$\sum_{u \in N(v)} (f(u) - f(v)) = \begin{cases} 1 & \text{if } v = b, \\ -1 & \text{if } v = a, \\ 0 & \text{otherwise.} \end{cases} \tag{6}$$

and

$$\sum_{u} f(u) = 0. \tag{7}$$

We denote this function by π_{ab}. If $e = ab$ is an edge, we also denote this function by π_e.

Expression (4) is equivalent to saying that the function $\delta\pi$ satisfies the flow condition at node i if and only if π is harmonic at i. Not every flow can be obtained from a harmonic function: for example, a non-zero circulation (a flow without sources and sinks) would correspond to a non-constant harmonic function with no poles, which cannot exist. In fact, the flow obtained by (1) satisfies, for every cycle C, the following condition:

$$\sum_{e \in C} f_\pi(e) = 0, \tag{8}$$

where the edges of C are oriented in a fixed direction around the cycle. We could say that the flow is "rotation-free", but we'll reserve this phrase for a slightly weaker notion in section 5.

3.2. Random walks, electrical networks, and rubber bands. Harmonic functions play an important role in the study of random walks: after all, the averaging in the definition can be interpreted as expectation after one move. They also come up in the theory of electrical networks, and in statics. This provides a connection between these fields, which can be exploited. In particular, various methods and results from the theory of electricity and statics, often motivated by physics, can be applied to provide results about random walks. We only touch upon these connections; see [**7, 23**] for much more.

Let a nonempty subset $S \subseteq V$ and a function $\pi_0 : S \to \mathbb{R}$ be given. We describe three constructions, one in each of the fields mentioned, that extend π_0 to a function $\pi : V \to \mathbb{R}$ so that the extension is harmonic at the nodes in $V \setminus S$.

EXAMPLE 1. Let $\pi(v)$ be the expectation of $\pi_0(s)$, where s is the (random) node where a random walk on the graph G starting at v first hits S.

We can re-state this construction as a discrete version of the Poisson Formula. Let $S \subseteq V(G)$. For every $i \in V(G) \setminus S$ and $j \in S$, let $K(i,j)$ denote the probability that a random walk started at i hits j before any other node in S. Then for every function f on $V(G)$ that is harmonic on $V \setminus S$, and every $i \in V \setminus S$

$$f(i) = \sum_{j \in S} K(i,j) f(j).$$

EXAMPLE 2. Consider the graph G as an electrical network, where each edge represents a unit resistance. Keep each node $s \in S$ at electric potential $\pi_0(s)$, and let the electric current flow through G. Define $\pi(v)$ as the electric potential of node v.

EXAMPLE 3. Consider the edges of the graph G as ideal springs with unit Hooke constant (i.e., it takes h units of force to stretch them to length h). Nail each node $s \in S$ to the point $\pi_0(s)$ on the real line, and let the graph find its equilibrium. The energy is a positive definite quadratic form of the positions of the nodes, and so there is a unique minimizing position, which is the equilibrium. Define $\pi(v)$ as the position of node v on the line.

More generally, fix the positions of the nodes in S (in any dimension), and let the remaining nodes find their equilibrium. Then every coordinate function is harmonic at every node of $V \setminus S$.

A consequence of the uniqueness property is that the harmonic functions constructed (for the case $|S| = 2$) in examples 1, 2 and 3 are the same. As an application of this idea, we show the following interesting connections (see Nash-Williams [**22**], Chandra at al. [**6**]). Let G be a graph with n nodes and m edges. Considering G as an electrical network, let R_{st} denote the effective resistance between nodes s and t. Considering the graph G as a spring structure in equilibrium, with two nodes s and t nailed down at 1 and 0, let F_{ab} denote the force pulling the nails. Doing a random walk on G, let $\kappa(a,b)$ denote the commute time between nodes a and b (i.e., the expected time it takes to start at a, walk until you first hit b, and then walk until you first hit a again).

THEOREM 3.3. $R_{ab} = \dfrac{1}{F_{ab}} = \dfrac{\kappa(a,b)}{2m}$.

Using the "topological formulas" from the theory of electrical networks for the resistance, we get a further well-known characterization of these quantities:

COROLLARY 3.4. *Let G' denote the graph obtained from G by identifying a and b, and let $\mathcal{T}(G)$ denote the number of spanning trees of G. Then*

$$R_{ab} = \frac{\mathcal{T}(G)}{\mathcal{T}(G')}.$$

4. Analytic functions on the grid

4.1. Definition and variations. Suppose that we have an analytic function g on the complex plane, and we can consider its restriction f to the set of lattice points (Gaussian integers) (say, for the purpose of numerical computation). Suppose that we want to "integrate" this function f along a path, which now is a polygon $v_0v_1 \dots v_n$ where $v_{k+1} - v_k \in \{\pm 1, \pm i\}$. A reasonable guess is to use the formula

$$\sum_{k=0}^{n-1} (v_{k+1} - v_k) \frac{f(v_{k+1}) + f(v_k)}{2}. \tag{9}$$

Unfortunately, this sum will in general depend on the path, not just on its endpoints. Of course, the dependence will be small, since the sum approximates the "true" integral.

Can we modify our strategy by defining f not as the restriction of g to the lattice, but as some other discrete approximation of g, for which the discrete integral (9) is independent from the path? To answer this question, we have to understand the structure of such discrete functions. Independence from the path means that the integral is 0 on closed paths, which in turn is equivalent to requiring that the integral is 0 on the simplest closed paths of the form $(z, z+1, z+1+i, z+i, z)$. In this case, the condition is

$$\frac{f(z+1)+f(z)}{2} + i\frac{f(z+1+i)+f(z+1)}{2} + (-1)\frac{f(z+i)+f(z+1+i)}{2} + (-i)\frac{f(z)+f(z+i)}{2} = 0.$$

By simple rearrangement, this condition can be written as

$$\frac{f(z+i+1)-f(z)}{i+1} = \frac{f(z+1)-f(z+i)}{1-i}. \tag{10}$$

This latter equation can be thought of as discrete version of the fact that the derivative is unique, or (after rotation), as a discrete version of the Cauchy-Riemann equation.

Let Ω be a subset of the plain that is the union of lattice squares. A function satisfying (10) for every square in Ω is called a *discrete analytic function* on Ω. This notion was introduced by Ferrand [**11**] and developed by Duffin [**8**]. There are several variations, some of which are equivalent to this, others are not (see e.g. Isaacs [**12**]).

The following version is essentially equivalent. The lattice of Gaussian integers can be split into "even" and "odd" lattice points ($a + bi$ with $a + b$ even or odd), and condition (10) only relates the differences of even values to the differences of odd values. We can take the even sublattice, rotate by 45°, and rescale it to get the standard lattice. We can think of the odd sublattice as the set of fundamental squares of the even lattice. This way a discrete analytic function can be thought of as a pair of complex-valued functions f_1 and f_2 defined on the lattice points and on the lattice squares, respectively. These are related by the following condition:

Discrete Cauchy-Riemann, complex version *Let ab be an edge of the lattice graph (so $b = a+1$ or $b = a+i$), and let p and q be the square to bordering ab from the left and right, respectively. Then*

$$f_1(b) - f_1(a) = i(f_2(p) - f_2(q)).$$

We call such a pair (f_1, f_2) a *complex discrete analytic pair.*

This form suggests a further simplification: since this equation relates the real part of f_1 to the imaginary part of f_2, and vice versa, we can separate these. So to understand discrete analytic functions, it suffices to consider pairs of *real valued* functions g_1 and g_2, one defined on the standard lattice, one on the lattice squares, related by the following condition:

Discrete Cauchy-Riemann, real version *Let ab be an edge of the lattice graph (so $b = a+1$ or $b = a+i$), and let p and q be the square to bordering ab from the left and right, respectively. Then*

$$g_1(b) - g_1(a) = g_2(p) - g_2(q).$$

To do computations, it is convenient to label each square by its lower left corner. This way a discrete analytic function can be thought of as two functions f_1 and f_2 defined on the lattice points, related by the equations

$$\begin{aligned} f_1(z+1) - f_1(z) &= -i(f_2(z) - f_2(z-i)), \\ f_1(z+i) - f_1(z) &= i(f_2(z) - f_2(z-1)). \end{aligned} \tag{11}$$

In the real version, we get the equations

$$\begin{aligned} g_1(x+1,y) - g_1(x,y) &= g_2(x,y) - g_2(x,y-1) \\ g_1(x,y+1) - g_1(x,y) &= g_2(x-1,y) - g_2(x,y). \end{aligned} \tag{12}$$

Both functions g and g_2 are harmonic on the infinite graph formed by lattice points, with edges connecting each lattice point to its four neighbors. Indeed,

$$\begin{aligned} &[g_1(x+1,y) - g_1(x,y)] + [g_1(x-1,y) - g_1(x,y)] \\ &\quad + [g_1(x,y+1) - g_1(x,y)] + [g_1(x,y-1) - g_1(x,y)] \\ &= [h(x,y) - h(x,y-1)] + [h(x-1,y-1) - h(x-1,y)] \\ &\quad + [h(x-1,y) - h(x,y)] + [h(x,y-1) - h(x-1,y-1)] \\ &= 0. \end{aligned}$$

Conversely, if we are given a harmonic function g_1, then we can define a function g_2 on the squares such that (g_1, g_2) satisfy (12). We define g_2 on one square arbitrarily, and then use (12) to extend the definition to all squares. The assumption that g_1 is harmonic guarantees that we don't run into contradiction by going around a lattice point; since the plane is simply connected, we don't run into contradiction at all.

So we see that *a discrete analytic function can be identified with a single complex valued harmonic function on the even sublattice, which in turn can be thought of a pair of real valued harmonic functions on the same lattice.* To each (real or complex) harmonic function we can compute a *conjugate* using (11) or (12). It turns out that both ways of looking at these functions are advantageous in some arguments.

4.2. Integration and differentiation. We defined discrete analytic functions so that integration should be well defined: Given a discrete analytic function f and two integer points a, b, we can define the integral from a to b by selecting a lattice path $a = z_0, z_1, \dots, z_n$, and defining

$$\int_a^b f\,dz = \sum_{k=0}^{n-1} \frac{f(z_{k+1}) + f(z_k)}{2}(z_{k+1} - z_k).$$

The main point in our definition of discrete analytic functions was that this is independent of the choice of the path. It is not obvious, but not hard to see, that the integral function

$$F(u) = \int_a^u f\,dz$$

is a discrete analytic function.

A warning sign that not everything works out smoothly is the following. Suppose that we have two discrete analytic functions f and g defined on Ω. It is natural to try to define the integral

$$\int_a^b f\,dg = \sum_{k=0}^{n-1} \frac{f(z_{k+1}) + f(z_k)}{2}(g(z_{k+1}) - g(z_k)).$$

It turns out that this integral is again independent of the path, but it is not an analytic function of the upper bound in general.

There are several ways one could try to define the derivative. The function defined by

$$(\nabla_a f)(z) = \frac{f(z+a) - f(z)}{a},$$

is discrete analytic for any Gaussian integer a ($a = i + 1$ seems the most natural choice in view of (10)). Unfortunately, neither one if these is the converse of integration. If $F(u) = \int_a^u f\,dz$, then for $a \in \{\pm 1, \pm i\}$,

$$(\nabla_a F)(z) = \frac{f(z+a) + f(z)}{2}. \tag{13}$$

There is in fact no unique converse, since adding $c \cdot (-1)^{x+y}$ to the function value at $x + iy$ does not change the integral along any path. This also implies that the converse of integration cannot be recovered "locally". But if we fix the value arbitrarily at (say) 0, then (13), applied with $a = 1$ and $a = i$, can be used to recover the values of f one-by one. This also can be expressed by integration (see [**8**]).

4.3. Constructions.

Example 4 (Extension). To see that there is a large variety of discrete analytic functions, we mention the following fact: *if we assign a complex number to every integer point on the real and imaginary axes, there is a unique discrete analytic function with these values.* Indeed, we can reconstruct the values of the function at the other integer points z one by one by induction on $|z|^2$, using (10).

EXAMPLE 5 (Discrete polynomials). The restriction of a linear or quadratic polynomial to the lattice points gives a discrete analytic function, but this is not so for polynomials of higher degree. But there are sequences of discrete analytic functions that can be thought of as analogues of powers of z.

One of these is best described in terms of an analytic pair. For $n \geq 1$, consider the functions

$$g_1^{\{n\}}(x,y) = n! \sum_{j=0}^{\lfloor n/2 \rfloor} (-1)^j \binom{x-j}{n-2j} \binom{y+j}{2j},$$

and

$$g_2^{\{n\}}(x,y) = n! \sum_{j=0}^{\lfloor (n-1)/2 \rfloor} (-1)^j \binom{x-j}{n-2j-1} \binom{y+j+1}{2j+1}$$

Then $g_1^{\{n\}}$ and $g_2^{\{n\}}$ satisfy the conditions (12). (To explain these formulas, note that if we replace $\binom{u}{k}$ by $u^k/k!$, then we get the real and imaginary parts of $(x+iy)^n$.) Taking linear combinations, we get "polynomials". These functions are polynomials in x and y, or (after a change of coordinates) in the complex numbers z and $\overline{z}$, but not necessarily polynomials in z.

Integration offers another way to define "pseudo-powers" of z:

$$z^{(0)} = 1, z^{(n)} = n \int_0^z w^{(n-1)}\, dw.$$

These functions are not the same as the analytic functions defined by the pairs $(g_1^{\{n\}}, g_2^{\{n\}})$ defined above, but they give rise to the same linear space of discrete polynomials. These functions approximate the true powers of z quite well: Duffin proves that $z^{(n)} - z^n$ is a polynomial in Z and $\overline{z}$ of degree at most $n-2$. Hence

$$z^{(n)} = z^n(1 + O(|z|^{-2})) \tag{14}$$

EXAMPLE 6 (Discrete exponentials). Once we have analogues of powers of z, we can obtain further discrete analytic functions by series expansion. As an example, we can define the exponential function by the formula

$$e(z) = \sum_{n=0}^{\infty} \frac{1}{n!} z^{(n)}.$$

More generally, one can introduce a continuous variable z, and define (at least for $|t| < 2$)

$$e(z,t) = \sum_{n=0}^{\infty} \frac{1}{n!} z^{(n)} t^n.$$

For this function, Ferrand proved the explicit formula

$$e(z,t) = \left(\frac{2+t}{2-t}\right)^x \left(\frac{2+it}{2-it}\right)^y.$$

This function is discrete analytic for every fixed $t \neq \pm 2, \pm 2i$.

4.4. Approximation. Let go back to the remark we used to motivate discrete analytic functions: that we want to use discrete analytic function to approximate a "true" analytic function by a function on a discrete set of points, in a more sophisticated way than restricting it.

One way to construct such an approximation is to first approximate $f(z)$ by a polynomial $p(z)$ (which could be a partial sum of the Taylor expansion), and then replace z^n by $z^{(n)}$ in the polynomial. It follows from (14) that by this, we introduce a relative error of $1 + O(|z|^{n-2})$. If we do this not on the lattice L of Gaussian integers, but on the lattice δL with $\delta \to 0$, then we get an approximation with relative error $1 + O(\delta^2)$. See Duffin and Peterson [**10**] for details.

While the space of polynomials is well-defined, which polynomials we want to call "powers of z" is a matter of taste, and expansion in terms of other sequences of polynomials may have better properties. For example, Zeilberger [**26**] constructs another sequence $(p_n(z))$ for which the series $\sum_n a_n p_n(z)$ converges absolutely to a discrete analytic function in the quadrant $x, y \geq 0$ whenever $|a_n|^{1/n} \to 0$.

Many results from complex analysis can be extended (mutatis mutandis) to discrete analytic functions. Besides Cauchy's integral formulas and the Maximum Principle, these include the Phragmén-Lindelöf Theorem, the Paley–Wiener–Schwartz theorem, and more. See also [**27, 28**] for details.

5. Holomorphic forms on maps

While no function can be harmonic at all nodes of a finite graph, the notion of *holomorphic forms* can be extended to any finite graph embedded in an orientable surface.

5.1. Preliminaries about maps. Let S be a 2-dimensional orientable manifold. By a *map* on S we mean graph $G = (V, E)$ embedded in S so that

(i) each face is an open topological disc, whose closure is compact,

(ii) every compact subset of S intersects only a finite number of edges.

We in fact will need mainly two cases: either S is the plane or S is compact. In the first case, G is necessarily infinite; in the second, G is finite.

We can describe the map by a triple $G = (V, E, \mathcal{F})$, where V is the set of nodes, E is the set of edges, and $\mathcal{F}$ is the set of faces of G. We set $n = |V|$, $m = |E|$, and $f = |\mathcal{F}|$.

We call an edge e *one-sided*, if it is incident with one and the same face on both sides, and *two-sided* otherwise.

For every map G, we can construct its *universal cover map* $\hat{G} = (\hat{V}, \hat{E}, \hat{\mathcal{F}})$ in the usual way. This is an infinite graph embedded in the plane, invariant under the action of an appropriate discrete group of isometries of the euclidean plane (in the case of the torus) or of the hyperbolic plane (in the case of higher genus).

Fixing any reference orientation of G, we can define for each edge a *right shore* $r_e \in \mathcal{F}$, and a *left shore* $l_e \in \mathcal{F}$. Recall that an edge $e = ij$ has a head $h_e = j$ and a tail $t_e = i$.

The embedding of G defines a *dual map* $G^* = (V^*, E^*, \mathcal{F}^*)$. Geometrically, we create a new node inside each face of G, to get V^*. For each edge $e \in E$, we connect the two faces bordering this edge by an edge e^* that crosses e exactly once. (If the same face is incident with e from both sides, then e^* is a loop.) It is not hard to arrange these curves so that these new edges give a map G^*. Combinatorially, we can think of G^* as the map where "node" and "face" are interchanged, "tail" is replaced "right shore", and "head" is replaced by "left shore". So $|E^*| = |E|$, and there is an obvious bijection $e \leftrightarrow e^*$. Note that "right shore" is replaced by "head" and "left shore" is replaced by "tail". So $(G^*)^*$ is not G, but G with all edges reversed.

Sometimes it is useful to consider another map $G_\diamond$ associated with a map G, called the *diamond map*. This is defined on the node set $V(G_\diamond) = V \cup V^*$, where the edges are those pairs xy where $x \in V$, $y \in V*$, and x is incident with the face F corresponding to y. (If F has t corners at x, we connect y to x by t edges, one through each of these corners.) Clearly G^* is a bipartite map where each face has 4 edges.

For every face $F \in \mathcal{F}$, we denote by $\partial F \in \mathbb{R}^E$ the boundary of F:

$$(\partial F)_e = \begin{cases} 1 & \text{if } r_e = F, \\ -1 & \text{if } l_e = F, \\ 0 & \text{otherwise.} \end{cases}$$

Then $d_F = |\partial F|^2$ is the length of the cycle bounding F.

Let e and f be two consecutive edges along the boundary of a face F, meeting at a node v. We call the quadruple (F, v, e, f) a *corner* (at node v on face F). If both edges are directed in or directed out of b, we call the corner *sharp*; else, we call it *blunt*.

5.2. Circulations, homology and discrete Hodge decomposition. If G is a map on a surface S, then the space of circulations on G has an important additional structure: for each face F, the vector ∂F is circulation. Circulations that are linear combinations of these special circulations ∂F are called *0-homologous*. Two circulations ϕ and ϕ' are *homologous* if $\phi - \phi'$ is 0-homologous.

Let ϕ be a circulation on G. We say that ϕ is *rotation-free*, if for every face $F \in \mathcal{F}$, we have

$$(\partial F)^\mathsf{T} \phi = \sum_{e:\ r_e = F} \phi(e) - \sum_{e:\ l_e = F} \phi(e) = 0.$$

This is equivalent to saying that ϕ is a circulation on the dual map G^*.

The following linear spaces correspond to the *Hodge decomposition*. Let $\mathcal{A} \subseteq \overrightarrow{\mathbb{R}}^E$ be the subspace generated by the vectors δv $(v \in V)$ and $\mathcal{B} \subseteq \mathbb{R}^E$, the subspace generated by the vectors ∂F $(F \in \mathcal{F})$. Vectors in $\mathcal{A}$ are sometimes called *tensions* or *potentials*. Vectors in $\mathcal{B}$ are 0-homologous circulations. It is easy to see that $\mathcal{A}$ and $\mathcal{B}$ are orthogonal to each other. The orthogonal complement $\mathcal{A}^\perp$ is the space of all circulations, while $\mathcal{B}^\perp$ consists of rotation-free vectors on the edges. The intersection $\mathcal{C} = \mathcal{A}^\perp \cap \mathcal{B}^\perp$ is the space of rotation-free circulations.

PROPOSITION 5.1. *For every map G on a surface S with genus G, the space of all 1-chains has a decomposition*

$$\overleftarrow{\mathbb{R}^E} = \mathcal{A} \oplus \mathcal{B} \oplus \mathcal{C}$$

into three mutually orthogonal subspaces, where $\mathcal{A}$ is the space of 0-homologous circulations, $\mathcal{B}$ is the space of all potentials, and $\mathcal{C}$ is the space of all rotation-free circulations.

If the map G is not obvious from the context, we denote these spaces by $\mathcal{A}(G), \mathcal{B}(G)$ and $\mathcal{C}(G)$.

From this proposition we conclude the following.

COROLLARY 5.2. *Every circulation is homologous to a unique rotation-free circulation.*

It also follows that $\mathcal{C}$ is isomorphic to the first homology group of S (over the reals), and hence we get the following:

COROLLARY 5.3. *The dimension of the space $\mathcal{C}$ of rotation-free circulations is $2g$.*

Indeed, we have

$$\dim(\mathcal{A}) = f - 1 \quad \text{and} \quad \dim(\mathcal{B}) = n - 1$$

by elementary graph theory, and hence

$$\dim(\mathcal{B}) = m - \dim(\mathcal{A}) - \dim(\mathcal{B}) = m - f - n + 2 = 2g$$

by Euler's Formula.

Figure 1 shows the (rather boring) situation on the toroidal grid: for every choice of a and b we get a rotation-free circulation, and by Corollary 5.3, these are all.

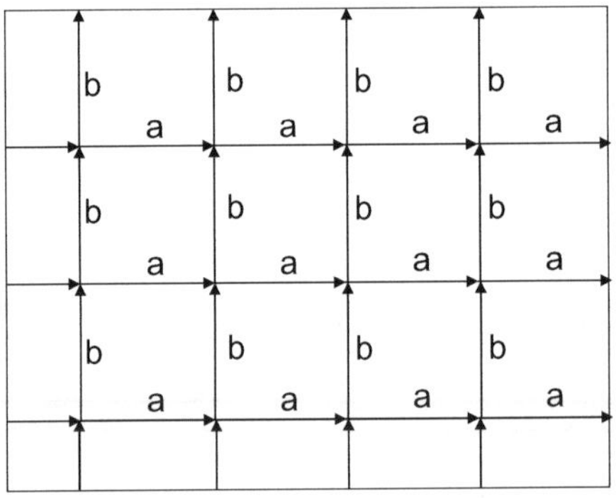

FIGURE 1. Rotation-free circulation on the toroidal grid.

5.3. Discrete analytic functions on a map. To explain the connection between rotation-free circulations and discrete analytic functions, let ϕ be a rotation-free circulation on a map G in the plane. Using that ϕ is rotation-free, we can construct a function $\pi : \hat{V} \to \mathbb{R}$ such that

$$\phi(e) = \pi(t_e) - \pi(h_e) \tag{15}$$

for every edge e. Similarly, the fact that ϕ is a circulation implies that there exists a function $\sigma : \hat{\mathcal{F}} \to \mathbb{R}$ such that

$$\phi(e) = \sigma(r_e) - \sigma(l_e) \tag{16}$$

for every edge e.

It is easy to see that π is harmonic at all nodes of $\hat{G}$ and σ is harmonic at all nodes of the dual map. Furthermore, π and σ are related by the following condition:

$$\sigma(l_e) - \sigma(r_e) = \pi(h_e) - \pi(t_e). \tag{17}$$

for every edge e (since both sides are just $\phi(e)$). We can think of π and σ as the real and imaginary parts of a (discrete) analytic function. The relation (17) is then a discrete analogue of the Cauchy–Riemann equations.

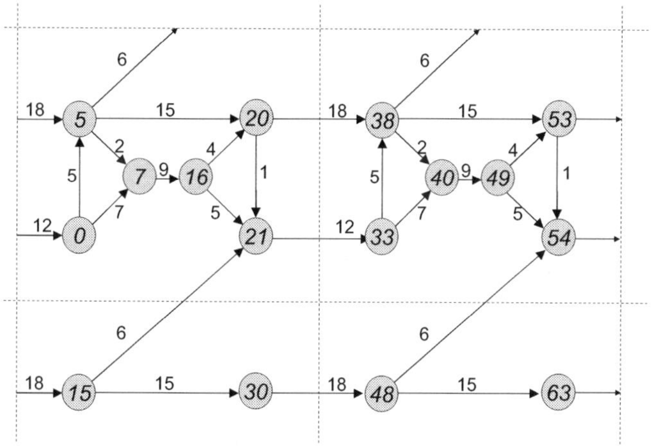

FIGURE 2. A rotation-free circulation on the torus, and a corresponding harmonic function on the universal cover.

Figures 2 and 3 show a rotation-free circulation on a graph embedded in the torus. The first figure shows how to obtain it from a harmonic function on the nodes of the universal cover map, the second, how to obtain it from a harmonic function on the faces.

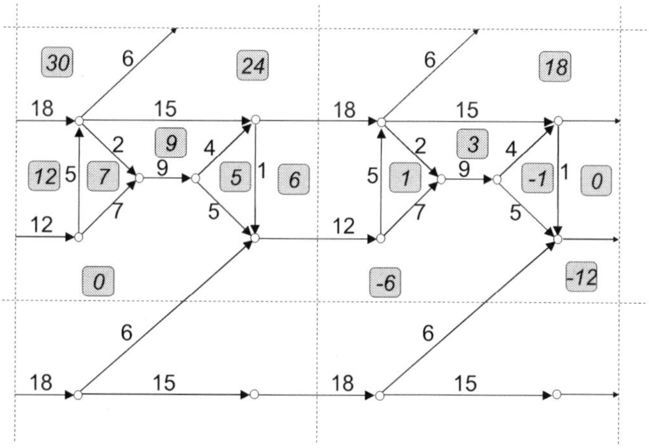

FIGURE 3. A harmonic function on the faces of the universal cover associated with the same rotation-free circulation.

As we mentioned in the introduction, discrete analytic functions and holomorphic forms on general maps were introduced by Mercat [**18**]. His definition is more general than the one above on two counts. First, he allows weighted edges; we'll come back to this extension a bit later. Second, he allows complex values. Let's have a closer look on this.

Let $G = (V, E, \mathcal{F})$ be a discrete map in the plane, and let $G^* = (V^*, E*, \mathcal{F}^*)$ be its dual map. Let $f: V \cup V^* \to \mathbf{C}$. We say that f is *analytic*, if

$$f(l_e) - f(r_e) = i(f(h_e) - f(t_e)). \tag{18}$$

for every edge e. Relation (18) implies a number of further properties; for example, the f to V is harmonic. Indeed, let $y_1, \dots, y_d$ be neighbors of x and let $(xy_k)* = p_k p_{k+1}$ (where $p_{d+1} = p_1$. Then

$$\sum_{i=1}^{d} (f(y_k) - f(x)) = \sum_{k=1}^{d} i(f(p_{k+1}) - f(p_k)) = 0.$$

It follows that if f is analytic, then the function $\phi: E \to \mathbf{C}$ defined by

$$\phi(e) = f(h_e) - f(t_e),$$

is a complex valued rotation-free circulation on G, which we call a *holomorphic form* on G.

Conversely, for any complex-valued function $\phi: E \to \mathbf{C}$, we define $\phi^*: E^* \to \mathbf{C}$ by

$$\phi^*(e^*) = i \cdot \phi(e) \tag{19}$$

Then ϕ is rotation-free if and only if ϕ^* is a circulation. So if both ϕ and ϕ^* are circulations, then they are both rotation-free. Similarly as in the real case, we can represent both ϕ and ϕ^* as differentials of functions on the nodes and faces, respectively. It is convenient to think of the two primitive functions as a single function f defined on $V \cup V^*$. So we have

$$\phi(e) = f(h_e) - f(t_e), \quad \text{and} \quad \phi^*(e) = f(l_e) - f(r_e)$$

for every edge e, and hence f is analytic on the whole map.

However, just like we saw it in the case of discrete analytic functions on a grid, the complex version is not substantially more general than the real. Indeed, a complex-valued function on the edges is a rotation-free circulation if and only if its real and imaginary parts are; and (19) only relates the real part of ϕ to the imaginary part of ϕ^* and vice versa. So a holomorphic form is just a pair of two rotation-free circulations, with no relation between them. In some cases (like in the topological considerations in section 6) the real format is more convenient, in others (like defining integration in section 8), the complex format is better.

Let f be a (complex-valued) discrete analytic function on a map in the plane; this corresponds to four real harmonic functions: $\pi_1 = \Re(f)$ on G, $\pi_2 = \Im(f)$ on G, $\sigma_2 = \Re(f)$ on G^* and $\sigma_1 = \Im(f)$ on G^*. Here π_1 and σ_1 are related by the Cauchy-Riemann equations, and so are π_2 and σ_2; but there is no relation between the pair (π_1, σ_1) and the pair (π_2, σ_2).

This way of looking at analytic functions explains the following construction, which does not seem to correspond to any classical notion. Let us multiply the second pair by -1; we get another 4-tuple of harmonic functions satisfying the Cauchy-Riemann equations, which correspond to a complex-valued discrete analytic function $f^\dagger$. In other words, we define the *conjugate* of f by

$$f^\dagger(i) = \begin{cases} \overline{f(i)}, & \text{if } i \in V(G), \\ -\overline{f(i)}, & \text{if } i \in V(G^*). \end{cases} \tag{20}$$

This construction was introduced in the special case of grids by Duffin, and it plays an important role in defining derivatives.

5.4. The weighted case. We can consider the following more general setup (Mercat [**18**]). Suppose that every edge e in the graph as well as in the dual graph has a positive weight ℓ_e associated with it, which we call its *length*. This assignment is symmetric, so that $\ell_{ij} = \ell_{ji}$. We can think of the length of the dual edge e^* as the "width" of the edge e.

For $\phi : E \to \mathbf{C}$, we define the flow condition as

$$\sum_e \delta v(e)\ell_{e^*}\phi(e) = 0,$$

and the rotation-free condition for a face F as

$$\sum_e \partial F(e)\ell_e\phi(e) = 0.$$

(In terms of hydrodynamics, we think of $\phi(e)$ as the speed of flow along the edge e.) This means that

$$\phi^*(e^*) = \phi(e)$$

defines a rotation-free circulation on the dual graph.

Similarly as before, we can consider real or complex valued circulations, and one complex rotation-free circulation will be equivalent to a pair of real ones.

Consider a complex valued rotation-free circulation ϕ on a map in the plane. Then there is a function f on $V \cup V^*$ so that for every edge e

$$\phi(e) = \frac{f(h_e) - f(t_e}{\ell_e} = i\frac{f(l_e) - f(r_e)}{\ell_{e^*}}.$$

Such a function f is called a *primitive function* of ϕ.

In most of this paper we'll not consider the weighted case, because it would not amount to much more than inserting 'ℓ_e' or 'ℓ_{e^*}' at appropriate places in the equations. In section 8, however, choosing the right weighting will be an important issue.

5.5. Constructing holomorphic forms. We give a more explicit construction of rotation-free circulations in the compact case, using electrical currents. For $e = ab \in E$, consider the harmonic function π_e with poles a and b (as defined in section 3.1). The function $\delta\pi_e$ is certainly rotation-free, but it is not a circulation: a is a source and b is a sink (all the other nodes satisfy the flow condition). We could try to repair this by sending a "backflow" along the edge ab; in other words, we consider $\pi_e - \chi_e$. This is now a circulation, but it is not rotation-free around the faces r_e and l_e.

The trick is to also consider the dual map G^*, the dual edge $(ab)^* = pq$, and the harmonic function π_{e^*}. We carry out the same construction as above, to get $\delta^*\pi_{e^*}$. Then we can combine these to repair the flow condition without creating rotation: We define

$$\eta_e = \delta\pi_e - \delta^*\pi_{e^*} + \chi_e. \tag{21}$$

The considerations above show that η_e is a rotation-free circulation. In addition, it has the following description:

LEMMA 5.4. *The circulation η_e is the orthogonal projection of χ_e to the space $\mathcal{C}$ of rotation-free circulations.*

PROOF. It suffices to show that

$$\chi_e - \eta_e = \delta\pi_e - \delta^*\pi_{e^*}$$

is orthogonal to every $\phi \in \mathcal{C}$. But $\delta\pi_e \in \mathcal{A}$ by (2, and similarly, $\delta^*\pi_{e^*} \in \mathcal{B}$. So both are orthogonal to $\mathcal{C}$. □

This lemma has some simple but interesting consequences. Since the vectors χ_e span $\mathbb{R}^E$, their projections η_e generate the space of rotation-free circulations. Since η_e is a projection of χ_e, we have

$$\eta_e(e) = \eta_e \cdot \chi_e = |\eta_e|^2 \geq 0. \tag{22}$$

Let R_e denote the effective resistance between the endpoints of an edge e, and let R_{e^*} denote the effective resistance of the dual map between the endpoints of the dual edge e^*. Then we get by Theorem 3.3 that

$$\eta_e(e) = 1 - R_e - R_{e^*}. \tag{23}$$

If we work with a map on the sphere, we must get 0 by Theorem 5.3. This fact has the following consequence (which is well known, and can also be derived e.g. from Corollary 3.4): *for every planar map, $R_e + R_{e^*} = 1$.* For any other underlying surface, we get $R_e + R_{e^*} \leq 1$. It will follow from theorem 6.6 below that strict inequality holds here, as soon as the map satisfies some mild conditions.

6. Topological properties

6.1. Combinatorial singularities. We need a simple combinatorial lemma about maps on compact surfaces. For every face F, let a_F denote the number of sharp corners of F. For every node v, let b_v denote the number of blunt corners at v. So a_F is the number of times the orientation changes if we move along the boundary of F, while b_v is the number of times the orientation changes in the cyclic order of edges as they emanate from v.

LEMMA 6.1. *Let $G = (V, E, \mathcal{F})$ be any digraph embedded on a surface S with Euler characteristic χ. Then*

$$\sum_{F\in\mathcal{F}} (a_F - 2) + \sum_{v\in V} (b_v - 2) = 4g - 4. \tag{24}$$

PROOF. Counting sharp corners, we get

$$\sum_F a_F = \sum_v (d_v - b_v),$$

and so by Euler's formula,

$$\sum_F a_F + \sum_v b_v = \sum_v d_v = 2m = 2n + 2f + 4g - 4.$$

Rearranging, we get the equality in the lemma. □

Suppose that the orientation of the map is such that there are no sources and sinks (so each node has at least one edge going out and at least one edge going in), and no face boundary is a directed cycle. Then $b_v \geq 2$ for each node and $a_F \geq 2$ for each face, and so every term in (24) is nonnegative. Lemma 6.1 says that all but at most $2g - 2$ nodes will have $b_v = 2$, which means that both the incoming edges and the outgoing edges are consecutive in the cyclic ordering around the node. Similarly, all but at most $2g - 2$ faces will have $a_F = 2$, which means that the face boundary consists of two directed paths.

6.2. Zero sets. Some useful nondegeneracy properties of rotation-free circulations were proved in [**2, 3**]. In this section we present these in a more general form.

An analytic function cannot vanish on a open set, unless it is identically 0. What is the corresponding statement for finite graphs? For which subgraphs of a map can we find a discrete holomorphic form that vanishes on all edges of the subgraph? In other words, what do we know about the support subgraph H_ϕ of a rotation-free circulation?

Let H be a subgraph of a finite connected graph G. Consider the connected components of $G \setminus V(H)$. A *bridge* of H is defined as a subgraph of G that consists of one of these components, together with all edges connecting it to H, and their endpoints in H. We also consider edges not in H but connecting two nodes in H as *trivial bridges* of H. Let $\mathcal{B}(H)$ denote the set of bridges of H. For every bridge $B \in \mathcal{B}(H)$, we call its nodes in H its *terminals.* The other nodes of the bridge are called *internal.*

If we look at a small neighborhood of a terminal v of B, then the edges of H incident with this node divide this neighborhood into "corners". The bridge B may have edges entering v through different corners. The number of corners B uses is the *multiplicity* of the terminal. We denote the sum of multiplicities of all terminals of B by $\tau(B)$.

THEOREM 6.2. *Let H be a subgraph of a map G on an orientable surface S with genus g.*

(a) *If H is the support of a rotation-free circulation ϕ, then*

$$\sum_{\substack{B\in\mathcal{B}(H)\\ \tau(B)\geq 2}} (\tau(B) - 2) \leq 4g - 2.$$

(b) *If*

$$\sum_{B\in\mathcal{B}(H)} (\tau(B) - 1) \leq 2g - 1,$$

then there is a rotation-free circulation with support contained in H.

Let us make some remarks in connection with this theorem.

1. Part (b) of the theorem implies that a rotation-free circulation can vanish on a rather large part of the graph, which could contain even the majority of the nodes. It is not the size of a set S that matters, but rather how well connected S is to the non-trivial parts of the graph.

2. It would be nice to replace $\tau(B) - 2$ by $\tau(B) - 1$ in (a); but no such result can be stated, since we cannot control the number of trivial bridges. As an example, consider the rotation-free circulation in Figure 1 on a toroidal grid with $a = 1,\ b = 0$.

3. If H satisfies (a) but not (b), then whether or not there exists a nonzero rotation-free circulation supported on H may depend on finer properties. An interesting case is a node of degree 6 in a map on the double torus ($g = 2$). One can construct examples where no rotation-free circulation can vanish on all 6 edges, and other examples where it can.

Let us formulate some corollaries of this theorem. The following theorem was proved in [**2, 3**]. We say that a connected subgraph H is *k-separable* in G, if G can be written as the union of two graphs G_1 and G_2 so that $|V(G_1) \cap V(G_2)| \leq k$, $V(H) \cap V(G_2) = \emptyset$, and G_2 contains a non-0-homologous cycle.

COROLLARY 6.3. *Let G be a map on an orientable surface S of genus $g > 0$, and let H be a connected subgraph of G. If H is not $(4g-1)$-separable in G, then every rotation-free circulation vanishing on all edges of $\overline{H}$ is identically 0.*

COROLLARY 6.4. *Assume that for every set X of $4g-1$ nodes, all but one of the components of $G - X$ are plane and have fewer than k' nodes. Let H be a connected subgraph with k nodes. Then every rotation-free circulation vanishing on all edges incident with H is identically 0.*

A map is called *k-representative*, if every non-contractible Jordan curve on the surface intersects the map in at least k points.

COROLLARY 6.5. *Let G be a $(4g-1)$-representative map on an orientable surface S of genus $g > 0$. Then every rotation-free circulation vanishing on all edges of a non-0-homologous cycle, and on all edges incident with it, is identically 0.*

6.3. Identically zero-sets. Most of the time, the motivation for the study of discrete analytic functions is to transfer the powerful methods from complex analysis to the study of graphs. In this section we look at questions that are natural for graphs. It would be interesting to find analogues or applications in the continuous setting.

Recall that for every oriented edge e, we introduced the rotation-free circulation η_e. We want to give a sufficient condition for this projection to be non-zero. The fact that η_e is the orthogonal projection of χ_e to $\mathcal{C}$ implies that the following three assertions are equivalent: $\eta_e \neq 0$; $\eta_e(e) \neq 0$; there exists a rotation-free circulation ϕ with $\phi(e) \neq 0$.

THEOREM 6.6. *Let G be a 3-connected simple map an orientable surface with genus $g > 0$. Then $\eta_e \neq 0$ for every edge e.*

The toroidal graphs in Figure 4 (where the surrounding area can be any graph embedded in the torus) show that the assumption of 3-connectivity and the exclusion of loops and parallel edges cannot be dropped[1]

COROLLARY 6.7. *If G is a 3-connected simple graph on a surface with positive genus, then there exists a nowhere-0 rotation-free circulation.*

Another corollary gives an explicit lower bound on the entries of η_{ab}.

COROLLARY 6.8. *If G is a 3-connected simple graph on a surface with positive genus, then for every edge e, $\eta_e(e) \geq n^{2-n} f^{2-f}$.*

Indeed, combining Theorem 6.6 with (22), we see that $\eta_e(e) > 0$ if $g > 0$. But $\eta_e(e) = 1 - R_e - R^*_{e^*}$ is a rational number, and from Theorem 3.4 it follows that its denominator is not larger than $n^{n-2} f^{f-2}$.

COROLLARY 6.9. *If G is a 3-connected simple graph on a surface with positive genus, then for every edge e, $R_e + R^*_{e^*} < 1$.*

[1]This condition was erroneously omitted in [**3**].

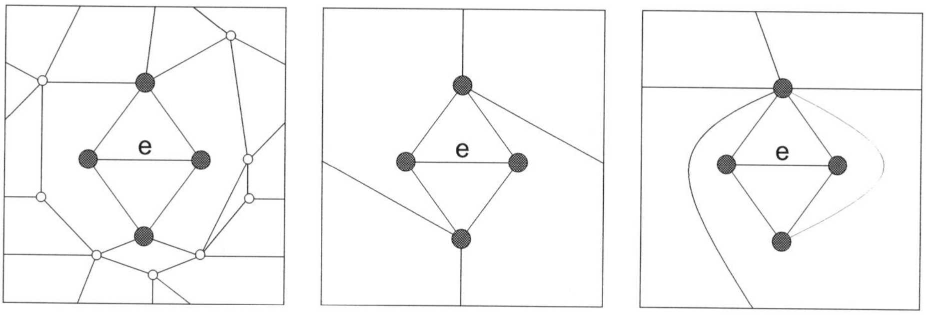

FIGURE 4. Every rotation-free circulation is 0 on the edge e.

6.4. Generic independence. The question whether every rotation-free circulation vanishes on a given edge is a special case of the following: given edges $e_1, \dots, e_k$, when can we independently prescribe the values of a rotation-free circulation on them? Since the dimension of $\dim(\mathcal{C}) = 2g$, we must have $k \le 2g$. There are other obvious conditions, like the set should not contain the boundary of a face or the coboundary of a node. But a complete answer appears to be difficult.

We get, however, a question that *can* be answered, if we look at the "generic" case: we consider the weighted version, and assume that there is no numerical coincidence, by taking (say) algebraically independent weights. Using methods from matroid theory, a complete characterization of such edge-sets can be given [**17**]. For example, the following theorem provides an NP-coNP characterization and (through matroid theory) a polynomial algorithm in the case when $k = 2g$. We denote by $c(G)$ the number of connected components of the graph G.

THEOREM 6.10. *Let $W \subseteq E$ be any set of $2g$ edges of a map on an orientable surface with genus g, with algebraically independent weights. Then the following are equivalent:*

(a) *Every set of prescribed values on W can be extended to a rotation-free circulation in a unique way.*

(b) *$E(G) - W$ can be partitioned into two sets T and T^* so that T forms a spanning tree in G and T^* forms a spanning tree in G^*.*

(c) *For every set $W \subseteq S \subseteq E(G)$ of edges*

$$c(G \setminus S) + c(G^* \setminus S) \le |S| + 2 - 2g.$$

As another special case, one gets that for a given edge e, there is a rotation-free circulation that is non-zero on e for some weighting (equivalently, for almost all weightings) of the edges if and only if the map contains a non-zero-homologous cycle through e.

7. Geometric connections

7.1. Straight line embeddings. We can view a (complex-valued) analytic function f on a map G as a mapping of the nodes into the complex plane. We can extend this to the whole graph by mapping each edge uv on the segment connecting $f(u)$ and $f(v)$. It turns out that under rather general conditions, this mapping is an embedding. To formulate the condition, note that on the nodes of G we can define a distance $d_G(u, v)$ as the minimum length of path in G connecting u and v.

THEOREM 7.1. *Let G be a simple 3-connected map in the plane and let f be an analytic function on G. Suppose that there exist a constant c such that*

$$\frac{1}{c} \leq \frac{|f(u) - f(v)|}{d_G(u,v)} \leq c. \tag{25}$$

for every pair of distinct nodes $u, v \in V \cup V^*$. Then f defines an embedding. Furthermore, this embedding has the additional property that every face is a convex polygon, and every node is in the center of gravity of its neighbors.

One case when the conditions in the theorem are automatically fulfilled is when the map G is the universal cover map of a toroidal map H and f is the primitive function of a holomorphic form on H. Then the embedding defined by f can be "rolled up" to the torus again. So we obtain the following corollary:

COROLLARY 7.2. *Every holomorphic form on a simple 3-connected toroidal map defines an embedding of it in the torus such that all edges are geodesic arcs.*

7.2. Square tilings. A beautiful connection between square tilings and rotation-free flows was described in the classic paper of Brooks, Smith, Stone and Tutte [**5**]. They considered tilings of squares by smaller squares, and used the connection with flows to construct a tiling of a square with squares whose edge-lengths are all different. For our purposes, periodic tilings of the whole plane are more relevant.

Consider tiling of the plane with squares, whose sides are parallel to the coordinate axes. Assume that the tiling is discrete, i.e., every bounded region contains only a finite number of squares. We associate a map in the plane with this tiling as follows. Represent any maximal horizontal segment composed of edges of the squares by a single node (say, positioned at the midpoint of the segment). Each square "connects" two horizontal segments, and we can represent it by an edge connecting the two corresponding nodes, directed top-down. We get an (infinite) directed graph G (Figure 5).

It is not hard to see that G is planar. If we assign the edge length of each square to the corresponding edge, we get a circulation: If a node v represents a segment I, then the total flow into v is the sum of edge length of squares attached to I from the top, while the total flow out of v is the sum of edge length of squares attached to I from the bottom. Both of these sums are equal to the length of I (let's ignore the possibility that I is infinite for the moment). Furthermore, since the edge-length of a square is also the difference between the y-coordinates of its upper and lower edges, this flow is rotation-free.

Now suppose that the tiling is double periodic with period vectors $a, b \in \mathbb{R}^2$ (i.e., we consider a square tiling of the torus). Then so will be the graph G, and so factoring out the period, we get a map on the torus. Since the tiling is discrete, we get a finite graph. This also fixes the problem with the infinite segment I: it will become a closed curve on the torus, and so we can argue with its length on the torus, which is finite now. The flow we constructed will also be periodic, so we get a rotation-free circulation on the torus.

We can repeat the same construction using the vertical edges of the squares. It is not hard to see this gives the dual graph, with the dual rotation-free circulation on it.

A little attention must be paid to points where four squares meet. Suppose that A, B, C, D share a corner p, where A is the upper left, and B, C, D follow clockwise. In this case, we must consider the lower edges of A and B to belong to a single

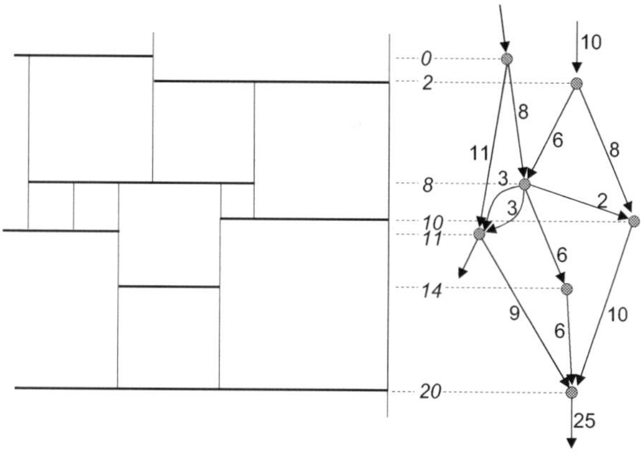

FIGURE 5. The Brooks–Smith–Stone–Tutte construction

horizontal segment, but interrupt the vertical segment at p, or vice versa. In other words, we can consider the lower edges of A and C "infinitesimally overlapping".

This construction can be reversed. Take a toroidal map G^* and any rotation-free circulation on it. Then this circulation can be obtained from a doubly periodic tiling of the plane by squares, where the edge-length of a square is the flow through the corresponding edge. (We suppress details.)

If an edge has 0 flow, then the corresponding square will degenerate to s single point. Luckily, we know (Corollary 6.7) that for a simple 3-connected toroidal map, there is always a nowhere-zero rotation-free circulation, so these graphs can be represented by a square tiling with no degenerate squares.

7.3. Rubber bands. Another important geometric method to represent planar graph was described by Tutte [**25**]. Tutte used it to obtain drawings of planar graphs, but we apply the method to toroidal graphs.

Let G be a toroidal map. We consider the torus as $\mathbb{R}^2/\mathbf{Z}^2$, endowed with the metric coming from the euclidean metric on $\mathbb{R}^2$. Let us replace each edge by a rubber band, and let the system find its equilibrium. Topology prevents the map from collapsing to a single point. In mathematical terms, we are minimizing

$$\sum_{ij\in E(G)} \ell(ij)^2, \tag{26}$$

where the length $\ell(ij)$ of the edge ij is measured in the given metric, and we are minimizing over all continuous mappings of the graph into the torus homomorphic to the original embedding.

It is not hard to see that the minimum is attained, and the minimizing mapping is unique up to isometries of the torus. We call it the *rubber band mapping.* Clearly, the edges are mapped onto geodesic curves. A nontrivial fact is that if G is a simple 3-connected toroidal map, then the rubber band mapping is an embedding. This follows from Theorem 7.1.

We can lift this optimizing embedding to the universal cover space, to get a planar map which is doubly periodic, and the edges are straight line segments. Moreover, *every node is at the center of gravity of its neighbors.* This follows simply from the minimality of (26). This means that both coordinate functions are harmonic and periodic, and so their coboundaries are rotation-free circulations on the original graph. Since the dimension of the space $\mathcal{C}$ of rotation-free circulations on a toroidal map is 2, this construction gives us the whole space $\mathcal{C}$.

This last remark also implies that if G is a simple 3-connected toroidal map, then selecting any basis ϕ_1, ϕ_2 in $\mathcal{C}$, the primitive functions of ϕ_1 and ϕ_2 give a doubly periodic straight-line embedding of the universal cover map in the plane.

7.4. Circle representations. Our third geometric construction that we want to relate to discrete holomorphic forms are circle representations. A celebrated theorem of Koebe [**15**] states that the nodes of every planar graph can be represented by openly disjoint circular discs in the plane, so that edges correspond to tangency of the circles. Andre'ev [**1**] improved this by showing that there is a simultaneous representation of the graph and its dual. Thurston [**24**] extended this to the toroidal graphs, and this is the version we need.

It is again best to go to the universal cover map $\hat{G}$. Then the result says that for every 3-connected toroidal graph G we can construct two (infinite, but discrete) families $\mathcal{F}$ and $\mathcal{F}^*$ of circles in the plane so that they are double periodic modulo a lattice $L = \mathbf{Z}a + \mathbf{Z}b$, $\mathcal{F}$ (mod L) corresponds to the nodes of G, $\mathcal{F}^*$ (mod L) corresponds to the faces of G, and for ever edge e, there are two circles C, C' representing h_e and t_e, and two circles D and D' representing r_e and l_e so that C, C' are tangent at a point p, D, D' are tangent at the same point p, and C, D are orthogonal.

If we consider the centers the circles in $\mathcal{F}$ as nodes, and connect two centers by a straight line segment if the circles touch each other, then we get a straight line embedding of the universal cover map in the plane (appropriately periodic modulo L). Let $f(i)$ denote the point representing node i of the universal cover map. or of its dual.

To get a holomorphic form out of this representation, consider the plane as the complex plane, and define $\phi(ij) = \rho(j) - \rho(i)$ for every edge of $\hat{G}$ or $\hat{G}^*$. Clearly ϕ is invariant under L, so it can be considered as a function on $E(G)$. By the orthogonality property of the circle representation, $\phi(e)/\phi(e^*)$ is a positive multiple of i. In other words,

$$\frac{\phi(e)}{|\phi(e)|} = i\frac{\phi(e^*)}{|\phi(e^*)|}$$

It follows that if we consider the map G with weights

$$\ell_e = |\phi(e)|, \qquad \ell_{e^*} = |\phi(e^*)|,$$

then ϕ is a discrete holomorphic form on this weighted map.

It would be nice to be able to turn this construction around, and construct a circle representation using discrete holomorphic forms.

8. Operations

8.1. Integration. Let f and g be two functions on the nodes of a discrete weighted map in the plane. Integration is easiest to define along a path $P =$

$(v_0, v_1, \dots, v_k)$ in the diamond graph $G_\diamond$ (this has the advantage that it is symmetric with respect to G and G^*). We define

$$\int_P f\,dg = \sum_{i=0}^{k-1} \frac{1}{2}(f(v_{i+1}) + f(v_i))(g(v_{i+1}) - g(v_i)).$$

The nice fact about this integral is that for analytic functions, it is independent of the path P, depends on the endpoints only. More precisely, let P and P' be two paths on $G_\diamond$ with the same beginning node and endnode. Then

$$\int_P f\,dg = \int_{P'} f\,dg. \tag{27}$$

This is equivalent to saying that

$$\int_P f\,dg = 0 \tag{28}$$

if P is a closed path. It suffices to verify this for the boundary of a face of $G_\diamond$, which only takes a straightforward computation. It follows that we can write

$$\int_u^v f\,dg$$

as long as the homotopy type of the path from u to v is determined (or understood).

Similarly, it is also easy to check the rule of integration by parts: If P is a path connecting $u, v \in V \cup V^*$, then

$$\int_P f\,dg = f(v)g(v) - f(u)g(u) - \int_P g\,df. \tag{29}$$

Let P be a closed path in $G_\diamond$ that bounds a disk D. Let f be an analytic function and g an arbitrary function. Define $\hat{g}(e) = g(h_e) - g(t_e) - i(g(l_e) - g(r_e))$ (the "analycity defect" of g on edge e. Then it is not hard to verify the following generalization of (28):

$$\int_P f\,dg = \sum_{e \subset D} (f(h_e) - f(t_e))\hat{g}(e). \tag{30}$$

This can be viewed as a discrete version of the Residue Theorem. For further versions, see [**18**].

Kenyon's ideas in [**13**] give a nice geometric interpretation of (28). Let G be a map in the plane and let g be an analytic function on G. Let us assume that g satisfies the conditions of Theorem 7.1, so that it gives a straight-line embedding of G in the plane with convex faces, and similarly, a straight-line embedding of G^* with convex faces. Let P_u denote the convex polygon representing the face of G (or G^*) corresponding to $u \in V^*$ (or $u \in V$)). Shrink each F_u from the point $g(u)$ by a factor of 2. Then we get a system of convex polygons where for every edge $uv \in G_\diamond$, the two polygons P_u and P_v share a vertex at the point $(g(u) + g(v))/2$ (Figure 6(a)). There are two kinds of polygons (corresponding to the nodes in V and V^*, respectively. It can be shown that the interiors of the polygons P_u will be disjoint (the point $g(u)$ is not necessarily in the interior of P_u). The white areas between the polygons correspond to the edges of G. They are rectangles, and the sides of the rectangle corresponding to edge e are $g(h_e) - g(t_e)$ and $g(l_e) - g(r_e)$.

FIGURE 6. Representation of an analytic function by touching polygons, and a deformation given by another analytic function.

Now take the other analytic function f, and construct the polygons $f(u)P_u$ (multiplication by the complex number $f(u)$ corresponds to blowing up and rotating). The resulting polygons will not meet at the appropriate vertices any more, but we can try to translate them so that they do. Now equation (28) tells us that we can do that (Figure 6(b)). Conversely, every "deformation" of the picture such that the polygons P_u remain similar to themselves defines an analytic function on G.

8.2. Critical analytic functions. These have been the good news. Now the bad part: for a fixed starting node u, the function

$$F(v) = \int_u^v f\,dg$$

is uniquely determined, but it is *not* analytic in general. In fact, a simple computation shows that for any edge e,

$$\begin{aligned} \hat{F}(e) &= \frac{F(h_e) - F(t_e)}{\ell_e} - i\frac{F(l_e) - F(r_e)}{\ell_{e^*}} \\ &= i\frac{f(h_e) - f(t_e)}{\ell_e}\Big[g(t_e) + g(h_e) - g(r_e) - g(l_e)\Big]. \end{aligned} \tag{31}$$

So we want an analytic function g such that the factor in brackets in (31) is 0 for every edge:

$$g(t_e) + g(h_e) = g(r_e) + g(l_e) \tag{32}$$

Let us call such an analytic function *critical.* What we found above is that $\int_u^v f\,dg$ is an analytic function of v for every analytic function f if and only if g is critical.

This notion was introduced in a somewhat different setting by Duffin [**9**] under the name of *rhombic lattice.* Mercat [**18**] defined *critical maps*: these are maps which admit a critical analytic function.

Geometrically, this condition means the following. Consider the function g as a mapping of $G \cup G^*$ into the complex plane $\mathbf{C}$. This defines embeddings of G, G^* and $G\diamondsuit$ in the plane with following (equivalent) properties:

(a) The faces of $G^\diamondsuit$ are rhomboids.

(b) Every edge of $G_\diamondsuit$ has the same length.

(c) Every face of G is inscribed in a unit circle.

(d) Every face of G^* is inscribed in a unit circle.

Criticality can be expressed in terms of holomorphic forms as well. Let ϕ be a (complex valued) holomorphic form on a weighted map G. We say that ϕ is *critical* if the following condition holds: Let $e = xy$ and $f = yz$ be two edges of G bounding a corner at y, with (say) directed so that the corner is on their left, then

$$\ell_e\phi(e) + \ell_f\phi(f) = \ell_{e^*}\phi(e^*) - \ell_{f^*}\phi(f^*). \tag{33}$$

Note that both f and f^* are directed into h_f, which explains the negative sign on the right hand side. To digest this condition, consider a plane piece of the map and a primitive function g of ψ. Then (33) means that

$$g(y') - g(y) = g(q) - g(q'),$$

which we can rewrite in the following form:

$$g(x) + g(y) - g(p) - g(q) = g(x) + g(y') - g(p) - g(q').$$

This means that $g(h_e) + g(t_e) - g(l_e) - g(r_e)$ is the same for every edge e, and since we are free to add a constant to the value of g at every node in V^* (say), we can choose the primitive function g so that g is critical.

Whether or not a weighted map in the plane has a critical holomorphic form depends on the weighting. Which maps can be weighted this way? A recent paper of Kenyon and Schlenker [**14**] answers this question. Consider any face F_0 of the diamond graph $G_\diamond$, and a face F_1 incident with it. This is a quadrilateral, so there is a well-defined face F_2 so that F_0 and F_2 are attached to F_1 along opposite edges. Repeating this, we get a sequence of faces $(F_0, F_1, F_2 \dots)$. Using the face attached to F_0 on the opposite side to F_1, we can extend this to a two-way infinite sequence $(\dots, F_{-1}, F_0, F_1, \dots)$. We call such a sequence a *track*.

THEOREM 8.1. *A planar map has a rhomboidal embedding in the plane if and only if every track consists of different faces and any two tracks have at most one face in common.*

8.3. Polynomials, exponentials, derivation and approximation. Once we can integrate, we can define polynomials. More exactly, let G be a map in the plane, and let us select any node to be called 0. Let Z denote a critical analytic function on G such that $Z(0) = 0$. Then we have

$$\int_0^x 1\, dZ = Z(x).$$

Now we can define higher powers of Z by repeated integration:

$$Z^{:n:}(x) = n \int_0^x Z^{:n-1:} dZ.$$

We can define a *discrete polynomial* of degree n as any linear combination of $1, Z, Z^{:2:}, \dots, Z^{:n:}$. The powers of Z of course depend on the choice of the origin, and the formulas describing how it is transformed by shifting the origin are more complicated than in the classical case. However, the space of polynomials of degree see is invariant under shifting the origin [**19**]).

Further, we can define the exponential function $exp(x)$ as a discrete analytic function $\mathrm{Exp}(x)$ on $V \cap V^*$ satisfying

$$d\mathrm{Exp}(x) = \mathrm{Exp}(x) dZ.$$

More generally, it is worth while to define a 2-variable function $\mathrm{Exp}(x, \lambda)$ as the solution of the difference equation

$$d\mathrm{Exp}(\lambda, x) = \lambda\mathrm{Exp}(\lambda, x)dZ.$$

It can be shown that there is a unique such function, and there are various more explicit formulas, including

$$\mathrm{Exp}(\lambda, x) = \sum_{n=0}^{\infty} \frac{Z^{:n:}}{n!},$$

(at least as long as the series on the right hand side is absolute convergent).

We can also define derivation, using the notion of the conjugate function defined in (20). Given a (complex) analytic function on (say) a map in the plane, we define

$$f'(j) = \left(\int_0^j f^{\dagger} dg\right)^{\dagger}.$$

Then it is not hard to see that f' is analytic, and

$$\int_0^v f'\, dg = f - f(0).$$

Mercat [**19, 20**] uses these tools to show that exponentials form a basis for all discrete analytic functions, and to generalize results of Duffin, Zeilberger and others about approximability of analytic functions by discrete analytic functions.

9. An application in computer science: Global information from local observation

Suppose that we live in a (finite) map on a compact orientable surface with genus g (we assume the embedding is reasonably dense). On the graph, a random process is going on, with local transitions. Can we determine the genus g, by observing the process in a small neighborhood of our location?

Discrete analytic functions motivate a reasonably natural and simple process, called the *noisy circulator*, which allows such a conclusion. This was constructed by Benjamini and the author [**2**]. Informally, this can be described as follows. Each edge carries a real weight. With some frequency, a node wakes up, and balances the weights on the edges incident with it, so that locally the flow condition is restored. With the same frequency, a face wakes up, and balances the weights on the edges incident with it, so that the rotation around the face is cancelled. Finally, with a much lower frequency, an edge wakes up, and increases or decreases its weight by 1.

To be precise, we consider a finite graph G, embedded on an orientable surface S, so that each face is a disk bounded by a simple cycle. We fix a reference orientation of G, and a number $0 < p < 1$. We start with the vector $x = 0 \in \mathbb{R}^E$. At each step, the following two operations are carried out on the current vector $x \in \mathbb{R}^E$:

(a) [Node balancing.] We choose a random node v. Let $a = (\delta v)^{\mathsf{T}} x$ be the "imbalance" at node v (the value by which the flow condition at v is violated). We modify f by subtracting $(a/d_v)\delta v$ from x.

(b) [Face balancing.] We choose a random face F. Let $r = (\partial F)^{\mathsf{T}} x$ be the rotation around F. We modify f by subtracting $(r/d_F)\partial F$ from x.

In addition, with some given probability $p > 0$, we do the following:

(c) [Excitation.] We choose a random edge e and a random number $X \in \{-1, 1\}$, and add X to x_e.

Rotation-free circulations are invariant under node and face balancing. Furthermore, under repeated application of (a) and (b), any vector converges to a rotation-free circulation.

Next we describe how we observe the process. Let U be a connected subgraph of G, which is not $(4g-1)$-separable in G. Our observation window is the set E_0 of edges incident with U (including the edges of U). Let $x(t) \in \mathbb{R}^E$ be the vector of edge-weights after t steps, and let $y(t)$ be the restriction of $x(t)$ to the edges in E_0. So we can observe the sequence random vectors $y(0), y(1), \dots$.

The main result of [**2**] about the noisy circulator is the following (we don't state the result in its strongest form).

THEOREM 9.1. *Assume that we know in advance an upper bound N on $n+m+f$. If $p = O(N^{-8})$, then observing the Noisy Circulator for $O(N^8/p)$ steps, we can determine the genus g with high probability.*

The idea behind the recovery of the genus g is that if the excitation frequency p is sufficiently small, then most of the time $x(t)$ will be essentially a rotation-free circulation. The random excitations guarantee that over sufficient time we get $2g$ linearly independent rotation-free circulations. Corollary 6.3 implies that even in our small window, we see $2g$ linearly independent weight assignments $y(t)$.

Acknowledgement

I am indebted to Oded Schramm, Lex Schrijver and Miki Simonovits for many valuable discussions while preparing this paper.

Appendix

Proof of Theorem 6.2. (a) Consider an edge e with $\phi(e) = 0$. There are various ways e can be eliminated. If e is two-sided, then we can delete e and get a map on the same surface with a rotation-free flow on it. If e is not a loop, then we can contract e and get a map on the same surface with a rotation-free flow on it. If e is a one-sided loop, we can change $\phi(e)$ to any non-zero value and still have a rotation-free circulation.

Of course, we don't want to eliminate all edges with $\phi = 0$, since then we don't get anything. We eliminate two-sided edges that constitute trivial bridges (this does not change the assertion in (a)). We contract edges that connect two different internal nodes in the same bridge, so that we may assume that every bridge has exactly one internal node. If there are two edges with $\phi = 0$ that together bound a disc (which necessarily connect the internal node of a bridge to a terminal), we delete one of them. We delete any two-sided loop with $\phi = 0$. Finally, if we have a one-sided loop with $\phi = 0$ attached at a node of H, then we send non-zero flow through it arbitrarily.

Let v_B denote the internal node of bridge B. The node v_B has $\tau(B)$ edges connecting it to H (there may be some one-sided loops left that are attached to v_B). For every face F, let $\beta(F)$ denote the number of times we switch between H and the rest when walking along the boundary.

We need some additional terminology. We call a corner *unpleasant* if it is at a node of H, and both bounding edges are outside H. Note that these edges necessarily belong to different bridges. Let $u(v)$ and $u(F)$ denote the number of unpleasant corners at node v and face F, respectively. Clearly $\sum_v u(v) = \sum_F u(F)$ is the total number of unpleasant corners.

Re-orient each edge of H in the direction of the flow ϕ. Re-orient each edge not in H randomly, independently of each other, with probability 1/2 in either direction. We evaluate the expectation of various terms in Lemma 6.1. A corner that has at most one edge of H will be sharp with probability 1/2 and blunt with probability 1/2. Hence for the internal nodes we have

$$\mathsf{E}(b_{v_B} - 2) = \frac{\tau(B)}{2} - 2. \tag{34}$$

We claim that for each node $v \in V(H)$,

$$\mathsf{E}(b_v - 2) \geq \frac{u(v)}{2}. \tag{35}$$

Trivially, v is never a source or a sink, and so $b_v - 2 \geq 0$ for any of the random orientations. So we may assume that $u(v) \geq 1$. H has at least two blunt corners at v. If all the bridges attached at v come in through the same corner, then one of these blunt corners is left intact in every orientation, and the $u(v)$ unpleasant corners give rise to $u(v) + 2$ corners at v bordered by at least one edge with $\phi = 0$. This means that

$$\mathsf{E}(b_v - 2) \geq 1 + \frac{u(v) + 2}{2} - 2 = \frac{u(v)}{2}.$$

If the bridges attached at v come in through at least two corners, then the $u(v)$ unpleasant corners give rise to at least $u(v) + 4$ corners at v bordered by at least one edge with $\phi = 0$, and so

$$\mathsf{E}(b_v - 2) \geq \frac{u(v) + 4}{2} - 2 = \frac{u(v)}{2}.$$

Finally, we claim that for any face F

$$\mathsf{E}(a_F - 2) \geq \frac{\beta(F) - u(F)}{2}. \tag{36}$$

The number of internal points and their neighbors on the boundary of F is at least $3\beta(F) - u(F)$, and so

$$\mathsf{E}(a_F - 2) \geq \frac{3\beta(F) - u(F)}{2}.$$

So if $\beta(F) \geq 2$, then we are done. Suppose that $\beta(F) = 1$. If the boundary contains edges with $\phi \neq 0$, then it must contain two such edges that are oriented in the opposite direction (since ϕ is rotation-free). There is an expected number of at least 3/2 orientation changes on the arc between these two edges that contains an internal node, and at least one on the other. This gives that

$$\mathsf{E}(a_F - 2) \geq 1 + \frac{3}{2} - 2 = \frac{1}{2} = \frac{\beta(F) - u(F)}{2}.$$

Finally, if $\beta(F) = 0$, then all we have to show is that the face is not oriented. If the face has at least one edge with $\phi \neq 0$ then this is obvious, since ϕ is rotation-free. If all its edges have $\phi = 0$, then these cannot connect H to v_B, which means that they must be one-sided. But then the face boundary passes through them twice in the opposite direction, so the face is not oriented.

Now we have, by Lemma 6.1,

$$\begin{aligned}
4g-4 &= \mathsf{E}\left(\sum_F (a_F-2)+\sum_v (b_v-2)\right) \\
&= \sum_F \mathsf{E}(a_v-2)+\sum_v \mathsf{E}(b_v-2) \\
&\geq \sum_F \frac{\beta(F)-u(F)}{2}+\sum_{v\in V(H)} \frac{u(v)}{2}+\sum_{B\in\mathcal{B}(H)}\left(\frac{\tau(B)}{2}-2\right) \\
&= \sum_F \frac{\beta(F)}{2}+\sum_{B\in\mathcal{B}(H)}\left(\frac{\tau(B)}{2}-2\right) \\
&= \sum_{B\in\mathcal{B}(H)} (\tau(B)-2),
\end{aligned}$$

as claimed.

(b) Contract the internal nodes of every non-trivial bridge B to a single node v_B. If we have two edges in $E(G)\setminus E(H)$ that bound a disc, we delete one of them. So we will have $\tau(B)$ edges connecting v_B to H. Let G denote the resulting map.

Let S be a set of edges that contains all the trivial bridges and also all but one edge between v_B and H for every bridge B. Then

$$|S| = \sum_B (\tau(B)-1) \leq 2g-1.$$

Since the space $\mathcal{C}(G')$ of rotation-free circulations has dimension $2g$, there is a $\phi \in \mathcal{C}(G')$ that is 0 on all edges of S. From the flow condition it follows that ϕ is 0 on the remaining edges between the nodes v_B and H. Now we can construct a $\psi \in \mathcal{C}(G)$ by keeping ϕ on H and extending it by 0 values to the rest of G.

Proof of Theorem 6.6. Suppose that $\eta_e = 0$. Then by the definition (21) of η_e, we have

$$\pi_e(h_f)-\pi_e(t_f)=\pi_{e^*}(r_f)-\pi_{e^*}(l_f) \tag{37}$$

for every edge $f \neq e$, but

$$\pi_e(h_e)-\pi_e(t_e)=\pi_{e^*}(r_e)-\pi_{e^*}(l_e)-1. \tag{38}$$

It will be convenient to set $\alpha(f)=\pi_e(h_f)-\pi_e(t_f)$. We may choose the reference orientation so that $\alpha(f)\geq 0$ for every edge.

Let U_c denote the union of faces F with $\pi_{e^*}(F) > c$. The boundary of U_c is an eulerian subgraph, and so it can be decomposed into edge-disjoint cycles $D_1,\dots,D_t$. For every edge $f\in E(D_j)$, $f\neq e$ we have $\alpha(f)>0$ by (37), and all these edges are oriented in the same way around the cycle. So π_e strictly increases as we traverse the cycle D_j. This is clearly a contradiction unless $t=1$ and e is an edge of D_1. Let $D(c)$ denote this unique boundary cycle of U_c. It also follows that all the values of π_e on this cycle are different.

Let G_0 denote the subgraph formed by those edges f for which $\alpha(f)=0$, and G_1 the subgraph formed by the other edges. Clearly π_e is constant on every connected component of G_0. Hence (i) *a cycle $D(c)$ meets a component of G_0 at most once.*

Next we show that (ii) *every node is in* G_1. Suppose $u \notin V(G_1)$. Since $D(c) \subseteq G_1$ for any c, and there are no parallel edges, the subgraph G_1 must have at least 3 nodes. By 3-connectivity, there are 3 paths connecting u to three nodes $v_1, v_2, v_3 \in V(G_1)$, which are disjoint from each other except for u and from G_1 except for the v_i. All edges incident with inner points of these paths have $\alpha(f) = 0$, and hence all faces F incident with inner points of these paths have the same $\pi_{e^*}(F) = c$. On the other hand, each v_i must be incident with an edge with $\alpha(f) \neq 0$, and hence also with a face F_i with $\pi_{e^*}(F_i) \neq c$. We may assume by symmetry that $\pi_{e^*}(F_1) < c$ and $\pi_{e^*}(F_2) < c$. But then $D(c)$ passes through v_1 and v_2, which belong to the same component of G_0, a contradiction.

Essentially the same argument shows that (iii) *no two edges of* G_0 *form a corner.*

We apply Lemma 6.1. Clearly $b_{h_e} = b_{t_e} = 0$. We claim that

$$b_v \leq 2 \tag{39}$$

for every other node. Suppose not, then there are four edges e_1, e_2, e_3, e_4 incident with v in this clockwise order, so that $h_{e_1} = h_{e_3} = v$ and $t_{e_2} = t_{e_3} = v$. Choose these edges so that as many of them as possible belong to G_1. Then (iv) *no two consecutive edges of these four can be in* G_0; indeed, by (iii) there would be an edge f of G_1 between them, and we could replace one of them by f.

First, suppose that all four of these edges belong to G_1. We may assume that there is no edge of G_1 between e_1 and e_2, nor between e_3 and e_4. Then all faces F between e_1 and e_2 have the same $\pi_{e^*}(F) = c$, and all faces F between e_3 and e_4 have the same $\pi_{e^*}(F) = c'$. Let (say) $c < c'$, then $D(c)$ passes through v twice, which is impossible.

Second, suppose that $e_1 \in E(G_0)$. Then $e_2, e_4 \in E(G_1)$ by (iv). The edge f that follows e_1 in the clockwise order around v must be in G_1 by (iii). This edge cannot be directed into v, since then we could replace e_1 by it and decrease the number of G_0-edges among the four. So we can replace e_2 by f. Thus we may assume that e_1 and e_2 form a corner, and similarly for e_1 and e_4. Let $\pi_{e^*}(l(e_1)) = \pi_{e^*}(l(e_1)) = c$, then $\pi_{e^*}(r(e_2)) > c$ and $\pi_{e^*}(rl(e_4)) < c$. Let $u = t(e_1)$, then (ii) implies that there is a face F with $\pi_{e^*}(F) \neq c$ incident with u. Let, say, $\pi_{e^*}(F) < c$, then $D(c)$ passes through both endpoints of e_1, which contradicts $(*)$. This completes the proof of (39).

A similar argument shows that

$$a_F \leq 2 \tag{40}$$

for every face. So substituting in Lemma 6.1 yields $-4 \geq 4g - 4$, or $g \leq 0$, a contradiction.

Proof of Theorem 7.1. The assertion that every node is in the center of gravity of its neighbors is just a restatement of the fact that every analytic function is harmonic. This also shows that (assuming that that f gives an embedding) no face can have a concave angle, and so the faces are convex polygons. So the main step is to show that f defines an embedding.

We start with observing that the image of every edge of G is a segment of length at most c, and for any two nodes u and v of G (adjacent or not) the distance of $f(u)$ and $f(v)$ is at least $1/c$.

We start with some lemmas.

LEMMA 9.2. *The image of G is not contained in a single line.*

PROOF. Suppose that a line ℓ, say the x-axis, contains the image of G. Let $V' = \{v \in V(G) : f(v) \geq 0$ and $V'' = \{v \in V(G) : f(v) < 0\}$. It follows from (25) that $f(V)$ is a discrete set, and the fact that each node is mapped to the center of gravity of its neighbors implies that both these sets must be infinite. It is not hard to see that there must be infinitely many disjoint edges $u_k v_k$ ($k = 1, 2, \ldots$) of G connecting V' and V''. But then $|f(u_k) - f(v_k)|$ could not remain bounded, which contradicts (25). □

LEMMA 9.3. *Let H be an open halfplane an v a node of G such that $f(v) \in H$. Then there exists an infinite path $P_v = (v_0 = v, v_1, \ldots$ starting at v such that $f(P_v) \subset H$ and the distance of $f(v_k)$ from the boundary of H tends to ∞.*

PROOF. Let G' be the subgraph spanned by those nodes that can be reached from v on a path Q such that $f(Q)$ lies in H. It suffices to show that the distance of points of $f(G')$ from the boundary of H is unbounded, since then the existence of P_v follows by simple compactness.

Suppose that $f(G')$ lies in a strip S of finite width, and let U denote the connected component of $\mathbf{C} \setminus f(G')$ containing the halfplane $H \setminus S$. The boundary of U is a polygon, whose vertices are points $f(i)$, $i \in V(G')$ and intersection points of images of edges. Neither type of vertices can give a concave angle, and so U is convex; since U contains a halfplane, it follows that U is a halfplane. So its boundary is a line ℓ.

Clearly G has nodes whose image is on ℓ (infinitely many such nodes, in fact). Furthermore, if a node v of G is mapped onto ℓ, then so are all its neighbors (otherwise, v could not be their center of gravity), so it follows that all nodes of G are mapped onto ℓ. But this contradicts Lemma 9.2. □

LEMMA 9.4. *For every (open or closed) halfplane H, the set $S = \{i \in V(G) : f(i) \in H\}$ induces a connected subgraph.*

PROOF. We may assume that H is the halfplane $\{y \geq 0\}$. Let u and v be two nodes in S, we want to show that they can be connected by a path whose image stays in the halfplane H. Consider any path P in G connecting u and v. We may assume that P is not just an edge, and that all the inner nodes of this path are outside H. Let w be a node on P which is lowest.

By Lemma 9.3, we can find three infinite paths P_u, P_v and P_w, starting at u, v and w, respectively, such that $f(P_u)$ and $f(P_v)$ lie in the upper halfplane, $f(P_w)$ lies in the lower halfplane, and the distance from the x-axis tends to infinity along each of these paths. Clearly P_w is node-disjoint from P_u and P_v. If P_u and P_v intersect, then the conclusion is trivial, so assume that they are node-disjoint.

Consider $P \cup P_u \cup P_v \cup P_w$ in the original planar embedding of G. This subgraph splits the plane into three infinite regions; let Ω be the region bounded by $P_u \cup P \cup P_v$. It is easy to see that there are infinitely many disjoint paths $Q_1, Q_2, \ldots$ connecting P_u and P_v inside Ω. We claim that if k is large enough, the image of Q_k must stay in H, proving that u and v can be connected by a path in S.

Let D denote the diameter of $f(P)$. Let u_1 be the last node on the path P_u such that the distance of $f(u_1)$ from the x-axis is at most $D + 2c^3$, let P'_u be the piece of P_u between u and u_1, and let $P''_u = P_u \setminus P'_u$. We define P'_v, P'_w etc. analogously.

Suppose that $f(P''_u)$ intersects $f(P''_v)$, say edge ij of $f(P''_u)$ intersects edge ab of $f(P''_v)$. By (25), the length of the image of any edge is at most c, so $|f(i)-f(a)| \leq 2c$, and so $d_G(i,a) \le 2c^2$. Thus there exists a path R of length at most $2c^2$ in G connecting i to a. Again by (25), the diameter of $f(R)$ is at most $2c^3$. By the definition of P''_u, the distance of $f(i)$ from the x-axis is more than $2c^3$, so $f(R)$ cannot cross the x-axis. It follows that u and v can be connected by a path whose image stays in the upper halfplane, using paths P_u, R, and P_v.

So we may assume that $f(P''_u)$ and $f(P''_v)$ are disjoint. Let T be the set of all nodes in G at a graph-distance at most c_2 from $P \cup P'_u \cup P'_v \cup P'_w$. Since T is a finite set, there is a k for which Q_k does not intersect T. By (25) we get that for every node s of Q_k and every node t of P,

$$|f(s) - f(t)| \ge \frac{d_G(s,t)}{c} \ge \frac{c_1}{c} = c_2.$$

In particular, $f(s)$ cannot be in the convex hull of $f(P)$.

If $f(Q_k)$ does not intersect the lower halfplane, then we are done. Suppose it does, then either it intersects $f(P'_w)$ or else it contains a subpath Q'_k such that $f(Q'_k)$ lies in the upper halfplane and intersects both $f(P''_u)$ and $f(P''_v)$.

Suppose that $f(Q_k)$ intersects $f(P_w)$. Similarly as above, we find a path R of length at most $2c^2$ in G connecting a node a on P_w to a node i on Q_k. This path must intersect the path $P \cup P_u \cup P_v$ at some node z; this means that $d_G(z,a) \le 2c^2$, and so $|f(z) - f(a)| \le 2c^3$. But $f(z)$ is either in the upper halfplane or at a distance at most D from it, and so $f(a)$ is at a distance at most $2c^3 + D$ from the upper halfplane. So $a \in V(P'_u)$ and hence $i \in T$, a contradiction, since Q_k avoids T.

Finally, if there is a path Q such that $f(Q)$ lies in the upper halfplane and intersects both $f(P''_u)$ and $f(P''_v)$, then similarly as above, we find two paths R_u and R_v of length at most $2c^2$ connecting Q to P''_u and P''_v, respectively. Similarly as above, these paths must stay in the upper halfplane, and so again we find that u and v can be connected by a path staying in the upper halfplane through P_u, R_u, Q, R_v and P_v. □

Now we turn to the proof that f defines an embedding. Let us triangulate each face of G arbitrarily, to get a new graph G_1. Let us draw the images of these new edges as straight segments. We claim that even with these new edges, f defines an embedding.

It is enough to show that (a) the images of every triangular face is a triangle (it does not degenerate), (b) two triangular faces of G_1 sharing an edge xy are mapped onto triangles on different sides of the line $f(x)f(y)$, and (c) the images of triangular faces incident with the same node x cover a neighborhood of $f(x)$ exactly once. We describe the proof of (b); the proof of (a) and (c) is similar.

So suppose that xyz and xyw are two triangular faces of G_1, and that $f(z)$ and $f(w)$ are on the same side of the line ℓ through $f(x)$ and $f(y)$, say on the right side. By Lemma 9.4, there is a path P in G connecting z and w whose image under f stays on the right side of ℓ. Since x is mapped to the center of gravity of its neighbors, there is a node x' adjacent to x in G such that $f(x')$ lies on the left side of ℓ, and similarly, y has a neighbor y' such that $f(y')$ lies on the left side of ℓ. Again by Lemma 9.4, there is a path Q in G connecting x' to y' such that the image of Q stays on the left side of ℓ. Extend Q with the edges xx' and yy' to get a path Q'.

Now obviously P and Q' are node-disjoint paths. But if we consider them in the planar embedding of G, it is clear that they must cross each other. This contradiction proves the theorem.

References

[1] E. Andre'ev: On convex polyhedra in Lobachevsky spaces, *Mat. Sbornik*, Nov. Ser. **81** (1970), 445–478.

[2] I. Benjamini and L. Lovász: Global Information from Local Observation, *Proc. 43rd Ann. Symp. on Found. of Comp. Sci.* (2002), 701–710.

[3] I. Benjamini and L. Lovász: Harmonic and analytic frunctions on graphs, *Journal of Geometry* **76** (2003), 3–15.

[4] I. Benjamini and O. Schramm: Harmonic functions on planar and almost planar graphs and manifolds, via circle packings, *Invent. Math.* **126** (1996), 565–587.

[5] R.L. Brooks, C.A.B. Smith, A.H. Stone, W.T. Tutte: The dissection of rectangles into squares, *Duke Math. J.* **7** (1940), 312–340.

[6] A.K. Chandra, P. Raghavan, W.L. Ruzzo, R. Smolensky and P. Tiwari: The electrical resistance of a graph captures its commute and cover times, *Proc. 21st ACM STOC* (1989), 574–586.

[7] P.D. Doyle and J.L. Snell: *Random walks and electric networks,* Math. Assoc. of Amer., Washington, D.C. 1984.

[8] R.J. Duffin: Basic properties of discrete analytic functions, *Duke Math. J.* **23** (1956), 335–363.

[9] R.J. Duffin: Potential theory on the rhombic lattice, *J. Comb. Theory* **5** (1968) 258–272.

[10] R.J. Duffin and E.L. Peterson: The discrete analogue of a class of entire functions, *J. Math. Anal. Appl.* **21** (1968) 619–642.

[11] J. Ferrand: Fonctions préharmoniques et fonctions préholomorphes, *Bull. Sci. Math.* **68** (1944), 152–180.

[12] R. Isaacs, Monodiffric functions, *Natl. Bureau Standards App. Math. Series* **18** (1952), 257–266.

[13] R. Kenyon: The Laplacian and Dirac operators on critical planar graphs, *Inventiones Math* **150** (2002), 409-439.

[14] R. Kenyon and J.-M. Schlenker: Rhombic embeddings of planar graphs with faces of degree 4 (math-ph/0305057).

[15] P. Koebe: Kontaktprobleme der konformen Abbildung, *Berichte über die Verhandlungen d. Sächs. Akad. d. Wiss.*, Math.–Phys. Klasse, **88** (1936) 141–164.

[16] R. Kenyon and J.-M. Schlenker: Rhombic embeddings of planar graphs (2003) (preprint)

[17] L. Lovász and A. Schrijver (unpublished)

[18] C. Mercat: Discrete Riemann surfaces and the Ising model, *Comm. Math. Phys.* **218** (2001), 177–216.

[19] C. Mercat: Discrete polynomials and discrete holomorphic approximation (math-ph/0206041).

[20] C. Mercat: Exponentials form a basis for discrete holomorphic functions (math-ph/0210016).

[21] B. Mohar and C. Thomassen: Graphs on surfaces. Johns Hopkins Studies in the Mathematical Sciences. Johns Hopkins University Press, Baltimore, MD, 2001.

[22] C. St.J. A. Nash-Williams: Random walks and electric currents in networks, *Proc. Cambridge Phil. Soc.* **55** (1959), 181–194.

[23] P.M. Soardi: *Potential Theory on Infinite Networks*, Lecture notes in Math. **1590**, Springer-Verlag, Berlin–Heidelberg, 1994.

[24] W.P. Thurston: *Three-dimensional Geometry and Topology*, Princeton Mathematical Series **35**, Princeton University Press, Princeton, NJ, 1997.

[25] W.T. Tutte: How to draw a graph, *Proc. London Math. Soc.* **13** (1963), 743-768.

[26] D. Zeilberger: A new basis for discrete analytic polynomials, *J. Austral. Math. Soc.* Ser. A **23** (1977), 95–104.

[27] D. Zeilberger: A new approach to the theory of discrete analytic functions, *J. Math. Anal. Appl.* **57** (1977), 350–367.

[28] D. Zeilberger and H. Dym: Further properties of discrete analytic functions, *J. Math. Anal. Appl.* **58** (1977), 405–418.

Microsoft Research, One Microsoft Way, Redmond, WA 98052, USA

Conformal properties in classical minimal surface theory

William H. Meeks III and Joaquín Pérez

Abstract. This is a survey of recent developments in the classical theory of minimal surfaces in $\mathbb{R}^3$ with an emphasis on the conformal properties of these surfaces such as recurrence and parabolicity. We cover the maximum principle at infinity for properly immersed minimal surfaces in $\mathbb{R}^3$ and some new results on harmonic functions as they relate to the classical theory. We define and demonstrate the usefulness of universal superharmonic functions. We present the compactness and regularity theory of Colding and Minicozzi for limits of sequences of simply connected minimal surfaces and its application by Meeks and Rosenberg in their proof of the uniqueness of the helicoid. Finally, we discuss some recent deep results on the topology and on the index of the stability operator of properly embedded minimal surfaces and give an application of the classical Shiffman Jacobi function to the classification of minimal surfaces of genus zero.

Contents

2000 *Mathematics Subject Classification.* Primary 53A10, Secondary 49Q05, 53C42.

Key words and phrases. Minimal surface, conformal structure, harmonic function, recurrence, parabolicity, harmonic measure, universal superharmonic function, Jacobi function, stability, index of stability, curvature estimates, maximum principle at infinity, limit tangent plane at infinity, minimal lamination, locally simply connected, blow-up on the scale of topology, parking garage, minimal planar domain.

Research of the first author was partially supported by NSF grant DMS - 0104044 and NSF DMS 9803206.

Research of the second author was partially supported by a MCYT/FEDER grant no. BFM2001-3318.

1. Introduction.

In this report on recent developments in the classical theory of minimal surfaces, we will focus on some of the spectacular progress that the subject has experienced in the past decade. The theme here is to explore relationships between the function theory and the conformal structure of complete embedded minimal surfaces. Embeddedness influences both local and global properties of these surfaces. This is because embeddedness allows one to use the minimal surface as a barrier against itself to construct stable minimal surfaces in its two complements in $\mathbb{R}^3$, see Subsection 4.2. These stable minimal surfaces act as guide posts for deciphering the structure and geometry of the surfaces and their complements.

The recent ground breaking work by Colding and Minicozzi in [**24, 18, 17, 19, 23, 14**] has been especially influential. Their theorems have been essential for recent progress made in understanding the local and global structure of embedded minimal surfaces which are simply connected or which have finite genus. Their results on the compactness and (partial) regularity of limits of sequences of uniformly locally simply connected minimal surfaces in Riemannian three-manifolds is a major story which we will only briefly touch on here; we refer the reader to our forthcoming survey [**73**] for a more complete discussion on the Colding-Minicozzi theory and its applications.

We now briefly outline the material of the survey. In Section 2 we cover most of the basic results in minimal surface theory. This Section offers a quick introduction to the main definitions, examples and classical results, and should provide the needed background to beginners in the subject. In Section 3 we go into some of the recent advances on the conformal structure of minimal surfaces with boundary. Section 4 is devoted to the important classical results on stable minimal surfaces. We include here a proof of the beautiful estimate of Colding-Minicozzi on the area of a stable minimal disk, a result on which we base the proofs of the other main theorems of the Section. In Section 5 we study properly embedded minimal surfaces with more than one end and discuss the basic ordering theorem of the ends of such surfaces. In Section 6 we explain how universal superharmonic functions (a concept developed in Section 3) can be used to obtain quadratic area growth estimates for the middle ends of properly embedded minimal surfaces, a result which implies that such surfaces can have at most two limit ends; this is a theorem of Collin, Kusner, Meeks and Rosenberg [**26**]. Another important application of universal superharmonic functions in this Section shows that when the minimal surface has exactly two limit ends, then it is recurrent for Brownian motion. In Section 7 we outline the recent proof of the general maximum principle at infinity for properly immersed minimal surfaces by Meeks and Rosenberg. Previous maximum principles at infinity have had a unifying effect on the theory and this final version will likely play a similar important role. A deep application of this principle appears in the proof that a properly embedded minimal surface in $\mathbb{R}^3$ with absolute Gaussian curvature at most 1 has an open regular neighborhood of radius 1 and so the surface has cubical area growth, which we explain as well in this Section. In Section 8 we discuss several Theorems and Conjectures related to flux, including for a properly embedded minimal surface the Flux Conjecture of Meeks and Rosenberg and a related flux conjecture for harmonic functions on parabolic Riemannian manifolds. In Section 9 we cover the aforementioned results of Colding-Minicozzi and the application of their theory to give a sketch of their alternative proof of Collin's Theorem that states that a properly embedded minimal surface of finite topology and at least two ends has finite total curvature. We also explain the important technique of blowing-up a sequence of embedded minimal surfaces on the scale of topology, and how this procedure yields a local-global parking garage picture for any embedded genus zero minimal surface in a neighborhood of a point of concentrated topology and more concentrated curvature (this technique was first developed as a tool in [**74**]). The existence and theory of minimal surfaces with a periodic parking garage structure with a finite number of columns appears in [**124**]. An important uniqueness for certain parking garage structures was recently given by Weber and Wolf [**126**], who proved that the columns of these parking garages project to the zeros of Hermite polynomials on the real line; this uniqueness result plays a crucial role in their proposed existence proof of an embedded genus g helicoid for every positive integer g. In Section 10 we cover some topological aspects of properly embedded minimal surfaces. These results include a discussion of the recent Topological Classification Theorem for Minimal Surfaces by Frohman and Meeks [**38**] and the recent topological obstructions of Meeks, Pérez and Ros [**75, 76**] for properly embedded minimal surfaces of finite genus. A particular consequence of these new topological obstructions is that every properly embedded minimal surface of finite genus in $\mathbb{R}^3$ is recurrent for Brownian motion. Another important theoretical consequence of these topological obstructions is that a properly embedded minimal surface in $\mathbb{R}^3$ of finite topology, at least two ends and with a given bound on its genus, has a related bound on its index of stability. We also sketch the proof of the recent Theorem by Meeks and Rosenberg [**80**] of the uniqueness of the plane and helicoid as the only properly embedded simply connected minimal surfaces in $\mathbb{R}^3$, including their general classification Theorem for minimal laminations of $\mathbb{R}^3$ that plays an important theoretical role in many recent advances in the theory. In the final Section 11, we give some partial results on the Genus Zero Conjecture. This Conjecture asserts that a properly embedded minimal surface of genus zero is a plane, a helicoid, a catenoid or an example in the 1-parameter family of minimal surfaces called the Riemann minimal examples (see Subsection 2.5), which are foliated by circles and straight lines in horizontal planes. The solution of this Conjecture reduces to a study of bounded Jacobi functions on properly embedded minimal surfaces M of genus zero with two limit ends. In this Section we define and study the classical Shiffman Jacobi function which is defined on such an M in order to obtain some partial results on the Genus Zero Conjecture.

The authors would like to thank the referee for helpful comments.

2. Basic results in classical minimal surface theory.

We will devote this Section to give a fast tour through the foundations of the theory, enough to understand and apply the results to be explained in future Sections.

2.1. Definitions and Theorems on equivalent properties. One can define a minimal surface from different points of view. The equivalences between these starting points give insight into the richness of the classical theory of minimal surfaces and its connections with other branches of Mathematics.

DEFINITION 1. Let $X\colon M \to \mathbb{R}^3$ be an isometric immersion of a Riemannian surface into space. X is said to be *minimal* if the coordinate functions x_1, x_2, x_3 are harmonic functions on M.

Very often, it is useful to identify a Riemannian surface M with its image under an isometric embedding. Since harmonicity is a local concept, the notion of minimality can be applied to a surface $M \subset \mathbb{R}^3$ (with the underlying induced Riemannian structure by the inclusion). Let H be the mean curvature function of X and $N\colon M \to \mathbb{S}^2 \subset \mathbb{R}^3$ its Gauss map[1]. The well-known formula $\Delta X = 2HN$, valid for an isometric immersion $X\colon M \to \mathbb{R}^3$, leads us to the following equivalent definition of minimality.

DEFINITION 2. A surface $M \subset \mathbb{R}^3$ is *minimal* if and only if its mean curvature vanishes identically.

Recall that any (regular) surface can be locally expressed as the graph of a function $u = u(x,y)$. The condition on the mean curvature to vanish identically can be expressed as a quasilinear elliptic second order partial differential equation,

$$(1) \qquad (1+u_x^2)u_{yy} - 2u_x u_y u_{xy} + (1+u_y^2)u_{xx} = 0.$$

DEFINITION 3. A surface $M \subset \mathbb{R}^3$ is *minimal* if and only if it can be locally expressed as the graph of a solution of the equation (1).

Let Ω be a relatively compact subdomain in a surface $M \subset \mathbb{R}^3$. If we perturb normally the inclusion map on Ω by a compactly supported smooth function $u \in C_0^\infty(\Omega)$, then $X + tuN$ is again an immersion for any $|t| < \varepsilon$ with ε sufficiently small. The mean curvature function of M is closely related to the infinitesimal variation of the area for compactly supported normal variations by means of the *first variation of area* (see for instance [**97**]):

$$(2) \qquad A'(0) = \left.\frac{d}{dt}\right|_{t=0} \text{Area}((X+tuN)(\Omega)) = -2\int_\Omega uH\, dA,$$

where dA stands for the area element of M. This variational formula lets us state a third equivalent definition of minimality.

DEFINITION 4. A surface $M \subset \mathbb{R}^3$ is *minimal* if and only if it is a critical point of the *area functional* for all compactly supported variations.

In fact, a consequence of the *second variation of area* (Subsection 2.8) is that any point in a minimal surface has a neighborhood with least area relative to its boundary. This property justifies the word "minimal" for this kind of surfaces. It

[1]Throughout the paper, all surfaces will be assumed to be *orientable*.

should be noted that the global minimization of area on any compact subdomain is a strong condition for a complete orientable minimal surface to satisfy; in fact, it forces the surface to be a plane (Theorem 8).

DEFINITION 5. A surface $M \subset \mathbb{R}^3$ is *minimal* if and only if every point $p \in M$ has a neighborhood with least area relative to its boundary.

Definitions 4 and 5 establish minimal surfaces as the 2-dimensional analogy to geodesics in Riemannian Geometry, and connect the theory of minimal surfaces with one of the classical important branches of Mathematics: the Calculus of Variations. Besides the area functional A, another well-known functional in the Calculus of Variations is the *Dirichlet energy,*

$$E = \int_{\Omega} |\nabla X|^2 dA,$$

where again $X : M \to \mathbb{R}^3$ is an isometric immersion and $\Omega \subset M$ is a subdomain with compact closure. These functionals are related by the inequality $E \geq 2A$, with equality if and only if the immersion $X : M \to \mathbb{R}^3$ is conformal. The classical formula $K - e^{2u}\overline{K} = \Delta u$ that relates the Gaussian curvature functions $K, \overline{K}$ for two conformally related metrics $g, \overline{g}$ on a 2-dimensional manifold (Δ stands for the Laplacian with respect to g) together with the existence of solutions of the Laplace equation $\Delta u = K$ for a relatively compact subdomain in a Riemannian manifold, guarantee the existence of local *isothermal* or *conformal coordinates* for any 2-dimensional Riemannian manifold, modeled on domains of $\mathbb{C}$. The relation between area and energy together with the existence of isothermal coordinates, allow us to give two further characterizations of minimality.

DEFINITION 6. A conformal immersion $X : M \to \mathbb{R}^3$ is *minimal* if and only if it is a critical point of the Dirichlet energy for compactly supported variations, or equivalently if any point $p \in M$ has a neighborhood with least energy relative to its boundary.

From a physical point of view, the mean curvature function of a homogeneous membrane separating two media is equal, up to a nonzero multiplicative constant, to the difference between the pressures on the two sides of the surface. When this pressure difference is zero, then the membrane has zero mean curvature. Therefore, soap films in space are physical realizations of the ideal concept of a minimal surface.

DEFINITION 7. A surface $M \subset \mathbb{R}^3$ is *minimal* if and only if every point $p \in M$ has a neighborhood D_p which is equal to the unique idealized soap film with boundary ∂D_p.

If $N : M \to \mathbb{S}^2$ is the Gauss map of M, then the tangent space T_pM of M at $p \in M$ identifies as subspace of $\mathbb{R}^3$ under parallel translation with the tangent space $T_{N(p)}\mathbb{S}^2$ to the sphere at $N(p)$, from where one can view the differential $A_p = -dN_p$ as an endomorphism of T_pM, called the *shape operator.* A_p is a symmetric linear transformation whose orthogonal eigenvectors are the *principal directions* of M at p, and the corresponding eigenvalues are the *principal curvatures* of M at p. Since the mean curvature function H of M equals the arithmetic mean of such principal curvatures, we deduce that minimality reduces to the expression

$$A_p = -dN_p = \begin{pmatrix} a & b \\ b & -a \end{pmatrix}$$

in an orthonormal tangent basis. After identification of N with its stereographic projection, the Cauchy-Riemann equations give the next last characterization of minimality.

DEFINITION 8. A surface $M \subset \mathbb{R}^3$ is *minimal* if and only if its stereographically projected Gauss map $g\colon M \to \mathbb{C} \cup \{\infty\}$ is a meromorphic function.

Definition 1 and the maximum principle for harmonic functions imply that no compact minimal surfaces in $\mathbb{R}^3$ exist. Although the study of compact minimal surfaces with boundary has been intensively developed and dates back to famous problems as the well-known *Plateau Problem,* in this survey we will focus on the study of *complete* minimal surfaces (possibly with boundary), in the sense that all geodesics can be indefinitely extended up to the boundary of the surface. A stronger global hypothesis, whose relationship with completeness is an active field of research in minimal surface theory, is presented in the following definition.

DEFINITION 9. A map $f\colon X \to Y$ between topological spaces is *proper* if $f^{-1}(C)$ is compact in X for any compact set $C \subset Y$. A minimal surface $M \subset \mathbb{R}^3$ is proper when the inclusion map is proper.

The Gaussian curvature function K of a surface $M \subset \mathbb{R}^3$ is the product of its principal curvatures. If M is minimal, then its principal curvatures are oppositely signed and thus, K is nonpositive. Another interpretation of K is the determinant of the shape operator A_p, or equivalently $|K|$ is the absolute value of the Jacobian for the Gauss map N. Adding up the curvature at all points of M (note that this integral may be $-\infty$ or a nonpositive number) we will obtain the same quantity as when computing the negative of the spherical area of M through its Gauss map, counting multiplicities. This quantity is called the *total curvature* of the minimal surface:

$$C(M) = \int_M K\, dA = -\text{Area}(N\colon M \to \mathbb{S}^2). \tag{3}$$

2.2. Weierstrass Representation. Recall that the Gauss map of a minimal surface M can be viewed as a meromorphic function on the underlying Riemann surface. Furthermore, the harmonicity of the third coordinate function x_3 lets us define (at least locally) its harmonic conjugate function x_3^*; hence, the so called *height differential*[2] $dh = dx_3 + i dx_3^*$ is a holomorphic differential on M. The pair (g, dh) is usually referred to as the *Weierstrass data* of the minimal surface, and the minimal immersion $X\colon M \to \mathbb{R}^3$ can be expressed up to translations solely in terms of these data as

$$X(p) = \Re \int_{p_0}^{p} \left(\frac{1}{2}\left(\frac{1}{g} - g\right), \frac{i}{2}\left(\frac{1}{g} + g\right), 1 \right) dh, \tag{4}$$

where $\Re$ stands for real part [**58, 99**]. The pair (g, dh) satisfies certain compatibility conditions, stated in assertions *i), ii)* below. The key point is that the procedure has the following converse, which gives a cook-book type recipe for analytically defining a minimal surface.

THEOREM 1 (Osserman [**98**]). *Let M be a Riemann surface, $g\colon M \to \mathbb{C} \cup \{\infty\}$ a meromorphic function and dh a holomorphic one-form on M. Assume that:*

[2] Note that the height differential might not be exact since x_3^* needs not to be globally well-defined on M. Nevertheless, the notation dh is commonly accepted and we will also make use of it here.

i) The zeros of dh coincide with the poles and zeros of g, with the same order.
ii) For any closed curve $\gamma \subset M$,

$$\overline{\int_\gamma g\, dh} = \int_\gamma \frac{dh}{g}, \qquad \Re \int_\gamma dh = 0. \tag{5}$$

Then, the map $X\colon M \to \mathbb{R}^3$ given by (4) is a conformal minimal immersion with Weierstrass data (g, dh).

Condition *i)* above expresses the non degeneracy of the induced metric by X on M, so by replacing it with the condition that the zeros and poles of g coincide with the zeros of dh with at most the same order, we allow the conformal X to be a *branched minimal surface.* Condition *ii)* deals with the independence of (4) on the integration path, and it is usually called the *period problem.* By Cauchy's Theorem, it suffices to consider the period problem on homology classes in M.

All geometric invariants of a minimal surface M can be expressed in terms of its Weierstrass data. For instance, the first and second fundamental forms are respectively (see [**44, 99**]):

$$ds^2 = \left(\frac{1}{2}(|g| + |g|^{-1})|dh|\right)^2, \qquad II(v,v) = \Re\left(\frac{dg}{g}(v)\cdot dh(v)\right), \tag{6}$$

where v is a tangent vector to M, and the Gaussian curvature is

$$K = -\left(\frac{4\,|dg/g|}{(|g| + |g|^{-1})^2|dh|}\right)^2. \tag{7}$$

If (g, dh) is the Weierstrass pair of a minimal surface $X : M \to \mathbb{R}^3$, then for each $\lambda > 0$ the pair $(\lambda g, dh)$ satisfies condition *i)* of Theorem 1 and the second equation in (5). The first equation in (5) holds for these new Weierstrass data if and only if $\int_\gamma g dh = \int_\gamma \frac{dh}{g} = 0$ for all homology classes γ in M, which in turn is equivalent to the fact that the *flux of M along γ* is vertical for all such γ. In general, the flux vector is defined as

$$F(\gamma) = \int_\gamma (\nabla x_1, \nabla x_2, \nabla x_3) = \Im \int_\gamma \left(\frac{1}{2}\left(\frac{1}{g} - g\right), \frac{i}{2}\left(\frac{1}{g} + g\right), 1\right) dh, \tag{8}$$

where $\Im$ stands for imaginary part. Thus, for a minimal surface X with vertical flux, the Weierstrass data $(\lambda g, dh)$ produce a well-defined minimal surface $X_\lambda\colon M \to \mathbb{R}^3$. The family $\{X_\lambda\}_\lambda$ is a smooth deformation of $X_1 = X$, called the *López-Ros deformation.* Clearly, the conformal structure, height differential and the set of points in M with vertical normal vector are preserved throughout this deformation. Another important property of this deformation is that if a component of a horizontal section of X is convex, then the same holds for the related component at the same height for any X_λ, $\lambda > 0$.

2.3. Minimal surfaces with finite total curvature. Among the family of complete minimal surfaces in space, those with *finite total curvature* have been extensively studied. The principal reason for this is that they can be thought of as compact algebraic objects in a natural sense, which opens tremendously the number and depth of tools that can be applied to these kinds of surfaces.

THEOREM 2 (Huber [**52**], Osserman [**99**]). *Let $M \subset \mathbb{R}^3$ be a complete oriented immersed minimal surface with finite total curvature. Then,*

i) M is conformally a compact Riemann surface $\overline{M}$ with a finite number of points removed (called the ends of M).

ii) The Weierstrass data (g, dh) extend meromorphically to $\overline{M}$. In particular, the total curvature of M is a multiple of -4π.

In this setting, the Gauss map g has a well-defined finite degree on $\overline{M}$. A direct consequence of (3) is that the total curvature of an M as in Theorem 2 is -4π times the degree of its Gauss map g. It turns out that this degree can be computed in terms of the genus of the compactification $\overline{M}$ and the number of ends by means of the *Jorge-Meeks formula* [**53**]. Rather that stating here this general formula for an immersed surface M as in Theorem 2, we will emphasize the particular case when all the ends of M are embedded:

$$\deg(g) = \text{genus}(\overline{M}) + \#(\text{ends}) - 1.$$

The asymptotic behavior of a complete embedded minimal surface with finite total curvature is well understood. Schoen [**118**] demonstrated that each embedded end of a complete minimal surface with finite total curvature can be parametrized as a graph over the exterior of a disk in the (x_1, x_2)-plane with height function

$$x_3(x_1, x_2) = a \log r + b + \frac{c_1 x_1 + c_2 x_2}{r^2} + \mathcal{O}(r^{-2}), \tag{9}$$

where $r = \sqrt{x_1^2 + x_2^2}$, $a, b \in \mathbb{R}$ and $\mathcal{O}(r^{-2})$ denotes a function such that $r^2\mathcal{O}(r^{-2})$ is bounded as $r \to \infty$. The coefficient a in (9) is called the *logarithmic growth* of the end. When $a \neq 0$, the end is called a *catenoidal end;* if $a = 0$, we have a *planar end.* We use this language since a catenoidal end is asymptotic to one of the ends of a catenoid and a planar end is asymptotic to the end of a plane. In particular, complete embedded minimal surfaces with finite total curvature are always proper; in fact, an elementary analysis of the asymptotic behavior shows that the equivalence between completeness and properness still holds for immersed minimal surfaces with finite total curvature.

A key result, proved by Collin in 1997, reduces the study of properly embedded minimal surfaces with finite topology and at least two ends (see Subsection 2.7 for the general definition of end) to the family of surfaces with finite total curvature.

THEOREM 3 (Collin [**25**]). *If $M \subset \mathbb{R}^3$ is a properly embedded minimal surface with and more than one end, then each annular end of M is asymptotic to the end of a plane or a catenoid. In particular, if M has finite topology and more than one end, then M has finite total curvature.*

The understanding of generic properties of minimal surfaces with finite total curvature leads to existence and uniqueness results, as well as to study the moduli spaces of such surfaces with a fixed topology. Along these lines, Schoen [**118**] proved in 1983 that the catenoid is the unique complete immersed minimal surface with finite total curvature and two embedded ends. Eight years later, López and Ros [**63**] characterized the plane and the catenoid as the only complete embedded minimal surfaces in $\mathbb{R}^3$ with genus zero and finite total curvature. The most celebrated complete minimal surface with finite total curvature since the classical examples from the nineteenth century was discovered in 1982 by Costa [**27, 28**]. This is a thrice punctured torus with two catenoidal ends and one planar middle end. Costa demonstrated existence of this surface but only its embeddedness outside a ball in $\mathbb{R}^3$. Hoffman and Meeks [**48**] proved global embeddedness for the

Costa torus and generalized this example to any genus $g \geq 2$ and three ends [**49**]. Also Hoffman and Meeks found a 1-parameter deformation of the Costa surface and of all of the higher genus Hoffman-Meeks minimal surfaces, where the middle planar end becomes catenoidal (unpublished). In [**44**], Hoffman and Karcher showed that the Hoffman-Meeks deformation families of the Costa-Hoffman-Meeks surfaces exist for all parameter values and that for genus $g > 1$, the deformed surfaces are embedded (see Subsection 2.5 for explicit formulas). This embeddedness is equivalent to showing that the logarithmic growth rate of the middle end never catches up with that of the other ends. They also mention that the situation in the case $g = 1$ (for the deformation of the Costa torus) is slightly more delicate (and still true), because the aforementioned growth rate of the middle end achieves the growth rate of the extreme ends in the limit. They do not provide this last computation in [**44**]. Costa [**29, 30**] showed that any complete embedded minimal surface in $\mathbb{R}^3$ with genus one and three ends has to be either the Costa surface or lies in the Hoffman-Meeks deformation, with the moduli space of complete embedded minimal thrice punctured tori being diffeomorphic to a real interval. Concerning moduli spaces of minimal surfaces with finite total curvature and prescribed topology, Pérez and Ros [**106**] gave general conditions on the space $\mathcal{M}(g, r)$ whose elements are the complete embedded minimal surfaces with finite total curvature, genus g and r ends, to have a structure of real analytic manifold of dimension $r - 2$ around a given minimal surface $M \in \mathcal{M}(g, r)$. Such conditions are expressed in terms of the bounded Jacobi functions on M (see Subsection 2.8 for the definition of Jacobi function). They also identified the tangent space of $\mathcal{M}(g, r)$ at a minimal surface $M \in \mathcal{M}(g, r)$ with the set of Jacobi functions on M with at most logarithmic singularities at the ends. Other compactness results for moduli spaces of complete embedded minimal surfaces with finite total curvature have been given in Ros [**112**] and Traizet [**122**]. We will explain in Section 10.2 some further recent advances in this area, see specifically Theorem 29.

2.4. Periodic minimal surfaces. A properly embedded minimal surface M in $\mathbb{R}^3$ is called *singly, doubly* or *triply periodic* when it is invariant by a discrete infinite group G of isometries of $\mathbb{R}^3$ of rank $1, 2, 3$ (respectively) that acts properly and discontinuously. Very often, it is useful to study such an M as a minimal surface in the complete flat three manifold $\mathbb{R}^3/G$. Up to finite coverings, these 3-manifolds reduce to $\mathbb{R}^3/T$, $\mathbb{R}^3/S_\theta$, $\mathbb{T}^2 \times \mathbb{R}$ and $\mathbb{T}^3$, where T denotes a nontrivial translation, S_θ is the screw motion symmetry resulting from the composition of a rotation of angle θ around the x_3-axis with a translation in the direction of this axis, and $\mathbb{T}^2, \mathbb{T}^3$ are flat tori of dimensions 2 and 3 obtained as quotients of $\mathbb{R}^2, \mathbb{R}^3$ by 2 or 3 linearly independent translations.

All known periodic minimal surfaces turn out to have finite total curvature (hence finite topology) when seen as surfaces in the corresponding $\mathbb{R}^3/G$. Meeks and Rosenberg [**81, 84**] developed the theory of periodic minimal surfaces. For instance, they obtained in this setting similar conclusions as the ones in Theorem 2, except that the Gauss map g of a minimal surface in $\mathbb{R}^3/G$ is not necessarily well-defined (the Gauss map does not descend to the quotient for surfaces in $\mathbb{R}^3/S_\theta$, $\theta \in (0, 2\pi)$, and in this case the role of the Gauss map g is played by the well-defined differential form dg/g). An important fact, due to Meeks and Rosenberg [**81, 84**], is that for properly embedded minimal surfaces in $\mathbb{R}^3/G$, $G \neq \{\text{identity}\}$, the conditions of finite total curvature and finite topology are equivalent[3]. Later Meeks [**72**] proved that every properly embedded minimal surface in $\mathbb{T}^2 \times \mathbb{R}$ has a finite number of ends, hence in this setting finite genus implies finite total curvature. Analogous to the Jorge-Meeks formula, Meeks and Rosenberg [**81, 84**] proved an explicit relation between the total curvature and the topology of such a surface. They also studied the asymptotic behavior of complete embedded minimal surfaces with finite total curvature in $\mathbb{R}^3/G$. In this setting, there are three possibilities: all ends must be simultaneously asymptotic to planes (as in the Riemann minimal examples, see Subsection 2.5), to halfplanes (as in the singly or doubly periodic Scherk minimal surfaces; for this reason, such ends are called *Scherk-type ends*) or to ends of helicoids (helicoidal type ends). Recently Meeks [**66**] proved that a properly embedded minimal surface in $\mathbb{R}^3/S_\theta$, $\theta \neq 0, \pi$, has a finite number of ends and if it has at least two ends, then the surface has at most quadratic area growth.

Concerning classification theorems for periodic minimal surfaces, Meeks, Pérez and Ros [**78**] proved that the classical Riemann minimal examples are the unique periodic nonsimply connected genus zero properly embedded minimal surfaces in $\mathbb{R}^3$. Lazard-Holly and Meeks [**59**] characterized the doubly periodic Scherk surfaces as the unique properly embedded examples of genus zero in $\mathbb{T}^2 \times \mathbb{R}$.

2.5. Examples of minimal surfaces. We will use the Weierstrass representation for introducing some of the most celebrated complete minimal surfaces.

THE PLANE. $M = \mathbb{C}$, $g(z) = 1$, $dh = dz$. It is the only flat minimal surface.

THE CATENOID. $M = \mathbb{C} - \{0\}$, $g(z) = z$, $dh = \frac{dz}{z}$. It has genus zero, two ends and total curvature -4π. Together with the plane, the catenoid is the only minimal surface of revolution (Bonnet [**4**]) and the unique complete embedded minimal surface with genus zero and finite total curvature (López and Ros [**63**]). Schoen [**118**] also characterized the catenoid as the unique complete immersed minimal surface with finite total curvature and two embedded ends.

THE HELICOID. $M = \mathbb{C}$, $g(z) = e^z$, $dh = i\,dz$. It has genus zero, one end and infinite total curvature. Together with the plane, the helicoid is the only ruled minimal surface (Catalan [**8**]) and the unique properly embedded simply connected minimal surface (Meeks and Rosenberg [**80**], see also Theorem 30 below). The vertical helicoid can be also seen as a genus zero surface with two ends in a quotient of $\mathbb{R}^3$ by a vertical translation or by a screw motion. The catenoid and the helicoid are *conjugate* minimal surfaces, in the sense that the coordinate functions of one of these surfaces are the harmonic conjugates of the coordinate functions of the other one; in this case, we consider the catenoid to be defined on its universal cover $e^z \colon \mathbb{C} \to \mathbb{C} - \{0\}$ in order for the harmonic conjugate of x_3 to be well-defined. Equivalently, both surfaces share the Gauss map e^z and their height differentials differ in multiplication by $i = \sqrt{-1}$.

THE ENNEPER SURFACE. $M = \mathbb{C}$, $g(z) = z$, $dh = z\,dz$. It is nonembedded, has genus zero, one end and total curvature -4π. The catenoid and the Enneper surface are the unique complete minimal surfaces in $\mathbb{R}^3$ with finite total curvature -4π (see [**99**]).

Given $k \in \mathbb{N}$, $k \geq 1$ and $a \in \mathbb{R} - \{0, -1\}$, we define the compact genus k surface $\overline{M}_{k,a} = \{(z, w) \in (\mathbb{C} \cup \{\infty\})^2 \mid w^{k+1} = \frac{(z+1)(z-a)}{z}\}$. Let $M_{k,a} = \overline{M}_{k,a} -$

[3]This equivalence does not hold for properly embedded minimal surfaces in $\mathbb{R}^3$, as demonstrates the helicoid.

$\{(-1,0),(\infty,\infty),(a,0)\}$ and

$$g_{k,a,m,A}(z,w) = A\frac{zw}{mz+1}, \qquad dh_{k,a,m} = \frac{mz+1}{(z+1)(z-a)}dz,$$

where $A \in \mathbb{R} - \{0\}$. Given $k \in \mathbb{N}$ and $a \in (0,\infty)$, there exist $m = m(a) \in \mathbb{R}$ and $A = A(a) \in \mathbb{R} - \{0\}$ such that the pair $(g_{k,a,m(a),A(a)}, dh_{k,a,m(a)})$ is the Weierstrass data of a well-defined minimal surface $X\colon M_{k,a} \to \mathbb{R}^3$ with genus k and three ends (Hoffman, Karcher [**44**]). Moreover, $m(1) = 0$ for any $k \in \mathbb{N}$. With this notation, we have the following examples.

THE COSTA TORUS. $M = M_{1,1}$, $g = g_{1,1,0,A(1)}$, $dh = dh_{1,1,0}$. Costa [**28**] proved existence of this surface, while its embeddedness is due to Hoffman and Meeks [**48**].

THE COSTA-HOFFMAN-MEEKS SURFACES. For any $k \geq 2$, take $M = M_{k,1}$, $g = g_{1,1,0,A(1)}$, $dh = dh_{1,1,0}$. Both existence and embeddedness were given by Hoffman and Meeks [**49**].

THE DEFORMATION OF THE COSTA TORUS. For any $a \in (0,\infty)$, take $M = M_{1,a}$, $g = g_{1,a,m(a),A(a)}$, $dh = dh_{1,a,m(a)}$ (when $a = 1$ we find the Costa torus). Hoffman and Karcher [**44**] proved existence of these surfaces. A complete proof of their embeddedness has not been published yet, see also the last paragraph of Subsection 2.3 and [**45**]. Costa [**29, 30**] showed that any complete embedded minimal torus with three ends must lie in this family.

THE DEFORMATION OF THE COSTA-HOFFMAN-MEEKS SURFACES. For any $k \geq 2$ and $a \in (0,\infty)$, take $M = M_{k,a}$, $g = g_{k,a,m(a),A(a)}$, $dh = dh_{k,a,m(a)}$. When $a = 1$ we find the Costa-Hoffman-Meeks surface of genus k and three ends. A complete proof of existence and embeddedness for these surfaces is given in [**44**] by Hoffman and Karcher.

THE SINGLY PERIODIC SCHERK SURFACES. $M = (\mathbb{C} \cup \{\infty\}) - \{\pm e^{\pm i\theta}\}$, $g(z) = z$, $dh = \frac{iz\,dz}{\prod(z \pm e^{\pm i\theta})}$, for fixed $\theta \in (0,\pi/4]$. Discovered (at least, the case $\theta = \pi/4$) by Scherk [**116**] in 1835, they form a 1-parameter family of genus zero surfaces in a quotient of $\mathbb{R}^3$ by a translation, with four ends. Each surface can be thought of geometrically as a desingularization of two vertical planes forming an angle of 2θ.

THE DOUBLY PERIODIC SCHERK SURFACES. $M = (\mathbb{C} \cup \{\infty\}) - \{\pm e^{\pm i\theta}\}$, $g(z) = z$, $dh = \frac{z\,dz}{\prod(z \pm e^{\pm i\theta})}$, where $\theta \in (0,\pi/4]$. These are the conjugate surfaces to the singly periodic Scherk surfaces, and can be thought of geometrically as the desingularization of two families of equally spaced vertical parallel halfplanes in opposite halfspaces, with the halfplanes in the upper family making an angle of 2θ with the halfplanes in the lower family. These surfaces are doubly periodic with genus zero in their corresponding quotient $\mathbb{T}^2 \times \mathbb{R}$, and were characterized by Lazard-Holly and Meeks [**59**] as the unique properly embedded minimal surfaces in $\mathbb{T}^2 \times \mathbb{R}$ with genus zero.

THE RIEMANN MINIMAL EXAMPLES. $M = \{(z,w) \in (\mathbb{C} \cup \{\infty\})^2 \mid w^2 = z(z-\lambda)(\lambda z+1)\} - \{(0,0),(\infty,\infty)\}$, $g(z,w) = z$, $dh = A_\lambda \frac{dz}{w}$, for each $\lambda > 0$, where A_λ is a nonzero complex number satisfying $A_\lambda^2 \in \mathbb{R}$. Discovered by Riemann (and posthumously published, Hattendorf and Riemann [**110, 111**]), these surfaces are invariant by a translation T_λ, and in the quotient space $\mathbb{R}^3/T_\lambda$ have genus one and two planar ends. The conjugate surface of the Riemann minimal example for a given $\lambda > 0$ is the Riemann minimal example for the parameter value $1/\lambda$ (the case $\lambda = 1$ gives the only self-conjugate surface in the family). Riemann minimal examples were characterized by Meeks, Pérez and Ros [**78**] as the unique periodic nonsimply connected genus zero properly embedded minimal surfaces in $\mathbb{R}^3$.

2.6. Monotonicity formula and classical maximum principles. As we will see in Section 6, the conformal type of a minimal surface is strongly related with its area growth in balls. The first result along these lines comes from the coarea formula applied to the distance function to a given point $p \in \mathbb{R}^3$. The following statement of the coarea formula appears in [**9**], see [**33**] for a more general version.

PROPOSITION 1. *Let Ω be a domain with compact closure in a Riemannian manifold M and $f : \overline{\Omega} \to \mathbb{R}$ a function in $C^0(\overline{\Omega}) \cap C^\infty(\Omega)$ with $f|_{\partial\Omega} = 0$. For any regular value t of $|f|$, we let $\Gamma(t) = |f|^{-1}(t)$ and $A(t) = \text{Area}(\Gamma(t))$. Then for any function $\phi \in L^1(\Omega)$ we have*

$$\int_\Omega \phi |\nabla f| \, dV = \int_0^\infty \left(\int_{\Gamma(t)} \phi \, dA_t \right) dt,$$

where ∇f is the gradient of f in M and dV, dA_t are respectively the volume elements in M and $\Gamma(t)$.

THEOREM 4 (Monotonicity formula [**20, 55**]). *Let $X : M \to \mathbb{R}^3$ be a properly immersed connected minimal surface. Given $p \in \mathbb{R}^3$, let $A(R)$ be the area of the portion of $X(M)$ inside a ball of radius $R > 0$ centered at p. Then, $A(R)R^{-2}$ is nondecreasing. In particular, $\lim_{R\to\infty} A(R)R^{-2} \geq \pi$ with equality if and only if M is a plane.*

One of the consequences of the fact that minimal surfaces can be viewed locally as solutions of the partial differential equation (1) is the validity of certain maximum principles for minimal surfaces. We will state them for minimal surfaces in $\mathbb{R}^3$, but they also hold when the ambient space is a complete flat three-manifold.

THEOREM 5 (Interior maximum principle [**118**]). *Let M_1, M_2 be connected minimal surfaces in $\mathbb{R}^3$ and p a point interior to both surfaces, such that $T_pM_1 = T_pM_2 = \{x_3 = 0\}$. If M_1, M_2 are locally expressed as the graphs of functions u_1, u_2 around p and $u_1 \leq u_2$ in a neighborhood of p, then $M_1 = M_2$ in a neighborhood of p.*

THEOREM 6 (Maximum principle at infinity [**82, 56**]). *Let $M_1, M_2 \subset N^3$ be disjoint connected properly immersed minimal surfaces with compact (possibly empty) boundary in a complete flat 3-manifold N^3.*

i) If $\partial M_1 \neq \emptyset$ or $\partial M_2 \neq \emptyset$, then after possibly reindexing, there exist points $p \in \partial M_1$, $q \in M_2$ such that $\text{dist}(p, q) = \text{dist}(M_1, M_2)$.
ii) If $\partial M_1 = \partial M_2 = \emptyset$, then M_1 and M_2 are flat.

The maximum principle at infinity can be generalized to the case of noncompact boundaries, see Theorem 21 below. A beautiful application of Theorem 5 is the following result by Hoffman and Meeks.

THEOREM 7 (Halfspace Theorem [**50**]). *A proper, connected, possibly branched, nonplanar minimal surface $M \subset \mathbb{R}^3$ cannot be contained in a halfspace.*

For later reference in the survey, we next provide a sketch of the original direct proof of Theorem 7. Arguing by contradiction, suppose that $M \subset H = \{x_3 \geq 0\} \subset \mathbb{R}^3$ is a connected, properly immersed, nonflat minimal surface. By the interior maximum principle, $M \subset \text{Int}(H)$. After a suitable vertical translation, we can assume that $\text{dist}(M, \partial H) = 0$. Let $C = \{(x_1, x_2, x_3) \mid x_1^2 + x_2^2 = \cosh^2 x_3,\ x_3 < 0\}$ be the lower half of a vertical catenoid. Since M is proper and disjoint from the

(x_1, x_2)-plane, there exists $\varepsilon > 0$ such that $(M - \varepsilon e_3) \cap C = \emptyset$ and $(M - \varepsilon e_3) \cap (\mathbb{D} \times [-1, 0]) = \emptyset$, where $e_3 = (0,0,1)$ and $\mathbb{D}$ is the unit disk in ∂H. Now consider the homothetically shrunk halfcatenoids $C_t = tC$, $0 < t \leq 1$. As $t \to 0$, C_t converges smoothly to $\partial H - \{0\}$ away from the origin. It follows that for $t > 0$ sufficiently small, $(M - \varepsilon e_3) \cap C_t \neq \emptyset$. From here it is not difficult to prove that there exists a largest $t_0 \in (0, 1]$ such that $(M - \varepsilon e_3) \cap C_{t_0} \neq \emptyset$. Since the intersection $(M - \varepsilon e_3) \cap C_{t_0}$ occurs outside the cylinder $\mathbb{D} \times [-1, 0]$, we find $p \in (M - \varepsilon e_3) \cap C_{t_0}$ which is interior to both surfaces. Since the translated surface $M - \varepsilon e_3$ is above C_{t_0} around p, the interior maximum principle insures that $M - \varepsilon e_3 = C_{t_0}$, which is a contradiction.

2.7. Ends of properly embedded minimal surfaces. One of the fundamental problems in classical minimal surface theory is to describe the behavior of a properly embedded minimal surface $M \subset \mathbb{R}^3$ outside a large compact set in space. This problem is well understood if M has finite total curvature (see Theorem 2), because each of its ends is asymptotic to an end of a plane or a catenoid. A recent Theorem by Meeks and Rosenberg [**80**] proves that if M has finite topology but infinite total curvature, then M is asymptotic to a helicoid (Theorem 30). More complicated asymptotic behaviors can be found in periodic minimal surfaces in $\mathbb{R}^3$, although this asymptotic behavior is completely understood when the periodic minimal surface has finite topology (hence finite total curvature) in the corresponding quotient ambient space; in this setting, only planar, helicoidal or Scherk-type ends can occur (Meeks and Rosenberg [**81, 84**]).

A crucial notion in the understanding of the asymptotic geometry of a generic properly embedded minimal surface is the notion of topological end, which we now explain. Let M be a noncompact connected manifold. We define an equivalence relation in the set $\mathcal{A} = \{\alpha : [0, \infty) \to M \mid \alpha \text{ is a proper arc}\}$ by setting $\alpha_1 \sim \alpha_2$ if for every compact set $C \subset M$, α_1, α_2 lie eventually[4] in the same component of $M - C$.

DEFINITION 10. Each equivalence class in $\mathcal{E}(M) = \mathcal{A}/_\sim$ is called an *end* of M. If $e \in \mathcal{E}(M)$, $\alpha \in e$ is a representative proper arc and $\Omega \subset M$ is a proper subdomain containing α with compact boundary, then we say that the domain Ω *represents* the end e.

$\mathcal{E}(M)$ has the following natural Hausdorff topology. For each proper domain $\Omega \subset M$ with compact boundary, we define the basis open set $B(\Omega) \subset \mathcal{E}(M)$ to be those equivalence classes in $\mathcal{E}(M)$ which have representatives contained in Ω. With this topology, $\mathcal{E}(M)$ is a totally disconnected compact space which embeds topologically as a subspace of $[0, 1] \subset \mathbb{R}$. Since this result is not known, or at least its proof does not seem to appear in the literature, we give a short proof of it at the end of this Subsection.

DEFINITION 11. Any isolated point $e \in \mathcal{E}(M)$ is called a *simple end of* M. If $e \in \mathcal{E}(M)$ is not a simple end (equivalently, if it is a limit point of $\mathcal{E}(M) \subset [0, 1]$), we will call it a *limit end* of M.

When M has dimension 2, then an end $e \in \mathcal{E}(M)$ is simple if and only if it can be represented by a proper subdomain $\Omega \subset M$ with compact boundary which is

[4]Throughout the paper, *eventually* for proper arcs means outside a compact subset of the parameter domain $[0, \infty)$.

homeomorphic to $\mathbb{S}^1 \times [0, \infty)$ (this case is called an *annular end*) or to $\mathbb{S}^1 \times [0, \infty)$ connected sum with an infinite number of tori where the n-th connected sum occurs at the point $(1, n) \in \mathbb{S}^1 \times [0, \infty)$, $n \in \mathbb{N}$ (this is a *simple end of infinite genus*). For limit ends there are similar notions: a limit end $e \in \mathcal{E}(M)$ is said to have *genus zero* if it can be represented by a proper subdomain $\Omega \subset M$ with compact boundary and genus zero. If a limit end e does not have genus zero, then we say that it has *infinite genus*; in this case every proper subdomain with compact boundary representing e has infinite genus.

We will devote Section 5 to the Ordering Theorem for ends of properly embedded minimal surfaces; this Theorem is the starting point for the theory of properly embedded minimal surfaces with more than one end. Concerning one-ended minimal surfaces, the classical example in this family is the helicoid. In 1993, Hoffman, Karcher and Wei [**46, 47**] found a surprising example with genus one and one helicoidal end. Recently, Hoffman, Weber and Wolf [**51**] have given a proof of the embeddedness of a genus one helicoid, and there are computational indications that point to the existence of a unique embedded example with one helicoidal end for any positive genus (the first computer graphics images of a higher genus helicoid are due to Traizet —unpublished—, see also Bobenko [**2**], Bobenko and Schmies [**3**]). From the theoretical point of view, a recent result by Meeks and Rosenberg [**80**] insures that any properly embedded one-ended minimal surface with finite topology must be necessarily asymptotic to a helicoid with finitely many handles and it can be described analytically by meromorphic data $(dg/g, dh)$ on a compact Riemann surface by means of the classical Weierstrass representation, see Theorem 30. Regarding one-ended surfaces with infinite topology, Callahan, Hoffman and Meeks [**6**] showed that any nonflat doubly or triply periodic minimal surface in $\mathbb{R}^3$ must have infinite genus and only one end.

We finish this Subsection by proving that the space $\mathcal{E}(M)$ of ends of a noncompact connected manifold M is a totally disconnected compact space that embeds in the unit interval $[0, 1]$. Our proof generalizes to many other spaces including finite dimensional simplicial complexes. If $\mathcal{E}(M)$ is a finite set, then the embedding property is obvious. Assume now that M has an infinite number of ends. Let $\Omega_1 \subset \ldots \subset \Omega_n \subset \ldots$ be a compact exhaustion of M. It is not difficult to inductively modify this exhaustion to produce a new exhaustion (denoted in the same way) which satisfies the following properties:

1. Ω_n is a manifold with boundary.
2. $\Omega_n \subset \text{Int}(\Omega_{n+1})$.
3. For $n \geq 2$, every boundary component of Ω_n separates M into two closed noncompact regions.
4. The number of boundary components of Ω_n is n.
5. Each end $e \in \mathcal{E}(M)$ has a representative α which begins in Ω_1 and intersects transversely each boundary component of Ω_n at most once point.

Consider the closure in M of the two components of $\Omega_3 - \Omega_2$. One of these components, which we label M_2, has exactly 2 boundary components, while the other component M_3 has 3 boundary components. This process can be inductively continued to label the closure in M of the n closed components of $\Omega_{n+1} - \Omega_n$ as follows. Assume that $\{M_{a_1,\ldots,a_{n-2}}\}$ is the collection of the components of the previous stage. We label a closed component of $\Omega_{n+1} - \Omega_n$ by $M_{a_1,\ldots,a_{n-1}}$ with

$a_{n-1} \in \{3,4,5,6,7\}$ so that $M_{a_1,\ldots,a_{n-1}}$ attaches to $M_{a_1,\ldots,a_{n-2}}$ and one of the following possibilities holds:

1. $a_{n-1} = 3$ if $M_{a_1,\ldots,a_{n-1}}$ has three boundary components.
2. $a_{n-1} = 4$ if $M_{a_1,\ldots,a_{n-1}}$ has two boundary components and $M_{a_1,\ldots,a_{n-2}}$ has also two boundary components.
3. $a_{n-1} = 5$ if $M_{a_1,\ldots,a_{n-1}}$ has two boundary components, $M_{a_1,\ldots,a_{n-2}}$ has three boundary components and the other component attaching to $M_{a_1,\ldots,a_{n-2}}$ has three boundary components.
4. $a_{n-1} = 6,7$ if $M_{a_1,\ldots,a_{n-1}}$ has two boundary components and it is not in the previous cases.

We now explain how to embed the totally disconnected compact Hausdorff space $\mathcal{E}(M)$ into $[0,1]$ where we consider every point in $[0,1]$ to be expressed as decimal point followed by an infinite sequence of digits. Let $e \in \mathcal{E}(M)$ and take a representative $\alpha \in e$ satisfying property 5 above. Then α eventually lies in $M_{a_1} \cup M_{a_1,a_2} \cup \ldots \cup M_{a_1,\ldots,a_n} \cup \ldots$ for a unique infinite sequence $S(\alpha) = (a_1, a_2, \ldots, a_n, \ldots) \in [0,1]$. Note that $S(\alpha)$ is independent of the choice of the proper arc $\alpha \in e$ satisfying property 5, so we can denote this decimal number as $S(e)$. It is straightforward to prove that the map $e \mapsto S(e)$ is a topological embedding. In the case that M is a properly embedded minimal surface in $\mathbb{R}^3$ with more than one end, there is a more natural topological embedding of $\mathcal{E}(M)$ into $[0,1]$ that uses the relative heights of the ends of M, see the Ordering Theorem 18 in Section 5.

2.8. Second variation of area and Jacobi functions. Let $M \subset \mathbb{R}^3$ be a minimal surface and $\Omega \subset M$ a subdomain with compact closure. Any compactly supported, smooth, normal deformation of the inclusion $X\colon M \to \mathbb{R}^3$ on Ω can be written as $X + tuN$, where N is the Gauss map of M and $u \in C_0^\infty(M)$. By (2), the area functional $A = A(t)$ for this deformation has $A'(0) = 0$. The *second variation of area* can be easily shown to be (see [**97**])

$$A''(0) = -\int_\Omega u(\Delta u - 2Ku)\, dA, \tag{10}$$

where K is the Gaussian curvature function of M and Δ its Laplace operator. Formula (10) can be seen as the classical bilinear form associated to the linear elliptic L^2-selfadjoint operator $L = \Delta - 2K = \Delta + |\nabla N|^2$, which is usually called the *Jacobi operator.*

DEFINITION 12. A C^2-function $u\colon M \to \mathbb{R}$ satisfying $\Delta u - 2Ku = 0$ on M is called a *Jacobi function.* We will let $\mathcal{J}(M)$ denote the space of Jacobi functions on the minimal surface M.

Classical elliptic theory implies that for a given subdomain $\Omega \subset M$ with compact closure, the Dirichlet problem for the Jacobi operator in Ω has an infinite discrete sequence $\{\lambda_k\}_{k\in\mathbb{N}\cup\{0\}}$ of eigenvalues with $\lambda_k \nearrow +\infty$ as k goes to infinity, and each eigenspace is a finite dimensional linear subspace of $C^\infty(\Omega) \cap H_0^1(\Omega)$, where $H_0^1(\Omega)$ denotes the usual Sobolev space of L^2 functions with L^2 weak partial derivatives and trace zero. Since any normal variation through minimal surfaces has vanishing second derivative of the area functional, it follows that the normal parts of variational fields coming from Killing or dilatation vector fields of $\mathbb{R}^3$ produce elements in $\mathcal{J}(M)$. For instance, translations give rise to the so called *linear Jacobi functions* $\langle N, v\rangle$ with $v \in \mathbb{R}^3$, rotations produce the Jacobi functions $\det(p, N, v)$ (where p denotes the position vector) and homotheties give the *support function* $\langle p, N\rangle \in \mathcal{J}(M)$. A particularly interesting Jacobi function, which is defined when the minimal surface is transverse to a family of horizontal planes, is the *Shiffman function,* which will be studied in Section 11.

DEFINITION 13. Let $\Omega \subset M$ be a subdomain with compact closure. The *index of stability* of Ω is the number of bounded states of L in such domain, i.e. the number of negative eigenvalues of the Dirichlet problem associated to L in Ω. The *nullity* of Ω is the dimension of $\mathcal{J}(\Omega) \cap H_0^1(\Omega)$. Ω is called *stable* if its index is zero, and *strictly stable* if both its index and nullity are zero.

Elliptic theory also implies that Ω is strictly stable provided that it is sufficiently small, which justifies the Definition 5 of minimal surface as a local minimum of area. Another consequence of elliptic theory is that Ω is stable if and only if it carries a positive Jacobi function. Since the Gauss map N of a graph defined on a domain in a plane Π has image set contained in an open halfsphere, the inner product of N with the unit normal to Π provides a positive Jacobi function, from where we conclude that any minimal graph is stable. Stability makes sense in the large, as we next explain.

DEFINITION 14. A minimal surface $M \subset \mathbb{R}^3$ is called *stable* if any relatively compact subdomain $\Omega \subset M$ is stable. For orientable minimal surfaces, stability is equivalent to the existence of a positive Jacobi function (Proposition 1 in [**34**]). M is said to have *finite index* if outside of a compact subset it is stable. The *index* of stability of M is the supremum of the indices of relatively compact subdomains in M.

By definition, stable surfaces have index zero. The following Theorem explains how restrictive is the property of stability for complete minimal surfaces. It was proved independently by Fischer-Colbrie and Schoen [**35**], do Carmo and Peng [**31**], and Pogorelov [**109**].

THEOREM 8. *If $M \subset \mathbb{R}^3$ is a complete (orientable) immersed stable minimal surface, then M is a plane.*

We will provide a short elementary proof of Theorem 8 in Section 4.1. If we weaken the stability hypothesis to finite index, then completeness also leads to a well-known family of minimal surfaces.

THEOREM 9 (Fischer-Colbrie [**34**]). *If $M \subset \mathbb{R}^3$ is a complete (orientable) minimal surface with possibly empty compact boundary, then M has finite index if and only if it has finite total curvature. In this case, the index and nullity of M coincides with the index and nullity of the meromorphic extension of its Gauss map to the compactification $\overline{M}$ obtained from M after attaching its ends, see Theorem 2.*

By the conformal invariance of the Dirichlet integral, both the index and nullity of the Jacobi operator $L = \Delta + |\nabla N|^2$ remain constant under a conformal change of metric. On the other hand, Osserman's Theorem implies that every complete immersed minimal surface $M \subset \mathbb{R}^3$ with finite total curvature is conformally equivalent to a finitely punctured compact Riemann surface $\overline{M}$. It can be shown (Pérez and Ros [**106**]) that there exists a smooth metric $d\overline{s}^2$ on the compactification $\overline{M}$ such that the metric ds^2 on M induced by the inner product of $\mathbb{R}^3$ can be expressed as $ds^2 = \mu\, d\overline{s}^2$, where μ is a positive smooth function that blows up at the ends of M. In this setting, both the index and nullity of L can be computed as the index and nullity of the operator $\overline{L} = \overline{\Delta} + |\overline{\nabla} N|^2$ on $\overline{M}$ minus the ends, where a bar means that the corresponding object is computed with respect to $d\overline{s}^2$. Also by Osserman's Theorem, N extends as a meromorphic function to $\overline{M}$, thus $\overline{L}$ is nothing more than the classical Schrödinger operator associated to such a meromorphic extension. The subspace $\mathcal{K}(M)$ of *bounded* Jacobi functions on M can be identified with the eigenspace associated to the eigenvalue 0 of the operator $\overline{L}$. Inside $\mathcal{K}(M)$ we have the subspace of linear functions $\mathcal{L}(M) = \{\langle N, v\rangle \mid v \in \mathbb{R}^3\}$. If additionally all the ends of M are horizontal, then $\det(p, N, e_3) \in \mathcal{K}(M)$, where $e_3 = (0, 0, 1)$. In particular, $\mathcal{K}(M)$ has dimension at least 4 for any complete embedded minimal surface of finite total curvature in $\mathbb{R}^3$ except for the catenoid where $\det(p, N, e_3)$ vanishes.

Montiel and Ros [**91**] stated a beautiful relationship between bounded Jacobi functions and branched minimal immersions. For a complete minimal surface $M \subset \mathbb{R}^3$ with finite total curvature, let $B(N) \subset \overline{M}$ be the set of branch points of the extended Gauss map and $\mathcal{M}(N)$ the linear space of all complete branched minimal immersions (including the constant maps) of $M - B(N)$ into $\mathbb{R}^3$ with the same Gauss map N as M.

THEOREM 10 (Montiel, Ros [**91**]). *Let $M \subset \mathbb{R}^3$ be a complete immersed minimal surface with finite total curvature*[5]. *Then, there exists a linear map $u \in \mathcal{K}(M) \mapsto X_u \in \mathcal{M}(N)$ such that the support function of X_u is u, and $u \in \mathcal{L}(M)$ if and only if X_u is constant. Furthermore, this linear map gives rise to an isomorphism between the quotient spaces $\mathcal{K}(M)/\mathcal{L}(M)$ and $\mathcal{M}(N)/\{constants\}$.*

Among the admissible conformal metrics which can be used to express questions related with the Jacobi operator, a particularly interesting choice comes from consideration of the pullback metric ds_N^2 through the Gauss map from the standard spherical metric on $\mathbb{S}^2$. The metric ds_N^2 has singularities at the branch points of N and the Jacobi operator transforms into $L_N = \Delta_N + 2$, where Δ_N is the Laplacian of ds_N^2. Eigenvalues and eigenfunctions of L_N are well-defined by a variational approach (Tysk [**125**]). In particular, the index of stability of a relatively compact subdomain $\Omega \subset M$ is equal to the number of eigenvalues of Δ_N which are strictly less than 2, and the nullity of Ω is the multiplicity of 2 as an eigenvalue of Δ_N. Using these ideas, Montiel and Ros [**91**] gave some estimates for the index and nullity under different geometrical assumptions, of which we emphasize the following one.

THEOREM 11. *Let $M \subset \mathbb{R}^3$ be a complete immersed minimal surface with finite total curvature*[5]. *If all the branch values of the Gauss map of M lie on a equator of $\mathbb{S}^2$, then the dimension of $\mathcal{K}(M)$ is* 3.

3. Conformal questions on minimal surfaces.

3.1. Recurrence and parabolicity for manifolds. The conformal structure of a complete minimal surface has a strong influence on its global properties. In particular, an important question is to decide the so called *type problem* for a minimal surface M, in the sense of classical Riemann surfaces: i.e. whether M

[5]Theorems 10 and 11 remain valid for complete minimal surfaces in any quotient of $\mathbb{R}^3$ where the Gauss map makes sense, and which have finite total curvature in the quotient.

is hyperbolic or parabolic[6] (as we have already noticed, the elliptic or compact case is impossible for a minimal surface). It turns out that the parabolicity for Riemann surfaces without boundary is equivalent to the *recurrence* of Brownian motion of such surfaces. This field lies in the borderline between several branches of Mathematics such as Riemannian Geometry, Stochastic Analysis, Partial Differential Equations and Potential Theory. A particularly interesting source where the reader can find an excellent introduction to these questions is the survey of recurrence and Brownian motion on Riemannian manifolds by Grigor'yan [**41**]. The goal of this Subsection is to introduce some key concepts that are useful in dealing with these conformal questions. In order to avoid concepts closely related with probability (such as random walks or Brownian motion), in this paper we will follow an alternative way to define recurrence and parabolicity that is slightly different from Grigor'yan's approach; this approach is well-known and is explained in greater detail in the notes by the second author [**101**]. However, we will briefly explain the connection between these two approaches. We will not provide proofs for most of the results stated in this Subsection, but the proofs can be found in [**41**] or [**101**].

DEFINITION 15. Let (M^n, g) be a n-dimensional Riemannian manifold with nonempty boundary. M is *parabolic* if every bounded harmonic function on M is determined by its boundary values.

DEFINITION 16. Let (M^n, g) be a n-dimensional Riemannian manifold without boundary. M is *recurrent* if for any nonempty open set $O \subsetneqq M$ with smooth boundary, $M - O$ is parabolic.

Given a Riemannian manifold (M, g) with boundary $\partial M \neq \emptyset$ and a point $p \in \text{Int}(M)$, the *harmonic measure* μ_p *with respect to* p can be defined as follows. Let $I \subset M$ be a nonempty open set with smooth boundary. Consider a compact exhaustion $I \subset \partial M_1 \subset M_1 \subset M_2 \subset \ldots$ of M. Given $k \in \mathbb{N}$, let $h_k \colon M \to [0,1]$ be the (bounded) harmonic function on M_k with boundary values 1 on $\text{Int}(I)$ and 0 on $\partial M_k - I$. After extending h_k by zero to M, we can see $\{h_k\}_k$ as an increasing sequence of harmonic functions, bounded from above by 1. Hence h_k limits to a unique bounded harmonic function $h \colon M \to [0,1]$. In this situation, we define $\mu_p(I) = h(p)$. It turns out that μ_p extends to a Borel measure μ_p on ∂M.

Another interpretation of μ_p, developed in [**41**], is that $\mu_p(I)$ is the probability of a Brownian path beginning at p, of hitting ∂M the first time somewhere on the interval I, and for this reason the harmonic measure of M is also called the *hitting measure with respect to* p. We now explain how to computationally calculate the hitting measure μ_p at an interval I contained in the boundary of a smooth domain $\Omega \subset \mathbb{R}^2$, where $p \in \text{Int}(\Omega)$. For $n \in \mathbb{N}$ and $\varepsilon > 0$, define the set $\Gamma(p, n, \varepsilon)$ to be the n-step orthogonal random ε-walks starting at p, i.e. continuous mappings $\sigma \colon [0, n\varepsilon] \to \mathbb{R}^2$ which begin at $\sigma(0) = p$ and for any integer $k = 0, \ldots, n-1$,

$$\left(\sigma|_{[k\varepsilon,(k+1)\varepsilon]}\right)(t) = \sigma(k\varepsilon) \pm t e_i,$$

[6]Classically, a Riemann surface without boundary is called *hyperbolic* if it carries a nonconstant positive superharmonic function, and *parabolic* if it is neither elliptic (i.e. compact) nor hyperbolic. The reader should be aware that we will use the concept of parabolicity for Riemannian manifolds with boundary (see Definition 15) and reserve the word *recurrent* for manifolds without boundary (Definition 16). For Riemannian manifolds, the relationship between parabolicity and recurrence will become clear soon.

where e_i is one of the unit vectors $(1,0),(0,1)$. We define $\mu_p(n,\varepsilon)(I)$ to be the probability that some $\sigma \in \Gamma(p,n,\varepsilon)$ crosses $\partial\Omega$ a first time in I. As $n \to \infty$, $\mu_p(n,\varepsilon)(I)$ converges quickly to a number $\mu_p(\varepsilon)(I) \in [0,1]$. Similarly, as $\varepsilon \to 0$, the $\mu_p(\varepsilon)$ converge to a measure μ_p on ∂M, which is equal to the hitting measure obtained from Brownian motion starting at p.

For an interval $I \subset \partial\Omega$, consider the function $P_I(n,\varepsilon) : \mathrm{Int}(\Omega) \to [0,1]$ defined as $P_I(n,\varepsilon)(p) = \mu_p(n,\varepsilon)(I)$ for $p \in \mathrm{Int}(\Omega)$. Note that for any $p \in \mathrm{Int}(\Omega)$ and for ε smaller that the distance from p to $\partial\Omega$, the following formula holds

$$P_I(n,\varepsilon)(p) = \frac{1}{4}\left(\sum_{i=1}^{2} P_I(n-1,\varepsilon)(p+e_i) + \sum_{i=1}^{2} P_I(n-1,\varepsilon)(p-e_i)\right),$$

and so, the limiting function $P_I(\varepsilon)$ satisfies an infinitesimal 4 point mean value property. As $\varepsilon \to 0$, $P_I(\varepsilon)$ converges to a function P_I on $\mathrm{Int}(\Omega)$ which satisfies the usual mean value property. Therefore, the function $p \mapsto P_I(p)$, which is the probability of a Brownian path starting at p of exiting Ω a first time on I, is a harmonic function which takes its values in $[0,1]$. Note that P_I has limiting values 1 on the interior of I and 0 on the interior of $\partial\Omega - I$ and so, it is the unique bounded harmonic function on $\mathrm{Int}(\Omega)$ whose boundary values correspond almost everywhere to the characteristic function of $I \subset \partial\Omega$. By definition of P_I, the hitting measure is $\mu_p(I) = P_I(p)$, which gives the desired equivalence between hitting and harmonic measure for planar domains.

The above discussion generalizes easily to a Riemannian manifold M with boundary. We will briefly explain this generalization in the case of dimension 2 and when M lies in the interior of a bigger complete manifold $\overline{M}$. Let M be a Riemannian surface with boundary and $p \in \mathrm{Int}(M)$. Given a unit tangent vector v_p, $n \in \mathbb{N}$ and $\varepsilon > 0$, we let $\Gamma(v_p,n,\varepsilon)$ denote the set of n-step orthogonal random ε-walks $\sigma\colon [0,n\varepsilon] \to \overline{M}$ such that $\sigma|_{[0,\varepsilon]}$ is the unit speed geodesic beginning at p in one of the directions $\pm v_p, \pm J v_p$, where J is a local almost complex structure in a neighborhood of p, and $\sigma|_{[k\varepsilon,(k+1)\varepsilon]}$ is the unit speed geodesic in $\overline{M}$ beginning at $\sigma(k\varepsilon)$ in one of the directions $\pm\sigma'(k\varepsilon), \pm J\sigma'(k\varepsilon)$, $0 \le k \le n-1$. These sets of random walks produce, as in the planar domain case, a limiting hitting measure on ∂M, which is independent of the initial choice of v_p, and by the previous arguments, is equal to the harmonic measure μ_p.

From the above discussion, it easily follows that a Riemannian manifold without boundary is recurrent (see Definition 16) precisely when almost all Brownian paths are dense in the manifold. Also, parabolicity and harmonic measure are closely related, as states the following result.

PROPOSITION 2. *Let (M,g) be a Riemannian manifold with $\partial M \neq \emptyset$. Then, the following statements are equivalent:*

1. *M is parabolic.*
2. *There exists a point $p \in \mathrm{Int}(M)$ such that the harmonic measure μ_p is full, i.e. $\int_{\partial M} \mu_p = 1$.*
3. *Given any $p \in \mathrm{Int}(M)$ and any bounded harmonic function $f\colon M \to \mathbb{R}$, then $f(p) = \int_{\partial M} f\,\mu_p$.*
4. *There exists a proper nonnegative superharmonic function on M.*

$\mathbb{R}^n$ is recurrent for Brownian motion if and only if $n \le 2$. The parabolicity of a Riemannian manifold with boundary is not affected by adding compact sets or by removing interiors of compact sets, and if a manifold M can be decomposed as the union of two parabolic domains with compact intersection, then M is parabolic (or recurrent, depending on if ∂M is empty or not).

Note that if $h\colon M \to \mathbb{R}$ is a nonconstant positive harmonic function on a recurrent Riemannian manifold, then for any positive regular value $t \in \mathbb{R}$ of h, the closed subset $M_t = h^{-1}((0,t])$ is parabolic and $h|_{M_t}$ is a bounded harmonic function with constant boundary value t. By Proposition 2, $h|_{M_t}$ is the constant function t, which contradicts that t is a regular value of h. This contradiction completes the proof of the following well-known result.

PROPOSITION 3 (Liouville Theorem). *Every positive harmonic function on a recurrent Riemannian manifold is constant.*

3.2. Recent results on parabolicity for minimal surfaces. As we have said before in this survey, a knowledge of the conformal type of a minimal surface M is crucial when tackling uniqueness questions. Sometimes it is useful to decompose M in pieces and study the conformal structure of each piece as a Riemann surface with boundary. For instance, the proof by Meeks and Rosenberg of the uniqueness of the helicoid [**80**] uses the fact that a simply connected properly embedded minimal surface $M \subset \mathbb{R}^3$ must admit a plane which intersects M transversely in a single proper arc γ. Each of the two closed complements of γ in M is contained in a closed halfspace, hence both are parabolic as follows from Theorem 12 below and ones then proves that M is conformally $\mathbb{C}$. This argument introduces one of the main open questions concerning minimal surfaces with boundary, which we now state after a definition.

DEFINITION 17. Let $W \subset \mathbb{R}^3$ be a connected region of space which is either open or the closure of an open set. We say that W is a *universal region for surfaces* if every complete, connected, properly immersed minimal surface $M \subset W$ is either recurrent (when $\partial M = \emptyset$) or a parabolic surface with boundary. W is called a *universal region for graphs* if every proper minimal graph $M \subset W$ is a parabolic surface with boundary.

QUESTION 1. *Which regions $W \subset \mathbb{R}^3$ are universal for surfaces or for graphs?*

Obviously, any universal region for surfaces is also universal for graphs. By an ingenious application of the classical Runge's Theorem, Rosenberg and Toubiana [**115**] gave an example of a nonflat minimal annulus without boundary which is properly immersed in an open slab. By Liouville's Theorem (Proposition 3), this example proves that an open slab is not universal for surfaces. A smart refinement of the ideas used by Nadirashvili [**94**] in his proof of the existence of a complete immersed minimal surface in a ball in $\mathbb{R}^3$, allows one to construct a minimal immersion of the open unit disk that is proper in $\mathbb{R}^3$ (Morales [**92**]), which shows that $\mathbb{R}^3$ is not universal for surfaces. This result demonstrates the necessity of the embeddedness assumption in Meeks and Rosenberg's proof that if a simply connected, proper minimal surface is embedded, then it will be conformally $\mathbb{C}$. Recently, Martín and Morales [**65**] have constructed a complete conformal minimal immersion of the open unit disk that is proper in an open ball, which implies that open balls are not universal regions for surfaces. More recently [**64**], they have generalized their result to prove that the interior of any convex region of $\mathbb{R}^3$ (including noncompact and nonsmooth ones) admits a proper complete minimal immersion of the unit disk, which implies that such regions are not universal for surfaces. Theorem 12 and Corollary 1 below imply respectively that any closed halfspace is universal for surfaces, and the region above a negative halfcatenoid is universal for graphs.

Part *5* in Proposition 2 gives a method for showing that a given region $W \subset \mathbb{R}^3$ is universal for surfaces, which consists of finding a proper nonnegative *universal superharmonic function* on W, a concept that we now define.

DEFINITION 18. Given a region $W \subset \mathbb{R}^3$, a function $h\colon W \to \mathbb{R}$ is said to be a *universal superharmonic function on* W if its restriction to any minimal surface $M \subset W$ is superharmonic.

Examples of universal superharmonic functions on all of $\mathbb{R}^3$ include coordinate functions such as x_1 or the function $-x_1^2$. Collin, Kusner, Meeks and Rosenberg proved the useful inequality (Lemma 2.2 in [**26**]) valid for any immersed minimal surface in $\mathbb{R}^3$:

$$|\Delta \ln r| \leq \frac{|\nabla x_3|^2}{r^2}, \tag{11}$$

where $r = \sqrt{x_1^2 + x_2^2}$ and ∇, Δ denote the intrinsic gradient and laplacian on M. Using the estimate (11), it is straightforward to check the following statement.

LEMMA 1.
i) The function $\ln r - x_3^2$ *is a universal superharmonic function in the region* $\{r^2 \geq \frac{1}{2}\}$.
ii) The function $\ln r - x_3 \arctan x_3 + \frac{1}{2}\ln(x_3^2 + 1)$ *is a universal superharmonic function in the region* $\{r^2 \geq x_3^2 + 1\}$.

With the above Lemma, we now prove that any closed halfspace is a universal region for surfaces. If M is a properly immersed minimal surface in a closed half-space and $\partial M = \emptyset$, then M is planar by the Halfspace Theorem (Theorem 7); in particular, any closed halfspace is a universal region for surfaces without boundary. The desired property of being planar also follows directly from the following general Theorem and the fact that on recurrent surfaces positive harmonic functions are constant (Proposition 3).

THEOREM 12 (Collin, Kusner, Meeks, Rosenberg [**26**]). *Let M be a connected properly immersed minimal surface in $\mathbb{R}^3$, possibly with boundary. Then, every component of the intersection of M with a closed halfspace is a parabolic surface with boundary. In particular, if M has empty boundary and intersects some plane in a compact set, then M is recurrent.*

PROOF. Up to a rotation, it suffices to check that any component C of $M(+) = M \cap \{x_3 \geq 0\}$ is parabolic. For fixed $n \in \mathbb{N}$, let $C_n = C \cap x_3^{-1}([0,n])$. By part *i)* of Lemma 1, the function $h = \ln r - x_3^2$ is superharmonic and proper when restricted to $C_n \cap \{r^2 \geq \frac{1}{2}\}$. Furthermore, h is positive outside a compact domain on C_n, which by part *4* of Proposition 2 implies that $C_n \cap \{r^2 \geq \frac{1}{2}\}$ is parabolic. Since M is proper and $\{r^2 \leq \frac{1}{2}\} \cap \{0 \leq x_3 \leq n\}$ is compact, we deduce that $C_n - \{r^2 \geq \frac{1}{2}\}$ is a compact subset of C_n. Since parabolicity is not affected by adding compact subsets, it follows that C_n is parabolic.

We now check that C is parabolic. Fix a point $p \in C$ with $x_3(p) > 0$ and let μ_p^C be the harmonic measure of C with respect to p. For n large enough, p lies in the interior of C_n. Since x_3 is a bounded harmonic function on the parabolic surface

C_n, part *3* of Proposition 2 insures that

$$x_3(p) = \int_{\partial C_n} x_3\, \mu_p^n \geq n \int_{\partial C_n \cap x_3^{-1}(n)} \mu_p^n,$$

where μ_p^n is the harmonic measure of C_n with respect to p. Since μ_p^n is full on ∂C_n, it follows that

$$\int_{\partial C_n - x_3^{-1}(n)} \mu_p^n = 1 - \int_{\partial C_n \cap x_3^{-1}(n)} \mu_p^n \geq 1 - \frac{x_3(p)}{n} \overset{(n\to\infty)}{\longrightarrow} 1.$$

Suppose that M and N are Riemannian manifolds with $M \subset N$, ∂ a component of $\partial M \cap \partial N$, $p \in \text{Int}(M)$ and μ_p^M and μ_p^N are the respective harmonic measures. Then it follows immediately from the definition of harmonic measure that $\int_\partial \mu_p^M \leq \int_\partial \mu_p^N \leq 1$. By letting $M = C_n$, $N = C$ and $\partial = \partial C_n - x_3^{-1}(n)$, the above displayed inequality implies $\lim_n \int_{\partial C_n - x_3^{-1}(n)} \mu_p^C \geq 1$, from where we conclude that $\int_{\partial C} \mu_p^C = 1$ and the proof is complete. □

An open subset Ω of a Riemann surface without boundary is called *hyperbolic* if Ω carries a nonconstant positive superharmonic function, or equivalently if for any point $q \in \Omega$, the Green's function with singularity at q exists (this function is the smallest positive harmonic function in $\Omega - \{q\}$ with a logarithmic singularity at q, see [**32**] for details about Green's functions and hyperbolicity). Suppose now that $X\colon M \to \mathbb{R}^3$ is a proper minimal immersion whose Gauss map g has image set contained in a hyperbolic open subset $\Omega \subset \mathbb{S}^2$. Since such an Ω does not have logarithmic capacity zero, a result by Osserman [**99**] implies that if M has no boundary, then the immersion is flat. Assume that X is not flat. A careful analysis of the role that x_3 plays in the last proof lets us exchange this coordinate function by the composition $G \circ g$, where G is the Green's function in Ω with singularity at a given point $q \in \Omega$. After a suitable choice of a universal superharmonic function h which constrains the region where $X(M)$ is contained, arguments not too different from the ones in the proof of Theorem 12 lead to the following statement.

THEOREM 13 (López, Pérez [**62**]). *Given $\alpha \in (0,1)$, let $W_\alpha = \{x_3 > -(x_1^2 + x_2^2)^{\alpha/2}\}$ and $X\colon M \to \mathbb{R}^3$ a proper nonflat minimal immersion with $X(M) \subset W_\alpha$. If, up to removing a compact set of M, the Gauss map of X has image contained in a hyperbolic open subset of the sphere, then M is a parabolic surface with nonempty boundary.*

The region above a vertical negative halfcatenoid and outside a certain compact set is contained in W_α for any $\alpha \in (0,1)$. Since the Gauss map image of a graph is contained in a closed halfsphere, it is also contained in an hyperbolic open subset of $\mathbb{S}^2$. Thus the following result is a direct consequence of Theorem 13.

COROLLARY 1. *Any proper minimal graph lying in the closed region above a vertical negative halfcatenoid is parabolic.*

We would like to finish this Section by mentioning an unpublished example by Collin of a complete stable minimal submersion of a disk minus a Cantor set of positive measure in its boundary into a plane, which we can view as a degenerate multigraph.

Let $W = \mathbb{R}^2 - \{0,1\}$ and let $h\colon \mathbb{D} \to W$ be the universal cover by the open unit disk $\mathbb{D} \subset \mathbb{C}$. Note that $\mathbb{D}$ is incomplete in the pulled-back flat metric through h.

Consider the simply connected subdomain $W(+) = \{(x_1, x_2) \in W \mid x_2 \geq 0\}$ and one of its lifts D_1 to $\mathbb{D}$. Let p_1, q_1 denote points in $\partial\mathbb{D}$ corresponding to the end points of D_1, which under completion would map to 0 and 1 in $\mathbb{R}^2$, respectively. Consider small disjoint arcs α_1, β_1 in $\partial\mathbb{D}$ centered at p_1, q_1 respectively and each of length $l(1) < \frac{1}{16}$. Consider the open "equilateral" triangles $\Delta(1,1), \Delta(1,2)$ in $\mathbb{D}$ with circle bases α_1, β_1, and sides being straight line segments of length $l(1)$. Let $D_1, D_2, \ldots, D_n, \ldots$ denote an ordering of the set of lifts of $W(+)$ to $\mathbb{D}$. For each D_k we find similar points p_k, q_k and arcs α_k, β_k with lengths at most $l(k) < \frac{1}{16^k}$. After removing from $\mathbb{D}$ all of the similarly defined open "equilateral" triangles $\Delta(k,1), \Delta(k,2)$, one obtains a flat surface M and an induced map $h \colon M \to \mathbb{R}^2$. Since the boundary of the closure $\overline{M}$ of M in the closed unit disk $\overline{\mathbb{D}}$ is a Lipschitz curve parametrized by the argument θ, the Riemann mapping Theorem preserves sets of positive Lebesgue measure on $\partial\overline{M}$. By construction, the set $\partial\overline{M} - \partial M$ has positive measure and so ∂M does not have full harmonic measure. On the other hand, it can be checked that the flat metric on M induced by the submersion is complete, which finishes our construction of the example of Collin.

4. Stable minimal surfaces.

Very often, stable minimal surfaces play the role of planes separating disjoint pieces of minimal surfaces. How to produce such separating stable surfaces will be the goal of Subsection 4.2. Before explaining this, we will show how to give bounds on the area and curvature for stable surfaces, results which in turn have important consequences to the global theory.

4.1. Area and curvature estimates for stable minimal surfaces. Let $D \subset M$ be an embedded geodesic disk of radius $r_0 > 0$ contained in a minimal surface $M \subset \mathbb{R}^3$. Since the Gaussian curvature of D is nonpositive, classical comparison with the Euclidean disk of the same radius gives $\pi r_0^2 \leq \text{Area}(D)$. Colding and Minicozzi (Theorem 1.2 in **[22]**) gave the following useful upper estimate for Area(D) by assuming stability.

THEOREM 14. *Let $D \subset M$ be a geodesic disk of radius r_0 inside a minimal surface $M \subset \mathbb{R}^3$. If D is stable, then*

$$\text{Area}(D) \leq \frac{4}{3}\pi r_0^2.$$

PROOF. The argument of Colding-Minicozzi uses a beautiful application of the stability inequality with a particular choice of a radial test function $f(r,\theta) = \eta(\theta)$ (here (r,θ) are polar geodesic coordinates on D), where $\eta \in C^1([0, r_0])$ with $\eta(r_0) = 0$. By stability, Green's formula and the coarea formula, we obtain:

(12)

$$0 \leq \int_D \left(|\nabla f|^2 + 2Kf^2\right) dA = \int_0^{r_0} (\eta'(r))^2 l(r)\, dr + 2\int_0^{r_0} \left(\int_{\partial D(r)} K ds_r\right) \eta^2(r)\, dr,$$

where K is the Gaussian curvature function on M, $D(r)$ is the disk of geodesic radius $r \in [0, r_0]$ concentric with D and $ds_r, l(r)$ are respectively the length element and the total length of $\partial D(r)$. Let $K(r) = \int_{D(r)} K\, dA$. As $K'(r) = \int_{\partial D(r)} K\, ds_r$ and $K(0) = \eta(r_0) = 0$, integration by parts gives $\int_0^{r_0} \left(\int_{\partial D(r)} K\, ds_r\right) \eta^2(r)\, dr = -\int_0^{r_0} K(r)(\eta^2(r))'\, dr$. By the Gauss-Bonnet formula and the first variation of length (see e.g. [**10**]), $K(r) = 2\pi - \int_{\partial D(r)} \kappa_g \, ds_r = 2\pi - l'(r)$, where κ_g denotes geodesic curvature. Substituting this formula for $K(r)$ into (12), we obtain:

$$0 \leq \int_0^{r_0} (\eta'(r))^2 l(r) \, dr - 2 \int_0^{r_0} (2\pi - l'(r))(\eta^2(r))' \, dr.$$

Now taking $\eta(r) = 1 - \frac{r}{r_0}$ in the last expression, we have

$$-\frac{1}{r_0^2} \int_0^{r_0} l(r) \, dr + \frac{4}{r_0} \int_0^{r_0} l'(r) \left(1 - \frac{r}{r_0}\right) dr \leq \frac{8\pi}{r_0} \int_0^{r_0} \left(1 - \frac{r}{r_0}\right) dr = 4\pi. \tag{13}$$

Finally, integration by parts gives $\frac{4}{r_0} \int_0^{r_0} l'(r) \left(1 - \frac{r}{r_0}\right) dr = \frac{4}{r_0^2} \int_0^{r_0} l(r) \, dr$. Plugging this into (13) and using that $\int_0^{r_0} l(r) \, dr = \text{Area}(D)$, we finish the proof. □

To see the usefulness of the above area estimate, we now give a short proof of Theorem 8.

PROOF OF THEOREM 8. Let $M \subset \mathbb{R}^3$ be a complete orientable immersed minimal surface which is stable. Recall that for orientable minimal surfaces, stability is equivalent to the existence of a positive Jacobi function (Proposition 1 in [**34**]). After lifting such a function on M to a Jacobi function on the universal cover of M, we can assume that M is simply-connected. Since the Gaussian curvature of M is nonpositive, Hadamard's Theorem implies that the intrinsic distance function to a given point $p_0 \in M$ is smooth outside p_0 without critical points and that the geodesic disk $D(r)$ centered at p_0 with radius $r > 0$ is embedded. Let $A(r)$ be the area of $D(r)$ (which is a smooth function of r) and $l(r)$ the length of its boundary. Then, $A'(r) = l(r)$ and thus, the first variation of length and the Gauss-Bonnet formula give $A''(r) = \int_{\partial D(r)} \kappa_g \, ds_r = 2\pi - \int_{D(r)} K \, dA$, where κ_g is the geodesic curvature of $\partial D(r)$, ds_r is its length element and K is the Gaussian curvature of M. The last equality implies that $A''(r)$ is monotonically increasing in r. This property together with $A(r) \leq \frac{4}{3}\pi r^2$ (Theorem 14) imply that $A''(r) \leq \frac{8}{3}\pi$, which shows that $-\int_{D(r)} K \, dA \leq \frac{2}{3}\pi$. Since r is arbitrary, we conclude that M has finite total absolute Gaussian curvature at most $\frac{2}{3}\pi$. By Theorem 2, the total curvature of a complete orientable nonplanar minimal surface is infinite or a positive integer multiple of -4π, and so we deduce that M must be a plane. This finishes the proof of Theorem 8. □

A crucial fact in minimal surface theory is that orientable minimally immersed stable surfaces with boundary in $\mathbb{R}^3$ have curvature estimates up to their boundary. These curvature estimates were firstly obtained by Schoen [**117**] and later proved in another way by Ros [**113**]. Here we give a different approach, using Theorem 8 and a blow-up argument. As a consequence, these curvature estimates are in fact easily derived from the area estimate in Theorem 14.

THEOREM 15. *There exists a universal constant $c > 0$ such that for any stable orientable minimally immersed surface M in $\mathbb{R}^3$, its absolute curvature function times the squared distance function to the boundary of M is bounded above by c.*

PROOF. Suppose on the contrary that there exist stable orientable minimally immersed surfaces M_n in $\mathbb{R}^3$ and points $p_n \in M_n$ in the interior of such surfaces such that $|K_{M_n}|(p_n) d_{M_n}(p_n, \partial M_n)^2 > n$ for all n, where K_{M_n}, d_{M_n} are respectively the Gaussian curvature and the intrinsic distance in M_n. By passing to the universal covering, we may assume that M_n is simply connected for each n. Let D_n be the embedded geodesic disk centered at p_n with radius $d_{M_n}(p_n, \partial M_n)$ and $q_n \in D_n$ a point where the function $d_{M_n}(\cdot, \partial D_n)^2 |K_{M_n}| \colon D_n \to [0, \infty)$ has a maximum value. We denote by $\widetilde{D}_n$ the minimal disk in $\mathbb{R}^3$ obtained by first translating by $-q_n$ the intrinsic disk in M_n with center q_n and radius $\frac{1}{2} d_{M_n}(q_n, \partial D_n)$ and then homothetically expanding this translated disk by the scaling factor $\sqrt{|K_{M_n}(q_n)|}$. Thus we have a sequence $\{\widetilde{D}_n\}_n$ of orientable stable minimal disks in $\mathbb{R}^3$, all passing through the origin with $d_{\widetilde{M}_n}(0, \partial \widetilde{D}_n)^2 \geq n$ and with Gaussian curvature function $K_{\widetilde{D}_n}$ satisfying $K_{\widetilde{D}_n}(0) = -1$ and $K_{\widetilde{D}_n} \geq -4$ on $\widetilde{D}_n$. A standard compactness result (see for example [**80**] or [**107**]) shows that a subsequence of the $\widetilde{D}_n$ converges uniformly on compact sets of $\mathbb{R}^3$ to an orientable, complete, simply-connected, immersed minimal surface $\widetilde{D}_\infty$ passing through the origin, with bounded Gaussian curvature $K_{\widetilde{D}_\infty}$, $K_{\widetilde{D}_\infty}(0) = -1$ and empty boundary. Since $\widetilde{D}_\infty$ is stable (because the smooth limit of stable minimal surfaces is stable), Theorem 8 implies that $\widetilde{D}_\infty$ is a plane, which is a contradiction. This finishes our proof of Theorem 15. □

REMARK 1. *Another blow-up argument implies that Theorem 15 also holds for stable minimal surfaces in a Riemannian three-manifold N^3 with injectivity radius bounded from below and which is uniformly locally quasi-isometric to balls in Euclidean space (in particular, it holds on any compact N^3).*

4.2. Barrier constructions. Barrier constructions allow one to construct compact and noncompact stable minimal surfaces in $\mathbb{R}^3$ that are constrained to lie in subdomains of $\mathbb{R}^3$ whose boundaries have nonnegative mean curvature. For example, consider two connected properly embedded disjoint minimal surfaces M_1, M_2 in $\mathbb{R}^3$ and the closed connected region W of $\mathbb{R}^3$ with $\partial W = M_1 \cup M_2$. We now show how to produce compact stable embedded minimal surfaces in W. First note that W is a complete flat manifold with boundary having zero mean curvature. Meeks and Yau [**89**] proved that W embeds isometrically in a complete homogeneously regular[7] Riemannian manifold $\widetilde{W}$ diffeomorphic to the interior of W and with metric g. Morrey [**93**] proved that in a homogeneously regular three-manifold one can solve the classical Plateau problem or other area minimizing problems. In particular, if Γ is an embedded 1-cycle in $\widetilde{W}$ which bounds an orientable chain in $\widetilde{W}$, then Γ is the boundary of a compact least area embedded surface $\Sigma_\Gamma(g) \subset \widetilde{W}$. Meeks and Yau prove that their metric g on $\widetilde{W}$ can be approximated by a family of homogeneously regular metrics $\{g_n\}_{n \in \mathbb{N}}$ which converge smoothly on compact subdomains to g and each g_n satisfies a convexity condition outside of $W \subset \widetilde{W}$, which forces the least area surface $\Sigma_\Gamma(g_n)$ to lie in W if Γ lies in W. A subsequence of the $\Sigma_\Gamma(g_n)$ converges to a smooth minimal surface Σ_Γ of least area in W with respect to the original flat metric. We now use this barrier construction to prove the Strong Halfspace Theorem (also see Theorem 7).

THEOREM 16. [**50, 85**] *If M_1 and M_2 are two disjoint properly immersed minimal surfaces in $\mathbb{R}^3$, then M_1 and M_2 are parallel planes.*

[7]A complete Riemannian manifold $(\widetilde{W}, g)$ is *homogeneously regular* if there exists an $\varepsilon > 0$ such that the injectivity radius of $\widetilde{W}$ is at least ε and ε-balls in $\widetilde{W}$ are uniformally quasi-isometric to ε-balls in $\mathbb{R}^3$. For example, every compact Riemannian three-manifold satisfies this rather weak property.

PROOF. Let W be the closed complement of $M_1 \cup M_2$ in $\mathbb{R}^3$ that has portions of both M_1 and M_2 on its boundary. The surface ∂W is a good barrier for solving Plateau-type problems in W. Let $M_1(1) \subset \ldots \subset M_1(n) \subset \ldots$ be a compact exhaustion of M_1 and let $\Sigma_1(n)$ be a least area surface in W with boundary $\partial M_1(n)$. Let α be a compact arc in W which joins a point in $M_1(1)$ to a point in $\partial W \cap M_2$. By elementary intersection theory, α intersects every least area surface $\Sigma_1(n)$. By compactness of least area surfaces, a subsequence of the surfaces $\Sigma_1(n)$ converges to a properly embedded area minimizing surface Σ in W with a component Σ_0 which intersects α. Since Σ_0 separates $\mathbb{R}^3$, Σ_0 is orientable and so by Theorem 8, Σ_0 is a plane. Hence, M_1 and M_2 lie in closed halfspaces of $\mathbb{R}^3$ and so, by Theorem 12, M_1 and M_2 are recurrent. But, then the height of M_1, M_2 over their separating plane is a positive harmonic function which must be constant (Proposition 3). Hence, M_1 and M_2 must be planes (instead of using Theorem 12, one could apply the Halfspace Theorem (Theorem 7) to conclude that M_1 and M_2 are parallel planes.) □

Another useful application of the barrier construction is the following. Suppose Γ is an extremal simple closed curve in $\mathbb{R}^3$, i.e. Γ lies on the boundary of its convex hull B. We first assume that ∂B is smooth. By the Jordan curve Theorem, Γ is the boundary of two disks $D_1, D_2 \subset \partial B$. Assume Γ bounds two different branched minimal immersions and let Σ denote their union. Let W_1, W_2 be the geodesic completions of the two components of $B - \Sigma$ which contain the disks D_1, D_2. In this case ∂W_1 and ∂W_2 consist of smooth pieces with zero mean curvature and convex corners. Meeks and Yau [**89**] proved that such boundaries are good barriers for solving least area problems. In fact, in this case they prove that Γ bounds a least area embedded disk $\widetilde{D}_1 \subset W_1$ and a different least area embedded disk $\widetilde{D}_2 \subset W_2$. Similarly, if Γ bounds a unique branched minimal surface which is not an embedded stable minimal disk, then with this barrier argument we produce two different embedded minimal disks with boundary Γ, which is a contradiction. If ∂B is not assumed to be smooth, then one can use an approximation argument by convex smooth boundaries (see e.g. [**88**]) to have the same conclusion.

On the other hand, Nitsche [**96**] proved that a regular analytic Jordan curve in $\mathbb{R}^3$ whose total curvature is at most 4π bounds a unique minimal disk. The hypothesis of analyticity for the boundary curve in Nitsche's Theorem comes from consideration of boundary branch points. When Γ is C^2 and extremal, there are never boundary branch points as shown in [**89**].

THEOREM 17. [**89**] *If Γ is a C^2-extremal curve with total curvature at most 4π, then Γ is the boundary of a unique compact branched minimal surface and this surface is a smooth embedded minimal disk of least area.*

5. The Ordering Theorem for the ends of properly embedded minimal surfaces.

The study of the ends of a properly embedded minimal surface $M \subset \mathbb{R}^3$ with more than one end has been extensively developed. Callahan, Hoffman and Meeks [**7**] showed that in one of the closed complements of M in $\mathbb{R}^3$ there exists a noncompact properly embedded minimal surface $\Sigma \subset \mathbb{R}^3 - M$ with compact boundary and finite total curvature. By the discussion following Theorem 2, the ends of Σ are of catenoidal or planar type, and the embeddedness of Σ forces its ends to have parallel normal vectors at infinity.

DEFINITION 19. In the above situation, the *limit tangent plane at infinity* of M is the plane in $\mathbb{R}^3$ passing through the origin whose normal vector equals (up to sign) the limiting normal vector at the ends of Σ. Such a plane is unique [**7**], in the sense that it does not depend on the finite total curvature minimal surface $\Sigma \subset \mathbb{R}^3 - M$.

The limit tangent plane at infinity is a key notion for studying the way in which a minimal surface with more than one end embeds properly in space.

THEOREM 18 (Ordering Theorem [**39**]). *Let $M \subset \mathbb{R}^3$ be a properly embedded minimal surface with more than one end and horizontal limit tangent plane at infinity. Then, the space $\mathcal{E}(M)$ of ends of M is linearly ordered geometrically by the relative heights of the ends over the (x_1, x_2)-plane, and embeds topologically in $[0,1]$ in an ordering preserving way. Furthermore, this ordering has a topological nature in the following sense: If M is properly isotopic to a properly embedded minimal surface M' with horizontal limit tangent plane at infinity, then the associated ordering of the ends of M' either agrees with or is opposite to the ordering coming from M.*

The linear ordering on the set of ends $\mathcal{E}(M)$ given by Theorem 18 lets us define the *top* end e_T of M as the unique maximal element in $\mathcal{E}(M)$ in the ordering (recall that $\mathcal{E}(M) \subset [0,1]$ is compact, hence e_T exists). Analogously, the *bottom* end e_B of M is the unique minimal element in $\mathcal{E}(M)$. If $e \in \mathcal{E}(M)$ is neither the top nor the bottom end of M, then it is called a *middle* end of M.

Rather than sketching the proof of the Ordering Theorem, we will be content to explain how one obtains the linear ordering. Suppose $M \subset \mathbb{R}^3$ is a minimal surface in the hypotheses of Theorem 18 and let $\mathcal{A} \subset \mathcal{E}(M)$ be the set of annular ends of M. By Theorem 1.1 in [**83**], each end $e \in \mathcal{A}$ is either planar or its third coordinate function is proper (since we are assuming that the limit tangent plane at infinity of M is horizontal). In this setting, Collin's Theorem (Theorem 3) insures that e has finite total curvature and thus, it is asymptotic to a horizontal plane or to a halfcatenoid. Since the ends in $\mathcal{A}$ are all graphs over complements of compact subdomains in the (x_1, x_2)-plane, we see that $\mathcal{A}$ has a natural linear ordering by relative heights of its ends over the (x_1, x_2)-plane. Hence the Ordering Theorem is proved when $\mathcal{A} = \mathcal{E}(M)$.

By Theorem 9, any end of M which can be represented by a proper stable subdomain can be also represented by a surface of finite total curvature and so, it can be represented by an annulus. Let $e_1 = [\alpha_1] \in \mathcal{E}(M)$ be an end which is not annular. Such an end can always be represented by a proper subdomain E_1 which is unstable and where ∂E_1 is connected and $\overline{M - E_1}$ is also unstable and noncompact (we are assuming M has at least 2 ends). Let W_1, W_2 be the two closed complements of M in $\mathbb{R}^3$. Note that we can consider E_1 to lie on the boundary of both of these complete flat 3-manifolds W_1, W_2, and that their boundaries $\partial W_1, \partial W_2$ are good barriers for solving Plateau-type problems. Since E_1 and $\overline{M - E_1}$ are both noncompact, elementary separation properties for surfaces in $\mathbb{R}^3$ imply that ∂E_1 is not homologous to zero in one of the domains W_1, W_2; suppose that ∂E_1 is not homologous to zero in W_1. Since ∂E_1 bounds the locally finite 2-chain E_1 in ∂W_1, the barrier argument in Subsection 4.2 shows that ∂E_1 is the boundary of a properly embedded orientable least area surface Σ_1 in W_1, which is noncompact since ∂E_1 is not homologous to zero in W_1. Similarly, ∂E_1 is the boundary of a least area (possibly compact) surface Σ_2 in W_2. Since E_1 and $M - E_1$ are unstable, the maximum principle implies that $(\Sigma_1 \cup \Sigma_2) \cap M = \partial E_1$.

Let R_1 be the closed complement of $\Sigma_1 \cup \Sigma_2$ in $\mathbb{R}^3$ which contains E_1 and let R_2 be the other closed complement. Since Σ_1 and $\overline{M - E_1}$ are both noncompact and M is properly embedded in $\mathbb{R}^3$, R_1 and R_2 are both noncompact. It follows that outside a large ball containing ∂E_1, the boundary of R_1, which equals $\Sigma_1 \cup \Sigma_2 = \partial R_2$, consists of a finite positive number of graphical ends which are asymptotic to the ends of horizontal planes and vertical catenoids. Let $e_2 = [\alpha_2] \in \mathcal{E}(M)$ be an end with a representative E_2 which is disjoint from E_1 (note that any two distinct ends can be chosen to have disjoint representatives). The proper arc α_1 eventually lies in R_1 and so, it eventually lies between two successive graphical ends of ∂R_1 or α_1 eventually lies in the region above the top graphical end of ∂R_1 or below the bottom graphical end of ∂R_1. A similar statement holds for the proper arc $\alpha_2 \subset R_2$. In particular there is a topological graphical plane P over the (x_1, x_2)-plane whose end is one of the ends of ∂R_1 and eventually α_1 and α_2 lie on opposite sides of P. If α_1 eventually lies below P and α_2 eventually lies above P, then $[\alpha_1] < [\alpha_2]$ in the linear ordering given by the Ordering Theorem; otherwise, $[\alpha_2] < [\alpha_1]$. The ordering we have just described can be proven to be a well-defined linear ordering, see [**39**] for more details.

6. Quadratic area growth and recurrence.

In this section we will sketch the proof of a Theorem that constrains both the geometry and the topology of properly embedded minimal surfaces in $\mathbb{R}^3$ with more than one end. This Theorem has been used in an essential way by Meeks, Perez and Ros in the proofs of their classification results in Theorem 29, Theorem 31 and Theorem 32 below, as well as by Frohman and Meeks in their proof of the Topological Classification of Minimal Surfaces (Theorem 28); all of these Theorems are discussed in Section 10 of this survey.

The Ordering Theorem in the previous Section represents the first step in trying to understand the geometry of properly embedded minimal surfaces with more than one end. By the proof of the Ordering Theorem, a middle end of a properly embedded minimal surface M with horizontal limit tangent plane at infinity can be represented by a proper subdomain $E \subset M$ with compact boundary such that E "lies between two catenoids." This means that E is contained in a neighborhood S of the (x_1, x_2)-plane, S being topologically a slab, whose width grows at most logarithmically with the distance from the origin.

Suppose for the moment that E is in fact contained in the region $W = \{(x_1, x_2, x_3) \mid r \geq 1,\ 0 \leq x_3 \leq 1\}$, where $r = \sqrt{x_1^2 + x_2^2}$. In Section 3 we defined the universal superharmonic function $\ln r - x_3^2$ in W. In particular, the restriction $f\colon E \to \mathbb{R}$ is superharmonic and proper. Suppose $f(\partial E) \subset [-1, c]$ for some $c > 0$. Replace E by $f^{-1}[c, \infty)$ and let $E(t) = f^{-1}[c, t]$ for $t > c$. Assuming that both c, t are regular values of f, the Divergence Theorem gives

$$\int_{E(t)} \Delta f \, dA = -\int_{f^{-1}(c)} |\nabla f| \, ds + \int_{f^{-1}(t)} |\nabla f| \, ds,$$

where ∇, Δ are the intrinsic gradient and laplacian on M, and dA, ds denote the corresponding area and length elements.

Since f is superharmonic, the function $t \mapsto \int_{E(t)} \Delta f \, dA$ is monotonically decreasing and bounded from below by $-\int_{f^{-1}(c)} |\nabla f| \, ds$. In particular, Δf lies in $L^1(E)$. Furthermore, $|\Delta f| = |\Delta \ln r - 2|\nabla x_3|^2| \geq -|\Delta \ln r| + 2|\nabla x_3|^2$. By estimate (11) in Section 3, $|\Delta \ln r| \leq \frac{|\nabla x_3|^2}{r^2}$, hence $|\Delta f| \geq \left(2 - \frac{1}{r^2}\right) |\nabla x_3|^2$. Since $r^2 \geq 1$ in W, it follows $|\Delta f| \geq |\nabla x_3|^2$ and thus, both $|\nabla x_3|^2$ and $\Delta \ln r$ are in $L^1(E)$. This implies that outside of a subdomain of E of finite area, E can be assumed to be as close to being horizontal as one desires and in particular for the radial function r on this horizontal part of E, $|\nabla r|$ is almost equal to 1.

Let $r_0 = \max r|_{\partial E}$. With a slight abuse of notation, redefine $E(t)$ to be the subdomain of E that lies inside the region $\{r_0^2 \leq x_1^2 + x_2^2 \leq t^2\}$. Since $\int_{E(t)} \Delta \ln r \, dA = -\int_{r=r_0} \frac{|\nabla r|}{r} ds + \int_{r=t} \frac{|\nabla r|}{r} ds = \text{const.} + \frac{1}{t} \int_{r=t} |\nabla r| \, ds$ and $\Delta \ln r \in L^1(E)$, then the following limit exists:

$$\lim_{t \to \infty} \frac{1}{t} \int_{r=t} |\nabla r| \, ds = C \tag{14}$$

for some positive constant C. Thus, $t \mapsto \int_{r=t} |\nabla r| \, ds$ grows at most linearly as $t \to \infty$. By the coarea formula, for t_1 fixed and large

$$\int_{E \cap \{t_1 \leq r \leq t\}} |\nabla r|^2 \, dA = \int_{t_1}^{t} \left(\int_{r=\tau} |\nabla r| \, ds \right) d\tau; \tag{15}$$

hence, $t \mapsto \int_{E \cap \{t_1 \leq r \leq t\}} |\nabla r|^2 \, dA$ grows at most quadratically as $t \to \infty$. Finally, since outside of a domain of finite area E is arbitrarily close to horizontal and $|\nabla r|$ is almost equal to one, we conclude that the area of $E \cap \{r \leq t\}$ grows at most quadratically as $t \to \infty$. In fact, from (14) and (15) it follows that

$$\int_{E \cap \{r \leq t\}} dA = \frac{C}{2} t^2 + o(t),$$

where $t^{-2} o(t) \to 0$ as $t \to \infty$. We now check that the constant C must be an integer multiple of 2π. The locally finite minimal integral varifolds associated to the homothetically shrunk surfaces $\frac{1}{n} E$ converge as $n \to \infty$ to a locally finite minimal integral varifold with empty boundary which is contained in the (x_1, x_2)-plane. Since this limit varifold must is an integer multiple of the (x_1, x_2)-plane, C must be an integer multiple of 2π.

In the case that the end E "lies between catenoids", a similar analysis (see [**26**] for details) using the universal superharmonic function $\ln r - c(x_3 \arctan(x_3) - \frac{1}{2} \ln(x_3^2 + 1))$, for some $c > 0$, shows that $E \cap \{r \leq t\}$ has area growth $n\pi t^2$ for some $n \in \mathbb{N}$. This in turn implies that E has area growth $n\pi R^2$ where $R = \sqrt{x_1^2 + x_2^2 + x_3^2}$, in the sense that $\lim_{t \to \infty} \frac{\text{Area}(M \cap \{R \leq t\})}{\pi t^2} = n$.

By the monotonicity formula for area (Theorem 4), every end representative of a minimal surface must have area growth at least equal to the area growth of a plane which is πR^2. Since we just checked that middle ends of a properly embedded minimal surface have quadratic area growth, then we can find representatives for middle ends which have exactly one end, which means they are never limit ends. This discussion gives the next Theorem and we refer the reader to [**26**] for further details.

THEOREM 19 (Collin, Kusner, Meeks, Rosenberg [**26**]). *Let $M \subset \mathbb{R}^3$ be a properly embedded minimal surface with more than one end and horizontal tangent* *plane at infinity. Then, any limit end of M must be a top or bottom end of M. In particular, M can have at most two limit ends, each middle end is simple and the number of ends of M is countable. Furthermore, each middle end E of M has limiting quadratic area growth $m(E)\pi R^2$ as $R \to \infty$, where $m(E)$ is a positive integer. The parity of $m(E)$ is called the parity of the middle end E.*

Collin, Kusner, Meeks and Collin [**26**] were also able to use universal superharmonic functions to control the geometry of properly embedded minimal surfaces with exactly two limit ends. Their proof of the following Theorem is motivated by the proof of a similar theorem by Callahan, Hoffman and Meeks [**7**] in the classical 1-periodic setting.

THEOREM 20. *Let $M \subset \mathbb{R}^3$ be a properly embedded minimal surface with two limit ends and horizontal limit tangent plane at infinity. Then there exists a proper collection $\{P_n \mid n \in \mathbb{Z}\}$ of horizontal planes in $\mathbb{R}^3$ such that every plane intersects M transversely in a finite number of simple closed curves. Furthermore, the closed slab S_n bounded by $P_n \cup P_{n+1}$ intersects M in a noncompact domain which represents the n-th end of M. In particular, by Theorem 12, M is recurrent.*

7. Maximum principle at infinity for properly immersed minimal surfaces.

In this Section we will present a sketch of the proof of a general maximum principle at infinity. As an important application of it, we will also show the existence of tubular neighborhoods for any properly embedded minimal surface with bounded Gaussian curvature, which in turn implies that the area growth of such a surface in balls is not more than cubical in the radial function R. The next statement generalizes the maximum principle at infinity described in Theorem 6 to the case of noncompact boundaries.

THEOREM 21 (Maximum principle at infinity [**79, 121**]). *Let $M_1, M_2 \subset N^3$ be disjoint connected properly immersed minimal surfaces with possibly empty boundaries in a complete flat 3-manifold N^3.*

i) If $\partial M_1 \neq \emptyset$ or $\partial M_2 \neq \emptyset$, then

$$\mathrm{dist}(M_1, M_2) = \min\{\mathrm{dist}(\partial M_1, M_2), \mathrm{dist}(\partial M_2, M_1)\}.$$

ii) If $\partial M_1 = \partial M_2 = \emptyset$, then M_1 and M_2 are flat.

SKETCH OF THE PROOF. The only real difference between the statements of Theorem 6 and of the above Theorem is that the boundaries of the surfaces in Theorem 21 are allowed to be noncompact. This noncompactness property presents some serious technical difficulties that are not easy to overcome. The most difficult of these problems arises from the unknown conformal structure of a stable orientable complete minimal surface Σ with boundary, which we now explain. By Theorem 9, when Σ has compact boundary, then it has finite total curvature and so it is conformally a finitely punctured compact Riemann surface with compact boundary. In particular, when Σ has compact boundary, then it is parabolic. On the other hand, at the end of Section 3 we discussed an example of a complete orientable stable minimal surface with noncompact boundary which was not parabolic. The fact that a stable minimal surface need not be parabolic makes the proof of the general maximum principle at infinity quite delicate. We now give the sketch of the proof and will eventually indicate how one circumvents this problem with the conformal structure. Note that by lifting the surfaces to the universal cover $\mathbb{R}^3$ of N^3, we may assume $N^3 = \mathbb{R}^3$.

Assume that Theorem 21 fails for some properly immersed minimal surfaces M_1, M_2. The first step in the proof of Theorem 21 is to reduce to the case where M_1, M_2 are properly embedded and stable. To do this one uses $M_1 \cup M_2$ together with small ε-neighborhoods of their boundary curves as barriers to obtain stable disjoint minimal surfaces M_1', M_2', where $\partial M_i'$ lies in the boundary of the ε-neighborhood of ∂M_i for $i = 1, 2$. This barrier argument is similar to one described in Subsection 4.2. Furthermore, the new stable surfaces are also constructed to be at least as close to each other as the previous ones and so by choosing ε-sufficiently small, M_1', M_2' are new counterexamples to the Theorem. So assume from now on that M_1, M_2 are properly embedded and stable.

Next we show how to complete the proof in the case that ∂M_1 and ∂M_2 are both compact (therefore proving Theorem 6). In this case, Theorem 9 implies that M_1, M_2 are two properly embedded connected minimal surfaces in $\mathbb{R}^3$ which have compact boundary and finite total curvature. Since the Theorem fails for M_1, M_2, then, after a fixed translation of M_1, we may assume that the distance from M_1 to M_2 is zero but the distances from ∂M_1 to M_2 and from ∂M_2 to M_1 are both positive. The only way that this can happen is that an end E_1 of M_1 has distance zero to some end E_2 of M_2. Since the ends of embedded finite total curvature surfaces can be taken to be graphs which are planar (zero logarithmic growth) or of catenoid-type (bounded logarithmic growth), we may assume after a rigid motion of $\mathbb{R}^3$ that E_1 and E_2 are nonnegative graphs over an annular domain $A = \{(x_1, x_2) \mid x_1^2 + x_2^2 \geq R_0^2\}$ with some fixed nonnegative logarithmic growth. In this case the estimate in formula (9) implies E_1 and E_2 are both asymptotic to the end of a common plane or to the end of a fixed catenoid. In particular, E_1 and E_2 are asymptotic to each other.

Suppose that the E_1 lies below E_2. After a small upward translation of E_1, we obtain a surface E_1' whose end lies above the end of E_2 and whose boundary lies below the boundary of E_2. An application of the usual maximum principle implies that $\Gamma = E_1' \cap E_2$ is a simple closed homotopically nontrivial curve on both surfaces. Let $E_1'(\Gamma) \subset E_1'$ and $E_2(\Gamma) \subset E_2$ be the annular ends of these surfaces with common boundary Γ. Note that the third component of the conormal to $E_1'(\Gamma)$ along Γ is pointwise greater in norm than the corresponding third component of the conormal to $E_2(\Gamma)$ along Γ, since $E_1'(\Gamma)$ lies above $E_2(\Gamma)$ along Γ. It follows that the vertical component of the flux of E_1' is greater in norm than the vertical component of the flux of E_2. But these vertical fluxes of E_1' and E_2 depend only on their logarithmic growths, which are equal. This contradiction completes the proof of the case where ∂M_1 and ∂M_2 are both compact.

We now continue with our proof in the case ∂M_1 or ∂M_2 is not compact. By our previous arguments, M_1 and M_2 can be supposed stable and properly embedded and, after a possible translation, have $\text{dist}(M_1, M_2) = 0$ but with both $\text{dist}(\partial M_1, M_2)$ and $\text{dist}(\partial M_2, M_1)$ being positive. By curvature estimates for stable minimal surfaces (Theorem 15), after removing small regular neighborhoods of their boundaries, the Gaussian curvature of each of the surfaces is bounded. Since M_1 has bounded Gaussian curvature, there exists $\varepsilon > 0$ such that the natural exponential map from the normal bundle $N(M_1)$ of M_1 into $\mathbb{R}^3$ restricts as a submersion to the subbundle $N_\varepsilon(M_1) = \{\xi \in N(M_1) \mid \|\xi\| < \varepsilon\}$. Note that M_1 can be seen as the zero section of $N_\varepsilon(M_1)$. Since some points of the interior of M_2 approach points of the interior of M_1, then after pulling-back portions of M_2 to $N_\varepsilon(M_1)$ we can find proper domains $\Omega_1 \subset M_1$, $\Omega_2 \subset M_2 \cap N_\varepsilon(M_1)$ such that Ω_2 is a normal graph over Ω_1 with boundary values ε and $\text{dist}(\Omega_1, \Omega_2) = 0$.

If Ω_1 is parabolic, then using the minimal surface equation for such a normal graph for ε small, one can produce a positive proper superharmonic function on Ω_1 which does not have its minimum on the boundary of Ω_1, which contradicts the maximum principle. Hence, Ω_1 is not parabolic. In fact, after removing any sufficiently small fixed size regular neighborhood of $\partial\Omega_1$ from Ω_1, this argument shows that we still obtain a surface which cannot be parabolic. This problem motivates the following definition.

DEFINITION 20. A Riemannian surface M with boundary is *ε-parabolic* if for any $\varepsilon > 0$, $M(\varepsilon) = \{p \in M \mid \text{dist}(p, \partial M) \geq \varepsilon\}$ is a parabolic surface.

Hence, to complete the proof of Theorem 21 it is sufficient to prove that Ω_1 is ε-parabolic. While it must certainly be the case that Ω_1 is ε-parabolic, the proof of this property is not what is shown. Rather one proves that between the zero section Ω_1 and the graph Ω_2, there are normal minimal graphs $\Delta(t)$ over Ω_1 for every $0 < t < \varepsilon$. One of these graphs $\Delta(t_0)$ is a limit of the other graphs and such that the limit induces a positive Jacobi function u on $\Delta(t_0)$ which is bounded away from zero in a fixed sized neighborhood of its boundary. The arguments of Fischer-Colbrie in [**34**] in her proof of Theorem 9 now imply that if g is the metric for $\Delta(t_0)$, then $u \cdot g$ is a new metric which is complete and has nonpositive Gaussian curvature. One then proves that such surfaces are ε-parabolic, which easily implies that $\Delta(t_0)$ is ε-parabolic in the original metric. Once one obtains such a $\Delta(t_0)$, then using the previous arguments one rather easily arrives at a contradiction. This completes our sketch of the proof to Theorem 21. □

We now come to a beautiful and deep application of the general maximum principle at infinity. The next Corollary appears in [**79**] and a slightly weaker variant of it in [**121**].

COROLLARY 2. *Suppose $M \subset \mathbb{R}^3$ is a properly embedded minimal surface with absolute bounded Gaussian curvature at most* 1. *Let $N_1(M)$ be the subbundle of the normal bundle of M given by the vectors of length strictly less that* 1, *and let* $\exp\colon N_1(M) \to \mathbb{R}^3$ *be the corresponding exponential map. Then* exp *is a smooth embedding. In particular, M has a open embedded tubular neighborhood of radius 1.*

Instead of giving a proof of Corollary 2, we will only check that a minimal surface M under the hypotheses in Corollary 2 has some tubular neighborhood. Consider $X = \exp^{-1}(M) \subset N_1(M)$ and note that X contains the zero section M of $N_1(M)$. Suppose X contains a component Δ which is different from the zero section. In this case, the distance from $\partial\Delta$ to M is equal to 1, but the distance from Δ to M is strictly less that 1. Since $\exp\colon N_1(M) \to \mathbb{R}^3$ is a submersion, under the pulled-back metric we can view $N_1(M)$ as a flat three-manifold. Inside this flat three-manifold, Δ and M contradict the maximum principle at infinity, or rather the proof of Theorem 21 holds in this setting and so gives the desired contradiction. We conclude that $X = M$. But then a simple application of the triangle inequality implies the restricted map $\exp|_{N_{\frac{1}{2}}(M)}$ is injective and so M has a tubular neighborhood of radius $\frac{1}{2}$. A special barrier argument is used to obtain the optimal maximal radius 1 tubular neighborhood given in the statement of the Corollary.

8. Flux conjectures and some properties of harmonic functions in parabolic Riemannian manifolds.

Algebraic-analytic invariants associated to a properly embedded minimal surface $M \subset \mathbb{R}^3$ are a way of not only distinguishing M from other such examples but also can be used as theoretical tools for acquiring deeper information on the surface. One of the most natural of these invariants is the flux of M associated to a Killing vector field V of $\mathbb{R}^3$. Given such a V and its restriction V_M to M, let V_M^T denote the tangent vector field on M which is the tangential part of V_M; note that the normal projection $V_M^N = V_M - V_M^T$ is a Jacobi vector field (i.e. the inner product of V_M with the Gauss map N of M is a Jacobi function). Given two oriented homologous cycles $\gamma_1, \gamma_2 \subset M$, the Divergence Theorem implies that the flux $\int_{\gamma_1} \langle V_M^T, \eta \rangle \, ds$ of V_M^T across γ_1 is equal to its flux across γ_2 (here η stands for the unit conormal vector to M along the integration curve). When the Killing vector field V is one of the parallel vector fields e_1, e_2, e_3 induced by the standard coordinate functions in $\mathbb{R}^3$, then for $V = e_i$ is $V_M^T = \nabla x_i$ where ∇x_i is the intrinsic gradient of the coordinate function x_i on M. Thus, for each homology class $\gamma \in H_1(M, \mathbb{Z})$, one obtains the flux vector $F(\gamma)$ defined in equation (8).

There is a well-known conjecture of Meeks concerning precisely when the so called *flux map* $F\colon H_1(M,\mathbb{Z}) \to \mathbb{R}^3$ is the zero map, or equivalently, it describes precisely which surfaces M have zero flux.

CONJECTURE 1 (Flux Conjecture (I)). *If $M \subset \mathbb{R}^3$ is a properly embedded minimal surface with zero flux, then M is a plane or a helicoid.*

Because of the characterization by Meeks and Rosenberg (Theorem 30 below) of the plane and the helicoid as the unique properly embedded simply connected minimal surfaces in $\mathbb{R}^3$, Conjecture 1 can be equivalently stated as follows.

CONJECTURE 2 (Flux Conjecture (II)). *If $M \subset \mathbb{R}^3$ is a properly embedded nonsimply connected minimal surface, then M has nonzero flux.*

An isometric minimal immersion $X\colon (M, ds^2) \to \mathbb{R}^3$ is *rigid* if given another isometric minimal immersion $Y\colon (M, ds^2) \to \mathbb{R}^3$, there exists a rigid motion $R\colon \mathbb{R}^3 \to \mathbb{R}^3$ such that $R \circ X = Y$. On the other hand, if X has Weierstrass data (g, dh), then for each $\theta \in [0, 2\pi)$ the pair $(g, e^{i\theta} dh)$ defines via Theorem 1 a possibly multivalued[8] minimal surface X_θ, which is called an *associate surface* of X. The associate surface for $\theta = \pi/2$ is the conjugate surface. Equation (6) implies that two associate minimal surfaces are locally isometric. The converse is true by Calabi's Rigidity Theorem [**5**], which asserts that if $X, Y\colon M \to \mathbb{R}^3$ are isometric minimal immersions and they are locally isometric, then Y is congruent to an associate surface of X.

LEMMA 2. *Let $X\colon M \to \mathbb{R}^3$ be a isometric minimal immersion. Then, the following statements are equivalent:*

1. *X is rigid.*
2. *All associate surfaces to X are multivalued, except X and X_π.*
3. *There exists a cycle $\gamma \in H_1(M, \mathbb{Z})$ such that the flux $F(\gamma)$ is not zero.*

PROOF. If X is rigid and $\theta \in [0, 2\pi)$, then the pulled-back metrics by X, X_θ coincide, hence if X_θ is univalent then $R \circ X_\theta = X$ for certain rigid motion R. Since

[8] In the sense that X_θ may have real periods although X solves the period problem.

X, X_θ share the Gauss map, the linear part of R must be the identity I or $-I$ and so, $\theta = 0$ or π, which proves *1* $\Rightarrow$ *2*. The implication *2* $\Rightarrow$ *3* follows easily from the fact that the flux vector of X on a given cycle $[\gamma] \in H_1(M, \mathbb{Z})$ is a period for its conjugate surface. To see *3* $\Rightarrow$ *1*, take an isometric minimal immersion $Y: M \to \mathbb{R}^3$. By Calabi's Rigidity Theorem, there exists a rigid motion $R: \mathbb{R}^3 \to \mathbb{R}^3$ and an associate surface X_θ such that $R \circ X_\theta = Y$. In particular, X_θ is well-defined on M. By hypothesis, there exists $\gamma \in H_1(M, \mathbb{Z})$ such that the $F(\gamma) \neq 0$. Since the period of X_θ along γ equals $\sin\theta \cdot F(\gamma)$, it follows that $\theta = 0$ or π, thus X and Y are congruent. □

Lemma 2 insures that the helicoid is not minimally rigid. Meeks and Rosenberg [**81**] proved that every nonflat doubly periodic minimal surface $M \subset \mathbb{R}^3$ with finite topology in the quotient has a cycle with nonzero flux. Meeks [**71**] extended this property to all triply periodic minimal surfaces and to singly periodic minimal surfaces with finite topology in the quotient and planar or Scherk-type ends. By Lemma 2 all these surfaces are rigid, and the helicoid shows that we cannot expect to extend these results to singly periodic minimal surfaces with helicoidal type ends. Pérez [**104**] showed that no other nonrigid example can exist in this last family of surfaces. All these results together imply that if a M is a nonflat properly embedded periodic minimal surface in $\mathbb{R}^3$ with finite topology in the quotient, then M is rigid or it is the helicoid.

An easy consequence of the definition of rigidity is that in a rigid minimal surface, every intrinsic isometry extends to an ambient isometry of $\mathbb{R}^3$ (this property is sometimes referred to as *weak rigidity*). Since intrinsic isometries of the helicoid and the plane extend to ambient isometries, the validity of the equivalent Conjectures 1 or 2 would imply the following conjecture by Meeks:

CONJECTURE 3 (Isometry Conjecture). *If $M \subset \mathbb{R}^3$ is a properly embedded minimal surface, then every intrinsic isometry of M extends to an ambient isometry.*

By the above arguments, Conjecture 3 holds when M is periodic and has finite topology in the quotient. In 1990, Choi, Meeks and White [**12**] proved that the stronger Conjecture 2 holds when M has more than one end and very recently, Meeks and Rosenberg (Theorem 30) extended the validity of Conjecture 2 to the case of M having one end and finite genus. In summary, Conjectures 1, 2 and 3 can only fail if the surface M has exactly one end and infinite genus. By Lemma 2, one way to prove Conjecture 1, 2 and 3 would be to show that when M has one end and infinite genus, then there exists a plane in $\mathbb{R}^3$ that intersects M in a set that contains a simple closed curve.

Another important conjecture related to flux, which is also closely related to classification questions, is the following one due to Meeks, Pérez and Ros.

CONJECTURE 4 (Flux Conjecture (III)). *If $M \subset \mathbb{R}^3$ is a properly embedded minimal surface and the rank of the flux map is* 1, *then M is the catenoid, one of the Riemann minimal examples or one of the Scherk doubly-periodic examples defined in Subsection 2.5.*

For a minimal surface $M \subset \mathbb{R}^3$ whose flux map has rank 1 we have at our disposal a powerful tool: the López-Ros deformation introduced in Subsection 2.2. This tool together with the maximum principle for minimal surfaces were crucial when proving certain uniqueness and nonexistence results, see for instance [**63,**

104, 105]. The interested reader can find an exposition of both this tool and some of its applications in Pérez and Ros [**107**].

By Collin's Theorem 3 and the López-Ros [**63**] characterization of the catenoid, Conjecture 4 holds when M has finite topology and more than on end (the only possible surface in this case is the catenoid). On the other hand, Theorems 31 and 32 below imply that any nonsimply connected properly embedded minimal surface $M \subset \mathbb{R}^3$ of genus zero has flux map of rank 1, and so a positive solution to Conjecture 4 would complete the classification of genus zero minimal surfaces in $\mathbb{R}^3$.

We now describe a recently discovered flux invariant associated to each coordinate function of a properly immersed minimal surface M in $\mathbb{R}^3$. Suppose $x\colon M \to \mathbb{R}$ is a coordinate function and $t \in \mathbb{R}$. Consider the scalar flux (possibly infinite) $\int_{x^{-1}(t)} |\nabla x|\, ds$ of the intrinsic gradient ∇x across the level set $x^{-1}(t)$, which is oriented almost everywhere by the normal vector $(\nabla x)|_{x^{-1}(t)}$. If $t_1 \leq t_2$ are real numbers, then Theorem 12 implies that $M[t_1, t_2] = x^{-1}[t_1, t_2]$ is a parabolic manifold. Thus, by the following Lemma, the scalar flux of ∇x across a level set is independent of the level set; we will call this number the *flux of* ∇x.

LEMMA 3 (Flux Lemma, Meeks [**67**]). *Suppose M is a parabolic Riemannian manifold, $h\colon M \to [0,1]$ is a nonconstant harmonic function and $\partial M = h^{-1}(\{0,1\})$. Then, the flux of ∇h across $h^{-1}(0)$ equals the flux of ∇h across $h^{-1}(1)$.*

The above Lemma is a consequence of Green's Theorem and the definition of a parabolic manifold. Meeks has made the following related Conjecture.

CONJECTURE 5 (Finite Geometric Flux Conjecture). *Let M be a parabolic Riemannian manifold and $h\colon M \to [0,1]$ a nonconstant harmonic function with $\partial M = h^{-1}(\{0,1\})$. If ∇h has finite flux across $h^{-1}(0)$, then almost all integral curves of ∇h begin at $h^{-1}(0)$ and end at $h^{-1}(1)$.*

Meeks and Wolf [**87**] have been able to prove Conjecture 5 in the case of dimension two. We finish this Section with a related problem to Conjecture 5. This problem constitutes one of the deepest conjectures in classical minimal surface theory and it is due to Meeks and Rosenberg.

CONJECTURE 6 (Geometric Flux Conjecture). *Suppose $M \subset \mathbb{R}^3$ is a properly embedded minimal surface. Then, except for a countable subset, the integral curves of a given coordinate function begin at $-\infty$ and end at $+\infty$.*

By the results contained in Theorem 3, Theorem 30, Theorem 31 and Theorem 32, the Geometric Flux Conjecture holds when M has finite genus.

9. The Colding-Minicozzi curvature estimates, compactness and regularity of limit laminations and applications.

The famous open *spherical space-form problem* asks if any free action of a finite group Γ of diffeomorphisms of the sphere $\mathbb{S}^3$ is conjugate to an action by isometries in $O(4)$. According to Pitts and Rubinstein [**108**], the failure of this question to be solved in the affirmative would create by minimax methods a sequence of embedded minimal unknotted tori M_n in some Γ-invariant metric on $\mathbb{S}^3$ with the index of stability of M_n at least n. This approach motivated the following open question by Pitts and Rubenstein.

QUESTION 2. *Does there exist a bound for the stability index of all closed embedded minimal surfaces with fixed genus in a given closed 3-manifold N with respect to some large class of Riemannian metrics?*

Arguments in Choi and Schoen [**11**] imply that if N is a closed Riemannian 3-manifold with a generic metric, then there exists a bound for the stability index of all closed embedded minimal surfaces in N with fixed genus and a fixed area bound. Thus, the above question by Pitts and Rubinstein leads to the study of sequences of closed embedded minimal surfaces with fixed genus in such a N^3 but without area (or density) bounds. Viewing a sequence of embedded minimal surfaces in N as a sequence of Radon measures on N, compactness results from Geometric Measure Theory imply that a subsequence of the measures (renormalized to have fixed mass) converges weakly to another Radon measure. The problem of understanding the geometric structure of this limit starts with the analysis of the local structure in a fixed ball, which motivates the central problem tackled by Colding and Minicozzi in a recent series of papers: the local structure of a limit of compact embedded minimal surfaces M_n with fixed genus and no area bound in a ball $B \subset \mathbb{R}^3$, with boundaries $\partial M_n \subset \partial B$. The most important case of their structure Theorem is when the M_n are disks and their Gaussian curvature blows up near the origin.

The basic example in this setting is a sequence of rescaled helicoids $M_n = a_n H$ where H is a fixed vertical helicoid with axis the x_3-axis and $a_n \in \mathbb{R}^+$, $a_n \to 0$. The curvature of the sequence $\{M_n\}_n$ blows up along the x_3-axis and the M_n converge away from the axis to a foliation $\mathcal{L}$ of $\mathbb{R}^3$ by horizontal planes. The x_3-axis is the singular set of convergence $S(\mathcal{L})$ of M_n to $\mathcal{L}$, but each leaf of $\mathcal{L}$ extends smoothly across its intersection with $S(\mathcal{L})$ (i.e. $S(\mathcal{L})$ consists of removable singularities of $\mathcal{L}$). With this model in mind, the statement of the so called *Limit Lamination Theorem* (Theorem 0.1 of [**18**]) can be easily understood. Given $p \in \mathbb{R}^3$ and $R > 0$, we denote by $B(p, R) = \{x \in \mathbb{R}^3 \mid \|x - p\| < R\}$, $B(R) = B(0, R)$ and K_M the Gaussian curvature function of a surface M.

THEOREM 22 (Colding, Minicozzi [**18**]). *Let $M_n \subset B(R_n)$ be a sequence of embedded minimal disks with $\partial M_n \subset \partial B(R_n)$ and $R_n \to \infty$. If $\sup |K_{M_n \cap B(1)}| \to \infty$, then there exists a subsequence of the M_n (denoted in the same way) and a Lipschitz curve $S\colon \mathbb{R} \to \mathbb{R}^3$ such that up to a rotation of $\mathbb{R}^3$,*

1. *$x_3(S(t)) = t$ for all $t \in \mathbb{R}$.*
2. *For each compact domain $C \subset \mathbb{R}^3 - S$ and n large enough, $M_n \cap C$ consists of two multivalued graphs over a subdomain of $\{x_3 = 0\}$ with $\{M_n \cap C\}_n$ converging to $C \cap \mathcal{L}$ as $n \to \infty$ in the C^α-topology for any $\alpha \in (0,1)$, where $\mathcal{L} = \{x_3 = t\}_{t \in \mathbb{R}}$ is the foliation of $\mathbb{R}^3$ by horizontal planes.*
3. *$\sup |K_{M_n \cap B(S(t), r)}| \to \infty$ as $n \to \infty$, for any $t \in \mathbb{R}$ and $r > 0$.*

Theorem 22 has two main ingredients in its proof, which we explain very roughly. The first ingredient is that the embedded minimal disk M_n with large curvature at some interior point can be divided into building blocks, each one being a multivalued graph $u_n(\rho, \theta)$ defined on an annulus[9], and that these basic pieces fit together properly. In particular, they prove that the number of sheets of $u_n(\rho, \theta)$

[9]In polar coordinates (ρ, θ) with $\rho > 0$ and $\theta \in \mathbb{R}$, a *N-valued graph on an annulus of inner radius r and outer radius R,* is a single valued graph of a function $u(\rho, \theta)$ defined over $\{(\rho, \theta) \mid r \leq \rho \leq R,\ |\theta| \leq N\pi\}$, N being a positive integer. The *separation* between consecutive sheets is $w(\rho, \theta) = u(\rho, \theta + 2\pi) - u(\rho, \theta) \in \mathbb{R}$.

rapidly grows as the curvature blows up and at the same time the sheets do not accumulate in a halfspace. This is obtained by means of sublinear and logarithmic bounds for the *separation*[9] $w_n(\rho, \theta)$ as a function of $\rho \to \infty$. Another consequence of these bounds is that by allowing the *inner radius*[9] of the annulus where the multigraph is defined to go to zero, the sheets of this multigraph collapse (i.e. $|w_n(\rho, \theta)| \to 0$ as $n \to \infty$ for ρ, θ fixed), thus a subsequence of the u_n converges to a smooth minimal graph through $\rho = 0$. The fact that the R_n go to ∞ then implies this limit graph is entire and, by Bernstein Theorem [**1**], it is a plane.

The second main ingredient in the proof of Theorem 22 is the so called *one-sided curvature estimate,* a scale invariant bound for the Gaussian curvature of embedded minimal disks in a halfspace.

THEOREM 23 (Colding, Minicozzi [**18**]). *There exists $\varepsilon > 0$ such that the following holds. Given $r > 0$ and an embedded minimal disk $M \subset B(2r) \cap \{x_3 > 0\}$ with $\partial M \subset \partial B(2r)$, then for any component M' of $M \cap B(r)$ which intersects $B(\varepsilon r)$,*

$$\sup_{M'} |K_M| \leq r^{-2}.$$

The hypothesis on M to be simply-connected in Theorem 23 is necessary, as the catenoid demonstrates. Theorem 23 basically says that if an embedded minimal disk is close enough to (and lies at one side of) a plane, then reasonably large components of it are graphs over this plane. This result is needed in the proof of Theorem 22 in the following manner: once it has been proven that an embedded minimal disk M contains a 2-valued graph $\widetilde{M}$, then $\widetilde{M}$ plays the role of the plane in the one-sided curvature estimate which implies that reasonably large pieces of M consist of multivalued graphs away from a cone with axis "orthogonal" to the 2-valued graph. The proofs of Theorems 22 and 23 are long and delicate. References [**24, 13, 14**] by Colding and Minicozzi are reading guides for the complete proofs of these results, which go through various papers [**15, 16, 18**]; see also the forthcoming survey by the authors in [**73**].

Theorems 22 and 23 have been applied to obtain a number of results. For instance, with these results in hand Meeks and Rosenberg [**80**] proved that the helicoid and the plane are the unique simply connected properly embedded minimal surfaces in $\mathbb{R}^3$ (see Theorem 30) and Meeks, Pérez and Ros showed that no properly embedded minimal surfaces with finite genus and one limit end can exist (Theorem 31). We will discuss other applications of Colding-Minicozzi results in Theorems 29 and 32.

DEFINITION 21. A *lamination* of an open subset $O \subset \mathbb{R}^3$ is the union of a collection of pairwise disjoint connected complete injectively immersed surfaces. More precisely, it is a pair $(\mathcal{L}, \mathcal{A})$ satisfying

1. $\mathcal{L}$ is a closed subset of O;
2. $\mathcal{A} = \{\varphi_\alpha \colon \mathbb{D} \times (0,1) \to U_\alpha\}_\alpha$ is a collection of coordinate charts of $\mathbb{R}^3$ (here $\mathbb{D}$ is the open unit disk, $(0,1)$ the open unit interval and U_α an open subset of O);
3. For each α, there exists a closed subset C_α of $(0,1)$ such that $\varphi_\alpha^{-1}(U_\alpha \cap \mathcal{L}) = \mathbb{D} \times C_\alpha$.

We will simply denote laminations as $\mathcal{L}$, omitting the charts φ_α in $\mathcal{A}$. A lamination $\mathcal{L}$ is said to be a *foliation of* O if $\mathcal{L} = O$. Every lamination $\mathcal{L}$ naturally decomposes into a union of disjoint surfaces, called the *leaves* of $\mathcal{L}$. As usual, the regularity of $\mathcal{L}$ requires the corresponding regularity on the change of coordinate charts. A lamination is *minimal* if all its leaves are minimal surfaces.

Each leaf L of a minimal lamination $\mathcal{L}$ is smooth, and if $C \subset L$ is a compact subset of a limit leaf $L \in \mathcal{L}$, then the leaves of $\mathcal{L}$ converge smoothly to L over C (i.e. they converge uniformly in the C^k-topology on C for any k).

Theorem 22 does not hold if we exchange the hypothesis that the radii R_n of the balls go to ∞ by R_n equals a constant, as demonstrated by a counterexample in [**13**]. Colding and Minicozzi construct a sequence of embedded minimal disks $M_n \subset B(1)$ with $\partial M_n \subset \partial B(1)$ all passing through the origin, and with Gaussian curvature blowing up only at the origin. This sequence produces a limit lamination of $B(1) - \{(0,0,0)\}$ with an isolated singularity at the origin. The limit lamination consists of three leaves, one of them being the flat horizontal punctured disk (which extends through the origin) and the other two being nonproper multigraphs with this disk as limit set. In particular, both smoothness and properness of leaves of the limit lamination fail for this local example.

Theorem 22 deals with limits of sequences of disks and it is also useful when studying more general situations, as for instance, uniformly locally simply connected sequences of minimal surfaces, a notion which we now define.

DEFINITION 22. Suppose $\{M_n\}_n$ is a sequence of properly embedded nonsimply connected minimal surfaces in $\mathbb{R}^3$. Given $p \in \mathbb{R}^3$ and $n \in N$, we let $r_n(p) > 0$ be the largest radius of an open ball B centered at p such that B intersects M_n in simply connected components. If for any $p \in \mathbb{R}^3$ the sequence $\{r_n(p)\}_n$ is bounded away from zero, we say that $\{M_n\}_n$ is *locally simply connected*. If for all $p \in \mathbb{R}^3$, the radius $r_n(p)$ is bounded from below by a positive constant for all n large, we will say that $\{M_n\}_n$ is *uniformly locally simply connected* (ULSC).

We have exploited the technique of blowing-up a sequence of immersed minimal surfaces on the scale curvature (see for instance the proof of Theorem 15 where we carry this out explicitly). When the surfaces in the sequence are properly embedded in $\mathbb{R}^3$, this blowing-up process produces a limit which is a properly embedded, nonflat minimal surface with bounded Gaussian curvature, whose genus and rank of homology groups are bounded by the ones for the M_n. For example, if each M_n is a planar domain, then the same holds for the limit.

For useful applications of the concept of ULSC sequence, it is essential to consider sequences of properly embedded minimal surfaces which *a priori* may not satisfy the ULSC condition, and the modify them to produce a new sequence which satisfies that condition. We accomplish this by considering a blow-up argument on a geometric scale which, in general, is different from blowing-up on the scale of curvature. We call this procedure *blowing-up by the scale of topology*. This scale was defined and used in [**74, 75**] to prove that any properly embedded minimal surface of finite genus has bounded curvature and is recurrent for Brownian motion. We now explain the elements of this new scale.

Suppose $\{M_n\}_n$ is a sequence of nonsimply connected, properly embedded minimal surfaces which is not ULSC. Note that the Gaussian curvature of the collection M_n is not uniformly bounded, and so, one could blow-up these surfaces on the scale of curvature to obtain a properly embedded nonflat minimal surface which may or may not be simply connected. Also note that, after choosing a subsequence, there exists points $p_n \in \mathbb{R}^3$ such that $r_n(p_n) \to 0$ as $n \to \infty$, where r_n is the function appearing in Definition 22. Let $\widetilde{p}_n$ be a point in $B(p_n, 1)$ where the function

$x \mapsto d(x, \partial B(p_n, r_n(p_n))) \cdot r_n(x)$ attains its maximum. Then the translated and rescaled surfaces $\widetilde{M_n} = \frac{1}{r_n(\widetilde{p}_n)}(M_n - \widetilde{p}_n)$ intersect for all n the closed ball centered at the origin with radius 1 in at least one component which is not simply connected, and for n large they intersect any ball of radius less that 1 in simply connected components, see [**74**] for details.

For the sake of clarity, we now illustrate this blow-up procedure on certain sequences of Riemann minimal examples, defined in Subsection 2.5. Each of these surfaces is foliated by circles and straight lines in horizontal planes, with a vertical plane of symmetry that can be assumed to be the (x_1, x_3)-plane. After a translation and a homothety, we assume that these surfaces are normalized so that the corresponding function r defined above attains its minimum value of 1 at the origin. Under this normalization, any sequence of Riemann minimal examples is ULSC. The flux of each Riemann minimal example along a compact horizontal section has horizontal and vertical components which are not zero, and the ratio V of the norm of its horizontal component over the vertical one parametrizes the 1-parameter family of these surfaces, with $V \in (0, \infty)$. When $V \to 0$, the Riemann minimal examples converge smoothly to the vertical catenoid centered at the origin with waist circle of radius 1. When $V \to \infty$, the Riemann minimal examples converge smoothly to a foliation of $\mathbb{R}^3$ by horizontal planes away from the two vertical lines passing through $(0, -1, 0), (0, 1, 0)$. In a neighborhood of any compact arc on these lines, the limiting Riemann examples are arbitrarily closed to a high sheeted vertical helicoid with axis along the line. Since the (x_1, x_3)-plane is a plane of symmetry of the approximated surfaces, these limit helicoids are oppositely handed. Outside the vertical cylinders containing the highly sheeted helicoids, the Riemann minimal examples consist of two multivalued almost flat graphs, which outside of a bigger cylinder containing both helicoids, are univalent graphs representing each of the ends of the Riemann minimal examples. This picture describes one particular case of what we call a parking garage structure for a surface. Roughly speaking, a *parking garage structure with n columns* is a smooth embedded surface in a horizontal slab $S \subset \mathbb{R}^3$ that can be decomposed into 2 disjoint almost flat horizontal multigraphs over the exterior of n disjoint disks in the (x_1, x_2)-plane, together n topological strips each one contained in one of the solid cylinders (these are the columns), such that each strip lies in a small regular neighborhood of the intersection of a vertical helicoid with S. One can associate to each column a + or − sign depending on the handedness of the corresponding helicoid. Note that a vertical helicoid is the basic example of a parking garage with 1 column, and the Riemann surfaces with $V \to \infty$ have the structure of a parking garage with two columns oppositely handed in any fixed size horizontal slab. Other parking garage structures with varying numbers of columns and associated signs can be found for other minimal surfaces, see Traizet and Weber [**124**].

There are interesting cases where ULSC guarantees the convergence of a sequence of minimal surfaces in $\mathbb{R}^3$ to a parking garage structure. Typically one proves that the sequence converges (up to a subsequence and a rotation) to a foliation of $\mathbb{R}^3$ by horizontal planes with singular set of convergence consisting of a locally finite set of Lipschitz curves parametrized by heights. In fact, these Lipschitz curves are vertical lines and locally around the lines the surfaces in the sequence approximate by highly sheeted vertical helicoids. To obtain this additional information, one applies in a neighborhood of each Lipschitz singular curve a blow-up argument on the scale of curvature and the uniqueness of the helicoid to prove that the Lipschitz curves are vertical lines. Using the fact that the unit normal field to a codimension-one minimal foliation in a 3-manifold is Lipschitz (Solomon [**120**]), another blow-up argument has the following consequence.

THEOREM 24 (Regularity of $S(\mathcal{L})$, Meeks [**68**]). *Suppose* $\{M_n\}_n$ *is a* ULSC *sequence of properly embedded minimal surfaces in a 3-manifold that converges smoothly to a minimal lamination* $\mathcal{L}$ *outside a locally finite collection of Lipschitz curves* $S(\mathcal{L})$ *transverse to* $\mathcal{L}$*. Then* $S(\mathcal{L})$ *consists of* $C^{1,1}$*-curves orthogonal to the leaves of* $\mathcal{L}$*.*

Recently, Meeks and Weber [**86**] have shown that the above $C^{1,1}$-regularity of $S(\mathcal{L})$ is the best possible. They do this by showing that any $C^{1,1}$-curve Γ properly embedded in an open set in $\mathbb{R}^3$ is the singular set of convergence for some Colding-Minicozzi limit foliation of some neighborhood of itself. In the special case that Γ is the unit circle in the (x_1, x_2)-plane, Meeks and Weber have defined for any $n \in \mathbb{N}$ a complete minimal annulus H_n of finite total curvature which contains the circle Γ; like in the helicoid, the Gauss map of H_n turns at a constant rate $2\pi n$ along its "circle axis" Γ. Meeks and Weber call the surfaces H_n *bent helicoids*, which have the Weierstrass data: $g(z) = z\frac{z^n+i}{z^n-i}$, $dh = \frac{z^{2n}+1}{z^{n+1}}dz$ defined on $\mathbb{C} - \{0\}$. They prove that there are compact annuli $\bar{H}_n \subset H_n$ which are embedded and which converge to the foliation $\mathcal{L}$ of $\mathbb{R}^3 - (x_3\text{-axis})$ by vertical halfplanes containing the x_3-axis and with singular set of convergence $S(\mathcal{L}) = \Gamma$ (for $n = \frac{1}{2}$, $H_{1/2}$ is the double cover of the Meeks's minimal Möbius strip [**69**] with total curvature -6π).

The regularity Theorem 24 allows one to replace the Lipschitz curves in the next Theorem by vertical lines, which on large balls, yields a parking garage structure with two columns. A little analysis shows that this asymptotic parking garage structure is $(+,-)$-handed, just like in the Riemann minimal case.

THEOREM 25 (Colding, Minicozzi [**19**]). *Let* $M_n \subset B(R_n)$ *be a* ULSC *sequence of embedded minimal planar domains with* $\partial M_n \subset \partial B(R_n)$*,* $R_n \to \infty$ *and* $M_n \cap B(2)$ *contains a component which is not a disk for any* n*. If* $\sup |K_{M_n \cap B(1)}| \to \infty$ *, then there exists a subsequence of the* M_n *(denoted in the same way) and two Lipschitz curves* $S_1, S_2 \colon \mathbb{R} \to \mathbb{R}^3$ *such that after a rotation of* $\mathbb{R}^3$*:*

1. $x_3(S_k(t)) = t$ *for all* $t \in \mathbb{R}$*.*
2. *Each* M_n *is horizontally locally graphical away from* $S_1 \cup S_2$*.*
3. *For each* $\alpha \in (0,1)$*,* $M_n - (S_1 \cup S_2)$ *converges in the* C^α*-topology to the foliation* $\mathcal{L}$ *of* $\mathbb{R}^3$ *by horizontal planes.*
4. $\sup |K_{M_n \cap B(S_k(t),r)}| \to \infty$ *as* $n \to \infty$*, for all* $t \in \mathbb{R}$ *and* $r > 0$*.*

To finish this Section, we next indicate how the one-sided curvature estimate of Colding and Minicozzi can be used to solve the *Generalized Nitsche Conjecture.* In 1962, Nitsche [**95**] conjectured that if a minimal surface meets every horizontal plane in a Jordan curve, then it must be a catenoid (he also proved this conjecture with the additional assumption that every horizontal section of the surface is a star shaped curve). In 1993, Meeks and Rosenberg [**83**] showed that if a properly embedded minimal surface $M \subset \mathbb{R}^3$ has at least two ends, then any annular end $E \subset M$ either has finite total curvature or it satisfies the hypotheses of the following conjecture.

CONJECTURE 7 (Generalized Nitsche Conjecture, Collin's Theorem [**25**]). *Let* $E \subset \{x_3 \geq 0\}$ *be a properly embedded minimal annulus with* $\partial E \subset \{x_3 = 0\}$*, such* *that E intersects each plane $\{x_3 = t\}$, $t > 0$, in a simple closed curve. Then, E has finite total curvature.*

This problem was originally solved by Collin [**25**] before the Colding-Minicozzi results, with a beautiful and long proof. The argument we present here is a short application of Theorem 23 which can be found in detail in Colding-Minicozzi [**21**] and in a recent survey by Rosenberg [**114**]. Given $\varepsilon \in \mathbb{R}$, we denote by $\mathcal{C}_\varepsilon$ the conical region $\{x_3 > \varepsilon\sqrt{x_1^2 + x_2^2}\}$. Conjecture 7 follows directly from the next result.

THEOREM 26. [**21**] *There exists $\delta > 0$ such that any complete properly embedded minimal annular end $E \subset \mathcal{C}_{-\delta}$ has finite total curvature.*

OUTLINE OF PROOF. The argument starts by showing, for each $\delta > 0$, the existence of a sequence $\{y_j\}_j \subset E - \mathcal{C}_\delta$ with $|y_j| \to \infty$ (this is done by contradiction: if for a given $\delta > 0$ this property fails, then one use E together with the boundary of $\mathcal{C}_\delta$ as barriers to construct an end of finite total curvature contained in $\mathcal{C}_\delta$, which is clearly impossible). The next step consists of choosing suitable radii $r_j > 0$ such that the connected component M_j of $E \cap B(y_j, 2r_j)$ which contains y_j is a disk. Now if $\delta > 0$ is sufficiently small in terms of the ε appearing in the one-sided curvature estimate, one can apply Theorem 23 and conclude a bound for the supremum of the absolute Gaussian curvature of the component M_j^1 of $M_j \cap B(y_j, r_j)$ which contains y_j. A Harnack type inequality together with this curvature bound gives a bound for the length of the intrinsic gradient of x_3 in the intrinsic ball $\mathcal{B}_j$ in M_j^1 centered at y_j with radius $5r_j/8$, which in turn implies (by choosing ε sufficiently small) that $\mathcal{B}_j$ is a graph with small gradient over $x_3 = 0$, and one can control a bound by below of the diameter of this graph. This allows to repeat the above argument exchanging y_j by a point y_j^1 in $\mathcal{B}_j^1$ at certain distance from y_j, and the estimates are carefully done so that the procedure can be iterated to go entirely around a curve $\gamma_j \subset E$ whose projection to the (x_1, x_2) plane links once around the x_3-axis. The graphical property of γ_j implies that either γ_j can be continued inside E to spiral indefinitely or it closes up with linking number one with the x_3-axis. The first possibility contradicts that E is properly embedded, and in the second case the topology of E implies that $\partial E \cup \gamma_j$ bounds an annulus E_j. The above gradient estimate gives a linear growth estimate for the length of γ_j in terms of $|y_j|$, from where the isoperimetric inequality for doubly connected minimal surfaces by Osserman and Schiffer [**100**] gives a quadratic growth estimate for the area of E_j. Finally, this quadratic area growth implies the finite total curvature property of E, finishing the outline of proof. □

10. Topological aspects of the theory of minimal surfaces.

Two of the main challenges in the classical theory of minimal surfaces are to decide which noncompact topological types are admissible as properly embedded minimal surfaces in space (Subsection 10.2), and given an admissible topological type in the previous sense, to show that there exists a unique way (up to ambient isotopy) of properly embedding this topological type as a minimal surface in $\mathbb{R}^3$ (Subsection 10.1). In the last decade amazing advances have been achieved in both problems, including a final solution to the second problem.

10.1. Topological classification of properly embedded minimal surfaces.

DEFINITION 23. Two properly embedded surfaces in a 3-manifold N^3 are called *ambiently isotopic* if one can be deformed to the other by a 1-parameter family of diffeomorphisms of N^3.

The problem of the topological uniqueness (up to ambient isotopy) of properly embedded minimal surfaces in 3-manifolds has been classically tackled by different authors. In 1970, Lawson [**57**] showed that two embedded closed diffeomorphic minimal surfaces in $\mathbb{S}^3$ are ambiently isotopic. Meeks [**70**] generalized a key result of Lawson to the case of orientable closed minimal surfaces in a closed 3-manifold with nonnegative Ricci curvature and also proved that any two compact diffeomorphic minimal surfaces embedded in a convex body B in $\mathbb{R}^3$, each with boundary a simple closed curve on the boundary of B, are ambiently isotopic in B (this result fails for more than one boundary curve, as demonstrated by a counterexample by Hall [**42**]). Later Meeks, Simon and Yau [**85**] generalized Lawson's Theorem to the ambient case of $\mathbb{S}^3$ with a metric of nonnegative scalar curvature. Returning to $\mathbb{R}^3$, Frohman [**37**] proved in 1990 that two triply periodic minimal surfaces are always ambiently isotopic. Although published two years later, Meeks and Yau [**90**] a decade earlier had shown that if M_1, M_2 are properly embedded minimal surfaces in $\mathbb{R}^3$ with the same finite topological type, then they are ambiently isotopic.

The essential first point in the proofs of these topological uniqueness results is to obtain a good understanding of the closed complements of the surfaces in the ambient space. This problem of the topological classification of the closed complements and the related uniqueness of the surfaces up to isotopy are closely related to the concept of a Heegaard surface in a 3-manifold, although here we will only deal with the case of the ambient space being $\mathbb{R}^3$.

DEFINITION 24. A 3-manifold with boundary is a *handlebody* if it is homeomorphic to a closed regular neighborhood of a properly embedded one-dimensional CW-complex in $\mathbb{R}^3$. A properly embedded surface $M \subset \mathbb{R}^3$ is called a *Heegaard surface* if each of the closed complements of M in $\mathbb{R}^3$ are handlebodies.

In 1997, Frohman and Meeks [**40**] proved that every properly embedded one-ended minimal surface in $\mathbb{R}^3$ is a Heegaard surface. Additionally, they obtained a topological uniqueness result for Heegaard surfaces in $\mathbb{R}^3$: two Heegaard surfaces of the same genus (possibly infinite) are properly ambiently isotopic. Joining these two results they obtained the following statement.

THEOREM 27 (Frohman, Meeks [**40**]). *Two properly embedded one-ended minimal surfaces in $\mathbb{R}^3$ with the same genus are ambiently isotopic.*

Furthermore, the topological model for any minimal surface M given in the hypotheses of Theorem 27, as well as for each of its complements in $\mathbb{R}^3$, is easy to describe. Attach g trivial one-handles to the closed lower halfspace H^- in $\mathbb{R}^3$, where g is the genus of M. If $g = \infty$, this attaching is performed on neighborhoods of the integer points on the x_1-axis in H^-, to obtain a one-periodic Heegaard surface Σ in $\mathbb{R}^3$. Let W be the handlebody of $\mathbb{R}^3$ with boundary Σ such that $H^- \subset W$. Then, there exists a diffeomorphism $h\colon \mathbb{R}^3 \to \mathbb{R}^3$ such that $h(M)$ is the Heegaard surface Σ and a prescribed closed complement of M in $\mathbb{R}^3$ maps to the handlebody W.

To appreciate the power of Theorem 27, we can consider the singly and doubly periodic Scherk minimal surfaces defined in Subsection 2.5. By Theorem 27, there exists a diffeomorphism of $\mathbb{R}^3$ that takes one surface to the other, although they look very different in space. We should also mention here a result by Callahan,

Hoffman and Meeks [**7**] that insures that every connected doubly periodic minimal surface has infinite genus and one end. Joining this result with Theorem 27 it follows that any two doubly periodic minimal surfaces in $\mathbb{R}^3$ are ambiently isotopic (for example, this applies to the classical P-Schwarz minimal surface and the doubly periodic Scherk surface!).

All these topological uniqueness results are special cases of a recent Theorem by Frohman and Meeks, which represents the final solution to the topological classification problem. It is also based on a deep topological analysis of the complements of a properly embedded minimal surface in space, based on previous work by Freedman [**36**], and shows what are the topological roles of the ordering on the set of ends given by Theorem 18 and the parity of each middle end defined in Theorem 19.

THEOREM 28 (Topological Classification Theorem, Frohman, Meeks [**38**]). *Two properly embedded minimal surfaces in $\mathbb{R}^3$ are ambiently isotopic if and only if there exists a homeomorphism between the surfaces that preserves both the ordering of their ends and the parity of their middle ends.*

10.2. Admissible topological types. We will start by discussing noncompact finite topologies, i.e. surfaces with finite genus and finitely many ends. Until the early eighties, no properly embedded minimal surfaces of finite topology other than the plane, the helicoid (both with genus zero, one end) and the catenoid (genus zero, two ends) were known. For a long time, some geometers supported the conjecture that no other examples of finite topology would exist. The discovery in 1982 of a new genus one three-ended example (Costa [**28**], Hoffman and Meeks [**48**]) not only disproved this conjecture, but also revitalized enormously the interest of geometers in classical minimal surface theory. Since then, a number of different new examples have appeared, sometimes even coming in multiparameter families [**43, 44, 45, 49, 54, 123, 127**].

For properly embedded minimal surfaces with finite topology, there is an interesting dichotomy between the one-end case and those surfaces with more that one end: surfaces in this last case always have finite total curvature (Collin's Theorem 3). Only the simplest finite topologies with more than one end have been characterized: López and Ros [**63**] proved that the unique examples with genus zero and finite topology are the plane and the catenoid, Schoen [**118**] demonstrated that the catenoid is the unique example with finite genus and two ends, and Costa [**29**] showed that the examples with genus one and three ends lie inside the one-parameter family of surfaces $\{M_{1,a} \mid 0 < a < \infty\}$ that appear in Subsection 2.5. Today we know many more examples of higher finite topologies and more than one end, and up to this date all known examples support the following conjecture by Hoffman and Meeks.

CONJECTURE 8 (Finite Topology Conjecture). *A connected noncompact (orientable) surface of finite topology, genus g and r ends, $r \neq 2$, can be properly minimally embedded in $\mathbb{R}^3$ if and only if $r \leq g+2$.*

Recently, Meeks, Pérez and Ros [**76**] have given the following partial result on the above conjecture.

THEOREM 29. [**76**] *For every nonnegative integer g, there exists an integer $e(g)$ such that if $M \subset \mathbb{R}^3$ is a properly embedded minimal surface of finite topology and with genus g, then the number of ends of M is at most $e(g)$.*

SKETCH OF THE PROOF. The argument is by contradiction. The failure of the Theorem to hold would produce an infinite sequence $\{M_n\}_n$ of properly embedded minimal surfaces with fixed finite genus g and a strictly increasing number of ends. Then one analyzes the nonsimply connected limits of subsequences of $\{M_n\}_n$. The key idea used to achieve these limits is to normalize M_n by a translation and then a homothety on the scale of topology. What this means is that we assume each surface intersects the closed unit ball centered at the origin in some nonsimply connected component, but that every open ball of radius 1 intersects the surface in simply connected components. Using the uniformly locally simply connected property of $\{M_n\}_n$, we prove that its limits are properly embedded nonsimply connected minimal surfaces with genus at most g and possibly infinitely many ends. The infinite topology limits are discarded by an application of either a descriptive Theorem of the geometry of properly embedded minimal surfaces in $\mathbb{R}^3$ with finite genus and two limit ends (Meeks, Pérez and Ros [**74**], see also Theorem 32 below), or a nonexistence Theorem for properly embedded minimal surfaces in $\mathbb{R}^3$ with finite genus and one limit end (Meeks, Pérez and Ros [**75**] or Theorem 31 below). Hence, any possible limit M of a subsequence of $\{M_n\}_n$ must be a finite total curvature surface or a helicoid with positive genus at most g. A surgery argument allows one to modify the surfaces M_n by replacing compact pieces of M_n close to the limit M by a finite number of disks, obtaining a new surface $\widetilde{M_n}$ with strictly less topology than M_n and which is not minimal in the replaced part. A careful study of the replaced parts during the sequence allows one to iterate the process of finding nonsimply connected minimal limits of the modified surfaces. The fact that all the M_n have the same genus, allows one to arrive to a stage in the process of producing limits from which all subsequent limits are catenoids. From this point in the proof it is not difficult to find a large integer n such that M_n contains a noncompact planar domain $\Omega \subset M_n$ whose boundary consists of two convex planar curves Γ_1, Γ_2 in parallel planes that each separate M_n and whose fluxes are orthogonal to the planes that contain Γ_1, Γ_2. In this setting, the López-Ros deformation defined in Subsection 2.2 (see [**63, 107**]) applies to Ω giving the desired contradiction. □

Since the finite index of a complete minimal surface of finite total curvature can be estimated from above by a function of the degree of its Gauss map (Tysk [**125**]), Theorem 29 has the following important theoretical consequence.

COROLLARY 3. *For every nonnegative integer g, there exists an integer $i(g)$ such that if $M \subset \mathbb{R}^3$ is a properly embedded minimal surface with genus g and a finite number of ends greater than 1, then the index of stability of M is at most $i(g)$.*

Theorem 29, as well as the related Theorems 31 and 32 below, rely on results by Colding and Minicozzi (see Section 9) to describe the basic local geometry of limits of sequences of properly embedded minimal surfaces with bounded genus and unbounded area.

Concerning one-ended minimal surfaces with finite topology, the model in this class is the helicoid. Recently, Meeks and Rosenberg [**80**] have proven the following result.

THEOREM 30 (Meeks, Rosenberg [**80**]). *Every properly embedded minimal surface with finite topology and one end has the conformal structure of a compact genus*

g Riemann surface $\overline{M}_g$ minus one point, can be analytically represented by meromorphic data on $\overline{M}_g$ and is asymptotic to a helicoid. Furthermore, for $g = 0$ the only possible examples are the plane and the helicoid.

SKETCH OF THE PROOF. We will just consider the case where M is simply connected. Let M be a properly embedded simply connected minimal surface in $\mathbb{R}^3$ which is not a plane. Consider any sequence of positive numbers $\{\lambda_n\}_n$ which decays to zero and let $M(n) = \lambda_n M$ be the surface scaled by λ_n. By Theorem 22, a subsequence of these surfaces converges on compact subsets of $\mathbb{R}^3$ to a minimal foliation $\mathcal{L}$ of $\mathbb{R}^3$ by parallel planes with singular set of convergence $S(\mathcal{L})$. Part of the proof of Theorem 22 depends on a unique extension result for the forming multigraphs, which in our case implies that for n large the almost flat multigraph which starts to form on $M(n)$ near the origin extends all the way to infinity. From here one can deduce that limit foliation $\mathcal{L}$ is independent of the sequence $\{\lambda_n\}_n$. After a rotation of M and replacement of the $M(n)$ by a subsequence, we can suppose that the $M(n)$ converge to the foliation $\mathcal{L}$ of $\mathbb{R}^3$ by horizontal planes, with singular set of convergence $S(\mathcal{L})$. The property that all surfaces $M(n)$ are simply connected is now crucial in showing that $S(\mathcal{L})$ consists of a single Lipschitz curve Γ which intersects each horizontal plane exactly once.

Since the origin is a singular point of convergence, the Lipschitz curve Γ passes through the origin and is contained in the solid cone $\mathcal{C}_\varepsilon = \{x_3^2 \geq \varepsilon^2(x_1^2 + x_2^2)\}$, where $\varepsilon > 0$ only depends on the curvature estimate in Theorem 23. Let Δ be the solid cylinder $\{x_1^2 + x_2^2 \leq 1,\ |x_3| \leq \varepsilon\}$. The two flat horizontal multigraphs $M_1(n), M_2(n)$ referred to in Theorem 22 intersect the cylindrical sides of $\partial\Delta$ almost orthogonally in two long spiraling arcs which are multigraphs over the unit circle $\mathbb{S}^1$ in the (x_1, x_2)-plane, possibly together with open arcs starting and finishing at the top (resp. bottom) planar disks of $\partial\Delta$ which are graphs over their projections in $\mathbb{S}^1$. Both spirals lie on the main compact component $D(n)$ of $M(n) \cap \Delta$. After a small perturbation $\Delta(n)$ of Δ near the top and bottom boundary disks of $\partial\Delta$ and replacing Δ by $\Delta(n)$, it can be shown that the boundary of $D(n)$ consists of the two spiraling arcs on the boundary of the cylinder together with two arcs which connect them, one on each of the boundary disks in $\partial\Delta(n)$; in this replacement the new top and bottom disks in $\partial\Delta(n)$ are minimal. Without much difficulty, one can extend the top and bottom disks of $\Delta(n)$ to an almost horizontal minimal foliation of $\Delta(n)$ by graphical minimal disks such that each boundary circle of these disks intersects each spiral curve in $\partial D(n)$ at a single point. Morse theory implies that each leaf of the minimal disk foliation of $\Delta(n)$ intersects $D(n)$ transversely in a simple arc. When $n \to \infty$, these foliations converge to the restricted foliation $\mathcal{L} \cap \Delta$ by flat horizontal disks. An important consequence of this last statement and of the openness of the Gauss map of the original surface M is that M is transverse to $\mathcal{L}$. This means that the stereographical projection of the Gauss map $g\colon M \to \mathbb{C} \cup \{\infty\}$ can be expressed as $g(z) = e^{f(z)}$ for some holomorphic function $f\colon M \to \mathbb{C}$.

The next part of the proof is longer and more delicate, and depends in part on a finiteness result for the number of components of minimal graphs over proper domains in $\mathbb{R}^2$ with zero boundary values. Through a series of geometric and analytic arguments using the double multigraph convergence of the $M(n)$ to $\mathcal{L}$ outside the cone $\mathcal{C}_\varepsilon$, one eventually proves that every horizontal plane in $\mathcal{L}$ intersects M transversely in a single proper arc. Then a straightforward argument using Theorem 12 implies M is recurrent, and thus M is conformally $\mathbb{C}$. The nonexistence of points in M with vertical normal vector and the connectedness of its horizontal sections force the height differential to be $dh = dx_3 + idx_3^* = dz$ in a conformal parametrization of M. In particular, the third coordinate $x_3 \colon \mathbb{C} \to \mathbb{R}$ is the linear function real part of z.

Recall that we have already shown that $g(z) = e^{f(z)}$. If the holomorphic function $f(z)$ is a linear function of the form $az + b$, then the Weierstrass data $(e^{f(z)}, dz)$ for M shows that M is an associate surface to the helicoid (see Section 8 for the definition of associate surface). Since none of the nontrivial associate surfaces to the helicoid are injective as mappings, M is a helicoid. Thus, it remains to show that $f(z)$ is linear. The formula (7) for the Gaussian curvature K and a straightforward application of Picard's Theorem imply $f(z)$ is linear if and only if M has bounded curvature. This fact completes the proof of the Theorem in the special case that K is bounded. However, Theorem 22 and a clever blow-up argument on the scale of curvature reduces the proof that $f(z)$ is linear in the general case to the case where K is bounded, and so M is a helicoid. For further details, see [**80**]. □

Theorem 30 solves a long standing conjecture about the uniqueness of the helicoid among properly embedded simply connected minimal surfaces in $\mathbb{R}^3$. In 1993, Hoffman, Karcher and Wei [**46, 47**] found a torus with one helicoidal end, called since then the *genus one helicoid,* which has been proved recently to be embedded by Hoffman, Weber and Wolf [**51**]. Very little is known about genus g helicoids with $g \geq 2$. Computer graphics seem to indicate that higher genus examples probably exist (Traizet —unpublished—, Bobenko [**2**], Bobenko and Schmies [**3**]) and Traizet and Weber [**124**] have given an approach based on the Implicit Function Theorem that could be useful for rigorously proving the existence of an embedded genus g helicoid for every g. The Conjecture in this direction, due to Meeks and Rosenberg, is the following one.

CONJECTURE 9 (One-ended Conjecture). *For every nonnegative integer g, there exists a unique nonplanar properly embedded minimal surface in $\mathbb{R}^3$ with genus g and one end.*

Next we deal with properly embedded minimal surfaces with finite genus and infinite topology. Since the number of ends of such a surface $M \subset \mathbb{R}^3$ is infinite and the set of ends $\mathcal{E}(M)$ of M is compact (Subsection 2.7), M must have at least one limit end. Up to a rotation, we can assume that the limit tangent plane at infinity of M (see Section 5) is horizontal. A crucial result by Collin, Kusner, Meeks and Rosenberg [**26**] (Theorem 19) insures that M has no middle limit ends, hence either it has one limit end (this one being the top or the bottom limit end) or both top and bottom ends are the limit ends of M, like in a Riemann minimal example. Very recently, Meeks, Pérez and Ros [**75**] have discarded the one limit end case through the following result.

THEOREM 31. [**75**] *If $M \subset \mathbb{R}^3$ is a properly embedded minimal surface with finite genus, then M cannot have exactly one limit end. Furthermore, M is recurrent.*

SKETCH OF THE PROOF. Assume M is a properly embedded minimal surface with finite genus and exactly one limit end. After a rotation, we can suppose that M has horizontal tangent plane at infinity and its set of ends, linearly ordered by increasing heights (see the Ordering Theorem 18), is $\mathcal{E}(M) = \{e_1, e_2, \ldots, e_\infty\}$

with the limit end of M being its top end e_∞. One first shows that M has a nice asymptotic behavior: each nonlimit end $e_n \in \mathcal{E}(M)$ is asymptotic to a graphical annular end E_n of a vertical catenoid with negative logarithmic growth a_n satisfying $a_1 \leq \ldots \leq a_n \leq \ldots < 0$. This is the statement of Theorem 2 in [**74**]. The next step consists of a detailed analysis of the limits (after passing to a subsequence) of homothetic shrinkings $\{\lambda_n M\}_n$, where $\{\lambda_n\}_n \subset \mathbb{R}^+$ is any sequence of numbers decaying to zero; we first show that $\{\lambda_n M_n\}_n$ is locally simply connected in $H(*) = \{x_3 \geq 0\} - \{0\} \subset \mathbb{R}^3$. This is a difficult technical part of the proof where the results of Colding-Minicozzi again play a crucial role. We next prove the limits of subsequences of $\{\lambda_n M_n\}_n$ consist of (possibly singular) minimal laminations $\mathcal{L}$ of $H(*) = \{x_3 \geq 0\} - \{0\} \subset \mathbb{R}^3$ containing $\partial H(*)$ as a leaf. Then we check that the limit lamination $\mathcal{L}$ is always smooth and that the singular set of convergence $S(\mathcal{L})$ of $\lambda_n M$ to $\mathcal{L}$ is empty. In particular, taking $\lambda_n = \|p_n\|^{-1}$ where p_n is any divergent sequence on M, the fact that $S(\mathcal{L}) = \emptyset$ for the corresponding limit minimal lamination $\mathcal{L}$ insures that the Gaussian curvature of M decays at least quadratically in terms of the distance function to the origin. Since the Gaussian curvature function times the squared distance to the origin is scale invariant, any leaf L of a limit lamination of $\lambda_n M$ for arbitrary $\lambda_n \searrow 0$ must also have quadratic decay of its Gaussian curvature function K_L. Finally, a suitable choice of the sequence $\lambda_n \searrow 0$ produces a limit lamination with a nonflat minimal leaf L properly embedded in $H(*)$ of infinite total curvature and such that the scalar flux of the gradient ∇x_3 along a certain horizontal section of L is finite. The quadratic decay property for K_L together with the existence of a sequence of horizontal planes $\{\Pi_k\}_k$ with heights diverging to $+\infty$ such that $L \cap \Pi_k$ contains a point with vertical tangent plane (this comes from the fact that L has infinite total curvature) implies that the flux of ∇x_3 is unbounded on horizontal sections of L. This leads to a contradiction with the invariance of the flux of the divergence free vector field ∇x_3 on L. This finishes the outline of the proof of the first statement of Theorem 31.

In order to finish the proof, it only remains to check that M is recurrent. If M has exactly one end, then M is conformally a compact Riemann surface minus one point (Theorem 30) and so, M is recurrent. If M has a finite number of ends greater than one, then M has finite total curvature (Theorem 3). By Huber-Osserman's Theorem, M is conformally a compact Riemann surface minus a finite number of points thus it is again recurrent. Finally, if M has infinitely many ends, then M has exactly two limit ends, see the paragraph just before the statement of Theorem 31. In this situation, Theorem 20 asserts that M is recurrent. This completes our sketch of the proof of Theorem 31. □

In the above sketch of the proof of Theorem 31 we mentioned that any limit lamination $\mathcal{L}$ of $H(*)$ obtained as a limit of (a subsequence of) homothetic shrinkings $\{\lambda_n M\}_n$ with $\lambda_n \searrow 0$, has no singularities and empty singular set of convergence $S(\mathcal{L})$. To understand why this last property is true, we show that if $\mathcal{L}$ had singularities (in which case $S(\mathcal{L}) \neq \emptyset$) or if $S(\mathcal{L})$ were nonempty for a given sequence of shrinkings of M, then some smooth leaf of the limit lamination $\mathcal{L}$ that intersects $S(\mathcal{L})$ would be a limit leaf, and hence stable. Our difficulty in discarding this possibility lies in the fact that the stable leaves of $\mathcal{L}$, while perhaps proper in $H(*)$, may not be complete and so, we do not know they must be planes. It is not difficult to prove that the smooth stable leaves in $\mathcal{L}$ in fact satisfy the hypotheses of the next Proposition, and so are in fact planes. Once one has that the smooth stable leaves in $\mathcal{L}$ are planes, then the proof of Theorem 22 leads to a contradiction, therefore showing that $S(\mathcal{L})$ is in fact empty. We include the proof of the next Proposition mostly to demonstrate how one can obtain information on the conformal structure of possibly incomplete minimal surfaces by studying conformally related metrics and then how to apply such information to constrain their geometry. Also, the proof of Proposition 4 below gives some new techniques and insights for possibly solving the following famous Conjecture, due to Gulliver and Lawson.

CONJECTURE 10 (Isolated Singularities Conjecture). *There does not exist a properly embedded minimal surface in a punctured ball $B - \{(0,0,0)\}$ whose closure is not a surface at the origin.*

Given $\mu > 0$ small, we define the upper half cones $\widetilde{C} = \{(x_1, x_2, x_3) \mid x_3 = \mu\sqrt{x_1^2 + x_2^2}\}$ and $C = \widetilde{C} + (0, 0, -1)$. Finally, let W be the component of $\mathbb{R}^3 - C$ which lies below C.

PROPOSITION 4. *Let L be a stable orientable minimal surface embedded in $H(*)$, not equal to $\partial H(*)$ and such that $L \cap W$ consists of a nonempty collection of horizontal planar ends. If any proper arc $\alpha\colon [0, \infty) \to L$ of finite length satisfies $\lim_{t\to\infty} \alpha(t) = (0, 0, 0)$, then L is a horizontal plane.*

PROOF. Assume L is not a horizontal plane. By the maximum principle, $L \subset \{x_3 > 0\} \subset \mathbb{R}^3$. As L is stable, orientable and is not a plane, Theorem 8 implies that L is not complete. Thus the set $\mathcal{A}$ of proper arcs $\alpha\colon [0, \infty) \to L$ with finite length is nonempty. Proposition 4 will be proven if we show that there exists $\alpha \in \mathcal{A}$ such that $\lim_{t\to\infty} \alpha(t) \in H(*) - \{(0, 0, 0)\}$. Reasoning by contradiction, suppose that any $\alpha \in \mathcal{A}$ satisfies $\lim_{t\to\infty} \alpha(t) = (0, 0, 0)$.

Consider the complete conformally related metric $\widetilde{g} = \frac{(1+x_3)^2}{R^2} g$ on L, where g is the induced metric on L by the inner product of $\mathbb{R}^3$ and $R = \sqrt{x_1^2 + x_2^2 + x_3^2}$. The Laplace operators and Gaussian curvature functions of (L, g), $(L, \widetilde{g})$ are related by the equations

$$\widetilde{\Delta} = \frac{R^2}{(1+x_3)^2}\Delta, \qquad \widetilde{K} = \frac{R^2}{(1+x_3)^2}\left(K - \Delta \ln \frac{1+x_3}{R}\right),$$

where as usual, the notation $\widetilde{\bullet}$ means that the corresponding object $\bullet$ is computed with respect to $\widetilde{g}$ (otherwise it refers to g). As $\Delta \ln(R) = \frac{2(1-|\nabla R|^2)}{R^2} \geq 0$ and $\Delta \ln(1 + x_3) = \frac{-|\nabla x_3|^2}{(1+x_3)^2} \leq 0$, it follows that $\widetilde{K} = \frac{R^2}{(1+x_3)^2} K + P$ where P is a nonnegative function. As (L, g) is stable, the operator $-\Delta + 2K$ is positive semidefinite on (L, g). Since $-\Delta + K \geq -\Delta + 2K$ and $\frac{R^2}{(1+x_3)^2} > 0$, it follows that $\frac{R^2}{(1+x_3)^2}(-\Delta + K)$ is also positive semidefinite on (L, g). Since $-\widetilde{\Delta} + \widetilde{K} \geq \frac{R^2}{(1+x_3)^2}(-\Delta + K)$, we conclude that $-\widetilde{\Delta} + \widetilde{K}$ is positive semidefinite on $(L, \widetilde{g})$.

Let E be the collection of points in the completion of $(L, \widetilde{g})$ corresponding to the annular planar ends of L. We claim that the metric completion $\overline{L}$ of $(L, \widetilde{g})$ is exactly $L \cup E$. To see this, let $\alpha\colon [0, \infty) \to L$ be a proper arc of finite length on $(L, \widetilde{g})$. If we consider α to lie on $(L, g) \subset (\mathbb{R}^3, <,>)$, then the factor $\frac{(1+x_3)^2}{R^2}$ in the metric $\widetilde{g}$ shows that α eventually lies in the component W of $\mathbb{R}^3 - C$ below C defined just before the Proposition. Since α is proper, it must lie in one of the annular planar ends of L and so, α diverges to one point in E. This proves our claim.

Note also that the metric $\widetilde{g}$ on L extends to a smooth Riemannian metric $\overline{g}$ on $\overline{L}$ (because the annular ends of L in E are asymptotic to horizontal planes). Since $-\widetilde{\Delta}+\widetilde{K}$ is positive semidefinite on $(L,\widetilde{g})$ and $E\subset\overline{L}$ is a discrete set, we conclude that the extended operator $-\overline{\Delta}+\overline{K}$ is positive semidefinite on $\overline{L}$ (here $\overline{\bullet}$ refers to the metric $\overline{g}$). As a consequence, the universal covering of $\overline{L}$ is conformally $\mathbb{C}$ [**35**]. Finally, since $x_3|_L$ is a nonconstant positive harmonic function on $(L,\widetilde{g})$ (because it is harmonic on (L,g)) and $x_3|_L$ extends smoothly through the points in E, we can lift such an extension to the universal covering of $\overline{L}$ as a positive harmonic function on $\mathbb{C}$, hence constant by Liouville's Theorem. In particular, $x_3|_L$ is also constant, which is the desired contradiction. □

If a properly embedded minimal surface $M\subset\mathbb{R}^3$ has finite genus and infinite topology, then Theorems 19 and 31 imply M has two limit ends which are its top and bottom ends (after a rotation so that the limit tangent plane at infinity of M is horizontal). The classical model in this setting is any of the surfaces in the 1-parameter family of Riemann minimal examples (see Subsection 2.5). In 1998, Meeks, Pérez and Ros [**78**] proved that these are the unique properly embedded minimal surfaces in $\mathbb{R}^3$ with genus zero, infinite topology and infinitely many symmetries. The extension of this characterization by eliminating the hypothesis on the symmetry group constitutes the following conjecture by the same authors.

CONJECTURE 11 (Genus zero Conjecture). *If $M\subset\mathbb{R}^3$ is a properly embedded minimal surface with genus zero and infinitely many ends, then M is a Riemann minimal example.*

This conjecture is an active field of research in the last years by Meeks, Pérez and Ros, and in part it has motivated further developments as explained in Theorems 29 and 31 above. Strong partial results have been achieved, such as the following statement.

THEOREM 32. [**74**] *Let $M\subset\mathbb{R}^3$ be a properly embedded minimal surface with finite genus, two limit ends and horizontal limit tangent plane at infinity. Then:*

1. *The middle ends $\{e_n\mid n\in\mathbb{Z}\}$ of M are planar, have heights $\mathcal{H}=\{x_3(e_n)\mid n\in\mathbb{Z}\}$ with $x_3(e_n)<x_3(e_{n+1})$ for all $n\in\mathbb{Z}$ and $\lim_{n\to\infty}x_3(e_n)=\infty$, $\lim_{n\to-\infty}x_3(e_n)=-\infty$.*
2. *Every horizontal plane sufficiently high or low intersects M in a simple closed curve when its height is not in $\mathcal{H}$ and in a single properly embedded arc when its height is in $\mathcal{H}$.*
3. *The flux vector of M along any compact (not necessarily connected) horizontal section does not depend on the height of the section. Both the horizontal and the vertical components of this flux vector are not zero. In what follows, we will rescale M so that this vertical component of the flux of M equals 1.*
4. *M has bounded Gaussian curvature. Furthermore, the maximum asymptotic curvature outside balls $\lim_{r\to\infty}\sup|K_{M-B(r)}|$ is bounded from above in terms only of the horizontal component of its flux[10]. In the case M has genus zero,*

[10]This means that if $\{M_n\}_n$ is a sequence of properly embedded minimal surfaces with fixed finite genus, horizontal limit tangent plane at infinity, the vertical components of the fluxes of the M_n along compact horizontal sections are all 1 and the horizontal components of such fluxes are bounded by above, then there are compact balls B_n centered at the origin so that $\{M_n-B_n\}_n$ has uniformly bounded Gaussian curvature.

the bound on the Gaussian curvature of M depends only on a bound of its horizontal flux.

5. *The spacings $S(n) = x_3(e_{n+1}) - x_3(e_n)$ between consecutive ends are bounded from above and below (by Corollary 2 in Section 7, the lower bound $2|\lambda|$ depends only on the curvature estimate λ for M given in the last point 4).*
6. *M is quasiperiodic in the following sense. There exists a divergent sequence $V(n) \in \mathbb{R}^3$ such that the translated surfaces $M + V(n)$ converge to a properly embedded minimal surface of genus zero, two limit ends and horizontal limit tangent plane at infinity.*
7. *If M has genus zero, then M has no points with vertical normal vector and the conclusions in point 2 above hold for every horizontal plane in $\mathbb{R}^3$.*

Using the Shiffman Jacobi function, Meeks, Pérez and Ros have proven the uniqueness of the Riemann minimal examples among properly embedded genus zero minimal surfaces under the hypotheses of Theorem 32, assuming additionally that the horizontal component of the flux along a compact horizontal section is sufficiently small. We will devote Section 11 to a more detailed discussion of this result and to further analysis on the Shiffman Jacobi function.

We finish this Subsection with a brief comment about properly embedded minimal surfaces of infinite genus. The collection of such surfaces with one end is extremely rich. One reason for this is that there are many doubly periodic examples (note that any triply periodic example can be viewed as a doubly periodic one), and as mentioned in Subsection 10.1, every doubly periodic properly embedded minimal surface in $\mathbb{R}^3$ has one end (Callahan, Hoffman and Meeks [**7**]). Besides Theorem 19 on the nonexistence of middle limit ends, very little is known about properly embedded minimal surfaces with infinite genus and infinitely many ends. The first known examples arise from singly periodic surfaces with planar ends and positive genus (Callahan, Hoffman and Meeks [**6**]), but these examples are better studied as finite total curvature minimal surfaces in the corresponding quotient space. A tentative example of (truly) infinite genus and one limit end might be constructed as follows. Weber and Wolf [**127**] proved the existence of a sequence $M_n \subset \mathbb{R}^3$ of properly immersed minimal surfaces of odd genus and $n+2$ horizontal planar ends. Computer graphics pictures indicate that all these surfaces are embedded. Assuming this embeddedness property holds, a suitable normalization of these surfaces should give as a limit a properly embedded minimal surface with a bottom catenoid end, infinitely many middle planar ends and a top limit end. By Theorem 31, this limit surface could not have finite genus.

10.3. Completeness versus properness and minimal laminations. In his beautiful recent survey on minimal surfaces, Harold Rosenberg [**114**] introduces the subject of his paper through a question asked to him by Andre Haefliger about twenty years ago: "Is there a foliation of $\mathbb{R}^3$ by minimal surfaces, other than a foliation by parallel planes?"

Any leaf L of a minimal foliation of $\mathbb{R}^3$ is a complete limit leaf and it can be proved to be stable. Now Theorem 8 implies L is a plane (provided L is orientable; but this technical assumption can be supposed after passing to the universal covering, which also turns out to be stable, see the proof of Lemma 1.1 in [**80**]). Thus the answer to Haefliger's question is *no*. Immediately one is tempted to extend this question to minimal laminations.

QUESTION 3. *What are the minimal laminations of* $\mathbb{R}^3$*?*

There are only two known types of minimal laminations of $\mathbb{R}^3$: a lamination with exactly one leaf which is a properly embedded minimal surface, or a lamination consisting of a closed set of parallel planes. Meeks and Rosenberg [**80**] have conjectured that these are the unique possible examples. Since every leaf of a minimal lamination of $\mathbb{R}^3$ is complete, the above question is closely related to the following one.

QUESTION 4. *When is a complete embedded minimal surface* $M \subset \mathbb{R}^3$ *proper?*

Given a minimal lamination $\mathcal{L}$ of $\mathbb{R}^3$, the function that assigns to each point p in $\mathcal{L}$ the Gaussian curvature of the leaf $L \in \mathcal{L}$ passing through p is continuous in the subspace topology. Since the intersection of $\mathcal{L}$ with any closed ball is compact, we conclude that the intersection of any leaf $L \in \mathcal{L}$ with any ball has Gaussian curvature bounded from below by a constant that only depends on the ball (in other words, L has *locally bounded Gaussian curvature*). Reciprocally, if M is a complete embedded minimal surface in $\mathbb{R}^3$ with locally bounded Gaussian curvature, then the closure $\overline{M}$ of M is a minimal lamination of $\mathbb{R}^3$ (Lemma 1.1 in [**80**]). With this perspective, it is natural to study complete embedded minimal surfaces $M \subset \mathbb{R}^3$ with locally bounded Gaussian curvature, as a first stage for possible answers to Questions 3 and 4. If M is such a minimal surface and it is not proper, then $\overline{M} - M$ may or may not be nonempty; but since M has locally bounded curvature, $\mathcal{L} = \overline{M}$ is a nontrivial minimal lamination of $\mathbb{R}^3$ and some leaf $L \in \mathcal{L}$ must be a limit leaf, hence stable. Now an argument similar to the one we used to answer Haefliger's question at the beginning of this Subsection insures that L is a plane; so in this case, $\overline{M} - M$ is always nonempty. This can be stated as follows.

LEMMA 4 (Meeks, Rosenberg [**80**]). *Let* $M \subset \mathbb{R}^3$ *be a connected complete embedded minimal surface with locally bounded Gaussian curvature. Then exactly one of the following holds:*

1. M *is properly embedded in* $\mathbb{R}^3$.
2. M *is properly embedded in an open halfspace, with limit set the boundary plane of this halfspace.*
3. M *is properly embedded in an open slab, with limit set consisting of the boundary planes of the slab.*

It should be mentioned that in a previous work, Xavier [**128**] proved that a complete, immersed, nonflat minimal surface of bounded curvature in $\mathbb{R}^3$ cannot be contained in a halfspace. This result together with Lemma 4 gives a partial answer to Question 4.

COROLLARY 4. [**80**] *If* $M \subset \mathbb{R}^3$ *is a connected nonflat complete embedded minimal surface with bounded Gaussian curvature, then* M *is proper.*

The next step in the study of complete embedded nonproper minimal surfaces consists of understanding how they accumulate to the limit set described in Lemma 4.

LEMMA 5. [**80**] *Let* $M \subset \mathbb{R}^3$ *be a connected complete embedded minimal surface with locally bounded Gaussian curvature. Suppose that* M *is not proper and let* Π *be a limit plane of* M*. Then, for any* $\varepsilon > 0$*, the closed* ε*-neighborhood of* Π *intersects* M *in a path connected set.*

OUTLINE OF THE PROOF. The argument is by contradiction. Assuming its failure, one can produce a stable minimal surface Σ between two components of the intersection of M with the slab $\{0 < x_3 < \varepsilon\}$ (we do not loss generality by assuming that $\Pi = \{x_3 = 0\}$ and that M limits to Π from above) by the usual barrier construction argument. Since Σ satisfies curvature estimates away from its boundary (Theorem 15), we conclude that for sufficiently small $\delta > 0$, the orthogonal projection π to Π restricted to $\Sigma(\delta) = \Sigma \cap \{0 < x_3 < \delta\}$ is a local diffeomorphism. A topological argument shows that $\pi|_{\Sigma(\delta)}$ is in fact bijective, so it is a diffeomorphism. This implies $\Sigma(\delta)$ is properly embedded in the slab $\{0 \leq x_3 \leq \delta\}$. Now the argument in the sketch of the proof of Theorem 7 applies to give a contradiction and proves Lemma 5. □

A refinement of the argument in the previous paragraph shows that if $M \subset \mathbb{R}^3$ is a connected complete nonproper embedded minimal surface with locally bounded Gaussian curvature that limits to the plane $\Pi = \{x_3 = 0\}$ from above, then for any $\varepsilon > 0$ the Gaussian curvature of $M \cap \{0 < x_3 \leq \varepsilon\}$ cannot be bounded from below. In other words, there exists a sequence $\{p_n\}_n \subset M$ with $x_3(p_n) \searrow 0$ and $|K_M(p_n)| \to \infty$ as n goes to infinity. Such a sequence must diverge in space because K_M is locally bounded. If we additionally assume M has finite topology, then an application of the Colding-Minicozzi one-sided curvature estimate (Theorem 23) contradicts that $|K_M(p_n)| \to \infty$. This is a rough sketch of the proof of the following statement.

THEOREM 33. [**80**] *If $M \subset \mathbb{R}^3$ is a connected complete embedded minimal surface in $\mathbb{R}^3$ with finite topology and locally bounded Gaussian curvature, then M is proper.*

Meeks, Pérez and Ros (Theorem 5 in [**74**]) have combined the last statement with deeper arguments using the results of Colding and Minicozzi, which let us exchange the finite topology assumption by the weaker hypothesis of finite genus.

THEOREM 34. [**74**] *If $M \subset \mathbb{R}^3$ is a connected complete embedded minimal surface in $\mathbb{R}^3$ with finite genus and locally bounded Gaussian curvature, then M is proper.*

In conclusion, we can state the following descriptive result for minimal laminations of $\mathbb{R}^3$.

THEOREM 35 (Meeks and Rosenberg [**80**], Meeks, Pérez and Ros [**74**]). *For a given minimal lamination $\mathcal{L}$ of $\mathbb{R}^3$, one of the following possibilities hold.*

i) $\mathcal{L}$ has one leaf which consists of a properly embedded minimal surface in $\mathbb{R}^3$.

ii) $\mathcal{L}$ has more that one leaf and consists of the disjoint union of a nonempty closed set of parallel planes $\mathcal{P} \subset \mathcal{L}$ together with a collection of complete minimal surfaces of unbounded Gaussian curvature and infinite genus that are properly embedded in the open slabs and halfspaces of $\mathbb{R}^3 - \mathcal{P}$. Furthermore, each of the open slabs and halfspaces in $\mathbb{R}^3 - \mathcal{P}$ contains at most one leaf of $\mathcal{L}$, every plane parallel to but different from the planes in $\mathcal{P}$ intersects at most one of the leaves of $\mathcal{L}$ and separates such an intersecting leaf into exactly two components.

To conclude this Subsection, we would like to mention that all the above results depend heavily on the embeddedness and properness assumptions. One sees this in part because of the existence of complete immersed minimal surfaces in a ball of $\mathbb{R}^3$. The first such an example was a minimal disk constructed by Nadirashvili [**94**] by a smart application of Runge's Theorem together with the López-Ros deformation. Later on, Martín and Morales [**65**] gave a bounded complete minimal annulus in $\mathbb{R}^3$, and using these techniques together with the Implicit Function Theorem,

López, Martín and Morales [**61, 60**] generalized these complete bounded minimal surfaces to similar examples with any finite topology. As we mentioned in Subsection 3.2, Martín and Morales [**64**] have recently generalized these results to prove that the interior of any convex, possibly noncompact or nonsmooth, region of $\mathbb{R}^3$ admits a proper complete minimal immersion of the unit disk. An interesting question is whether or not such a minimal disk can be embedded.

11. The Shiffman Jacobi function on properly embedded planar domains.

Next we explain how an interesting Jacobi function, called the *Shiffman function*, can help in our attempt to solve Conjecture 11. Let $M \subset \mathbb{R}^3$ be a properly embedded minimal surface with genus zero and infinitely many ends. By Theorems 19 and 31, M has two limit ends. After a rotation, we will assume M has horizontal tangent plane at infinity. By Theorem 32, the Gauss map g of M (stereographically projected from the sphere) is a holomorphic function without zeros or poles on M. In particular, M intersects each horizontal plane transversally. For minimal surfaces with this last property, Shiffman [**119**] introduced in 1956 a function that incorporates the curvature variation of the horizontal sections of the surface.

The Shiffman function can be defined locally. Assume that $(g(z), dh = dz)$ is the Weierstrass pair of a minimal surface $M \subset \mathbb{R}^3$, where z is a local conformal coordinate in M (in particular, g has no zeros or poles and any minimal surface admits such a local representation around a point with nonvertical normal vector). By (6), the induced metric ds^2 by the inner product of $\mathbb{R}^3$ is $ds^2 = \Lambda^2|dz|^2$, where $\Lambda = \frac{1}{2}(|g| + |g|^{-1})$. The horizontal level curves $x_3 = c$ correspond to $z_c(y) = c + iy$ in the z-plane (here $z = x + iy$ with $x, y \in \mathbb{R}$ and $i^2 = -1$) and the planar curvature of this level curve can be computed as

$$\kappa_c(y) = \left[\frac{|g|}{1+|g|^2}\Re\left(\frac{g'}{g}\right)\right]\Bigg|_{z=z_c(y)}, \tag{16}$$

where the prime stands for derivative with respect to z.

DEFINITION 25. We define the *Shiffman function* of M as

$$u = \Lambda\frac{\partial \kappa_c}{\partial y} = \Im\left[\frac{3}{2}\left(\frac{g'}{g}\right)^2 - \frac{g''}{g} - \frac{1}{1+|g|^2}\left(\frac{g'}{g}\right)^2\right],$$

where $\Im$ stands for imaginary part.

Since Λ is a positive function, the zeros of u coincide with the critical points of $\kappa_c(y)$. Thus, $u = 0$ vanishes identically if and only if M is foliated by pieces of circles and straight lines in horizontal planes. In a posthumously published paper, B. Riemann [**110, 111**] classified all minimal surfaces with such a foliation property: they reduce to the plane, catenoid, helicoid and the 1-parameter family of surfaces which, since then, have been known as *Riemann minimal examples* (see Subsection 2.5).

Coming back to our properly embedded minimal surface $M \subset \mathbb{R}^3$ with genus zero, infinitely many ends and horizontal tangent plane at infinity, we now see that a way to solve Conjecture 11 consists of proving that the Shiffman function of M vanishes identically. A crucial property of the Shiffman function is that it satisfies $\Delta u - 2Ku = 0$ on M (here K is the Gaussian curvature of M), i.e. u is a Jacobi function. This observation allows one to weaken the condition $u = 0$ that characterizes the Riemann minimal examples to the following condition. Recall from Subsection 2.8 that a *linear Jacobi function* is any function of the type $\langle N, v\rangle$ with fixed $v \in \mathbb{R}^3$, where N is the Gauss map of M.

LEMMA 6. [**102, 103**] *Let $M \subset \mathbb{R}^3$ be a properly embedded planar domain with infinitely many ends and transverse to horizontal planes. If the Shiffman function of M is linear, then M is a Riemann minimal example.*

SKETCH OF PROOF. The argument uses the Montiel-Ros correspondence between Jacobi functions on M and branched minimal immersions with Gauss map N (Theorem 10) to conclude that if the Shiffman function $u = \Im(f)$ of M is linear, then its Jacobi-conjugate function[11] $u^* = \Re(f)$ is also linear, where $f = \frac{3}{2}\left(\frac{g'}{g}\right)^2 - \frac{g''}{g} - \frac{1}{1+|g|^2}\left(\frac{g'}{g}\right)^2$. Thus $f = \langle N, z_0\rangle$ for a certain $z_0 \in \mathbb{C}^3$, which in turns implies that g satisfies a differential equation of the type $(g')^2 = g(\alpha g^2 + \beta g + \delta)$ with $\alpha, \beta, \delta \in \mathbb{C}$. From here it is not difficult to deduce that M is an unbranched covering of a properly embedded minimal torus with two planar ends in a certain quotient of $\mathbb{R}^3$ by a translation. Now the classification Theorem by Meeks, Pérez, Ros [**78**] in the periodic setting applies to finish the proof of the Lemma. □

We devote the remainder of this Section to proving that the Shiffman function vanishes on a properly embedded planar domain M with two limit ends and horizontal tangent plane at infinity, provided that the ratio between the horizontal and vertical components of its flux is small enough. To prove this result, we first need to understand the global behavior of the Shiffman function u on such a surface M. Using that the Weierstrass pair of M around any of its middle ends is $(g(z) = z^2 t(z), dh = dz)$ where t is a holomorphic function of z with $t(0) \neq 0$ and $t'(0) = 0$ (here $z = 0$ corresponds to the puncture, we have assumed that the limit normal vector at the end points to the South Pole of $\mathbb{S}^2$ and $t'(0) = 0$ comes from the fact that the end has no period), a straightforward calculation shows that $u(0) = -\Im\left(\frac{t''(0)}{t(0)}\right)$, which implies that u extends smoothly through $z = 0$. A similar result holds at the ends where g has a pole. Thus u can be viewed as a continuous function on the conformal cylinder $\overline{M}$ obtained after attaching the middle ends to M. By elliptic regularity, u is smooth on $\overline{M}$.

LEMMA 7. *Let $\delta \in (0,1)$ and let $\Omega \subset \mathbb{R}^3$ be a complete noncompact minimal surface with nonempty compact boundary and finite total curvature, such that its Gauss map N satisfies $N_3 = \langle N, e_3\rangle \geq 1 - \delta$ in Ω. Then, for every bounded Jacobi function v on Ω,*

$$(1-\delta)\sup_{\Omega}|v| \leq \sup_{\partial\Omega}|v|.$$

[11] Two Jacobi functions v, v^* on M are called *Jacobi-conjugate* if there exists a globally defined complex solution f of the Jacobi equation $\Delta f - 2Kf = 0$ on M such that $v = \Re(f)$ and $v^* = \Im(f)$.

PROOF. Since Ω has finite total curvature, Ω compactifies after attaching its ends to a compact Riemann surface $\overline{\Omega}$ with boundary. Since v is bounded on Ω, v extends smoothly across the punctures to a Jacobi function on $\overline{\Omega}$. We will let $a = \sup_{\partial\Omega} |v|$. Since $N_3 \geq 1 - \delta > 0$ in Ω and N_3 is Jacobi, we conclude that Ω is strictly stable and so, $a > 0$. Now, $v + \frac{a}{1-\delta} N_3 \geq 0$ in $\partial\Omega$ and $v + \frac{a}{1-\delta} N_3$ is Jacobi on Ω, thus by stability $v + \frac{a}{1-\delta} N_3 \geq 0$ in Ω. Analogously, $v - \frac{a}{1-\delta} N_3 \leq 0$ in $\partial\Omega$, hence $v - \frac{a}{1-\delta} N_3 \leq 0$ in Ω. These inequalities together with $N_3 \leq 1$ give $|v| \leq \frac{a}{1-\delta}$ in Ω, as desired. □

Theorem 32 implies that each properly embedded minimal planar domain M with infinite topology and horizontal tangent plane at infinity has a well-defined *flux vector,* which is the flux of M along any compact horizontal section, and this vector is neither vertical nor horizontal. In the sequel, we will normalize M by a homothety so that its flux vector has the form $(F(M), 1) \in \mathbb{C} \times \mathbb{R} \equiv \mathbb{R}^3$. Let $\mathcal{S}$ be the space of all properly embedded minimal planar domains with infinitely many ends, horizontal tangent plane at infinity and vertical component of the flux equal to one.

THEOREM 36 (Meeks, Pérez, Ros [**102**]). *There exists $\varepsilon > 0$ such that if $M \in \mathcal{S}$ has $|F(M)| < \varepsilon$, then M is a Riemann minimal example.*

PROOF. We will present here a different proof from the one in [**102**]. By contradiction, assume we have a sequence $\{M_n\}_n \subset \mathcal{S}$ with $F(M_n) \to 0$, and none of the M_n is a Riemann minimal example. Point *4* in Theorem 32 insures that $\{M_n\}_n$ has uniformly bounded Gaussian curvature. A suitable modification of the arguments in the proof of Lemma 3 in [**78**] can be used to show that as $n \to \infty$, the surfaces M_n become arbitrarily close to an infinite discrete collection of larger and larger translated pieces of a vertical catenoid with flux $e_3 = (0, 0, 1)$ joined by flatter and flatter graphs containing the ends of M_n. For each n, let $\overline{M}_n$ be the conformal cylinder obtained by attaching the middle ends to M_n, and let u_n be the Shiffman function of M_n.

ASSERTION 1. *For $n \in \mathbb{N}$ large enough, u_n is bounded on $\overline{M}_n$.*

PROOF OF ASSERTION 1. Suppose that the Assertion fails. To simplify the notation, we will denote the surface M_n only by M, and think of M as being arbitrarily close to pieces of translated catenoids and flat graphs as above. The failure of the Assertion allows us to find a subsequence of points $p_k \in M$ such that $|u(p_k)| \to \infty$ as k goes to ∞ (here u is the Shiffman function of M). Note that $\{p_k\}_k$ must be a divergent sequence in height (otherwise we contradict that u stays bounded at any middle end of M). By Theorem 32, $\{M - p_k\}_k$ is a sequence of properly embedded minimal surfaces in $\mathbb{R}^3$ with uniform bounds for the Gaussian curvature and area. After passing to a subsequence, $M - p_k$ converges to a properly embedded minimal surface $M_\infty \in \mathbb{R}^3$ with $0 \in M_\infty$.

First suppose that the value of the Gauss map N of M at p_k does not converge to vertical as $k \to \infty$. Then, a suitable modification of the arguments in the proof of Lemma 2 in [**78**] insures that M_∞ cannot be flat. This implies the sequence $\{M - p_k\}_k$ converges smoothly to M_∞ with multiplicity one, and a lifting argument shows that M_∞ has genus zero. Since M_∞ has no points with vertical normal vector (which comes from the open mapping Theorem applied to the Gauss map of M_∞), M_∞ must have a well-defined Shiffman function u_∞, which is nothing but the limit of $u_k(q) = u(q + p_k)$, $q \in M - p_k$. In particular $u(p_k)$ converges to $u_\infty(0)$, which is a contradiction.

Now assume that, after extracting a subsequence, $N(p_k)$ converges to e_3 as $k \to \infty$. By taking k sufficiently large, we conclude that p_k lies in one of the almost flat graphs $\Omega \subset M = M_n$ that joins two consecutive catenoids forming. This contradicts Lemma 7, since $|u|$ is arbitrarily small in $\partial\Omega$ but $|u(p_k)| \to \infty$. Now Assertion 1 is proved.

In the sequel, we will work with n large so that Assertion 1 holds. Note that for fixed n, the function $|u_n|$ needs not attain its maximum on M_n, but in that case we can exchange each M_n by a limit of suitable translations of M_n so that the Shiffman function in absolute value reaches its maximum on this limit. Since the flux of a surface in $\mathcal{S}$ does not change under translations, we do not loss generality by assuming that for all n large, $|u_n|$ attains its maximum at a point $p_n \in M_n$. We now define $v_n = \frac{1}{|u_n(p_n)|} u_n$.

Take a sequence $\{\delta(n)\}_n \subset (0,1)$ converging to 1. For n large, let $C_n \subset M_n$ be one of the connected components of $\langle N_n, e_3\rangle^{-1}[-\delta(n), \delta(n)]$ which contains p_n or is adjacent to a horizontal graphical region containing p_n. By our previous arguments, C_n is arbitrarily close to a translated image of the intersection of a vertical catenoid C_∞ of vertical flux e_3 centered at the origin with a ball of arbitrarily large radius also centered at the origin.

ASSERTION 2. *$\{\sup_{C_n} |v_n|\}_n$ tends to zero as $n \to \infty$.*

PROOF OF ASSERTION 2. Since $\{v_n|_{C_n}\}_n$ is a bounded sequence of Jacobi functions on the C_n and suitable translations of the C_n converge to the catenoid C_∞, it is not difficult to check that a subsequence of $\{v_n|_{C_n}\}_n$ (denoted in the same way) converges to a bounded Jacobi function on C_∞. Since bounded Jacobi functions on a catenoid are linear, we conclude that $\{v_n|_{C_n}\}_n$ converges to a linear Jacobi function v on C_∞ (or by identifying C_∞ with the sphere $\mathbb{S}^2$ through its Gauss map, we can see v as a linear function on $\mathbb{S}^2$). We now check that v is identically zero on $\mathbb{S}^2$.

By contradiction, suppose v is not identically zero on $\mathbb{S}^2$. Recall that the Shiffman function $u_n|_{C_n}$ measures the derivative of the curvature of each planar section of C_n with respect to a certain parameter times a positive function. By the Four Vertex Theorem, each horizontal section of C_n contains at least four zeros of u_n and so, also at least four zeros of v_n. Since horizontal sections of the C_n (suitably translated) converge to horizontal sections of C_∞ and any nontrivial linear function on $\mathbb{S}^2$ has at most two zeros on each horizontal section (with a possible exceptional horizontal section if the linear function is the vertical coordinate, but this does not affect to our argument by taking a different horizontal section), we conclude that at least two zeros of v_n in a certain horizontal section must collapse into a zero of v, hence the gradient of v will vanish at such a collapsing zero. But the gradient of a nontrivial linear function on $\mathbb{S}^2$ never vanishes at a zero of the function. This contradiction proves Assertion 2.

Recall that $|v_n(p_n)| = 1$ for all n. By Assertion 2, $N_n(p_n)$ must converge to the vertical or equivalently, p_n must lie in one of the graphical components of the complement of all the catenoidal pieces in M_n, a noncompact minimal graph which we will denote by Ω_n. Note that Ω_n is a graph over an unbounded domain in the plane $\{x_3 = 0\}$, $\partial\Omega_n$ consists of two almost-circular almost-horizontal curves with $\langle N_n, e_3\rangle|_{\partial\Omega_n} = \pm\delta(n)$ and Ω_n contains exactly one end of M_n. Hence we can apply

Lemma 7 to the minimal surface Ω_n and to the bounded Jacobi function $v_n|_{\Omega_n}$, contradicting that $v_n|_{\partial\Omega_n}$ converges to zero (Assertion 2) but $|v(p_n)| = 1$. This contradiction finishes the proof of the Theorem. □

The proof of Theorem 36 that appears in [**102**] is based on the relationship between the Shiffman function and the index form $Q(v,v) = \int_\Omega(|\nabla v|^2 + 2Kv^2)$ on any connected complement Ω of $\{|g| = 1\}$ in a planar domain $M \in \mathcal{S}$. A similar technique with the nodal domains of a linear function can be applied to conclude the following description of all bounded Jacobi functions on any Riemann minimal example.

THEOREM 37 (Meeks, Pérez, Ros [**77**]). *Let $M \subset \mathbb{R}^3$ be a Riemann minimal example. Then any bounded Jacobi function on M is linear.*

The above Theorem plays a central role in our program to prove that any finite genus limit end of a properly embedded minimal surface converges exponentially quickly to a limit end of one of the Riemann minimal examples. The Theorem should also be useful in proving that the moduli space of genus 1 properly embedded minimal surfaces with horizontal limit tangent plane at infinity and two limit ends is an open interval parametrized by the ratio of the horizontal and vertical components of the flux, in the same way that the Riemann examples are parametrized, and for finite $g \geq 2$, the moduli space of the genus g examples with two limit ends has an infinite number of path components.

References

[1] S. Bernstein. Uber ein geometrisches theorem und seine anwendung auf die partiellen differentialglechungen vom elliptischen typus. *Math. Z.*, 26:551–558, 1927.

[2] A. I. Bobenko. Helicoids with handles and Baker-Akhiezer spinors. *Math. Z.*, (1):9–29, 1998.

[3] A. I. Bobenko and M. Schmies. Computer graphics experiments for helicoids with handles. Personal communication.

[4] O. Bonnet. Mémoire sur l'emploi d'un nouveau systeme de variables dans l'etude des surfaces courbes. *J. Mathemém. p. appl.*, 2:153–266, 1860.

[5] E. Calabi. Quelques applications de l'Analyse complexe aux surfaces d'Aire minima. In *Topics in Complex Manifolds*, pages 59–81. Les Presses de l'Université de Montréal, 1967. H. Rossi, editor.

[6] M. Callahan, D. Hoffman, and W. H. Meeks III. Embedded minimal surfaces with an infinite number of ends. *Invent. Math.*, 96:459–505, 1989.

[7] M. Callahan, D. Hoffman, and W. H. Meeks III. The structure of singly-periodic minimal surfaces. *Invent. Math.*, 99:455–481, 1990.

[8] E. Catalan. Sur les surfaces réglées dont l'aire est un minimum. *J. Mathem. p. appl.*, 7:203–211, 1842.

[9] I. Chavel. *Riemannian Geometry: a modern introduction.* Cambridge University Press, 1993.

[10] J. Cheeger and D. G. Ebin. *Comparison theorems in Riemannian Geometry.* North-Holland Mathematical Library, Vol. 9., 1975.

[11] H. I. Choi and R. Schoen. The space of minimal embeddings of a surface into a three-dimensional manifold of positive Ricci curvature. *Invent. Math.*, 81:387–394, 1985.

[12] T. Choi, W. H. Meeks III, and B. White. A rigidity theorem for properly embedded minimal surfaces in $\mathbb{R}^3$. *J. of Differential Geometry*, 32:65–76, 1990.

[13] T. H. Colding and W. P. Minicozzi II. Embedded minimal disks: proper versus nonproper - global versus local. *Transactions of A.M.S.* (to appear).

[14] T. H. Colding and W. P. Minicozzi II. An excursion into geometric analysis. *J. Diff. Geom.*

[15] T. H. Colding and W. P. Minicozzi II. The space of embedded minimal surfaces of fixed genus in a 3-manifold I; Estimates off the axis for disks. *Annals of Math.* (to appear).

[16] T. H. Colding and W. P. Minicozzi II. The space of embedded minimal surfaces of fixed genus in a 3-manifold II; Multi-valued graphs in disks. *Annals of Math.* (to appear).

[17] T. H. Colding and W. P. Minicozzi II. The space of embedded minimal surfaces of fixed genus in a 3-manifold III; Planar domains. *Annals of Math.* (to appear).

[18] T. H. Colding and W. P. Minicozzi II. The space of embedded minimal surfaces of fixed genus in a 3-manifold IV; Locally simply-connected. *Annals of Math.* (to appear).

[19] T. H. Colding and W. P. Minicozzi II. The space of embedded minimal surfaces of fixed genus in a 3-manifold V; Fixed genus. in preparation.

[20] T. H. Colding and W. P. Minicozzi II. *Minimal surfaces.* Courant Institute of Mathematical Sciences Lecture Notes **4**, 1999.

[21] T. H. Colding and W. P. Minicozzi II. Complete properly embedded minimal surfaces in $\mathbb{R}^3$. *Duke Math. J.*, 107:421–426, 2001.

[22] T. H. Colding and W. P. Minicozzi II. Estimates for parametric elliptic integrands. *Int. Math. Res. Not.*, (6):291–297, 2002.

[23] T. H. Colding and W. P. Minicozzi II. Multi-valued minimal graphs and properness of disks. *International Mathematical Research Notices*, (21):1111–1127, 2002.

[24] T. H. Colding and W. P. Minicozzi II. Disks that are double spiral staircases. *Notices of A.M.S.*, 50(3):327–339, 2003.

[25] P. Collin. Topologie et courbure des surfaces minimales de $\mathbb{R}^3$. *Annals of Math. 2nd Series*, 145–1:1–31, 1997.

[26] P. Collin, R. Kusner, W. H. Meeks III, and H. Rosenberg. The geometry, conformal structure and topology of minimal surfaces with infinite topology. Preprint.

[27] C. Costa. *Imersöes minimas en* $\mathbb{R}^3$ *de gênero un e curvatura total finita.* PhD thesis, IMPA, Rio de Janeiro, Brasil, 1982.

[28] C. Costa. Example of a complete minimal immersion in $\mathbb{R}^3$ of genus one and three embedded ends. *Bull. Soc. Bras. Mat.*, 15:47–54, 1984.

[29] C. Costa. Uniqueness of minimal surfaces embedded in $\mathbb{R}^3$ with total curvature 12π. *J. of Differential Geometry*, 30(3):597–618, 1989.

[30] C. Costa. Classification of complete minimal surfaces in $\mathbb{R}^3$ with total curvature 12π. *Invent. Math.*, 105(2):273–303, 1991.

[31] M. do Carmo and C. K. Peng. Stable minimal surfaces in $\mathbb{R}^3$ are planes. *Bulletin of the AMS*, 1:903–906, 1979.

[32] H. M. Farkas and I. Kra. *Riemann Surfaces.* Number 72 in Graduate texts in mathematics. Springer-Verlag, 1980.

[33] H. Federer. *Geometric measure theory.* Springer Verlag, Berlin-Heidelberg, New York, 1969.

[34] D. Fischer-Colbrie. On complete minimal surfaces with finite Morse index in 3-manifolds. *Invent. Math.*, 82:121–132, 1985.

[35] D. Fischer-Colbrie and R. Schoen. The structure of complete stable minimal surfaces in 3-manifolds of non-negative scalar curvature. *Comm. on Pure and Appl. Math.*, 33:199–211, 1980.

[36] M. Freedman. An unknotting result for complete minimal surfaces in $\mathbb{R}^3$. *Invent. Math*, 109(1):41–46, 1992.

[37] C. Frohman. The topological uniqueness of triply-periodic minimal surfaces in $\mathbb{R}^3$. *J. of Differential Geometry*, 31:277–283, 1990.

[38] C. Frohman and W. H. Meeks III. The topological classification of minimal surfaces in $\mathbb{R}^3$. Preprint.

[39] C. Frohman and W. H. Meeks III. The ordering theorem for the ends of properly embedded minimal surfaces. *Topology*, 36(3):605–617, 1997.

[40] C. Frohman and W. H. Meeks III. The topological uniqueness of complete one-ended minimal surfaces and Heegaard surfaces in $\mathbb{R}^3$. *J. of the Amer. Math. Soc.*, 10(3):495–512, 1997.

[41] A. Grigor'yan. Analytic and geometric background of recurrence and non-explosion of Brownian motion on Riemannian manifolds. *Bull. of A.M.S*, 36(2):135–249, 1999.

[42] P. Hall. Two topological examples in minimal surface theory. *J. of Differential Geometry*, 19:475–481, 1984.

[43] D. Hoffman. The computer-aided discovery of new embedded minimal surfaces. *Mathematical Intelligencer*, 9(3):8–21, 1987.

[44] D. Hoffman and H. Karcher. Complete embedded minimal surfaces of finite total curvature. In R. Osserman, editor, *Encyclopedia of Mathematics, Vol. 90, Geometry V*, pages 5–93. Springer Verlag, 1997.
[45] D. Hoffman, H. Karcher, and W. H. Meeks III. One-parameter families of embedded complete minimal surfaces of finite topology. *GANG preprint* in preparation.
[46] D. Hoffman, H. Karcher, and F. Wei. Adding handles to the helicoid. *Bulletin of the AMS, New Series*, 29(1):77–84, 1993.
[47] D. Hoffman, H. Karcher, and F. Wei. The genus one helicoid and the minimal surfaces that led to its discovery. In *Global Analysis and Modern Mathematics*. Publish or Perish Press, 1993. K. Uhlenbeck, editor, p. 119–170.
[48] D. Hoffman and W. H. Meeks III. A complete embedded minimal surface in $\mathbb{R}^3$ with genus one and three ends. *J. of Differential Geometry*, 21:109–127, 1985.
[49] D. Hoffman and W. H. Meeks III. Embedded minimal surfaces of finite topology. *Annals of Math.*, 131:1–34, 1990.
[50] D. Hoffman and W. H. Meeks III. The strong halfspace theorem for minimal surfaces. *Invent. Math.*, 101:373–377, 1990.
[51] D. Hoffman, M. Weber, and M. Wolf. The existence of the genus-one helicoid. Preprint.
[52] A. Huber. On subharmonic functions and differential geometry in the large. *Comment. Math. Helvetici*, 32:181–206, 1957.
[53] L. Jorge and W. H. Meeks III. The topology of complete minimal surfaces of finite total Gaussian curvature. *Topology*, 22(2):203–221, 1983.
[54] N. Kapouleas. Complete embedded minimal surfaces of finite total curvature. *J. Differential Geom.*, 47(1):95–169, 1997.
[55] N. Korevaar, R. Kusner, and B. Solomon. The structure of complete embedded surfaces with constant mean curvature. *J. of Differential Geometry*, 30:465–503, 1989.
[56] R. Langevin and H. Rosenberg. A maximum principle at infinity for minimal surfaces and applications. *Duke Math. J.*, 57(3):819–828, 1988.
[57] H. B. Lawson. The unknottedness of minimal embeddings. *Invent. Math.*, 11:183–187, 1970.
[58] H. B. Lawson, Jr. *Lectures on Minimal Submanifolds*. Publish or Perish Press, Berkeley, 1971.
[59] H. Lazard-Holly and W. H. Meeks III. The classification of embedded doubly–periodic minimal surfaces of genus zero. *Invent. Math.*, 143:1–27, 2001.
[60] F. J. López, F. Martín, and S. Morales. Complete nonorientable minimal surfaces in a ball of $\mathbb{R}^3$. Preprint.
[61] F. J. López, F. Martín, and S. Morales. Adding handles to Nadirashvili's surfaces. *Journal of Diff. Geom.*, 60:155–175, 2002.
[62] F. J. López and J. Pérez. Parabolicity and Gauss map of minimal surfaces. *Indiana J. of Math.*, 52(4):1017–1026, 2003.
[63] F. J. López and A. Ros. On embedded complete minimal surfaces of genus zero. *J. of Differential Geometry*, 33(1):293–300, 1991.
[64] F. Martin and S. Morales. Complete and proper minimal surfaces in convex bodies. Preprint.
[65] F. Martin and S. Morales. On the asymptotic behavior of a complete bounded minimal surface in $\mathbb{R}^3$. *Trans. of the A.M.S.* (to appear).
[66] W. H. Meeks III. The geometry and topology of singly-periodic minimal surfaces. to appear in the Asian J. of Math.
[67] W. H. Meeks III. Global problems in classical minimal surface theory. To appear in the Clay Minimal Surface Conference Proceedings.
[68] W. H. Meeks III. The regularity of the singular set in the Colding and Minicozzi lamination theorem. To appear in Duke Math. J.
[69] W. H. Meeks III. The classification of complete minimal surfaces with total curvature greater than -8π. *Duke Math. J.*, 48:523–535, 1981.
[70] W. H. Meeks III. The topological uniqueness of minimal surfaces in three-dimensional Euclidean space. *Topology*, 20:389–410, 1981.
[71] W. H. Meeks III. The theory of triply-periodic minimal surfaces. *Indiana Univ. Math. J.*, 39(3):877–936, 1990.
[72] W. H. Meeks III. The geometry, topology, and existence of periodic minimal surfaces. *Proceedings of Symposia in Pure Math.*, 54:333–374, 1993. Part I.

[73] W. H. Meeks III and J. Pérez. Recent advances in classical minimal surface theory. Work in progress.

[74] W. H. Meeks III, J. Pérez, and A. Ros. The geometry of minimal surfaces of finite genus I; curvature estimates and quasiperiodicity. To appear in J. Diff. Geom.

[75] W. H. Meeks III, J. Pérez, and A. Ros. The geometry of minimal surfaces of finite genus II; nonexistence of one limit end examples. To appear in Invent. Math.

[76] W. H. Meeks III, J. Pérez, and A. Ros. The geometry of minimal surfaces of finite genus III; bounds on the topology and index of classical minimal surfaces. Preprint.

[77] W. H. Meeks III, J. Pérez, and A. Ros. The geometry of minimal surfaces of finite genus IV; Jacobi fields and uniqueness for small flux. Work in progress.

[78] W. H. Meeks III, J. Pérez, and A. Ros. Uniqueness of the Riemann minimal examples. *Invent. Math.*, 131:107–132, 1998.

[79] W. H. Meeks III and H. Rosenberg. Maximum principles at infinity with applications to minimal and constant mean curvature surfaces. Preprint.

[80] W. H. Meeks III and H. Rosenberg. The uniqueness of the helicoid and the asymptotic geometry of properly embedded minimal surfaces with finite topology. To appear in Annals of Math.

[81] W. H. Meeks III and H. Rosenberg. The global theory of doubly periodic minimal surfaces. *Invent. Math.*, 97:351–379, 1989.

[82] W. H. Meeks III and H. Rosenberg. The maximum principle at infinity for minimal surfaces in flat three-manifolds. *Comment. Math. Helvetici*, 65:255–270, 1990.

[83] W. H. Meeks III and H. Rosenberg. The geometry and conformal structure of properly embedded minimal surfaces of finite topology in $\mathbb{R}^3$. *Invent. Math.*, 114:625–639, 1993.

[84] W. H. Meeks III and H. Rosenberg. The geometry of periodic minimal surfaces. *Comment. Math. Helvetici*, 68:538–578, 1993.

[85] W. H. Meeks III, L. Simon, and S. T. Yau. The existence of embedded minimal surfaces, exotic spheres and positive Ricci curvature. *Annals of Math.*, 116:221–259, 1982.

[86] W. H. Meeks III and M. Weber. Existence of bent helicoids and the geometry of the singular set in the Colding-Minicozzi lamination theorem. Preprint.

[87] W. H. Meeks III and M. Wolf. Solution to the finite flux conjecture in dimension two. Work in progress.

[88] W. H. Meeks III and S. T. Yau. The classical Plateau problem and the topology of three-dimensional manifolds. *Topology*, 21(4):409–442, 1982.

[89] W. H. Meeks III and S. T. Yau. The existence of embedded minimal surfaces and the problem of uniqueness. *Math. Z.*, 179:151–168, 1982.

[90] W. H. Meeks III and S. T. Yau. The topological uniqueness of complete minimal surfaces of finite topological type. *Topology*, 31(2):305–316, 1992.

[91] S. Montiel and A. Ros. Schrödinger operators associated to a holomorphic map. In *Global Differential Geometry and Global Analysis (Berlin, 1990)*, volume 1481 of *Lecture Notes in Mathematics*, pages 147–174. Springer-Verlag, 1990.

[92] S. Morales. On the existence of a proper minimal surface in $\mathbb{R}^3$ with the conformal type of a disk. To appear in GAFA.

[93] C. B. Morrey. The problem of Plateau in a Riemannian manifold. *Annals of Math.*, 49:807–851, 1948.

[94] N. Nadirashvili. Hadamard's and Calabi-Yau's conjectures on negatively curved and minimal surfaces. *Invent. Math.*, 126(3):457–465, 1996.

[95] J. C. C. Nitsche. A characterization of the catenoid. *J. of Math. Mech.*, 11:293–302, 1962.

[96] J. C. C. Nitsche. A new uniqueness theorem for minimal surfaces. *Arch. Rat. Mech. Anal.*, 52:319–329, 1973.

[97] J. C. C. Nitsche. *Lectures on Minimal Surfaces*, volume 1. Cambridge University Press, 1989.

[98] R. Osserman. Global properties of minimal surfaces in E^3 and E^n. *Annals of Math.*, 80(2):340–364, 1964.

[99] R. Osserman. *A Survey of Minimal Surfaces*. Dover Publications, New York, 2nd edition, 1986.

[100] R. Osserman and M. Schiffer. Doubly-connected minimal surfaces. *Arch. Rat. Mech. Anal.*, 58:285–307, 1975.

[101] J. Pérez. Parabolicity and minimal surfaces. To appear in the *Proceedings of the Conference on Global Theory of Minimal Surfaces,* Berkeley 2001.
[102] J. Pérez. Uniqueness of the Riemann minimal surfaces. To appear in the *Proceedings of the Conference on Global Theory of Minimal Surfaces,* Berkeley 2001.
[103] J. Pérez. On singly-periodic minimal surfaces with planar ends. *Trans. of the A.M.S.*, 6:2371–2389, 1997.
[104] J. Pérez. A rigidity theorem for periodic minimal surfaces. *Comm. in Analysis and Geom.*, 7(1):95–104, 1999.
[105] J. Pérez and A. Ros. Some uniqueness and nonexistence theorems for embedded minimal surfaces. *Math.Ann.*, 295(3):513–525, 1993.
[106] J. Pérez and A. Ros. The space of properly embedded minimal surfaces with finite total curvature. *Indiana Univ. Math. J.*, 45(1):177–204, 1996.
[107] J. Pérez and A. Ros. Properly embedded minimal surfaces with finite total curvature. In *The Global Theory of Minimal Surfaces in Flat Spaces-LN M-1775*, pages 15–66. Springer Verlag, 1999. G. P. Pirola, editor.
[108] J. Pitts and H. Rubenstein. Applications of minimax to minimal surfaces and the topology of three-manifolds. In *Proceedings for the Center for Mathematical Analysis*, pages 137–170, Canberra, Australia, 1987. Australian National University.
[109] A. V. Pogorelov. On the stability of minimal surfaces. *Soviet Math. Dokl.*, 24:274–276, 1981.
[110] B. Riemann. Über die Fläche vom kleinsten Inhalt bei gegebener Begrenzung. *Abh. Königl, d. Wiss. Göttingen, Mathem. Cl.*, 13:3–52, 1867. K. Hattendorf, editor.
[111] B. Riemann. *Ouevres Mathématiques de Riemann.* Gauthiers-Villars, Paris, 1898.
[112] A. Ros. Compactness of spaces properly embedded minimal surfaces with finite total curvature. *Indiana Univ. Math. J.*, 44(1):139–152, 1995.
[113] A. Ros. The Gauss map of minimal surfaces. In *Differential Geometry, Valencia 2001, Proceedings of the conference in honour of Antonio M. Naveira*, pages 235–252. World Scientific, 2002.
[114] H. Rosenberg. Some topics in minimal surface theory. Preprint.
[115] H. Rosenberg and E. Toubiana. A cylindrical type complete minimal surface in a slab of $\mathbb{R}^3$. *Bull. Sc. Math. III*, pages 241–245, 1987.
[116] H. F. Scherk. Bemerkungen über die kleinste Fläche innerhalb gegebener Grenzen. *J. R. Angew. Math.*, 13:185–208, 1835.
[117] R. Schoen. *Estimates for Stable Minimal Surfaces in Three Dimensional Manifolds*, volume 103 of *Annals of Math. Studies.* Princeton University Press, 1983.
[118] R. Schoen. Uniqueness, symmetry, and embeddedness of minimal surfaces. *J. of Differential Geometry*, 18:791–809, 1983.
[119] M. Shiffman. On surfaces of stationary area bounded by two circles, or convex curves, in parallel planes. *Annals of Math.*, 63:77–90, 1956.
[120] B. Solomon. On foliations of $\mathbb{R}^{n+1}$ by minimal hypersurfaces. *Comm. Math. Helv.*, 61:67–83, 1986.
[121] M. Soret. Maximum principle at infinity for complete minimal surfaces in flat 3-manifolds. *Annals of Global Analysis and Geometry*, 13:101–116, 1995.
[122] M. Traizet. A balancing condition for weak limits of minimal surfaces. Preprint.
[123] M. Traizet. An embedded minimal surface with no symmetries. *J. Differential Geometry*, 60(1):103–153, 2002.
[124] M. Traizet and M. Weber. Hermite polynomials and helicoidal minimal surfaces. Preprint.
[125] J. Tysk. Eigenvalue estimates with applications to minimal surfaces. *Pacific J. of Math.*, 128:361–366, 1987.
[126] M. Weber and M. Wolf. personal communication.
[127] M. Weber and M. Wolf. Teichmuller theory and handle addition for minimal surfaces. *Annals of Math.*, 156:713–795, 2002.
[128] F. Xavier. Convex hulls of complete minimal surfaces. *Math. Ann.*, 269:179–182, 1984.

Mathematics Department, University of Massachusetts, Amherst, MA 01003, USA
E-mail address: bill@gang.umass.edu

Universidad de Granada, Departamento de Geometria y Topologia, Fuentenueva s/n 18071 Granada, Spain
E-mail address: jperez@ugr.es

Surveys in Differential Geometry
IX, International Press

Analysis of the Cut Locus via the Heat Kernel

Robert Neel and Daniel Stroock

ABSTRACT. We study the Hessian of the logarithm of the heat kernel to see what it says about the cut locus of a point. In particular, we show that the cut locus is the set of points at which this Hessian diverges faster than t^{-1} as $t \searrow 0$. In addition, we relate the rate of divergence to the conjugacy and other structural properties.

1. Introduction

Our purpose here is to present some recent research connecting behavior of heat kernel to properties of the cut locus. The unpublished results mentioned below constitute part of the first author's thesis, where they will be proved in detail.

To explain what sort of results we have in mind, let M be a compact, connected Riemannian manifold of dimension n, and use $p_t(x,y)$ to denote the heat kernel for the heat equation $\partial_t u = \frac{1}{2}\Delta u$. As a special case of a well-known result due to Varadhan,

$$E_t(x,y) \equiv -t \log p_t(x,y) \to E(x,y) \quad \text{uniformly on } M \times M \text{ as } t \searrow 0,$$

where E is the energy function, given in terms of the Riemannian distance by $E(x,y) = \frac{1}{2}\operatorname{dist}(x,y)^2$. Thus, $(x,y) \rightsquigarrow E_t(x,y)$ can be considered to be a geometrically natural mollification of $(x,y) \rightsquigarrow E(x,y)$. In particular, given a fixed $x \in M$, we can hope to learn something about the cut locus $\operatorname{Cut}(x)$ of x by examining derivatives of $y \rightsquigarrow E_t(x,y)$.

Before going further, we present an example which, although somewhat trivial, may help explain what we have in mind. Namely, take $M = \mathbb{S}^1 \equiv \mathbb{R}/2\pi\mathbb{Z}$. In this case, the heat kernel is a theta function

$$p_t(0,\theta) = \frac{1}{\sqrt{2\pi t}} \sum_{n=-\infty}^{\infty} \exp\left[-\frac{(\theta + 2\pi n)^2}{2t}\right],$$

which is obtained by "wrapping" the heat kernel for $\mathbb{R}$ (i.e., the centered Gaussian kernel with variance t) around $\mathbb{S}^1$. It is clear that, as $t \searrow 0$, for any $m \geq 0$ and

The second author thanks the NSF for funds provided in DMS-0244991.

$\theta \in (-\pi, \pi)$:

$$\begin{aligned}\partial_\theta^m E_t &\simeq -\partial_\theta^m t \log\left(e^{-\frac{(\theta+\pi)}{2t}} + e^{-\frac{(\theta-\pi)}{2t}}\right)\\ &= \partial_\theta^m t\left(\frac{\theta^2}{2t} - \log\cosh\frac{\pi\theta}{t}\right),\end{aligned}$$

from which it follows that

$$\lim_{t\searrow 0} \partial_\theta^m E_t(0,\theta) = \partial_\theta^m \frac{(\pi - |\theta|)^2}{2} \quad \text{for } \theta \in (-\pi,\pi).$$

On the other hand,

$$\begin{aligned}&\lim_{t\searrow 0} \partial_\theta E_t(0,\theta)|_{\theta=\pi} = 0,\\ &\lim_{t\searrow 0} \partial_\theta^2 E_t(0,\theta)\big|_{\theta=\pi} = -\infty,\end{aligned}$$

but

$$-\lim_{t\searrow 0} t\partial_\theta^2 E_t(0,\theta)\big|_{\theta=\pi} = \pi^2,$$

and things get worse when $m > 2$.

The lessons to be learned from this example are

- The the behavior as $t \searrow 0$ of derivatives of E_t undergo dramatic change at the cut locus. For $\mathbb{S}^1$, this change is already evident in discontinuity which the first derivative has there, and it becomes even more dramatic in the second derivative, which goes from 1 when $\theta \neq 0$ to $-\infty$ when $\theta = 0$.
- The rate at which the Hessian of $E_t(x,y)$ explodes when $y \in \mathrm{Cut}(x)$ can be as high as t^{-1}.
- The direction in which the Hessian explodes is toward $-\infty$. In the case of $\mathbb{S}^1$, the intuitive explanation for this is easy: there are no strictly convex functions on a compact manifold and the Hessian of $E(0,\cdot)$ is 1 except at π. Hence, for $\mathbb{S}^1$, all the "concavity" of $E(0,\cdot)$ must live at π.

That the behavior of E_t for $\mathbb{S}^1$ is somewhat typical was proved in [**3**]. Namely, the following result was proved there.

THEOREM 1. [1] *Given M and some fixed x as above,*

$$\lim_{t\searrow 0} \nabla^2 E_t(x,y) = \nabla^2 E(x,y)$$

uniformly for y in compact subsets of $M\setminus \mathrm{Cut}(x)$. In addition, when $y \in \mathrm{Cut}(x)$ and there are multiple minimal geodesics from x to y, then, if the initial velocities of these geodesics have a sufficiently nice structure (e.g., if they form a submanifold of the tangent space to x), then there is a (strictly) positive definite, symmetric 2-tensor A to which $t\nabla^2 E_t(x,y)$ converges as $t \searrow 0$.

Some aspects of Theorem 1 are quite general. For example, if we think about the Hessian $\nabla^2 E(x,\cdot)$ of E as a distribution (in the sense of Laurent Schwartz), then it is relatively easy to show that its singular support is contained in $\mathrm{Cut}(x)$ and that, as a distribution, $\nabla^2 E(x,\cdot)$ can (cf. [**6**]) be estimated from above in terms of the uniform lower bound on the sectional curvature. Alternatively, the key ingredient in the proof of Theorem 1 is the use of pathspace integration to express

[1] An extension of this result to higher derivatives was given in [**7**], where it was shown that, at the cut locus, the mth order of E_t may explode as fast as $t^{-\frac{m}{2}}$.

$\nabla^2 E_t(x,y)$ as the sum of two terms, one of which stays bounded as $t \searrow 0$ and the other of which is $-t^{-1}$ times the variance of a random variable. Because when $y \notin \mathrm{Cut}(x)$ there is only one minimal geodesic and the first variation around this unique minimal geodesic is non-degenerate, an application of Laplace asymptotics to the pathspace integral allows one to show that the distribution of this random variable degenerates fast enough as $t \searrow 0$ to kill off the variance term. On the other hand, when $y \in \mathrm{Cut}(x)$ for the reason that there are multiple minimal geodesics, there can be residual variance, and this is what accounts for the final statement in Theorem 1.[2]

Unfortunately, the method employed in [**3**] is too cumbersome to encourage any attempt to extend it to more delicate situations or obtain more detailed information. For this reason, it seems wise to attempt seek alternative approaches. Perhaps the most geometrically natural alternative would be to see if one can mimic the computation in the preceding example writing M as the quotient of $\mathbb{R}^n$ by some sufficiently nice group of transformations. For example, it is not hard to analyze the flat torus in the same way as we did circle. More generally, one might hope to get something out of writing M as the quotient of its universal cover by the group of deck transformations. In particular, if M has non-positive sectional curvature, then the Cartan–Hadamard Theorem says that its universal cover will have no cut locus, and so everything should come down to an analysis of the way the deck transformations act on the heat kernel on the universal cover. However, except in very special case, like $\mathbb{S}^1$, such an analysis appears to be very difficult. Even worse, when M can have positive curvature, the structure of geodesics on the universal cover may not be essentially simpler than it is on M itself.

2. Another Approach

For the reasons alluded to toward the end of the preceding section, we will now discuss another, and enormously simpler, way to think about the sort of analysis on which Theorem 1 rests. Namely, it has been realized for some time (cf., for example, Pinsky [**4**]) that more precise asymptotics for $p_t(x,y)$ with $y \in \mathrm{Cut}(x)$ can be obtained by the following method, which we sketch here. By the Chapman–Kolmogorov equation, we can write $p_t(x,y)$ as the integral of $p_{\frac{t}{2}}(x,\cdot)p_{\frac{t}{2}}(\cdot,y)$ over M. Loosely speaking, as $t \searrow 0$,

$$p_{\frac{t}{2}}(x,z)p_{\frac{t}{2}}(z,y) \simeq \frac{1}{\sqrt{V(x,\frac{t}{2})V(y,\frac{t}{2})}} \exp\left(-\frac{h_{x,y}(z)}{2t}\right),$$

where $h_{x,y}(z) \equiv E(x,z) + E(z,y)$ is the *hinged energy function* and $V(p,r)$ denotes the volume of the ball of radius r centered at p. In particular, a naive Laplace asymptotic argument indicates the integral should be getting more and more concentrated on the set Γ of z's where $h_{x,y}(z)$ achieves its minimum value. Equivalently, $\Gamma = \{z : h_{x,y}(z) = E(x,y)\}$ and, as such, is the set of mid-points of minimal geodesics running from x to y. Thus, because Γ is a uniformly positive distance from both $\mathrm{Cut}(x)$ and $\mathrm{Cut}(y)$, one can apply the Pleijel expansion to each of the factors $p_{\frac{t}{2}}(x,\cdot)$ and $p_{\frac{t}{2}}(\cdot,y)$. The result is an expression for the asymptotics of $p_t(x,y)$ in terms of a Laplace integral of the asymptotics of the heat kernel near Γ, which is valid whether or not $y \in \mathrm{Cut}(x)$.

---

[2]In the extension in [**7**], the coefficient of $t^{-\frac{m}{2}}$ can be interpreted as the mth cumulant of the random variable whose variance appears in this discussion.

Using more recent results on the heat kernel, including Theorem 1 and the estimates in [**8**], it is possible to employ this method to study logarithmic derivatives of the heat kernel. In order to give a precise statement of what the method says when applied to the Hessian of $E_t(x,\cdot)$ at $\mathrm{Cut}(x)$, we need to introduce a little notation. We have already introduced Γ, the set of midpoints of minimal geodesics from x to y, and the hinged energy function $h_{x,y}(z) = E(x,z)+E(z,y)$. As we said, Γ is precisely the place where $z \rightsquigarrow h_{x,y}(z)$ achieves its minimum value $E(x,y)$ Now let Γ_ϵ be an ϵ-neighborhood of Γ, where we implicitly think of $\epsilon > 0$ as being strictly smaller than $\frac{1}{2}\operatorname{dist}(x,y)$. Further, given $z \in M\setminus \mathrm{Cut}(x)$, there is a unique $Z \in T_xM$ such that $s \in [0,1] \longmapsto \exp_x(sZ)$ is the minimal geodesic from x to z, and we will use $H(x,z)$ to denote the Jacobian of $\exp_x$ at Z. For fixed x, $z \in M \setminus \mathrm{Cut}(x) \longmapsto H(x,z) \in \mathbb{R}$ is a smooth function on $M\setminus \mathrm{Cut}(x)$.

THEOREM 2. *Let M be a smooth, compact, connected Riemannian manifold. Choose any two distinct points x and y on M any $A \in T_yM$. Then there exists a positive constant ϵ such that Γ_ϵ is a strictly positive distance from $\{x,y\}\cup\mathrm{Cut}(x)\cup \mathrm{Cut}(y)$ and*

$$\nabla^2_{A,A}E_t(x,y) = -\frac{4}{t}\left\{\frac{\int_{\Gamma_\epsilon}(\nabla_A E(z,y))^2\exp\left[-\frac{2}{t}h_{x,y}(z)\right]H(x,z)H(y,z)\,dz}{\int_{\Gamma_\epsilon}\exp\left[-\frac{2}{t}h_{x,y}(z)\right]H(x,z)H(y,z)\,dz}\right.$$
$$\left. - \left[\frac{\int_{\Gamma_\epsilon}\nabla_A E(z,y)\exp\left[-\frac{2}{t}h_{x,y}(z)\right]H(x,z)H(y,z)\,dz}{\int_{\Gamma_\epsilon}\exp\left[-\frac{2}{t}h_{x,y}(z)\right]H(x,z)H(y,z)\,dz}\right]^2\right\} + O(1),$$

where $\nabla_A E(z,y)$ stands for differentiation in the second variable, evaluated at y.

In many ways, the formula in Theorem 2 is an exact replica of the formula on which Theorem 1 was based. Indeed, here, like there, the coefficient of $-t^{-1}$ is a variance. In addition, as was the case there, all the integrals in this formula lend themselves to analysis via Laplace asymptotics as $t \searrow 0$. The difference is that here Laplace asymptotics is for finite dimensional integrals, whereas there it was for integrals in pathspace. Thus, everything should be simpler here. On the other hand, even though we are now working in finite dimensions, the asymptotics can be far from trivial. Indeed, the set Γ onto which the integral is being forced to collapse can be a very complicated and ugly!

3. Preliminary Conclusions

We begin our discussion of Theorem 2 by making it explicit that the coefficient of $-t^{-1}$ is a variance. For this purpose, set

$$\begin{aligned}\mu_t(dz) &= \frac{\mathbf{1}_{\Gamma_\epsilon}(z)}{Z_t}H(x,z)H(y,z)\exp\left(-\frac{2h_{x,y}(z)}{t}\right)dz\\ \text{where } Z_t &= \int_{\Gamma_\epsilon}H(x,z)H(y,z)\exp\left(-\frac{2h_{x,y}(z)}{t}\right)dz.\end{aligned} \tag{1}$$

Clearly, the coefficient of $-t^{-1}$ is the variance $\mathrm{Var}^{\mu_t}(\nabla_A E(\cdot,y))$ of $\nabla_A E(\cdot,y)$ with respect to μ_t. Moreover, because Γ_ϵ is compact, we know that $\{\mu_t : t>0\}$ is relatively compact in the weak topology, and it is clear that the set L_0 of limit points as $t \searrow 0$ consists of probability measures which are supported on Γ. In particular, if $\mu \in L_0$ comes from $t_i \searrow 0$, then we have that

$$\lim_{i\to\infty} t_i \nabla^2_{A,A} E_{t_i}(x,y) = 4\,\mathrm{Var}^{\mu}\left[\nabla_A E(\cdot,y)\right].$$

In order to get a more explicit expression for $\nabla_A E(z,y)$, let $z \in \Gamma$ be given and take $Y(z)$ be the (unique) unit vector at y such that $\exp_y\left[\mathrm{dist}(z,y)Y(z)\right] = z$. Then we know that

$$\nabla_A E(z,y) = -\frac{1}{2}\,\mathrm{dist}(x,y)\,\langle A, Y(z)\rangle .$$

Finally, use $\theta_A(z)$ to denote the angle between A and $Y(z)$. Then, for any $A \in T_yM$, we have

$$\lim_{i\to\infty} t_i \nabla^2_{A,A} E_{t_i}(x,y) = -|A|^2\,\mathrm{dist}(x,y)^2\,\mathrm{Var}^{\mu}\left[\cos\theta_A(z)\right]. \tag{2}$$

Remark: We digress here in order to expand on the connections between the approach which we are taking here and the one taken in [**3**]. In [**3**], the integrals were taken with respect to Brownian paths on M which start at x and are conditioned to arrive at y at time 1. Using the intuition which comes from the Feynman picture (cf. [**5**]) of Brownian integrals as being Gaussian integrals in which the weight is given by[3]

$$\exp\left(-\frac{1}{2}\int_0^1 |\dot{w}(t)|^2\,dt\right),$$

the heuristic expression for the heat kernel is

$$p_t(x,y) = \frac{1}{Z(t)} \int_{w(0)=x\ \&\ w(1)=y} \exp\left(-\frac{1}{2t}\int_0^1 |\dot{w}(t)|^2\,dt\right)\,dw,$$

where the "dw" is supposed to indicate that the integral is taken with respect to the (non-existent) Lebesgue measure on pathspace and the constant out in front is a (equally non-existent) normalizing factor. Fanciful as this expression may be, it strongly indicates that, as $t \searrow 0$, the overwhelming contribution to the integral will come from those paths w for whose energy is nearly minimal, and, in the limit, one should expect that the integral will be over minimal geodesics. Of course, this is exactly what (2) says. Namely, because Γ parameterizes the minimal geodesics from x to y, the measure μ can be thought of as a probability measure on the space of these minimal geodesics and the function $\cos\theta_A$ can be thought of a function there.

Some simple facts about the log Hessian follow immediately from equation (2). In the first place, if we homothetically scale M by a factor of $a > 0$,

$$\lim_{i\to\infty} t_i \nabla^2_{A,A} E_{t_i}(x,y)$$

is multiplied by a^2. Secondly, we have the inequality

$$0 \geq \limsup_{t\searrow 0} t\nabla^2_{A,A} E_t(x,y) \geq \liminf_{t\searrow 0} t\nabla^2_{A,A} E_t(x,y) \geq -|A|^2\,\mathrm{dist}(x,y)^2.$$

Before looking more closely at $h_{x,y}$ and its accompanying Laplace asymptotics, we take a moment to compute a specific example. Earlier, we observed that our explicit computation for $\mathbb{S}^1$ would not extend to higher dimensional spheres. Using Theorem 2, however, we can easily compute the leading term of the log Hessian of the heat kernel on the spheres. Choose any point $x \in \mathbb{S}^n$ (here n can be any

[3]The use of w to denote a generic path is in honor of Wiener.

integer greater than or equal to 2). Then $\mathrm{Cut}(x)$ consists of a single point, namely, the antipodal point to x. Thus, we may as well let x and y be the north and south pole, respectively, which we will denote N and S. In this case, Γ is the equatorial sphere $\mathbb{S}^{n-1}$. Further, by symmetry we see that μ_t converges to uniform probability measure on the equatorial sphere (with respect to the induced volume measure). Next, let A be any unit vector in $T_y\mathbb{S}^n$ (it doesn't matter which one, again by symmetry). The equatorial sphere decomposes nicely into level sets of $\theta_A(z)$. In particular, the level set for a given θ is $\mathbb{S}^{n-2}(\sin\theta)$.[4]

Given the preceding, we can compute the relevant integrals. If we let ω_m denote the volume of the unit sphere of dimension m, we then have that

$$\mathbb{E}^\mu\left[\cos^2\theta_A(z)\right] = \frac{1}{\omega_{n-1}}\int_{\theta=0}^{\pi}\left(\frac{\pi}{2}\cos\theta\right)^2(\sin\theta)^{n-2}\omega_{n-2}\,d\theta$$

and

$$\mathbb{E}^\mu\left[\cos\theta_A(z)\right]^2 = \left(\frac{\omega_{n-2}}{\omega_{n-1}}\right)^2\frac{\pi^2}{4}\left(\int_{\theta=0}^{\pi}\cos\theta(\sin\theta)^{n-2}\,d\theta\right)^2.$$

The expectation-squared term vanishes because $\cos\theta$ is anti-symmetric about $\pi/2$ while $\sin\theta$ is symmetric, and thus the relevant integral vanishes. Plugging this in gives

$$\begin{aligned}\lim_{t\searrow 0} t^2\left[\nabla^2_{A,A}\log p_t(N,S)\right] &= \frac{\omega_{n-2}\pi^2\int_{\theta=0}^{\pi}(\cos\theta)^2(\sin\theta)^{n-2}\,d\theta}{\omega_{n-1}}\\ &= \frac{\omega_{n-2}\pi^2\int_{\theta=0}^{\pi}(\cos\theta)^2(\sin\theta)^{n-2}\,d\theta}{\int_{\theta=0}^{\pi}\omega_{n-2}(\sin\theta)^{n-2}\,d\theta}\\ &= \frac{\pi^2}{n},\end{aligned}$$

where this last quotient of integrals can be evaluated using integration by parts. Thus, for any $n \geq 1$, we have now shown that $\nabla^2_{A,A}E_t(N,S) \sim -\frac{\pi^2}{nt}|A|^2$ as $t \searrow 0$ for any $A \in T_SM$.

4. Degenerate Minima and Conjugacy

In general it will not be so easy to determine the limiting measure, or measures, $\mu \in L_0$. Even when the set Γ is rather simple (e.g., a finite set of points), one needs information about the nature of the minima which $h_{x,y}$ has on Γ in order to understand L_0. That is, are some or all the minima degenerate and, if they are, how degenerate are they?

Before getting into a discussion of how degeneracy manifests itself in the asymptotics of $\nabla^2 E_t$, we take a moment to give, in terms of more familiar geometric quantities, an interpretation of what it means for $h_{x,y}$ to have a degenerate or non-degenerate minimum at a point $z \in \Gamma$. Namely, we want to show that the degeneracy of $h_{x,y}$ at $z \in \Gamma$ gives precise information about the conjugacy of the minimal geodesic from x to y which runs through z. That something of this sort ought to be true is clear. To wit, the most extreme degeneracy of $h_{x,y}$ occurs when z is one of a whole submanifold $M' \subseteq \Gamma$ having dimension $n' \geq 1$, as will be the case when $M = \mathbb{S}^n$ for some $n \geq 2$. Because, in this case, the exponential map will be constant as one moves away from z in any direction $A \in T_zM'$, the geodesic

[4] By $\mathbb{S}^n(a)$ we mean the standard n-dimensional sphere scaled by a factor of a, that is, $\mathbb{S}^n(a)$ is the set of points a distance a from the origin in $\mathbb{R}^{n+1}$.

through z will certainly be conjugate. A less extreme case occurs when z is an isolated point of Γ (equivalently, an isolated minimum of $h_{x,y}$). If we think about how $h_{x,y}$ behaves as one moves away from z in some direction, then high order vanishing of $h_{x,y}$ in that direction should indicate the presence of nearby points which are "almost" the midpoints of minimal geodesics from x to y. In other words, we should expect that in this case the minimal geodesic through z is conjugate, although now the conjugacy will usually be a consequence of finite order degeneracy of the exponential map.

To make the preceding precise, given a smooth, real-valued function f which is defined in a neighborhood of the origin in $\mathbb{R}^N$, we will say *f is constant to exactly order m at the origin in the direction $\xi \in \mathbb{S}^{N-1}$ if $(\partial_t)^i f(t\xi)\big|_{t=0}$ is zero for $1 \le i < m$ but is non-zero for $i = m$.* Now, let γ be a minimal geodesic connecting a point x and y in M, and take $(r, \theta_1, \ldots, \theta_{n-1})$ to be the polar coordinate system on T_xM such that $\gamma(r) = \exp_x(r, 0, \ldots, 0)$ for $r \in [0, \operatorname{dist}(x, y)]$. We then say that *$\gamma$ is conjugate to exactly order m in the direction $\xi \in \mathbb{S}^{n-2}$ if $\theta \rightsquigarrow \exp_x(r, \theta)$ is constant to exactly order m in the direction ξ.* Notice that this terminology has the annoying feature that geodesics which are conjugate of order 1 are *not* conjugate in the usual sense!

We can now make a precise statement about the relationship between the degeneracy of $h_{x,y}$ and conjugacy of geodesics.

LEMMA 3. *Choose distinct points x and y on M. Let $(r, \theta_1, \ldots, \theta_{n-1})$ and γ be as above. Then $h_{x,y}$ vanishes to exactly order $2m$ at $(\operatorname{dist}(x, y)/2, 0, \ldots, 0)$ in the direction ξ if and only if γ is conjugate to exactly order $2m - 1$ in that direction.*

Thus if $z \in \Gamma$, then z is a non-degenerate minimum of $h_{x,y}$ (i.e., $h_{x,y}$ vanishes to exact order 2 in all directions) if and only if x and y are not conjugate along the minimal geodesic γ passing through z. On the other hand, if z is a degenerate minimum, then x and y are conjugate, and furthermore, the index and orders of conjugacy can be determined from information about which partial derivatives of $h_{x,y}$ are zero.

5. More Refined Laplace Asymptotics when Γ is Discrete

Having related the degeneracy properties $h_{x,y}$ to geodesic geometry, we now return to problem of understanding the set L_0 of limits, as $t \searrow$, of (cf. equation (1)) $\{\mu_t : t > 0\}$, and we begin by considering the case when Γ consists of finitely many points, say $z_1, \ldots, z_m$. Obviously, by taking ϵ small, we can write the integrals with respect to the μ_t's as a sum of integrals over neighborhoods of the individual z_i's. Thus, we can study the asymptotics around each z_i separately.

In order to understand what is happening to μ_t near z_i as $t \searrow 0$, we must understand the structure of the Laplace asymptotics of integrals of the form

$$e^{-2h_{x,y}(z_i)/t} \int_{B_\epsilon(z_i)} \varphi(z) e^{-g(z)/t}\, dz \tag{3}$$

as $t \searrow 0$, where $g(z) \equiv 2h_{x,y}(z) - 2h_{x,y}(z_i)$ and φ is a smooth function. By assumption, g is non-negative and has a unique zero at z_i. Laplace determined the first term of the asymptotic expansion of this integral in the case when the region of integration is one-dimensional and where g has a non-degenerate minimum (that is, $g''(z_i) > 0$). In order to see what happens in n-dimensions, we first suppose that g can be diagonalized, in the sense that we can find coordinates $(u_1, \dots, u_n)$ around z_i so that

$$g(u_1, \dots, u_n) = \sum_{j=1}^{n} u_k^{2k_j} \tag{4}$$

for some positive integers $k_1 \le \cdots \le k_n$. Of course, at a non-degenerate minimum, the Morse Lemma guarantees the existence of such coordinates with $k_j = 1$ for each j. However, as we will discuss further below, in general diagonalizability represents a serious problem. Be that as it may, when g can be diagonalized at z_i, results of Estrada and Kanwal [**2**] allow us to give a complete expansion of (3). For the present, we will content ourselves with the first term. Namely,

$$\int_{B_\epsilon(z_i)} \varphi(z) e^{-g(z)/t}\, dz = t^{1/2k_1+\cdots+1/2k_n} \left[c_1 \operatorname{vol}_u(z_i)\varphi(z_i) + O\left(t^{1/k_n}\right)\right] \tag{5}$$

where vol_u is the volume element in the u coordinate chart and c_1 is a constant which depends only on n and the k_j's.

From (5), we see that a geodesic which is conjugate in many directions and/or to high order contributes more to the integral over Γ_ϵ than a "less conjugate" geodesic. In particular, suppose that g can be diagonalized around each of its minima and that z_i has associated to it the order of its leading term, $l_i = 1/2k_{1,i} + \cdots + 1/2k_{n,i}$. Then we see that, as $t \searrow 0$, μ_t converges to a limit μ which is supported on those z_i with $l_i = \min\{l_1, \dots, l_m\}$ and furthermore, the density at these points is given by the coefficient of the leading term of the expansion coming from (5), normalized to have total mass one. Therefore, not only does the limit exist, but we know we know what it is.

In more general situations, g may not be diagonalizable around some of the z_i. Nonetheless, in a series of papers (for example [**9**]) which culminate in the monograph [**1**], Arnold and his school have provided a fairly complete analysis of the asymptotic expansion of equation (3) . We give a very brief summary of the most relevant results. First, we need[5] to assume that g vanishes to finite order at z_1. This presents no problem if we work in the analytic category, but in the smooth category it need not be the case. Given that g vanishes to finite order at z_i, there exists a resolution of singularities (in the sense of Hironaka) from which one can compute the order of the leading term of the expansion around z_i, which we continue to denote l_i. Further, for generic g, l_i can be determined simply by looking at the Newton polytope, which is a finite combinatorial object depending only on finitely many terms in the Taylor expansion of g, associated to g. Thus, at least in principle, if we assume that g vanishes to finite order at each of the z_i's, we can associate to each z_i the leading order of the expansion l_i, and the conclusion about μ will be same as above.

6. Laplace Asymptotics when Γ is not Discrete

So far we've restricted our attention to cases in which Γ is composed of a finite number of points. Obviously, this will not always be true. In general, Γ can be quite complex, and we are very far from a result which covers all possibilities. Nonetheless, we will now discuss one fairly broad situation. Namely, suppose that Γ can be decomposed as a finite collection of isolated, smooth submanifolds (possibly with

[5] In fact, we know of no general results for the infinite order of vanishing case.

boundary) $N_1, \ldots, N_m$ of M. Then the integral over Γ_ϵ can be decomposed accordingly into integrals over the ϵ-neighborhood $(N_i)_\epsilon$ of the individual N_i. Further, by Fubini's Theorem, the integral over $(N_i)_\epsilon$ can be written as an iterated integral in which the horizontal and normal directions to N_i are segregated. But this means that, for each $z \in N_i$, we can, when it applies, use the analysis just discussed. In particular, if we assume that, for each N_i, the asymptotics of the integral in the normal direction has the same order l_i at all points, then, just as before, μ_t will converge to a measure μ which is supported on the union of those N_i's for which l_i is minimal. In fact, on each such N_i, μ will be absolutely continuous with respect to the induced Riemann measure on N_i.

7. A Cautionary Example and a Positive Result

In the preceding two sections, we dealt with relatively nice situations in which Γ was given by isolated smooth submanifolds. Here we construct an example which shows that this need not be the case.

Our example involves $\mathbb{S}^2$ with a metric somewhat deformed from the standard one. To be precise, start with the standard $\mathbb{S}^2$, the one which is embedded in $\mathbb{R}^3$ as the set of points of distance one from the origin, and let x be the north pole and y the south pole. Next, parameterize the equator by $\theta \in [-\pi, \pi)$. With the aid of C^∞ bump functions, we can increase the radius in a neighborhood of some sections of the equator so that, after this deformation, $\Gamma = \{0\} \cup \{2^{-m}\pi : m \geq 1\}$. Further we can achieve this in such a way that $h_{x,y}$ has a non-degenerate minimum at $2^{-m}\pi$ for each $m \geq 1$. On the other hand, it is clear that $h_{x,y}$ will have to vanish to infinite order at 0.

Obviously, we are well outside the situations considered here-to-fore, and the methods developed above do not apply. Nonetheless, we can argue as follows. Pick any point $\theta = \pi/2^{-m}$ for some $m \geq 1$, and choose an open set $U \cap \Gamma = \{\theta\}$. When we apply Laplace asymptotics to the integral over U, we see that

$$\int_U e^{-\frac{h_{x,y}(z)}{t}}\, dz \sim t^{-1} e^{-t^{-1}d} c_1 \operatorname{vol}_u(\theta),$$

where d is half the distance from x to y and c_1 is the constant which appeared earlier. After applying this line of reasoning to each $2^{-m}\pi$, we conclude that

$$\varlimsup_{t \searrow 0} \mu_t(U) \leq \frac{\det\big(\nabla^2 \mathrm{h_{x,y}}(2^{-\mathrm{m}}\pi)\big)}{\sum_{\ell=1}^{\infty} \det\big(\nabla^2 \mathrm{h_{x,y}}(2^{-\ell}\pi)\big)} = 0,$$

since the denominator is infinite. But this means that, as $t \searrow 0$, μ_t converges to the unit point mass at the point on the equator corresponding to $\theta = 0$.

Here, even though Γ had an accumulation point at which $h_{x,y}$ vanished to infinite order, we were still able to determine the limiting behavior of μ_t. One could easily imagine extending the above construction to produce more pathological examples where determining the limiting behavior of μ_t would be quite difficult, if it could be done at all.

In addition, this example demonstrates that multiplicity of minimal geodesics is no guarantee that μ is not a point mass. In terms of the asymptotics of $\nabla^2 E_t(x, y)$, L_0 contains some μ other than a point mass precisely when

$$\limsup_{t \searrow 0} t \nabla^2_{A,A} E_t(x, y) < 0 \quad \text{for some } A \in T_y M.$$

Thus, it is clear that the $-t^{-1}$ term in the asymptotics of $\nabla^2_{A,A}E_t(x,y)$ is not sufficient to determine when $y \in \mathrm{Cut}(x)$. Nonetheless we have the following positive result about the relationship between $\mathrm{Cut}(x)$ and the asymptotics of $\nabla^2 E_t(x,y)$.

THEOREM 4. *With the same notation as before, we have that $y \notin \mathrm{Cut}(x)$ if and only if*

$$\lim_{t\searrow 0} \nabla^2 E_t(x,y) = \nabla^2 E(x,y)$$

and $y \in \mathrm{Cut}(x)$ if and only if

$$\lim_{t\searrow 0} \|\nabla^2 E_t(x,y)\|_{\mathrm{op}} = \infty,$$

where $\|\cdot\|_{\mathrm{op}}$ is the operator norm.

In general, the qualitative result of Theorem 4 is the most we can say about the leading order of $\nabla^2 E_t$ for $y \in \mathrm{Cut}(x)$. However, in the special case where Γ contains only one point, say z_1, around which g is diagonalizable (in particular, we assume equation (4) holds), we can give more detail. Using the further terms in the expansion (5), one can show that, in this case, $\nabla^2_{A,A}E_t(x,y) \sim -Q(A)t^{1/k_n - 1}$ where $Q(A)$ is a symmetric, non-negative definite quadratic form on T_yM. Further, let $Q^{\perp}$ be the restriction of Q to the $n-1$ dimensional subspace perpendicular to γ, where γ is the unique minimal geodesic between x and y, and let d be the dimension of the kernel of $Q^{\perp}$. Then the number of i for which $k_i = k_n$ is given by $n-1-d$. In other words, the leading order tells us the highest order of conjugacy of γ, and knowing the leading coefficient as a function of the vector A tells us the number of (independent) directions in which this maximum order of conjugacy is achieved. First, this case shows that every rational of the form $-(m-1)/m$ for a positive integer m can be achieved as the leading order of the expansion of $\nabla^2 E_t$. Second, it indicates that coefficients in the expansion other than that of t^{-1} may have geometric significance. Unfortunately, these coefficients are hard to compute in general, and at the moment we know nothing more about them.

8. Mollification of the Energy Function

So far, we've been concerned with understanding the asymptotics of the log Hessian of the heat kernel for fixed points x and y. However, Theorem 2 can also be used to investigate the distributional Hessian of $E(x,y)$, where we think of x as a fixed base point and y as varying over M. A result of Stroock **[6]** implies that, in the compact case with which we are concerned, $\nabla^2 E(x,y)$, thought of as a distribution, is bounded above by a non-negative constant. It follows that the singular part can be at worst a negative measure, which for fixed base point x and smooth vector field A we denote by $\nu_{x,A}$. It is this measure which we will now investigate. From Varadhan's result, we know that $t\log p_t(x,y)$ gives a smooth mollifier of $-E(x,y)$ as $t \searrow 0$. Thus, computing the (distributional) limit of $-t\nabla^2_{A,A}\log p_t(x,y)$ as $t\searrow 0$ gives the distribution $\nabla^2_{A,A}E(x,y)$. With this in mind, we turn our attention to studying this limit.

We know that in a neighborhood of a point y not in $\mathrm{Cut}(x)$, the distribution $\nabla^2_{A,A}E(x,y)$ is just a smooth function, and Theorem 1 tells us that our mollifier converges to this limit pointwise. In particular, the singular support of $\nabla^2_{A,A}E(x,y)$ is contained in $\mathrm{Cut}(x)$. Looking at Theorem 2, we see that any terms not coming from the variance are $O(t)$ and thus don't contribute to the singular part $\nu_{x,A}$.

More concretely, let φ be a smooth function with support in an ϵ-neighborhood of $\mathrm{Cut}(x)$. Then we have

$$\langle \varphi, \nu_{x,A} \rangle = -\lim_{\epsilon \searrow 0} \lim_{t \searrow 0} \int_{B_\epsilon(\mathrm{Cut}(x))} \varphi(y) \frac{1}{t} \mathrm{Var}^{\mu_{t,y}}(\nabla_A E(\cdot, y))\, dy$$

where $\mu_{t,y}$ is the measure μ_t defined by equation (1) corresponding to the point y, with a slight modification. Namely, we need to enlarge the set Γ corresponding to a given y to include not only the midpoints of minimal geodesics to y, but also the midpoints of minimal geodesics from x to any point in a small neighborhood of y (this is so that the error term in Theorem 2 is bounded uniformly as y approaches $\mathrm{Cut}(x)$). We know the limit on the right exists precisely because, by the general theory of distributions, it must be equal to the quantity on the left.

Next, we will lift all of our considerations to the tangent space to M at x. First we recall that $\mathrm{Cut}(x)$ has measure zero, and thus we can ignore it in the preceding integral. Next, observe that $M \setminus \mathrm{Cut}(x)$ is contained in a single coordinate patch. Namely, let (r, θ) be (normal) polar coordinates around x, and let $d(\theta)$ be the distance to the cut locus along the geodesic corresponding to θ. Let $U_x = \{(r, \theta) : \theta \in \mathbb{S}^{n-1}, r \in (d(\theta) - \epsilon, d(\theta))\}$. Then the exponential map gives a diffeomorphism from U_x to $M \setminus \mathrm{Cut}(x)$. We have that

$$\langle \varphi, \nu_{x,A} \rangle =$$

$$-\lim_{\epsilon \searrow 0} \lim_{t \searrow 0} \int_{\mathbb{S}^{n-1}} \left[\int_{d(\theta)-\epsilon}^{d(\theta)} \varphi(r, \theta) \frac{1}{t} \mathrm{Var}^{\mu_{t,(r,\theta)}}(\nabla_A E(\cdot, (r, \theta)))\, \mathrm{vol}(r, \theta)\, dr \right] d\theta. \tag{6}$$

Note that $(d(\theta), \theta) = \partial U$ parameterizes the preimage of $\mathrm{Cut}(x)$ in the tangent space to x (which is commonly called the tangential cut locus). We can thus identify $\mathbb{S}^{n-1}$ with ∂U using polar coordinates, and then identify $\mathbb{S}^{n-1}$ with the set of minimal geodesics from x to points in $\mathrm{Cut}(x)$ (and also, for that matter, with the union of $\Gamma^{x,y}$ over all $y \in \mathrm{Cut}(x)$). It is this identification of $\mathbb{S}^{n-1}$ with the minimal geodesics from x to $\mathrm{Cut}(x)$ which we will have in mind when, for example, stating that some $\theta \in \mathbb{S}^{n-1}$ corresponds to a conjugate geodesic.

If we look at the integral in equation (6), we see that the central issue is the fact that, while $\mathrm{Var}^{\mu_{t,(r,\theta)}}(\nabla_A E(\cdot, (r, \theta)))$ converges pointwise at any given r in the region of integration, this convergence is not uniform as $r \nearrow d(\theta)$. Thus one must be able to estimate the variance as a function of t and r in such a way that one can first integrate with respect to r and only then let $t \searrow 0$. Here we will say only that this can be done, the central observation being that the the contribution to the variance from nearby geodesics is determined by the Jacobi fields along the geodesic corresponding to θ, as is the volume form. Carrying these ideas through allows one to prove the a pair of theorems about the singular part of $\nabla^2_{A,A} E(x, y)$.

First, one can show that, in a sense, lifting our concerns to the tangent space is the right thing to do. A priori, the right-hand side of equation (6) only makes sense for functions φ on the tangent space which are lifts of smooth functions on M. However, one can show that the relevant limits exist for almost every θ. In fact, we have the following theorem.

THEOREM 5. *Let M be a smooth, compact Riemannian manifold and let x be any point in M. Let A be any smooth vector field on M. Choose (normal) polar coordinates on T_xM and define U_x as above. Then the right-hand side of equation 6 defines a negative measure on ∂U, which is absolutely continuous with respect to* *the measure $d\theta$ on ∂U obtained by identifying it with $\mathbb{S}^{n-1}$ via polar coordinates. Denote the corresponding Radon-Nikodym derivative by $\rho(\theta)$; then $\rho(\theta)$ is bounded. Thought of as a distribution on M, $\nabla^2_{A,A}E(x,y))$ has as its singular part a negative measure $\nu_{x,A}$ supported on* $\mathrm{Cut}(x)$, *and further, $\nu_{x,A}$ is given by the pushforward of $\rho(\theta)\,d\theta$ under the exponential map.*

On M, there need not be any natural reference measure with which to compare $\nu_{x,A}$. Theorem 5 tells us that, on the tangent plane, there is such a natural reference measure, namely the measure induced by identifying the set of directions around x with $\mathbb{S}^{n-1}$.

This leads us to wonder what we can say about $\rho(\theta)$. We have the following result, which will require us to introduce a little notation. Let $C \subset \mathbb{S}^{n-1}$ be the set of all θ which correspond to conjugate geodesics. Next, say that the geodesics corresponding to θ and $\tilde{\theta}$ are *associated* if they lead to the same point in $\mathrm{Cut}(x)$ (that is, if $d(\theta) = d(\tilde{\theta})$ and $(d(\theta), \theta)$ and $(d(\tilde{\theta}), \tilde{\theta})$ are mapped to the same point under $\exp_x$). Let $P \subset \mathbb{S}^{n-1}$ be the set of $\theta \in \mathbb{S}^{n-1} \backslash C$ to which there is associated to precisely one other $\tilde{\theta} \in \mathbb{S}^{n-1}$ and such that $\tilde{\theta} \notin C$. Finally, let $R = \mathbb{S}^{n-1} \setminus (C \cup P)$ (so R consists of non-conjugate θ which are associated to more than one other geodesic or which are associated to a conjugate geodesic). The three sets C, P and R are disjoint and partition $\mathbb{S}^{n-1}$.

THEOREM 6. *Let the hypotheses be as in Theorem 5. If $\theta \in C$, then $\rho(\theta) = 0$. Also, R has measure zero as a subset of $\mathbb{S}^{n-1}$ with respect to $d\theta$, and ρ is continuous except possibly at points of R.*

In addition, we can give an explicit expression for ρ on P, although this requires introducing more notation. Let θ be in P, and let $\tilde{\theta}$ be the (one) associated geodesic. Let y be their common endpoint. Also, let z be the midpoint of the geodesic corresponding to θ. We know that $h_{x,y}$ has non-degenerate Hessian at z; let B denote this Hessian. Let $\tilde{z}$ and $\tilde{B}$ be the corresponding objects associated to $\tilde{\theta}$. Next, let $A_y \in T_yM$ be the value of the vector field A at y. Then let ψ be the angle between the geodesic given by θ and A_y, $\tilde{\psi}$ be the angle between the geodesic corresponding to $\tilde{\theta}$ and A_y, and φ the angle between the geodesics θ and $\tilde{\theta}$. Finally, recall that $H(x,z)$ is the Jacobian of $\exp_x$ at the vector Z corresponding to z. Then

$$\rho(\theta) = \mathrm{dist}(x,y)|A_y|^2 \left(\cos\psi - \cos\tilde{\psi}\right)^2 \mathrm{vol}(d(\theta),\theta) \times \left[4(1-\cos\varphi)\left(1 + \frac{H(x,z)H(y,z)\sqrt{\det B}}{H(x,\tilde{z})H(y,\tilde{z}\sqrt{\det \tilde{B}}}\right)\right]^{-1}.$$

Note that the volume element, all of the functions $H(\cdot,\cdot)$ appearing above, and both B and $\tilde{B}$ can be computed from the Jacobi fields along the geodesics given by θ and $\tilde{\theta}$.

There are a few things for us to observe in regard to Theorem 6. First of all, conjugate geodesics do not contribute to the singular part of the distribution. From the point of view of characterizing the cut locus, this means that simply looking at the singular part of $\nabla^2 E(x,y)$ is insufficient, in contrast to the pointwise limits of Theorem 4. On the other hand, in terms of understanding $\nabla^2 E(x,y)$, this says that its singular part is not too bad, in some sense. While the cut locus itself can be quite complicated (for example, it may not be triangulable), the only contributions to the singular part come from points in P, which on M are places where locally the cut locus looks like a smooth hypersurface and the singular part of $\nabla^2 E(x,y)$ is just given by the jump discontinuity of $\nabla E(x,y)$ across this hypersurface. We should point out, however, that even though there may not by any singular part of $\nabla^2 E(x,y)$ in a neighborhood of a conjugate point, in general one expects $\nabla^2 E(x,y)$ to be unbounded near a conjugate point. Finally, note that, for a given x, $\nabla^2 E(x,y)$ has no singular part if and only if $\mathrm{Cut}(x)$ and the first conjugate locus of x coincide.

References

[1] V. I. Arnol′d, S. M. Guseĭ n Zade, and A. N. Varchenko, *Singularities of differentiable maps. Vol. II: Monodromy and asymptotics of integrals*, Monographs in Mathematics, vol. 83, Birkhäuser Boston Inc., Boston, MA, 1988, Translated from the Russian by Hugh Porteous, Translation revised by the authors and James Montaldi.

[2] Ricardo Estrada and Ram P. Kanwal, *A distributional approach to asymptotics: Theory and applications*, second ed., Birkhäuser Advanced Texts: Basel Textbooks, Birkhäuser Boston Inc., Boston, MA, 2002.

[3] Paul Malliavin and Daniel W. Stroock, *Short time behavior of the heat kernel and its logarithmic derivatives*, J. Differential Geom. **44** (1996), no. 3, 550–570.

[4] Mark A. Pinsky, *Stochastic Riemannian geometry*, Probabilistic analysis and related topics, Vol. 1, Academic Press, New York, 1978, pp. 199–236.

[5] Daniel W. Stroock, *Gaussian measures in traditional and not so traditional settings*, Bull. Amer. Math. Soc. (N.S.) **33** (1996), no. 2, 135–155.

[6] ———, *Non-divergence form operators and variations on Yau's explosion criterion*, J. Fourier Anal. Appl. **4** (1998), no. 4-5, 565–574.

[7] Daniel W. Stroock and James Turetsky, *Short time behavior of logarithmic derivatives of the heat kernel*, Asian J. Math. **1** (1997), no. 1, 17–33.

[8] ———, *Upper bounds on derivatives of the logarithm of the heat kernel*, Comm. Anal. Geom. **6** (1998), no. 4, 669–685.

[9] B. A. Vasil′ev, *Asymptotic exponential integrals, Newton's diagram, and the classification of minimal points*, Functional Anal. Appl. **11** (1977), no. 3, 163–172.

Harvard University M.I.T.
E-mail address: neel@fas.harvard.edu & dws@math.mit.edu

Surveys in Differential Geometry
IX, International Press

Analysis on Riemannian co-compact covers

Laurent Saloff-Coste

ABSTRACT. This is a survey on analysis on non-compact co-compact Riemannian covers and how it relates to random walks on finitely generated groups. The focus is on the long time behavior of the heat kernel and related topics such as Liouville theorems, scale invariant Harnack inequalities and isoperimetric profiles.

CONTENTS

1. Introduction

This article surveys results concerning the large scale potential theory of non-compact co-compact Riemannian regular covers. Here, "large scale potential theory" means the large scale analysis of the Laplace operator and heat equation. This includes heat kernel estimates, Liouville theorems and related topics. The fundamental idea behind most of the results described in this survey is that the large scale potential theory of a co-compact Riemannian cover is determined by the behavior of some basic random walks on its deck transformation group. Thus this article also reviews the relevant results concerning random walks on finitely generated groups. Of course, what is reviewed in this survey reflects in parts the author's personal taste and limitations. For pointers to certain aspects (e.g., Novikov-Shubin invariants) that are not covered here, see [**20, 41, 81**].

2000 *Mathematics Subject Classification.* Primary 20F69,31C12,58J35,60G50.

Key words and phrases. co-compact regular covers, heat kernel, random walks, potential theory, isoperimetry.

Research supported in part by NSF Grant # 0102126.

1.1. Acknowledgments. This surveys owes a lot to contacts with colleagues and friends over the years. Many thanks to Alano Ancona, Philippe Bougerol, Yves Guivarc'h, V. Kaimanovich, François Ledrappier, Terry Lyons, and Wolfgang Woess. Thanks to T. Coulhon, A. Grigor'yan, W. Hebisch and C. Pittet for sharing their ideas with me. One of the well-known open questions recorded in this survey concerns the stability of the Liouville property, say, on Cayley graphs. I remember very well being introduced to this question and many related ideas present in this survey by N. Varopoulos, just after I finished my Ph.D, twenty years ago.

In August 2003, I sent a first draft of this survey to a few friends, including Martine Babillot, asking for help and comments. I learned that Martine would not respond. She had left us. I miss her.

1.2. Manifolds. Let (M, g) be a non-compact complete Riemannian manifold. Let $\mathcal{C}_0^\infty(M)$ be the space of smooth compactly supported functions on M. Denote by Δ the Laplace-Beltrami operator and by μ the Riemannian measure. Thus, in local coordinates,

$$\Delta f = \frac{1}{\sqrt{|g|}} \sum_{i,j} \left(\sqrt{|g|}\, g^{i,j} f_j\right)_i, \quad d\mu(x) = \sqrt{|g|(x)}dx,$$

where $|g|$ is the determinant of the metric tensor $(g_{i,j})$ and $(g^{i,j}) = (g_{i,j})^{-1}$. Let $B(x, r)$ denote the geodesic ball of radius r around x and set $V(x, r) = \mu(B(x, r))$.

A smooth function u is harmonic in an open set Ω if $\Delta u = 0$ in Ω. The heat kernel of M is the minimal positive solution $(t, x, y) \mapsto p(t, x, y)$ of

$$\begin{cases} \partial_t u = \Delta u \text{ on } (0, \infty) \times M \\ u(0, x) = \delta_y(x). \end{cases}$$

It is a smooth function of the three variables (t, x, y) on $(0, \infty) \times M \times M$. For instance, if $M = \mathbf{R}^n$ equipped with its Euclidean structure,

$$p(t, x, y) = \left(\frac{1}{4\pi t}\right)^{n/2} \exp\left(-\frac{\|x - y\|^2}{4t}\right).$$

Although we will not use it explicitly in the sequel, recall that there exists a Markov process $(X_t)_{t\geq 0}$ having continuous paths and such that, for any bounded continuous function f,

$$e^{t\Delta} f(x) = \int_M p(t, x, y) f(y) d\mu(y) = \mathbb{E}_x(f(X_t)).$$

In particular, $\int_M p(t, x, y) d\mu(y) \leq 1$ (in fact, for the class of manifolds of interest to us here, $\int_M p(t, xy) d\mu(y) = 1$). The process $(X_t)_{t\geq 0}$ is called the Brownian motion on M. For background information, see the excellent survey [**56**].

1.3. Graphs. For our purpose, a graph is a pair (X, E) where $E \subset X \times X$ is symmetric (i.e., $(x, y) \in E$ if and only if $(y, x) \in E$). The set X is the vertex set and the elements of E are called edges. Two vertices $x, y \in X$ are neighbors $(x \sim y)$ if and only if $(x, y) \in E$. Our graphs are oriented but symmetric so that orientation is merely a notational convention. There is a loop at x if $(x, x) \in E$. We do not consider multiple edges. For $x \in X$, set

$$N(x) = \#\{y \in X : (x, y) \in E\}.$$

The integer $N(x)$ is the degree of the vertex x (we will only consider locally finite graphs, i.e., $N < \infty$. In fact, most of our graphs will have uniformly bounded degree). For a subset $A \subset X$, set $N(A) = \sum_{x\in A} N(x)$. Thus the function N defines the "volume" on the graph (X,E). For any edge $e = (x,y)$ and any function f on X, set $df(e) = f(y) - f(x)$. The graph distance $d(x,y)$ between two points x and y in X is the minimal number of edges one has to cross to move from x to y. A graph is connected if $d(x,y) < \infty$ for all x,y. Set $B(x,r) = \{y : d(x,y) \le r\}$ and $V(x,r) = N(B(x,r))$. Denote by $\mathcal{C}_0(X)$ the set of all finitely supported functions on X.

The simple random walk on (X,E) is a stochastic process $(X_n)_0^\infty$ with values in X evolving as follows. If $X_n = x$ then X_{n+1} is one of the $N(x)$ neighbors of x chosen uniformly at random. This Markov process is associated with the kernel

$$K(x,y) = \begin{cases} 1/N(x) & \text{if } y \sim x \\ 0 & \text{otherwise.} \end{cases}$$

We can also view K as a self-adjoint bounded operator acting on $L^2(X,N)$ by

$$Kf(x) = \frac{1}{N(x)} \sum_{y\sim x} f(y).$$

The iterated kernel $K_n(x,y)$ is defined inductively by

$$K_1 = K, \quad K_n(x,y) = \sum_z K_{n-1}(x,z)K(z,y).$$

In this setting, a harmonic function in a set Ω is a function u defined on $\Omega_1 = \{x : d(x,\Omega) \le 1\}$ and such that $Ku = u$ in Ω.

1.4. Cayley graphs. Let Γ be a finitely generated group equipped with a finite symmetric generating set S. The Cayley graph (G,S) is the oriented symmetric graph with vertex set G and an edge from x to y if and only if $y = xs$ for some $s \in S$. Thus, the edge set E is

$$E = \{(x,xs) : x \in \Gamma,\ s \in S\}.$$

The distance $d(x,y)$ from x to y is the smallest k such that $y = xs_1 \dots s_k$ with $s_i \in S$, $i = 1,\dots,k$. Set $B(x,r) = \{y \in \Gamma : d(x,y) \le r\}$ and $V(x,r) = \#B(x,r)$. Obviously, these objects depend on the choice of S. Note that, to be consistent with the notation introduced above for general graphs, we should have set $V(x,r) = |S|\#B(x,r)$ instead.

Given a probability measure q on Γ, the left-invariant random walk on Γ driven by q is the discrete Markov process (X_n) which evolves as follows. If the position at time n is $X_n = x$ then pick s in Γ with probability $q(s)$ and move to $X_{n+1} = xs$. Thus, if the process starts from x at time 0, the probability that $X_n = y$ is $q^{(n)}(x^{-1}y)$ where $q^{(n)}$ is the n-fold convolution of q with itself (recall that $u * v(x) = \sum_y u(y)v(y^{-1}x)$). A function u is q-harmonic on Γ if $u * q = u$. When $q = q_S = (\#S)^{-1}\mathbf{1}_S$, the corresponding random walk is called the simple random walk on (G,S). For instance, if $\Gamma = \mathbb{Z}$ and $S = \{+1,-1\}$ then

$$q_S^n(x) = 2^{-n}\binom{n}{(n-x)/2}$$

if $n - x$ is even and $q_S^{(n)}(x) = 0$ otherwise. In particular,

$$q_S^{(2n)}(0) = 2^{-2n}\binom{2n}{n} \sim (\pi n)^{-1/2}.$$

On the d-dimensional square lattice $\mathbb{Z}^d$ with $S = \{\pm e_i, 1 \le i \le d\}$, we have

$$q_S^{(2n)}(0) \sim c_d n^{-d/2}, \quad c_d = 2(d/4\pi)^{d/2}.$$

For background on random walks, see Spitzer's book [**114**] and [**78, 131**].

1.5. Algebraic properties. Throughout this survey, we will encounter several classes of groups defined by certain algebraic properties. For the convenience of the reader, we gather here the main definitions. See, e.g., [**107**] for details.

Solvable groups. A group G is solvable if it admits a descending normal series $G = H_1 \supset H_2 \supset \cdots \supset H_{k+1} = \{e\}$ such that H_i/H_{i+1} is abelian. It is polycyclic if there is such a series with H_i/H_{i+1} cyclic. Polycyclic groups are always finitely generated hence countable. They are obviously solvable. One of the essential differences between polycyclic groups and general solvable groups is that subgroups of a polycyclic group are always finitely generated whereas a non-polycyclic finitely generated solvable groups always has (abelian) subgroups that are not finitely generated. In fact, polycyclic groups are exactly those solvable groups all of whose subgroups (equivalently, abelian subgroups) are finitely generated.

The lower central series of a group G is the non-increasing sequence of subgroups defined by $G = G_1$, $G_{i+1} = [G_i, G]$, $i = 2, \ldots$. A group is nilpotent if there is a k such that $G_{k+1} = \{e\}$. Finitely generated nilpotent groups are always polycyclic.

Examples of solvable groups. A typical finitely generated nilpotent group is the group $\mathrm{Up}_n(\mathbb{Z})$ of all $n \times n$ upper-triangular matrices with integer coefficients and diagonal entries all equal to 1 ("Up" stands for upper and for unipotent!).

The semidirect products $G = \mathbb{Z} \ltimes_A \mathbb{Z}^2$, $A \in SL_2(\mathbb{Z})$, where the product is given by

$$(k, x)(k', x') = (k + k', x + A^k x')$$

for $k, k' \in \mathbb{Z}$ and $x, x' \in \mathbb{Z}^2$ are polycyclic groups. For instance, $A = \begin{pmatrix} 2 & 1 \\ 1 & 1 \end{pmatrix}$ gives a polycyclic group having exponential volume growth. This group is isomorphic to a lattice in the three dimensional Lie group which defines the so-called "Sol" geometry in the theory of 3-manifolds [**116**].

Examples of solvable groups that are not polycyclic can be obtained as follows. Consider the affine group of the real line $\mathrm{Aff}(\mathbb{R})$, i.e., the group of all affine transformations $x \mapsto ax + b$, $a > 0$, $b \in \mathbb{R}$, under composition. Fix a real $\lambda > 1$ and consider the subgroup $\mathbb{A}_\lambda$ of $\mathrm{Aff}(\mathbb{R})$ generated by the transformations $x \mapsto x + 1$ and $x \mapsto \lambda x$. This group is solvable but, for most values of λ, it is not polycyclic. In fact, it is polycyclic if and only if the group T_λ of all translations contained in $\mathbb{A}_\lambda$ is finitely generated. For instance, if $\lambda = k > 1$ is an integer then T_λ is the ring $\mathbb{Z}[1/k] \subset \mathbb{Q}$ which, as a group, is not finitely generated.

Wreath products. The following construction is known to play an important role in the theory of solvable groups. Let A and B be two finitely generated groups. Consider the algebraic direct sum $A_B = \sum_{b \in B} A_b$ of a countable number of copies of A indexed by B. Thus A_B is the set of all sequences $(a_b)_B$ where all but a finite number of a_b are trivial (i.e., equal to the neutral element in A). The group law in A_B is product coordinate by coordinate. Now, define the wreath product $A \wr B$ as the semidirect product $A_B \rtimes_\tau B$ where the action τ is given by $\tau_c((a_b)) = (a_{c^{-1}b})$ for all $(a_b) \in A_B$ and $c \in B$. Thus the product in $A \wr B$ is given by

$$(u, c).(v, d) = (u\tau_c(v), cd), \quad u, v \in A_B, \ c, d \in B.$$

If A and B are finitely generated, so is $A \wr B$ although it contains the subgroup A_B which is not finitely generated unless B is finite. The reader should work out this definition for $A = \{0, 1\} = \mathbb{Z}_2$ and $B = \mathbb{Z}$. The group $\mathbb{Z}_2 \wr \mathbb{Z}$ (also known as the lamplighter group) is an example of a solvable group which is not polycyclic. When iterating this construction, parentheses should be used with care. Moreover, iterated wreath products obtained through the above definition are distinct from the groups obtained by the more general iterating procedure described in [**107**].

Amenability. Recall that a "mean" on a discrete group G is a continuous linear functional ν defined on the space of all bounded functions such that $f \geq 0 \Rightarrow \nu(f) \geq 0$ and $\nu(1) = 1$. A mean is invariant if for any $g \in G$, $\nu(f_g) = \nu(f)$ where $f_g(x) = f(gx)$. A group is amenable if it admits an invariant mean. All abelian groups are amenable and so are all solvable groups because amenability is preserved by quotient, extension, passage to a subgroup and increasing limit. Note however that the existence of a mean on $\mathbb{Z}$ requires the use of the axiom of choice. See e.g., [**95**, Chapter 0].

The free group $\mathbb{F}_r$ on $r \geq 2$ generators is an example of a group that is not amenable and so is $SL_2(\mathbb{Z})$. See [**95**, Chapter 0]. The surface group S_g, i.e., the fundamental group of a compact surface of genus $g > 1$ is non-amenable. May be the most surprising non-amenable groups are the Burnside groups $B(r, n)$ (exponent n and r generators) for large enough odd exponent n and $r \geq 2$ [**1**] (Adian's proof uses the co-growth criteria of Grigorchuk [**51, 53**]).

It should also be noted that it is not always easy to decide whether a group is amenable or not. One currently popular example is Thompson's group F defined by the presentation (see, e.g., [**19**])

$$\langle x_0, x_1, \dots | x_i^{-1} x_n x_i = x_{n+1} \text{ for } i < n \rangle.$$

The group F is generated by x_0, x_1. In fact, it is finitely presented and has several interesting realizations. It has exponential growth. Whether this group is amenable or not is an open problem.

1.6. Regular coverings. Let M be a complete Riemannian manifold. Assume that there exists a discrete subgroup Γ of the group of isometries of M such that $N = M/\Gamma$ is a compact Riemannian manifold (we write M/Γ even so we always think of Γ as acting on the left). Such a manifold M is called a regular covering (or cover) of N with deck transformation group Γ. If we consider the fundamental groups $\pi_1(M), \pi_1(N)$ then $\pi_1(M)$ is a normal subgroup of $\pi_1(N)$ and $\Gamma \cong \pi_1(N)/\pi_1(M)$. We can realize Γ as net in M by picking an arbitrary origin $o \in M$ and considering $\Gamma o = \{\gamma o : \gamma \in \Gamma\}$. One calls this "a net" because there are positive constants c, C such that any point $x \in M$ is at distance at most C of Γo and any two points in Γo are at least distance c apart.

By construction, the local geometry of a Riemannian co-compact regular cover is uniformly under control. More precisely, fix $r_0 > 1$. Then there are positive finite constants c, C such that $c \leq \mu(B) \leq C$ for all balls of radius between r_0 and $1/r_0$. Moreover, for all $x \in M$ and $0 < r \leq r_0$, $V(x, r) \leq CV(x, 2r)$. Any co-compact regular cover satisfies uniform local Harnack inequalities, both elliptic and parabolic. In particular, for all $t \geq 1/r_0$ and for all x, y, z with $d(y, z) \leq r_0$,

$$cp(t, x, x) \leq p(t, y, z) \leq Cp(t, x, x).$$

Of course, any co-compact regular cover has curvature bounded below and positive injectivity radius.

1.7. Quasi-isometries. The following definition is useful to capture in a very general form the idea that a co-compact regular cover and its deck transformation group are similar. See, e.g., [**22, 58, 59, 74**]. It owes its fame to the work and ideas of Gromov.

Definition 1.1. Let (X,d), (X',d') be two metric spaces. A map $\psi : X \to X'$ is a quasi-isometry from X to X' if there are constants $C_1,\dots,C_5$ such that:

(a) For all $x' \in X'$, there exists $x \in X$ such that $d'(x',\psi(x)) \le C_1$;
(b) For all $x,y \in X$, $C_2 d(x,y) - C_3 \le d'(\psi(x),\psi(y)) \le C_4 d(x,y) + C_5$.

Property (a) says that no points in X' are very far from $\psi(X)$. Property (b) says that distances are roughly preserved at large scale. If ψ is as in Definition 1.1, there is a quasi-isometry ψ' from X' to X such that $\sup_{x\in X} d(\psi'\circ\psi(x),x) < \infty$.

It is not hard to check that a regular Riemannian covering M of a compact manifold with deck transformation group Γ is quasi-isometric to any fixed Cayley graph (Γ,S). Also, any two Cayley graphs of a given finitely generated group are quasi-isometric (with the identity map as quasi-isometry). More generally, if (Γ_i,S_i), $i\in\{1,2\}$, are two Cayley graphs and Γ_2 is either a subgroup of Γ_1 with finite index or a quotient of Γ_1 by a finite subgroup then these two Cayley graphs are quasi-isometric.

Definition 1.2. Let u,v be two positive functions defined on a neighborhood of ∞ in $\mathbb{R}_+$ or $\mathbb{N}$. We say that u dominates v ($u \gg v$) if there are positive finite constants a,b,c, $a\le b$, such that, for all t large enough,

$$v(t) \le c \sup_{at\le s\le bt} u(t), \qquad \inf_{at\le s\le bt} v(t) \le cu(t).$$

We say that u and v are equivalent ($u \approx v$) if u dominates v and vice versa.

This definition is useful to construct quasi-isometric invariants. Note that it simplifies considerably if one of the two functions u,v is monotone. Indeed, if either u or v is monotone then $u \gg v$ is equivalent to say that there exist two positive finite constants c,b such that, for all t large enough, $v(t)\le cu(bt)$.

Example 1.3. For any metric space (X,d) equipped with a measure μ, define the volume growth of (X,d,μ) as the $\approx$-equivalence class of $v(r)=\mu(B(o,r))$ where o is an arbitrary fixed point in X. In general, the volume growth is not a quasi-isometric invariant. However, it is preserved under quasi-isometry if we restrict our attention to spaces such that

$$\sup_{x\in X}\frac{\mu(B(x,2r))}{\mu(B(x,r))} = C(r)$$

is finite for each $r>0$ and to quasi-isometries $\psi : (X,d,\mu)\to(X',d',\mu')$ such that

$$\forall x\in X,\ \ c \le \frac{\mu(B(x,1))}{\mu'(B(\psi(x),1))} \le C$$

for some finite positive constants c,C.

2. Some quasi-isometric invariants

2.1. Bounded geometry. It will be convenient to consider the collection **BG** (bounded geometry) of metric measure spaces which we now describe. By definition, any space in **BG** either is a complete Riemannian manifold equipped with its Riemannian distance and measure or is a connected graph equipped with the graph distance and the measure N ($N(x)$ is the degree of x). The manifolds in **BG** are exactly those with Ricci curvature bounded below and with a uniform lower bound on the volume of balls of radius 1. The graphs in **BG** are those with uniformly bounded degree.

If one thinks in terms of metric measure spaces, the above definition is obviously very narrow minded but it will serve our purpose for the present survey.

It is not hard to see that any manifold in **BG** is quasi-isometric to some graph(s) in **BG** (and vice-versa). From a local viewpoint, manifolds in **BG** are similar to regular coverings in that their local geometry is uniformly under control. The volume function is uniformly doubling as long as the radius stays bounded above. Balls of radius 1 all have comparable volume. Uniform local Harnack inequalities, elliptic and parabolic are satisfied thanks to the groundbreaking work of S-T. Yau and his collaborators on analysis on manifolds with Ricci curvature bounded below, see, e.g., [**24, 80, 134**] and [**22, 111**].

2.2. Volume growth. The following simple result goes back to [**44, 113**] and follows from the discussion in Example 1.3. Indeed, for manifolds with Ricci curvature bounded below, the Bishop–Gromov volume comparison theorem gives (see, e.g., [**22**, Theorem 3.10])

$$\sup_{x\in M} V(x,1) < \infty \quad \text{and} \quad \sup_{\substack{x\in M\\ 0<r\leq R}} \frac{V(x,2r)}{V(x,r)} < \infty.$$

PROPOSITION 2.1. *In* **BG**, *the volume growth is a quasi-isometric invariant. In particular, a regular Riemannian covering of a compact manifold has the same volume growth as its deck transformation group.*

Note in particular that the volume growth of a finitely generated group is independent of the choice of a symmetric finite generating set.

Example 2.2. The fundamental group of a compact manifold N with negative sectional curvature has exponential volume growth because its volume growth is comparable to that of the universal cover of N which is exponential (see, e.g., [**22**, Theorem 3.7]).

Example 2.3. The fundamental group of a compact n-manifold N with non-negative Ricci curvature has volume growth $\ll r^n$. Indeed, if $V(x,r)$ is the volume function on the universal cover of N and $V_n(r) = c_n r^n$ is the n-dimesnional Euclidean volume, Bishop's volume comparison theorem (e.g., [**22**, Theorem 3.9]) gives $V(x,r) \leq V_n(r)$.

Proposition 2.1 shows that the study of the volume growth of finitely generated groups is of great importance for the development of the large scale analysis on Riemannian co-compact covers. A good source of information on this subject is [**63**]. We recall the following results.

(0) Finitely generated groups have at most exponential volume growth.
(1) Non-amenable groups have exponential growth ([**48**]).
(2) For finitely generated discrete subgroups of almost connected Lie groups, the volume is either exponential or polynomial r^d for some $d = 0, 1, 2, \ldots$ ([**13, 61, 92, 117, 133**]).

(2) For finitely generated solvable groups, the volume is either exponential or polynomial r^d for some $d = 0, 1, 2, \dots$ ([**13, 61, 90, 133**]). This led Milnor to ask whether the growth function of any group is either exponential or polynomial. Grigorchuk surprised many by showing that the answer is no. See (4) below.

(3) A nilpotent group H has polynomial volume growth r^d with

$$d = \sum_1^k i \operatorname{rank}(H_i/H_{i+1}) \tag{2.1}$$

where $H_1 = H$, $H_i = [H_{i-1}, H]$, k is the smallest integer such that $H_{k+1} = \{e\}$, and $\operatorname{rank}(A)$ denotes the torsion free rank of the abelian group A. See [**13, 61**] and the discussion in [**63**].

(4) There are many finitely generated groups of intermediate growth, that is, whose volume growth is slower than exponential but is faster than polynomial. Such groups where first discovered by Grigorchuk. See [**52**] and the references therein.

(5) A group such that $\liminf_{r\to\infty} r^{-D}V(r) < \infty$ for some finite D contains a nilpotent subgroup of finite index ([**58, 118**]). This is Gromov's celebrated result on groups with polynomial growth.

Specialists of volume growth conjecture that there are no groups with volume growth faster than any polynomial but slower than $\exp(r^{1/2})$ (see, e.g., [**63**, p. 203] and [**52**]). Not only did Grigorchuk give examples of groups with intermediate volume growth, he also proved that the set of all possible growths has the cardinality of the continuum and that there are groups G_1, G_2 with volume growths v_1, v_2 such that neither $v_1 \ll v_2$ nor $v_2 \ll v_1$ hold true. See [**52**]. At this writing, there are no examples of groups of intermediate growth for which the ($\approx$-equivalence class of) the volume growth is precisely known.

2.3. Parabolicity. On any complete Riemannian manifold M, set

$$G(x,y) = \int_0^\infty p(t,x,y)dt.$$

If, for $x \neq y$, $G(x,y) < \infty$ then the function G is called the Green function on M. It is the minimal positive solution of $\Delta_y G(x,y) = -\delta_x(y)$. When $G(x,y) = \infty$ for some (equivalently any) $x \neq y$, one says that M is parabolic. A manifold M is parabolic if and only if the Brownian motion $(X_t)_{t\geq 0}$ on M is recurrent, i.e., almost surely, $(X_t)_{t\geq 0}$ visits any fixed non-empty relatively compact open set infinitely often (see, e.g., [**56**]). It is proved in [**24**] that Riemannian manifolds with $\liminf_{r\to\infty} r^{-2}V(x_0,r) < \infty$ are parabolic (see the discussion in [**33**] for further references).

On a connected graph (X,E), the simple random walk is recurrent (i.e., almost surely (X_n) visits any fixed point infinitely often) if and only if $\sum_n K^n(x,x) = \infty$. When $\sum_n K^n(x,x) = \infty$, we say that (X,E) is recurrent or, equivalently, parabolic. For a finitely generated group, we say that Γ is recurrent (equivalently parabolic) if and only if the Cayley graph (Γ, S) is recurrent for at least one symmetric finite generating set S. A walk that is not recurrent is called transient. For the following theorem, see [**3, 73, 119, 120, 125**].

THEOREM 2.4. *In* **BG**, *parabolicity is a quasi-isometric invariant. In particular, a co-compact regular cover is parabolic if and only if its deck transformation group is parabolic.*

SKETCH OF THE PROOF. It turns out that parabolicity can be described in many different ways. One useful way is that a graph is non-parabolic if and only if, for some/any point $o \in X$ there exists a constant $C(o)$ such that

$$\forall f \in \mathcal{C}_0(\Gamma), \ \ |f(o)|^2 \leq C(o) \sum_E |df|^2.$$

Similarly, a manifold is non-parabolic if and only if

$$\forall f \in \mathcal{C}_0^\infty(M), \ \ \int_U |f|^2 d\mu \leq C(U) \int_M |\nabla f|^2 d\mu$$

for one/any non-empty open precompact set U. These criteria go back to Deny [**39**]. See also [**83, 121**].

Consider for instance two distinct Cayley graphs (Γ, S_1), (Γ, S_2) of a given finitely generated group Γ. Then it is not hard to see that there are finite positive constants c, C such that

$$c \sum_{E_1} |df|^2 \leq \sum_{E_2} |df|^2 \leq C \sum_{E_1} |df|^2$$

where E_1, E_2 are the corresponding edge sets. It follows from the criterion above that if the simple random walk on a particular Cayley graph of a finitely generated group Γ is recurrent then the simple random walk on any of its Cayley graphs is recurrent. Thus recurrence (i.e., parabolicity) is a property of the group Γ, i.e., it does not depend on the choice of the finite symmetric generating set S. Theorem 2.4 extends this to show that parabolicity is a quasi-isometric invariant in **BG**. To see how to pass from a manifold to a graph and vice versa, see the proof of Theorem 2.8 below. □

Pòlya was the first to realize that the simple random walk on the square lattice $\mathbb{Z}^d$ is recurrent in dimension $d = 1, 2$ and transient otherwise. The question of finding exactly which finitely generated groups are recurrent emerged from the work of Kesten in the sixties and was solved around 1985 by Varopoulos. Varopoulos' proof uses Gromov's theorem on groups of polynomial volume growth. See Section 4.2 below.

THEOREM 2.5. *A finitely generated group is recurrent (i.e., parabolic) if and only if it is a finite extention of* $\{e\}$, $\mathbb{Z}$ *or* $\mathbb{Z}^2$, *that is, if and only if it contains one of group* $\{e\}$, $\mathbb{Z}$ *or* $\mathbb{Z}^2$ *as a subgroup with finite index.*

COROLLARY 2.6. *Let* M *be a regular Riemannian covering of a compact manifold and assume that* M *is parabolic. Then* M *covers a (possibly different) compact manifold with deck transformation group equal to either* $\{e\}, \mathbb{Z}$, *or* $\mathbb{Z}^2$.

PROOF. Let M cover the compact manifold N with deck transformation group Γ. If M is parabolic, so is Γ by Theorem 2.4. By Theorem 2.5, Γ contains a subgroup H equals to $\{e\}$, $\mathbb{Z}$ or $\mathbb{Z}^2$ and with finite index in Γ. Taking the quotient of M by H gives a compact Riemannian manifold covered by M with deck transformation group H. □

Let us quote another corollary (due to G. Mess). For the definition of quasi-conformal maps, we refer the reader to [**4, 65**] and the references therein.

COROLLARY 2.7. *Let M be a 2-dimensional Riemannian manifold in* **BG**. *Assume that M is quasi-isometric to a finitely generated group and quasi-conformally equivalent to the plane $\mathbb{R}^2$. Then M is quasi-isometric to $\mathbb{R}^2$.*

SKETCH OF THE PROOF. In dimension 2, parabolicity is a quasi-conformal invariant. Thus M is parabolic. Let Γ be the finitely generated group which is quasi-isometric to M. By Theorem 2.4, Γ is parabolic. By Theorem 2.5, Γ is a finite extention of $\{e\}$, $\mathbb{Z}$ or $\mathbb{Z}^2$. It is easy to rule out $\{e\}$ and $\mathbb{Z}$ because M is quasi-conformal to the plane. □

The question naturally arises of generalizing this result. For instance, the author asked several years ago the following question.

QUESTION 1. *Let M be a 3 dimensional manifold in* **BG** *quasi-isometric to a finitely generated group and quasi-conformally equivalent to $\mathbb{R}^3$. Is M quasi-isometric to $\mathbb{R}^3$?*

Recently, Maillot made interesting advances in this direction [**87**]. Let M_0 be a regular co-compact Riemannian covering. Say that M_0 is "large-scale conformally rigid" if, for any finitely generated group Γ which is quasi-isometric to a manifold conformally equivalent to M_0 and of bounded geometry, Γ is quasi-isometric to M_0 (Maillot's definition of bounded geometry is more restrictive than the one used in the definition of **BG**). It is proved in [**87**] that $\mathbb{R}^3$ (in fact, any 3-manifold M_0 which is conformally flat and homeomorphe to $\mathbb{R}^3$) is large scale conformally rigid. This gives a positive answer to Question 1 (up to the different definitions of bounded geometry, a technicality).

Say a manifold is not p-parabolic if for one/any relatively compact non-empty open set U, there is a constant $C(U)$ such that

$$\forall f \in \mathcal{C}_0^\infty(M), \quad \int_U |f|^p d\mu \le C(U) \int |\nabla f|^p d\mu.$$

This notion of p-parabolicity is equivalent to others found in the literature on quasi-regular maps and non-linear potential theory. See, e.g., [**33**]. By using p-parabolicity with $p = 3$, it is easy to see that the manifold M in Question 1 is quasi-isometric to $\mathbb{Z}^d$ with $d = 0, 1, 2$ or 3. See the sketch of the proof of Theorem 2.5 and [**128**, Chapter X]. As before, it is not very hard to eliminate $\{e\}$ and $\mathbb{Z}$. I did not know how to eliminate $\mathbb{Z}^2$ but Maillot is able to use the simple connectedness at infinity of $\mathbb{R}^3$ to conclude that $\mathbb{Z}^3$ is the only possibility.

One can ask the same question in higher dimension. If a manifold is quasi-isometric to a group and is n-parabolic then the group must be of polynomial growth of degree at most n. In dimension 4, the possible groups are (up to finite extentions) $\mathbb{Z}^d$ with $d = 0, 1, 2, 3, 4$ and the discrete Heisenberg group $\mathrm{Up}_3(\mathbb{Z})$.

2.4. Spectral gaps. On a Riemannian manifold M, for any $1 \le p < \infty$, let $\lambda_p(M)$ be the largest real λ such that

$$\lambda \int |f|^p d\mu \le \int |\nabla f|^p d\mu \tag{2.2}$$

for all smooth functions f with compact support. For $p = 2$, $\lambda_2(M)$ is the bottom of the L^2-spectrum of $-\Delta$ and, for any $x \in M$,

$$\lambda_2(M) = -\lim_{t\to\infty} t^{-1}\log p(t,x,x).$$

See [**37**, Prop.12] and the references therein. For $p = 1$, $\lambda_1(M)$ is an isoperimetric constant. Indeed, by the familiar co-area formula, (2.2) with $p = 1$ is equivalent to

$$\lambda\mu(A) \le \mu'(\partial A)$$

for all compact sets A with smooth boundary ∂A. Here μ' denotes the induced Riemannian measure on the hypersurface ∂A.

On a graph (X, E) with vertex set X and (symmetric oriented) edge set E, let $\lambda_p(X)$ be the largest real λ such that

$$2\lambda\sum_X |f|^p N \le \sum_E |df|^p \tag{2.3}$$

When $p = 2$, $\lambda_2(X)$ is the bottom of the L^2-spectrum of the operator $(I - K)$. Moreover, for any $x \in X$,

$$1 - \lambda_2(X) = \lim_{n\to\infty} K^{2n}(x,x)^{1/2n}.$$

If (X, E) is a Cayley graph (Γ, S) then $K(x,y) = q_S(x^{-1}y)$, the operator K is convolution by q_S and $K^{2n}(x,x) = q^{(2n)}(e)$.

On graphs, the discrete version of the co-area formula asserts that

$$\sum_E |df| = 2\int_0^\infty \#\partial\{f > t\}dt$$

for any function $f \ge 0$ with finite support. Thus, when $p = 1$, (2.3) is equivalent to

$$\lambda N(A) \le \#\partial A$$

for all finite sets $A \subset X$. Here $N(A) = \sum_A N$ and ∂A is the boundary of A defined as the set of those oriented edges $e = (x, y)$ in E such that $x \in A$ and $y \in X \setminus A$.

THEOREM 2.8. *In* **BG**, *for each* $p \in [1,\infty)$, *the property "*$\lambda_p > 0$*" is a quasi-isometric invariant. In particular, for a regular covering* M *of a compact manifold with deck transformation group* Γ, $\lambda_p(M) > 0$ *if and only if* $\lambda_p(\Gamma) > 0$.

SKETCH OF THE PROOF. (See, e.g., [**3, 36, 74**]) We only discuss the case of a co-compact regular cover M with deck transformation group Γ. Identify Γ with the subset $\{\gamma o : \gamma \in \Gamma\}$ of M and consider a partition of unity χ_γ, $\sum_\Gamma \chi_\gamma \equiv 1$, such that each χ_γ is a smooth compactly supported bump function centered around $\gamma o \in M$ and whose "profile" is essentially independent of γ. Then, we can consider the maps:

$$\mathbf{rst} \;:\; \mathcal{C}_0^\infty(M) \to \mathcal{C}_0(\Gamma),\;\; f \mapsto \mathbf{rst}(f) = \sum_\gamma \left(\int f\chi_\gamma d\mu\right)\mathbf{1}_\gamma$$

$$\mathbf{ext} \;:\; \mathcal{C}_0(\Gamma) \to \mathcal{C}_0^\infty(M),\;\; f \mapsto \mathbf{ext}(f) = \sum_\gamma f(\gamma)\chi_\gamma.$$

These two maps are sort of "inverse" of each others. The maps **rst** "restricts" smooth functions defined on M to functions on Γ whereas **ext** "extends" functions defined on Γ to smooth functions on M. Moreover, if we restrict our attention to non-negative functions, these two maps essentially preserve the size of the function and the size of its gradient. More precisely, for non-negative functions,

$$\sum_{\Gamma}|f|^p \le C\int_M |\mathbf{ext}(f)|^p d\mu, \quad \int_M |\mathbf{ext}(f)|^p d\mu \le C\sum_{\Gamma}|f|^p,$$

$$\int_M |\nabla(\mathbf{ext}(f))|^p d\mu \le C\sum_E |df|^p$$

and

$$\sum_{\Gamma}|\mathbf{rst}(f)|^p \le C\left(\int_M f^p d\mu + \int_M |\nabla f|^p d\mu\right),$$

$$\int_M f^p d\mu \le C\left(\sum_{\Gamma}|\mathbf{rst}(f)|^p + \int_M |\nabla f|^p d\mu\right),$$

$$\sum_E |d(\mathbf{rst}(f))|^p \le C\int_M |\nabla f|^p d\mu.$$

where $E = \{(\gamma, \gamma s) : \gamma \in \Gamma, s \in S\}$, S being a fixed symmetric finite generating set of Γ. As λ_p can be computed by using only non-negative functions, the theorem follows. The two crucial properties of manifolds in **BG** used in this discretization procedure are that, for any $r > 0$, (a) all balls of radius in $(r, 2r)$ have comparable volumes and (b) all balls of radius in $(r, 2r)$ satisfy the Poincaré inequality

$$\int_B |f - f_B|^p d\mu \le C(p,r)\int_{2B} |\nabla f|^p d\mu$$

where f_B is the mean value of f over B. □

PROPOSITION 2.9. *In* **BG**, *for* $1 \le p, q < \infty$, *the properties "*$\lambda_p > 0$*" and "*$\lambda_q > 0$*" are equivalent.*

SKETCH OF THE PROOF. (See, e.g., [**3, 36, 40, 74, 121**]) Using the change of function $f \to |f|^s$ for an appropriate $s > 1$ and Hölder inequality, it is easy to see that $\lambda_p > 0$ implies $\lambda_q > 0$ for all $q \ge p \ge 1$. For instance, on a Riemannian manifold, for non-negative functions and $s > 1$, we have $|\nabla f^s| = sf^{s-1}|\nabla f|$. Hence, if $q > p$, $s = q/p$ and $s' = s/(s-1)$ then

$$\begin{aligned}\int_M |\nabla f^s|^p d\mu &= s^p\int_M f^{(s-1)p}|\nabla f|^p d\mu \\ &\le s^p\left(\int f^{(s-1)ps'} d\mu\right)^{1/s'}\left(\int_M |\nabla f|^{ps} d\mu\right)^{1/s} \\ &\le s^p\left(\int f^q d\mu\right)^{1-1/s}\left(\int_M |\nabla f|^q d\mu\right)^{1/s}.\end{aligned}$$

It follows that $\lambda_q \ge (p/q)^q \lambda_p^{q/p}$. Up to technical details, the same proof works on graphs.

To prove the converse, we will work on graphs and appeal to Theorem 2.8 to conclude in the case of manifolds (the proof given for graphs does not easily adapt to manifolds). Let (X, E) be a graph with $\lambda_p(X) > 0$. According to what we just proved, it suffices to show that $\lambda_1(X) > 0$. Setting $f = \mathbf{1}_A$ in (2.3) we obtain

$$\lambda_p(X)N(A) \le \#\partial A \tag{2.4}$$

for any finite set $A \subset X$ is a finite set. By the discrete co-area formula, (2.4) is equivalent to

$$2\lambda_p(X)\sum |f|N \leq \sum_E |df|.$$

This shows that $\lambda_p(X) \leq \lambda_1(X)$ as desired. It follows from this discussion that, for graphs, $\lambda_p > 0 \Leftrightarrow \lambda_q > 0$ for all $1 \leq q, p < \infty$. By Theorem 2.8, this is also true in **BG**. □

Now we recall the following fundamental results of Følner [**48**] and Kesten [**76, 77**]. Without using the present terminology, Følner was working essentially with the isoperimetric constant λ_1. Kesten was interested in random walks and the spectral constant λ_2. At that time (1955-59), the relation between the two was not clearly established.

THEOREM 2.10. *A finitely generated group Γ satisfies $\lambda_p > 0$ for one/any $p \in [1, \infty)$ if and only if it is non-amenable.*

Together, the two theorems above give the following well-known statement due to Brooks [**18**] (see also [**3, 119**]; a different proof is in [**112**]).

THEOREM 2.11. *The covering M of a compact manifold satisfies $\lambda_p(M) > 0$ for one/any $p \in [1, \infty)$ if and only if its deck transformation group is non-amenable.*

Example 2.12. Let S_g be the fundamental group of a compact surface of genus $g > 1$. Then S_g is quasi-isometric to the hyperbolic plane. Hence $\lambda_p(S_g) > 0$ for all $p \in [1, \infty)$ and thus S_g is non-amenable. More generally, the fundamental group of a compact manifold admitting a metric of negative sectional curvature is non-amenable.

2.5. Isoperimetry. Let Ψ be an arbitrary positive function. We say that a graph (X, E) satisfies the Ψ-isoperimetric inequality if

$$N(A) \leq \Psi(N(A))\#\partial A. \tag{2.5}$$

If we set

$$I(t) = \inf\{\#\partial A : A \subset X,\ \ N(A) = t\}$$

then $t/I(t)$ is the best possible Ψ for the graph (X, E). These definitions should be handled with care because the so-called isoperimetric profile I is not an increasing function in general. Even for Cayley graphs, it is not known if I is $\approx$-equivalent to the monotone isoperimetric profile $I^\uparrow(k) = \inf\{\#\partial A : A \subset X,\ \ N(A) \geq k\}$. Moreover, $I^\uparrow$ is inappropriate for the discussion of inequalities of the form (2.5). Indeed, the smallest non-decreasing Ψ satisfying (2.5) is

$$J(k) = \sup\left\{\frac{N(A)}{\#\partial A} : N(A) \leq k\right\}.$$

Even for Cayley graphs, it is not known if $J(k) \approx k/I(k)$ nor if $J(k) \approx k/I^\uparrow(k)$.

The Følner function of (G, S) is defined by setting

$$F(t) = \min\{N(A) : \text{ there exists } A \subset X \text{ such that } \#\partial A < t^{-1}N(A)\}.$$

This is a non-decreasing function. It is related to J by

$$F(t) > k \Leftrightarrow J(k) \leq t.$$

The following is an elementary result.

PROPOSITION 2.13. *For graphs in* **BG**, $I, I^{\uparrow}, J, F$ *are all* $\approx$*-invariant under quasi-isometry.*

On a complete Riemannian manifold M, we define the isoperimetric functions $I, I^{\uparrow}, J$ of M and the Følner function F of M by replacing $N(A)$ by $\mu(A)$ and $\#\partial A$ by $\mu'(\partial A)$, working with precompact sets A with smooth boundary instead of finite sets. In particular, we say that M satisfies the Ψ-isoperimetric inequality if

$$\mu(A) \leq \Psi(\mu(A))\mu'(\partial A) \tag{2.6}$$

for all relatively compact open sets A with smooth boundary.

It is not entirely obvious how to distinguish between small scale and large scale isoperimetry on a manifold (see [**60**, Lemma 6.21] and [**28, 23**]). To state a precise and fairly general result, we will use Young conjugate functions and Orlicz norms (for a detailled treatment of Orlicz spaces, see [**75**]; for their use in the present context, see, e.g., [**9, 70, 89**]). Let P, Q be a pair of conjugate Young functions (this implies that P, Q are convexe increasing functions). Define the Orlicz P-norm of a function f by

$$\|f\|_P = \sup\left\{\int fgd\mu : \int Q(|g|)d\mu \leq 1\right\}.$$

The Orlicz P-norm of the caracteristic function of a set U is

$$\|\mathbf{1}_U\|_P = \mu(U)Q^{-1}(1/\mu(U))$$

where Q^{-1} is the inverse function of Q. It follows from the co-area formula that the isoperimetric inequality

$$\mu(A) \leq \Psi(\mu(A))\mu'(\partial A)$$

for relatively compact sets with smooth boundary is equivalent to the Sobolev inequality (See [**89**])

$$\forall f \in \mathcal{C}_0^\infty(M), \quad \|f\|_P \leq \|\nabla f\|_1$$

as long as Ψ and P are related by $\Psi(t) = 1/Q^{-1}(1/t)$. As $1/Q^{-1}(1/t)$ is increasing we can state this equivalence as

$$\forall f \in \mathcal{C}_0^\infty(M), \quad \|f\|_P \leq \|\nabla f\|_1 \quad \Leftrightarrow \quad J(t) \leq 1/Q^{-1}(1/t).$$

PROPOSITION 2.14. *Let P, Q be a pair of Young conjugate functions and assume that P satisfies $P(2t) \leq CP(t)$ for all $t > 0$. Set $\Psi(t) = 1/Q^{-1}(1/t)$. Let $M_1, M_2 \in$* **BG** *(thus each M_i is either a graph or a manifold with bounded geometry). Assume that M_1 and M_2 satisfy the local isoperimetric inequality*

$$\mu(A) \leq C\Psi(\mu(A))[\mu(A) + \mu'(\partial A)] \tag{2.7}$$

and that M_1 and M_2 are quasi-isometric. Then M_1 satisfies $J(t) \ll \Psi(t)$, that is,

$$\mu(A) \leq C\Psi(\mu(A))\mu'(\partial A)$$

if and only if M_2 does.

If M_i is a graph, $\mu(A)$ should be understood as $N(A)$ and $\mu'(A)$ as $\#\partial A$ and in this case the hypothesis that M_i satisfies $N(A) \leq C\Psi(\mu(A))[N(A) + \#\partial A]$ is trivially satisfied because ψ is increasing and $N(A)$ is bounded away form 0. It is also useful to note that the fact that P is doubling implies that P^{-1} and Q^{-1} (hence Ψ) are also doubling although Q may not be doubling.

SKETCH OF THE PROOF OF PROPOSITION 2.14. The proof of Proposition 2.14 is along the same lines as the proof of Theorem 2.8. If M_1 and M_2 are both graphs, Proposition 2.13 gives a more precise result. Thus its suffices to treat the case where $M_1 = X$ is a graph and $M_2 = M$ is a manifold. For simplicity, we outline the proof in the case when M is a regular covering of a compact manifold with deck transformation group Γ. We keep the notation introduced in the sketch of the proof of Theorem 2.8. In addition to the maps **rst** and **ext** introduced in that proof, we consider the regularization map $\mathbf{reg} : \mathcal{C}_0^\infty(M) \mapsto \mathcal{C}_0^\infty(M)$ defined by

$$\mathbf{reg}(f)(x) = \sum_\gamma \left(\frac{\int f\chi_\gamma d\mu}{\int \chi_\gamma d\mu} \right) \chi_\gamma.$$

It is plain that $\|\mathbf{reg}(f)\|_P \approx \|\mathbf{rst}(f)\|_P$. It is easy to show that a ψ-isoperimetric inequality on M implies the same type of inequality on Γ (more generaly, X). To go from Γ to M, write

$$\|f\|_P \leq \|\mathbf{reg}(f)\|_P + \|f - \mathbf{reg}(f)\|_P.$$

Observe that $\|\mathbf{reg}(f)\|_P \approx \|\mathbf{rst}(f)\|_P$ and estimate this term using the P-Sobolev inequality on Γ and $\sum_E |d(\mathbf{rst}(f))| \leq C \int_M |\nabla f| d\mu$. To estimate the second term $\|f - \mathbf{reg}(f)\|_P$, use the local P-Sobolev inequality $\|f\|_P \leq C[\|f\|_1 + \|\nabla f\|_1]$ (which is equivalent to (2.7)) and the inequality $\|f - \mathbf{reg}(f)\|_1 \leq C\|\nabla f\|_1$. □

Let $V(t)$ be the volume growth function of a finitely generated group Γ (for some finite symmetric set of generators) and set $W(t) = \inf\{s > 0 : V(s) > t\}$. It is proved in [**35**] that Γ satisfies the isoperimetric inequality $J(t) \ll W(t)$, that is

$$\#A \leq CW(C\#A)\#\partial A. \tag{2.8}$$

For instance, if $V(t) \gg t^d$ then $J(t) \ll t^{1/d}$ whereas $V(t) \gg \exp(t^\alpha)$ implies $J(t) \leq [\log(1+t)]^{1/\alpha}$. Comments are in order concerning (2.8) since in his wonderful and very influencial book [**60**, Chap. 6, E+], Gromov gives a very misleading and erroneous account of the discovery of this inequality. Varopoulos was the first to obtain valuable general isoperimetric inequalities on groups based on volume growth. He proved (2.8) for groups of polynomial growth [**122, 123, 124**] using several structure theorems such as Malcev's embedding of any torsion free finitely generated nilpotent group as a co-compact lattice in a nilpotent Lie group and Gromov's theorem on groups of polynomial volume growth. He also proved that groups of super-polynomial growth satisfy $J(t) \ll t^{1/d}$ for all $d > 0$ and that $J(t) \ll [\log(1+t)]^{2/\alpha}$ if $V(t) \gg \exp(t^\alpha)$ (see [**123, 126, 127**]). In [**126**], Varopoulos conjectured that $V(t) \gg \exp(t^\alpha) \implies J(t) \ll [\log(1+t)]^{1/\alpha}$ (as noted above, this follows immediately from (2.8)). The method used by Varopoulos to prove the partial results mentioned above are not suited to handle general volume growth functions and, as far as I know, they cannot be used to prove (2.8) or even the much simpler implication $V(t) \gg \exp(t^\alpha) \implies J(t) \ll [\log(1+t)]^{1/\alpha}$. The proof of (2.8) in [**35**] uses a variation on an idea used by D. W. Robinson in [**106**] to prove a Nash inequality on Lie groups.

By Proposition 2.14, we obtain the following result (see [**35**, Théorème 4] and [**96, 97, 128**]).

THEOREM 2.15. *Let M be a n-dimensional Riemannian regular covering of a compact manifold. Let $V(o,r)$ be the volume growth function from a fixed origin o.*

- *If $V(o,r) \gg r^d$ for some $d > 0$ then M satisfies the Ψ-isoperimetric inequality (2.6) where $\Psi(t)$ is a concave function satisfying $\Psi(t) \approx t^{1/n}$ at $t = 0$ and $\Psi(t) \approx t^{1/d}$ at $t = \infty$.*
- *If $V(o,r) \gg \exp(r^\alpha)$ with $\alpha \in (0,1]$ then M satisfies the Ψ-isoperimetric inequality (2.6) where $\Psi(t)$ is a concave function satisfying $\Psi(t) \approx t^{1/n}$ at $t = 0$ and $\Psi(t) \approx (\log t)^{1/\alpha}$ at $t = \infty$.*
- *If the deck transformation group of M contains a nilpotent group H of finite index then*

$$J(t) \approx \begin{cases} t^{1/n} & \text{at } 0 \\ t^{1/d} & \text{at } \infty \end{cases}$$

where d is given by (2.1) applied to the group H.
- *If the deck transformation group of M contains a polycyclic group H of finite index and exponential volume growth then*

$$J(t) \approx \begin{cases} t^{1/n} & \text{at } 0 \\ \log t & \text{at } \infty. \end{cases}$$

- *If the deck transformation group of M is not amenable then*

$$J(t) \approx \begin{cases} t^{1/n} & \text{at } 0 \\ 1 & \text{at } \infty. \end{cases}$$

For refined results concerning the very special case of periodic metrics on $\mathbb{R}^n$, see [**94**].

The results of [**45, 99**] and Proposition 2.14 give examples of regular coverings of compact manifolds whose volume growth is exponential and whose isoperimetric profile J at infinity is intermediate between the extreme case $J(t) \approx 1$ (non-amenable) and $J(t) \approx \log t$ (polycyclic of exponential growth). The simplest example is the wreath product $\mathbb{Z} \wr \mathbb{Z}$ which has $J(t) \approx (\log t)/(\log\log t)$. See [**45**] where there are also examples for which $J(t)$ behaves at infinity as $(\log t)^{1/k}$ with k an integer, or $(\log_m t)^{1/k}$ where m, k are integers and $\log_m$ denotes the m-iterated logarithm.

QUESTION 2 ([**103**]). *Let Γ be a solvable group with exponential volume growth and finite Prüfer rank (i.e., there exists k such that any finitely generated subgroup of Γ contains a generating set with atmost k elements). Prove that $J(t) \approx \log t$.*

Pittet and the author make the following conjecture.

CONJECTURE 1. *For any finitely generated torsion free solvable group with exponential growth, $J(t) \approx \log t$ if and only if the group has finite Prüfer rank.*

The group $\mathbb{Z}_2 \wr \mathbb{Z}$ is solvable with exponential growth. It is not of finite Prüfer rank and satisfies $J(t) \approx \log t$ but it is not torsion free.

3. Invariance of the heat decay

3.1. Questions concerning the invariance of the heat decay in BG.

On a manifold, define

$$\phi_M(x,t) = p(t,x,x), \quad \phi_M(t) = \sup_{x \in M} p(t,x,x).$$

On a graph set

$$\phi_X(x,n) = K_{2n}(x,x), \quad \phi_X(n) = \sup_{x \in X} K_{2n}(x,x).$$

In what follows we will often write M for either a graph or a manifold. In great generality, thanks to the local parabolic Harnack inequality, we have

$$0 < c(x,y) \le \lim_{t\to\infty} \frac{\phi_M(x,t)}{\phi_M(y,t)} \le C(x,y) < \infty.$$

Thus $\phi_M(x,t) \approx \phi_M(y,t)$ for any pair x, y. Note that it may easily happen that the $\phi_M(x,t)$ is not $\approx$ equivalent to $\phi_M(t)$. For instance, one may have $\phi_M(x,t) \approx t^{-d/2}$ whereas $\phi_M(t) \approx t^{-D/2}$ with $0 < d < D < \infty$ arbitrary integers. See, e.g., [**37, 57**].

In view of the results presented in the previous sections, one may ask the following question.

QUESTION 3. *In* **BG**, *is the $\approx$-equivalence class of the functions $\phi_M(x,\cdot)$, $x \in M$, a quasi-isometric invariant? Is the $\approx$-equivalence class of the function ϕ_M a quasi-isometric invariant?*

A more precise form of these questions is as follows.

QUESTION 4. *Let M_1, M_2 be two objects in* **BG** *(thus, each is either a manifold or a graph). Let ψ be a quasi-isometry from M_1 to M_2. Are there constants $0 < c \le 1 \le C < \infty$ such that, for all $x \in M$ and all $t \ge 1$,*

$$c\phi_{M_1}(x, Ct) \le \phi_{M_2}(\psi(x), t) \le C\phi_{M_1}(x, ct)?$$

3.2. Invariance for Cayley graphs and regular covers. For a Cayley graph (Γ, S), we have

$$\phi_\Gamma(x,n) = \phi_\Gamma(n) = q_S^{(2n)}(e).$$

For a cover of a compact manifolds, there is a positive finite constant c such that

$$\forall x \in M, \quad c\phi_M(t) \le \phi_M(x,t) \le \phi_M(t).$$

Thus, in these cases, we consider only the functions $\phi_\Gamma(t), \phi_M(t)$. We call "heat decay" of Γ (resp. M) the $\approx$-equivalence class of ϕ_Γ (resp. ϕ_M). The following result is taken from [**100**].

THEOREM 3.1. *For Cayley graphs and regular covers of compact manifolds the heat decay is a quasi-isometric invariant. In particular, a regular cover of a compact manifold and its deck transformation group share the same heat decay.*

In more concrete terms, if M covers a compact manifold with deck transformation group Γ, and if S is any fixed symmetric finite set generating Γ, there are positive finite constants c, C such that

$$\forall x \in M, \ \forall n \in \{1, 2, \dots\}, \quad cp(Cn, x, x) \le q_S^{(2n)}(e) \le Cp(cn, x, x).$$

What is perhaps remarkable about this result is that, in the present generality, we know very little about the behavior of $p(n, x, x) \approx q^{(2n)}(e)$. In the next section we review what is known about $q^{(2n)}(e) = \phi_\Gamma(n)$ under various algebraic assumptions on Γ.

4. The heat decay for random walks

This section reviews what is known about the heat decay for finitely generated groups. Of course, by Theorem 3.1, most of the results described below can be translated in the context of regular covers of compact manifolds.

Fix a finitely generated group Γ. We assume that Γ is equipped with a finite symmetric generating set S. Distance and volume growth on Γ are computed with respect to S. Theorem 3.1 shows that the $\approx$-equivalence class of $\phi_{(\Gamma,S)}$ is independent of the choice of S. By definition, it is the heat decay of Γ and we denote it by ϕ_Γ.

4.1. Further stability results. We start by collecting some results from [**100**] that complement Theorem 3.1.

THEOREM 4.1. *Let Γ be a finitely generated group. Let Γ' be a finitely generated subgroup of a quotient of Γ. Then $\phi_\Gamma \ll \phi_{\Gamma'}$.*

Note that this applies in particular to quotients and to subgroups of Γ.

THEOREM 4.2. *Let Γ be a finitely generated group. Let q be a symmetric probability measure on Γ.*

- *Assume that $\sum_\Gamma |\gamma|^2 q(\gamma) < \infty$. Then $\phi_\Gamma \ll \phi_q$.*
- *Assume that q has finite generating support. Then $\phi_q \ll \phi_\Gamma$.*

It may be surprising at first that such a wide variety of probability measures shares the same basic behavior. However, in $\mathbb{Z}^d$, it is well-known that the second moment condition $\sum_\gamma |\gamma|^2 q(\gamma) < \infty$ delimits the natural domain of application of the central limit theorem.

4.2. From volume growth to the heat decay. Suppose that M is a graph or manifold in **BG** and that we are given a uniform lower bound

$$\forall x, \forall r > 1, \;\; V(x,r) \geq v(r)$$

where v is a continuous positive increasing convex function. What can be said about the heat decay? The answer found in [**12**] is that ϕ_M is then bounded above by the inverse function of $sv^{-1}(s)$ where v^{-1} is the inverse function of v. Hence, if $v(r) = cr^d$ with $d \geq 1$ then $\phi_M(t) \leq Ct^{d/(1+d)}$. If $v(r) = \exp(cr)$ then $\phi_M(t) \leq Ct/\log t$. Moreover, examples show that these results are essentially sharp.

For Cayley graphs, the situation is much better thanks to the following result. The precise form given here is from [**27, 104**]. The two special cases emphasized in the statement below were obtained earlier in ground breaking works by Varopoulos. See [**127, 128**].

THEOREM 4.3. *Let Γ be a finitely generated group with finite symmetric generating set S. Let $w(t) = \inf\{n : V(n) > t\}$. Define ψ implicitely for t large enough by*

$$t = \int_1^{1/\rho(t)} \frac{w(s)^2 ds}{s}.$$

Then $\phi_\Gamma \ll \rho$. In particular:

- *if $V(n) \gg n^d$ then $\phi_\Gamma(t) \ll t^{-d/2}$.*
- *if $V(n) \gg \exp(n^\alpha)$ then $\phi_\Gamma(t) \ll \exp(-t^{\alpha/(\alpha+2)})$.*

The proof of this remarkable result although not very difficult is too long to be included here. The crucial ingredient that catches the difference between, say, general graphs of bounded degree and Cayley graphs is the following "Calculus-type" inequality. For any $f \in \mathcal{C}_0(\Gamma)$, we have

$$\forall y \in \Gamma, \; \sum_{x\in\Gamma} |f(xy) - f(x)| \leq |y|_S \sum_E |df|.$$

The proof of this inequality is the only place where the group structure is used in the proof of Theorem 4.3.

Example 4.4. Let Γ be a finitely generated group. Assume that Γ contains a infinite normal subgroup Z_1, the quotient Γ/Z_1 contains an infinite normal subgroup Z_2 and so on, k times. Then Γ has volume growth $V(n) \gg n^k$ (this simple result is noted in [**58**, p.59]). Thus $\phi_\Gamma(n) \ll n^{-k/2}$.

Theorem 4.3 is one of the keys to the proof of Theorem 2.5 (i.e., to Varopoulos' solution of Kesten's conjecture concerning recurrent groups). Indeed, Theorem 4.3 shows that a recurrent group must satisfy $\liminf r^{-\alpha}V_S(r) < \infty$ for each $\alpha > 2$ (if not, we would have $V(r) \gg r^\alpha$ for some $\alpha > 2$. By Theorem 4.3, this would give $\phi_\Gamma(n) \ll n^{-\alpha/2}$ and thus $\phi_\Gamma(n)$ would be summable, contradicting the hypothesis that Γ is recurrent). By Gromov's theorem on groups of polynomial volume growth, this implies that $V_S(n) \approx n^d$ with $d = 0, 1$ or 2 and it is not hard to show that the only groups with such growth are the finite extensions of $\{e\}$, $\mathbb{Z}$ and $\mathbb{Z}^2$.

Coulhon and Grigor'yan [**31**] give a simple and general proof that if $V_S(n) \approx r^d$ then $\phi_\Gamma(n) \gg n^{-d/2}$ (properly stated, their result is not restricted to groups). Alexopoulos [**2**] shows that polycyclic groups with exponential volume growth satisfy $\phi_\Gamma(n) \gg \exp(-n^{1/3})$. Thus the results stated in Theorem 4.3 are sharp on some examples.

4.3. Classical behavior and discrete linear groups. There are several equivalent definitions of polycyclic groups (see Section 1.5). In particular, polycyclic groups are exactly those countable solvable groups that can be realized as a closed subgroup of a Lie group with finitely many connected components. See [**105**]. Moreover, it turns out that any discrete subgroup of a connected Lie group is either non-amenable or contains a polycyclic subgroup of finite index. This follows from the work of Tits [**117**] and Mostow [**92**]. Thus, for discrete subgroups of connected Lie groups, we have the following (see, e.g., [**98**]).

THEOREM 4.5. *Let Γ be a discrete subgroup of a real Lie group with finitely many connected components. Then either (a) Γ is non-amenable or (b) Γ is finitely generated and contains a polycyclic subgroup of finite index. In the second case, either Γ has exponential volume growth and satisfies $\phi_\Gamma(n) \approx \exp(-n^{1/3})$ or Γ has polynomial volume growth $V(r) \approx r^d$ and satisfies $\phi_\Gamma(n) \approx n^{-d/2}$ for some $d = 0, 1, 2, \dots$.*

Thus for a finitely generated discrete subgroup Γ of a connected Lie group there are only three cases: $\phi_\Gamma \approx \exp(-n)$, $\phi_\Gamma \approx \exp(-n^{1/3})$, $\phi_\Gamma \approx n^{-d/2}$ for some integer d. We call these three behaviors the classical behaviors.

4.4. Exotic behaviors for solvable groups. Because of the great variety of groups of intermediate growth discovered by Grigorchuk [**52**], it seems unlikely that one can obtain a complete classification of the possible behaviors of ϕ_Γ in full generality. One might, however, dream of a classification of all possible behaviors of ϕ_Γ for finitely generated solvable groups. Indeed, this would be very easy is ϕ_Γ was determined by the volume growth for solvable groups but it is not as the following examples show.

Example 4.6 ([**99, 101, 102**]). Let λ be a real that is not algebraic. Then the subgroup of $\mathbb{A}_\lambda$ of $\mathrm{Aff}(\mathbb{R})$ generated by $x \mapsto x+1$ and $x \mapsto \lambda x$ is isomorphic to the wreath product $\mathbb{Z} \wr \mathbb{Z}$ and satisfies $\phi_{\mathbb{A}_\lambda}(n) \approx \exp(-n^{1/3}(\log n)^{2/3})$.

Not only is $\mathbb{Z} \wr \mathbb{Z}$ solvable, it is in fact metabelian and even abelian by cyclic. It is proved in [**102**] that, for any abelian by cyclic group Γ with exponential growth, either $\phi_\Gamma(n) \approx \exp(-n^{1/3})$ or $\mathbb{Z} \wr \mathbb{Z} \subset \Gamma$ and $\phi_\Gamma(n) \approx \exp(-n^{1/3}(\log n)^{2/3})$. Thus, the volume growth is not sufficient to determine the behavior of ϕ_Γ in the class of solvable groups. This fact is more dramatically illustrated by the following examples, all of exponential volume growth (each of the different classes of groups considered below contains solvable groups).

Example 4.7 ([**45, 97, 101, 104**]). Let A and B be two finitely generated groups and consider the wreath product $\Gamma = A \wr B$.

- Assume that B has polynomial growth of degree d (i.e., $V(r) \approx r^d$). Then

$$\phi_\Gamma(n) \approx \begin{cases} \exp(-n^{d/(d+2)}) \text{ if } A \text{ is finite and non trivial} \\ \exp(-n^{d/(d+2)}(\log n)^{2/(d+2)}) \text{ if } A \text{ is infinite with polynomial growth} \\ \exp(-n^{(d+1)/(d+3)}) \text{ if } A \text{ is polycyclic with exponential growth.} \end{cases}$$

- Assume that B is polycyclic with exponential volume growth. Then for any non-trivial A, finite or polycyclic, we have

$$\phi_{A \wr B}(n) \approx \exp(-n(\log n)^{-2}).$$

- Assume that $\Gamma = A_k \wr (A_{k-1} \wr (\cdots (A_1 \wr B) \cdots))$ where each A_i is non-trivial polycyclic. Set $\log_k(t) = \log(1 + \log_{k-1}(t))$ with $\log_0(t) = t$, $t > 1$.
  - If B is polycyclic with exponential growth then

$$\phi_\Gamma(n) \approx \exp(-n(\log_k n)^{-2}).$$

  - If $k \geq 2$ and B has polynomial growth of degree $d \geq 1$, let V_1 be the volume growth function of A_1. Then

$$\phi_\Gamma(n) \approx \begin{cases} \exp(-n(\log_{k-1} n)^{-2/(d+1)} & \text{if } V_1 \text{ is exponential} \\ \exp(-n[(\log_{k-1} n)/\log_k n]^{-2/d}) & \text{if } V_1 \text{ is polynomial}, V_1 \not\approx 1 \\ \exp(-n(\log_{k-1} n)^{-2/d}) & \text{if } V_1 \approx 1, \text{ i.e., } A_1 \text{ is finite} \end{cases}$$

More examples can be constructed from the results in [**45, 101**]. In particular, Erschler's results from [**45**] represent a breakthrough and provide a variety of examples. See also [**47**].

As mentioned above, for abelian by cyclic groups of exponential growth, only two behaviors of ϕ_Γ are possible. The papers [**101, 103**] give two other results in this direction. Recall that a group Γ has finite Prüfer rank if there exists an integer k such that any finitely generated subgroup of Γ can be generated by k elements or less.

THEOREM 4.8. *Let Γ be a finitely generated solvable group.*

- *If Γ is metabelian then $\phi_\Gamma(n) \gg \exp(-n^{1-\epsilon})$ for some $\epsilon \in (0,1)$.*
- *If Γ has finite Prüfer rank then $\phi_\Gamma(n) \gg \exp(-n^{1/3})$.*

Pittet and the author make the following conjecture.

CONJECTURE 2. *Let Γ be a finitely generated solvable torsion free group. Then $\phi_\Gamma(n) \gg \exp(-n^{1/3})$ if and only if Γ has finite Prüfer rank.*

The group $\mathbb{Z}_2 \wr \mathbb{Z}$ is solvable with exponential growth. It is not of finite Prüfer rank and satisfies $\phi_{\mathbb{Z} \wr \mathbb{Z}}(n) \approx \exp(-n^{1/3})$ but it is not torsion free.

QUESTION 5. *Is it true that if Γ is metabelian of exponential growth then there exist two rational numbers α, β, such that $\phi_\Gamma(n) \approx \exp(-n^\alpha(\log n)^\beta)$?*

QUESTION 6. *Let* $\mathbb{S}_d^2 = \mathbb{F}_d/\mathbb{F}_d''$ *be the free metabelian group with* d *generators* ($\mathbb{F}_d''$ *is the second derived group). What is the behavior of* $\phi_{\mathbb{S}_d^2}$*?*

The study of the group $\mathbb{S}_d^2$ is advocated in [**130**] where the Poisson boundary is studied (see also [**46**]). In preliminary work, Pittet and the author observed that

$$\exp(-n^{d/(d+2)}) \ll \phi_{\mathbb{S}_d^2}(n) \ll \exp(-n^{(d-1)/(d+1)}(\log n)^{2/(d+1)}).$$

QUESTION 7. *Is it true that for any step-three solvable group* Γ *there exists* $\epsilon > 0$ *such that*

$$\phi_\Gamma(n) \gg \exp(-n(\log n)^{-\epsilon})?$$

4.5. Isoperimetry and the heat decay on groups. It is well understood (e.g., [**30, 55, 104, 128**] and the different approach of [**91**]) that, in a very general setting, control of the isoperimetric profile yields control on the uniform heat kernel decay. In general, something is lost in this type of argument (see [**34**]).

In the case of a Cayley graph (Γ, S), one has the following precise statement, versions of which hold in much greater generality. Define $J_\Gamma = J$ by

$$J(k) = \sup\left\{\frac{\#A}{\#\partial A} : \#A \le k\right\}$$

and define the function $\rho = \rho_J$ implicitely by

$$t = \int_1^{1/\rho(t)} J(s)^2 \frac{ds}{s}.$$

THEOREM 4.9. *The heat decay* ϕ_Γ *of the Cayley graph* (Γ, S) *is bounded by*

$$\phi_\Gamma \ll \rho.$$

It is also possible to estimate J in terms the heat decay but the known results in this direction are less satisfactory and the following question is open.

QUESTION 8. *Let* (Γ_i, S_i) *be two Cayley graphs.*

- *Suppose* $J_{\Gamma_1} \approx J_{\Gamma_2}$. *Does it implies* $\phi_{\Gamma_1} \approx \phi_{\Gamma_2}$*?*
- *Suppose* $\phi_{\Gamma_1} \approx \phi_{\Gamma_2}$. *Does it implies* $J_{\Gamma_1} \approx J_{\Gamma_2}$*?*

That is, does J *and* ϕ *contain the same information about the underlying group?*

For further discussions and results concerning this question, see [**31, 32, 104**].

4.6. The heat decay and Sobolev inequality on regular covers. This section gather statements concerning the on-diagonal behavior of the heat kernel on co-compact Riemannian covers. These statements are simple consequences of the results reviewed earlier. First, as a consequence of Theorems 4.5 and 3.1, we have the following result.

THEOREM 4.10. *Let* M *be a co-compact Riemannian regular cover. Assume that the deck transformation group of* M *can be realized as a discrete subgroup of a real Lie group with finitely many connected components. Then the heat decay* $\phi_M(t)$, $t \gg 1$, *is given by one of the following three exclusive possibilities:*

1. $\phi_M(t) \approx e^{-t}$;
2. $\phi_M(t) \approx e^{-t^{1/3}}$;
3. *There exists an integer* d *such that* $\phi_M(t) \approx t^{-d/2}$.

Of course, the first case corresponds exactly to the case where the deck transformation group Γ is non-amenable, the second case to the case where Γ is amenable with exponential growth and the third case to the case where Γ has polynomial volume growth of degree d. The examples of Section 4.4 show that many other behaviors are possible when Γ cannot be realized as a discrete subgroup of a real Lie group with finitely many connected components. In such cases, the following corollary to Theorems 3.1, 4.3 may be useful.

THEOREM 4.11. *Let M be a co-compact Riemannian regular cover.*

- *Assume that $V(x,r) \geq cr^d$, $r > 1$. Then, for all $t > 1$, $\phi_M(t) \leq Ct^{-d/2}$.*
- *Assume that $V(x,r) \geq c\exp(cr^\alpha)$, $r > 1$. Then*

$$\forall t > 1, \quad \phi_M(t) \leq C_1 \exp(-c_1 t^{\alpha/(2+\alpha)}).$$

The next statement connects these heat kernel estimates to the most classical form of the Sobolev inequality. See, e.g., [**128**].

THEOREM 4.12. *Let M be a co-compact Riemannian regular cover of dimension n. Then M satisfies the Sobolev inequality*

$$\forall f \in \mathcal{C}_0(M), \quad \left(\int_M |f|^{2\nu/(\nu-2)} d\mu\right)^{(\nu-2)/2\nu} \leq C \int_M |\nabla f|^2 d\mu$$

if and only if $\nu \geq n$ and $V(x,r) \geq cr^\nu$, $r > 1$.

Note that the last volume condition above is satisfied if and only if the deck transformation group has volume growth at least r^ν. One can prove other versions of the above inequality (see, e.g., [**110, 111, 128**]). More precisely, assume that $V(x,r) \geq cr^\nu$ for all $r > 0$ (hence, $\nu > n$) and fix $1 \leq p < \nu$. Then there exists a constant C such that

$$\forall f \in \mathcal{C}_0^\infty(M), \quad \|f\|_{p\nu/(\nu-p)} \leq C\|\nabla f\|_p.$$

Assume in addition that the volume growth is subexponential, i.e., $\lim_{n\to\infty} V(n)^{1/n} < 1$. Then there exists a constant C such that for any smooth function f on M satisfying $|\nabla f| \in L^p$ (not necessarily compactly supported), there exists a constant $C(f)$ for which

$$\forall f \in \mathcal{C}_0^\infty(M), \quad \|f - C(f)\|_{p\nu/(\nu-p)} \leq C\|\nabla f\|_p.$$

In cases where Γ has polynomial volume growth or degree d with $d < n$, no Sobolev inequalities of the form stated above can possibly hold but there are several ways to cope with this situation. Assume indeed that Γ has polynomial volume growth or degree d and set $\nu = \max\{n, d\}$. Then the families of local Sobolev inequalities

$$\forall B = B(x,r), \quad \forall f \in \mathcal{C}_0^\infty(B), \quad \|f\|_{B,p\nu/\nu-p)} \leq C_p\, r\, \mu(B)^{-1/\nu} \|\nabla f\|_{B,p}$$

and

$$\forall B = B(x,r), \quad \forall f \in \mathcal{C}^\infty(B), \quad \|f - f_B\|_{B,p\nu/(\nu-p)} \leq C_p\, r\, \mu(B)^{-1/\nu} \|\nabla f\|_{B,p},$$

hold for all $1 \leq p < \nu$ (see, e.g., [**111**]). Here, $\|\cdot\|_{B,p}$ is the L^p norm restricted to the set B and f_B denotes the mean of f over B. Alternatively, one can use the Faber-Krahn type inequality (see [**55**])

$$\forall B = B(x,r), \quad \forall \Omega \subset B, \quad \lambda(\Omega) \geq \frac{c}{r^2}\left(\frac{\mu(\Omega)}{\mu(B)}\right)^{2/\nu}$$

where $\lambda(\Omega)$ denotes the lowest eigenvalue of minus the Laplacian with Dirichlet boundary condition in the open set Ω.

5. Harmonic functions

5.1. Liouville properties. A complete Riemannian manifold is Liouville if any bounded harmonic function is constant. It has the strong Liouville property if any positive harmonic function is constant. The same definitions apply to graphs. A measure q on a group is Liouville (resp. strong Liouvillle) if any bounded (resp. positive) solution of $u * q = u$ is constant.

The study of Liouville's properties is really only the iceberg's top of the deeper problem of studying the space of all bounded harmonic functions or the cone of all positive harmonic functions through the construction and understanding of the Poisson and Martin boundaries. The Liouville property is equivalent to the fact that the Poisson boundary is reduced to a singleton. In the class **BG**, the strong Liouville property is equivalent to the fact that the Martin boundary is reduced to a singleton. We will not discuss Poisson and Martin boundaries here. Pointers to the literature can be found in [**3, 71, 72, 85, 129, 131**].

Blackwell [**16**] appears to be the first to have observed that any bounded harmonic functions on the square lattice $\mathbb{Z}^d$ must be constant. This is a special case of the celebrated Choquet-Deny theorem [**25**] which implies that, under mild conditions on the probability measure q, bounded q-harmonic functions on a locally compact abelian group are constants. Dynkin and Maljutov [**43**] observed that finitely generated nilpotent groups have the Liouville property. Later, the following three satisfactory results concerning the Liouville property were obtained:

- If q is a probability measure with finite generating support on a group Γ and Γ has subexponential volume growth (i.e., $\lim_{n\to\infty} V(n)^{1/n} = 1$) then any bounded q-harmonic function is constant. See [**5**].
- If q is a symmetric finitely supported measure on a polycyclic group then any bounded q-harmonic function is constant. See [**2, 68**].
- If the support of q generates Γ and Γ is non-amenable then there are non-trivial bounded q-harmonic functions on Γ. See [**6**] and also [**50**].

Concerning the strong Liouville property, Margulis [**88**] proved that if q is symmetric and Γ is nilpotent then any positive q-harmonic function is constant. More recently, Bougerol and Elie [**17**] proved the following theorem and corollary (their results are slightly more general than stated here).

THEOREM 5.1. *Let Γ be a finitely generated group. Assume that there exists a continuous homomorphism h from Γ into a group G having finitely many connected components such that the closure of $h(\Gamma)$ has exponential volume growth. Assume that q is symmetric with generating support and has a third moment (i.e., $\sum |\gamma|^3 q(\gamma) < \infty$). Then there are non-constant positive q-harmonic functions on Γ.*

COROLLARY 5.2. *If q is symmetric, with finite generating support and Γ is polycyclic then the positive q-harmonic functions are constant if and only if Γ has polynomial volume growth.*

Example 5.3. Theorem 5.1 applies to the subgroups $\mathbb{A}_\lambda$ of the affine group of the line (recall that $\mathbb{A}_\lambda$ is the group generated by $x \mapsto x+1$ and $x \mapsto \lambda x$). For any

$\lambda > 1$, and for any symmetric probability measure q with finite generating support on $\mathbb{A}_\lambda$, there are non-constant positive q-harmonic functions.

5.2. Instability of the Liouville properties. In [**84**], Lyons gives examples of quasi-isometric pairs M_1, M_2 in **BG** such that M_1 admits non-constant bounded harmonic functions whereas M_2 has the strong Liouville property. He provides examples where both M_1 and M_2 are graphs, or both manifolds, or one is a graph and the other a manifold. Thus, each of the Liouville properties is unstable under quasi-isometry in **BG**. See also [**14**]. However, these works leave open the following basic questions.

QUESTION 9. *Let (Γ_1, S_1) and (Γ_2, S_2) be two quasi-isometric Cayley graphs. Is it true that (Γ_1, S_1) has the (strong) Liouville property if and only if (Γ_2, S_2) does?*

QUESTION 10. *Given two symmetric probability measure q_1, q_2 with finite generating supports on a group Γ, does the (strong) Liouville property for q_1 implies the same property for q_2?*

QUESTION 11. *Let M be a regular cocompact cover with deck transformation group Γ. Let g_1, g_2 be two Γ invariant Riemannian metrics on M. Is it true that (M, g_1) has the (strong) Liouville property if and only if (M, g_2) does?*

5.3. Passage from the deck transformation group to the cover. Despite the fact that we do not know if the Liouville properties are stable when passing from a covering to its deck transformation group (assuming the quotient is compact), there is a way to transport some information thanks to a discretization procedure introduced by Furstenberg [**49**] and refined successively by Lyons and Sullivan [**86**], Ancona [**3**], Kaimanovich [**69**], and Ballman and Ledrappier [**10**] (the result in [**10**] is more precise than the one stated below).

PROPOSITION 5.4. *Let M be a regular Riemannian covering of a compact manifold with deck transformation group Γ. There exists a symmetric probability measure q on Γ whose support is Γ, which satisfies $\sum_{\gamma\in\Gamma} e^{c|\gamma|}q(\gamma) < \infty$ for some $c > 0$, and such that:*

- *Any bounded harmonic function on M restricted to Γ is q-harmonic.*
- *any positive q-harmonic function on Γ can be extended to a positive harmonic function on M.*

Suppose we want to apply this proposition to show that a certain co-compact Riemannian covering has (does not have) the Liouville property by using information concerning the deck transformation group Γ. Then it is not enough to have information about a specific random walk on Γ, nor does it suffice to know about all random walks driven by finitely symmetric supported measures. In the form stated above, Proposition 5.4 requires dealing with symmetric measures μ having an exponential moment. This is in contrast with what happens when a property is stable under quasi-isometries.

Example 5.5. Let M be a regular co-compact cover with non-amenable deck transformation group. Then, by the aforementioned result of Azencott (any non-amenable group admits non trivial bounded harmonic functions for (essentially) any random walk) and Theorem 5.4, M admits non-constant bounded harmonic function.

Example 5.6. Let M be a regular covering of a compact manifold with deck transformation group $\Gamma = \mathbb{A}_\lambda$, $\lambda > 1$. (this group is isomorphic to $\mathbb{Z} \wr \mathbb{Z}$ when λ is not algebraic). Then, using Proposition 5.4 and Theorem 5.1, we obtain that M admits non-constant positive harmonic functions.

Example 5.7. Let M be a regular covering of a compact manifold with deck transformation group Γ having sub-exponential growth (i.e., $\lim V(n)^{1/n} < 1$). Then any bounded harmonic function on M is constant. See [**67**] and [**3**, Théorème 3.1].

Example 5.8. Consider the wreath products $\Gamma_k = \mathbb{Z}_2 \wr \mathbb{Z}^k$. Then the Cayley graphs of Γ_k have the Liouville property if and only if $k \leq 2$ [**71, 72**]. From Proposition 5.4 and [**71**, Theorem 3.3], it follows that a co-compact regular cover with deck transformation group Γ_k has the Liouville property if and only if $k \leq 2$.

Further examples are dicussed in [**46**] where, in particular, a compact Riemannian manifold with amenable fundamental group whose universal cover does not have the Liouville property is constructed.

The following result is from [**17**]. See also [**3, 8, 62, 69**].

THEOREM 5.9. *Assume that M is a regular co-compact cover whose deck transfomation group is a closed subgroup of a group having finitely many connected components. Then M has non-trivial positive harmonic functions if and only if Γ has exponential volume growth. It has non-trivial bounded harmonic functions if and only if Γ is non-amenable.*

5.4. Harmonic functions with finite energy. On a manifold, consider the energy form $D(f) = \int_M |\nabla f|^2 d\mu$ whereas, on a graph, set $D(f) = \sum_{e \in E} |df(e)|^2$. Royden work [**108**] sparkled off interest on whether or not harmonic functions with finite energy are constant. Say that the "finite energy Liouville property" holds if any harmonic function u with $D(u) < \infty$ is constant.

The following result of Holopainen and Soardi [**66**] may be a little surprising in view of the instability of the Liouville and strong Liouville properties.

THEOREM 5.10. *The finite energy Liouville property is invariant under quasi-isometries in* **BG**. *In particular, a co-compact Riemannain cover has the finite energy Liouville property if and only if its deck transformation group does.*

There are several simple but not completely obvious facts concerning harmonic functions with finite eneregy that are worth noting (for graphs, a good reference is [**115**]). Any harmonic function of finite energy is the difference of two positive harmonic functions. Thus, if the strong Liouville property holds, so does the finite energy Liouville property. In fact, the finite energy Liouville property holds already if bounded harmonic functions with finite energy are constant. In particular, it holds if the Liouville property holds (see, e.g., [**66**] which treats p-harmonic functions).

Example 5.11. Let M be a regular Riemannian co-compact cover with deck transformation group Γ. Assume that Γ as subexponential volume growth. Then M has the finite energy Liouville property.

5.5. Harnack inequalities. Assume that M is a metric space equipped with a family of harmonic functions (for our purpose, it suffices to think of the examples above where M is a Riemannian manifold or a Cayley graph). We say that M

satisfies the elliptic Harnack inequality if there is a constant C such that for any ball B and any positive function u harmonic in $2B$, we have

$$\sup_B\{u\} \le C \inf_B\{u\}. \tag{5.1}$$

Let us emphasize that C is independent of the ball B and of the function u. In particular, this is in an obvious sense a scale invariant property.

The Harnack inequality implies the strong Liouville property (the converse is false as will become clear from some of the examples discussed below). Classical harmonic functions in Euclidean space satisfy the Harnack inequality. For manifolds with non-negative Ricci curvature, (5.1) was obtained in [**24**] under the stronger form $|\nabla u| \le \frac{C}{r} u$ in B.

QUESTION 12. *Is the elliptic Harnack inequality* (5.1) *a quasi-isometric invariant in* **BG***?*

For simplicity, we define the parabolic Harnack inequality only on manifolds (see, e.g., [**29, 38**] for the graph version). Given a real s, $r \in (0,\infty)$ and $x \in M$, set $Q = (s, s+4r^2) \times B(x,2r)$ and

$$Q_- = (s+r^2, s+2r^2) \times B(x,r), \quad Q_+ = (s+3r^2, s+4r^2) \times B(x,r).$$

Thus, Q_- and Q_+ are two sub-cylinders contained in Q with Q_+ sitting well above Q_- and Q_- floating in the bottom part of Q.

We say that a manifold M satisfies the parabolic Harnack inequality if there exists a constant C such that for any non-negative solution u of $(\partial_t + \Delta)u = 0$ in Q we have

$$\sup_{Q_-}\{u\} \le C \inf_{Q_+}\{u\}. \tag{5.2}$$

It is proved in [**36**] that the parabolic Harnack inequality (5.2) is stable under quasi-isometries in **BG**. This is because it can be characterized by the volume doubling property and the scale invariant Poincaré inequality. See [**54, 109, 111**].

5.6. A necessary condition for Harnack inequality in BG. The following useful result is due to Barlow [**11**].

PROPOSITION 5.12. *Let M be a graph or a manifold in* **BG***. If M satisfies the Harnack inequality* (5.1) *then the volume growth of M is bounded above by a power function.*

SKETCH OF THE PROOF. In [**11**], Barlow proves this result (in a more general form) for graphs. Here, we adapt Barlow's proof to manifolds. In M, consider a maximal set of point $X = \{x_i, i = 1, 2, \dots\}$ such that $d(x_i, x_j) > 16$ if $i \neq j$. Obviously, the (closed) ball $B(x_i, 8)$ are pairwise disjoint whereas the (closed) balls $B(x_i, 16)$ cover M. Fix $o \in M$. Let $F_n = \{i : B(x_i, 1) \cap B(o,n) \setminus B(o, n-1) \neq \emptyset\}$. We claim that $B(o,n)$ is covered by the balls $B(x_i, 64)$, $i \in F_n$. Indeed, let z be in $B(o,n)$. Let j be such that $z \in B(x_j, 16)$. Along a shortest path from x_j to o, let y be the first point such that $d(o,y) \le n - 16$ and let j_1 be such that $y \in B(x_{j_1}, 16)$. Then $B(x_{j_1}, 1)$ intersects $B(o,n)$. As $d(o, x_j) \le n + 16$, we must have $d(x_j, y) \le 32$ hence $d(x_j, x_{j_1}) \le 48$. Thus $d(x_{j_1}, z) \le 64$. As balls of a fixed radius all have comparable volume (because $M \in$ **BG**), there exists a constant C

$$\mu(B(o,n)) \le \sum_{k=0}^{n} \sum_{i \in F_k} \mu(B(x_i, 64)) \le C \sum_0^n \#F_k.$$

We now claim that there exist $C, \alpha > 0$ such that

$$\#F_k \leq Ck^\alpha. \tag{5.3}$$

If this is the case then $\mu(B(o,n)) \leq C' n^{\alpha+1}$ which is the desired result. To prove this claim, we use the Brownian motion $(X_t)_{t>0}$ on M. Fix k and set

$$A = A_k = \cup_{i\in F_k} B(x_i, 1),$$

$$\tau = \inf\{t > 0 : X_t \in A_k\}$$

and

$$h_i(x) = \mathbb{P}_x(X_\tau \in B(x_i, 1)).$$

Thus $h_i(x)$ is the probability that, starting from x, one first enters A by entering $B(x_i, 1)$. By the strong Markov property, each h_i is harmonic on $M \setminus A$.

Assertion 5.13. *There exists a positive constant $c_0 > 0$ such that, for every k, $i \in F_k$, and x such that $d(x, x_i) = 2$, we have $h_i(x) \geq c_0$.*

Proof. This assertion is satisfied because M has bounded geometry (i.e., $M \in$ **BG**). To see this, for $x \in B(x_i, 8)$, observe that

$$h_i(x) \geq u_i(x) = \mathbb{P}_x(X_{\tau_i} \in B(x_i, 1) \text{ and } \tau_i < \sigma_i)$$

where τ_i is the first hitting time of $B(x_i, 1)$ and σ_i is the first exit time from $B(x_i, 8)$. Then, for $x \in B(x_i, 8)$, we have

$$u_i(x) \approx \frac{g_i(x_i, x)}{g_i(x_i, y)}$$

where g_i is the Green function with Dirichlet boundary condition in $B(x_i, 8)$ and y is an arbitrary point on $\partial B(x_i, 1)$. By well-known Dirichlet heat kernel estimates, for z such that $1 \leq d(x_i, z) \leq 7$, we have $g_i(x_i, z) \approx 1$. □

Assertion 5.14. *There are finite positive constants c, α such that, for every k and every $i \in F_k$, $h_i(o) \geq ck^{-\alpha}$.*

Proof. Fix $i \in F_k$. Let γ_t, $0 \leq t \leq T$ be a shortest curve from x_i to o parametrized by the distance from x_i so that $d(x_i, \gamma_t) = t$, $0 \leq t \leq T$. We need to control de distance from γ_t to A. By construction, if $t \leq 8$, $d(\gamma_t, A) = d(\gamma_t, B(x_i, 1)) = t - 1$. Moreover, the triangle inequality shows that for any $j \in F_k$ with $j \neq i$, we have

$$d(\gamma_t, B(x_j, 1)) \geq d(o, B(x_j, 1)) - d(o, \gamma_t) \geq t - 3.$$

Thus, for $t \in [2, T]$,

$$d(\gamma_t, A) \geq t/2.$$

Now, consider the points $z_j = \gamma_{t_j}$ with $t_j = 2(1 + 1/8)^{j-1}$, $j = 1, 2, \ldots, j_0$ where j_0 is such that $d(o, x_{j_0}) \leq k/4$. It follows that $j_0 \approx \log k$. For each $j = 1, \ldots, j_0 - 1$, $d(x_j, x_{j+1}) = t_j/8$ and the function h_i is harmonic in $B(x_j, t_j/2)$. Thus the Harnack inequality (5.1) gives $h_i(z_{j+1}) \geq C^{-1} h_i(z_j)$ (C being the constant from (5.1)). It follows that

$$h_i(z_{j_0}) \geq C^{-j_0} h_i(z_1).$$

Applying (5.1) one more time to pass from z_{j_0} to o and using Assertion 5.13 to estimate $h_i(z_1)$, we obtain $h_i(o) \geq c_0 C^{-j_0} \geq ck^{-\alpha}$ for some $c, \alpha > 0$, as desired. □

To finish the proof of Proposition 5.12, we show that Assertion 5.14 implies (5.3). Indeed, as the balls $B(x_i, 1)$, $i \in F_k$, form a partition of A_k, $\sum_{i \in F_k} h_i(x) = 1$. Hence

$$1 \geq \sum_{i \in F_k} h_i(o) \geq c \# F_k \, k^{-\alpha}$$

which gives (5.3). □

5.7. Analysis on regular covers having polynomial volume growth. Using Proposition 5.12, Gromov's theorem on groups of polynomial growth, and the results of [**54, 109**], we can describe those co-compact regular covers satisfying the elliptic (and parabolic) Harnack inequality.

THEOREM 5.15. *Let M be a Riemannian co-compact regular cover with deck transformation group Γ. The following properties are equivalent.*

- *The elliptic Harnack inequality* (5.1) *is satisfied.*
- *The parabolic Harnack inequality* (5.2) *is satisfied.*
- *The group Γ has polynomial volume growth.*
- *The heat kernel $p(t, x, y)$ satisfies the two-sided estimate*

$$\frac{c_1}{V(x, \sqrt{t})} e^{-c_2 d(x,y)^2/t} \leq p(t, x, y) \leq \frac{c_3}{V(x, \sqrt{t})} e^{-c_4 d(x,y)^2/t}$$

for all $t > 0$ and $x, y \in M$.

If M is a co-compact Riemannian regular cover and its deck transformation group Γ has polynomial growth then M satisfies the Poincaré inequality

$$\forall f \in \mathcal{C}^\infty(B), \quad \int_B |f - f_B|^p d\mu \leq C_p \, r(B)^p \int_B |\nabla f|^p d\mu \tag{5.4}$$

for each $p \in [1, \infty)$. Here $r(B)$ denotes the radius of that ball, f_B is the mean of f over the ball B, and the constant C_p is independent of B (see, e.g., [**36**]). Note that, by the results of [**54, 109, 111**], the two sided Gaussian heat kernel inequality of Theorem 5.15 holds on a co-compact cover M having polynomial volume growth as soon as the Riemannian metric is uniformly comparable to a metric lifted from the compact quotient. This type of global heat kernel estimate is similar to the estimates obtained by Li and Yau [**80**] for manifolds with non-negative Ricci curvature. Note however that, in general, the manifolds in Theorem 5.15 do not have non-negative Ricci curvature. For instance, Γ may have volume growth r^d with d strictly larger than the dimension of M, preventing the existence of a Riemannian metric of non-negative Ricci curvature on M.

As a corollary of Theorem 5.15, we see that a regular co-compact Riemannian cover M with deck transformation group of polynomial volume growth of degree $d \geq 3$ admits a Green function $G(x, y)$ which satisfies

$$c d(x, y)^{2-d} \leq G(x, y) \leq C d(x, y)^{2-d}$$

for all $x, y \in M$ with $d(x, y) \geq 1$. For the case of abelian cover, [**7**] gives the following more precise result (obtained by Fourier transform techniques).

THEOREM 5.16. *Let M be a regular Riemannian co-compact cover with deck transformation group equal to $\mathbb{Z}^d$, $d \geq 3$. Then there exists an Euclidean norm $\|\cdot\|$ on $\mathbb{R}^d$ such that the green function G satisfies*

$$\lim_{\gamma \to \infty} \|\gamma\|^{d-2} G(x, \gamma(y)) = c_d > 0$$

for all $x, y \in M$ (in the limit, γ is an element of $\Gamma = \mathbb{Z}^d$ and $\gamma(y)$ is the image of y under the action of γ).

A result that is similar in spirit to the above large scale asymptotic of the Green function and concerns the heat kernel is obtained in [**82**].

In a recent work, N. Dungey [**42**] has obtained gradient estimates that complement the results of Theorem 5.15.

THEOREM 5.17. *Let M be a co-compact Riemannian cover with deck transformation group Γ of polynomial volume growth. Let S be a symmetric finite generating set of Γ. Let $p(t,x,y)$ be the heat kernel on M_0. Then, for all $t>0$, and $x, y \in M_0$,*

$$|\nabla_y p(t,x,y)| \le \frac{C}{t^{1/2}V(x,\sqrt{t})} \exp\left(-cd(x,y)^2/t\right). \tag{5.5}$$

Dungey uses (5.5) and further estimates obtained in [**42**] to prove that the Riesz transforms are bounded on $L^p(M,\mu)$, $1<p<\infty$, that is,

$$\forall f \in \mathcal{C}_0^\infty(M), \quad c_p\|\nabla f\|_p \le \|(-\Delta)^{1/2} f\|_p \le C_p \|\nabla f\|_p.$$

Again, Dungey's gradient estimate is similar to the global gradient estimates that follows from [**80**] in the case of manifolds with non-negative Ricci curvature.

Another result worth mentioning here concerns harmonic functions of at most polynomial growth. See [**26, 79**].

THEOREM 5.18. *Let M be a co-compact Riemannian cover with deck transformation group Γ of polynomial volume growth of degree d. Then there exists a constant C such that the space of all harmonic functions on M satisfying*

$$\sup_{r>1} \sup_{d(o,x)\le r} \{r^{-\alpha}|u(x)|\} < \infty$$

has finite dimension bounded by $C\alpha^{d-1}$.

An interesting and non-trivial generalization of Theorem 5.15 concerns quotients of regular covering by subgroups that are not necessarily normal. See [**64, 109**].

THEOREM 5.19. *Let M be a Riemannian co-compact regular cover with deck transformation group Γ. Assume that Γ has polynomial volume growth. Let Γ_0 be a subgroup of Γ (not necessarily a normal subgroup) and let $M_0 = M/\Gamma_0$ be the Riemannian manifold obtained by taking the quotient of M by Γ_0. Then M_0 satisfies the elliptic and parabolic Harnack inequalities* (5.1), (5.2) *and the heat kernel $p(t,x,y)$ on M_0 satisfies the two-sided estimate*

$$\frac{c_1}{V(x,\sqrt{t})} e^{-c_2 d(x,y)^2/t} \le p(t,x,y) \le \frac{c_3}{V(x,\sqrt{t})} e^{-c_4 d(x,y)^2/t}$$

for all $t>0$ and $x, y \in M_0$.

Note that in the setting of this theorem the volume function $V(x,r)$ will often exhibit a non-uniform behavior as x varies. The manifolds in Theorem 5.19 satisfy the Poincaré inequality (5.4). Dungey's gradient estimate (5.5) and the boundedness of the Riesz transforms also hold in this context.

References

[1] S.I. Adian *Random walks on free periodic groups.* Math. USSR Izvestiya **21** (1983), 425-434.

[2] G. Alexopoulos *A lower estimate for central probabilities on polycyclic groups.* Canadian Math. J., **44** (1992), 897-910.

[3] A. Ancona *Théorie du potentiel sur les graphes et les variétés.* Lecture Notes in Math. **1427** (1990), Springer.

[4] G. Anderson, M. Vamanamurthy and M. Vuorinen *Conformal invariants, inequalities, and quasiconformal maps.* Wiley & Sons, 1997.

[5] A. Avez *Harmonic functions on groups.* In "Differential Geometry and Relativity" (1976), 27–32. Reidel, Dordrecht.

[6] R. Azencott *Espace de Poisson des groupes localement compacts.* Lecture Notes in Math. **148** (1970), Springer.

[7] M. Babillot *Théorie du renouvelement pour des chaines semi-markoviennes transientes.* Ann. Inst. H. Poincaré, Probab. Stat. **24** (1988), 507–569.

[8] M. Babillot, P. Bougerol and L. Elie *On the strong Liouville property for co-compact Riemannain covers.* Rend. Sem. Mat. Fis. Milano **64** (1994), 77–84.

[9] D. Bakry, T. Coulhon, M. Ledoux and L. Saloff-Coste *Sobolev inequalities in disguise.* Indiana Univ. Math. J. **44** (1995), 1033–1074.

[10] W. Ballman and F. Ledrappier *Discretization of positive harmonic functions on Riemannian manifolds and Martin boundaries.* In "Actes de la table ronde en l'honneur de Marcel Berger", Séminaires et Congrés **1**, Soc. Math. France, (1996).

[11] M. Barlow *Some remarks on the elliptic Harnack inequality for graphs.* Preprint, 2003.

[12] M. Barlow, T. Coulhon and A. Grigor'yan *Manifolds and graphs with slow heat kernel decay.* Inventiones Math. **144** (2001), 609–649.

[13] H. Bass *The degree of polynomial growth of finitely generated nilpotent groups.* Proc. London Math. Soc., **25** (1972), 603-614.

[14] I. Benjamini *Instability of the Liouville property for quasi-isometric graphs and manifolds of polynomial volume growth.* J. Theoret. Probab. **4** (1991), 631–637.

[15] I. Benjamini and J. Cao *Examples of simply connected Liouville manifolds with positive spectrum.* Differential Geom. Appl. **6** (1996), 31-50.

[16] D. Blackwell *On transient markov processes with a countable number of states and stationary transition probabilities.* Ann. Math. Stat. **26** (1955), 654-658.

[17] P. Bougerol and L. Elie *Existence of positive harmonic functions on groups and on covering manifolds.* Ann. Inst. H. Poincaré **31** (1995), 54–80.

[18] R. Brooks *The fundamental group and the spectrum of the Laplacian.* Comment. Math. Helv. **56** (1981), 581–598.

[19] K. Brown and R. Geoghegan *An infinite dimensional torsion free FP_∞ group.* Invent. Math. **77** (1984), 367–381.

[20] A. Carey, T. Coulhon, V. Mathai, J. Phillips *Von Neumann spectra near the spectral gap.* Bull. Sci. Math. **122** (1998), 203-242.

[21] G. Carron *Inégalités isopérimétriques de Faber-Krahn et conséquences.* In Actes de la table ronde de géométrie différentielle (Luminy, France, 1992), Séminaires et Congrés **1**, Soc. Math. France, 1996, pp. 205–232.

[22] I. Chavel *Riemannian geometry: a modern introduction.* Cambridge University Press, 1993.

[23] I. Chavel *Isoperimetric inequalities. Differential geometric and analytic perspectives.* Cambridge Tracts in Mathematics, 145. Cambridge University Press, Cambridge, 2001.

[24] S.Y. Cheng and S-T. Yau *Differential equations on Riemannian manifolds and their geometric applications.* Comm. Pure Appl. Math. **28** (1975), 333–354.

[25] G. Choquet and J. Deny *Sur l'équation de convolution $\mu = \mu * \sigma$* C. R. Acad. Sci. Paris **250** (1960), 799-801.

[26] T. Colding and W. Minicozzi *Harmonic functions on manifolds.* Ann. of Math. **146** (1997), 725–747.

[27] T. Coulhon *Ultracontractivity and Nash type inequalities.* J. Funct. Anal. **141** (1996), 510–539.

[28] T. Coulhon *Noyau de la chaleur et discrétisation d'une variété riemannienne.* Israel J. Math. **80** (1992), 289–300.

[29] T. Coulhon *Analysis on infinite graphs with regular volume growth.* In "Random walks and discrete potential theory, Cortona 1997" Sympos. Math. XXXIX (1999), 165–187, Cambridge University Press.

[30] T. Coulhon *Heat kernel and isoperimetry on non-compact Riemannian manifolds.* Contemporary Math. to appear.

[31] T. Coulhon and A. Grigor'yan *On-diagonal lower bounds for heat kernels and Markov chains.* Duke Math. J **89** (1997), 133–199.

[32] T. Coulhon, A. Grigor'yan and C. Pittet *A geometric approach to on diagonal heat kernel lower bounds on groups.* Ann. Inst. Fourier **51** (2001), 1763–1827.

[33] T. Coulhon, I. Holopainen, L. Saloff-Coste *Harnack inequality and hyperbolicity for subelliptic p-Laplacians with applications to Picard type theorems.* Geom. Funct. Anal. **11** (2001), 1139–1191.

[34] T. Coulhon and M. Ledoux *Isopérimétrie, décroissance du noyau de la chaleur et transformations de Riesz: un contre exemple.* Ark. Mat. **32** (1994), 63–77.

[35] T. Coulhon and L. Saloff-Coste *Isopérimétrie sur les groupes et les variétés.* Rev. Math. Iberoamericana **9** (1993), 293–314.

[36] T. Coulhon and L. Saloff-Coste *Variétés riemanniennes isométriques à l'infini.* Rev. Mat. Iberoamericana **13** (1995), 687–726.

[37] E.B. Davies *Non-Gaussian aspects of heat kernel behaviour.* J. London Math. Soc. **55** (1997), 105–125.

[38] T. Delmotte *Parabolic Harnack inequality and estimates of Markov chains on graphs.* Rev. Mat. Iberoamericana **15** (1999), 181–232.

[39] J. Deny *Méthodes hilbertiennes en théorie du potentiel.* C.I.M.E., 1 ciclo, Stresa 1969 (1970), 121–201, Edizioni Cremonese, Rome.

[40] J. Dodziuk *Difference equations, isoperimetric inequality and transience of certain random walks.* Trans. Amer. Math. Soc. **284** (1984), 787–794.

[41] J. Dodziuk and V. Mathai *Approximating L^2 invariants of amenable covering spaces: a combinatorial approach.* J. Funct. Anal. **154** (1998), 359–378.

[42] N. Dungey *Heat kernel estimates and Riesz transforms on some Riemannian covering manifolds.* To appear in Math. Zeit.

[43] E. Dynkin and M. Maljutov *Random walks on roups with a finite number of generators.* (Russian) Dokl. Akad. Nauk SSSR **137** (1961), 1042–1045.

[44] V. A. Efremovic *The proximity geometry of Riemannian manifolds.* Uspekhi Math. Nauk. **8** (1953), 189.

[45] A. Erschler (Dyubina) *On isoperimetric profiles of finitely generated groups.* Geometriae Dedicata **100** (2003), 157–171.

[46] A. Erschler (Dyubina) *Liouville property for groups and manifolds.* Invent. Math. **155** (2004), 55–80.

[47] A. Erschler (Dyubina) *Isoperimetry for wreath products of Markov chains and multiplicity of the selfintersections of randowm walks.* Preprint, 2004.

[48] E. Følner *On groups with full Banach mean value.* Math. Scand. **3** (1955), 243–254.

[49] H. Furstenberg *Random walks and discrete subgroups of Lie groups.* 1971 Advances in Probability and Related Topics, Vol. 1, 1–63 Dekker, New York.

[50] H. Furstenberg *Boundary theory and stochastic processes on homogeneous spaces.* In "Harmonic analysis on homogeneous spaces" (Proc. Sympos. Pure Math., Vol. XXVI, Williams Coll., Williamstown, Mass., 1972), 193–229. Amer. Math. Soc., Providence, R.I., 1973

[51] R. Grigorchuk *Symmetrical random walks on discrete groups.* In "Multicomponent random systems" (Dobrushin et al, eds), Advances in Probability and Related Topics **6** (1980), 285–325, Dekker.

[52] R. Grigorchuk *On growth in group theory.* In "Proceeding of the International Congress of Mathematicians" (Kyoto, 1990), Math. Soc. Japan, 1991, 325–338.

[53] R. Grigorchuk and P. de la Harpe *On problems related to growth, entropy, and spectrum in group theory.* J. Dynam. Control Systems **3** (1997), 51–89.

[54] A. Grigor'yan *The heat equation on non-compact Riemannian manifolds.* Matem Sbornik **182** (1991), 55-87. Engl. Transl. Math. USSR Sb. **72** (1992), 47-77.

[55] A. Grigor'yan *Heat kernel upper bounds on a complete non-compact manifold.* Rev. Mat. Iberoamericana **10** (1994), 395–452.

[56] A. Grigor'yan *Analytic and geometric background of recurrence and non-explosion of the Brownian motion on Riemannian manifolds.* Bull. Amer. Math. Soc. **36** (1999), 135–249.

[57] A. Grigor'yan and L. Saloff-Coste *Heat kernel on connected sums of Riemannian manifolds.* Math. Res. Letters **6** (1999), 307–321.

[58] M. Gromov *Groups of polynomial growth and expanding maps.* Publ. Math. I.H.E.S. **53** (1981), 53–73.

[59] M. Gromov *Asymptotic invariants of infinite groups in Geometric Group Theory.* LMS Lecture notes series 182(II), Cambridge University Press, 1993.

[60] M. Gromov *Metric structures for Riemannian and non-Riemannian spaces.* Birkhäuser, 1999.

[61] Y. Guivarc'h *Croissance polynômiale et périodes des fonctions harmoniques.* Bull. Soc. Math. France **101** (1973), 333–379.

[62] Y. Guivarc'h *Application d'un théorème limite local à la transience et à la récurrence de marches de Markov.* In "Théorie du Potentiel" Lecture Notes in Math. **1096** (1984), 301-332, Springer.

[63] P. de la Harpe *Topics on geometric group theory.* Chicago Lectures in Mathematics. University of Chicago Press, 2000.

[64] W. Hebisch and L. Saloff-Coste *Gaussian estimates for Markov chains and random walks on groups.* Ann. Prob. **21** (1993), 673–709.

[65] J. Heinonen and P. Koskela *Quasiconformal maps in metric spaces with controlled geometry.* Acta Math. **181** (1998), 1–61.

[66] I. Holopainen and P. Soardi *p-harmonic functions on graphs and manifolds.* Manuscripta Math. **94** (1997), 95–110.

[67] V. Kaimanovich *Brownian motion and harmonic functions on covering manifolds. An entropic approach.* (Russian) Dokl. Akad. Nauk SSSR **288** (1986), 1045–1049.

[68] V. Kaimanovich *Boundaries of random walks on polycyclic groups and the law of large numbers for solvable Lie groups.* Vestnik Leningrad University: Math. **20** (1987), 49–52.

[69] V. Kaimanovich *Discretization of bounded harmonic functions on Riemannian manifolds and entropy.* In "Proceedings of the International Conference on Potential Theory, Nagoya" (M. Kishi, ed.), de Gruyter (1992), 213–223.

[70] V. Kaimanovich *Dirichlet norms, capacities and generalized isoperimetric inequalities for Markov operators.* Pot. Anal. **1** (1992), 61-82.

[71] V. Kaimanovich *Poisson boundaries of random walks on discrete solvable groups.* In "Probability measures on groups X" (Oberwolfach 1990, H. Heyer, ed) (1991), 205–238, Plenum.

[72] V. Kaimanovich and A. Vershik *Random walks on discrete groups: boundary and entropy* Ann. Probab. **11** (1983), 457–490.

[73] M. Kanai *Rough isometries and the parabolicity of Riemannian manifolds.* J. Math. Soc. Japan **38** (1986), 227–238.

[74] M. Kanai *Rough isometries and the combinatorial approximations of geometries of noncompact Riemannian manifolds.* J. Math. Soc. Japan **37** (1985), 391–413.

[75] M. Krasnoselsky and Y. Rutitsky *Convex functions and Orlicz spaces.* Noordhof, 1961.

[76] H. Kesten *Symmetric random walks on groups.* Trans. Amer. Math. Soc. **92** (1959), 336–354.

[77] H. Kesten *Full Banach mean values on countable groups.* Math. Scand. **7** (1959), 146–156.

[78] G. Lawler *Intersections of random walks* Bikhäuser, 1991.

[79] P. Li *Curvature and function theory on Riemannian manifolds.* In "Surveys in Differential Geometry, VII" International Press (2000), 375–432.

[80] P. Li and S-T. Yau *On the parabolic kernel of the Schrödinger operator* Acta Math. **156** (1986), 153–201.

[81] J. Lott *Heat kernels on covering spaces and topological invariants* J. Diff. Geom. **35** (1992), 471–510.

[82] J. Lott *Remarks about heat diffusion on periodic spaces.* Proc. Amer. Math. Soc. **127** (1997), 1243–1249.

[83] T. Lyons *A simple criterion for transience of a reversible Markov chain.* Ann. Probab. **11** (1983), 393–402.

[84] T. Lyons *Instability of the Liouville property for quasi-isometric Riemannian manifolds and reversible Markov chains.* J. Diff. Geom. **26** (1981), 33-66.

[85] T. Lyons *Random thoughts on reversible potential theory.* In "Summer School in Potential Theory, Joensuu 1990" (I. Laine, ed), University of Joensuu (1992), 71–114.

[86] T. Lyons and D. Sullivan *Function theory, random paths and covering spaces.* J. Diff. Geom. **19** (1984), 299–323.

[87] S. Maillot *Large-scale conformal rigidity in dimension three.* Preprint, 2002.

[88] G. Margulis *Positive harmonic functions on nilpotent groups.* (Russian) Dokl Akad. Nauk SSSR **166** (1966), 1054–1057. Engl. Transl. Soviet Math. Dokl. **7** (1966), 241–244.

[89] V. Maz'ja *Sobolev Spaces.* Springer, 1985.

[90] J. Milnor *Growth in finitely generated solvable groups.* J. Diff. Geom. **2** (1968), 447–449.

[91] B. Morris and Y. Peres *Evolving sets, mixing and heat kernel bounds.* Probab. Theory Related Fields, to appear.

[92] G.D. Mostov *Fundamental groups of homogeneous spaces.* Ann. of Math. **60** (1954), 1–27.

[93] Nash J. *Continuity of solutions of parabolic and elliptic equations.* American Math. J. **80** (1958), 931–954.

[94] P. Pansu *Profil isopérimétrique, métriques périodiques et formes d'équilibre des cristaux.* ESAIM Control Optim. Calc. Var. **4** (1999), 631–665

[95] A. L. Paterson *Amenability.* Mathematical Surveys and Monographs, Vol. 29. American Mathematical Society, 1988.

[96] Ch. Pittet *Følner sequences on polycyclic groups.* Rev. Math. Iberoamericana **11** (1995), 675–685.

[97] Ch. Pittet *The isoperimetric profile of homogeneous Riemannian manifolds.* J. Diff. Geom. **54** (2000), 255–302.

[98] Ch. Pittet and L. Saloff-Coste *Isoperimetry and random walk on discrete subgroups of connected Lie groups.* In Random Walk and Discrete Potential Theory, Cortona, (M. Picardello and W. Woess, eds.) Cambridge University Press, 1999, pp 306–319.

[99] Ch. Pittet and L. Saloff-Coste *Amenable groups, isoperimetric profile and random walks.* In "Geometric Group Theory Down Under" (J. Cossey, Ch. F. Miller III, W. Neumann, and M. Shapiro, eds.), Walter de Gruyter, 1999, pp. 293–316.

[100] Ch. Pittet and L. Saloff-Coste *On the stability of the behavior of random walks on groups.* J.Geom. Anal. **10** (2001), 701–726.

[101] Ch. Pittet and L. Saloff-Coste *On random walks on wreath products.* Ann. Probab. **30** (2002), 948–977.

[102] Ch. Pittet and L. Saloff-Coste *Random walks on abelian-by-cyclic groups.* Proc. Amer. Math. Soc. **131** (2002), 1071-1079.

[103] Ch. Pittet and L. Saloff-Coste *Random walks on finite rank solvable groups.* J. Europ. Math. Soc. **5** (2003), 313–342.

[104] Ch. Pittet and L. Saloff-Coste *A survey on the relationships between volume growth, isoperimetry, and the behavior of simple random walks on Cayley graphs, with examples.* Unpublished manuscript.

[105] M.S. Raghunathan *Discrete subgroups of Lie groups.* Ergebnisse der Mathematik und ihrer Grenzgebiete, Band 68. Springer, 1972.

[106] [43] D. W. Robinson *Elliptic operators and Lie groups.* Oxford Mathematical Monographs. Oxford University Press, New York, 1991.

[107] D.J.S. Robinson *A Course in the Theory of Groups.* Graduate texts in Mathematics, Springer, 1993.

[108] H. Royden *On the ideal boundary of a Riemann surface.* In "Contributions to the theory of Riemann surfaces" Annals of Mathematics Studies, no. 30 (1953), 107–109, Princeton University Press.

[109] L. Saloff-Coste *A note on Poincaré, Sobolev and Harnack inequalities.* Duke Math. J. **65** (1992), IMRN 27-38.

[110] L. Saloff-Coste *On global Sobolev inequalities.* Forum Math. **6** (1994), 271–286.

[111] L. Saloff-Coste *Aspects of Sobolev Inequalities.* Cambridge university Press, 2001.

[112] L. Saloff-Coste and W. Woess *Transition operators on co-compact G-spaces.* Preprint (2003).

[113] A. S. Schwarzc *A volume invariant of coverings* (In Russian) Dokl. Ak. Nauk. **105** (1955), 32-34.

[114] F. Spitzer *Principles of Random Walk.* Van Nostrand, (1972).

[115] P. Soardi *Potential theory on infinite networks.* Lect. Notes in Math. 1590. Springer, 1994.

[116] W. Thurston *Three-Dimensional Geometry and Topology.* Princeton University Press, 1997.

[117] J. Tits *Free subgroups in linear groups.* J. Algebra, **20** (1972), 250–270.

[118] L. Van den Dries and A. Wilkie *Gromov's theorem on groups of polynomial growth and elementary logic.* J. Alg. **89** (1984), 349–374.

[119] N. Varopoulos *Brownian motion and transient groups.* Ann. Inst. Fourier **33** (1983), 241–261.

[120] N. Varopoulos *Brownian motion and random walks on manifolds.* Ann. Inst. Fourier **34** (1984), 243–269.

[121] N. Varopoulos *Isoperimetric inequalities and Markov chains.* J. Funct. Anal. **63** (1985), 215–239.

[122] N. Varopoulos *Théorie du potentiel sur les groupes nilpotents.* C. R. Acad. Sci. Paris Sr. I Math. **301** (1985), 143–144.

[123] N. Varopoulos *Théorie du potentiel sur des groupes et des variétés.* C. R. Acad. Sci. Paris Sr. I Math. **302** (1986), 203–205.

[124] N. Varopoulos, *Analysis on nilpotent groups.* J. Funct. Anal. **66** (1986), 406–431.

[125] N. Varopoulos *Random walks and Brownian motion on manifolds.* In "Harmonic Analysis, Symmetric Spaces and Probability Theory, Cortona 1984" Sympos. Math. XXIX (1987), 97-109, Academic Press.

[126] N. Varopoulos *Groups of superpolynomial growth.* In "Harmonic analysis (Sendai, 1990)", 194–200, ICM-90 Satell. Conf. Proc., Springer, Tokyo, 1991.

[127] N. Varopoulos *Analysis and geometry on groups.* In "Proceeding of the International Congress of Mathematicians" (Kyoto, 1990), Math. Soc. Japan, 1991, 951–957..

[128] N. Varopoulos, L. Saloff-Coste and T. Coulhon *Analysis and geometry on groups.* Cambridge University Press, (1993).

[129] A. M. Vershik *Dynamic theory of growth in groups: entropy, boundaries, examples.* (Russian. Russian summary) Uspekhi Mat. Nauk 55 (2000), 59–128; translation in Russian Math. Surveys 55 (2000), 667–733.

[130] A. M. Vershik *Geometry and dynamics on the free solvable groups.* arXiv:math.GR/0006177, June 2000.

[131] W. Woess *Random walks on infinite graphs and groups — A survey on selected topics.* Bull. London Math. Soc. **26** (1994), 1–60.

[132] W. Woess *Random walks on infinite graphs and groups.* Cambridge University Press, 2000.

[133] J. Wolf *Growth of finitely generated solvable groups and curvature of Riemannian manifolds.* J. Diff. Geom. **2** (1968), 424–446.

[134] S-T. Yau *Harmonic functions on complete Riemannian manifolds* Comm. Pure Appl. Math. **28** (1975), 201–228.

Department of Mathematics, Cornell University, Ithaca NY 14853
E-mail address: lsc@math.cornell.edu

Surveys in Differential Geometry
IX, International Press

Functoriality and Small Eigenvalues of Laplacian on Riemann Surfaces

Freydoon Shahidi

ABSTRACT. The purpose of this article is to survey the recent progress made on estimating positive eigenvalues of Laplacian on hyperbolic Riemann surfaces in the case of congruence subgroups in connection with the Selberg conjecture, as well as certain related ones. The results are obtained as consequences of establishing certain important cases of Langlands' functoriality conjecture.

CONTENTS

1. Introduction

Hyperbolic Riemann surfaces are one dimensional complex connected manifolds whose universal coverings are the upper half plane H (conformally equivalent to the open disc Δ of radius 1). They comprise most Riemann surfaces and are uniformized as $M = \Gamma\backslash H$, where Γ is a freely acting Fuchsian group, i.e., a discrete subgroup of $PSL_2(\mathbb{R})$, the group of conformal automorphisms of H, acting without fixed points on H through fractional linear transformations

$$z \longrightarrow \gamma \cdot z = (az+b)/(cz+d) \quad (\gamma = \begin{pmatrix} a & b \\ c & d \end{pmatrix} \in \Gamma). \tag{1.1}$$

Then $\Gamma = \pi_1(M)$, the fundamental group of M.

2000 *Mathematics Subject Classification.* Primary 11F70, 30F35; Secondary 11R39, 11R42, 34B25.

Key words and phrases. Hyperbolic Riemann surfaces, Laplace operators, Selberg and Ramanujan conjectures, Langlands functoriality conjecture.

The author was partially supported by NSF grant DMS-0200325.

Throughout this paper we are interested only in those M which are parametrized by Fuchsian groups of the first kind. These are simply Fuchsian groups for which $\mathrm{Vol}(\Gamma\backslash H) < \infty$ and they are usually called *"finite volume type groups."* Here the volume is calculated with respect to the hyperbolic measure $dxdy/y^2$ on H.

Let us recall that a *congruence subgroup* is a subgroup of $SL_2(\mathbb{Z})$ containing a *principal congruence subgroup* $\Gamma(N)$, the subgroup of all $\gamma \in SL_2(\mathbb{Z})$ satisfying $\gamma \equiv I (\mathrm{mod}\ N)$ for some positive integer N. In view of the fact that the *congruence subgroup problem* (cf. [**60**]) is not valid for $SL_2(\mathbb{R})$, not every arithmetic subgroup (i.e., finite index) of $SL_2(\mathbb{Z})$ is a congruence subgroup. This was first observed in 1887 by Fricke[**16**] and Pick [**52**]. We refer to [**44**], page 251, for a discussion of this and examples of non–congruence arithmetic subgroups of $SL_2(\mathbb{Z})$.

The Laplace operator on $M = \Gamma\backslash H$ is simply

$$\Delta = -y^2\left(\frac{\partial^2}{\partial x^2} + \frac{\partial^2}{\partial y^2}\right). \tag{1.2}$$

This is a symmetric and non–negative operator which has a self–adjoint extension to all of $L^2(\Gamma\backslash H)$. Spectral decomposition of Δ on $L^2(\Gamma\backslash H)$ decomposes $L^2(\Gamma\backslash H)$ to the direct sum of its discrete and continuous spectrum.

It is traditional to write $\lambda = s(1-s)$, $s \in \mathbb{C}$, to denote an eigenvalue for Δ. We note that the continuous part $s = \frac{1}{2} + it$, $t \in \mathbb{R}$, giving $\lambda = \frac{1}{4} + t^2$. For the discrete spectrum, we denote the distinct eigenvalues as $0 = \lambda_0 < \lambda_1 < \lambda_2 < \ldots$. If $0 < \lambda_i < 1/4$, we call λ_i exceptional; meaning in particular, that they are of a different nature than those greater than or equal to $1/4$ (cf. Selberg's conjecture below). They are finite in number.

One of the central problems in the theory of Riemann surfaces is how small λ_1 can get. When Γ is cocompact, there are many examples in which $\lambda_1 < \frac{1}{4}$. In fact, if $\Gamma\backslash H$ is of "signature $(g,0,0)$", then Schoen, Wolpert and Yau [**58**] have shown that λ_{2g-3} can be made as small as one wishes, allowing exceptional eigenvalues below $1/4$, while Buser (cf. [**3**]) has proved

$$\lambda_{4g-2} \geq 1/4,$$

i.e., λ_{4g-2} is never exceptional, but λ_{2g-3} can be exceptional for arbitrary $g \geq 2$. Remarkably enough λ_1 has a universal upper bound due to Yang and Yau [**72**]:

$$\lambda_1 \leq 2\,\frac{g+1}{g-1} \leq 6.$$

As for non–compact Riemann surfaces, one can find examples where $\lambda_1(\Gamma\backslash H) < 1/4$ ([**59, 73**]). But they are not congruence subgroups. In fact, for a congruence subgroup Selberg made the following remarkable conjecture.

CONJECTURE 1.1 (SELBERG [**59**], 1965). *There are no exceptional eigenvalues for congruence subgroups, i.e., one has*

$$\lambda_1 \geq 1/4.$$

In fact, in [**59**], Selberg proved:

THEOREM 1.2. *For any congruence subgroup* Γ

$$\lambda_1 = \lambda_1(\Gamma\backslash H) \geq 3/16.$$

With all the examples discussed earlier, it is clear that the conjecture is very much of arithmetic nature and quite deep. In fact, most of the progress in improving these lower bounds have come from the theory of automorphic forms. The purpose of this article is to report on recent progress on this important conjecture and other related ones such as that of Ramanujan. We refer to [**1, 4, 5, 6, 20, 25, 27, 55, 56, 69, 70**] for some recent excellent expository articles on different aspects of these conjectures and more.

I would like to thank the editors of "Surveys in Differential Geometry" and in particular Professor S.T. Yau for their invitation that I prepare such a report for inclusion in the volume. I would also like to thank the referees for their comments towards the improvement of this exposition.

2. Ramanujan Conjecture

To study the eigenvalues of Δ one needs to concentrate on eigenfunctions. These are now functions on $\Gamma\backslash H$ and those for eigenvalue $\lambda = s(1-s)$ can be written as

$$f(x+iy) = f_0(y) + \sum_{n\neq 0} a_n(|n|y)^{1/2} K_{s-\frac{1}{2}}(2\pi|n|y)e^{2\pi inx}, \tag{2.1}$$

where $K_\nu(z)$ is the Whittaker–Bessel function bounded at infinity, i.e., the solution to the differential equation

$$t^2K_\nu'' + tK_\nu' - (t^2+\nu^2)K_\nu = 0 \tag{2.2}$$

satisfying

$$K_\nu(t) \sim \sqrt{\frac{\pi}{2t}}\, e^{-t} \tag{2.3}$$

as t goes to $+\infty$.

The complex numbers a_n are the corresponding Fourier coefficients. Moreover, when Γ is a congruence subgroup the eigenfunctions for exceptional eigenvalues, if any, will all be cuspidal, i.e., $f_0(y) \equiv 0$.

When $f_0(y) = 0$, the functions defined by (2.1) are the so called Maass cusp forms and although they exist in plenty (Weyl's Law), no explicit examples are known, unless one resorts to congruence subgroups in which case there are explicit constructions of these forms using Galois–Weil representations [**47, 54**].

There is an analogue of the Selberg conjecture for the Fourier coefficients a_n when f is an eigenfunction for all the Hecke operators (cf. [**17, 47, 71**]). Let us further normalize f by assuming $a_1 = 1$. Then

Conjecture 2.1 (Ramanujan-Petersson). *For every Maass cusp form f which is an eigenfunction for all the Hecke operators with $a_1 = 1$, and every prime number p,*

$$|a_p| \leq 2p^{-1/2}.$$

Henceforth, following the traditional terminology, we shall call this the Ramanujan conjecture.

Both Ramanujan and Selberg conjectures are quite hard and are still out of reach. But remarkable progress has been recently made in finding better bounds [**32, 34, 35, 36, 37**] which we shall now begin to explain. Before taking on this task, let us explain a problem of interest to both number theorists and geometers whose solution requires only partial improvements of existing bounds towards these conjectures, and which has now been completely resolved as a consequence of this progress [**37**]. These new bounds are consequences of the recent progress in establishing new cases of Langlands' functoriality conjecture all of which were considered out of reach even a few years ago.

Remark 2.2. *There is an analogue of the Ramanujan–Petersson conjecture for normalized holomorphic cuspidal eigenforms of weight k, stating $|a_p| \leq 2p^{(k-1)/2}$. This was proved by Deligne in 1973, as a consequence of his proof of Weil's conjecture* [**14**]. *The holomorphic structure of these forms play a central role in his proof, something that Maass forms lack.*

3. The Hyperbolic Circle Problem

The hyperbolic distance function $\rho(z,w)$ on H is simply given by

$$\rho(z,w) = \log \frac{|z-\overline{w}| + |z-w|}{|z-\overline{w}| - |z-w|} \quad (z, w \in H) \tag{3.1}$$

from which one can deduce a function $u(z,w)$

$$u(z,w) = \frac{|z-w|^2}{4 Im z Im w} \tag{3.2}$$

through

$$\cosh \rho(z,w) = 1 + 2u(z,w) \tag{3.3}$$

which is easier to work with. Given a positive real number X, the hyperbolic circle problem demands an estimate or asymptotic for the number of lattice points inside a hyperbolic circle of radius X centered at a point $w \in H$ and generated by the Γ–orbit of another point $z \in H$ (Γ–lattice points). More precisely, one wants to estimate

$$P(X) = \# \{\gamma \in \Gamma | 4u(\gamma \cdot z, w) + 2 \leq X\}. \tag{3.4}$$

Since $\Gamma\backslash H$ has a negative constant curvature ($\kappa = -1$), the euclidean circle packing arguments cannot be used. In fact, as the area and the length are of the same order of magnitude, Gauss's circle problem estimate which appeals to an area calculation for the circle to estimate the lattice points inside it, fails. Instead one uses spectral theory of Δ on $L^2(\Gamma\backslash H)$ by cleverly choosing a kernel function for counting these points, and then estimating it by using spectral theory [**26**]. We refer to [**43**] for the first published version of these results. They have been known to Selberg in our setting, i.e., with comparable error estimates, but were never published (cf. [**13**]). We refer to [**51**] for an earlier result on this problem. The error estimates in [**51**] are not as good as those in [**13, 43**]. We should finally mention [**3**] for a detailed discussion of the problem and its history in the compact case.

To explain the result, let $\{u_j(z)\}$ be a complete set of orthonormal eigen–cusp forms for Δ attached to complex parameters $1/2 < s_j \leq 1$, i.e., $0 \leq \lambda_j < 1/4$, then one can show [**26**]:

Let Γ be a congruence subgroup. Then for $X \geq 2$, we have:

$$P(X) = \sum_{1/2 < s_j \leq 1} c\pi^{1/2} \frac{\Gamma(s_j - \frac{1}{2})}{\Gamma(s_j + 1)} u_j(z) \overline{u}_j(w) X^{s_j} + O(X^{2/3}), \tag{3.5}$$

where $c = 2$ or 1 according as -1 is in Γ or not.

We refer to [**26**] for a detailed discussion of this problem and many other issues in the theory of Maass forms.

Remark 3.1. *This is in fact true for any finite volume group Γ, if we include also the residual eigenfunctions of Δ, i.e., the non–cuspidal discrete eigenfunctions. They can be constructed as residues of Eisenstein series.*

One can then quickly show that if $F = \mathrm{Vol}(\Gamma\backslash H)$, then

$$P(X) = \sum_{\frac{2}{3}<s_j<1} c\pi^{1/2} \frac{\Gamma(s_j - \frac{1}{2})}{\Gamma(s_j+1)} u_j(z)\overline{u}_j(w)X^{s_j} + c\pi F^{-1}X + O(X^{2/3}) \qquad (c = 1 \text{ or } 2) \tag{3.6}$$

and conclude:

Proposition 3.2. *Suppose there are no exceptional eigenvalues in the range $\frac{2}{3} < s_j < 1$, i.e., $\lambda_1 \geq 2/9 = 0.222\ldots$ Then*

$$P(X) = \frac{c\pi}{F}\, X + O(X^{2/3}) \qquad (c = 1 \text{ or } 2). \tag{3.7}$$

It is amusing to see that when this is applied to $\Gamma = SL_2(\mathbb{Z})$ and $z = w = \sqrt{-1}$ for which $c = 2$ and $F = \pi/3$ one gets

$$\mathrm{Card}\{(a,b,c,d) \in \mathbb{Z}^4 | ad - bc = 1 \text{ and } a^2 + b^2 + c^2 + d^2 \leq X\} = 6X + O(X^{2/3}). \tag{3.8}$$

The fact that there are no exceptional eigenvalues for $SL_2(\mathbb{Z})$ is well known and old (cf. Section 11.3 of [**26**]).

Many other arithmetic approximations can be deduced from this for which we refer to Section 12 of [**26**]. For example, if $r(m)$ is the number of integral points inside a Euclidean circle of radius $\sqrt{m}$, $m \in \mathbb{N}$, then by applying the proposition to an appropriate conjugate of $\Gamma_0(2)$, one can show

$$\sum_{m \leq X} r(m)r(m+1) = 8X + O(X^{2/3}). \tag{3.9}$$

While for such large congruence subgroups Selberg's conjecture has been known for sometime (in fact for subgroups of level ≤ 7), the general case remains unavailable and it is for this reason and for the complete resolution of certain problems such as the hyperbolic circle problem (cf. [**26, 37**] for other examples), that partial improvements are quite desirable and highly appreciated by experts.

As for the Selberg conjecture the best result established so far is:

Theorem 3.3. (Kim–Sarnak [**34**]).

$$\lambda_1 \geq \frac{1}{4} - \left(\frac{7}{64}\right)^2 = \frac{975}{4096} \simeq 0.2380371. \tag{3.10}$$

While this is more than enough to remove the assumption on λ_1 in Proposition 3.2, i.e., $\lambda_1 \geq 0.222\ldots$, the unconditional resolution of the hyperbolic circle problem was first proved in [**37**] as:

Proposition 3.4. (Kim–Shahidi [**37**]). *Suppose $X \geq 2$. Then*

$$P(X) = \frac{c\pi}{F} X + O(X^{2/3}), \tag{3.11}$$

where $c = 2$ or 1 according as $-1 \in \Gamma$ or not.

The proposition is a consequence of the following estimate proved in [**37**].

THEOREM 3.5 [**37**]. $\lambda_1 \geq \frac{1}{4} - (\frac{5}{34})^2 = \frac{66}{289} = 0.2283737\ldots$.

As for the Ramanujan conjecture the best estimate at present is [**34**]:

$$p^{-\frac{1}{2}}(p^{7/64} - p^{-7/64}) \leq |a_p| \leq p^{-1/2}(p^{7/64} + p^{-7/64}) \tag{3.12}$$

4. Functoriality and Maass Forms

The recent striking improvements towards the Ramanujan and the Selberg conjectures are consequences of certain special cases of Langlands' functoriality conjecture [**1, 42**] proved recently [**32, 35, 37**].

The functoriality conjecture is one of the central components of a vast program (the Langlands program) which deals with automorphic forms on general reductive groups and considering our limitation it is better to only discuss it in the present context. We refer to [**18**] for an elementary introduction to the Langlands program.

A Maass (cusp) form f is simply a real analytic eigenfunction of Δ in $L^2(\Gamma\backslash H)$ orthogonal to all the Eisenstein series and their residues. We will further assume that f is an eigenfunction for all the Hecke operators (cf. [**17, 26, 47, 71**]) and normalize it so that $a_1 = 1$.

To explain functoriality it is best to introduce the *adeles*. This is simply a ring defined as the restricted product of all the completions of the field $\mathbb{Q}$ of rational numbers, with respect to their ring of integers. In fact, if $\mathbb{Q}_p$ and $\mathbb{Z}_p$ are respectively the field of p–adic numbers and its ring of integers, i.e., those whose p–adic absolute values are less than or equal to 1, then

$$x = (x_p) \in \prod_{p \leq \infty} \mathbb{Q}_p$$

is an adele if and only if $x_p \in \mathbb{Z}_p$ for almost all finite primes p. Here by convention $\mathbb{Q}_\infty = \mathbb{R}$, the field of real numbers for which $\mathbb{Z}_\infty$ is not a ring! The ring of adeles of $\mathbb{Q}$ is denoted by $\mathbb{A}_\mathbb{Q}$. It is a locally compact ring under the direct limit topology. The group $\mathbb{A}_\mathbb{Q}^* = \mathbb{I}_\mathbb{Q}$ is called the group of *ideles*. It is a locally compact group if one takes the topology induced by

$$\mathbb{I}_\mathbb{Q} \simeq \{(x, x^{-1}) | x \in \mathbb{I}_\mathbb{Q}\} \subset \mathbb{A}_\mathbb{Q} \times \mathbb{A}_\mathbb{Q}.$$

This whole notion can be extended to any number field F, i.e., a finite field extension of $\mathbb{Q}$; and one defines its ring of adeles $\mathbb{A}_F$ and group of ideles $\mathbb{I}_F = \mathbb{A}_F^*$ in the same manner. One notes that F (resp. F^*) is a discrete subgroup of $\mathbb{A}_F$ (resp. $\mathbb{I}_F$).

It is quite standard [**17**] to attach to an eigen–cusp Maass form f an irreducible subrepresentation π of

$$L^2(GL_2(\mathbb{Q})\backslash GL_2(\mathbb{A}_\mathbb{Q}), \chi),$$

the space of square integrable functions on $\mathbb{A}_\mathbb{Q}^* GL_2(\mathbb{Q})\backslash GL_2(\mathbb{A}_\mathbb{Q})$ transforming under $\mathbb{A}_\mathbb{Q}^*$ according to a grössencharacter χ, i.e., a complex (unitary) character of $\mathbb{Q}^*\backslash\mathbb{A}_\mathbb{Q}^*$. There is a non–unique way of factorizing π to $\pi = \otimes_p' \pi_p$, where each π_p is an irreducible preunitary representation of $GL_2(\mathbb{Q}_p)(GL_2(\mathbb{R})$ if $p = \infty)$. But the class of each π_p is unique. Moreover, almost all of them are unramified or spherical, i.e., each have a vector fixed by $GL_2(\mathbb{Z}_p)$. One can then realize such π_p as the full space of all the locally constant complex functions

$$\varphi \colon GL_2(\mathbb{Q}_p) \to \mathbb{C}$$

satisfying

$$\varphi(\begin{pmatrix} a & x \\ 0 & b \end{pmatrix} g) = \mu_p(a)\nu_p(b)|a/b|_p^{1/2}\varphi(g), \tag{4.1}$$

$a, b \in \mathbb{Q}_p^*$, $x \in \mathbb{Q}_p$, where μ_p and ν_p are a pair of unramified characters of $\mathbb{Q}_p^*$.

The class of the representation π_p is then determined by the conjugacy class of the diagonal element

$$t_p = \begin{pmatrix} \mu_p(p) & 0 \\ 0 & \nu_p(p) \end{pmatrix} \tag{4.2}$$

in $GL_2(\mathbb{C})$. Let $\alpha_p = \mu_p(p)$ and $\beta_p = \nu_p(p)$. Then the Ramanujan conjecture demands

$$|\alpha_p| = |\beta_p| = 1. \tag{4.3}$$

For a Maass form, π_∞ is also fully induced from a pair of characters (μ_∞, ν_∞) of $\mathbb{R}^*$ with

$$\mu_\infty/\nu_\infty = |\ \ |^{s_\infty} \qquad (s_\infty \in \mathbb{C}), \tag{4.4}$$

and the corresponding eigenvalue λ, defined by $\Delta f = \lambda f$, is given by

$$\lambda = \frac{1 - s_\infty^2}{4}, \tag{4.5}$$

in which $s_\infty \in (-1, 1) \cup i\mathbb{R}$, $i = \sqrt{-1}$.

The Selberg conjecture is then equivalent to

$$s_\infty \in i\mathbb{R}. \tag{4.6}$$

REMARK 4.1. *Representations of these forms satisfying (4.3) and (4.6) are among a class of representations which are called "tempered", and the generalized Ramanujan conjecture* [**57**] *requires that for a cuspidal representation $\pi = \otimes'_p \pi_p$, every π_p be tempered (not necessarily only of the above forms). Therefore the Selberg conjecture is a special case of the generalized Ramanujan conjecture at $p = \infty$, while the Ramanujan is one at $p < \infty$.*

The partial estimates in [**34**] are that

$$p^{-7/64} \leq |\alpha_p|,\ |\beta_p| \leq p^{7/64} \tag{4.7}$$

and

$$s_\infty \in [-7/32, 7/32] \cup i\mathbb{R} \tag{4.8}$$

or

$$\lambda \geq \lambda_1 \geq \frac{1}{4} - (\frac{7}{64})^2 = \frac{975}{4096} \simeq 0.2380371. \tag{4.9}$$

Langlands' functoriality conjecture can be formulated for any pair of arbitrary connected reductive groups. But for our purposes we will restrict ourselves to the case of general linear groups.

Given $m \in \mathbb{N}$, a natural number, we define a homomorphism

$$\mathrm{Sym}^m \colon GL_2(\mathbb{C}) \longrightarrow GL_{m+1}(\mathbb{C}) \tag{4.10}$$

as follows. For a form $P(x, y)$ (homogeneous polynomial) of degree m in variables x and y, and a $g \in GL_2(\mathbb{C})$, we let $\mathrm{Sym}^m g \in GL_{m+1}(\mathbb{C})$ be the matrix which expresses the coefficients of $P((x, y)g)$ in terms of those of $P(x, y)$. Then Sym^m defines a homomorphism from $GL_2(\mathbb{C})$ into $GL_{m+1}(\mathbb{C})$.

For $\pi = \otimes'_p \pi_p$ a cuspidal representation of $GL_2(\mathbb{A}_{\mathbb{Q}})$, we let, as before,

$$t_p = \begin{pmatrix} \alpha_p & 0 \\ 0 & \beta_p \end{pmatrix} \in GL_2(\mathbb{C}) \tag{4.11}$$

parametrize π_p for almost all $p < \infty$. Observe that

$$\mathrm{Sym}^m t_p = \mathrm{diag}(\alpha_p^m,\ \alpha_p^{m-1}\beta_p, \ldots, \beta_p^m) \in GL_{m+1}(\mathbb{C}) \tag{4.12}$$

can be used to define a semisimple conjugacy class in $GL_{m+1}(\mathbb{C})$.

It is well–known that each $\mathrm{Sym}^m t_p$ determines an irreducible admissible representation of $GL_{m+1}(\mathbb{Q}_p)$, denoted by $\mathrm{Sym}^m \pi_p$, which has a vector fixed by $GL_{m+1}(\mathbb{Z}_p)$.

Langlands' functoriality conjecture in this case then demands the existence of representations $\mathrm{Sym}^m \pi_p$ at all other places p such that

$$\mathrm{Sym}^m \pi = \otimes'_{p \leq \infty} \mathrm{Sym}^m \pi_p \tag{4.13}$$

is an automorphic representation of $GL_{m+1}(\mathbb{A}_{\mathbb{Q}})$, i.e., it appears in

$$L^2(GL_{m+1}(\mathbb{Q}) \backslash GL_{m+1}(\mathbb{A}_{\mathbb{Q}}),\ \omega_\pi^{m(m+1)/2}), \tag{4.14}$$

where ω_π is the central character of π, i.e., $\pi|\mathbb{A}_{\mathbb{Q}}^*$.

The appearance of $\mathrm{Sym}^m \pi$ does not need to be as a discrete subspace. It could appear through an Eisenstein series as part of the continuous spectrum (by means of incomplete Eisenstein series [**26, 41, 48**]). But the fact that it appears is very deep as it is evident from the consequences of the validity of the conjecture even for $m = 3$ and 4 (c.f. [**32, 34, 36, 37, 53, 54, 67**]).

Theorem 4.2 ([**32, 37**]). *Let π be a cuspidal representation of $GL_2(\mathbb{A}_{\mathbb{Q}})$. Then $\mathrm{Sym}^m \pi$ is automorphic for not only $m = 1$ and 2 but also for $m = 3$ and 4. (The case $m = 1$ is trivial and the case $m = 2$ is due to Gelbart–Jacquet* [**19**].*)*

Remark 4.3. *The theorem is valid for cuspidal representations of $GL_2(\mathbb{A}_F)$ for any number field F.*

There are many consequences of these results in automorphic forms and number theory. But let us only point out how (4.7), (4.8) and (4.9) are obtained from Theorem 4.2 here.

It follows from a result of Luo–Rudnick–Sarnak [**46**] that one can bound α_p^m and β_p^m by

$$p^{-(\frac{1}{2} - \frac{1}{(m+1)^2+1})} \leq |\alpha_p|^m,\ |\beta_p|^m \leq p^{\frac{1}{2} - \frac{1}{(m+1)^2+1}}, \tag{4.15}$$

or

$$p^{-\frac{1}{m}(\frac{1}{2} - \frac{1}{(m+1)^2+1})} \leq |\alpha_p|,\ |\beta_p| \leq p^{\frac{1}{m}(\frac{1}{2} - \frac{1}{(m+1)^2+1})}, \tag{4.16}$$

As for the Selberg conjecture, using [**46**] one gets

$$s_\infty \in [-\ell, \ell] \cup i\mathbb{R}, \tag{4.17}$$

where

$$\ell = \frac{2}{m}\Big(\frac{1}{2} - \frac{1}{(m+1)^2+1}\Big). \tag{4.18}$$

Remark 4.4. *The result in* [**46**] *is quite foundational and must be considered as a breakthrough. It provides us with the first non–local estimates for Hecke–eigenvalues of cuspidal representations of* $GL_m(\mathbb{A}_F)$. *Most of the improvements on the Selberg and Ramanujan Conjectures are consequences of combining functorial transfers of forms on* $GL_2(\mathbb{A}_F)$ *to appropriate* $GL_m(\mathbb{A}_F)$, *which we will discuss in the next section, with these estimates.*

Using case $m = 3$ of Theorem 4.2 we have

COROLLARY 4.5. $s_\infty \in [-5/17, 5/17] \cup i\mathbb{R}$ *or equivalently*

$$\lambda_1 \geq \frac{1}{4} - (\frac{5}{34})^2 = \frac{66}{289} \simeq 0.2283737 > 2/9. \tag{4.19}$$

Consequently, if Γ *is a congruence subgroup*

$$P(X) = \frac{c\pi}{F} X + O(X^{2/3}), \tag{4.20}$$

where $c = 2$ *or 1 according as* $-1 \in \Gamma$ *or not, i.e., the hyperbolic circle estimates are valid.*

A direct appeal to (4.15) and (4.16) for $m = 4$ will provide us with the estimates (cf. [**32**])

$$p^{-3/26} \leq |\alpha_p|,\ |\beta_p| \leq p^{3/26} \tag{4.21}$$

and

$$s_\infty \in [-3/13, 3/13] \cup i\mathbb{R} \tag{4.22}$$

or

$$\lambda_1 \geq \frac{1}{4} - (\frac{3}{26})^2 = \frac{40}{169} \simeq 0.2366839, \tag{4.23}$$

which although still quite striking, are even weaker than

$$p^{-1/9} < |\alpha_p|,\ |\beta_p| < p^{1/9} \tag{4.24}$$

and

$$s_\infty \in (-2/9, 2/9) \cup i\mathbb{R} \tag{4.25}$$

or equivalently

$$\lambda_1 > \frac{1}{4} - \frac{1}{81} = \frac{77}{324} \simeq 0.23765432 \tag{4.26}$$

obtained in[**36**] (cf. [**33**] for the archimedean estimate), a result which although slightly weaker than (3.10) and (3.12), is valid over every number field.

The estimates (3.10) and (3.12) proved in [**34**] require further appeal to the theory of automorphic L–functions (that of $L(s, \mathrm{Sym}^4\pi, \mathrm{Sym}^2)$, to be precise, which is now available using the same machinery [**32, 34, 63, 64, 66**] since $\mathrm{Sym}^4\pi$ is automorphic) and methods of analytic number theory [**2, 15**]. We refer to [**34**] for details.

Remark 4.6. *It is important to put the recent progress in perspective. The earlier estimates on both conjectures were in the range of exponent 1/5 (over arbitrary number fields) and the slightly better exponent* $\frac{5}{28} + \varepsilon$ *for all* $\varepsilon > 0$, *over* $\mathbb{Q}$. *They were obtained in* [**65**] *(cf.* [**68**] *for an exposition) and* [**2**], *respectively. For the Selberg conjecture* $\frac{1}{5} + \varepsilon$, *all* $\varepsilon > 0$, *was the best one proved in* [**45**]. *It gave the lower bound* $\lambda_1 \geq 0.21$. *They were all consequences of the automorphy of* $Sym^2\pi$

proved in [**19**] *and techniques and estimates such as those in* [**15, 46**] *(cf. Remark 4.4).*

For quite a long time experts struggled to obtain an estimate in the range of 1/6 as that would already simplify and improve a number of results in number theory and automorphic forms. It was therefore quite surprising when the estimate $\frac{5}{34} + \varepsilon$ *over any number field, which was only slightly weaker than 1/7, was announced by the author and Kim. The striking estimates (3.10), (3.12), (4.24) and (4.25) were obtained a few months after that* [**33, 34, 36**] *as soon as the automorphy of* $Sym^4\pi$ *was also ready at hand* [**32**]. *The techniques in establishing (3.10), (3.12), (4.24) and (4.25) using the automorphy of* $Sym^3\pi$ *and* $Sym^4\pi$ *are similar to those for* $5/28 + \varepsilon$ *and 1/5.*

5. The Method

It is now clear that these estimates are consequences of the existence of $\mathrm{Sym}^3\pi$ and $\mathrm{Sym}^4\pi$ as automorphic forms on $GL_4(\mathbb{A}_\mathbb{Q})$ and $GL_5(\mathbb{A}_\mathbb{Q})$, respectively. In this section we will try to briefly explain the machinery behind it.

The automorphy of $\mathrm{Sym}^3\pi$ and $\mathrm{Sym}^4\pi$ are consequences of applying converse theorems of Cogdell and Piatetski–Shapiro [**10, 11**] to analytic properties of certain L–functions proved by the Langlands–Shahidi method [**21, 31, 40, 41, 62, 63, 64, 65, 66**]. At present, they cannot be obtained from other methods which have been developed to prove functoriality.

To start off, let us note that in our approach, and more generally in the Langlands program, the choice of the number field is of no concern and one may assume π is a cuspidal representation of $GL_2(\mathbb{A}_F)$, where $\mathbb{A}_F$ is the ring of adeles of an arbitrary number field. Beside the fact that functoriality needs to be proved for groups over all global fields (as well as local ones), many applications, even those in algebraic number theory [**6, 12**], require estimates and results over arbitrary number fields. The reader should now appreciate that in this approach and context there will no longer be any need to study Hilbert modular forms as a different entity, at least when these questions arise, than those over $\mathbb{Q}$ or any other number field. We may and will therefore assume π is an infinite dimensional irreducible admissible subrepresentation of $L^2(GL_2(F)\backslash GL_2(\mathbb{A}_F), \chi)$ for some grössencharacter χ. It will automatically be cuspidal.

We can again write $\pi = \otimes'_v \pi_v$, where each π_v is an irreducible unitary representation of $GL_2(F_v)$, where F_v is the completion of F in a place v. There are obvious generalizations of all the notions visited in the previous section such as the fact that almost all π_v are spherical, i.e., have a vector fixed by $GL_2(O_v)$, and that they are given by a pair of unramified characters of F_v^*. (Here O_v is the ring of integers of F_v.) In particular, for almost all v, π_v, or more specifically its class, is given by the conjugacy class of a diagonal element $t_v = \mathrm{diag}(\alpha_v, \beta_v) \in GL_2(\mathbb{C})$.

As explained earlier, the automorphy of $\mathrm{Sym}^2\pi = \otimes'_v \mathrm{Sym}^2\pi_v$ was established more than 25 years ago [**19**]. When π_v is spherical, it is attached to the diagonal element $\mathrm{Sym}^2 t_v \in GL_3(\mathbb{C})$ whose entries define the unramified characters which determine the class of $\mathrm{Sym}^2\pi_v$. The diagonal element $t_v \otimes \mathrm{Sym}^2 t_v \in GL_6(\mathbb{C})$ will then determine a spherical representation $\pi_v \boxtimes \mathrm{Sym}^2\pi_v$ of $GL_6(F_v)$. The operation $\boxtimes$ mirrors that of tensor products when one parametrizes these representations by means of two and three dimensional representations of W'_{F_v}, the corresponding Deligne–Weil group. The parameterization problem for irreducible admissible representations of $GL_n(F_v)$ by means of n–dimensional complex representations of W'_{F_v} has now been completely resolved by Harris–Taylor [**23**] and Henniart [**24**]. This is a particularly deep and important result and of particular interest to us. In fact, it allows us to extend the operation to any pair of representations and not only spherical ones (cf. [**42**] for archimedean places). In particular, one can define $\pi_v \boxtimes \mathrm{Sym}^2\pi_v$ for all v. One of the special cases of the main result of [**37**] is

THEOREM 5.1. *a)* $\pi \boxtimes Sym^2\pi = \otimes'_v(\pi_v \boxtimes Sym^2\pi_v)$ *is an automorphic representation of* $GL_6(\mathbb{A}_F)$.

b) $\pi \boxtimes Sym^2\pi$ *appears in the continuous spectrum (cf.* [**30**]*) of* $GL_6(\mathbb{A}_F)$ *and is defined by an "Eisenstein series of type* $(2,4)$*" attached to* $(\pi \otimes \omega_\pi,\ Sym^3\pi)$*, where* ω_π *is the central character of* π*. In particular,* $Sym^3\pi$ *is an automorphic representation of* $GL_4(\mathbb{A}_F)$.

The main result of [**37**] (Theorem 5.1) proves the automorphy of $\pi_1 \boxtimes \pi_2 = \otimes'_v(\pi_{1v} \boxtimes \pi_{2v})$ for any pair of automorphic cuspidal representations π_1 and π_2 of $GL_2(\mathbb{A}_F)$ and $GL_3(\mathbb{A}_F)$, respectively.

This is a very fine result, incorporating both local and global functoriality, and consequently its proof is quite complicated. Using the Langlands–Shahidi method, one attaches a triple product L–function $L(s, \pi_1 \times \pi_2 \times (\tau \otimes \eta))$ to π_1, π_2 and τ cuspidal representations of $GL_2(\mathbb{A}_F)$, $GL_3(\mathbb{A}_F)$ and $GL_n(\mathbb{A}_F)$, $1 \le n \le 4$, respectively. Here σ is assumed to be unramified at every place where $v < \infty$ and either π_{1v}or π_{2v} is ramified. Moreover, η is a grössencharacter of $\mathbb{A}_F^*$, i.e., one of $F^*\backslash\mathbb{A}_F^*$, which we shall assume to be highly ramified at least at one place where one of π_{iv}'s, $i = 1, 2$, are ramified. Observe that now all the archimedean places at which the Selberg conjecture is rooted, are put outside the bad primes, enabling us to conclude deep results on them.

To utilize the method one notes that these L–functions appear in constant terms of certain Eisenstein series [**22, 40, 41, 48, 50, 65, 66**] defined on exceptional groups Spin(10), E_6 and E_7, granting the case $n = 1$ as a special case of Rankin–Selberg product L–functions on $GL_2(\mathbb{A}_F) \times GL_3(\mathbb{A}_F)$. The Langlands–Shahidi method then proves the necessary analytic properties of these L–functions. The twist by the highly ramified grössencharacter destroys all the possible symmetries that could lead to the existence of poles, proving they are entire (cf. [**31**]). One can also conclude that these entire functions are bounded in vertical strips of finite width [**21**]. Moreover, the technology developed in [**63, 64, 65, 66**] allows us to define root numbers $\varepsilon(s, \pi_1 \times \pi_2 \times (\sigma \otimes \eta))$ by means of which one establishes the functional equation

$$L(s, \pi_1 \times \pi_2 \times (\sigma \otimes \eta)) = \varepsilon(s, \pi_1 \times \pi_2 \times (\sigma \otimes \eta))L(1 - s, \tilde{\pi}_1 \times \tilde{\pi}_2 \times (\tilde{\sigma} \otimes \eta^{-1})), \tag{5.1}$$

where $\sim$ signifies appropriately defined duals.

Finally, we have

$$L_S(s, \pi_1 \times \pi_2 \times (\sigma \otimes \eta)) = L_S(s, (\pi_1 \boxtimes \pi_2) \times (\sigma \otimes \eta)), \tag{5.2}$$

where the L–function on the right is that of Rankin–Selberg studied in [**28, 29, 30, 49, 62, 63, 64**] defined by an infinite product of local factors in which for all $v \notin S$, S a finite set of places of F, either π_v is unramified or $v = \infty$.

It is to the L–functions on the right hand side of (5.2) that one can apply converse theorems of Cogdell and Piatetski–Shapiro [**10, 11**] which states that *if they are entire, bounded in vertical strips of finite width and satisfying (5.1),* *then there exists an automorphic representation* $\Pi = \otimes'_v \Pi_v$ *of* $GL_6(\mathbb{A}_F)$ *for which* $\Pi_v \simeq \pi_{1v} \boxtimes \pi_{2v}$ *whenever both* π_{iv} *are unramified or* v *is archimedean.*

While we know nothing about the analytic properties of $L_S(s, (\pi_1 \boxtimes \pi_2) \times (\sigma \otimes \eta))$ (as they are defined only by means of an infinite product of local L–functions), their equality with $L_S(s, \pi_1 \times \pi_2 \times (\sigma \otimes \eta))$ provides us with the knowledge needed for applying the converse theorem as explained earlier.

Quite a bit more technical work is needed to show that in fact $\Pi_v \simeq \pi_{1v} \boxtimes \pi_{2v}$ for all v and therefore $\pi_1 \boxtimes \pi_2$ is automorphic. We refer to [**37**] for details and appropriate references to both local and global difficulties.

The converse theorem [**11**] of Cogdell and Piatetski–Shapiro that we have used is non–standard and fairly recent (1999). It is designed precisely to handle L–functions for which holomorphy is obtained only up to a twist, something that one can provide from our method as it was observed by Kim in [**31**].

The automorphy of $\mathrm{Sym}^4\pi$ is proved again inductively. One applies the same machinery to certain Eisenstein series on groups of type $\mathrm{Spin}(2n)$, $4 \leq n \leq 7$, to transfer $\mathrm{Sym}^3\pi$ to an automorphic form [**30**] on $GL_6(\mathbb{A}_F)$ of type $(1, 5)$ attached to $(\omega_\pi^3,\ \mathrm{Sym}^4\pi \otimes \omega_\pi)$ from which automorphy of $\mathrm{Sym}^4\pi$ follows. We refer the reader to [**32**] for details.

Having established the automorphy of $\mathrm{Sym}^3\pi$ and $\mathrm{Sym}^4\pi$, one can then apply techniques of [**65**] (Section 5 and in particular Lemma 5.8 of [**65**]) to prove:

THEOREM 5.2. [**36**] *Let* $\pi = \otimes'_v \pi_v$ *be a cuspidal representation of* $GL_2(\mathbb{A}_F)$. *For each unramified* π_v *let* $t_v = \mathrm{diag}(\alpha_v, \beta_v) \in GL_2(\mathbb{C})$ *represent the corresponding semisimple conjugacy class. Then*

$$q_v^{-1/9} < |\alpha_v|,\ |\beta_v| < q_v^{1/9}. \tag{5.3}$$

The archimedean analogue of this result is also valid. It can be formulated as in (4.25) for the corresponding s_∞ defined by (4.4) at each archimedean place of F. We observe that unless $F = \mathbb{Q}$, this is the best bound available for $|\alpha_v|$ and $|\beta_v|$ at present (cf. (3.10) and (3.12) for $F = \mathbb{Q}$). We refer to [**7, 12**] for an application of this result to number theory.

There are many other applications of the existence of $\mathrm{Sym}^3\pi$ and $\mathrm{Sym}^4\pi$. But covering them all will become a bulky task and outside the scope and purpose of this paper. We refer the reader to [**27, 32, 36, 37, 53, 54**] for some of these applications. As has been the case with $\mathrm{Sym}^2\pi$ for the past 25 years we expect many consequences of the automorphy of $\mathrm{Sym}^3\pi$ and $\mathrm{Sym}^4\pi$ in several important subjects, ranging from number theory to automorphic forms and arithmetic geometry. At present the automorphy of $\mathrm{Sym}^5\pi$ remains out of reach!

REMARK 5.3. *As explained in Remark 4.1 the Ramanujan and Selberg conjectures are equivalent to each* π_v *being a tempered representation, i.e., one appearing in* $L^2(F_v^* \backslash GL_2(F_v))$, *either discretely or continuously, the latter case implying that* μ_v *and* ν_v *are unitary for every* v, *finite or infinite.*

A generalization of these conjectures to the effect that all the local components of a cuspidal representation of a general reductive group are tempered is one of the central problems in the theory of automorphic forms. Although this is false in general, one expects its validity for general linear groups.

There are instances of functoriality through which the study of automorphic forms on (split) classical groups $SO(2n)$, $SO(2n+1)$ *and* $Sp(2n)$ *is reduced to that of* $GL(N)$ *with* $N = 2n$, *except in the symplectic case* $Sp(2n)$ *for which* $N =$

$2n+1$. *For the generic spectrum of these classical groups, i.e., those with a non-zero Fourier coefficient of highest rank, the functoriality has now been established in* [**8, 9**] *and consequently the Ramanujan conjecture for these groups is reduced to that for* $GL(m)$, *for all* $m \leq N$. *We refer to* [**8, 9**] *where the same techniques as those used to establish existence of symmetric powers are used to prove these new results.*

When our method is fully developed (converse theorems already are developed for any global field) and these transfers are established for the generic spectrum of cuspidal representstions of classical groups over function fields, this should lead to a proof of the Ramanujan conjecture for generic cuspidal representations of these groups. This can be done using the validity of the conjecture for $GL(N)$ *which has been proved by Lafforgue* [**38**] *for general linear groups over function fields. The case of number fields even for* $GL(2)$ *is still out of reach.*

References

[1] J. Arthur, *The principle of functoriality*, Bull. Amer. Math. Soc. (N.S.) **40**(2002), no. 1, 39–53 (electronic), Mathematical challenges of the 21st century (Los Angeles, CA, 2000).

[2] D. Bump, W. Duke, J. Hoffstein, and H. Iwaniec, *An estimate for the Hecke eigenvalues of Maass forms*, IMRN **4** (1992), 75–81.

[3] P. Buser, *Geometry and Spectra of Compact Riemann Surfaces*, Progress in Mathematics, Vol. 106, Birkhäuser, Boston–Basel–Berlin, 1992.

[4] L. Clozel, *Spectral theory of automorphic forms*, IAS/Park City Lecture Notes, Park City, Utah, 2002.

[5] J.W. Cogdell, *L–functions and Converse Theorems for* $GL(n)$, IAS/Park City Lecture Notes, Park City, Utah, 2002.

[6] J.W. Cogdell, *Converse theorems, functoriality, and applications to number theory*, Proceedings of the International Congress of Mathematicians, Vol. II (Beijing, 2002), Higher Ed. Press, Beijing, 2002, pp. 119–128.

[7] J.W. Cogdell,*On sums of three squares*, Journal de Théorie des Nombres de Bordeaux **15** (2003), 33–44.

[8] J.W. Cogdell, H. Kim, I.I. Piatetski–Shapiro and F. Shahidi, *On lifting from classical groups to* GL_N, Publ. Math. Inst. Hautes Études Sci. **93** (2001), 5–30.

[9] J.W. Cogdell, H. Kim, I.I. Piatetski–Shapiro and F. Shahidi, *Functoriality for the classical groups*, Publ. Math. Inst. Hautes Études Sci., to appear.

[10] J.W. Cogdell and I.I. Piatetski–Shapiro, Converse theorems for GL_n, Publ. Math. Inst. Hautes Études Sci. **79** (1994), 157–214.

[11] J.W. Cogdell and I.I. Piatetski–Shapiro, *Converse theorems for* GL_n *II*, J. Reine Angew. Math. **507** (1999), 165–188.

[12] J.W. Cogdell, I.I. Piatetski–Shapiro and P. Sarnak, *Estimates on the critical line for Hilbert modular L–functions and applications*, in preparation.

[13] P. Cohn and P. Sarnak, *Discontinuous Groups and Harmonic Analysis*, unpublished lecture notes.

[14] P. Deligne, *La conjecture de Weil. I*, Inst. Publ. Math. Hautes Études Sci. **43** (1974), 273–307.

[15] W. Duke and H. Iwaniec, *Estimates for coefficients of L–functions, I*, Automorphic forms and analytic number theory (Montreal, PQ, 1989), Univ. Montréal, Montreal, QC, 1990, pp. 43–47.

[16] R. Fricke, *Die Elliptische Funktionen und ihre Anwendungen*, Vol. 1, 2nd ed., Teubner, Leipzig, 1930; First reprint by Johnson Reprint Corporation, New York, 1972.

[17] S. Gelbart, *Automorphic forms on adéle groups*, Princeton University Press, Princeton, NJ, 1975, Annals of Mathematics Studies, no. 83.

[18] S. Gelbart, *An elementary introduction to Langlands program*, Bulletin of AMS **10** (1984), no. 2, 177–219.

[19] S. Gelbart and H. Jacquet, *A relation between automorphic representations of* $GL(2)$ *and* $GL(3)$, Ann. Sci. École Norm. Sup. **11** (1978), no. 4, 471–542.

[20] S. Gelbart and S. Miller, *Riemann's zeta function and beyond*, Bulletin of AMS **41** (2003), no. 1, 59–112.

[21] S. Gelbart and F. Shahidi, *Boundedness of automorphic L–functions in vertical strips*, J. Amer. Math. Soc. **14** (2001), no. 1, 79–107.

[22] Harish-Chandra, *Automorphic forms on semisimple Lie groups*, Notes by J.G.M. Mars. Lecture Notes in Mathematics, no. 62, Springer–Verlag, Berlin, 1968.

[23] M. Harris and R. Taylor, *The geometry and cohomology of some simple Shimura varieties*, Annals of Mathematics Studies, vol. 151, Princeton University Press, Princeton, NJ, 2001, with an appendix by Vladimir G. Berkovich.

[24] G. Henniart, *Une preuve simple des conjectures de Langlands pour $GL(n)$ sur un corps p–adique*, Invent. Math. **139** (2000), 439–455.

[25] G. Henniart, *Progrés récents en fonctorialité de Langlands*, Seminaire Bourbaki, Vol. 2000/2001, Astérisque No. 282 (2002), Exp. 890, 301–322.

[26] H. Iwaniec, *Spectral methods of automorphic forms*, 2nd ed., Graduate Studies in Mathematics, vol. 53, American Mathematical Society, Providence, RI, 2002.

[27] H. Iwaniec and P. Sarnak, *Perspectives on the analytic theory of L–functions*, Geom. Funct. Anal. (2000), no. Special Volume, 705–741, GAFA 2000 (Tel Aviv, 1999).

[28] H. Jacquet, I.I. Piatetski–Shapiro, and J.A. Shalika, *Rankin–Selberg convolutions*, Amer. J. Math. **105** (1983), no. 2, 367–464.

[29] H. Jacquet and J.A. Shalika, *On Euler products and the classification of automorphic representations, I*, Amer. J. Math. **103** (1981), no. 3, 499–558.

[30] H. Jacquet and J.A. Shalika, *On Euler products and the classification of automorphic representations, II*, Amer. J. Math. **103** (1981), 777–815.

[31] H. Kim, *Langlands–Shahidi method and poles of automorphic L–functions: application to exterior square L–functions*, Canad. J. Math. **51** (1999), no. 4, 835–849.

[32] H. Kim, *Functoriality for the exterior square of GL_4 and the symmetric fourth of GL_2*, J. Amer. Math. Soc. **16** (2002), no. 1, 139–183, with appendix 1 by D. Ramakrishnan and appendix 2 by H. Kim and P. Sarnak.

[33] H. Kim, *On local L–functions and normalized intertwining operators*, Canad. J. Math., to appear.

[34] H. Kim and P. Sarnak, *Refined estimates towards the Ramanujan and Selberg conjectures*, Appendix 2 to [29], J. Amer. Math. Soc. **16** (2002), no. 1, 175–181.

[35] H. Kim and F. Shahidi, *Functorial products for $GL_2 \times GL_3$ and functorial symmetric cube for GL_2*, C.R. Acad. Sci. Paris Sér. I Math. **331** (2000), no. 8, 599–604.

[36] H. Kim and F. Shahidi, *Cuspidality of symmetric powers with applications*, Duke Math. J. **112** (2002), no. 1, 177–197.

[37] H. Kim and F. Shahidi, *Functorial products for $GL_2 \times GL_3$ and the symmetric cube for GL_2*, Ann. of Math. **155** (2002), no. 3, 837–893, with an appendix by C.J. Bushnell and G. Henniart.

[38] L. Lafforgue, *Chtoucas de Drinfeld et correspondance de Langlands*, Invent. Math. **147** (2002), no. 1, 1–241.

[39] R.P. Langlands, *Problems in the theory of automorphic forms*, in Lecture Notes in Math. 170, Springer–Verlag, Berlin–Heidelberg–New York, 1970, 18–86.

[40] R.P. Langlands, *Euler products*, a James K. Whittemore Lecture in Mathematics given at Yale University, 1967; Yale Mathematical Monographs, 1, Yale University Press, New Haven, CT, 1971.

[41] R.P. Langlands, *On the functional equations satisfied by Eisenstein series*, Lecture Notes in Mathematics, Vol. 544, Springer–Verlag, Berlin, 1976.

[42] R.P. Langlands, *On the classification of irreducible representations of real algebraic groups*, Representation theory and harmonic analysis on semisimple Lie groups, Math. Surveys Monogr., vol. 31, Amer. Math. Soc., Providence, RI, 1989, pp. 101–170.

[43] P.D. Lax and R.S. Phillips, *The asymptotic distribution of lattice points in Euclidean and non–Euclidean spaces*, J. Funct. Ana. 46 (1982), 280–350.

[44] J. Lehner, *Discontinuous Groups and Automorphic Functions*, Math. Surveys, Number VIII, AMS, Providence, Rhode Island, 1964.

[45] W. Luo, Z. Rudnick, and P. Sarnak, *On Selberg's eigenvalue conjecture*, Geom. Funct. Anal. **5** (1995), no. 2, 387–401.

[46] W. Luo, Z. Rudnick and P. Sarnak, *On the generalized Ramanujan conjecture for $GL(n)$*, Automorphic forms, automorphic representations, and arithmetic (Fort Worth, TX, 1996), Proc. Sympos. Pure Math., vol. 66, Amer. Math. Soc., Providence, RI, 1999, pp. 301–310.

[47] H. Maass, *Über eine neue Art von nichtanalytischen automorphen Funktionen und die Bestimmung Dirichletscher Reihen durch Funktionalgleichungen*, Math. Ann. **121** (1949), 141–183.

[48] C. Moeglin and J.-L. Waldspurger, *Spectral decomposition and Eisenstein series*, Cambridge Tracts in Mathematics, vol. 113, Cambridge University Press, Cambridge, 1995.

[49] C. Moeglin and J.-L. Waldspurger, *Le Spectre residuel de $GL(n)$*, Ann. Scient. Ec. Norm. Sup. **22** (1989), 605–674.

[50] W. Müller, *The trace class conjecture in the theory of automorphic forms*, Ann. of Math. **130** (1989), no. 3, 473–529.

[51] S.J. Patterson, *A lattice point problem in hyperbolic space*, Mathematika 22 (1975), 81-88.

[52] G. Pick, *Ueber gewisse ganzzahlige lineare substitutionen, welche sich nicht durch algebraische congruenzen erklären lassen*, Math. Ann. **28** (1887), 119–124.

[53] D. Ramakrishnan and S. Wang, *On the exceptional zeros of Rankin-Selberg L-functions*, Comp. Math. **135** (2003), 211–244.

[54] P. Sarnak, *Maass cusp forms with integer coefficients*, A Panorama of Number Theory or The View from Baker's Garden (G. Wüstholz, ed.), Cambridge University Press, 2002, pp. 121–128.

[55] P. Sarnak, *Spectra of hyperbolic surfaces*, Bull. Amer. Math. Soc. (N.S.) **40** (2003), 441–478.

[56] P. Sarnak, *Notes on the generalized Ramanujan conjectures*, Fields Institute Lectures, June 2003.

[57] I. Satake, *Spherical functions and Ramanujan Conjecture*, Proc. Sympos. Pure Math., Vol. IX, Amer. Math. Soc., Providence, RI, 1966, pp. 258–264.

[58] R. Schoen, S. Wolpert and S.T. Yau, *Geometric bounds to the low eigenvalues of a compact surface*, in Geometry of the Laplace Operators, Proc. Sympos. Pure Math., Vol. **36** (1980), AMS, Providence, R.I., pp. 279–285.

[59] A. Selberg, *On the estimation of Fourier coefficients of modular forms*, Proc. Sympos. Pure Math., Vol. VIII, Amer. Math. Soc., Providence, RI, 1965, pp. 1–15.

[60] J-P. Serre, *Le probléme des groupes de congruence pour SL_2*, Ann. of Math. **92** (1970), 489–527.

[61] J-P. Serre, *Abelian l–adic Representations and Elliptic Curves*, W.A. Benjamin, New York, 1968.

[62] F. Shahidi, *Fourier transforms of intertwining operators and Plancherel measures for $GL(n)$*, Amer. J. Math. **106** (1984), 67-111.

[63] F. Shahidi, *On certain L–functions*, Amer. J. Math. **103** (1981), no. 2, 297–355.

[64] F. Shahidi, *Local coefficients as Artin factors for real groups*, Duke Math. J. **52** (1985), no. 4, 973–1007.

[65] F. Shahidi, *On the Ramanujan conjecture and finiteness of poles for certain L–functions*, Ann. of Math. **127** (1988), no. 3, 547–584.

[66] F. Shahidi, *A proof of Langlands' conjecture on Plancherel measures; complementary series for p–adic groups*, Ann. of Math. **132** (1990), no. 2, 273–330.

[67] F. Shahidi, *Symmetric power L–functions for $GL(2)$*, Elliptic curves and related topics, CRM Proc. Lecture Notes, vol. 4, Amer. Math. Soc., Providence, RI, 1994, pp. 159–182, with an appendix by J-P. Serre.

[68] F. Shahidi, *Intertwining Operators, L–functions, and Representation Theory*, Lecture Notes of the Elevent KAIST Mathematics Workshop (Ja Kyung Koo, ed.), 1996, pp. 1–63.

[69] F. Shahidi, *Automorphic L–functions and functoriality*, Proceedings of the International Congress of Mathematicians, Vol. II (Beijing, 2002), Higher Ed. Press, Beijing, 2002, pp. 655–666.

[70] F. Shahidi, *Langlands–Shahidi Method*, IAS/Park City Lecture Notes, Park City, Utah, 2002.

[71] G. Shimura, *Introduction to the Arithmetic Theory of Automorphic Functions*, Princeton University Press, 1994.

[72] P.C. Yang and S.T. Yau, *Eigenvalues of the Laplacian of Compact Riemann surfaces and minimal submanifolds*, Annali Scuola Normale Superiore–Pisa, Serie 4, **7** (1980), no. 1, 55–63.

[73] P.G. Zograf, *On the spectrum of automorphic Laplacians in spaces of parabolic functions*, Soviet Math. Dokl. **27** (1983), no. 2, 420–422.

Department of Mathematics, Purdue University, West Lafayette, Indiana, USA 47907

E-mail address: shahidi@math.purdue.edu

Surveys in Differential Geometry
IX, International Press

The inverse spectral problem

Steve Zelditch

CONTENTS

1. Introduction

The inverse spectral problem on a Riemannian manifold (M, g), possibly with boundary, is to determine as much as possible of the geometry of (M, g) from the spectrum of its Laplacian Δ_g (with some given boundary conditions). The special inverse problem of Kac is to determine a Euclidean domain $\Omega \subset \mathbb{R}^n$ up to isometry from the spectrum $Spec_B(\Omega)$ of its Laplacian Δ_B with Dirichlet, Neumann or more general boundary conditions B. The physical motivation is to identify physical objects from the light or sound they emit, which may be all that is observable of remote objects such as stars or atoms.

The inverse spectral problem is just one among many kinds of inverse problems whose goal is to determine a metric, domain or scatterer from physically relevant invariants. A comparison with other inverse problems shows just how small a set of invariants the spectrum is. For instance, the boundary inverse problem asks to determine the metric g on a fixed bounded domain $\Omega \subset M$ of a Riemannian manifold (M, g) from the spectrum of the Dirichlet Laplacian on $L^2(\Omega)$, and from

Research partially supported by NSF grant #DMS-0071358.

the Cauchy data $\frac{\partial \phi_\lambda}{\partial \nu}|_{\partial \Omega}$ of its eigenfunctions, or equivalently from its Dirichlet-to-Neumann operator [**BK, KKL, LU**]. The inverse scattering problem seeks to determine an obstacle from its scattering amplitude [**Ma, Ma2**], or from its scattering length spectrum [**St, St2**], a set of lengths parametrized by $S^{n-1} \times S^{n-1}$. By comparison, the inverse spectral problem has to make do with just a discrete 'unformatted' set of eigenvalues (or resonance poles in the open case).

It is probably a consequence of this relative poverty of invariants that most of the results in inverse spectral theory over the last two decades (especially since the appearance of Sunada's article [**Su**]) are 'negative results', showing that one cannot determine metrics or domains by their spectra. The collection of non-isometric isospectral pairs of Riemannian manifolds would require a lengthy survey of its own and for that we refer to the lectures of C. Gordon (cf. [**Gor, Gor2**]). By comparison, the number of 'positive results' showing that one can indeed recover a domain or metric is rather small.

This survey is devoted to the positive results. The emphasis is on relations between the spectrum of the Laplacian and the dynamics of the geodesic flow $G^t : S^*M \to S^*M$. Most of the new material concerns wave trace invariants and their applications to solving concrete inverse spectral problems. The wave group is the quantization of the geodesic flow, and so wave trace methods often 'reduce' inverse spectral problems to inverse dynamical problems. Because of their relevance, we have attempted to describe inverse dynamical problems and results.

1.1. Some basic inverse spectral problems. Let us introduce some basic terminology. The spectrum of a compact Riemannian manifold defines a map

$$Spec : \mathcal{M} \to \mathbb{R}_+^{\mathbb{N}}, \quad (g, B) \to Spec(\Delta_{g,B}) = \{\lambda_0 < \lambda_1^2 \le \lambda_2^2 \le \cdots\}$$

from some class of metrics $\mathcal{M}$ on a manifold M to the spectrum of its Laplacian,

$$\begin{cases} \Delta \phi_j = \lambda_j^2 \phi_j, \ \ \langle \phi_i, \phi_j \rangle = \delta_{ij} \\ \\ B\phi_j = 0 \ \text{ on } \partial M, \end{cases}$$

with boundary conditions $B : C^\infty(M) \to C^\infty(\partial M)$ if $\partial M \neq \emptyset$. Here, Δ denotes the *positive* Laplacian

$$\Delta = -\frac{1}{\sqrt{g}} \sum_{i,j=1}^n \frac{\partial}{\partial x_i} g^{ij} \sqrt{g} \frac{\partial}{\partial x_j}$$

of a Riemannian manifold (M, g), where $g_{ij} = g(\frac{\partial}{\partial x_i}, \frac{\partial}{\partial x_j})$, $[g^{ij}]$ is the inverse matrix to $[g_{ij}]$ and $g = \det[g_{ij}]$. We will only consider Dirichlet $Bu = u|_{\partial M}$ and Neumann $Bu = \partial_\nu u|_{\partial M}$. Eigenvalues are repeated according to their multiplicities.

Two metrics or domains are called isospectral if they have the same spectrum. The main problem in inverse spectral theory is to describe the possible spectra $\Lambda \subset \mathbb{R}^{\mathbf{N}}$ of Laplacians and, for each possible spectrum, to describe the metrics or domains in the spectral class

$$Spec^{-1}(\Lambda). \tag{1.1}$$

Somewhat simpler is to describe the possible smooth curves in the isospectral class, since it apriori eliminates irregular subsets. An isospectral deformation of a Riemannian manifold (possibly with boundary) is one-parameter family of metrics satisfying $Spec(M, g_t) = Spec(M, g_0)$ for each t. Similarly, an isospectral deformation of a domain with a fixed background metric g_0 and boundary conditions

B is a family Ω_t with $Spec_B(\Omega_t) = Spec_B(\Omega)$. One could also pose the inverse spectral problems for boundary conditions (while holding the other data fixed) as in [**GM2, PT**].

The inverse spectral and isospectral deformation problems are difficult because the map $Spec$ is highly nonlinear. The linearization of the problem is to find infinitesimal isospectral deformations, i.e. deformations for which the eigenvalue variations vanish to first order. By first order perturbation theory, the variations of the eigenvalues under a variation of the metric are given by

$$(1.2) \qquad \dot{\lambda}_j = \frac{d}{dt}\lambda_j(t)|_{t=0} = \langle \dot{\Delta}\phi_j, \phi_j \rangle,$$

where $\dot{\Delta}$ is the variation of the Laplacian and where $\phi_j = \phi_j(0)$ is an orthonormal basis of eigenfunctions which varies smoothly in t (such a basis exists by the Kato-Rellich theory). As will be recalled below, lengths L_γ of closed geodesics γ are spectral invariants (at least, when there is at most one closed geodesic of each length), so infinitesimal iso-length-spectral deformations are those where

$$(1.3) \qquad \dot{L}_\gamma = 0, \quad \forall \gamma.$$

The deformation of the metric is a symmetric 2-tensor $\dot{g}$, so the linearized problem is to determine the space of $\dot{g} \in S^2T^*M$ (modulo tensors arising from diffeomorphisms $\phi_t^*(g)$) for which

$$(1.4) \qquad \int_\gamma \dot{g} ds = 0 \, \forall \gamma \iff \langle \dot{\Delta}\phi_j, \phi_j \rangle = 0, \quad (\forall j).$$

The operator $\dot{\Delta} = Op(\dot{g})$ is the differential operator with symbol $\dot{g}$. The linearized problem is still very difficult because it requires a study of the asymptotic behavior of the expressions (1.4) as the lengths or eigenvalues tend to infinity. This is tantamount to the study of the equidistribution theory of closed geodesics and eigenfunctions.

The basic distinctions in inverse spectral theory are the following. We say that

- a metric or domain is *spectrally determined* (within $\mathcal{M}$) if it is the unique element of $\mathcal{M}$ with its spectrum;
- it is *locally spectrally determined* if there exists a neighborhood of the metric or domain in $\mathcal{M}$ on which it is spectrally determined;
- a metric or domain is *spectrally rigid* in $\mathcal{M}$ if it does not admit an isospectral deformation within the class;
- the inverse spectral problem is solvable in $\mathcal{M}$ if $Spec|_{\mathcal{M}}$ is $1-1$, i.e. if any other metric or domain in $\mathcal{M}$ with the same spectrum is isometric to it. If not, one has found a counterexample.

There are analogous problems for Laplacians on non-compact Riemannian manifolds, which often have continuous spectra as well as discretely occurring eigenvalues. In place of eigenvalues, one considers the *resonances* $Res(\Delta)$ of Δ, i.e. the poles of the analytic continuation of its resolvent

$$R(z) = (\Delta + z^2)^{-1}.$$

Depending on whether the dimension is odd or even, $Res(\Delta)$ is a discrete subset of $\mathbb{C}$ or of the logarithmic plane. The inverse spectral problems above have natural analogues for resonance poles. We refer to Zworski's expository articles [**Zw2, Zw3**] for background.

To illustrate the current state of knowledge, we note that even the simplest special metrics are not known to be spectrally determined at the present time (at least to the author's knowledge). It is not known:

- if the standard metric g_0 on S^n is determined by its spectrum (in dimensions ≥ 7), i.e. if (M, g) (or even (S^n, g)) is isospectral to (S^n, g_0) then it is isometric to it. This has been proved in dimensions ≤ 6 [**T**].
- if ellipses in the plane are determined by their Dirichlet spectra, or even if they are spectrally rigid, i.e. if there exist isospectral deformations of ellipses (with Dirichlet boundary conditions).
- if hyperbolic manifolds are determined by their spectra in dimensions ≥ 3. I.e. if (M_0, g_0) is hyperbolic and (M, g) is isospectral to it, then is (M, g) hyperbolic? This is of course true in dimension 2. Is (M, g) isometric to (M_0, g_0)? In dimension 2, this is known to be false for some hyperbolic surfaces.
- if flat metrics are determined by their spectra in the sense that if (M, g_0) is flat and (M, g) is isospectral to it, then (M, g) is flat (it is known that this is true in dimensions ≤ 6 or in all dimensions if additionally g is assumed to lie in a sufficiently small neighborhood of g_0 [**Ku3**]); it is also classical that there are non-isometric flat tori with the same spectra.

These special cases are tests of the strength of the known methods. Another test is given by the two-dimensional inverse spectral problem. One-dimensional problems are comparatively well understood because the eigenvalue problems are ordinary differential equations and the underlying dynamics consists of just one orbit (an interval)! Two-dimensional problems are already rich in spectral and dynamical complexities, as illustrated by the classical dynamics of twist maps or geodesic flows on Riemannian surfaces. In general, the inverse spectral problem grows rapidly in difficulty with the dimension, and is already quite open for analytic surfaces and domains in two dimensions. This motivates our concentration on two-dimensional problems for much of the survey. The following simple-sounding problems are still apparently beyond the reach of known methods:

- Are convex analytic domains determined by their spectra among other such domains? Are they spectrally rigid?
- Are convex analytic surfaces of revolution determined by their spectra among all metrics on S^2? Are they spectrally rigid?

These problems are in some ways analogous to each other in that the unknown is a function of one variable (the boundary or the profile curve), and that is limit of what wave trace invariants at one orbit can hope to recover. Surfaces of revolution are simpler than plane domains since the geodesic flow is integrable, while billiards on plane domains could have any dynamical type. On the other hand, in the domain problem the class $\mathcal{M}$ only consists of convex plane domains, whereas in the second problem we allow any other metric. If we similarly restricted the class of surfaces of the first problem entirely to analytic convex surfaces of revolution, then the answer is known to be 'yes' [**Z2**].

The inverse problem for surfaces of revolution might sound reasonably simple since the geodesic flow is completely integrable, and the feeling arises that one should be able to detect this property from the spectrum. This is one of many problems which relate spectral invariants to dynamics of the geodesic flow.

The relations between Laplace spectrum and dynamics have been at the center of at least the positive results in inverse spectral theory in the last thirty years, by comparison with the emphasis on heat invariants in the earlier period. Before going into the technical relations between spectral (in particular, wave) invariants and dynamics, it might be helpful to give some heuristic principles which suggest the relevance of dynamical inverse spectral problems to Laplace inverse spectral problems. The first is the relation of classical to quantum mechanics. Two Laplacians are isospectral if

$$\Delta_{g_1} = U\Delta_{g_2}U^*, \tag{1.5}$$

where $U : L^2(M_1, g_1) \to L^2(M_2, g_2)$ is a unitary operator. In the Dirac dictionary of analogies, the classical analogue of this similarity is the symplectic conjugacy

$$|\xi|_{g_1} = \chi^*|\xi|_{g_2} \iff G^t_{g_1} = \chi \circ G^t_{g_2} \circ \chi^{-1}$$

of the corresponding geodesic flows (cf. (3.8)). Here, $\chi : T^*M_1\backslash 0 \to T^*M_2\backslash 0$ is a homogeneous symplectic diffeomorphism.) This analogy should not be taken too literally, but it is useful in suggesting conjectures. In modern language, the analogue would hold if U were a unitary Fourier integral operator quantizing χ. It would clearly be difficult to prove, even in special cases, that isospectral Laplacians are conjugate by unitary Fourier integral operators, though it was observed independently by Uribe and the author (see [**Z6**]) and by P. Bèrard ([**Be, Be2**]) that the Sunada counterexamples [**Su**] have this property (the resulting Fourier integral operators were termed 'transplantations' by Bèrard). It was also observed in [**Z6**] that such Fourier integral intertwining operators need not be quantizations of symplectic diffeomorphisms, but could be (and indeed they are, in the Sunada examples) quantizations of multi-valued symplectic correspondences.

But the analogy is suggestive and is fruitful on a local (or more accurately, formal local) level. One of the main results in inverse spectral theory in recent years is the theorem due to V. Guillemin [**G, G3**] (see also [**Z3, Z4**]) that isospectrality (with a simple length spectrum assumption) implies the *formal local* symplectic equivalence of the geodesic flows around corresponding pairs of closed geodesics, i.e. it implies the equality of their *Birkhoff normal forms.* This implies local symplectic equivalence around hyperbolic orbits, although not around elliptic orbits (see Problem 3.4). It is perhaps the closest that the above heuristic principle comes to being valid in a general setting.

A further heuristic principal is that much of the dynamics of a flow is encoded in the structure of the flow near closed orbits. To the extent that this is true, local equivalence would be a powerful fact and one should be able to obtain strong information about the metric g from studying the wave trace expansion around closed geodesics. This raises the possibility that the zeta functions of the flows might determine their dynamical type (see Problem 3.5).

- Can one determine the dynamical type of the geodesic flow from the spectrum of Δ, i.e. whether the geodesic flow is integrable, ergodic or of some other type? The Ruelle zeta function of the geodesic flow is generically a Laplace spectral invariant. Can one determine dynamical type from its analytic properties.
- In the above problems, we almost always assume in addition that the length spectrum is multiplicity-free: i.e. that the set of closed geodesics of a fixed length, or more generally the fixed point sets of the geodesic flow, should

contain at most two components interchanged by the time reversal involution $(x, \xi) \to (x, -\xi)$. Are any of the known counterexamples, i.e. non-isometric isospectral pairs, multiplicity free?

The motivation for the second problem is that wave trace methods cannot get off the ground, and in particular the dynamical zeta function need not be a spectral invariant, unless the length spectrum is simple. The issue is that there could exist complicated cancellations among invariants of closed geodesics of the same length.

1.2. Strategies for solving the inverse spectral problem. As mentioned above, we are concerned here primarily with the positive results, ones which prove that certain geometric data is determined by the spectra. Roughly speaking, the strategy for obtaining positive results has long consisted of the the following steps:

(A) Define a lot of spectral invariants;
(B) Calculate them in terms of geometric or dynamical invariants;
(C) Try to determine the metric or domain from the invariants.

A crucial limitation arises in step (B), which accounts for the relative paucity of positive results compared to negative results. It is easy to define a complete set of spectral invariants, namely the 'special values' of any one of

(1.6)
$$\begin{cases} \text{The heat trace,} & Z(t) = Tr e^{-t\Delta} = \sum_{j=0}^{\infty} e^{-\lambda_j^2 t} \;\; (t > 0), \\ \text{The zeta function} & \zeta(s) = Tr \Delta^{-s} = \sum_{j=0}^{\infty} \lambda_j^{-2s} \;\; (\Re s > n) \\ \text{The wave trace} & S(t) = Tr e^{it\sqrt{\Delta}} = \sum_{j=0}^{\infty} e^{i\lambda_j t}, \text{ or } S_{ev}(t) = Tr \cos t\sqrt{\Delta}. \end{cases}$$

Of course, $\zeta(s)$ must be meromorphically continued to $\mathbb{C}$ and $S(t)$ is a distribution rather than a function. But the point we are making is that special values are rarely computable in terms of the geometry and are therefore of limited use for positive results on the inverse spectral problem. By comparison, special values can be used to prove negative results by showing that the traces of any of the above operators are the same for two non-isometric (M, g). A key step in obtaining positive results is to find computable invariants, and to give efficient algorithms for computing them in terms of the simplest possible geometric invariants. It should be mentioned that there exists spectral invariants such as λ_1 and $\log \det \Delta = -\zeta'(0)$ which are useful although not computable in the above sense. To maintain our theme of wave invariants and dynamics, we will not discuss such invariants.

Oversimplifying a bit, the computable invariants arise from the singularity (or related) asymptotic expansions of the traces defined above or, in another language, from non-commutative residues of functions of the Laplacian (we re refer to [**G, G2, Z9**] for discussion of non-commutative residues). In fact, all of the computable invariants known to the author are wave invariants, i.e. arising from the singularities of the distribution trace or residues of the wave operator $U(t) = e^{it\sqrt{\Delta}}$ at times t in the length spectrum of (M, g) (including $t = 0$).

And it should not be forgotten that the goal of inverse spectral theory is step (C). There now exist a number of rather abstract results showing that various dynamical or quantum mechanical invariants (e.g. Birkhoff normal forms) are spectral invariants [**G, GM2, ISZ, Z1, Z3, Z4**]. But this only trades one inverse problem for another, and there is relatively little work on the subsequent inverse problem of determining the domain or metric from these invariants. For instance, it is not hard to see that the classical Birkhoff normal form of the Poincare map of a bouncing ball orbit *does not* determine all of the Taylor coefficients of the boundary at the endpoints of the orbit, even if the domain has one symmetry. Often step (C) is the deepest, requiring a separate study of inverse dynamical problems.

1.3. Contents of the survey. The focus of this survey is on the use of wave trace formula to derive information about the metric and geodesic flow around closed geodesics and to determine metrics or domains from the information. We survey in some detail the relation between wave invariants and Birkhoff normal form invariants on general Riemannian manifolds, initiated by V. Guillemin, and developed by the author and by Iantchenko-Sjöstrand-Zworski [**G, G3, ISZ, SjZ, Z3, Z4**]. In an appendix to this article J. Sjöstrand and M. Zworski describe in more detail how their general results on quantum monodromy apply to the Laplacian.

But as mentioned above, to succeed with step (C) we need to be able to determine a metric or domain from such invariants. In the end, the crucial problem is to compute wave trace invariants in the simplest and most efficient way, to analyze them in detail, and to reconstruct the domain or metric. This is most feasible in dimension two, so we review in some detail the articles which have succeeded in determining special families of domains or metrics from wave invariants, to wit, bounded plane domains [**CdV, P, Z1, Z2, Z5, Z7, Z10, ISZ, GM, MM, S1**] and surfaces of revolution [**Z2**]. One of our aims is to describe a new method for calculating wave invariants from [**Z5**] which so far has achieved better results than the Birkhoff normal form approach.

In addition, we provide a fair amount of background that hopefully puts the special problems in context. There already exist a number of surveys on the inverse spectral problem (e.g. [**Ber, Be3, C3, Me, Gor, Gor2**]) including our own expository articles [**Z9, Z10**], and we have tried to avoid duplication of material which already appears elsewhere. However, to make the survey more self-contained we quote from a number or prior surveys, including our own. We also follow the lecture notes of Melrose [**Me**] in our discussion of the Lifshits (Penrose mushroom) example of two domains with the same wave invariants, and also the (much better known) examples of domains with the same heat invariants.

We also omit a number of topics as being too far from our focus on wave invariants and dynamics. As mentioned above, we do not discuss counterexamples and negative results (cf. [**Gor, Gor2**]). We also omit discussion of compactness results of isospectral sets, of which there are many since the (unpublished) work of Melrose and the work of Osgood-Phillips-Sarnak [**OPS**] on isospectral sets of plane domains (for the resonance analogue, see [**HZel2**]). To avoid dissipation of energy, we do not discuss the inverse resonance problem in detail, but only mention some recent results closely related to the inverse spectral problems covered in this survey. We refer to [**Z10**] for further discussion of inverse resonance problems for exterior domains and to [**BJP, BP**] for geometric scattering settings.

The author would like to thank V. Baladi, G. Besson, Y. Colin de Verdière, G. Courtois, R. Kuwabara, R. de La Llave, G. Lebeau, M. Rouleux, K. F. Siburg and M. Zworski for informative remarks on the contents of this survey. Of course, errors and omissions are the author's responsibility.

2. Expansions at $t = 0$

The most classical spectral invariants are the heat invariants, namely the coefficients of the expansion at $t = 0$ of the trace of the heat kernel. They are closely related to the coefficients of the trace of the wave group at $t = 0$, although it should be noted that the powers $t^{-\frac{n}{2}+m}$ of the heat kernel expansion at $t = 0$ are not singular if m is even and if $m \geq \frac{n}{2}$ (where $n = \dim M$). Hence, the heat kernel expansion contains more information than the singularity expansion of the wave trace at $t = 0$.

2.1. Boundaryless case. The earliest work in inverse spectral was based on calculations of heat invariants in terms of curvature invariants, and the recovery of special metrics or domains from these curvature invariants [**Ber, Pa, T, T2, Ku2, Ku3**].

We recall that the heat trace expansion in dimension n on a boundaryless manifold has the asymptotic expansion,

$$Tre^{t\Delta_g} \sim t^{-n/2} \sum_{j=0}^{\infty} a_j\, t^j. \tag{2.1}$$

The coefficients a_j are the heat invariants. We note that when n is odd, the powers of t are singular and hence the expansion may be viewed as a singularity expansion in which the terms become more regular. When n is even, the terms with $-n/2+j<0$ are singular but the rest are smooth and hence are not residual. Rather one may view the expansion as a Taylor expansion at $t = 0$ of $t^{n/2}Tre^{t\Delta_g}$. But just like the singular terms, the coefficients are spectral invariants given by integrals of curvature invariants.

The first four heat invariants in the boundaryless case are given by [**T, T2**]

$$\begin{aligned}
a_0 &= Vol(M) = \textstyle\int dVol_M \\
a_1 &= \tfrac{1}{6} \textstyle\int S dVol_M \\
a_2 &= \tfrac{1}{360} \textstyle\int \{2|R|^2 \;-2|Ric|^2 \;+5S^2]dVol_M \\
a_3 &= \tfrac{1}{6!} \textstyle\int \{-\tfrac{1}{9}|\nabla R|^2 - \tfrac{26}{63}|\nabla Ric|^2 - \tfrac{143}{63}|\nabla S|^2 \\
&\quad -\tfrac{8}{21} R^{ij}_{k\ell} R^{kl}_{rs} R^{rs}_{ijk\ell} - \tfrac{8}{63} R^{rs} R^{jk\ell}_{r} R_{sjk\ell} + \tfrac{2}{3} S|R|^2 \\
&\quad -\tfrac{20}{63} R^{ik} R_{j\ell} R_{ijk\ell} - \tfrac{4}{7} R^{i}_{j} R^{j}_{k} R^{k}_{i} - \tfrac{2}{3} S|Ric|^2 + \tfrac{5}{9} S^3\}dVol_M.
\end{aligned} \tag{2.2}$$

Here, S is the scalar curvature, Ric is the Ricci tensor and R is the Riemann tensor. In general the heat invariants are integrals of curvature polynomials of various weights in the metric. We refer to [**Ber**] for background.

The heat invariants are complicated and it is difficult to detect meaningful patterns in the curvature polynomials. Nevertheless, they have been successfully used to obtain inverse spectral results.

- Spheres: Tanno [**T2**] used a_0, a_1, a_2, a_3 to prove that the round metric g_0 on S^n for $n \leq 6$ is determined among all Riemannian manifolds by its spectrum, i.e. any isospectral metric g is necessarily isometric to g_0. He also used a_3 [**T2**] to prove that canonical spheres are locally spectrally determined (hence spectrally rigid) in all dimensions. Patodi proved that round spheres are determined by the spectra $\text{Spec}^0(M, g)$ and $\text{Spec}^1(M, g)$ on zero and 1 forms.

- Complex projective space: Let (M, g, J) be a compact Kähler manifold and let $(CP^n(H), g_0, H_0)$ be a complex n-dimensional projective space with the Fubini-Study metric of constant holomorphic sectional curvature H. Tanno [**T2**] proves that if the complex dimension $n \leq 6$ and if

$$\text{Spec}(M, g, J) = \text{Spec}(CP^n(H), g_0, J_0)$$

, then (M, g, J) is holomorphically isometric to $(CP^n(H), g_0, J_0)$. He also proves that $(CP^n(H), g_0, H_0)$ is locally spectrally determined in all dimensions [**T3**].
- Flat manifolds: Patodi [**Pa**] and Tanno [**T, T2**] used the heat invariants to prove in dimension ≤ 5 that if (M, g) is isospectral to a flat manifold, then it is flat. More precisely, they showed that if $a_j = 0$ for $j \geq 1$, and if $n \leq 5$ then (M, g) is flat. The result is sharp, as Patodi (loc. cit.) showed that $a_j = 0$ for $j \geq 1$ for the product of a 3-dimensional sphere with a 3-dimensional space of constant negative curvature. In fact, Tanno showed that if $a_2 = a_3 = 0$, then (M, g) is either E^6/Γ_1, where Γ_1 is some discontinuous group of translations of the Euclidean space E^6, or (2) $[S^3(C) \times H^3(-C)]/\Gamma_2$, where $S^3(C)[H^3(-C)]$ is the 3-sphere [hyperbolic 3-space] with constant curvature $C > 0[-C < 0]$ and Γ_2 is some discontinuous group of isometries of $S^3(C) \times H^3(-C)$. Kuwabara [**Ku2, Ku3**] used the invariants to prove that flat manifolds are locally spectrally determined, hence spectrally rigid.

2.1.1. *The boundary case.* When $\partial\Omega \neq 0$, the heat trace has the form

$$Tre^{t\Delta_g} \sim t^{-n/2} \sum_{j=0}^{\infty} a_j \, t^{j/2}. \tag{2.3}$$

The coefficients have been calculated for a variety of boundary conditions (see [**BG**] and its references).

The formulae are simplest for plane domains, where the only invariant is the curvature κ of the boundary. Using a nicely adapted calculus of pseudodifferential operators, L. Smith obtained the first five heat kernel coefficients in the case of Dirichlet boundary conditions. They are given by :

$$\begin{cases} a_0 = \text{ area of } \Omega, \\ a_1 = -\sqrt{\frac{\pi}{2}} \, |\partial\Omega| \text{ (the length of the boundary)}, \\ a_2 = \frac{1}{3} \int_{\partial\Omega} \kappa ds, \\ a_3 = \frac{\sqrt{\pi}}{64} \int_{\partial\Omega} \kappa^2 ds, \\ a_4 = \frac{4}{315} \int_{\partial\Omega} \kappa^3 ds, \\ a_5 = \frac{37\sqrt{\pi}}{2^{13}} \int_{\partial\Omega} \kappa^4 ds - \frac{\sqrt{\pi}}{2^{10}} \int_{\partial\Omega} (\kappa')^2 ds, \\ . \end{cases} \tag{2.4}$$

Here, ds is arclength and κ' is the derivative with respect to arclength. Certain useful patterns in the heat coefficients were used by R. B. Melrose to prove a compactness result (later improved by Osgood-Phillips-Sarnak). We refer to [**Me**] for the details. In higher dimensions, one still has

$$a_0 = C_n \, Vol_n(\Omega), \quad a_1 = C_n' \, Vol_{n-1}(\partial\Omega). \tag{2.5}$$

There exist a few inverse inverse spectral results using heat invariants:

- Euclidean balls in all dimensions are spectrally determined among simply connected bounded Euclidean domains by their Dirichlet or Neumann spectra. This follows from (2.5) and from the fact that isoperimetric hypersurfaces in $\mathbb{R}^n$ are spheres.
- The exterior of the unit ball $B_3 \subset \mathbb{R}^3$ in dimension 3 is uniquely determined among exterior domains of simply connected compact obstacles by its resonance poles. [**HZ**].

2.2. Domains and metrics with the same heat invariants. It was soon realized that heat invariants are insufficient to determine smooth metrics or domains. This is due to the fact that they are integrals of local invariants of the metrics. Pairs of non-isometric metrics with the same heat invariants can be obtained by putting two isometric bumps. The bumped spheres will not be isometric if the distances between the bumps are different, but the heat invariants will be the same. There are many variations on this well-known example. But heat invariants might be quite useful for analytic metrics and domains, and have also been used in compactness results.

3. Dynamics and dynamical inverse problems

We now turn to the more dynamical theory of the wave group in inverse spectral theory. The trace of the wave group expresses spectral invariants in terms of the dynamics of the geodesic flow, and often 'reduces' inverse Laplace spectral problems to inverse problems in dynamics. We therefore begin by recalling the relevant dynamical notions and inverse problems.

3.1. Geodesic flow on boundaryless manifolds. We denote by

$$(T^*M, \sum_j dx_j \wedge d\xi_j)$$

the cotangent bundle of M equipped with its natural symplectic form. Given a metric g, we define the metric Hamiltonian in a standard notation by

$$H(x,\xi) = |\xi| := \sqrt{\sum_{ij=1}^{n+1} g^{ij}(x)\xi_i\xi_j} \tag{3.1}$$

and define the energy surface to be the unit sphere bundle $S^*_g M = \{(x,\xi)| : |\xi|_g = 1\}$. The spectral theorists geodesic flow is the Hamiltonian flow

(3.2)

$$G^t = \exp t\Xi_H : T^*M\backslash \to T^*M\backslash 0, \quad \Xi_H = \text{ the Hamiltonian vector field of } H.$$

It is homogeneous of degree 1 with respect to the dilation $(x,\xi) \to (x, r\xi), r > 0$, so nothing is lost by restricting G^t to $S^*_g M$. We also denote its generator by Ξ, the Hamiltonian and metric being understood.

The periodic orbits of G^t of period T of the geodesic flow are the fixed points of G^T on S^*M. Equivalently they are the critical points of the length functional on the free loop space of M of length T.

3.2. Billard flow and billiard map on domains with boundary.

To define the geodesic or billiard flow G^t on a domain Ω with boundary $\partial\Omega$, we need to specify what happens when a geodesic intersects the boundary. The definition is dictated by the propagation of singularities theorem for solutions of the wave equation (due to R. B. Melrose and J. Sjöstrand [**MS**]), and is therefore not purely geometric or dynamical. A billiard trajectory is defined to be the path along which a singularity of a solution of the wave equation propagates. We give a quick, informal review of the flow; for more details, the reader might consult [**MS**], [**PS**]. For simplicity we assume that there are no points of infinite order tangency.

We denote by $S^*\Omega$ the unit tangent vectors to the interior of Ω and by $S^*_{in}\partial\Omega$ the manifold with boundary of inward pointing unit tangent vectors to Ω with footpoints in $\partial\Omega$. The boundary consists of unit vectors tangent to $\partial\Omega$. The billiard flow G^t is a flow on $S^*\Omega \cup S^*_{in}\partial\Omega$, defined as follows: When an interior geodesic of Ω intersects the boundary $\partial\Omega$ transversally, it is reflected by the usual Snell law of equal angles. Such trajectories are called (transversal) reflecting rays. The complications occur when a geodesic intersects the boundary tangentially in $S^*\partial\Omega$.

Convex domains are simpler than non-convex domains, since interior rays cannot intersect the boundary tangentially. Dynamical studies of billiards (see e.g. [**MF**]) often restrict to convex domains. It should be noted that geodesics of $\partial\Omega$ with the induced metric are important billiard trajectories. They are limits of 'creeping rays', i.e. rays with many small links (interior segments) which stay close to $\partial\Omega$. In particular, the boundary of a convex plane domain is a closed billiard trajectory. In higher dimensions, closed geodesics on the boundary which are limits of interior creeping rays are closed billiard trajectories. (Are all closed geodesics on the boundary limits in this sense?)

On a non-convex domain, a general billiard trajectory is divided into segments which are either geodesic segments in the interior or geodesic segments of the boundary. Geodesic segments of the boundary only occur where the boundary is convex. Intuitively, the boundary segments are limits of creeping rays along the convex parts of the boundary. Trajectories enter and exist the boundary at inflection points. In particular, the boundary of a non-convex plane domains is not a closed billiard trajectory. If a trajectory intersects the boundary tangentially at a non-inflection point of a non-convex domain, it goes straight past the point of intersection into the interior.

When (Ω, g) is non-trapping (i.e. if there is no geodesic ray which remains forever in the interior), the set $S^*_{in}\partial\Omega$ behaves like a global cross section to the billiard flow. It is then natural to reduce the dimension by defining the *billiard ball map* $\beta : B^*(\partial\Omega) \to B^*(\partial\Omega)$, where $B^*(\partial\Omega)$ is the ball bundle of the boundary. We first identify $S^*_{in}\partial\Omega \simeq B^*(\partial\Omega)$ by adding to a tangent (co)vector $\eta \in B^*_q\partial\Omega$ of length < 1 a multiple $c\nu_q$ of the inward point unit normal ν_q to form a covector in $S^{in}_{\partial\Omega}\Omega$. The image $\beta(q, v)$ is then defined to be the tangential part of the first intersection of $G^t(q, \eta + c\nu_q)$ with $\partial\Omega$. The billiard map is symplectic with respect to the natural symplectic form on $B^*(\partial\Omega)$.

An equivalent description of the billiard map of a plane domain is as follows. Let $q \in \partial\Omega$ and let $\phi \in (0, \pi)$. The point (q, ϕ) corresponds to an inward pointing unit vector making an angle ϕ with the tangent line, with $\phi = 0$ corresponding to a fixed orientation (say counter-clockwise). The billiard map is then $\beta(q, \phi) = (q', \phi')$ where (q', ϕ') are the parameters of the reflected ray at the first point of intersection with the boundary. The map β is then area preserving with respect to $\sin \phi ds \wedge d\phi$ (see e.g. [**MF**]).

3.3. Closed orbits and their Poincaré maps. Closed orbits (or periodic orbits) γ of flows are orbits of points $(x, \xi) \in T^*M$ satisfying $G^T(x, \xi) = (x, \xi)$ for some $T \neq 0$ (the period). They project to closed geodesics on the Riemannian manifold or domain.

We recall the definition of the nonlinear Poincare map $\mathcal{P}_\gamma$: in S^*M one forms a symplectic transversal S_γ to γ at some point m_0. One then defines the first return map, or nonlinear Poincar map,

$$\mathcal{P}_\gamma(\zeta) : S_\gamma \to S_\gamma$$

by setting $\mathcal{P}_\gamma(\zeta) = G^{T(\zeta)}(\zeta)$, where $T(\zeta)$ is the first return time of the trajectory to S_γ. This map is well-defined and symplectic from a small neighborhood of $\gamma(0) = m_0$ to a larger neighborhood. By definition, the linear Poincare map is its derivative, $P_\gamma = d\mathcal{P}_\gamma(m_0)$.

Closed geodesics are classified by the spectral properties of the symplectic linear map P_γ. Its eigenvalues come in 4-tuples $\lambda, \bar{\lambda}, \lambda^{-1}, \bar{\lambda}^{-1}$. A closed geodesic γ is called:

- *non-degenerate* if $\det(I - P_\gamma) \neq 0$;
- *elliptic* if all of its eigenvalues are of modulus one and not equal to ± 1, in which case they come in complex conjugate pairs $e^{i \pm \alpha_j}$.
- *hyperbolic* if all of its eigenvalues are real, in which case they come in inverse pairs $\lambda_j \lambda_j^{-1}$
- *loxodromic* or *complex hyperbolic* in the case where the 4-tuple consists of distinct eigenvalues as above.

There are other possibilities (parabolic) in the degenerate case. In the case of Euclidean domains, or more generally domains where there is a unique geodesic between each pair of boundary points, one can specify a billiard trajectory by its successive points of contact $q_0, q_1, q_2, \dots$ with the boundary. The *n-link periodic reflecting rays* are the trajectories where $q_n = q_0$ for some $n > 1$. The point $q_0, \dots, q_n$ is then a critical point of the length functional

$$\mathcal{L}(q_0, \dots, q_n) = \sum_{i=0}^{n-1} |q_{i+1} - q_i|$$

on $(\partial\Omega)^n$.

Among the periodic orbits, bouncing ball orbits often have special applications in inverse spectral theory. By a *bouncing ball* orbit γ one means a periodic 2-link reflecting ray, i.e. $q_0 = q_2$. Convex domains always have at least two bouncing ball orbits, one of which is its diameter. There exist non-convex domains without any bouncing ball orbits. For geometric aspects of bouncing ball orbits, we refer to [**Gh**].

The projection to Ω consists of a segment $\overline{q_0 q_1}$ which is orthogonal to the boundary at both endpoints, i.e. $\overline{q_0 q_1}$ is an extremal diameter. The period is of course twice the length of the segment, which we denote by L. We write $q_0 = A, q_1 = B$. We orient the domain so that the links of γ are vertical and so that the midpoint is at the origin of $\mathbb{R}^2$. The top, resp. bottom, of the boundary $\partial\Omega$ is then the graph of a function $y = f_+(x)$ resp. $y = f_-(x)$ over the x-axis. It is clear that the wave invariants depend only on the Taylor coefficients of $f_\pm$ at A, B.

We denote by R_A, resp. R_B the radius of curvature of the boundary at the endpoints A, resp. B of an extremal diameter. The bouncing ball orbit is elliptic if $L < \min\{R_A, R_B\}$ or if $\max\{R_A, R_B\} < L < R_A + R_B$, and is hyperbolic if $L > R_A + R_B$ or if $\min\{R_A, R_B\} < L < \max\{R_A, R_B\}$ (see [**KT**]). If $\bar{AB}$ is a local minimum diameter than $L < R_A + R_B$ while if it is a local maximum diameter then $L > R_A + R_B$. So a (non-degenerate) local maximum diameter must be hyperbolic; a local minimum diameter is elliptic if it satisfies the additional inequalities above.

When γ is elliptic, the eigenvalues of P_γ are of the form $\{e^{\pm i\alpha}\}$ while in the hyperbolic case they are of the form $\{e^{\pm\lambda}\}$. The explicit formulae for them are:

$$(3.3)\qquad \begin{aligned} \cos\alpha/2 &= \sqrt{(1 - \tfrac{L}{R_A})(1 - \tfrac{L}{R_B})} \text{ (elliptic case)}, \\ \cosh\lambda/2 &= \sqrt{(1 - \tfrac{L}{R_A})(1 - \tfrac{L}{R_B})} \text{ (hyperbolic case)}. \end{aligned}$$

We note that $f''_\pm(0) = \frac{1}{R_\pm} = \kappa_\pm$, where $\kappa_\pm$ denotes the curvature at $(A = -, B = +)$.

3.4. The length spectrum and the marked length spectrum. The *length spectrum* of a boundaryless manifold (M, g) is the discrete set

$$(3.4)\qquad Lsp(M, g) = \{L_{\gamma_1} < L_{\gamma_2} < \cdots\}$$

of lengths of closed geodesics γ_j. In the boundary case, the length spectrum $Lsp(\Omega)$ is the set of lengths of closed billiard trajectories in the sense of §3.2 and is no longer discrete, but rather has points of accumulation at lengths of trajectories which have intervals along the boundary. In the case of convex plane domains, e.g., the length spectrum is the union of the lengths of periodic reflecting rays and multiples of $|\partial\Omega|$. According to the standard terminology, $Lsp(M, g)$ is the set of distinct lengths, not including multiplicities, and one refers to the the length spectrum repeated according to multiplicity as the *extended length spectrum.*

In the notation for $Lsp(M, g)$ we wrote L_{γ_j} as if the closed geodesics of this length were isolated. But in many examples (e.g. spheres or flat tori), the geodesics come in families, and the associated length T is the common length of closed geodesics in the family. In place of closed geodesics, one has components of the fixed point sets of G^T at this time. The fixed point sets could be quite messy, so it is also common to assume that they are clean, i.e. that the fixed point sets are manifolds, and that their tangent spaces are fixed point sets of dG^T. It is equivalent that the length functional is Bott-Morse on the free loop space.

The set of lengths is unformatted in the sense that one does not know which lengths in the list correspond to which closed geodesics. A formatted notion in which lengths are assigned to topogically distinct types of closed geodesics is the *marked length spectrum* ML_g. On a manifold without boundary, it assigns to each free homotopy class of closed loops the length of the shortest closed geodesic in its class. On a convex plane domain with boundary, topologically distinct closed geodesics correspond to different rotation numbers $\frac{m}{n} = \frac{\text{winding number}}{\text{number of reflections}}$. The marked length spectrum then associates to each rational rotation number the *maximal length* of the closed geodesics having n reflection points and winding number m. We refer to §8.4 (based on [**S1, S2**]) for further discussion and applications to inverse spectral theory.

3.5. Birkhoff normal forms. Birkhoff normal forms are approximations to Hamiltonians (or symplectic maps) near equilibria by completely integrable Hamiltonians (or symplectic maps). We now briefly consider the simplest kind of equilibrium occurring for geodesic flows, a closed geodesic γ.

Let us first consider the Birkhoff normal form of the metric Hamiltonian (3.1) near a non- degenerate elliptic closed geodesic γ. To put H into normal form is to conjugate it (approximately) to a function of locally defined action variables $(\sigma, I_1, \dots, I_n)$ on the model space $T^*(S^1 \times \mathbb{R}^n)$, where σ is the momentum coordinate in T^*S^1 and I_j are transversal action variables which depend on the type (elliptic, hyperbolic, loxodromic) of γ. More precisely, the normal form algorithm defines a sequence of canonical transformations χ_M at γ which conjugate H to the normal forms

$$(3.5)\quad \chi_M^* H \equiv \sigma + \frac{1}{L}\sum_{j=1}^{n} \alpha_j I_j + \frac{p_1(I_1, ..., I_n)}{\sigma} + ... + \frac{p_M(I_1, ..., I_n)}{\sigma^M} \qquad mod \quad O^1_{M+1}$$

where p_k is homogeneous of order k+1 in $I_1, \dots, I_n$, and where O^1_{M+1} is the space of germs of functions homogeneous of degree 1 which vanish to order $M+1$ along γ. Note that all the terms in (3.5) are homogenous of degree 1 in $(\sigma, I_1, ..., I_n)$, and that the order of vanishing at $|I| = 0$ equals one plus the order of decay in σ. The coefficients of the monomials in the $p_j(I_1, \dots, I_n)$ are known as the classical Birkhoff normal form invariants.

The algorithm for putting a symplectic map (or Hamiltonian) into Birkhoff normal form around a fixed point (or equilibrium point) can be found in many places (see e.g. [**AKN**]). For geodesic flows, we need Birkhoff normal forms for homogeneous Hamiltonians around periodic orbits. An algorithm for putting a metric Hamiltonian into Birkhoff normal form around a closed geodesic is described in [**G**] and also in the appendix to [**Z3**], among other places. To carry out the algorithm to infinite order, one needs to assume that no eigenvalues of γ are roots of unity, which is of course a stronger condition than non-degeneracy.

There are also a Birkhoff normal forms for the geodesic flow and for the Poincaré map. The normal form is simpler for $\mathcal{P}_\gamma$ since we have eliminated the directions along γ and therefore may express $\mathcal{P}_\gamma$ just in terms of local action-angle variables (I, ϕ). The normal form is as follows:

$$(3.6)\qquad \mathcal{P}(I, \phi) = (I, \phi + \nabla_I G_M(I)), \quad mod \quad O^1_{M+1},$$

where $G_M(I)$ is a polynomial of degree M in the I variables. Thus, to order $M+1$, the Poincare map leaves invariant the level sets of the actions (ellipses, hyperbolas etc. according to the type of γ) and 'rotates' the angle along them.

The well-known question arises whether the full (infinite series) Birkhoff normal form converges and whether the formal symplectic map conjugating the Hamiltonian to its normal form converges. The latter is sometimes called the Birkhoff transformation. Clearly, if the Birkhoff normal form diverges, then so must the Birkhoff transformation. According to a recent article of Perez-Marco [**PM**], there are no known examples of analytic Hamiltonians having divergent Birkhoff normal forms, and we refer to that article for further background and history on the problem. If both the normal form and the transformation converge, the Hamiltonian must be integrable in a neighborhood of the orbit. The Birkhoff normal form is then the expression of the Hamiltonian in local action-angle variables. In the generic non-resonant case, it was proved by H. Ito that the Birkhoff transformation converges [**I1**] (see also [**I2**] for the resonant case).

For non-integrable systems, the Birkhoff normal form (3.5) and the conjugating map an around elliptic orbit γ are only approximations in small neighborhoods of γ, which shrink at M increases. The hyperbolic case is simpler. In the case of hyperbolic orbits of analytic symplectic maps in two degrees of freedom, it was proved by J. Moser [**Mo**] that the Birkhoff normal form and transformation do converge. Convergence is more complicated in higher dimensional, and we refer to Banyaga-de La Llave-Wayne [**BLW**], Perez-Marco [**PM**] and to Rouleux [**R**] for recent results. In Rouleux, results of [**BLW**] are used to prove the existence of a local smooth symplectic conjugacy $\kappa^* H = q(I)$ of the metric Hamiltonian H near a hyperbolic fixed point (or orbits) to a smooth normal form $q(I)$, obtained by Borel summation of the formal Birkhoff normal form of H. (In the elliptic case, one only has a conjugacy $\kappa^* H = q(I) + r$ up to a remainder r which vanishes to infinite order at the orbit.) It follows that two metric Hamiltonians with the same Birkhoff normal form at respective hyperbolic closed geodesics have locally symplectically equivalent geodesic flows near those orbits. (The author thanks M. Rouleux for corroborating this point).

3.6. Livsic cohomology. The cohomology problem asks whether a function (cocycle) $F \in C^\infty(S^*M)$ satisfying $\int_\gamma F ds = 0$ for every closed geodesic of the metric g is necessarily a co-boundary, $F = \Xi(f)$ where Ξ is the generator of the geodesic flow G^t and f is a function with some degree of regularity (see [**Ll**] and other works of de la Llave and others for regularity results).

It is relevant to the inverse length spectral problem for the following reason, first observed by Guillemin-Kazhdan [**GK**]: Under a deformation g_ϵ of a metric $g = g_0$ preserving the extended $Lsp(M, g)$ (including multiplicities), one has

$$\int_\gamma \dot{g} ds = 0, \quad \forall \gamma. \tag{3.7}$$

When the cohomology is trivial, one can therefore write $\dot{g} = \Xi(f)$ for some f with the given regularity. Then one can study the harmonic analysis on S^*M to see whether the symmetric 2-tensor $\dot{g}$ can be expressed as $\Xi(f)$. The answer is no for negatively curved surfaces, and that gave the rigidity result of [**GK**]. Since then their result has been improved (see [**CS**] for the most general result on rigidity for negatively curved manifolds), but the general strategy retains some potential for other kinds of metrics.

It is known that the Livsic cohomology problem is always solvable for an Anosov flow on a closed manifold. We refer to [**Ll, Wa**] for recent results. One might ask for the analogous result for hyperbolic billiard flows on bounded domains. It is difficult to formulate the analogous regularity result since the dynamics are not smooth. For instance, the billiards are not even defined at the corners of a domain with piecewise smooth boundary and with concave boundary faces. One studies billiards on the domain by puncturing out the measure zero set of orbits which ever run into the corners. The resulting phase space is then not a closed manifold and there would be complicated issues about regularity of solutions as one approached the punctured set. Perhaps the simplest setting would be that of Sinai billiards, i.e. the exterior of a convex obstacle in a compact manifold (e.g. the exterior of a disc in a two-dimensional torus). The billiards are hyperbolic and there are no glancing orbits or corner orbits to puncture out. But we are not aware of any work on the Livsic equation in this setting.

The Livsic equation can be studied for for non-hyperbolic flows, but there do not seem to exist many studies of it other than manifolds all of whose geodesics are closed (in which case a simple Fourier analysis suffices). One study in the near-integrable case is the article [**L2**] of de la Llave. One does not expect the cohomology to be trivial in general settings, but the results on this equation might have interesting implications for the length spectral deformation problem. As will be explained in §4.3, in certain cases such as integrable systems, the relevant homological equation is not the Livsic equation but an analogue where one integrates over non-degenerate critical manifolds of geodesics.

3.7. Dynamical inverse problems. As mentioned above, inverse spectral theory often proceeds by showing that certain dynamical invariants are invariants of the Laplace spectrum. This gives rise to inverse dynamical problems. In this section, we briefly survey some of the problems and results. A recent survey of rigidity and conjugacy problems has been written by C. Croke [**C3**]. We are grateful to V. Baladi, G. Besson, G. Courtois, C. Gordon, R. de La Llave, and K. F. Siburg for advice on this section.

One of the important problems in dynamics is the following inverse spectral problem in dynamics.

PROBLEM 3.1. *For which (M, g) does the marked length spectrum of (M, g) determine (M, g) up to isometry?*

In the positive direction, V. Bangert [**B**] proved that the marked length spectrum of a flat two-torus determines the flat metric up to isometry. J. P. Otal [**O1**] and C. Croke [**C**] (see also [**CFF, C3**]) independently proved that the marked length spectrum determines surfaces of negative curvature; for more general results see [**C3**]. U. Hamenstädt [**H**] proved that a locally symmetric manifold is determined by its marked length spectrum. An example due to F. Bonahon shows that metric structures more general than Riemannian metrics are not always determined by their marked length spectra.

A related problem concerns the conjugacy rigidity of Riemannian manifolds. The geodesic flows G_j^t of Riemannian manifolds (M_j, g_j) are called C^k conjugate if there exists a time-preserving C^k conjugacy between them, i.e. a C^k homeomorphism $\chi : S^*_{g_1} M_1 \to S^*_{g_2} M_2$ satisfying

$$\chi \circ G_1^t \circ \chi^{-1} = G_2^t. \tag{3.8}$$

If χ extends to a homogeneous symplectic diffeomorphism of the cotangent bundles, the flows are called symplectically conjugate. Two billiard maps are conjugate if there exists a symplectic diffeomorphism $\chi : B^*\Omega_1 \to B^*\Omega_2$ such that

$$\chi \circ \beta_1 \circ \chi^{-1} = \beta_2. \tag{3.9}$$

PROBLEM 3.2. *When does existence of a C^k conjugacy between two geodesic flows (for a given k) imply isometry of the metrics? Does symplectic equivalence of billiard maps of convex domains imply their isometry?*

In some cases, equality of marked length spectra implies C^0 conjugacy of geodesic flows. It was proved independently by J. P. Otal [**O1**] and by C. Croke [**C**] that negatively curved surfaces with the same marked length spectrum have C^0-conjugate geodesic flows, and that such surfaces must be isometric. The corresponding statement in higher dimensions still appears to be open in general. It is stated in [**C3**] that compact flat manifolds are C^∞ conjugacy rigid. In [**H2**], Hammenstädt constructed a time-preserving C^0 conjugacy between negatively curved manifolds with the same marked length spectrum. When this conjugacy is C^1 the work of Besson, Courtois and Gallot [**BCG**] proves that the manifolds must be isometric. The method of [**H**] avoids this regularity issue. The conjugacy problem for certain nilmanifolds was studied by C. Gordon, D. Schueth and Y. Mao in [**GorM, GMS**]. They proved for special classes of 2-step nilmanifolds that C^0 conjugacy of the geodesic flow implies isometry. These manifolds occur in a non-trivial isospectral deformation and have the same marked length spectrum, so these examples also show that equality of marked length spectra does not necessarily imply C^0 conjugacy of the geodesic flows.

There is at least one known example where symplectic conjugacy does not imply isometry, namely for surfaces all of whose geodesics are closed. The geodesic flow of such a Zoll surface was shown by Weinstein to be symplectically equivalent to that of the standard 2-sphere. Aside from related examples, all of the other results known to the author (see [**C3**]) are rigidity results showing that conjugacy implies isometry. The relations between marked length spectral equivalence, conjugacy of geodesic flows and isometry do not seem to have been studied in other settings.

One could pose local versions of the symplectic conjugacy problem, where there should be much less rigidity. A local version (on the level of germs around closed orbits) of symplectic conjugacy is symplectic conjugacy between Poincare maps:

$$\chi : S_{\gamma_1} \subset S^*_{g_1} M_1 \to S_{\gamma_2} \subset S^*_{g_2} M_2, \quad \chi \mathcal{P}_{\gamma_1} \chi^{-1} = \mathcal{P}_{\gamma_2}. \tag{3.10}$$

One could ask:

PROBLEM 3.3. *When does local symplectic conjugacy of Poincare maps at a closed geodesic (or local symplectic conjugacy of geodesic flows) imply local isometry?*

As a special case of the local equivalence problem, suppose that (M, g) is a hyperbolic manifold, and let γ be a closed geodesic. Let g' be a second real analytic metric on M. Suppose that there exists a closed geodesic for g' for which the Poincare maps $\mathcal{P}_\gamma$ are symplectically conjugate. Must g' be a hyperbolic metric?

Even less rigid is the Birkhoff normal form, since in general there only exists a formal (power series) canonical transformation conjugating the germs of the metrics.

PROBLEM 3.4. *To what extent can the germ of a metric g at γ be determined by its Birkhoff normal form at γ?*

The sceptic might suspect that when the formal conjugacy does not converge, the Birkhoff normal form gives little beyond a list of numerical invariants, similar to and no simpler than the wave invariants. When it does converge, there might be too little rigidity to determine very much about the metric.

For instance, it was observed by Y. Colin de Verdière [**CdV**] the Birkhoff normal form of the billiard map around a bouncing ball orbit of a domain with the symmetries of an ellipse determines all of the Taylor coefficients of the domain at the endpoints of the orbit. Hence a real analytic domain with two symmetries is determined by the Birkhoff normal form of the Poincare map at a bouncing ball orbit.

If the domain has only one symmetry, then the calculation of [**CdV**] shows that the Birkhoff normal form at a bouncing ball orbit does not contain enough information to determine the Taylor coefficients at both the top and bottom of the domain. This simple, 2-dimensional example shows how weak an invariant the Birkhoff normal form is.

3.7.1. *Zeta functions of geodesic flows.* We add a final inverse problem which to our knowledge has not been studied before. Let us consider two zeta functions associated to the geodesic flow. The first is the Ruelle zeta function defined by

$$\log\, Z_{(M,g)}(s) = \sum_{\gamma \in \mathcal{P}} \sum_{k=1}^{\infty} \frac{e^{-skL(\gamma)}}{k|det(I - P_\gamma^k)|}, \tag{3.11}$$

Here, we are assuming that the metric is bumpy so that the fixed point sets of the geodesic flow consist of isolated, non-degenerate closed orbits. We denote by $\mathcal{P} = \{\gamma\}$ the set of primitive periodic orbits of the geodesic flow Φ^t and by P_γ the linear Poincare map of γ. As will be seen below, it is a spectral invariant as long as the (extended) length spectrum is simple (multiplicity free) or if the metric has no conjugate points.

The zeta function $\log\, Z_{(M,g)}(s)$ arises as the Laplace transform of the so-called flat trace of the unitary Koopman operator

$$W_t : L^2(S^*M) \to L^2(S^*M), \;\; W_t f(x,\xi) := f(G^t(x,\xi)).$$

This operator does not possess a distribution trace in the usual sense, but the integral of its diagonal part $\int_{S^*M} W(t,\omega,\omega) dV(\omega)$ is well-defined in the sense of pushing forward and pulling back distributions. In this sense, it was observed by V. Guillemin [**G2**] that

$$TrW(t) = C_n Vol(M,g)\delta^{(n)}(t) + \sum_{\gamma \in \mathcal{P}} \sum_{k=1}^{\infty} \frac{\delta(t - L_\gamma)}{|det(I - P_\gamma^k)|}. \tag{3.12}$$

The Laplace transform may be viewed as the flat trace of the resolvent of Ξ, so that

$$\log\, Z_{(M,g)}(s) = Tr(\Xi - s)^{-1}. \tag{3.13}$$

In a formal sense, the Ruelle zeta function is a trace of a spectral function of the geodesic flow and it is natural to wonder how much of the spectrum of W_t can be determined from the traces. In particular, ergodicity, weak mixing and mixing are spectral properties of geodesic flows, i.e. can be read off from the spectrum of W_t. For instance, ergodicity is equivalent to the statement that the multiplicity of the eigenvalue 1 for W_t equals one.

But the trace is not a distribution trace on L^2 in the sense that if $\rho \in C_0^\infty$, $\int_{\mathbb{R}} \rho(t) W_t dt$ is usually not a trace class operator on $L^2(S^*M)$. Hence, it is far from clear that one can read off spectral properties of W_t from these traces. At this time of writing, almost nothing seems to be known about zeta functions except in the case of hyperbolic flows, in which case one knows many properties of the spectrum of the geodesic flow, e.g. that it is mixing. We refer to papers of V. Baladi (see e.g. [**Ba**] for her survey) for background on spectral interpretations of zeta functions and to C. Liverani [**BKL**] et. al. for some possible future extensions. C. Deninger and others (see e.g. [**DS**]) have recently studied the flat trace in non-hyperbolic dynamical settings.

It might be interesting to develop the theory of zeta functions for broader classes of flows. Indeed we are interested in the inverse problem of determining the dynamical properties from the trace.

A related zeta function is defined for $\Re s$ large by

$$\log \zeta_{(M,g)}(s) = \sum_{\gamma \in \mathcal{P}} \sum_{k=1}^{\infty} \frac{e^{-skL(\gamma)}}{k|\det(I - P_{\gamma}^{k})|^{1/2}}, \tag{3.14}$$

It is a Laplace spectral invariant of metrics without conjugate points, even if (as with hyperbolic quotients) the extended length spectrum has multiplicity.

Problem 3.5. *What are the analytical properties of these zeta functions when the geodesic flow has some given dynamical signature. Can one determine apriori from $Z_{(M,g)}(s)$ whether the geodesic flow is ergodic, weak mixing or mixing?*

The first step is to see how far the zeta function can be analytically continued, and whether one can determine the multiplicity of the pole at $s = 0$. We then ask if the multiplicity has a spectral interpretation as the dimension of the invariant L^2-functions under the flow.

We observe that the zeta functions $\zeta_{(M,g)}(s)$ are Laplace spectral invariant for metrics with multiplicity free length spectra as well as for those without conjugate points. Hence this dynamical inverse result would immediately imply that one can determine spectral properties of the geodesic flow from the Laplace spectrum.

4. Wave invariants

We now return to the Laplace spectrum and introduce the wave invariants, which are a more discriminating set of spectral invariants than heat invariants. They arise from the trace of the wave group $U(t) = e^{it\sqrt{\Delta}}$ of (M, g). One forms the (distribution) trace

$$TrU(t) = \sum_{\lambda_j \in Sp(\sqrt{\Delta})} e^{it\lambda_j}. \tag{4.1}$$

It is a tempered distribution on $\mathbb{R}$. We denote its singular support (the complement of the set where it is a smooth function) by Sing Supp $TrU(t)$.

The first result on the wave trace is the Poisson relation on a manifold without boundary,

$$\text{Sing Supp} TrU(t) \subset \; Lsp(M, g), \tag{4.2}$$

proved by Y. Colin de Verdière [**CdV2, CdV3**], Chazarain [**Ch2**], and Duistermaat-Guillemin [**DG**] (following non-rigorous work of Balian-Bloch [**BB2**] and Gutzwiller [**Gutz**]). The generalization to manifolds with boundary was proved by Anderson-Melrose [**AM**] and Guillemin-Melrose [**GM**]. As above, we denote the length of a closed geodesic γ by L_γ. For each $L = L_\gamma \in Lsp(M, g)$ there are at least two closed geodesics of that length, namely γ and γ^{-1} (its time reversal). The singularities due to these lengths are identical so one often considers the even part of $TrU(t)$ i.e. $TrE(t)$ where $E(t) = \cos(t\sqrt{\Delta})$.

We emphasize that (4.2) is only known to be a containment relation. As will be seen below, cancellations could take place if a length $L \in Lsp(M, g)$ is multiple, so that $TrU(t)$ might be smooth at $L \in Lsp(M, g)$.

PROBLEM 4.1. *Is $Lsp(M, g)$ a spectral invariant? Are there any examples where $TrU(t)$ is C^∞ at $t = L_\gamma \in Lsp(M, g)$?*

No example of this seems to be known. Y. Colin de Verdière has pointed out that it is even unknown whether $TrU(t)$ could be smooth on all of $\mathbb{R}\backslash\{0\}$. Moreover, $Lsp(M, g)$ does not include information about multiplicities of lengths. Sunada-type isospectral pairs always have multiple length spectra, and for many (presumably, generic) examples, the length spectra have different multiplicities. The reason is that the geodesics on the base manifold of a Sunada quadruple can split in different ways in the covers. Other examples of isospectral pairs (both in the sense of eigenvalue and length spectra) with different multiplicities of lengths have been constructed by R. Gornet, C. Gordon and others.

The indeterminacy of multiplicities raises a natural problem. Let us recall that the topological entropy of the geodesic flow of (M, g) is the exponential growth rate of the length spectrum:

$$h_{top} := \liminf_{L \to \infty} \log \#\{\gamma : L_\gamma \leq L\}. \tag{4.3}$$

Here, lengths are counted with multiplicity. In the case where geodesics come in families, we count components of the fixed point sets.

PROBLEM 4.2. *Is h_{top} a spectral invariant?*

The question is only non-trivial when multiplicities in the length spectrum grow as fast as the length spectrum, as occurs for compact hyperbolic manifolds in dim $\geq$ 3 and for arithmetic hyperbolic quotients in dimension 2 . As observed by Besson-Courtois-Gallot [**BCG**], an affirmative answer would show that that hyperbolic metrics are spectrally determined among other negatively curved metrics, since they are the unique minimizers of h_{top}. A closely related problem is:

PROBLEM 4.3. *(see [**BCG**]) Suppose that M with $\dim M \geq 3$ possesses a hyperbolic metric, and let $\mathcal{M}_-$ denote the class of negatively curved metrics on M. Is the hyperbolic metric g determined by its spectrum among metrics in $\mathcal{M}_-$? I.e. can there exist another non-isometric metric in this class which is isospectral to the hyperbolic metric?*

4.0.2. *Singular support versus analytic singular support.* While discussing the Poisson relation, we pose the following question concerning the analytic Poisson relation:

PROBLEM 4.4. *Can one tell from $Spec(\Delta)$ if a metric (or the underlying manifold or domain) is real analytic?*

The idea is to calculate the analytic wave front set of $TrU(t)$, i.e.

$$WF_a(\sum_{\lambda_j \in Sp(\sqrt{\Delta})} e^{it\lambda_j}).$$

The analytic wave front set is the complement of the set where the trace is real analytic. When (M, g) is a C^ω (real analytic) Riemannian manifold, the analytic wave front set $WF_aTrU(t)$ is also the set $Lsp(M, g)$, that is, $TrU(t)$ is a real analytic function outside of this discrete set. If (M, g) is a C^∞ but not a C^ω Riemannian manifold, it is plausible that $WF_a TrU(t)$ could contain an interval or be all of $\mathbb{R}$. Thus simply the discreteness of $WF_a TrU(t)$ would say that (M, g) is real analytic.

4.1. Singularity expansions. Much more is true than the Poisson relation: $TrU(t)$ has a singularity expansion at each $L \in Lsp(M, g)$:

$$TrU(t) \equiv e_0(t) + \textstyle\sum_{L \in Lsp(M,g)} e_L(t) \;\; mod \;\; C^\infty, \tag{4.4}$$

where e_0, e_L are Lagrangian distributions with singularities at just one point, i.e. $singsuppe_0 = \{0\}, singsuppe_L = \{L\}$. When the length functional on the loopspace of M is a Bott-Morse functional, the terms have complete asymptotic expansions. In the Morse case (i.e. bumpy metrics), the expansions take the form

$$e_0(t) = a_{0,-n}(t + i0)^{-n} + a_{0,-n+1}(t + i0)^{-n+1} + \cdots \tag{4.5}$$

$$\begin{array}{rcl} e_L(t) & = & a_{L,-1}(t - L + i0)^{-1} + a_{L,0} \log(t - (L + i0)) \\ & + & a_{L,1}(t - L + i0) \log(t - (L + i0)) + \cdots \;, \end{array} \tag{4.6}$$

where $\cdots$ refers to homogeneous terms of ever higher integral degrees ([DG]). The wave coefficients $a_{0,k}$ at $t = 0$ are essentially the same as the singular heat coefficients, hence are given by integrals over M of $\int_M P_j(R, \nabla R, ...)$dvol of homogeneous curvature polynomials. The wave invariants for $t \neq 0$ have the form:

$$a_{L,j} = \sum_{\gamma : L_\gamma = L} a_{\gamma,j}, \tag{4.7}$$

where $a_{\gamma,j}$ involves on the germ of the metric along γ. Here, $\{\gamma\}$ runs over the set of closed geodesics, and where L_γ, $L_\gamma^\#$, m_γ, resp. P_γ are the length, primitive length, Maslov index and linear Poincaré map of γ. (The primitive length of a closed orbit is the least non-zero period of the orbit, i.e the length once around). For instance, the principal wave invariant at $t = L$ in the case of a non-degenerate closed geodesic is given by

$$a_{L,-1} = \sum_{\gamma : L_\gamma = L} \frac{e^{\frac{i\pi}{4} m_\gamma} L_\gamma^\#}{|\det(I - P_\gamma)|^{\frac{1}{2}}}. \tag{4.8}$$

The same formula for the leading singularity is valid for periodic reflecting rays of compact smooth Riemannian domains with boundary and with Neumann boundary conditions, while in the Dirichlet case the numerator must be multiplied by $(-1)^r$ where r is the number of reflection points (see [**GM, PS**]).

The wave invariants for $t \neq 0$ are both less global and more global than the heat invariants. First, they are more global in that they are not integrals of local invariants, but involve the semi-global first return map $\mathcal{P}_\gamma$. One could imagine different local geometries producing the same first return map. Second, they are less global because they are determined by the germ of the metric at γ and are unchanged if the metric is changed outside γ.

Thus, associated to any closed geodesic γ of (M, g) is the sequence $\{a_{\gamma^r,j}\}$ of wave invariants of γ and of its iterates γ^r. These invariants depend only on the germ of the metric at γ. The principal question of this survey may be stated as follows:

PROBLEM 4.5. *How much of the local geometry of the metric g at γ is contained in the wave invariants $\{a_{\gamma^r,j}\}$? Can the germ of the metric g at γ be determined from the wave invariants? At least, can the symplectic equivalence class of its germ be determined?*

As will be discussed in the next section §6, one of the principal results on this problem is the theorem (due to V. Guillemin [**G, G2**], with some input and generalizations by the author [**Z3, Z4**]) that the classical Birkhoff normal form of the metric (or the Poincaré map $\mathcal{P}_\gamma$) at γ is determined by the wave trace invariants. On a global level:

PROBLEM 4.6. *How much of the global geometry (M, g) is contained in the entire set of wave invariants $\{a_{\gamma^r,j}\}$?*

As these questions suggest, one may divide the potential use of wave invariants into two classes: (i) those which use all of the closed geodesics, and (ii) those which involve one or a few closed geodesics. Obviously, (i) is more powerful if one can combine information from all of the geodesics, since it adds precisely the global feature which is lost by studying just one closed geodesic. But it seems very difficult in general to combine information coming from different geodesics.

To the author's knowledge, the global problem of combining wave invariants of all closed geodesics has only been led to successful results on isospectral deformations. We will briefly survey the methods and results in the next section §4.3. Otherwise, the main results use only one or two closed geodesics, and this cannot possibly succeed unless the metrics or domains are assumed real analytic. We will survey the results in the analytic case in §7.5.

Another dichotomy in the use of wave invariants is whether one uses only 'principal term' information at each geodesic, i.e. the invariant $a_{\gamma,-1}$ (4.8), or whether one uses all of the terms. Before the latter is possible, one needs to calculate the lower order terms to some degree. We will discuss the possible calculations in detail in later sections.

4.2. Inverse spectral results using wave invariants. We now consider results which use wave invariants to specify a class of metrics or domains which are spectrally determined in the class or which admit no isospectral deformations. We will discuss the proofs of some of the results in later sections to illustrate the methods.

The positive results based on wave invariant analysis are as follows.

- Negatively curved compact manifolds are spectrally rigid [**GK, C**].
- Simple real analytic surfaces of revolution of 'simple type' (with one critical distance from the axis) are spectrally determined within the class of such surfaces [**Z6**]. Any other surface which is isospectral to a simple surface of revolution must be C^0-integrable [**Z7, S2**]. Smooth surfaces of revolution with a mirror symmetry through the $x-y$ plane are spectrally determined among metrics of this kind [**BH**].
- Simply connected analytic plane domains with two symmetry axes (i.e. with the symmetries of an ellipse) and with a bouncing ball orbit of fixed length L are spectrally determined within this class ([**Z1, ISZ, GM2**], see also [**CdV**] for an earlier result proving spectral rigidity of domains in this class). A closely related result is that convex analytic domains with two symmetry axes are spectrally determined within this class [**Z1**]. The shortest orbit

is necessarily a bouncing ball orbit and of course its length is a spectral invariant [**Gh**].

- Simply connected analytic plane domains with one symmetry, and with a bouncing ball orbit whose orientation of a fixed length L which is reversed by the symmetry, are spectrally determined within this class [**Z5**]. (This implies the preceding result, but we state it separately since it is a new result based on different methods).
- There is a spectrally determined class of convex plane domains (ellipses?) for which each element is spectrally determined among all convex plane domains [**MM**]. (Other isolated spectrally determined examples have recently been given in [**W**]).
- The *mean minimal action* of a convex billiard table is invariant under isospectral deformations [**S1**] (see §8.4 for further discussion).
- The exterior of a two-component analytic obstacle with two symmetries around a bouncing ball orbit between the components is determined by its resonances (poles of its scattering matrix) among other such exterior domains [**Z10**]. The proof is essentially the same as in the interior case, once some known facts on resonance poles (explained to the author by M. Zworski) are added.

4.2.1. *Domains and metrics with the same wave invariants.* A Penrose mushroom type example due to Michael Lifshits (see [**Me**], [**Rau**] for pictures and background) shows that wave invariants are not sufficient to discriminate between all pairs of smooth billiard tables. Indeed, Lifshitz constructs (many) pairs of smooth domains (Ω_1, Ω_2) which have the same length spectra and the same wave invariants at corresponding pairs of closed billiard orbits γ_j of Ω_j $(j = 1, 2)$. It follows that the Poincaré maps $\mathcal{P}_{\gamma_j}$ have the same Birkhoff normal forms as well.

The idea is to exploit the complete integrability of the billiard flow on an ellipse, i.e the fact that it is foliated by caustics. Caustics are curves in the domain with the property that any billiard trajectory which starts off tangent to the caustic will remain tangent to it, as with the confocal ellipses and hyperbolae of an ellipse. The billiard trajectories fall into two families separated by the bouncing ball orbit between the foci. One family consists of trajectories which remain tangent to confocal ellipses (which degenerate to the segment between the foci). The second family consists of trajectories which are tangent to confocal hyperbolae that pass through the segment between the foci. Now divide the major axis into two parts, the segment between the foci and the other two segments. Remove the segment between the foci and replace it with any simple curve with the foci as endpoints, e.g. a 'tongue' below the segment. Next, remove the outer segments of the axis between the foci and replace them with any 'bumps' below the segment. The trajectories which start in the outer bumps never intersect the segment between the foci and therefore never go into the tongue. Similarly trajectories which come into the elliptical part from the tongue pass through the segment between the foci and never go into the outer bumps.

It follows that the closed billiard orbits fall into two families: those which never intersect the segment between the foci and therefore bounce back and forth between the bumps; or those which do intersect this segment and never go into the outer bumps.

To obtain non-isometric domains with the same wave invariants it suffices to reverse the relative orientations of the two families, either by reflecting the segment between the foci at the center (i.e. mirror reversing the tongue) or equivalently by reversing the outer bumps. Since the reversal involution is an isometry on each 'half' of the domain, it does not change the wave invariants for each half. But there is no 'interaction' between the halves (i.e. no closed geodesic intersecting both halves), so the involution preserves all wave invariants. We refer to [**Me, Rau**] for pictures.

The question may occur whether such a domain and its reversal are isospectral or not. This seems dubious, but we are not aware of a proof of it. As mentioned above, equality of wave invariants implies equality of Birkhoff normal form invariants at corresponding pairs. Are the billiard maps of each pair symplectically conjugate? Again dubious, but again we don't know a proof.

4.3. Isospectral deformations and spectral rigidity. One of the first uses of wave invariants and dynamics was the proof by Guillemin-Kazhdan [**GK**] that negatively curved surfaces are spectrally rigid. In the negatively curved case, the Maslov indices are always zero and no cancellation takes place in the wave trace formula as one sums over closed geodesics of the same length. Hence, $Lsp(M, g)$ is a spectral invariant of negatively curved manifolds. An isospectral deformation therefore preserves the length spectrum.

If γ is an isolated, non-degenerate closed geodesic of g, then for any deformation g_t of g, γ deforms smoothly as a closed geodesic γ_t of g_t and one may define its variation

$$\dot{L_\gamma} = \frac{d}{dt}|_{t=0} L_{\gamma_t}. \tag{4.9}$$

It is not hard to compute that

$$\dot{L_\gamma} = \int_\gamma \dot{g} ds, \tag{4.10}$$

where $\dot{g}$ is viewed as a quadratic function on TM and γ is viewed as the curve $(\gamma(s), \gamma'(s))$ in TM.

It follows that whenever the closed geodesics are non-degenerate and of multiplicity one in the length spectrum, we have

$$\int_\gamma \dot{g} ds = 0, \quad \forall \gamma. \tag{4.11}$$

Thus, the Livsic cohomology problem enters. Since $Lsp(M, g)$ is generically simple, this equation holds for generic isospectral deformations. We note that only the principal level of the wave invariants was used here.

In the case of negatively curved surfaces, the geodesic flow is Anosov and it is known that the cohomology is trivial, i.e. that $\dot{g} = \Xi(f)$ for some smooth f (the regularity problem is a separate issue, but for the sake of brevity we will not consider it). The next step is to study harmonic analysis on the unit sphere bundle S^*M to determine if there actually can exist f when $\dot{g}$ is a quadratic form. For surfaces of negative curvature, it was proved in [**GK**] that there cannot exist a smooth solution, and hence there exist no isospectral deformations of negatively curved surfaces. After a series of partial results by Guillemin-Kazhdan, Min-Oo and others, the result was extended to higher dimensions by Croke-Sharafutdinov [**CS**]. In [**SU**], Sharafutdinov- Uhlmann prove infinitesimal spectral rigidity for the more general class of closed two-dimensional manifold without focal points whose geodesic flow is of Anosov type.

Isospectral deformations also lead to the inverse marked length spectral problem and hence (in some cases) to the C^0 conjugacy rigidity problem for the geodesic flows. An isospectral deformation preserves lengths of closed geodesics in the generic case where the multiplicities all equal one (or more generally where no length is cancelled in the wave trace formula), and therefore marks the length spectrum, i.e. gives a correspondence between closed geodesics and lengths, by the length spectrum of the initial metric. More precisely, as one deforms the metric or domain, each one-parameter family of closed geodesics γ_ϵ stays in a fixed free homotopy class of the fundamental group of M. Therefore, an isospectral deformation gives rise to a one-parameter family of metrics or domains with the same marked length spectrum. As discussed above, the marked length spectrum determines the metric for surfaces of negative curvature (Croke, Otal), which gave a different proof of the result of [**GK**].

There do not seem to exist many studies of the Livsic cohomology equation for non-hyperbolic systems. It was proved by Kuwabara using heat invariants that flat tori are spectrally rigid. It might be interesting to review the result in terms of the Livsic equation, and possibly generalizing them to a broader class of completely integrable systems. As will be discussed in §8.1, closed geodesics of integrable systems (such as surfaces of revolution or Liouville tori) come in families on invariant tori $\mathcal{T}$, and the analogue of (4.11) would say that

$$\int_{\mathcal{T}} \dot{g} dx = 0 \tag{4.12}$$

for each invariant torus of the geodesic flow. This is an analogue of the Livsic cohomology equation when geodesics come in non-degenerate critical manifolds. In the case of integrable systems, it could be studied using Fourier analysis relative to the foliation by invariant tori of a completely integrable system. However, it is weaker than the Livsic equation. At least for some Riemannian surfaces, isospectral deformations of metrics with integrable geodesic flows must be through metrics whose geodesic flows are C^0 integrable (see §8.1), and the study of the Livsic type equation (4.12) might combine to form a more powerful tool. The simplest case to consider is whether simple surfaces of revolution on S^2 are spectrally rigid (see §8.1 for background).

To give an extreme example of an open isospectral deformation problem, consider the case of Zoll manifolds. It is simple to see that isospectral deformations of Zoll manifolds must be Zoll [**Z8**], but it is an open problem whether any non-trivial isospectral deformations exist even for $M = S^2$. One of the main problems is that all principal symbol level spectral invariants of Zoll manifolds are the same, so one has to dig further into the wave trace expansions to find obstructions to isospectral deformability.

4.3.1. *Isospectral deformations of domains.* A few words on the analogues for Euclidean domains. To the author's knowledge, there are no spectral rigidity results even for Euclidean plane domains except in the case of a disc. It is not even known whether ellipses are spectrally rigid. The length spectrum of a bounded domain is a spectral invariant for both Dirichlet and Neumann boundary conditions, so an isospectral deformation must preserve the length spectrum and even the marked length spectrum. We will discuss some results of K.-F. Siburg on the marked length spectrum problem in §8.4.

Since the length spectrum is generically an isospectral invariant, it is natural to consider the analogue of the Guillemin-Kazhdan rigidity result for domains with hyperbolic billiards. The billiard flow is hyperbolic, so presumably the Livsic equation can be solved (although the author is not aware of a specific reference). It is not clear how the harmonic analysis would change from the case of negatively curved surfaces.

There is another approach which reduces the wave group to the boundary. Suppose that Ω_t is a smooth one parameter family of smooth compact plane domains such that the Dirichlet (resp. Neumann) spectrum $Spec(\Delta_D^{\Omega(0)})$ (resp. $Spec(\Delta_N^{\Omega(0)})$) is constant in t. Since the area and length of the boundary are spectral invariants, both must be fixed under the deformation. With no loss of generality, we assume that the variation is generated by a normal vector field $\rho\nu$, where ν is the outward unit normal to $\Omega(0)$ and where $\rho \in C^\infty(\partial\Omega(0))$.

The variations of the eigenvalues are given by Hadamard's variational formulae [**O1**]:

$$\dot{\lambda}_j(0) = \begin{cases} \int_{\partial\Omega(0)} \rho\, |\partial_\nu \phi_j(0)|_{\partial\Omega}|^2 dA, & \textit{Dirichlet} \\ \\ \int_{\partial\Omega} \rho\, |\phi_j(0)|_{\partial\Omega(0)}|^2\, dA, & \textit{Neumann}. \end{cases} \tag{4.13}$$

Hence, the infinitesimal deformation condition is that the right hand sides are zero for all j. To normalize the problem, we assume with no loss of generality that the deformation is volume preserving, which implies that

$$\int_{\partial\Omega} \rho dA = 0. \tag{4.14}$$

Any such ρ defines a volume preserving deformation of Ω.

Thus, the infinitesimal deformation is orthogonal to all boundary traces of eigenfunctions:

$$\dot{\lambda}_j(0) = 0 \forall j, \iff \begin{cases} \int_{\partial\Omega(0)} \rho\, |\partial_\nu \phi_j(0)|_{\partial\Omega}|^2 dA = 0, & \textit{Dirichlet} \\ \\ \int_{\partial\Omega} \rho\, |\phi_j(0)|_{\partial\Omega(0)}|^2\, dA = 0, & \textit{Neumann}. \end{cases} \tag{4.15}$$

We may rewrite these conditions in terms of the boundary values of the wave kernel:

$$E^b(t,x,x) := \begin{cases} \partial_{\nu_x}\partial_{\nu_y} U(t,x,x)|_{x\in\partial\Omega}, & \textit{Dirichlet} \\ \\ U(t,x,x)|_{x\in\partial\Omega}, & \textit{Neumann} \end{cases} \tag{4.16}$$

as saying that

$$\int_{\partial\Omega} E^b(t,x,x)\rho(x) dA(x) = 0, \forall t. \tag{4.17}$$

Using the calculations in [**O1**], one can obtain expansions for (4.17) at $t = 0$ similar to a heat kernel expansion in terms of integrals of ρ against polynomials in the (extrinsic) curvature invariants of the boundary. The singularities at $t \neq 0$ in turn give integrals over closed orbits of the billiard map. In the case of an ellipse, one can directly study the boundary traces of the eigenfunctions, and it appears that $\int_\Gamma \rho ds = 0$ for every caustic Γ for the billiard map. It would be interesting to explore this further.

5. Formulae for wave invariants

When studying the inverse spectral problem for individual pairs of metrics rather than isospectral deformations, it is difficult in general to bring in the global dynamics of the geodesic flow. Nothing better is known than to concentrate on a single closed geodesic. So to make progress on the inverse spectral problem, it is crucial to have useful formulae for the wave invariants $a_{\gamma^r,j}$ associated to a closed geodesic γ and its iterates γ^r. The purpose of this section is to survey the known methods of calculation and the formulae which they bring.

There are several potential approaches:

(i) Construct a microlocal parametrix for $e^{it\sqrt{\Delta}}$ at γ and apply a stationary phase method to calculate the wave invariants (cf. [**Z9**]).

(ii) Construct a Birkhoff normal form for $e^{it\sqrt{\Delta}}$ at γ and relate normal form invariants to wave invariants (cf. [**G, Z1, Z3, Z4, Z9**].

(iii) Construct a Birkhoff normal form for the monodromy operator $\mathcal{M}$ on the microlocal solution space $\ker_{m_0}(\Delta - \lambda^2)$ at a fixed initial point m_0 of γ and relate it to the wave invariants (cf. [**SjZ, ISZ**]).

(iv) For bounded domains, apply the Balian-Bloch or Calderon projector methods (cf. [**Z5**]§7).

We now display the formulae and give a brief discussion of the methods. In the following sections, we describe the methods in much more detail.

5.1. The parametrix method. In boundaryless manifolds one can construct a Hadamard parametrix for $\cos t\sqrt{\Delta}(x,x)$ for small times. This operator solves the initial value problem

$$\left\{ \begin{array}{ll} (\frac{\partial}{\partial t}^2 - \Delta)u = 0 & \\ u|_{t=0} = f & \frac{\partial}{\partial t}u|_{t=0} = 0 \end{array} \right. , \tag{5.1}$$

and has the form of the real part of the oscillatory integral

$$\int_0^\infty e^{i\theta(r^2-t^2)} \sum_{j=0}^\infty U_j(x,y)\theta_{reg}^{\frac{n-1}{2}-j} d\theta \mod C^\infty \tag{5.2}$$

where the Hadamard-Riesz coefficients U_j are determined inductively by the transport equations

$$\begin{array}{l} \frac{\Theta'}{2\Theta}U_0 + \frac{\partial U_0}{\partial r} = 0 \\ \\ 4ir(x,y)\{(\frac{k+1}{r(x,y)} + \frac{\Theta'}{2\Theta})U_{k+1} + \frac{\partial U_{k+1}}{\partial r}\} = \Delta_y W U_k. \end{array} \tag{5.3}$$

The solutions are given by:

$$\begin{array}{l} U_0(x,y) = \Theta^{-1/2}(x,y) \\ \\ U_{j+1}(x,y) = \Theta^{-1/2}(x,y) \int_0^1 s^j \Theta(x,x_s)^{1/2} \Delta_2 U_j(x,x_s) ds \end{array} \tag{5.4}$$

where x_s is the geodesic from x to y parametrized proportionately to arc-length, where $\Theta(x,y)$ denotes the volume density in normal coordinates centered at x, and where Δ_2 operates in the second variable.

The simplest case is where the metric is without conjugate points, in which case the parametrix is global in time on the universal cover. It may then be projected to M in the usual way by summing over the deck transformation group. One then takes the trace and computes the coefficients by a stationary phase method.

This was first done by Donnelly [**D**] for negatively curved surfaces and then for all manifolds without conjugate points by the author in [**Z9**]. To get an impression of the complexity of the result, here is the formula for the subprincipal wave invariant (the coefficient of the logarithmic singularity):

$$\begin{aligned} a_{\gamma o} &= \frac{1}{|\det(I-P_\gamma)|^{\frac{1}{2}}}\{C^o_{n,o,o,o}\int_\gamma \Theta_\gamma^{\frac{1}{2}}(\sigma)\{\Theta_\gamma^{-\frac{1}{2}}\mathrm{Hess}(f_\gamma)_\sigma^{-1}(J_\gamma)((0,0),(\sigma,0) \\ &+ \mathrm{Hess}(f_\gamma)_\sigma^{-1}(\Theta_\gamma^{-\frac{1}{2}}((0,0),(\sigma,0)d\sigma \\ &+ C^o_{n,o,o,1}\int_0^{L_\gamma}\int_0^{L_\gamma}\Theta_\gamma^{\frac{1}{2}}(s\sigma_1)\{\mathrm{Hess}(f_\gamma)_\sigma^{-1}\}^2(g_\gamma\Theta_\gamma^{-\frac{1}{2}}J_{[\gamma]})|_{\nu=0}ds \\ &+ C^o_{n,o,o,1}\int_0^{L_\gamma}\int_0^{L_\gamma}\Theta_\gamma^{\frac{1}{2}}(s\sigma_1)\{\mathrm{Hess}(f_\gamma)_\sigma^{-1}\}^3(f_\gamma)\Theta_\gamma^{-\frac{1}{2}}|_{\nu=0}ds \\ &+ C^o_{n,o,1}\int_0^{L_\gamma}\int_0^{L_\gamma}\Theta_\gamma^{\frac{1}{2}}(\sigma_1)(\Delta_2\Theta^{-\frac{1}{2}})((\sigma_1,0),(\sigma_2,0))\cdot d\sigma_1 d\sigma_2\}\,. \end{aligned}$$

Here, $\Theta_\gamma(x) = \Theta(x, \gamma x)$, J_γ is a similar quantity using geodesic polar coordinates, $r = r(x, y)$ denote the distance function of the universal cover, $\gamma \in \Gamma$, the fundamental group, $f_\gamma(x) = r(x, \gamma x)^2$ denotes the displacement function. See [**Z9**] for details. The higher wave invariants are of course exponentially more complicated.

In the case of bounded domains, the parametrix approach is even messier. A microlocal parametrix for $\cos t\sqrt{\Delta}$ near a transversal reflecting ray was constructed by J. Chazarain [**Ch**] (see also [**GM, PS**] for more details). In principle it could be used in the same way as a microlocal parametrix in the boundaryless case, but in the boundary case there are almost always conjugate points and additionally the transport equations and phase are more complicated. Applying the method of stationary phase to all orders to such a parametrix descends into a jungle of formulae. To our knowledge, no concrete inverse spectral results have been proved using the parametrix method.

5.2. Wave invariants and quantum Birkhoff normal forms. Evidently, some guiding principle is needed to civilize the wilderness of formulae. Two such principles have emerged, at least in the case of bounded domains. The first is the method of quantum Birkhoff normal forms. As for classical Birkhoff normal forms, one conjugates the Laplacian to a model normal form on a model space. The wave invariants are then expressed in terms of the coefficients of the normal form. The method is described in detail in the next section. For the moment, we are interested in the formula it gives for the wave invariants.

To state the result we will need some notation. The wave invariants at a closed geodesic γ are invariants of the germ of the metric at γ. For simplicity, we assume that the geodesic is isolated and non-degenerate ($\det(I - P_\gamma) \neq 0$). We assume $\dim M = n + 1$ and introduce Fermi normal coordinates (s, y) along γ. We denote the corresponding metric coefficients by g_{ij}, and refer to the coordinate vector fields $\frac{\partial}{\partial s}, \frac{\partial}{\partial y_j}$ $(j = 1, \ldots, n)$ and their real linear combinations as Fermi normal vector fields along γ. We denote the Riemannian connection, resp. curvature tensor, by ∇ resp. R and refer to contractions of tensor products of the $\nabla^m R$'s with the Fermi normal vector fields as *Fermi curvature polynomials*. Such polynomials will be called *invariant* if they are invariant under the action of $O(n)$ in the normal spaces.

We consider the Jacobi equation $Y'' + R(Y, \gamma')\gamma' = 0$ along γ. The space of complex Jacobi fields along γ is denoted $\mathcal{J}_\gamma$. The linear Poincare map is the monodromy map $Y(t) \to Y(t + L_\gamma)$ on this space. We refer to its eigenvectors $Y_j, \overline{Y}_j$ as Jacobi eigenvectors. We denote by y_{jk} their components relative to the Fermi normal vector fields.

We then define *Fermi–Jacobi polynomials* to be invariant contractions of tensor products of $\nabla^m R$'s against $\frac{\partial}{\partial s}$ and against the Jacobi eigenvectors, with coefficients given by invariant polynomials in the components y_{jk}. We will also use this term repeated integrals over γ of such polynomials. Finally, FJ polynomials whose coefficients are given by polynomials in the Floquet invariants

$$\beta_j = (1 - e^{i\alpha_j})^{-1}$$

are *Fermi–Jacobi–Floquet* polynomials. Here, we assume for simplicity that the closed geodesic is elliptic and that the eigenvalues of its Poincaré map are $\{e^{i\alpha_j}\}$. (The Floquet invariants in the general non-degenerate case are analogous; see [**Z4**] for details).

We define the 'weight' of the data going into a Fermi-Jacobi-Floquet polynomial in terms of its scaling behavior under $g \to \epsilon^2 g$. Thus, the variables g_{ij}, $D^m_{s,y} g_{ij}$ (with $m := (m_1, \ldots, m_{n+1})$), $L := L_\gamma, \alpha_j, y_{ij}, \dot{y}_{ij}$ have the following weights:

$$\text{wgt}(D^m_{s,y} g_{ij}) = -|m|,$$

$\text{wgt}(L) = 1$, $\text{wgt}(\alpha_j) = 0$, $\text{wgt}(y_{ij}) = \frac{1}{2}, \text{wgt}(\dot{y}_{ij}) = -\frac{1}{2}$. A polynomial in this data is homogeneous of weight s if all its monomials have weight s under this scaling.

The general result is [**Z4**]:

THEOREM 5.1. *Let γ be a strongly non-degenerate closed geodesic. Then $a_{\gamma k} = \int_\gamma I_{\gamma;k}(s; g) ds$ where:*
(i) $I_{\gamma;k}(s; g)$ is a homogeneous Fermi–Jacobi–Floquet polynomial of weight $-k-1$ in the data $\{y_{ij}, \dot{y}_{ij}, D^m_{s,y} g\}$ with $m = (m_1, \ldots, m_{n+1})$ satisfying $|m| \leq 2k+4$;
(ii) The degree of $I_{\gamma;k}$ in the Jacobi field components is at most $6k+6$;
(iii) At most $2k+1$ indefinite integrations over γ occur in $I_{\gamma;k}$;
(iv) The degree of $I_{\gamma;k}$ in the Floquet invariants β_j is at most $k+2$.

The formula is simplest in dimension 2, where there is only one Floquet invariant β. We use the notation τ for the scalar curvature, τ_ν for its unit normal derivative along γ, $\tau_{\nu\nu}$ for the Hessian $\text{Hess}(\tau)(\nu, \nu)$. We denote by Y the unique normalized Jacobi eigenvector along γ and by $\dot{Y}$ its time-derivative.

The subprincipal wave invariant $a_{\gamma 0}$ is then given by:

$$a_{\gamma 0} = \frac{a_{\gamma,-1}}{L^\#}[B_{\gamma 0;4}(2\beta^2 - \beta - \frac{3}{4}) + B_{\gamma 0;0}] \tag{5.5}$$

where:
(a) $a_{\gamma,-1}$ is the principal wave invariant (4.8);
(b) $L^\#$ is the primitive length of γ; σ is its Morse index; P_γ is its Poincaré map;
(c) $B_{\gamma 0;j}$ has the form:

$$B_{\gamma 0;j} = \frac{1}{L^\#}\int_o^{L^\#} [a\, |\dot{Y}|^4 + b_1\, \tau |\dot{Y}\cdot Y|^2 + b_2\, \tau \text{Re}\, (\bar{Y}\dot{Y})^2 + c\, \tau^2 |Y|^4 + d\, \tau_{\nu\nu} |Y|^4 + e\, \delta_{j0}\tau] ds$$

$$+\frac{1}{L^{\#}} \sum_{0 \leq m,n \leq 3; m+n=3} C_{1;mn} \frac{\sin((n-m)\alpha)}{|(1-e^{i(m-n)\alpha})|^2} \left| \int_o^{L^{\#}} \tau_\nu(s) \bar{Y}^m \cdot Y^n(s) ds \right|^2$$

$$+\frac{1}{L^{\#}} \sum_{0 \leq m,n \leq 3; m+n=3} C_{2;mn} \operatorname{Im} \left\{ \int_o^{L^{\#}} \tau_\nu(s) \bar{Y}^m \cdot Y^n(s) \left[\int_o^s \tau_\nu(t) \bar{Y}^n \cdot Y^m(t) dt \right] ds \right\}$$

for various universal coefficients.

The formula is very complicated, and it is only the second term in the expansion! That is why one hopes to directly use the information that these coefficients determine the Birkhoff normal form.

In the case of surfaces of revolution, it is possible to simplify the formulae to the point where one can determine the metric along the invariant geodesics. The point is to recover the Taylor expansion of the curvature and its normal derivatives along invariant geodesics. In [**Z2**], a variant of this method was used to show that 'simple' surfaces analytic of revolution (those with just one invariant geodesic) are determined by their wave invariants, hence are spectrally determined among such surfaces. The variant was to use the existence of a global Birkhoff normal form for the metric and Laplacian, which greatly simplified the calculations.

The wave invariants can also be calculated by the method of Birkhoff normal forms in the boundary case, but the method becomes appreciably more difficult. We therefore use a different approach which is inspired by the Balian-Bloch approach to the Poisson formula.

5.3. Balian-Bloch approach in the boundary case. We now consider wave invariants associated to periodic reflecting rays of a bounded plane domain Ω. For simplicity and because they are often useful in applications, we consider wave invariants at a a 2-link periodic reflecting ray γ of length rL_γ (a 'bouncing ball orbit'). We refer to the §3.3 for the basic definitions.

We introduce some further notation: we write $y = f_{\sigma(j)}(x_j)$ where $\sigma(2j+1) = -, \sigma(2j) = +$. The length functional in Cartesian coordinates for a given assignment σ of signs is given by

$$\mathcal{L}_\sigma(x_1, \ldots, x_{2r}) = \sum_{j=1}^{2r-1} \sqrt{(x_{j+1}-x_j)^2 + (f_{\sigma(j+1)}(x_{j+1}) - f_{\sigma(j)}(x_j))^2}. \tag{5.6}$$

We will need formulae for the entries of its Hessian H_{2r} in Cartesian coordinates at the critical point corresponding to the rth repetition of the bouncing ball orbit with a given orientation. Thus, $x_j = 0$ for all j. We will assume, with no essential loss of generality, that $q(x_{odd}) = A, q(x_{ev}) = B$. We put: $a = 1 - Lf_+''(0), b = 1 - Lf_-''(0)$.

The Hessian is given in either angular or Cartesian coordinates by:

$$H_{2r} = \frac{1}{L}\left\{\begin{array}{cccccc} a & 1 & 0 & \dots & 1 \\ 1 & b & 1 & \dots & 0 \\ 0 & 1 & a & 1 & 0 \\ 0 & 0 & 1 & b & 1\dots \\ \dots & \dots & \dots & \dots & \dots \\ 1 & 0 & 0 & \dots & b \end{array}\right\}, \tag{5.7}$$

where $a = 2(1-\frac{L}{R_A})$, $b = 2(1-\frac{L}{R_B})$. A well known formula relates the determinant of this Hessian to that of the Poincaré map:

$$\det(I - P_{\gamma^r}) = L^{2r} \det H_{2r}; \quad (\gamma\ 2-\text{link.}) \tag{5.8}$$

We denote the matrix elements of the inverse Hessian H_{2r}^{-1} at the bouncing ball orbit by h^{ij}. To be precise, the bouncing ball orbit has two possible orientations, one (which we denote by $+$ which starts at the top graph $y = f_+$ and proceeds to the bottom $y = f_-$ and the other – which in the reverse order. We then have two (closely related) length functions and Hessians. We denote the matrix elements of their inverses by $h^{ij}_{\pm}$. One has $h^{pq}_- = h^{p-1,q-1}_+$.

We now state the formulae for the wave invariants.

THEOREM 5.2. [**Z5**] *Let γ be a 2-link reflecting ray of length L_γ, and let γ^r, resp. γ^{-r} be the rth iterate of γ, resp. γ^{-1}. Then there exist polynomials*

$$p_{2,r,j}(\xi_1, \dots, \xi_{2j}; \eta_1, \dots, \eta_{2j}),$$

which are homogeneous of degree $-j$ under the dilation $f \to \lambda f$, which are invariant under the substitutions $\xi_j \iff \eta_j$ and under $f(x) \to f(-x)$, and which have degree $j+1$ in the Floquet data $e^{i\alpha r}$, such that

$$a_{\gamma^r,j} = p_{2,r,j}(f_-^{(2)}(0), f_-^{(3)}(0), \cdots, f_-^{(2j+2)}(0); f_+^{(2)}(0), f_+^{(3)}(0), \cdots, f_+^{(2j+2)}(0)).$$

The leading order term in derivatives of $f_\pm$ has the form

$$r\{(h_+^{22})^j \textstyle\sum_{q=1;q\equiv 0}^{2r} h_+^{2q} + (h_+^{22})^{j-2} \sum_{q;q\equiv 0}^{2r} (h_+^{2q})^3\} f_+^{(2j-1)}(0) f_+^{(3)}(0)$$

$$+r\{(h_+^{11})^j \textstyle\sum_{q;q\equiv 1}^{2r} h_-^{1q} + (h_+^{11})^{j-2} \sum_{q;q\equiv 1}^{2r} (h_+^{1q})^3\} f_-^{(2j-1)}(0) f_-^{(3)}(0)$$

$$-r\{(h_+^{22})^{j-1} h_+^{11} \textstyle\sum_{q=1;q\equiv 1}^{2r} h_\pm^{2q} + (h_+^{22})^{j-2} \sum_{q=1;q\equiv 1}^{2r} (h_+^{2q})^3\} f_+^{(2j-1)}(0) f_-^{(3)}(0)$$

$$-r\{(h_+^{11})^{j-1} h_+^{22} \textstyle\sum_{q=1;q\equiv 0}^{2r} h_\pm^{1q} + (h_+^{11})^{j-2} \sum_{q=1;q\equiv 0}^{2r} (h_+^{1q})^3\} f_-^{(2j-1)}(0) f_+^{(3)}(0)$$

$$+R_{2r}(j^{2j-2} f_+(0), j^{2j-2} f_-(0)),$$

where the remainder $R_{2r}(j^{2j-2} f_+(0), j^{2j-2} f_-(0))$ is a polynomial in the designated jet of $f_\pm$.

In the case where the domain has a symmetry interchanging the top and bottom, so that $f_\pm$ are mirror images of the graph of $y = f(x)$. The length functions for γ^r and γ^{-r} (and hence their Hessians) are equal, so we may drop the subscripts $\pm$. The formula simplifies as follows:

COROLLARY 5.3. *Suppose that γ (as above) is invariant under an isometric involution σ. Then, modulo the error term $R_{2r}(j^{2j-2}f(0))$, we have:*

$$la_{\gamma^r,j-1} = r\{2(h^{11})^j f^{(2j)}(0) + \{2(h^{11})^j \frac{1}{2 - 2\cos\alpha/2}$$

$$+ (h^{11})^{j-2} \sum_{q=1}^{2r} (h^{1q})^3\} f^{(3)}(0) f^{(2j-1)}(0)\}\}.$$

Here, we sum over repeated indices.

The inverse spectral problem is then reduced to analyzing Hessian coefficients. It is not evident to the author how this reduction of the inverse spectral problem would be visible using Birkhoff normal forms.

5.3.1. *Melrose-Marvizi invariants.* In [**MM**], Melrose-Marvizi introduce further spectral invariants of a convex domain which could be interpreted as quantum normal form invariants, namely the normal form of Δ around the closed geodesic $\partial\Omega$. The calculation in [**MM**] involves Melrose's normal form for glancing hypersurfaces and may be viewed as a normal form construction, but one might also follow Lazutkin's construction of whispering gallery quasi-modes to put Δ into normal form as a function of a so-called Airy operator. This would presumably give a new way to calculate the invariants. J. Toth and the author have partial results in this direction (unpublished), and from discussions with G. Popov it appears that they are known to others.

The first two Melrose-Marvizi integrals have the form [**MM**]:

$$I_1 = -2\int_{\partial\Omega} \kappa^{2/3} ds, \quad 1080 I_2 = \int_{\partial\Omega} (9\kappa^{4/3} + 8\kappa^{-8/3}\dot{\kappa}^2) ds.$$

6. Calculation of wave invariants I: Birkhoff normal forms

We now outline the construction of the Birkhoff normal form of Δ in both the boundaryless and boundary cases, and indicate how it is used to calculate the wave invariants. Our goals are to explain how the calculations in Theorem 5.1 are done.

Let γ be a non-degenerate closed geodesic on an n-dimensional Riemannian manifold, and at first let us assume it to be elliptic. During the 70's, various authors (Babic, Lazutkin, Ralston, Guillemin-Weinstein) constructed a series of quasi-modes and quasi-eigenvalues associated to γ. Roughly speaking, the results showed that for each transversal quantum number $q \in \mathbb{Z}^{n-1}$, there was an approximate eigenvalue of the form

$$\lambda_{kq} \equiv r_{kq} + \frac{p_1(q)}{r_{kq}} + \frac{p_2(q)}{r_{kq}^2} + \ldots \tag{6.1}$$

where

$$r_{kq} = \frac{1}{L}(2\pi k + \sum_{j=1}^{n}(q_j + \frac{1}{2})\alpha_j)$$

where the coefficients are polynomials of specified degrees and parities. We refer to [**BB**] for a clear exposition and for details.

A natural question is whether the wave invariants $a_{\gamma j}$ of (4.6) can be determined from the quasi-eigenvalues (6.1), i.e. from the polynomials $p_j(q)$, which could be computed explicitly from the quasimode constructions described in [**BB**]. This question was difficult to answer or even to formulate precisely until the article of V. Guillemin [**G**] on quantum Birkhoff normal forms for the Laplacian around elliptic closed orbits. This article does not refer to quasi-modes or quasi-eigenvalues but it effectively proves that the wave invariants may indeed be expressed in terms of the polynomials $p_j(q)$ and conversely that these polynomials are spectral invariants. A somewhat different proof of this result, which constructed quantum Birkhoff normal forms by adapting the algorithm in [**BB**] for constructing quasi-modes, was given by the author for elliptic orbits in [**Z2**], and for general orbits in [**Z3**].

The results of [**G**] are stated in the terms of quantum Birkhoff normal forms. A number of expositions now exist which explain this notion in detail (see for instance [**Z9**] in addition to the original articles), so we will only briefly review the notion. To put Δ into normal form, is first to conjugate it into a distinguished maximal abelian algebra $\mathcal{A}$ of pseudodifferential operators on a model space, the cylinder $S^1_L \times \mathbb{R}^n$, where S^1_L is the circle of length L. The algebra is generated by the tangential operator $D_s := \frac{\partial}{i\partial s}$ on S^1_L together with the transverse action operators. The nature of these action operators depends on γ. When γ is elliptic, the action operators are harmonic oscillators

$$I_j = I_j(y, D_y) := \frac{1}{2}(D^2_{y_j} + y_j^2),$$

while in the real hyperbolic case they have the form

$$I_j = y_j D_{y_j} + D_{y_j} y_j.$$

When γ is non-degenerate, they involve some mixture of these operators (and also complex hyperbolic actions) according to the spectral decomposition of P_γ. For notational simplicity we restrict to the elliptic case and put

$$H_\alpha := \frac{1}{2}\sum_{k=1}^{n} \alpha_k I_k$$

where $e^{\pm i\alpha_k}$ are the eigenvalues of the Poincare map P_γ.

To put Δ into normal form is to conjugate it to the model space and algebra as a function of D_s and the action operators. The conjugation is only defined in a neighborhood of γ in $T^*M - 0$, i.e. one constructs a microlocally elliptic Fourier Integral operator W from the conic neighborhood of $\mathbb{R}^+\gamma$ in $T^*N_\gamma - 0$ to a conic neighborhood of $T^*_+ S^1_L$ in $T^*(S^1_L \times R^n)$ such that:

(6.2)
$$W\sqrt{\Delta_\psi}W^{-1} \equiv \mathcal{D} + \frac{\tilde{p}_1(\hat{I}_1, \ldots, \hat{I}_n)}{L\mathcal{D}} + \frac{\tilde{p}_2(\hat{I}_1, \ldots, \hat{I}_n)}{(L\mathcal{D})^2} + \cdots + \frac{\tilde{p}_{k+1}(\hat{I}_1, \ldots, \hat{I}_n)}{(L\mathcal{D})^{k+1}} + \ldots$$

where the numerators $p_j(\hat{I}_1, ..., \hat{I}_n), \tilde{p}_j(\hat{I}_1, ..., \hat{I}_n)$ are polynomials of degree j+1 in the variables $\hat{I}_1, ..., \hat{I}_n$, where W^{-1} denotes a microlocal inverse to W. Here, $\mathcal{D} = D_s + \frac{1}{L}H_\alpha$. The kth remainder term lies in the space $\oplus_{j=o}^{k+2} O_{2(k+2-j)}\Psi^{1-j}$, where Ψ^s denotes the pseudo-differential operators on the model space of order s and where $O_{2(k+2-j)}$ denotes the operators whose symbols vanish to the order $2(k+2-j)$ along γ. We observe that (6.2) is an operator version of (6.1).

The inverse result of [**G**] (see also [**Z3, Z4**]) is:

Theorem *Let γ be a non-degenerate closed geodesic. Then the quantum Birkhoff normal form around γ is a spectral invariant; in particular the classical Birkhoff normal form is a spectral invariant.*

In other words, one can determine the polynomials $p_j(\hat{I}_1, ..., \hat{I}_n)$ from the wave trace invariants of Δ at γ. When γ is a non-elliptic (e.g. hyperbolic) geodesics, the action operators have continuous spectra and there are no approximate eigenvalue expansions as in (6.1). Thus the quantum Birkhoff normal form approach is conceptually clearer.

We give a brief review of how wave invariants and normal form invariants for the Laplacian are calculated by the algorithm in [**Z2, Z3, Z4, Z5**] in the case of an elliptic closed geodesic. The algorithm is similar in the general non-degenerate case. We then consider the algorithm for calculating the normal form of the monodromy operator in [**ISZ**].

6.1. Manifolds without boundary. Our algorithm for conjugating Δ to normal form along a closed geodesic γ is inspired by the constructions due to Lazutkin [**L**] and Babich-Buldyrev [**BB**] for constructing quasimodes associated to elliptic closed orbits. The same conjugation method works in the general non-degenerate case.

Since we have recently written an exposition of the method [**Z9**], we only include only a few formal aspects of the calculations, in the hope they provide sufficient explanation of how the wave invariants are calculated. In particular, we wish to emphasize:

- The *semiclassical normal form*, and how it arises from a a semi-classical scaling of the Laplacian along γ. Our method differs from the others [**G, ISZ**] by working entirely on the quantum level and inductively on the Taylor expansion of the metric around γ. The homogeneous normal form (6.2) is obtained from the semiclassical normal form by (roughly speaking) replacing the large parameter by $|D_s|$ along γ.
- The homological equations and the obstructions to their solvability along a closed orbit. The obstructions determine the normal form.

6.2. Quasi-mode heuristics. We wish to conjugate the Laplacian to a normal form on a model space, namely the normal bundle N_γ of γ, or equivalently, the cylinder $S^1_L \times \mathbb{R}^n$, where as above $S^1_L = \mathbb{R}/L\mathbb{Z}$. On the phase space level, the model is $T^*(S^1_L \times \mathbb{R}^n)$ or more precisely the cone $|\eta| \leq \epsilon\sigma, |y| \leq \epsilon$ in the natural symplectic coordinates (s, σ, y, η) corresponding to Fermi normal coordinates along γ.

In the construction of quasi-modes associated to γ, one uses the WKB ansatz

$$\Phi_{kq}(s, \sqrt{r_{kq}}y) = e^{ir_{kq}s} U_{kq}(s, \sqrt{r_{kq}}y, r_{kq}^{-1}),$$

where $r_{kq} = \frac{2\pi}{L}(k + (q + \frac{1}{2})\alpha)$ and where

$$U_{kq}(s, \sqrt{r_{kq}}y, r_{kq}^{-1}) \sim \sum_{j=0}^{\infty} r_{kq}^{-\frac{j}{2}} U_q^{\frac{j}{2}}(s, \sqrt{r_{kq}}y, r_{kq}^{-1}).$$

One then solves asymptotically the eigenvalue

$$\Delta_y e^{ir_{kq}s} U_{kq}(s, \sqrt{r_{kq}}y, r_{kq}^{-1}) \sim \lambda_{kq} e^{ir_{kq}s} U_{kq}(s, \sqrt{r_{kq}}y, r_{kq}^{-1}),$$

satisfying the periodicity condition of being well defined on the cylinder. The intertwining operator W_γ to normal form may be thought of as the operator taking the model eigenfunctions to the quasimodes,

$$W_\gamma \phi_{kq}(s, y) = \Phi_{kq}(s, \sqrt{r_{kq}}y).$$

We now change our point of view and concentrate on the construction of the intertwining operator.

6.3. Semiclassically scaled Laplacian. In view of the form of the quasi-eigenvalue problem,

$$\Delta_u e^{\frac{i}{hL}s} U(s, h^{-\frac{1}{2}}u, h) = \lambda(h) e^{\frac{i}{hL}s} U(s, h^{-\frac{1}{2}}u, h),$$

it is natural to rescale the Laplacian before conjugating it to normal form. In fact, this rescaling is all we retain from the quasi-mode construction. We therefore introduce the unitary operators T_h and M_h on $\mathcal{H}_T$ or equivalently on the 1/2-density version $L^2_T(\mathbb{R}^1 \times \mathbb{R}^n, \Omega_{1/2})$ given by

$$T_h(f(s,u)|ds|^{1/2}|du|^{1/2}) := h^{-n/2} f(s, h^{-\frac{1}{2}}u)|ds|^{1/2}|du|^{1/2}$$

$$M_h(f(s,u)|ds|^{1/2}|du|^{1/2}) := e^{\frac{i}{hL}s} f(s,y)|ds|^{1/2}|du|^{1/2}.$$

Definition *The rescaling of an operator $A_u = a(s, D_s, u, D_u)$ of the adapted model is the operator*

$$A_h := T_h^* M_h^* A T_h M_h.$$

We now rescale the Laplacian in Fermi normal coordinates. It is convenient to first conjugate to the unitarily equivalent 1/2-density Laplacian

$$\Delta_{1/2} := J^{1/2} \Delta J^{-1/2},$$

which can be written in the form:

$$-\Delta_{1/2} = J^{-1/2} \partial_s g^{oo} J \partial_s J^{-1/2} + \sum_{ij=1}^{n} J^{-1/2} \partial_{u_i} g^{ij} J \partial_{u_j} J^{-1/2}$$

$$\equiv g^{oo} \partial_s^2 + \Gamma^o \partial_s + \sum_{ij=1}^{n} g^{ij} \partial_{u_i} \partial_{u_j} + \sum_{i=1}^{n} \Gamma^i \partial_{u_i} + \sigma_o.$$

Here, $\partial_x := \frac{\partial}{\partial x}$, and $J = J(s,u) = \sqrt{g}$ is the volume density in these coordinates.

We then have:

$$-M_h^* \Delta M_h = -(hL)^{-2} g^{oo} + 2i(hL)^{-1} g^{oo} \partial_s + i(hL)^{-1} \Gamma^o + \Delta$$

Conjugation with T_h then gives

$$\begin{aligned} -\Delta_h \quad &= -(hL)^{-2} g^{oo}_{[h]} + 2i(hL)^{-1} g^{oo}_{[h]} \partial_s + i(hL)^{-1} \Gamma^o_{[h]} \\ &+ h^{-1}(\textstyle\sum_{ij=1}^n g^{ij}_{[h]} \partial_{u_i} \partial_{u_j}) + h^{-\frac{1}{2}}(\textstyle\sum_{i=1}^n \Gamma^i_{[h]} \partial_{u_i}) + (\sigma)_{[h]}, \end{aligned}$$

the subscript $[h]$ indicating to dilate the coefficients of the operator in the form, $f_h(s,u) := f(s, h^{\frac{1}{2}}u)$.

Expanding the coefficients in Taylor series at $h = 0$, we obtain the asymptotic expansion

$$\Delta_h \sim \sum_{m=0}^{\infty} h^{(-2+m/2)} \mathcal{L}_{2-m/2}$$

where $\mathcal{L}_2 = L^{-2}$, $\mathcal{L}_{3/2} = 0$ and where

$$\mathcal{L}_1 = 2L^{-1}[i\frac{\partial}{\partial s} + \frac{1}{2}\{\sum_{j=1}^{n} \partial_{u_j}^2 - \sum_{ij=1}^{n} K_{ij}(s) u_i u_j\}].$$

6.4. Conjugation of scaled Laplacian to semi-classical normal form. We now conjugate the semi-classically scaled Laplacian to normal form. The conjugating operators are s-dependent operators acting on the transverse space to γ (they are sometimes called tangential semi-classical Fourier integral operators).

The first step is to conjugate the principal term into quadratic normal form on the transverse space. Since it is quadratic, there exists a metaplectic conjugation $\mu(s)$ depending on s on the transverse space which conjugates it to the operator $\mathcal{D}$ of (6.2). That is, for each $s \in [0, L]$ there exists an operator $\mu(s)$ in the metaplectic representation of the metaplectic group $ML(n, \mathbb{R}) \rightarrow Sp(n, \mathbb{R})$ which acts on the transverse $\mathbb{R}^n$ of the model space. For background on metaplectic operators and details on the conjugation we refer to [**Z3, Z4**]. This conjugates the other operators $\mathcal{L}_j$ to new operators $\mathcal{D}_j$.

Once the principal term is in normal form, we continue by conjugating the lower order terms to functions of the actions by using perturbation theory, i.e. formal series in h. Thus we wish to put the formal series Δ_h into a *semi-classical normal form* by an infinite sequence of conjugations, each one putting one new order in h into normal form. We carry out the procedure to two orders to illustrate the main points.

The first step is to construct $Q_{\frac{1}{2}}(s, x, D_x)$ such that

$$e^{-ih^{\frac{1}{2}} Q_{\frac{1}{2}}} \mathcal{D}_h e^{ih^{\frac{1}{2}} Q_{\frac{1}{2}}}|_o = [-h^{-2} + 2h^{-1}\mathcal{D} + \mathcal{D}_o^{\frac{1}{2}} + \dots]|_o$$

where

(6.3) $\quad |_o$ denotes the restriction of the operator to functions of y,

and where the dots ... indicate higher powers in h. Introduction of the $|_o$ operator is motivated by the construction of quasi-modes, since the normal form is only being applied to the amplitude (a function of y). A more conceptual explanation (suggested by the work of Iantchenko-Sjöstrand-Zworski [**ISZ**]) is that the correct Hilbert space on which to define the operators is the microlocal solution space of $\Delta - \lambda^2$ along γ. These are essentially the space of quasi-modes along the open arc $(0, L)$ of γ. Thus, $|_o$ restricts the operator to the microlocal solution space.

Expanding the exponential, we find that the operator $Q_{\frac{1}{2}}$ then must satisfy the *homological equation*

$$\{[L^{-1}\mathcal{D}, Q_{\frac{1}{2}}] + \mathcal{D}_{\frac{1}{2}}\}|_o = 0.$$

One may solve this equation explicitly by further conjugating $\mathcal{D}$ to D_s. When $\mathcal{D} = D_s$, the homological equation becomes

$$\{[L^{-1}D_s, Q_{\frac{1}{2}}] + \mathcal{D}_{\frac{1}{2}}\}|_o = 0,$$

that is,

$$L^{-1}\partial_s\, Q_{\frac{1}{2}}|_o = -i\{\mathcal{D}_{\frac{1}{2}}\}|_o.$$

Here, $\partial_s A$ is the Weyl operator whose complete symbol is the s-derivative of that of A. To solve this equation we rewrite it in terms of complete Weyl symbols. We will use the notation $A(s,x,\xi)$ for the complete Weyl symbol of the operator $A(s,x,D_x)$. We then arrive at the homological equation

$$L^{-1}\partial_s\tilde{Q}_{\frac{1}{2}}(s,x,\xi) = -i\mathcal{D}_{\frac{1}{2}}|_o(s,x,\xi)$$

We solve with the Weyl symbol

$$\tilde{Q}_{\frac{1}{2}}(s,x,\xi) = \tilde{Q}_{\frac{1}{2}}(0,x,\xi) + L\int_0^s -i\mathcal{D}_{\frac{1}{2}}|_o(u,x,\xi)du$$

where $\tilde{Q}_{\frac{1}{2}}(0,x,\xi)$ is determined by the consistency condition

$$\tilde{Q}_{\frac{1}{2}}(L,x,\xi) - \tilde{Q}_{\frac{1}{2}}(0,x,\xi) = L\int_0^L -i\mathcal{D}_{\frac{1}{2}}|_o(u,x,\xi)du.)$$

To solve the equation, we invoke the fact (which is not obvious) that $\mathcal{D}_{\frac{1}{2}}|_o(u,x,\xi)$ is a polynomial of degree 3 in (x,ξ). We also switch to complex coordinates $z_j = x_j + i\xi_j$ and $\bar{z}_j = x_j - i\xi_j$ in which the action of $r_\alpha(L)$ is diagonal. The homological equation becomes

$$\tilde{Q}_{\frac{1}{2}}(0,e^{i\alpha}z,e^{-i\alpha}\bar{z}) - \tilde{Q}_{\frac{1}{2}}(0,z,\bar{z}) = L\int_0^L -i\mathcal{D}_{\frac{1}{2}}|_o(u,z,\bar{z})du.$$

We put:

$$\tilde{Q}_{\frac{1}{2}}(s,z,\bar{z}) = \sum_{|m|+|n|\le 3} q_{\frac{1}{2};mn}(s)z^m\bar{z}^n$$

and

$$\mathcal{D}_{\frac{1}{2}}|_o(s,z,\bar{z})du = \sum_{|m|+|n|\le 3} d_{\frac{1}{2};mn}(s)z^m\bar{z}^n$$

then the homological equation becomes

$$\sum_{|m|+|n|\le 3}(1-e^{(m-n)\alpha})q_{\frac{1}{2};mn}(0)z^m\bar{z}^n = -iL^2\sum_{|m|+|n|\le 3}\bar{d}_{\frac{1}{2};mn}z^m\bar{z}^n.$$

Since there are no terms with $m=n$ in this (odd-index) equation, and since the α_j's are independent of π over $\mathbb{Z}$, there is no obstruction to the solution. Thus we can simply eliminate the term $\mathcal{D}_{\frac{1}{2}}$.

6.4.1. *The normal form coefficients.* We now consider the second step, where the first obstruction occurs. We thus seek a pseudodifferential operator $\tilde{Q}_1(s,x,D_x)$ and a quadratic polynomial $f_o(I_1,...,I_n)$ in the action operators so that

$$e^{-ih\tilde{Q}_1}\mathcal{D}^{\frac{1}{2}}e^{ih\tilde{Q}_1} = h^{-2}L^{-2} + h^{-1}L^{-1}D_s + h^{-\frac{1}{2}}\mathcal{D}^{\frac{1}{2}}_{\frac{1}{2}} + \mathcal{D}^1_o(s,D_s,x,D_x) + \ldots$$

with

$$\mathcal{D}^1_o(s,D_s,x,D_x)|_o = f_o(I_1,...,I_n).$$

As usual, the dots signify terms of higher order in h. The homological equation is then

$$\{[D_s,\tilde{Q}_1] + \mathcal{D}^{\frac{1}{2}}_o\}|_o = f_o(I_1,...,I_n),$$

or equivalently

$$\partial_s\tilde{Q}_1|_o = \{-\mathcal{D}^{\frac{1}{2}}_o + f_o(I_1,...,I_n)\}|_o.$$

We rewrite the equation in terms of the complete Weyl symbols, to obtain

$$L^{-1}\partial_s\tilde{Q}_1(s,z,\bar{z}) = -i\{\mathcal{D}_o^{\frac{1}{2}}|_o(s,z,\bar{z}) - f_o(|z_1|^2,\ldots,|z_n|^2)\}$$

or equivalently

$$\tilde{Q}_1(s,z,\bar{z}) = \tilde{Q}_1(0,z,\bar{z}) - iL\int_0^s [\mathcal{D}_o^{\frac{1}{2}}|_o(u,z,\bar{z}) - f_o(|z_1|^2,\ldots,|z_n|^2)]du,$$

or again,

$$\tilde{Q}_1(0,e^{i\alpha}z,e^{-i\alpha}\bar{z}) - \tilde{Q}_1(0,z,\bar{z}) = -iL\{\int_0^L \mathcal{D}_o^{\frac{1}{2}}|_o(u,z,\bar{z})du - Lf_o(|z_1|^2,\ldots,|z_n|^2)\}.$$

By construction, $\mathcal{D}_o^{\frac{1}{2}}|_o(u,z,\bar{z})$ is a polynomial of degree 4, so if we put

$$\tilde{Q}_1(s,z,\bar{z}) = \sum_{|m|+|n|\leq 4} q_{1;mn}(s)z^m\bar{z}^n, \qquad f_o(|z_1|^2,\ldots,|z_n|^2) = \sum_{|k|\leq 2} c_{ok}|z|^{2k}$$

and

$$\mathcal{D}_o^{\frac{1}{2}}|_o(s,z,\bar{z})du := \sum_{|m|+|n|\leq 4} d_{o;mn}^{\frac{1}{2}}(s)z^m\bar{z}^n, \qquad \bar{d}_{o;mn}^{\frac{1}{2}} := \frac{1}{L}\int_o^L d_{o;mn}^{\frac{1}{2}}(s)ds,$$

we can solve for the off-diagonal coefficients,

$$q_{1;mn}(0) = -iL^2(1-e^{i(m-n)\alpha})^{-1}\bar{d}_{o;mn}^{\frac{1}{2}}.$$

We also must set the diagonal coefficients equal to zero, and this determines the c_{ok} coefficients:

$$c_{ok} = \bar{d}_{1;kk}^{\frac{1}{2}}.$$

These coefficients are the quantum normal form coefficients. Since we used only algebraic operations on the rescaled Laplacian, it is clear that the coefficients can be calculated in terms of the data described in Theorem 5.1. We refer to [**Z4, Z9**] for further details.

6.5. Bounded domains. There are two serious differences in the normal form for the Laplacian around a bouncing ball orbit of a bounded domain. First, we need to straighten out the domain to define a model space. Secondly, we need the conjugation to normal form to incorporate the boundary conditions.

6.5.1. *The model domain Ω_o.* The configuration space of the model is the infinite strip Ω_o. We denote the coordinate on $[0,L]$ by s and that on $\mathbb{R}$ by y , with dual cotangent coordinates σ,η on $T^*[0,L]\times T^*\mathbb{R}$. As in the boundaryless case, we would like to view the model space as the normal bundle of the orbit. In the case of a bouncing ball orbit in the boundary case, the normal bundle and exponential map are ill-defined at the reflection points but Lazutkin has constructed a nice replacement for them. Namely, he constructs a formal power series map from Ω_ϵ to Ω_0,

$$\Phi:\Omega_\epsilon\to U_\epsilon, \qquad \Phi(s,y) = (\bar{s},\bar{y})$$

of the form

$$\bar{s} = s + \textstyle\sum_{m=2}^{\infty}\kappa_{mm}(s)y^m$$

$$\bar{y} = \textstyle\sum_{p=1}^{\infty}\psi_{pp}(s)y^p$$

with real valued analytic coefficients in s extending analytically to a neighborhood of $[0, L]$ in $\mathbb{C}$ and satisfying the boundary conditions:

$$\{\bar{s} = 0\} \cup \{\bar{s} = L\} = \Phi(\partial\Omega_\epsilon).$$

LEMMA 6.1. *Suppose that $\overline{AB}$ is a bouncing ball orbit. Then there exists a transversal power series map $\Phi : (\Omega_\epsilon, \partial\Omega_\epsilon) \to (\Omega_o, \partial\Omega_o)$ which straightens the domain and puts $\bar{\Delta}$ in the form:*

$$\bar{\Delta} \sim D_{\bar{s}}^2 + B(\bar{s}, \bar{y})(\bar{y}^2 D_{\bar{s}}^2 + D_{\bar{y}}^2) + \Gamma'_s D_s + \Gamma'_y D_y \quad \text{elliptic case}$$

$$\bar{\Delta} \sim D_{\bar{s}}^2 + B(\bar{s}, \bar{y})(D_{\bar{y}}^2 - \bar{y}^2 D_{\bar{s}}^2) + \Gamma'_s D_s + \Gamma'_y D_y \quad \text{hyperbolic case}$$

in the sense that the left and right sides agree to infinite order at $y = 0$. Here, $B(s, y)$ is a transversal power series.

By choosing Φ carefully, one can arrange that the Laplacian $(\Phi^{*-1}\Delta\Phi^*)_{\frac{1}{2}}$ in the transformed coordinates has the principal term $D_{\bar{s}}^2 + (a_{22}(s)\bar{y}^2 D_{\bar{s}}^2 + e_{11}(s)\bar{y}D_{\bar{y}} + b_{00}(s)D_{\bar{y}}^2$. We can further put the principal part into the normal form

$$(6.4) \quad \begin{array}{ll} D_{\bar{s}}^2 + \frac{1}{2}\dot{b}_{00}(s)[\bar{y}^2 D_{\bar{s}}^2 + D_{\bar{y}}^2] & \text{elliptic case} \\ D_{\bar{s}}^2 + \frac{1}{2}\dot{b}_{00}(s)[D_{\bar{y}}^2 - \bar{y}^2 D_{\bar{s}}^2] & \text{hyperbolic case} \end{array} .$$

6.6. Semi-classical scaling of the Laplacian. Having straightened the domain, and hence having transferred the information about the boundary into the metric, we now follow the approach in the boundaryless case by scaling the Laplacian. Following [**L, Z2**], we use the notation $N = h^{-1}$.

We define operators T_N, M_N on the model space $L^2(\Omega_o)$ by

$$(6.5) \quad \begin{array}{l} \bullet \quad T_N f(\bar{s}, \bar{y}) := N f(\bar{s}, N\bar{y}) \\ \bullet \quad M_N f(\bar{s}, \bar{y}) := e^{iN^2\bar{s}} f(\bar{s}, \bar{y}) \end{array}$$

DEFINITION 6.2. *The semiclassically scaled Laplacian is the operator on Ω_o defined by*

$$\bar{\Delta}_N = M_N^* T_N^* \bar{\Delta} T_N M_N.$$

In the straightened form, the rescaled Weyl symbol of $\bar{\Delta}$ has the form:

$$(6.6) \quad \sigma_{\bar{\Delta}_N}^w \sim (\sigma + N)^2 + B(s, \frac{1}{N}y)(\pm y^2 N^{-2}(\sigma + N^2)^2 + N^2\eta^2) + K(s, \frac{1}{N}y).$$

6.7. Conjugation to semiclassical normal form. As in the boundaryless case, we begin by conjugating the principal term (which again is quadratic)

$$(6.7) \quad \sigma_{\bar{\Delta}_N}^w = N^4 + 2N^2[D_{\bar{s}} + \dot{b}_{00}(s)\hat{I}] \mod N$$

into quadratic normal form by an s-dependent metaplectic conjugation on the transverse space to the bouncing ball orbit. Here, $\hat{I}$ denotes the quantum action operator: $\hat{I}^e = \frac{1}{2}(D_{\bar{y}}^2 + y^2)$ in the elliptic case and $\hat{I}^h = \frac{1}{2}(D_{\bar{y}}^2 - y^2)$ in the hyperbolic case.

In [**Z2**], we prove that there exists an $SL(2, \mathbb{R})$-valued function $a_\alpha(\bar{s})$ so that

$$\mu(a_\alpha)^*[D_{\bar{s}} + b_{00}(\bar{s})\hat{I}]\mu(a_\alpha) = D_s + \tfrac{\alpha}{L}\hat{I}, \quad \mu(a_\alpha)(0) = \mu(a_\alpha)(L) = Id,$$

where $\alpha = \int_0^L b_{00}(s)ds$. We then conjugate the resulting operator

$$(6.8) \quad \mathcal{R}_N := \mu(a_\alpha)^* \bar{\Delta}_N \mu(a_\alpha)$$

to the semiclassical normal form

$$F_N(\hat{I})^2 \sim N^4 + N^2 \frac{\alpha \hat{I}}{L} + p_1(\hat{I}) + N^{-2} p_2(\hat{I}) + \cdots \tag{6.9}$$

by conjugations as in the boundaryless case, but now additionally preserving the boundary conditions.

We assume for simplicity that the bouncing ball orbit is elliptic, but the same argument and result hold in the hyperbolic case. We again use the notation of (6.3): we denote by $A|_o$ the restriction of A to functions of y only. In the boundary case, we need to construct complex valued Weyl symbols $P_{j/2}, Q_{j/2}$ so that iterated composition with $e^{N^{-j}(P+iQ)^w_{j/2}}$ will successively remove the lower order terms in $\mathcal{R}_N$ after restriction by $|_o$ and so that the boundary condition is satisfied. It turns out that the boundary condition on the conjugating operator involves only the odd part of the real part and the even part of the imaginary part:

$$P^o_{j/2}(0, \bar{y}, \bar{\eta}) = P^o_{j/2}(L, \bar{y}, \bar{\eta}) = 0, \qquad Q^e_{j/2}(0, \bar{y}, \bar{\eta}) = Q^e_{j/2}(L, \bar{y}, \bar{\eta}) = 0. \tag{6.10}$$

We emphasize that there is no condition on the odd part $Q^o_{j/2}$ of the imaginary part or the even part $P^e_{j/2}$ of the real part.

6.7.1. *A Sturm-Liouville homological equation.* As before, we see what happens in the first and second steps. We first find $P_{\frac{1}{2}}(\bar{s}, \bar{y}, D_{\bar{y}}),\ Q_{\frac{1}{2}}(\bar{s}, \bar{y}, D_{\bar{y}})$ so that the boundary conditions are satisfied and so that

$$e^{-N^{-1}(P+iQ)_{\frac{1}{2}}} \mathcal{R}_N e^{N^{-1}(P+iQ)_{\frac{1}{2}}}|_o = \{N^4 + N^2[D_s + \frac{\alpha}{L}\hat{I}] + \cdots\}|_o. \tag{6.11}$$

Expanding the exponential, we get the *homological equation*:

$$\{[D_s + \frac{\alpha}{L}\hat{I}, (P+iQ)_{\frac{1}{2}}] + \mathcal{R}_{\frac{1}{2}}\}|_o = 0. \tag{6.12}$$

Taking the complete symbol of both sides we get the symbolic homological equation:

$$i\{\sigma + \frac{\alpha}{L} I, P_{\frac{1}{2}} + iQ_{\frac{1}{2}}\} + \mathcal{R}_{\frac{1}{2}}|_o = 0. \tag{6.13}$$

The equation may be rewritten in the form:

$$\partial_{\bar{s}}(P+iQ)_{\frac{1}{2}}(\bar{s}, r_\alpha(\bar{s})(\bar{y}, \bar{\eta})) = -i\mathcal{R}_{\frac{1}{2}}|_o, \tag{6.14}$$

whose solution is given by

$$(P+iQ)_{\frac{1}{2}}(\bar{s}, r_\alpha(\bar{s})(\bar{y}, \bar{\eta})) = (P+iQ)_{\frac{1}{2}}(0) - i\int_0^{\bar{s}} \mathcal{R}_{\frac{1}{2}}(u, \bar{y}, \bar{\eta}) du. \tag{6.15}$$

We thus need to determine $(P+iQ)_{\frac{1}{2}}(0)$ so that the boundary conditions at $\bar{s} = 0$ and $\bar{s} = L$ are satisfied. Thus, we must solve the system

$$\begin{cases} \partial_{\bar{s}} P^e_{\frac{1}{2}} + \frac{\alpha}{L}\{I, P^o_{\frac{1}{2}}\} = \Im \mathcal{R}^e_{\frac{1}{2}}|_o & \partial_{\bar{s}} P^o_{\frac{1}{2}} + \frac{\alpha}{L}\{I, P^e_{\frac{1}{2}}\} = \Im \mathcal{R}^o_{\frac{1}{2}}|_o \\ \partial_{\bar{s}} Q^e_{\frac{1}{2}} + \frac{\alpha}{L}\{I, Q^o_{\frac{1}{2}}\} = -\Re \mathcal{R}^e_{\frac{1}{2}}|_o & \partial_{\bar{s}} Q^o_{\frac{1}{2}} + \frac{\alpha}{L}\{I, Q^e_{\frac{1}{2}}\} = -\Re \mathcal{R}^o_{\frac{1}{2}}|_o \end{cases} \tag{6.16}$$

with the boundary conditions

$$P^o_{\frac{1}{2}}(0) = P^o_{\frac{1}{2}}(L) = 0, \qquad Q^e_{\frac{1}{2}}(0) = Q^e_{\frac{1}{2}}(L) = 0. \tag{6.17}$$

As in the boundaryless case, $\mathcal{R}^{\pm}_{\frac{1}{2}}(u, \bar{y}, \bar{\eta}) = \bar{y} \circ (ar_\alpha^{-1}(u))I$ is a polynomial of degree 3 in $(\bar{y}, \eta)$ in which every term is of odd degree in $(\bar{y}, \eta)$. It follows that $P_{\frac{1}{2}}, Q_{\frac{1}{2}}$ are also odd polynomials of degree 3. Assuming we are in the elliptic case, we change coordinates to the complex cotangent variables $z = \bar{y} + i\bar{\eta}, \bar{z} = \bar{y} - i\bar{\eta}$ and write

$$(6.18)\quad \begin{cases} P^e_{\frac{1}{2}}(\bar{s}, z, \bar{z}) = \sum_{m,n:m+n\leq 3} p^e_{\frac{1}{2}mn}(\bar{s})(z^m\bar{z}^n + \bar{z}^m z^n), \\ P^o_{\frac{1}{2}}(\bar{s}, z, \bar{z}) = \sum_{m,n:m+n\leq 3} p^o_{\frac{1}{2}mn}(\bar{s})(z^m\bar{z}^n - \bar{z}^m z^n). \\ Q^e_{\frac{1}{2}}(\bar{s}, z, \bar{z}) = \sum_{m,n:m+n\leq 3} q^e_{\frac{1}{2}mn}(\bar{s})(z^m\bar{z}^n + \bar{z}^m z^n), \\ Q^o_{\frac{1}{2}}(\bar{s}, z, \bar{z}) = \sum_{m,n:m+n\leq 3} q^o_{\frac{1}{2}mn}(\bar{s})(z^m\bar{z}^n - \bar{z}^m z^n). \end{cases}$$

Then (6.16) may be rewritten in terms of these coordinates. In the elliptic case, we have:

$$(6.19)\quad \begin{cases} \frac{d}{ds}p^e_{\frac{1}{2}mn}(s) + i\frac{\alpha}{L}(m-n)p^o_{\frac{1}{2}mn}(s) = \Im\mathcal{R}^e_{\frac{1}{2}}|_o \\ \frac{d}{ds}p^o_{\frac{1}{2}mn}(s) + i\frac{\alpha}{L}(m-n)p^e_{\frac{1}{2}mn}(s) = \Im\mathcal{R}^o_{\frac{1}{2}}|_o \\ \frac{d}{ds}q^e_{\frac{1}{2}mn}(s) + i\frac{\alpha}{L}(m-n)q^o_{\frac{1}{2}mn}(s) = -\Re\mathcal{R}^e_{\frac{1}{2}}|_o \\ \frac{d}{ds}q^o_{\frac{1}{2}mn}(s) + i\frac{\alpha}{L}(m-n)q^e_{\frac{1}{2}mn}(s) = -\Re\mathcal{R}^o_{\frac{1}{2}}|_o \end{cases}$$

Since $m \neq n$, we can eliminate $p^e_{\frac{1}{2}mn}, q^o_{\frac{1}{2}mn}$ and reduce the homological equation to (uncoupled) second order Sturm-Liouville boundary value problems

$$(6.20)\quad \begin{cases} -\frac{d^2}{ds^2}p^o_{\frac{1}{2}mn}(s) - [\frac{\alpha}{L}(m-n)]^2 p^o_{\frac{1}{2}mn}(s) = \frac{L}{\alpha(m-n)}\frac{d}{ds}\Im\mathcal{R}^e_{\frac{1}{2}}|_o + \Im\mathcal{R}^e_{\frac{1}{2}}|_o \\ -\frac{d^2}{ds^2}q^e_{\frac{1}{2}mn}(s) - [\frac{\alpha}{L}(m-n)]^2 q^e_{\frac{1}{2}mn}(s) = -\frac{L}{\alpha(m-n)}\frac{d}{ds}\Re\mathcal{R}^e_{\frac{1}{2}}|_o + \Re\mathcal{R}^e_{\frac{1}{2}}|_o \\ p^o_{\frac{1}{2}mn}(0) = 0,\ p^o_{\frac{1}{2}mn}(L) = 0;\ \ q^e_{\frac{1}{2}mn}(0) = 0, q^e_{\frac{1}{2}mn}(L) = 0 \end{cases}$$

for the independent variable $p^o_{\frac{1}{2}mn}, q^e_{\frac{1}{2}mn}(s)$. The boundary value problem (6.20) is always solvable unless 0 is an eigenvalue of the operator $D^2_{\bar{s}} - [\frac{\alpha}{L}(m-n)]^2$ (elliptic case). The eigenfunction would have to have the form $\sin(\frac{\alpha}{L}(m-n)\bar{s})$, hence a sufficient condition for solvability is that $\alpha/\pi \notin \mathbb{Q}$. The analogous boundary problem in the hyperbolic case is always solvable.

Thus, the conjugation eliminated the sub-principal term $\mathcal{R}_{\frac{1}{2}}$ and put our operator in normal form up to order N. The exponents $P_{\frac{1}{2}}, Q_{\frac{1}{2}}$ are odd polynomial differential operators of degree 3 with smooth coefficients defined in a neighborhood of $[0, L]$.

6.7.2. *The normal form coefficients emerge.* We carry the process forward one more step because, as in the boundaryless case, the even steps produce the non-trivial normal form coefficients. In outline, the conjugated operator in the second step, $\mathcal{R}^{\frac{1}{2}}_N$, is further conjugated with an exponential of the form $e^{N^{-2}(P+iQ)_1}$. The homological equation is again a Sturm-Liouville boundary problem, with the new feature that diagonal terms with $m = n$ do occur in (6.19). In the elliptic case, a diagonal term is a function of $|z|^2$, hence is even under the involution $z \to \bar{z}$. We write the diagonal terms in the form:

$$(6.21)\qquad \begin{cases} P_j^d(\bar{s},|z|^2) = \sum_{m:m\leq j} p_{jm}^d(\bar{s})|z|^{2m} \\ Q_j^d(\bar{s},|z|^2) = \sum_{m:m\leq j} q_{jm}^d(\bar{s})|z|^{2m} \end{cases}$$

It is impossible to solve the inhomogeneous boundary problem (6.19) for the diagonal terms as in the odd case with zero boundary conditions. To obtain a solvable system, we must add new terms $f_j(|z|^2)$ to the right side:

$$(6.22)\qquad \begin{cases} \frac{d}{ds}p_{jm}^d(s) = \Im \mathcal{R}_{2-j}^d|_o - \Im f_j(|z|^2) \\ \frac{d}{ds}q_{jm}^d(s) = -\Re \mathcal{R}_{2-j}^d|_o + \Re f_j(|z|^2) \\ p_{jm}^d(0) = p_{jm}^d(L) = 0 \\ q_{jm}^d(0) = q_{jm}^d(L) = 0. \end{cases}$$

The right hand sides are determined by the condition that equations with

$$(6.23)\qquad \begin{cases} p_{jm}^d(s) = \int_0^s \{-\Im \mathcal{R}_{2-j}^{j,d}|_o(u,|z|^2) + \Re f_j(|z|^2)\} \\ q_{jm}^d(s) = \int_0^s \{\Re \mathcal{R}_{2-j}^{j,d}|_o(u,|z|^2) + \Im f_j(|z|^2)\}du, \end{cases}$$

satisfies the boundary conditions. This forces

$$(6.24)\qquad f_j(|z|^2) = -\frac{1}{L}\int_0^L \mathcal{R}_{2-j}^{j,d}|_o(u,|z|^2)du.$$

In the case $j=1$, we have then conjugated $\mathcal{R}_N^+$ to $N^4 + N^2\mathcal{R} + Op^w(f_1(I)) + O(N^{-1})$. The coefficients of the polynomial $f_1(I)$ are the first Birkhoff invariants. We note that $Op^w(f_1(I))$ is a function of $\hat{I}$, so we have conjugated to normal form to fourth order. We now repeat the process to complete the normal form.

6.8. Monodromy operator. We now outline a different approach to normal forms due to Iantchenko-Sjöstrand-Zworski [**SjZ, ISZ**]. (See also the appendix). Their goal is to construct the normal form of a so-called *quantum monodromy operator* rather than of the Laplacian or wave group. The normal form of the monodromy operator is apparently the same as the semi-classical normal form of the wave group that we constructed in §6.4 and §6.7. However, it is presented in a conceptually clearer way which works equally well in the boundary or boundaryless case. On the other hand, the method presented above also gives an algorithm for calculating the normal form, which is at present lacking in the approach of [**ISZ**].

The new concepts introduced in [**SjZ, ISZ**] following earlier one-dimensional definitions in works of Helffer-Sjöstrand and Colin de Verdire- Parisse) are the following:

- The monodromy operator $\mathcal{M}(\lambda) : \ker_{m_0}(\sqrt{\Delta}-\lambda) \to \ker_{m_0}(\sqrt{\Delta}-\lambda)$ acting on the space $\ker_{m_0}(\sqrt{\Delta}-\lambda)$ of microlocal solutions of the equation $(\sqrt{\Delta}-\lambda)u(h) = O(\lambda^{-\infty})$ near m_0, where m_0 is is an arbitrarily chosen base point of the given geodesic γ.
- The *flux norm* on the microlocal solution space $\ker_{m_0}(\sqrt{\Delta}-\lambda)$, with respect to which $\mathcal{M}(\lambda)$ is unitary;

- The Grushin reduction of the wave operator to the monodromy operator and the expression of the wave trace as a trace of the monodromy operator.

The microlocal solution space is essentially the space of quasi-modes along arcs of γ, so the monodromy approach is not in essence so different from the approach we outline above. We give a brief exposition of the key ideas, which are discussed in more detail in the appendix by Sjöstrand-Zworski.

Assuming for simplicity that γ projects to an embedded curve in M, we again work on the model space $S^1 \times \mathbb{R}^{n-1}$ given by the normal bundle N_γ to M. We then consider the universal cover $\mathbb{R} \times \mathbb{R}^{n-1}$, fix the base point $m = (0,0)$, and consider microlocal solutions along the geodesic $\mathbb{R} \times \{0\}$. In the simply connected space $\mathbb{R} \times \mathbb{R}^{n-1}$, there is no obstruction to constructing a global quasimode along $\mathbb{R} \times \{0\}$. So we may define

$$\mathcal{M}(\lambda) : \ker_0(\sqrt{\Delta} - \lambda) \to \ker_L(\sqrt{\Delta} - \lambda) \tag{6.25}$$

where $L = L_\gamma$ by

$$\mathcal{M}(\lambda)[u_\lambda]_0 = [u_\lambda]_{(L,0)} \tag{6.26}$$

where $[u_\lambda]_x$ is the germ of the quasimode at the point x. We obtain a quasi-mode when $(I - \mathcal{M}(\lambda))[u_\lambda]_0 = 0$.

To make contact with the wave group, we quote the following observation of Sjöstrand-Zworski (see the appendix):

PROPOSITION 6.3. *We have:*

$$\mathcal{M}(\lambda) = \exp(-iL(\sqrt{\Delta} - \lambda)) : \ker_0(\sqrt{\Delta} - \lambda) \to \ker_L(\sqrt{\Delta} - \lambda).$$

Sketch of Proof (See the appendix for more details) It is obvious that $\exp(-iL(\sqrt{\Delta} - \lambda))$ takes microlocal solutions into themselves since it commutes with $\sqrt{\Delta}$. The rest of the proof is to verify that the complete symbol of the quasimode $\exp(-iL(\sqrt{\Delta} - \lambda))u_\lambda(s,y)$ at $(0,0)$ is the same as that of $u_\lambda(s,y)$ at $s = L$. QED

6.8.1. *Flux norm.* We now describe the *flux norm* $\langle \cdot, \cdot \rangle$ on $\ker_m(\sqrt{\Delta} - \lambda)$. It is motivated by the properties of the probability current density $\mathbf{j}$ of a solution $\psi(x,t)$ of the Schrödinger equation in quantum mechanics (see e.g. [**LL**], §19). It is defined by

$$\mathbf{j} = \frac{\hbar}{i}(\psi \nabla \overline{\psi} - \overline{\psi} \nabla \psi).$$

The integral $\int_S \mathbf{j} \cdot dA$ over a surface measures the probability that the particle described by ψ will cross the surface S in a unit of time. As is pointed out in [**LL**], if $\psi = Ae^{\frac{i}{\hbar}S}$ then $\mathbf{j} = |A|^2 \nabla S$.

The flux norm on microlocal solutions is an invariantly defined version of this. We consider the microlocal solution space $\ker_{m_0}$ for $P = \hbar^2\Delta - 1$ along an initial arc of a closed geodesic γ. We let Y denote a transversal to γ in M. If ν_Y is the unit normal to Y, then

$$||u||^2_{QF} = \int_Y \mathbf{j} \cdot \nu_Y dvol_Y, \quad u \in \ker_{m_0}(P).$$

More precisely and generally,

$$||u||^2_{QF} = \frac{i}{h} \langle [P, \chi]u, u \rangle,$$

where we use the semi-classical notation $P = -h^2\Delta - 1$ with $h = \lambda^{-1}$, and where χ is a microlocal cutoff defined in the universal cover of a small tube around γ, whose complete symbol equals 0 before m_0 and 1 after m_0 in a somewhat smaller tube. We use the semi-classical notation to conform to the notation of [**ISZ, SjZ**] and also because their results apply to much more general semi-classical operators $P(h)$. For further details we refer to [**ISZ, SjZ**].

PROPOSITION 6.4. [**ISZ**] *The monodromy operator* $\mathcal{M}$ *is a unitary operator on* $\ker_{m_0}(\sqrt{\Delta} - \lambda)$ *with respect to the flux norm* $||\cdot||_{QF}$.

6.8.2. *Grushin problem.* One of the basic steps of [**SjZ**] is the proof of a trace formula which relates the trace of the wave group at γ to the traces involving the monodromy operator. Formally, the trace formula says roughly that

$$tr\rho(P/h)\chi(P)A = \sum_{\pm} tr \int_{\mathbb{R}} \rho(z/h)\frac{d}{dz}\log det(I + \mathcal{M}(z \pm i0))\chi(z)dz, \tag{6.27}$$

where $\hat{\rho} \in C_0^\infty(L_\gamma - \epsilon, L_\gamma + \epsilon)$ (compare to (7.1)), and χ is a cutoff to the sphere bundle. The precise statement is given in Theorem 1 of [**SjZ**]:

$$tr\rho(P/h)\chi(P)A = \frac{1}{2\pi}\sum_{-N-1}^{N+1} tr \int_{\mathbb{R}} \rho(z/h)\mathcal{M}(z,h)^k\frac{d}{dz}\mathcal{M}(z,h)\chi(z)dz + O(h^\infty). \tag{6.28}$$

This formula is analogous to one in the case of domains with boundary, where the boundary replaces the transversal, see Proposition 7.1.

To prove the formula, one sets up a microlocal Grushin problem near the closed trajectory γ. One begins by forming a microlocally invertible system

$$\mathcal{P}(z) = \begin{pmatrix} \frac{i}{h}P(z) & R_-(z) \\ R_+(z) & 0 \end{pmatrix} : \mathcal{D}'(M) \times \mathcal{D}'(\mathbb{R}^{n-1}) \to \mathcal{D}'(M) \times \mathcal{D}'(\mathbb{R}^{n-1}),$$

where $R_\pm(z)$ are defined microlocally near γ. The microlocal inverse of $\mathcal{P}(z)$ is given by:

$$\mathcal{E}(z) = \begin{pmatrix} E(z) & E_+(z) \\ E_-(z) & E_{-+}(z) \end{pmatrix} \tag{6.29}$$

We thus obtain the key formula

$$(\sqrt{\Delta} - \lambda)^{-1} \sim E(\lambda) - E_+(\lambda)(I - \mathcal{M})^{-1}E_-(\lambda), \tag{6.30}$$

which reduces the eigenvalue problem to the microlocal invertibility of $(I - \mathcal{M})$. Using this formula one proves (6.27). This equation is analogous to one which occurs in the Fredholm-Neumann reduction of the Dirichlet problem to the boundary of this type (see 7.4).

The authors of [**ISZ, SjZ, ISj**] then put $\mathcal{M}(\lambda)$ into quantum Birkhoff normal form. In view of Proposition 6.3, the normal form of the monodromy operator is just what we called the semi-classical normal form of the wave group. The operation $|_o$ thus has been interpreted as restriction to the microlocal solution space, or as part of a Grushin reduction to the transversal. In [**Z4, Z5**], we proved that the semi-classical normal form was a spectral invariant by first turning into the homogeneous normal form, by proving that the homogeneous normal form was a spectral invariant and then by relating the semi-classical and homogeneous normal forms. The formula (6.27) eliminates this latter step and directly shows that the semi-classical normal form is a spectral invariant by using the Grushin reduction.

At the present time, the only application of the monodromy method to inverse spectral results for plane domains is for the case of bouncing ball orbits of analytic domains with two symmetries as in [**Z2**]. This does not involve any new calculations since the principal symbol of the normal form of the monodromy operator is the classical Birkhoff normal form of $\mathcal{P}_\gamma$ and, as mentioned above, this normal form determines Ω when it is analytic with two symmetries (as proved in [**CdV**]).

7. Calculation of wave invariants II: Balian-Bloch approach

In this section, we describe a method for calculating wave invariants of bounded domains which does not use normal forms or parametrices, but rather is based on an exact formula for the Dirichlet (or Neumann) resolvent in terms of the 'free resolvent'. Our goal is to explain how the calculations of Theorem 5.2 and Corollary 5.3 are done.

For simplicity, we will confine ourselves to calculating wave invariants at a bouncing ball orbit of a bounded simply connected domain $\Omega \subset \mathbb{R}^2$. The method extends with no difficulty to higher dimensions, more general metrics and domains (in particular, non-convex domains) and more general periodic orbits, but the special case is already sufficiently rich and difficult.

The method is based on the use of a classical formula due to C. Neumann and I. Fredholm for the resolvent $R_B^\Omega(z)$ for the boundary problem in Ω with boundary conditions B in terms of layer potentials and the boundary integral operators they induce. This approach was first used by the physicists Balian-Bloch [**BB1, BB2**] in their well-known work on the Poisson relation on three dimensional Euclidean domains, and we will refer to it as the Balian-Bloch approach. The crucial advantage of this approach over the use of normal forms or monodromy operators is its computability. For the first time, one can calculate wave invariants of all orders explicitly and find out what information they contain.

7.1. Wave invariants as semi-classical resolvent trace invariants. First, we explain how wave trace expansions are equivalent to resolvent expansions. We fix a non-degenerate bouncing ball orbit γ of length L_γ, and let $\hat{\rho} \in C_0^\infty(L_\gamma - \epsilon, L_\gamma + \epsilon)$ be a cutoff, equal to one on an interval $(L_\gamma - \epsilon/2, L_\gamma + \epsilon/2)$ which contains no other lengths in $\mathrm{Lsp}(\Omega)$ occur in its support. We then define the regularized resolvent by

$$R_\rho^\Omega(k + i\tau) := \int_{\mathbb{R}} \rho(k - \mu)(\mu + i\tau) R_B^\Omega(\mu + i\tau) d\mu. \tag{7.1}$$

From the resolvent identity (e.g.)

$$R_B^\Omega(\mu + i\tau) = \frac{1}{\mu + i\tau} \int_0^\infty e^{i(\mu + i\tau)t} E_B^{\Omega^c}(t) dt,$$

it follows that

$$R_{\rho B}^\Omega(k + i\tau) = \int_0^\infty \hat{\rho}(t) e^{i(k + i\tau)t} E_B^\Omega(t) dt. \tag{7.2}$$

When γ, γ^{-1} are the unique closed orbits of length L_γ, it follows from the Poisson relation for manifolds with boundary (see §3 and [**GM, PS**]) that the trace

$Tr 1_\Omega R_\rho((k+i\tau))$ of the regularized resolvent on $L^2(\Omega)$ admits a complete asymptotic expansion of the form:

$$Tr 1_\Omega R_\rho(k+i\tau) \sim e^{(ik-\tau)L_\gamma} \sum_{j=1}^{\infty} (B_{\gamma,j} + B_{\gamma^{-1},j}) k^{-j}, \quad k \to \infty \tag{7.3}$$

with coefficients $B_{\gamma,j}, B_{\gamma^{-1},j}$ determined by the jet of Ω at the reflection points of γ. The coefficients $B_{\gamma,j}, B_{\gamma^{-1},j}$ are thus essentially the same as the wave trace coefficients at the singularity $t = L_\gamma$.

7.2. Reduction to the boundary. To calculate $Tr 1_\Omega R_\rho(k+i\tau)$ asymptotically, we use the Fredholm-Neumann reduction of the Dirichlet or Neumann problems in a bounded domain to the boundary. The key formula is similar to (6.30):

$$R_\Omega(k+i\tau) = R_0(k+i\tau) - \mathcal{D}\ell(k+i\tau)(I + N(k+i\tau))^{-1}\gamma \mathcal{S}\ell^{tr}(k+i\tau), \tag{7.4}$$

where $R_0(k+i\tau)$ is the free resolvent $-(\Delta_0 + (k+i\tau)^2)^{-1}$ on $\mathbb{R}^2$. Here, $\gamma : H^s(\Omega) \to H^{s-1/2}(\partial\Omega)$ is the restriction to the boundary, and $\mathcal{D}\ell(k+i\tau)$ (resp. $\mathcal{S}\ell(k+i\tau)$) is the double (resp. single) layer potential is the operator from $H^s(\partial\Omega) \to H^{s+1/2}_{loc}(\Omega)$ defined by

$$\begin{cases} \mathcal{S}\ell(k+i\tau)f(x) = \int_{\partial\Omega} G_0(k+i\tau, x, q) f(q) ds(q), \\ \\ \mathcal{D}\ell(k+i\tau)f(x) = \int_{\partial\Omega} \frac{\partial}{\partial \nu_y} G_0(k+i\tau, x, q) f(q) ds(q), \end{cases} \tag{7.5}$$

where $ds(q)$ is the arc-length measure on $\partial\Omega$, where ν is the interior unit normal to Ω, and where $\partial_\nu = \nu \cdot \nabla$. Also, $\mathcal{S}\ell^{tr}$ is its transpose (from the interior to the boundary), and $G_0(z, x, y)$ is the free Green's function, i.e. the kernel of $R_0(z)$. Further,

$$N(k+i\tau) f(q) = 2 \int_{\partial\Omega} \frac{\partial}{\partial \nu_y} G_0(k+i\tau, q, q') f(q') ds(q') \tag{7.6}$$

is the boundary integral operator induced by $\mathcal{D}\ell$. It is classical that this operator satisfies

$$\begin{array}{ll} (i) & N(k+i\tau) \in \Psi^{-1}(\partial\Omega), \\ \\ (ii) & (I + N(k+i\tau)) : H^s(\partial\Omega) \to H^s(\partial\Omega) \text{ is an isomorphism.} \end{array} \tag{7.7}$$

The reader might now compare the formulae (7.4) and (6.30) to see how $N(\lambda)$ is analogous to $\mathcal{M}(\lambda)$. Moreover, it is used in an analogous way to reduce the calculation of the (distribution) trace of $R_\Omega(k+i\tau)$, formula (7.4) to the boundary. In doing so, we found it [**Z10**] more convenient to combine the interior and exterior problems as follows: We write $L^2(\mathbb{R}^2) = L^2(\Omega) \oplus L^2(\Omega^c)$ and let R_N^Ω, resp. $(k + i\tau), R_D^{\Omega^c}(k+i\tau)$ denote the Neumann resolvent on the exterior domain, resp. the Dirichlet resolvent on the interior domain. We then regard $R_D^{\Omega^c}(k+i\tau) \oplus R_N^\Omega(k+i\tau)$ as an operator on this space. The reason for combining the interior/exterior problems is that we can cycle around the layer potentials in (7.4) when taking the trace and simplify the formula to:

$$Tr_{\mathbb{R}^2}[R_D^{\Omega^c}(k+i\tau) \oplus R_N^\Omega(k+i\tau) - R_0(k+i\tau)] = \frac{d}{dk} \log \det \Big(I + N(k+i\tau) \Big), \tag{7.8}$$

where the determinant is the usual Fredholm determinant. We refer to [**Z10**] for a proof of this apparently well-known formula. This gives:

PROPOSITION 7.1. *Suppose that L_γ is the only length in the support of $\hat{\rho}$. Then,*

$$\int_{\mathbb{R}} \rho(k-\lambda) \frac{d}{d\lambda} \log\det(I + N(\lambda + i\tau))d\lambda \sim \sum_{j=0}^{\infty} B_{\gamma;j} k^{-j},$$

where $B_{\gamma;j}$ are the wave invariants of γ.

7.3. Semi-classical analysis of $N(k + i\tau)$. The next step is to analyse the operator $N(k + i\tau)$ and the geometric series expansion of $(I + N(k + i\tau)^{-1}$. The operator $N(k+i\tau)$ has the singularity of a homogeneous pseudodifferential operator of order -1 on the diagonal (in fact, it is of order -2 in dimension 2) and that is the way it is normally described in potential theory. However, away from the diagonal, it has a WKB approximation which exhibits it as a semi-classical Fourier integral operator with phase $d_{\partial\Omega}(q, q') = |q - q'|$ on $\partial\Omega \times \partial\Omega$, the boundary distance function of Ω. Indeed, the free Green's function in dimension two is given by:

$$G_0(k + i\tau, x, y) = H_0^{(1)}((k + i\tau)|x - y|) = \int_{\mathbb{R}^2} e^{i\langle x-y,\xi\rangle}(|\xi|^2 - (k + i\tau)^2)^{-1} d\xi.$$

Here, $H_0^{(1)}(z)$ is the Hankel function of index 0. It has the near diagonal and off-diagonal asymptotics,

$$(7.9) \qquad (i)\ H_0^{(1)}(z) \sim \begin{cases} -\frac{1}{2\pi} \ln|z| \text{ as } |z| \to 0, \\ (ii)\ e^{i(z|x-y| - \pi/4)} \frac{1}{|z|^{1/2}} \text{ as } |z| \to \infty. \end{cases}$$

By the explicit formula we have:

$$\begin{aligned} \tfrac{1}{2}N(k + i\tau, q(\phi), q(\phi')) &= \partial_{\nu_y} G_0(\mu, q(\phi), q(\phi')) \\ &= -\ (k + i\tau) H_1^{(1)}(k + i\tau |q(\phi) - q(\phi')|) \\ &\quad \times\ \cos\angle(q(\phi) - q(\phi'), \nu_{q(\phi)}). \end{aligned}$$

Combining with (7.9(ii)), we see that when $|q(\phi) - q(\phi')| \geq |k|^{1-\epsilon}$ for some $\epsilon < 1$, , $N(k + i\tau, q(\phi), q(\phi'))$ is a semi-classical Fourier integral kernel whose phase is the boundary distance function. For a convex domain, the boundary distance function generates the billiard map of $\partial\Omega$ and hence we view $N(k + i\tau)$ as a global quantization of the billiard map. For non-convex domains, the boundary distance function additionally generates 'ghost orbits' which in part exit the domain, but these only present a mild complication. We refer to [**HZel**] for discussion of these orbits and to [**Z2, HZel**] for further discussion of $N(k)$.

We now explain how to use Proposition (7.1) to calculate the wave trace coefficients $B_{\gamma,j}$ at a closed geodesic. When calculating these coefficients we first compose with a special kind of semiclassical cutoff operator $\chi(x, k^{-1}D_x)$ to a neighborhood of γ on both sides of (7.4). Since we are not dealing with conventional Fourier integral operators, it must be proved that composition with such a cutoff does in fact microlocalize to γ. We will suppress this issue until the end.

At least formally, we expand $(I+N(k+i\tau))^{-1}$ in a finite geometric series plus remainder:
(7.10)
$$(I+N(\lambda+i\tau))^{-1}=\sum_{M=0}^{M_0}(-1)^M\; N(\lambda)^M+(-1)^{M_0+1}\; N(\lambda)^{M_0+1}(I+N(\lambda+i\tau))^{-1}.$$

To calculate a given wave invariant, we need to show that, for each order k^{-J} in the trace expansion of Corollary (7.1), there exists $M_0(J)$ such that
(7.11)
$$(i)\quad \sum_{M=0}^{M_0}(-1)^M Tr\int_{\mathbb{R}}\rho(k-\lambda)\; N(\lambda)^M N'(k+i\tau)d\lambda$$
$$=\sum_{j=0}^{J} B_{\gamma;j}k^{-j}+O(k^{-J-1}),$$
$$(ii)\quad Tr\int_{\mathbb{R}}\rho(k-\lambda)N(\lambda)^{M_0+1}(I+N(\lambda+i\tau))^{-1}N'(k+i\tau)d\lambda=O(k^{-J-1}).$$

We outline the method for obtaining the wave trace asymptotics at a closed geodesic γ from the first term. Since N has singularities on the diagonal, one cannot just apply the stationary phase method to the trace. Rather one has to regularize the operator. We do this by separating out the tangential and transversal parts of N by introducing a cutoff of the form $\chi(k^{1-\delta}|q-q'|)$ to the diagonal, where $\delta>1/2$ and where $\chi\in C_0^\infty(\mathbb{R})$ is a cutoff to a neighborhood of 0. We then put

$$N(k+i\tau)=N_0(k+i\tau)+N_1(k+i\tau),\quad \text{with} \tag{7.12}$$

$$\begin{cases} N_0(k+i\tau,q,q')=\chi(k^{1-\delta}|q-q'|)\; N(k+i\tau,q,q'),\\ \\ N_1(k+i\tau,q,q')=(1-\chi(k^{1-\delta}|q-q'|))\; N(k+i\tau,q,q'). \end{cases} \tag{7.13}$$

The term N_1 is a semiclassical Fourier integral kernel quantizing the billiard map, while N_0 behaves like an Airy operator close to the diagonal with the singularity of a homogeneous pseudodifferential operator on the diagonal.

Now consider the powers $N(k+i\tau)^M$ in the first term (i) of (7.11). We write

$$(N_0+N_1)^M=\sum_{\sigma:\{1,\dots,M\}\to\{0,1\}} N_{\sigma(1)}\circ N_{\sigma(2)}\circ\cdots\circ N_{\sigma(M)}. \tag{7.14}$$

We regularize N^M by eliminating the factors of N_0 from each of these terms. This is obviously not possible for the term N_0^M but it is possible for the other terms. By explicitly writing out the composition in terms of Hankel functions and using the basic identities for these special functions, we prove that $N_0\circ N_1\circ\chi_0(k+i\tau,\phi_1,\phi_2))$ is a semiclassical Fourier integral operator on $\partial\Omega$ of order -1 associated to the billiard map. Thus, composition with N_0 lowers the order. The remaining terms N_0^M, when composed with a cut-off to γ, do not contribute asymptotically to the trace.

The successive removal of the factors of N_0 thus gives a semi-classical quantization of the billiard map near γ. We then calculate the traces of each term by the stationary phase method and obtain the result stated in (5.2). The terms displayed there with the maximum number of derivatives of the defining function of $\partial\Omega$ come only from the N_1^M terms. Thus, although the removal of the N_0 factors does change the amplitude of the N_1 factors and does contribute to the wave trace, it turns out to be negligible in the inverse spectral problem. This is one of the principal virtues of the Balian-Bloch approach.

We will explain how the explicit formulae for the wave invariants are derived in the next section.

7.3.1. *Remainder estimate.* We now address some sketchy remarks to the remainder estimate, in the hopes of convincing the reader that it is plausible to expect the remainder trace to be small. The ability to insert a microlocal cutoff to γ is crucial here. The main obstacle to the remainder is that the norm of $N(k+i\tau)$ fails to decrease with increasing τ due to the Airy part associated to creeping rays. This again is a difference to the monodromy operator.

To obtain a small remainder we set the spectral parameter in N equal to $k + i\tau \log k$. The presence of the $\log k$ in the imaginary part changes the wave trace expansions by k^{-CrL_γ}, but this does not hurt the expansions such the remainder estimates will be of lower order. We then estimate the remainder

$$Tr \int_{\mathbb{R}} \rho(k-\lambda)N(\lambda)^{M_0+1}(I+N(\lambda+i\tau))^{-1}\, \chi_\gamma\, N'(k+i\tau)d\lambda$$

by applying the Schwarz inequality for the Hilbert-Schmidt inner product, and using the relation $(I+N(\lambda+i\tau))^{-1}$ to the Poisson kernel to estimate this factor. These estimates leave a trace of $N^M N^{*M}$ for fixed M, microlocalized to γ. We regularize these traces as above and obtain oscillatory integrals whose phases have critical points corresponding to M-link closed circuits which being at some point q, bounce along the boundary until q' and then return to q by traversing the links in reverse order. However, the cutoffs to γ force the links in critical paths to point in the direction of γ and hence to be of length roughly ML_γ. The imaginary part $i\tau \log k$ of the semiclassical parameter then contributes a damping factor of $e^{-\tau ML_\gamma \log k}$ for each link. The links correspond to the N_1 factors. Thus, for each string, we have one k^{-1} for each N_0 factor and one $e^{-\tau ML_\gamma \log k}$ for each N_1 factor. For sufficiently large τ these combine to give a factor of k^{-R} for any prescribed R.

7.3.2. $N(\lambda)$ *versus* $\mathcal{M}(\lambda)$. We digress momentarily to compare $N(\lambda)$ and the monodromy operator $\mathcal{M}(\lambda)$. Both operators arise as a reduction of the wave group to the boundary (or a transversal) and are quantizations of the billiard map, $N(\lambda)$ globally and $\mathcal{M}(\lambda)$ microlocally at a closed orbit γ. (An earlier reduction also occurs in [**MM, P, P3**] but in a rather different way.) In applications to boundary problems, they are quite similar, as evidence by the comparisons of (6.27) with Proposition 7.1 and of (6.30) with (7.4).

But there do exist rather important differences. The operator $N(\lambda)$ is not a standard Fourier integral operator on $\partial\Omega$, while $\mathcal{M}(\lambda)$ is one. The diagonal singularities of $N(\lambda)$ require a complicated regularization procedure. Moreover it is a global invariant of $\partial\Omega$, not a microlocal one at γ; On the other hand, $N(\lambda)$ is just the restriction of a canonically defined free Green's kernel to $\partial\Omega \times \partial\Omega$, and thus is an elementary and computable object. By comparison, $\mathcal{M}(\lambda)$ must be constructed by some kind of parametrix method. It was precisely the complexity of microlocal parametrices for the wave group, even at periodic reflecting rays, which motivated our turning to the Balian-Bloch approach.

7.4. Stationary phase expansion. After regularizing the traces, we end up with oscillatory integrals in the standard sense and obtain expansions by applying stationary phase. The amplitudes and phases are canonical, since we began with canonical amplitudes and phases and since the regularization procedure is essentially the same for all domains. So most of the complexity of the expansion is due to the stationary phase method.

The key point in inverse spectral theory is to identify data in the stationary phase term of order k^{-j} which represents 'new data' not contained in the previous terms. Terms of the coefficient of k^{-j} which contain the maximum number of derivatives of the phase are the most important ones. Thus we face the combinatorial problem of locating such terms in the stationary phase expansion. The Feynman diagram method of assigning labelled graphs to each term in the expansion proves to be very effective for this purpose.

Consider a general oscillatory integral $Z_k = \int_{\mathbb{R}^n} a(x)e^{ikS(x)}dx$ where $a \in C_0^\infty(\mathbb{R}^n)$ and where S has a unique critical point in supp$a$ at 0. Let us write H for the Hessian of S at 0. The stationary phase expansion takes the form:

$$Z_k = \left(\frac{2\pi}{k}\right)^{n/2} \frac{e^{i\pi sgn(H)/4}}{\sqrt{|detH|}} e^{ikS(0)} Z_k^{h\ell},$$
$$\text{where } Z_k^{h\ell} = \sum_{j=0}^\infty k^{-j} \left\{ \sum_{(\Gamma,\ell):\chi_{\Gamma'}=j} \frac{I_\ell(\Gamma)}{S(\Gamma)} \right\}.$$

Here, the sum runs over the set $\mathcal{G}_{V,I}$ of labelled graphs (Γ, ℓ) with V closed vertices of valency ≥ 3 (each corresponding to the phase), with one open vertex (corresponding to the amplitude), and with I edges. Further, the graph Γ' is defined to be Γ minus the open vertex, and $\chi_{\Gamma'} = V - I$ equals its Euler characteristic. We note that there are only finitely many graphs for each χ because the valency condition forces $I \geq 3/2V$. Thus, $V \leq 2j, I \leq 3j$.

The function ℓ 'labels' each end of each edge of Γ with an index $j \in \{1, \ldots, n\}$. Also, $S(\Gamma)$ denotes the order of the automorphism group of Γ, and $I_\ell(\Gamma)$ denotes the 'Feynman amplitude' associated to (Γ, ℓ). By definition, $I_\ell(\Gamma)$ is obtained by the following rule: To each edge with end labels j, k one assigns a factor of $\frac{-1}{ik}h^{jk}$ where $H^{-1} = (h^{jk})$. To each closed vertex one assigns a factor of $ik\frac{\partial^\nu S(0)}{\partial x^{i_1}\cdots\partial x^{i_\nu}}$ where ν is the valency of the vertex and $i_1 \ldots, i_\nu$ at the index labels of the edge ends incident on the vertex. To the open vertex, one assigns the factor $\frac{\partial^\nu a(0)}{\partial x^{i_1}\ldots\partial x^{i_\nu}}$, where ν is its valence. Then $I_\ell(\Gamma)$ is the product of all these factors. To the empty graph one assigns the amplitude 1. In summing over (Γ, ℓ) with a fixed graph Γ, one sums the product of all the factors as the indices run over $\{1, \ldots, n\}$.

7.4.1. *The data* $\boldsymbol{f_\pm^{2j}(0)}$. An analysis of the diagrams and amplitudes shows that the jth even Taylor coefficients $f_\pm^{(2j)}(0)$ of the boundary defining functions appear first in the k^{-j+1} term.

When the domain has one symmetry axis, which we visualize as an up/down symmetry, the terms with this data have the form

$$2rL(h^{11})^j f^{(2j)}(0) + \cdots,$$

where $\cdots$ refers to terms with $\leq 2j - 1$ derivatives.

When the domain has two symmetries, a left/right symmetry in addition to an up/down symmetry, the odd Taylor coefficients vanish and we see immediately that the even Taylor coefficients can be determined inductively from the wave trace invariants. This gives a new proof that analytic domains with two symmetries can be determined from the wave trace invariants at a bouncing ball orbit which is one of the symmetry axes.

This does not quite prove that such domains are spectrally determined among other analytic domains with two symmetries, since the length of the bouncing ball orbit must be known in order to obtain the wave invariants. This length is a spectral invariant if the domains are additionally convex [**Gh**]. For non-convex domains we needed to add as an assumption that the bouncing ball symmetry axis had a fixed length L in [**Z1**].

7.4.2. *The data* $\boldsymbol{f_{\pm}^{(2j-1)}(0)}$. When we do not assume a left/right symmetry, the odd Taylor coefficients are non-zero in general, and the problem arises whether there is sufficient information in the wave invariants to determine all of the even and odd Taylor coefficients of the boundary defining function (or curvature function) of a domain with one symmetry. We assume the axis of symmetry is a bouncing ball orbit whose orientation is reversed by the symmetry.

An analysis of the diagrams and amplitudes for the odd Taylor coefficients $f_{\pm}^{(2j-1)}(0)$ show that they appear first in the term of order k^{-j+1}. It turns out that five diagrams contain this data, but the amplitudes of three automatically vanish. They two amplitudes have the following forms:

(i) $(h_{\pm}^{pp})^{j-1} h_{\pm}^{qq} h_{\pm}^{pq} f^{(2j-1)}(0) f^{(3)}(0)$.

(ii) $(h_{\pm}^{pp})^{j-2} (h_{\pm}^{pq})^3 f^{(2j-1)}(0) f^{(3)}(0)$.

To decouple them, we need to analyze the behavior of power sums of columns the Hessian matrix elements. In [**Z5**] we proved that cubic column sums are linearly independent from linear ones as $r \to \infty$.

7.5. Positive results for analytic domains and metrics. We now review the proof in [**Z5**] that the Taylor coefficients $f^{2j-1}(0), f^{2j}(0)$ can be determined from the wave invariants $B_{\gamma^r,j}$ as r varies over $r = 1, 2, 3, \ldots$.

It suffices to separately determine the two terms

$$(7.15)\qquad \begin{array}{l} 2(h_{2r}^{11})^2 \left\{ f^{(2j)}(0) + \frac{1}{2-2\cos\alpha/2}\, f^{(3)}(0) f^{(2j-1)}(0) \right\}, \\ \text{and} \quad \left\{ \sum_{q=1}^{2r} (h_{2r}^{1q})^3 \right\} f^{(3)}(0) f^{(2j-1)}(0). \end{array}$$

It is easy to see that the terms decouple as r varies if and only if the cubic sums $\sum_{q=1}^{2r} (h_{2r}^{1q})^3$ are non-constant in $r = 1, 2, 3, \ldots$. By the explicit calculation in [**Z2**], we have:

$$\sum_{q=1}^{2r} (h^{pq})^3 =$$

$$2r\sum_{k_1,k_2=0}^{2r} \frac{1}{(\cosh\alpha/2+\cos\frac{k_1\pi}{r})(\cosh\alpha/2+\cos\frac{k_2\pi}{r})(\cos\alpha/2+\cosh\frac{(k_1+k_2)\pi}{r})}.$$

It is obvious that the sum is strictly increasing as r varies over even integers.

This independence of the linear and cubic sums permits us to use an inductive argument when the domain has the given symmetry. From the $j = 0$ term we determine $f''(0)$. Indeed, $(1 - Lf^{(2)}(0) = \cos(h)\alpha/2$ and α is a wave trace invariant. From the $j = 2$ term we recover $f^3(0), f^4(0)$. The induction hypothesis is then that the Taylor polynomial of f of degree $2j-2$ has been recovered by the $j-1$st stage. By the decoupling argument we can determine $f^{2j}(0), f^{2j-1}(0)$. (Strictly speaking, there is the minor annoyance of the factor of $f^{(3)}(0)$, which is resolved in [**Z5**]).

7.5.1. *Final comments on the Balian-Bloch invariants.* It would of course be desirable to remove all the symmetry assumptions, if possible. Thus one would need to recover the Taylor coefficients $f_{\pm}^{(2j)}(0), f_{\pm}^{(2j-1)}(0)$ when $f_+ \neq f_-$ from the wave invariants. The problem is that the induction does not work because the term $R_{2r}(j^{2j-2}f(0))$, which we do not need to know in the symmetric case, might have a different dependence on the Taylor coefficients of $f_{\pm}$ than the 'principal term', i.e. the one with the highest derivatives. The latter is basically a sum of the coefficients of $f_\pm$ while $R_{2r}(j^{2j-2}f(0))$ could involve any symmetric polynomial in these coefficients. At the present time, we do not even know whether the set $\{f_+^{(2)}(0), f_-^{(2)}(0)\}$ of second derivatives can be determined from the second wave coefficient.

Some hope has been provided by computer calculations of C. Hillar [**Hil**]. The results suggest that Hessian column power sums with different powers are linearly independent as $r \to \infty$. This would give quite a large supply of Taylor coefficients.

8. Surfaces and domains with integrable dynamics

As mentioned above, Birkhoff normal forms at a periodic orbit (on both the classical and quantum level) are approximations to the Hamiltonian by an integrable one. When the Hamiltonian is completely integrable, its dynamics and Birkhoff normal forms are very special and the metrics might be rather spectrally rigid, at least in low dimensions. This is the case with flat metrics [**Ku3**]. In this section, we consider some model inverse spectral problems on two-dimensional surfaces or domains with integrable dynamics.

Let us recall that the geodesic flow of an n-dimensional Riemannian manifold (M, g) is *completely integrable* if it commutes with a Hamiltonian action of $\mathbb{R}^n$ on $T^*M - 0$ ($n = \dim M$). That is, the metric Hamiltonian $|\xi|_g$ satisfies

$$\{|\xi|_g, p_j\} = 0 = \{p_i, p_j\}, \;\; i, j = 1, \ldots, n$$

where $p_j : T^*M - 0 \to \mathbb{R}$ are homogeneous of degree one and are independent in the sense that

$$dp_1 \wedge dp_2 \wedge \cdots \wedge dp_n \neq 0 \;\text{ on a dense open subset } U \subset T^*M.$$

The orbits of an $\mathbb{R}^n$ action give a (usually singular) foliation of S_g^*M (the unit sphere bundle for the metric), called the Liouville foliation, by affine manifolds of the form $\mathbb{R}^n \cdot (x, \xi) \equiv \mathbb{R}^n/\Gamma$ where Γ is the isotropy group at (x, ξ). If it is a lattice of full rank, the orbit is a torus T^n of dimension n. If it has less than full rank, the orbit type is $T^k \times \mathbb{R}^{n-k}$. The isotropy group might have the form $\mathbb{R}^k \times \mathbb{Z}^{n-k}$ in which case the orbit becomes a singular torus of dimension $n - k$. The orbits which contain periodic geodesics are sometimes called 'periodic tori', and they furnish the components $\mathcal{T}$ in the trace formula.

In the case of bounded plane domains, complete integrability of the billiard map β means that $B^*\partial\Omega$ is foliated by invariant curves. The well-known conjecture of Birkhoff is that ellipses are the only example of compact smooth Euclidean plane domains with integrable billiard map. There are more examples if curved metrics are allowed, see Popov-Topalov [**PT**].

8.1. Trace formulae for integrable systems. So far, we have mainly considered the trace of the wave group around non-degenerate periodic orbits. For integrable systems, the periodic orbits usually come in families filling out invariant tori. We now consider the appropriate notions of non-degeneracy in this context.

We will always assume that the closed geodesics come in clean families in the following sense:

DEFINITION 8.1. *A metric g on a compact manifold M will be said to have a* simple clean length spectrum *if the length function L_g on the loop space $Map(S^1, M)$*

is a Bott-Morse function which takes distinct values at distinct components of its critical set, $Crit(L_g)$.

The term Bott-Morse means that each component of $Crit(L_g)$ is a manifold, whose tangent space is the kernel of dL_g. Equivalently, each component is a clean fixed point set for G^t_g. One needs the clean (Bott-Morse) condition to get a nice wave trace expansion and one needs the simple length spectrum condition to determine geometric information from the expansion.

Under this assumption, the trace of its wave group has the form:

$$Tr e^{it\sqrt{\Delta_g}} = e_0(t) + \sum_{\mathcal{T}} e_{\mathcal{T}}(t) \tag{8.1}$$

where the singular term $e_0(t) = C_n\ Vol(M,g)\ (t+i0)^{-n} + \dots$ at $t=0$ is the same as in the non-degenerate case, where $\{\mathcal{T}\}$ runs over the critical point components of L_g, and where

$$e_{\mathcal{T}} = c_{\mathcal{T},d_{\mathcal{T}}}(t - L_{\mathcal{T}} + i0)^{-d_{\mathcal{T}}/2} + c_{\mathcal{T},d_{\mathcal{T}}-1}(t - L_{\mathcal{T}} + i0)^{(-d_{\mathcal{T}}/2+1)} + \dots, \quad d_{\mathcal{T}} = \dim \mathcal{T}.$$

Here, $d_{\mathcal{T}}$ is the dimension of the symplectic cone formed by the family of closed geodesics within $T^*_g M$ and $L_{\mathcal{T}}$ is the common length of the closed geodesic in $\mathcal{T}$. For instance, in the non-degenerate case, $\mathcal{T} = \mathbb{R}_+\gamma$ is the symplectic cone generated by $\gamma \subset S^*_g M$ and $d_{\mathcal{T}} = 2$.

8.2. Spectral determination of simple surfaces of revolution. The simplest example of what can be done with integrable systems is given by 'simple' analytic surfaces of revolution. We first review the proof in [**Z2**] that they are spectrally determined among other simple analytic surfaces of revolution and then sketch a result observed independently by the author and G. Forni and by K.F. Siburg regarding their spectral determination among all metrics on S^2.

The precise class of metrics we consider are those metrics g on S^2 which belong to the class $\mathcal{R}^*$ of real analytic, rotationally invariant metrics on S^2 with simple length spectrum in the above sense and satisfying the following 'simplicity' condition

- $g = dr^2 + a(r)^2 d\theta^2$
- $\exists! r_0 : a'(r_0) = 0;$
- The Poincare map $\mathcal{P}_0$ of $r = r_0$ is elliptic of twist type

Convex analytic surfaces of revolution are examples, but of course there are others. The unique isolated closed geodesic (at distance r_0) is an elliptic orbit. Peanut surfaces are obviously non-simple, since they possess three isolated closed geodesics of which one is hyperbolic. The standard round metric and other Zoll metrics of revolution are non-simple since the Poincare maps are not twist maps.

The following results shows that one can solve the inverse spectral problem in the class of analytic simple surfaces of revolution.

THEOREM 8.2. *([**Z2**]) Suppose that g_1, g_2 are two real analytic metrics on S^2 such that (S^2, g_i) are simple surfaces of revolution with simple length spectra. Then $Sp(\Delta_{g_1}) = Sp(\Delta_{g_2})$ implies $g_1 = g_2$.*

The proof is based on quantum Birkhoff normal forms for the Laplacian Δ. But the special feature of simple surfaces of revolution is that there exists a global

Birkhoff normal form as well as local ones around the critical closed orbit or the invariant tori. This is because Δ is a toric integrable Laplacian in the following sense: there exist commuting first order pseudo-differential operators $\hat{I}_1, \hat{i}_2$ such that:

- the joint spectrum is integral, i.e. $Sp(\mathcal{I}) \subset \mathbb{Z}^2 \cap \Gamma + \{\mu\}$ where Γ is the cone $I_2 \geq |I_1|$ in $\mathbb{R}^2$.
- The square root of Δ is a first order polyhomogeneous function $\sqrt{\Delta} = \hat{H}(\hat{I}_1, \hat{I}_2)$ of the action operators.

By polyhomogeneous, we mean that $\hat{H}$ has an asymptotic expansion in homogeneous functions of the form:

$$\hat{H} \sim H_1 + H_o + H_{-1} + \dots, \qquad H_j(rI) = r^j H_j(I).$$

The principal symbols I_j of the $\hat{I}_j$'s generate a classical Hamiltonian torus action on $T^*S^2 - 0$. Analysis of the normal form shows that $H_o = 0$.

It follows that

$$Sp(\sqrt{\Delta_g}) = \{\hat{H}(N + \mu) : N \in \mathbb{Z}^2 \cap \Gamma_o\},$$

where the eigenvalues have expansions

$$\lambda_N \sim H_1(N + \mu) + H_{-1}(N + \mu) + \dots.$$

THEOREM 8.3. *([**Z2**]) Let (S^2, g) be an analytic simple surface of revolution with simple length spectrum. Then the normal form $\hat{H}(\xi_1, \xi_2)$ is a spectral invariant.*

The normal form and the proof are very different from the non-degenerate case in [**G, Z3, Z4**], although the philosophy of the proof is similar.

To complete the proof of Theorem 8.2, we need to show that $\hat{H}$ determines a metric in $\mathcal{R}$. As in the bounded domain case outline above, the crucial point is to calculate the normal form invariants. It turns out to be sufficient to calculate $H_1 = H$ and H_{-1} in terms of the metric (i.e. in terms of $a(r)$) and then to invert the expressions to determine $a(r)$.

The method given in [**Z2**] for calculating H and H_{-1} was to study the spectral asymptotics of $\sqrt{\Delta} = \hat{H}(\hat{I}_1, \hat{I}_2)$ along 'rays of representations' of the quantum torus action, i.e. along multiples of a given lattice point (n_o, k_o). The lattice points $(n_o, k_o + \frac{1}{2})$ parametrize tori T_{n_o, k_o} satisfying so-called Bohr-Sommerfeld quantization conditions, which imply that one can construct associated joint eigenfunctions ϕ_{n_o, k_o} of $(\hat{I}_1, \hat{I}_2)$ by the WKB method. This is reminiscent of the quasi-mode method mentioned above around non-degenerate closed geodesics, but there we studied the conjugation to normal form rather than the asymptotics of quasi-modes. Here, the existence of a global torus action makes the quasi-modes easier to study, and indeed the $\{\phi_{n,k}\}$ are actually modes (eigenfunctions) of Δ with complete asymptotic expansions along rays. By studying the eigenvalue problem as $|(n,k)| \to \infty$ we determine the H_{-j}'s.

Once it is known that the global Birkhoff normal for $\hat{H}$ is a spectral invariant, it follows that for each $n_0 \in \mathbb{Z}$, the function $\hat{H}(n_o, \hat{I}_2)$ is a known function of the variable $I := \hat{I}_2$. Its principal symbol $H_{n_o}(I) := H_1(n_o, I)$ is then a known function and its inverse function

$$I_{n_o}(E) = \int_{r_-(E)}^{r_+(E)} \sqrt{E - \frac{1}{a(r)^2}} dr$$

is also known. Here, $r_\pm(E)$ are the upper and lower values of the radius of the projection of the torus of energy E and angular momentum n_0 to S^2. We may write the integral in the form

$$\int_{\mathbb{R}} (E-x)_+^{\frac{1}{2}} d\mu(x)$$

where μ is the distribution function $\mu(x) := |\{r : \frac{1}{a(r)^2} \le x\}|$ of $\frac{1}{a^2}$, with $|\cdot|$ the Lebesgue measure. This Abel transform is invertible and hence

$$d\mu(x) = \sum_{r: \frac{1}{a(r)^2}=x} |\frac{d}{dr}\frac{1}{a(r)^2}|^{-1} dx$$

and therefore

$$J(x) := \sum_{r:a(r)=x} \frac{1}{|a'(r)|}$$

are spectral invariants. By the simplicity assumption on a, there are just two solutions of $a(r) = x$; the smaller will be written $r_-(x)$ and the larger, $r_+(x)$. Thus, the function

$$J(x) = \frac{1}{|a'(r_-(x))|} + \frac{1}{|a'(r_+(x))|}$$

is a spectral invariant.

By studying H_{-1}, we find in a somewhat similar way that

$$K(x) = |a'(r_-(x))| + |a'(r_+(x))|$$

is a spectral invariant. It follows that we can determine $a'(r_+(x))$ and $a'(r_-(x))$.

Since both metrics g_1 and g_2 are assumed to belong to $\mathcal{R}^*$, they are determined by their respective functions $a_j(r)$. We conclude that $a_1 = a_2$ and hence $g_1 = g_2$.

8.3. Unconditional spectral determination. We now ask whether we can remove the assumption that $g_2 \in \mathcal{R}^*$ in Theorem 8.2? The question is, if SpecΔ_g = SpecΔ_h and $g \in \mathcal{R}^*$, then is $h \in \mathcal{R}^*$? An affirmative answer would give a large class of metrics which are spectrally determined. To the author's knowledge, the only metric on S^2 known to be spectrally determined is the canonical round one.

We cannot answer this question, but we will give a partial result suggesting that it is true. The following theorem was stated in [**Z7**] and was worked out in a conversation with G. Forni in 1997. A similar but somewhat stronger conclusion was drawn by K. F. Siburg (Theorem 5.2 of [**S2**]), under the stronger hypothesis of an isospectral (or length-spectral) deformation. After sketching our proof, we will also sketch his, which has other applications.

THEOREM 8.4. *Let $g \in \mathcal{R}^*$ and let h be any metric on S^2 with simple clean length spectrum for which SpecΔ_g = SpecΔ_h. Then h has the following properties:*

(i) *It has just one isolated non-degenerate closed geodesic γ_h (up to orientation); all other closed geodesics come in one-parameter families lying on invariant tori in $S^*_h S^2$;*

(ii) *The Birkhoff normal form of G^t_h at γ_h is identical to that of G^t_g at its unique non-degenerate closed orbit. Hence it is convergent.*

(iii) *the geodesic flow G^t_h of h is C^0- integrable. That is, $S^*_h S^2$ has a C^0-foliation by 2-tori invariant under G^t_h.*

If we knew in (ii) that the Birkhoff transformation conjugating G_h^t to its Birkhoff normal form was convergent, then it would follow that G_h^t is completely integrable with global action-angle variables, and that it would commute with a Hamiltonian torus action. We conjecture that this is the case. Statement (iii) shows that it is at least integrable in the C^0 sense. At the present time, metrics on S^2 whose geodesic flows commute with Hamiltonian torus actions have not been classified. For a result which classifies such metrics on the torus (they must be flat), see [**LS**].

8.3.1. *Wave invariants for $g \in \mathcal{R}^*$.* The proof is based on a study of the wave trace formula in this setting.

PROPOSITION 8.5. *Suppose that $g \in \mathcal{R}^*$. Then the trace of its wave group has the form:*

$$Tre^{it\sqrt{\Delta_g}} = e_0(t) + e_{\gamma_g}(t) + \sum_{\mathcal{T}} e_{\mathcal{T}}(t)$$

where

$$e_0(t) = C_n area(M, g)(t + i0)^{-n} + \dots$$

is singular only at $t = 0$, *where*

$$e_{\gamma_g}(t) = c_\gamma(t - L_\gamma + i0)^{-1} + a_{\gamma 0}\log(t - L_\gamma + i0) + \dots$$

and where

$$e_{\mathcal{T}} = c_{\mathcal{T}}(t - L_{\mathcal{T}} + i0)^{-3/2} + \dots.$$

Suppose now that h is any other metric with Spec $\Delta_h = \text{Spec}\Delta_g$ and with simple length spectrum. Then the wave trace of h has precisely the same singularities as the wave trace of g. Since there is only one singularity of the order $(t+i0)^{-1}$, there can exist only one non-degenerate closed geodesic, proving (i). By Guillemin's inverse result, the Birkhoff normal form of the metric and Poincaré map for γ_h is the same as for γ_g. In particular, the Poincare map $\mathcal{P}_h$ of γ_h is elliptic of twist type. From the fact that all other critical components have the singularity of a three-dimensional cone, it follows that in $S_h^*S^2$ the other closed geodesics come in one-parameter families. They weep out a surface foliated by circles, which can only be a two dimensional torus, proving (ii)

It is the third statement (iii) which requires a new idea. So far we only know that periodic orbits lie on invariant tori, but we do not know what lies between these tori. Aubry-Mather theory will now close the gaps.

We recall that Aubry-Mather theory is concerned with an area-preserving diffeomorphism ϕ of an annulus $A = S^1 \times (a, b)$ [**KH, MF, Ka, M, S1, S2**]. Let $\tilde{\phi}$ denote a lift to $\mathbb{R} \times (a, b)$ with $\tilde{\phi}(x + 1, y) = \tilde{\phi}(x, y) + (1, 0)$. The map ϕ is called a *monotone twist mapping* if it preserves the orientation of A, if it preserves the boundary components and if the lift $\tilde{\phi}(x_0, y_0) = (x_1, y_1)$ satisfies:

- The twist condition: $\frac{\partial x_1}{\partial y_0} > 0$;
- The exactness condition: $y_1 dx_1 - y_0 dx_0 = dh(x_0, x_1)$.

If a, b are finite, $\tilde{\phi}$ extends continuously to the boundary as a pair of 'rotations':

$$\tilde{\phi}(x, a) = (x + \omega_-, a),; \quad \tilde{\phi}(x, b) = (x + \omega_+, b).$$

The interval (ω_-, ω_+) is called the *twist interval* of ϕ.

Let $\{(x_i, y_i)\}$ be an orbit of $\tilde{\phi}$. Its *rotation number* is defined to be

$$\lim_{|i|\to\infty} \frac{x_i - x_0}{i}.$$

A curve $C \subset A$ is called an *invariant circle* if it is an invariant set which is homeomorphic to the circle and which separates boundary components. According to Birkhoff's invariant circle theorem, an invariant circle is a Lipschitz graph over the factor S^1 of A. Any invariant circle has a well-defined rotation number (the common rotation number of orbits in the circle) and the rotation number belongs to the twist interval.

An orbit $\{(x_i, y_i)\}$ is determined by the sequence $\{x_i\}$ of its x-coordinates. It is called *minimal* if every finite segment is action-minimizing with fixed endpoints:

$$\sum_{i=k}^{n-1} h(x_i, x_{i+1}) \leq \sum_{i=k}^{n-1} h(\xi_i, \xi_{i+1}), \quad \forall(\xi_k, \dots, \xi_n) \text{ with } \xi_k = x_k, \xi_n = x_n.$$

The corresponding orbit orbit (x_i, y_i) is called a *minimal orbit.* The Aubrey-Mather theorem states (cf. [**Ka**], Theorem 1;) :

A monotone twist map possesses minimal orbits for each rotation number $\omega \in (\omega_1, \omega_+)$ in its twist interval. Every minimal orbit lies on a Lipschitz graph over the x-axis. For each rational rotation number $\omega = \frac{p}{q}$, there exists a periodic minimal orbit of rotation number $\frac{p}{q}$. When ω is irrational, there exists either an invariant circle with rotation number ω, or an invariant Cantor set E.

The theorem also describes three possible orbit types in both the rational or irrational case.

We now return to our problem on simple surfaces of revolution. We fix local transverse discs (Poincaré sections) S_g, resp. S_h to the geodesic flows G_g^t, resp. G_h^t at the orbits γ_g, resp. γ_h. Concretely, the transversals can be taken to be small variations of γ_g', resp. γ_h' moved up and down a small orthogonal geodesic arc to γ_g', resp. γ_h'. Since γ_g is non-degenerate elliptic, its Poincaré map $\mathcal{P}_g$ defines area-preserving map of the symplectic disc S_g with a non-degenerate elliptic fixed point corresponding to γ_g. Since the Birkhoff normal forms of $\mathcal{P}_g$ and $\mathcal{P}_h$ are the same, $\mathcal{P}_h$ defines the same kind of map of S_h. To obtain a twist map of an annulus, we puncture out the fixed point of $\mathcal{P}_g$, resp. $\mathcal{P}_h$. We define the rotation angle ω_0 corresponding to this orbit by continuity from nearby orbits, which can be read off from the Birkhoff normal form. Indeed, the Poincaré maps have the form

$$\begin{pmatrix} x \\ y \end{pmatrix} \to \begin{pmatrix} \cos 2\pi\theta & -\sin 2\pi\theta \\ \sin 2\pi\theta & \cos 2\pi\theta \end{pmatrix} \cdot \begin{pmatrix} x \\ y \end{pmatrix} + O(|x|^2 + |y|^2),$$

with $\theta \sim \omega_0 + \beta(x^2 + y^2) + \cdots$, as $x^2 + y^2 \to 0$. For further discussion of area-preserving maps around elliptic fixed points and twist maps, see [**S2**].

The foliation of $S^*S_g \backslash \gamma_h$ by 2-tori intersects S_g in a foliation by invariant circles converging to the fixed point. Circles with rational rotation numbers contain only periodic orbits, and conversely all periodic orbits belong to invariant circles with rational rotation numbers.

In the case of h, we know that once γ_g is punctured out, all periodic orbits come in two-tori which project to invariant curves in S_h which are diffeomorphic to circles. The Birkhoff normal form shows that they are invariant circles in the sense that they are also homotopically non-trivial in the punctured S_h (indeed, the action-angle variables are essentially polar coordinates on this disc).

Let $C \subset S_h$ be an invariant circle for $\mathcal{P}_h$. Set:

- $I(C)$ = area enclosed by C.
- C_I = invariant circle enclosing an area I. (It is clearly unique).
- ω_I = rotation number of $\mathcal{P}_H|_{C_I}$.

Since $\mathcal{P}_h$ is a twist map, the rotation number ω_I is a monotone increasing function of I.

Let I_+ denote the area of S_h with respect to the symplectic form. By shrinking S_h we may assume that the boundary of S_h is a periodic circle C_{I_+}.

Since the origin is a fixed point of $\mathcal{P}_h$ and since the rotation number is increasing, it is clear that the set of rotation numbers lie in the interval $[0, \omega_{I_+}]$. It follows by the Aubry-Mather theorem that every rational number $p/q \in [0, \omega_{I_+}]$ is the rotation number of a periodic circle $C_I \subset S_h$.

LEMMA 8.6. *Let $\alpha \in [0, \omega_{I_+}]$. Then there exists an invariant circle for $\mathcal{P}_h$ of rotation number α.*

Proof This follows from Corollary 6.1 of [**Ka**], which shows that as soon as Birkhoff periodic points of all rational rotation numbers are constructed, then there exist orbits of each irrational rotation number which are dense in an invariant circle. Alternatively one could let $p_n/q_n \to \alpha$ and let C_{p_n/q_n} be the corresponding periodic circles. Each C_{p_n/q_n} is a Lipschitz circle and the sequence of these circles tends monotonically to a limit circle. It is Lipschitz and its rotation number is α.

To complete the proof, we make the observation:

LEMMA 8.7. *S_h is foliated in the C^0 sense by invariant circles for $\mathcal{P}_h$.*

Proof: If not there exists an annulus $A \subset S_h$ with boundary consisting of two invariant circles and containing no invariant circles in its interior. But by the Aubrey-Mather theorem, there must exist a periodic point in A. Since the periodic points come in circles, there must exist a periodic circle, contradicting the non-existence of invariant circles in A. QED

This completes the proof of Theorem 8.4.

8.3.2. *Isospectral class of an ellipse.* Instead of surfaces in $\mathcal{R}^*$, one can apply this reasoning to the Dirichlet (or Neumann) problem for an ellipse $E_{a,b} = \{(x, y) : \frac{x^2}{a^2} + \frac{y^2}{b^2} = 1\}$. It is well-known that ellipses have integrable billiards.

The ellipse has three distinguished periodic billiard orbits:

- The bouncing ball orbit along the minor axis, which is a non-degenerate elliptic orbit;
- The bouncing ball orbit along the major axis, which is a non-degenerate hyperbolic orbit;
- Its boundary.

All other periodic orbits come in one-parameter families. The existence of a hyperbolic orbit means that the billiard flow is very different from geodesic flows of metrics in $\mathcal{R}^*$.

The wave trace formula shows that any domain Ω with $Spec(\Omega) = Spec(E_{a,b})$ has precisely one isolated elliptic orbit, one isolated hyperbolic orbit. The accumulation points in the length spectrum must be multiples of the perimeter of the domain (a spectral invariant), so the boundary must be a closed geodesic as well.

One can apply the twist map theory either to the boundary orbit or to the unique non-degenerate elliptic orbit with isolated length in the length spectrum. The argument above shows that there exists a C^0 foliation by invariant circles at least near these two orbits.

8.4. Marked length spectral rigidity of domains. We now review a result due to K.F. Siburg [**S1, S2**] on isospectral deformations of integrable systems which has interesting applications to metrics in $\mathcal{R}^*$ and to bounded plane domains.

The key invariant is the *mean minimal action*

$$\alpha : [\omega_-, \omega_+] \to \mathbb{R}, \tag{8.2}$$

of a twist map ϕ, which associates to a rotation number ω in the 'twist interval' the mean action

$$\alpha(\omega) = - \lim_{N\to\infty} \frac{1}{2N} \sum_{i=-N}^{N} h(x_i, x_{i+1}) \tag{8.3}$$

of a minimal orbit (q_i, η_i) of ϕ of rotation number ω. It is a strictly convex function which is differentiable at all irrational numbers. If $\omega = p/q$, then α is differentiable at ω if and only if there exists an invariant circle of rotation number p/q consisting entirely of periodic minimal orbits. If a monotone twist map possesses an invariant circle of rotation number ω, then every orbit on the circle is minimal ([**MF**], Theorem 17.4). In the case of a bounded plane domain, $h(q, q') = -|q - q'|$.

It is observed by K. F. Siburg [**S1**] (Theorem 4.1) that the marked length spectrum is essentially the same invariant as the mean minimal action. The mean minimal action is therefore an isospectral deformation invariant. He used this to give a proof that there cannot exist isospectral deformations within $\mathcal{R}^*$ (or within more general classes of deformations, see [**S2**], Theorem 4.5). Indeed, the mean minimal actions α would all be the same. Hence if that α_0 of the original surface is differentiable at all rationals, so are they all. But this implies that they all have invariant circles of rational rotation numbers. By taking limits, one obtains invariant circles of all rotation numbers. (We remark that the trace formula already shows that there existed invariant circles of all rational rotation numbers).

Siburg further connects the mean minimal to the Melrose-Marvizi invariants: Let $\alpha^* : [-1, 1] \to \mathbb{R}$ denote its convex conjugate of α. Then Siburg shows (loc. cit. p. 300) that the Melrose-Marvizi invariants are algebraically equivalent to the Taylor coefficients of $(\alpha^*)^{2/3}$ at -1.

It appears that the only explicitly known mean minimal action is that of the disc D, where it is given by $\alpha(\omega) = \frac{-1}{\pi} \sin \pi\omega$ ([**S1**]); α is only smooth when $\Omega = D$ ([**S1**], Theorem 4.6) It is likely that it is computable in the case of an ellipse (perhaps in terms of elliptic functions). Perhaps it can be proved that ellipses are the unique domains with these particular minimal action functions (which could be simpler than the long outstanding problem of proving that they are the unique integrable billiard systems). If so, this would prove that ellipses are spectrally rigid.

PROBLEM 8.1. *Is the map from curvature functions κ of convex plane domains to the mean minimal action α of the associated convex domain injective or finitely many to one? at least near ellipses or under some additional analyticity or discrete symmetry condition?*

References

[AM] K. G. Andersson and R. B. Melrose, The propagation of singularities along gliding rays. Invent. Math. 41 (1977), no. 3, 197–232.

[AKN] V. I. Arnol'd, V.V. Kozlov, and A. I. Neushtadt, *Mathematical aspects of classical and celestial mechanics. Dynamical systems, III*, pp. vii–xiv and 1–291, Encyclopaedia Math. Sci., 3, Springer, Berlin, 1993.

[BB] V.M.Babic, V.S. Buldyrev: *Short-Wavelength Diffraction Theory*, Springer Series on Wave Phenomena 4, Springer-Verlag, New York (1991).

[Ba] V. Baladi, Periodic orbits and dynamical spectra. Ergodic Theory Dynam. Systems 18 (1998), no. 2, 255–292.

[BB1] R. Balian and C. Bloch, Distribution of eigenfrequencies for the wave equation in a finite domain I: three-dimensional problem with smooth boundary surface, Ann. Phys. 60 (1970), 401-447.

[BB2] R. Balian and C. Bloch, Distribution of eigenfrequencies for the wave equation in a finite domain. III. Eigenfrequency density oscillations. Ann. Physics 69 (1972), 76–160.

[B] V. Bangert, Geodesic rays, Busemann functions and monotone twist maps. Calc. Var. Partial Differential Equations 2 (1994), no. 1, 49–63.

[BLW] A. Banyaga, R. de la Llave, and C. E. Wayne, Cohomology equations near hyperbolic points and geometric versions of Sternberg linearization theorem. J. Geom. Anal. 6 (1996), no. 4, 613–649 (1997).

[BGR] C. Bardos, J. C. Guillot, and J. Ralston, La relation de Poisson pour l'équation des ondes dans un ouvert non-born, Comm. Partial Differential Equations 7 (1982), 905–958

[BK] M. I. Belishev and Y. V. Kurylev, To the reconstruction of a Riemannian manifold via its spectral data (BC-method). Comm. Partial Differential Equations 17 (1992), no. 5-6, 767–804

[Be] P. Bérard, Transplantation et isospectralité. Math. Ann. 292, 547–560 (1992)

[Be2] P. Bèrard, Transplantation et isospectralit. II. J. London Math. Soc. (2) 48 (1993), no. 3, 565–576.

[Be3] P. Bèrard, *Spectral geometry: direct and inverse problems.* With appendixes by Grard Besson, and by Bèrard and Marcel Berger. Lecture Notes in Mathematics, 1207. Springer-Verlag, Berlin, 1986.

[Ber] M. Berger, P. Gauduchon, and E. Mazet, *Le spectre d'une variètè riemannienne.* Lecture Notes in Mathematics, Vol. 194 Springer-Verlag, Berlin-New York 1971

[BCG] G. Besson, G. Courtois, and S. Gallot, Entropies et rigidités des espaces localement symtriques de courbure strictement ngative. Geom. Funct. Anal. 5 (1995), no. 5, 731–799.

[Bia] M. Bialy,Convex billiards and a theorem by E. Hopf. Math. Z. 214 (1993), no. 1, 147–154

[BKL] M. Blank, G.Keller, and C. Liverani, Ruelle-Perron-Frobenius spectrum for Anosov maps Authors: (arXiv preprint nlin.CD/0104031).

[BJP] D. Borthwick, C. Judge, and P. A. Perry, Determinants of Laplacians and isopolar metrics on surfaces of infinite area. Duke Math. J. 118 (2003), no. 1, 61–102.

[BG] T.P. Branson and P. Gilkey, The asymptotics of the Laplacian on a manifold with boundary. Comm. Partial Differential Equations 15 (1990), no. 2, 245–272.

[BGKV] T. Branson, P. B. Gilkey, K. Kirsten and D. Vassilevich, Heat kernel asymptotics with mixed boundary conditions, Nuclear Phys. B563 (1999), 603–626.

[BP] R. Brooks and P. A. Perry, Isophasal scattering manifolds in two dimensions. Comm. Math. Phys. 223 (2001), no. 3, 465–474.

[BH] J.Bruning and E.Heintze, Spektrale starrheit gewisser Drehflachen, Math.Ann. 269 (1984), 95-101.

[Ch] J. Chazarain, Construction de la paramtrix du problme mixte hyperbolique pour l'quation des ondes. C. R. Acad. Sci. Paris Sr. A-B 276 (1973), A1213–A1215.

[Ch2] J. Chazarain, Formule de Poisson pour les varits riemanniennes. Invent. Math. 24 (1974), 65–82

[CdV] Y. Colin de Verdière, Sur les longueurs des trajectoires priodiques d'un billard, in: Dazord, Desolneux (eds.): *Géométrie symplectique et de contact*, Sem. Sud-Rhod. Gom. (1984), 122 139.

[CdV2] Y. Colin de Verdière, Spectre du laplacien et longueurs des godsiques priodiques. C. R. Acad. Sci. Paris Sr. A-B 275 (1972), A805–A808.

[CdV3] Y. Colin de Verdière, Spectre du laplacien et longueurs des godsiques priodiques. I, II. Compositio Math. 27 (1973), 83–106; ibid. 27 (1973), 159–184.

[CdV4] Y. Colin de Verdière, Spectre conjoint d'opérateurs pseudo-diffrentiels qui commutent. II. Le cas intégrable. Math. Z. 171 (1980), no. 1, 51–73.

[C] C. Croke, Rigidity for surfaces of nonpositive curvature. Comment. Math. Helv. 65 (1990), no. 1, 150–169.

[C2] C. Croke, Rigidity and the distance between boundary points. J. Differential Geom. 33 (1991), no. 2, 445–464.

[C3] C. Croke, Rigidity theorems in Riemannian geometry, to appear in IMA Volume 137: *Geometric Methods in Inverse Problems and PDE Control.* C.B. Croke, I. Lasiecka, G. Uhlmann, and M. S.Vogelius eds.

[CFF] C. Croke, A. Fathi and J. Feldman, The marked length-spectrum of a surface of nonpositive curvature. Topology 31 (1992), no. 4, 847–855.

[CS] C. B. Croke and V. Sharafutdinov, Spectral rigidity of a compact negatively curved manifold. Topology 37 (1998), no. 6, 1265–1273.

[DS] C. Deninger and W. Singhof, A note on dynamical trace formulas. Dynamical, spectral, and arithmetic zeta functions (San Antonio, TX, 1999), 41–55, Contemp. Math., 290, Amer. Math. Soc., Providence, RI, 2001.

[D] H. Donnelly, On the wave equation asymptotics of a compact negatively curved surface. Invent. Math. 45 (1978), no. 2, 115–137.

[DG] J.J.Duistermaat and V.Guillemin, The spectrum of positive elliptic operators and periodic bicharacteristics, Inv.Math. 24 (1975), 39-80.

[Gh] M. Ghomi, Shortest periodic billiard trajectories in convex bodies, to appear in Geom. Funct. Anal.

[Gor] C. Gordon,Survey of isospectral manifolds. *Handbook of differential geometry*, Vol. I, 747–778, North-Holland, Amsterdam, 2000.

[Gor2] C. Gordon, CBMS Lecture Notes (to appear).

[GorM] C. Gordon and Y. Mao, Geodesic conjugacies of two-step nilmanifolds. Michigan Math. J. 45 (1998), no. 3, 451–481.

[GMS] C. Gordon, Y. Mao, and D. Schueth, Symplectic rigidity of geodesic flows on two-step nilmanifolds. Ann. Sci. cole Norm. Sup. (4) 30 (1997), no. 4, 417–427.

[GWW] C. Gordon, D. Webb, and S. Wolpert, Isospectral plane domains and surfaces via Riemannian orbifolds. Invent. Math. 110 (1992), no. 1, 1–22.

[G] V. Guillemin, Wave-trace invariants. Duke Math. J. 83 (1996), no. 2, 287–352.

[G2] V. Guillemin, Lectures on spectral theory of elliptic operators. Duke Math. J. 44 (1977), no. 3, 485–517.

[G3] V. Guillemin, Wave-trace invariants and a theorem of Zelditch. Internat. Math. Res. Notices 1993, no. 12, 303–308

[GK] V. Guillemin and D. Kazhdan, Some inverse spectral results for negatively curved 2-manifolds. Topology 19 (1980), no. 3, 301–312.

[GM] V. Guillemin and R. B. Melrose, The Poisson summation formula for manifolds with boundary, Adv. Math. 32 (1979), 204–232.

[GM2] V. Guillemin and R. B. Melrose, The wave equation on manifolds with boundary, Colloque Boutet de Monvel (2003).

[GM3] V. Guillemin and R. B. Melrose, An inverse spectral result for elliptical regions in R^2. Adv. in Math. 32 (1979), no. 2, 128–148.

[GM4] V. Guillemin and R. B. Melrose, A cohomological invariant of discrete dynamical systems. E. B. Christoffel (Aachen/Monschau, 1979), pp. 672–679, Birkhuser, Basel-Boston, Mass., 1981

[Gutz] M. C. Gutzwiller, J. Math. Phys. 12, 343 (1971).

[H] U. Hamenstädt, Cocycles, symplectic structures and intersection. Geom. Funct. Anal. 9 (1999), no. 1, 90–140.

[H2] U. Hamenstädt, Time-preserving conjugacies of geodesic flows. Ergodic Theory Dynam. Systems 12 (1992), no. 1, 67–74.

[HZel] A. Hassell and S. Zelditch, Quantum ergodicity of boundary values of eigenfunctions (2003), arxiv preprint math.SP/0211140.

[HZel2] A. Hassell and S. Zelditch, Determinants of Laplacians in exterior domains. Internat. Math. Res. Notices 1999, no. 18, 971–1004.

[HZ] A. Hassell and M. Zworski, Resonant rigidity of S^2, J. Funct. Anal. 169 (1999), no. 2, 604–609.

[Hil] C. Hillar (private communication).

[ISj] A. Iantchenko and J. Sjöstrand, Birkhoff normal forms for Fourier integral operators. II. Amer. J. Math. 124 (2002), no. 4, 817–850.

[ISZ] A. Iantchenko, J. Sjöstrand and M. Zworski, Birkhoff normal forms in semi-classical inverse problems. Math. Res. Lett. 9 (2002), no. 2-3, 337–362.

[I1] H.Ito, Convergence to Birkhoff normal forms for integrable systems, Comment.Math.Helv. 64 (1989), 412-461.

[I2] H.Ito, Integrability of Hamiltonian systems and Birkhoff normal forms in the simple resonance case, Math.Ann.292 (1992), 411-444.

[K] W. Klingenberg, *Lectures on Closed Geodesics*, Grundlehren der. math. W. **230**, Springer-Verlag (1978).

[KKL] A. Katchalov, Y. Kurylev, and Lassas, *Inverse boundary spectral problems.* Chapman and Hall/CRC Monographs and Surveys in Pure and Applied Mathematics, 123. Boca Raton, FL, 2001 Ruishi On isospectral deformations of Riemannian metrics. Compositio Math. 40 (1980), no. 3, 319–324.

[Ka] A. Katok, Periodic and quasiperiodic orbits for twist maps. *Dynamical systems and chaos* (Sitges/Barcelona, 1982), 47–65, Lecture Notes in Phys., 179, Springer, Berlin, 1983.

[KH] A. Katok and B. Hasselblatt, *Introduction to the modern theory of dynamical systems.* Encyclopedia of Mathematics and its Applications, 54. Cambridge University Press, Cambridge, 1995.

[KT] V.V. Kozlov and D. V. Treshchë v, *Billiards*, Trans. Math. Mono. 89, AMS, Providence R.I. (1991).

[Ku2] R. Kuwabara, On isospectral deformations of Riemannian metrics. II. Compositio Math. 47 (1982), no. 2, 195–205.

[Ku3] R. Kuwabara, On the characterization of flat metrics by the spectrum. Comment. Math. Helv. 55 (1980), no. 3, 427–444.

[LL] L. D. Landau and E.M. Lifshitz, *Quantum Mechanics*, Course of Theoretical Physics Vol. 3, Third Edition, Pergmanon Press (1977).

[LU] M. Lassas and G. Uhlmann, On determining a Riemannian manifold from the Dirichlet-to-Neumann map. Ann. Sci. cole Norm. Sup. (4) 34 (2001), no. 5, 771–787

[L] V. F. Lazutkin, Construction of an asymptotic series for eigenfunctions of the "bouncing ball" type. In *Asymptotic methods and stochastic models in wave propagation problems.* (Russian) Trudy Mat. Inst. Steklov. 95 (1968), 106–118, 215

[La2] V. F. Lazutkin, Existence of caustics for the billiard problem in a convex domain. Math. USSR-Izv. 7 (1973), 185–214.

[Ll] R. de la Llave, Analytic regularity of solutions of Livsic's cohomology equation and some applications to analytic conjugacy of hyperbolic dynamical systems, Ergodic Theory Dynam. Systems 17 (1997), 649–662.

[L2] R. de la Llave, On necessary and sufficient conditions for uniform integrability of families of Hamiltonian systems. International Conference on Dynamical Systems (Montevideo, 1995), 76–109, Pitman Res. Notes Math. Ser., 362, Longman, Harlow, 1996.

[LS] E. Lerman and N. Shirokova, Completely integrable torus actions on symplectic cones. Math. Res. Lett. 9 (2002), no. 1, 105–115.

[Ma] A. Majda, High frequency asymptotics for the scattering matrix and the inverse problem of acoustical scattering. Comm. Pure Appl. Math. 29 (1976), no. 3, 261–291.

[Ma2] A. Majda, A representation formula for the scattering operator and the inverse problem for arbitrary bodies. Comm. Pure Appl. Math. 30 (1977), no. 2, 165–194.

[M] J. D. Meiss, Symplectic maps, variational principles, and transport. Rev. Modern Phys. 64 (1992), no. 3, 795–848.

[MM] Sh. Marvizi and R. B. Melrose, Spectral invariants of convex planar regions. J. Differ. Geom. 17, 475–502 (1982).

[MF] J.N. Mather and G. Forni, Action minimizing orbits in Hamiltonian systems, in: Graffi (ed.): *Transition to Chaos in Classical and Quantum Mechanics*, Springer LNM 1589 (1992), 92–186.

[Me] R. B. Melrose, The inverse spectral problem for planar domains. *Instructional Workshop on Analysis and Geometry*, Part I (Canberra, 1995), 137–160, Proc. Centre Math. Appl. Austral. Nat. Univ., 34, Austral. Nat. Univ., Canberra, 1996.

[Me2] R. B. Melrose, "Geometric Scattering Theory," Cambridge Univ. Press, Cambridge/New York/Melbourne, 1995

[MS] R. B. Melrose and J. Sjöstrand, Singularities of boundary value problems. I. Comm. Pure Appl. Math. 31 (1978), no. 5, 593–617.

[M1] R. Michel, Sur la rigidité imposée par la longueur des géodésiques, Invent. Math. 65 (1981), 71–83.

[M2] R. Michel, Restriction de la distance géodésique à un arc et rigidité, Bull. Soc. Math. France 122 (1994), 435–442.

[Mo] J. Moser, On invariant curves of area-preserving mappings of an annulus. Nachr. Akad. Wiss. Gttingen Math.-Phys. Kl. II 1962 1962 1–20.

[Mo1] J.Moser, Convergent series expansions for quasi-periodic motions, Math.Annalen 169 (1967), 136-176.

[Mo.2] J.Moser, *Stable and Random Motions in Dynamical Systems*, Ann.Math.Studies 77, Princeton U Press, Princeton (1973).

[Mu] R. G. Mukhometov, The reconstruction problem of a two-dimensional Riemannian metric, and integral geometry, (Russian), Dokl. Akad. Nauk SSSR 232 (1977), 32–35.

[OPS] B. Osgood, R. Phillips, and P. Sarnak, Moduli space, heights and isospectral sets of plane domains. Ann. of Math. (2) 129 (1989), no. 2, 293–362.

[O1] J. P. Otal, Le spectre marqu des longueurs des surfaces courbure ngative. Ann. of Math. (2) 131 (1990), no. 1, 151–162.

[O2] J. P. Otal, Sur les longuer des géodésiques d'une métrique a courbure négative dans le disque, Comment. Math. Helv. 65 (1990), 334–347.

[Pa] V.K. Patodi, Curvature and the fundamental solution of the heat operator. J. Indian Math. Soc. 34 (1970), no. 3-4, 269–285 (1971).

[PM] R. Pérez-Marco, Convergence or generic divergence of the Birkhoff normal form. Ann. of Math. (2) 157 (2003), no. 2, 557–574.

[PS] V.M.Petkov and L.N.Stoyanov, *Geometry of Reflecting Rays and Inverse Spectral Problems*, John Wiley and Sons, N.Y. (1992).

[P] G. Popov, Invariants of the length spectrum and spectral invariants for convex planar domains. Commun. Math. Phys. 161, 335–364 (1994)

[P2] G. Popov, Length spectrum invariants of Riemannian manifolds. Math. Z. 213 (1993), no. 2, 311–351.

[P3] G. Popov, On the contribution of degenerate periodic trajectories to the wave-trace. Comm. Math. Phys. 196 (1998), no. 2, 363–383.

[PT] G. Popov and P. Topalov, Liouville billiard tables and an inverse spectral result. Ergodic Theory Dynam. Systems 23 (2003), no. 1, 225–248.

[Rau] J. Rauch, Illumination of bounded domains. Amer. Math. Monthly 85 (1978), no. 5, 359–361.

[R] M. Rouleux, Semi-classical integrability, hyperbolic flows and the Birkhoff normal form, to appear in Canadian J. Math (see also arXiv preprint math.DS/0207026).

[SU] V. Sharafutdinov and G. Uhlmann, On deformation boundary rigidity and spectral rigidity of Riemannian surfaces with no focal points. J. Differential Geom. 56 (2000), no. 1, 93–110.

[S1] K. F. Siburg, Aubry-Mather theory and the inverse spectral problem for planar convex domains. Israel J. Math. 113 (1999), 285–304.

[S2] K. F. Siburg, Symplectic invariants of elliptic fixed points. Comment. Math. Helv. 75 (2000), no. 4, 681–700.

[S3] K. F. Siburg, The minimal action in geometry and dynamics (to appear).

[S.M] C.L.Siegel and J.Moser, *Lectures on Celestial Mechanics*, Grundlehren der math Wiss.Einz. 187, Springer-Verlag, Berlin (1971).

[SjZ] J. Sjöstrand and M. Zworski, Quantum monodromy and semi-classical trace formulæ, J. Math. Pure Appl. 81 (2002), 1–33.

[Sm] L. Smith, The asymptotics of the heat equation for a boundary value problem. Invent. Math. 63 (1981), 467–493.

[StU] P. Stefanov and G. Uhlmann, Rigidity for metrics with the same lengths of geodesics. Math. Res. Lett. 5 (1998), no. 1-2, 83–96.

[St] L. Stoyanov, Rigidity of the scattering length spectrum. Math. Ann. 324 (2002), no. 4, 743–771.

[St2] L. Stoyanov, On the scattering length spectrum for real analytic obstacles. J. Funct. Anal. 177 (2000), no. 2, 459–488.

[Su] T. Sunada, Riemannian coverings and isospectral manifolds. Ann. of Math. (2) 121 (1985), no. 1, 169–186.

[T] S. Tanno, A characterization of the canonical spheres by the spectrum. Math. Z. 175 (1980), no. 3, 267–274.

[T2] S. Tanno, Eigenvalues of the Laplacian of Riemannian manifolds. Tŏhoku Math. J. (2) 25 (1973), 391–403.

[T3] S. Tanno, A characterization of a complex projective space by the spectrum. Kodai Math. J. 5 (1982), no. 2, 230–237

[Wa] C.P. Walkden, C. P. Livsic theorems for hyperbolic flows. Trans. Amer. Math. Soc. 352 (2000), no. 3, 1299–1313

[W] K. Watanabe, Plane domains which are spectrally determined. Ann. Global Anal. Geom. 18 (2000), no. 5, 447–475.

[Z1] S. Zelditch, Spectral determination of analytic bi-axisymmetric plane domains. Geom. Funct. Anal. 10 (2000), no. 3, 628–677.

[Z2] S. Zelditch, The inverse spectral problem for surfaces of revolution. J. Differential Geom. 49 (1998), no. 2, 207–264.

[Z3] S. Zelditch, Wave invariants at elliptic closed geodesics. Geom. Funct. Anal. 7 (1997), no. 1, 145–213.

[Z4] S. Zelditch, Wave invariants for non-degenerate closed geodesics. Geom. Funct. Anal. 8 (1998), no. 1, 179–217.

[Z5] S. Zelditch, The inverse spectral problem for analytic plane domains. I: Balian-Bloch trace formula (http://arXiv.org/abs/math.SP/0111077, to appear in Comm. Math. Phys.); and II: domains with symmetry, (http://arXiv.org/abs/math.SP/0111078.)

[Z6] S. Zelditch, Isospectrality in the FIO category. J. Differential Geom. 35 (1992), no. 3, 689–710.

[Z7] S. Zelditch, Normal forms and inverse spectral theory, Journes Équations aux dérivées partielles, 2–5 June 1998, GDR 1151 (CNRS).

[Z8] S. Zelditch, Maximally degenerate Laplacians. Ann. Inst. Fourier (Grenoble) 46 (1996), no. 2, 547–587.

[Z9] S. Zelditch, Lectures on wave invariants. *Spectral theory and geometry* (Edinburgh, 1998), 284–328, London Math. Soc. Lecture Note Ser., 273, Cambridge Univ. Press, Cambridge, 1999.

[Z10] S. Zelditch Inverse resonance problem for Z_2 symmetric analytic obstacles in the plane, IMA Volume 137: *Geometric Methods in Inverse Problems and PDE Control.* C.B. Croke, I. Lasiecka, G. Uhlmann, and M. S.Vogelius eds.

[Z11] S. Zelditch, Kuznecov sum formulae and Szegö limit formulas on manifolds. Comm. P.D.E. 17, 221–260 (1992)

[Zw2] M. Zworski, Resonances in physics and geometry. Notices Amer. Math. Soc. 46 (1999), no. 3, 319–328.

[Zw3] M. Zworski, Quantum resonances and partial differential equations. Proceedings of the International Congress of Mathematicians, Vol. III (Beijing, 2002), 243–252, Higher Ed. Press, Beijing, 2002.

9. Appendix: Quantum monodromy revisited (by Johannes Sjöstrand and Maciej Zworski)

We present a few simplifications of the presentation of the *quantum monodromy* operator in [**1**] and [**2**].

We first repeat the comment made in [**1**, §4]: the trace formula of [**2**] is formulated in terms of a general Hamiltonian, $P(z)$ (for instance an effective Hamiltonian with a non-linear dependence on z). However, the proof can be reduced to the case of $P - z$. In fact, the assumptions of [**2**, Theorem 2], the implicit function theorem, and the usual symbolic iteration, imply that

$$P(z) = A(z)^*(P - z)A(z)\,,$$

with $A(z) \in \Psi_h^{0,k/2}(X)$ elliptic near $\gamma(0)$, and $P \in \Psi_h^{0,0}(X)$ self-adjoint. Replacing $P(z)$ by $P - z$ in [**2**, Theorem 2] changes the trace by $\mathcal{O}(h^\infty)$.

In the special case of $P(z) = P - z$ the monodromy operator can be written quite simply (though we still believe that it is interesting to consider $M(z)$ for the non-linear $P(z)$ as done in [**2**]). Let us recall that at a point on an integral curve of $P - z$, $\gamma(z)$, $m_0(z) \in \gamma$, we can define the microlocal kernel of $P - z$ at $m_0(z)$, to be the set of families $u(h)$, such that $u(h)$ are microlocally defined near m_0 and

$$(P - z)u(h) = \mathcal{O}(h^\infty) \text{ near } m_0.$$

We denote it by $\ker_{m_0(z)}(P - z)$. Since microlocally, near a given point, the operator $P - z$ can be reduced to hD_{x_1} any solution can be continued microlocally along $\gamma(z)$ and we denote the corresponding forward and backward continuations by $I_\pm(z)$. We can also define the propagator $\exp(-it(P - z)/h)$ and we see that

$$\exp(-it(P - z)/h) \;:\; \ker_{m_0(z)}(P - z) \longrightarrow \ker_{\exp(tH_p)m_0(z)}(P - z)\,.$$

This follows from the fact that $(P - z)\exp(-it(P - z)/h) = \exp(-it(P - z)/h)(P - z)$, and propagation of semi-classical wave fronts: $WF_h(\exp(-it(P - z)/h)u)$ is contained in a neighbourhood of $\exp(tH_p)m_0(z)$ if $WF_h(u)$ is contained in a neighbourhood of $m_0(z)$. Hence we have

$$I_\pm(z) = \exp(\mp it(P - z)/h) \tag{9.1}$$

microlocally near $(m_0(z), \exp(tH_p)(m_0(z)))$.

To define the quantum monodromy we take $m_1(z) \neq m_0(z)$ be another point on $\gamma(z)$ and put

$$\begin{aligned} &I_-(z)\mathcal{M}(z) = I_+(z)\,, \quad \text{near } m_1, \\ &\mathcal{M}(z) \;:\; \ker_{m_0(z)}(P - z) \longrightarrow \ker_{m_0(z)}(P - z)\,. \end{aligned} \tag{9.2}$$

In view of (9.1) we now have

$$\mathcal{M}(z) = \exp(-iT(z)(P - z)/h) \;:\; \ker_{m_0(z)}(P - z) \longrightarrow \ker_{m_0(z)}(P - z)\,, \tag{9.3}$$

where $T(z)$ is the period of $\gamma(z)$ but for z small we can replace it by a fixed period, $T(0)$.

The operator $P(z)$ is assumed to be self-adjoint with respect to some inner product $\langle \bullet, \bullet \rangle$, and we define the *quantum flux* norm on $\ker_{m_0(z)}(P - z)$ as follows: let χ be a microlocal cut-off function supported near γ and equal to one near the part of γ between m_0 and m_1, in the positive direction determined by H_p. We denote by $[P, \chi]_+$ the part of the commutator supported near m_0, or more generally, near the left end point (using the orientation determined by H_p) of the support of $\chi|_\gamma$. We then put

$$\langle u, v \rangle_{\mathrm{QF}} \stackrel{\mathrm{def}}{=} \langle [(i/h)P, \chi]_+ u, v \rangle\,, \quad u, v \in \ker_{m_0(z)}(P - z)\,.$$

It is easy to check that this norm is independent of the choice of χ: if $\tilde{\chi}$ agrees with χ near m_1 we see that $[P, \tilde{\chi} - \chi]_+ = [P, \tilde{\chi} - \chi]$ and clearly $\langle [P, \tilde{\chi} - \chi]u, v \rangle = 0$

(see [**2**, Lemma 4.4] for more details). This independence leads to the unitarity of $\mathcal{M}(z)$:

$$\begin{aligned}\langle \mathcal{M}(z)u, \mathcal{M}(z)u\rangle_{\mathrm{QF}} &= \langle (i/h)[P,\chi]_+ e^{-iT(z)(P-z)/h}u, e^{-iT(z)(P-z)/h}v\rangle \\ &= \langle (i/h)[P, e^{iT(z)(P-z)/h}\chi e^{-iT(z)(P-z)/h}]_+ u, v\rangle \\ &= \langle (i/h)[P,\chi]_+ u, v\rangle = \langle u, v\rangle_{\mathrm{QF}}\end{aligned} \tag{9.4}$$

As already recalled above the operator $P - z$ can be reduced to hD_{x_1} we can identify $\ker_{m_0(z)}(P-z)$ with $\mathcal{D}'(\mathbb{R}^n)$. This is done microlocally near $(0,0)$, and we can choose the identification, $K(z)$, so that

$$K^*(z)(i/h)[P,\chi]_+ K(z) = Id\,.$$

This guarantees that the corresponding monodromy operator,

$$M(z) \stackrel{\mathrm{def}}{=} K(z)^{-1}\mathcal{M}(z)K(z)^{-1} \;:\; \mathcal{D}'(\mathbb{R}^n) \longrightarrow \mathcal{D}'(\mathbb{R}^n)\,,$$

microlocally defined near $(0,0)$, is unitary (microlocally near $(0,0)$. Here $(0,0)$ corresponds to the closed orbit intersecting a transversal identified with $T^*\mathbb{R}^n$. We easily see that $M(z)$ is a semi-classical Fourier integral operator which quantizes the Poincaré map of $\gamma(z)$.

Using (9.3),

$$M(z) = K(z)^{-1} \circ \exp(-iT(0)(P-z)/h) \circ K(z)\,. \tag{9.5}$$

This expression trivializes the proof of [**2**, Lemma 6.2] in the case $P(z) = P - z$. For $P = hD_{x_1}$, $K(z)u(x) = e^{izx_1/h}u(x')$, $x = (x_1, x')$, $x' \in \mathbb{R}^n$, and hence the complexification of z in $K(z)$ produces growth of size $\mathcal{O}(e^{\epsilon|\operatorname{Im} z|/h})$. Then (9.5) shows that (for z close to 0)

$$\begin{aligned}\|M(z)\| &\leq e^{-(T(0)-\epsilon)\operatorname{Im} z/h}\,, \quad 0 < \operatorname{Im} z\,, \\ \|M(z)^{-1}\| &\leq e^{(T(0)-\epsilon)\operatorname{Im} z/h}\,, \quad \operatorname{Im} z < 0\,.\end{aligned} \tag{9.6}$$

The rather subtle [**2**, Lemma 6.1] is altogether unnecessary (unless we want the results for the general $P(z)$; they are however not needed for the trace formula). The estimates (9.6) also give a slight improvement in [**2**, Theorem 1]: we can make the conditions on the support of $\hat{f}$ there optimal: $\hat{f} \in \mathcal{C}_{\mathrm{c}}^\infty(\mathbb{R})$, supp $\hat{f} \subset (-NT, NT) \setminus \{0\}$,

For a discussion of the quantum monodromy operator in a concrete setting and a relation of the quantum flux norm to the more standard objects, see the appendix to [**1**].

References

[1] A. Iantchenko, J. Sjöstrand, and M. Zworski, *Birkhoff normal forms in semi-classical inverse problems,* Math. Res. Lett. **9**(2002), 337-362.

[2] J. Sjöstrand and M. Zworski, *Quantum monodromy and semi-classical trace formulæ*, J. Math. Pure Appl. **81**(2002), 1-33.

S.Zelditch, Department of Mathematics, Johns Hopkins University, Baltimore, MD 21218, USA

E-mail address: `zelditch@@math.jhu.edu`

J. Sjöstrand, Ecole Polytechnique, Centre de Mathematiques, UMR 7460, CNRS, F-91128 Palaiseau Cedex, France

E-mail address: `johannes@math.polytechnique.fr`

M. Zworski, University of California, Mathematics Department, Evans Hall, Berkeley, CA 94720, USA
E-mail address: zworski@math.berkeley.edu

Made in the USA
Monee, IL
08 November 2026

68055345R00262